SPARKING A WORLDWIDE ENERGY REVOLUTION

SOCIAL STRUGGLES IN THE TRANSITION TO A POST-PETROL WORLD

SPARKING A WORLDWIDE ENERGY REVOLUTION

SOCIAL STRUGGLES IN THE TRANSITION
TO A POST-PETROL WORLD

Edited by

KOLYA ABRAMSKY

Sparking a Worldwide Energy Revolution:
Social Struggles in the Transition to a Post-petrol World

ISBN-13: 9781849350051

Library of Congress Control Number: 2009933022

AK Press
674-A 23rd Street
Oakland, CA 94612
USA
www.akpress.org
akpress@akpress.org

AK Press
PO Box 12766
Edinburgh, EH8 9YE
Scotland
www.akuk.com
ak@akedin.demon.co.uk

The above addresses would be delighted to provide you with the latest AK Press distribution catalog, which features the several thousand books, pamphlets, zines, audio and video products, and stylish apparel published and/or distributed by AK Press. Alternatively, visit our web site for the complete catalog, latest news, and secure ordering.

Visit us at www.akpress.org *and* www.revolutionbythebook.akpress.org.

Printed in Canada on acid free, recycled paper with union labor.

Cover by John Yates (stealworks.com)

For my parents,
For being there when I needed you the most.
I am glad I could also be there when you needed me.

In loving memory of my granddad Chimen,
You never made it to see this book,
Although you asked about it right until the end.
I hope it would have made you proud.

TABLE OF CONTENTS

PART 4: COMMUNITY AND WORKER STRUGGLES OVER OWNERSHIP AND CONTROL IN THE FOSSIL FUEL SECTOR AND THEIR ROLE IN A TRANSITION TO A POST-PETROL WORLD

SECTION 2: FROM PETROL TO RENEWABLE ENERGIES: SOCIALLY PROGRESSIVE EFFORTS AT TRANSITION WITHIN THE CONTEXT OF EXISTING GLOBAL POLITICAL AND ECONOMIC RELATIONS

PART 5: LEADING THE WAY: A SAMPLE OF EMERGING "BEST PRACTICES"

SECTION 3: STRUGGLES OVER THE CHOICE OF FUTURE ENERGY SOURCES AND TECHNOLOGIES

PART 6: TECHNOFIXES

SECTION 4: POSSIBLE FUTURES: THE EMERGING STRUGGLE FOR CONTROL OF THE GLOBALLY EXPANDING RENEWABLE ENERGY SECTOR AND THE ROADS AHEAD

PART 11: EMERGING SOCIAL CONFLICTS IN THE RENEWABLE ENERGY SECTOR: THE EXAMPLE OF WIND

PART 12: TIME TO SPEED UP! RENEWABLE ENERGY AS A POSSIBLE WAY OUT OF THE WORLD ECONOMIC CRISIS?

PART 13: TOWARDS A TRANSITION BASED ON DECENTRALIZATION, COMMON OWNERSHIP, DIGNIFIED WORK, AND COMMUNITY AUTONOMY

ACKNOWLEDGEMENTS

Putting together this book has been a fun, learning, and tiring experience. I have met many interesting people along the way, and received a lot from them all. I feel uncomfortable writing acknowledgements. Not because I am unappreciative of people's help, but because it seems an overly formal process, and much of the input is highly informal. Nevertheless, let me give it a go ... and apologies in advance if I have forgotten anyone.

It is always a special pleasure and great learning experience to talk with my own personal gang, who are working on energy-related struggles—George, Silvia, Massimo, Sergio, Olivier, and Ramon. Your perspectives add many different angles to the discussion. Thanks to Massimo and Silvia, for insisting that I clarify the difference between public and common ownership, and that I emphasis the importance of the political aspects of democratic, participatory, and collective control and decision-making, as well as the economic aspects of ownership. To Sergio, for arguing and strongly disagreeing with me about the importance of ownership struggles in the fossil fuel sector, and labor struggles in the wind sector, for sharing with me a number of unpublished texts related to the Copenhagen mobilizations that were useful in writing the book's introduction and final chapter ... and, of course, for embarking on yet another mad leap towards creating an unknown future! And, to George for agreeing wholeheartedly with me about exactly the same issues that Sergio disagreed about. Thank you, Ramon, for the wonderful gazpacho, for excursions to the Sierra, and for taking the time to carefully discuss my work when you are in the middle of your own large book on energy. Thank you, Olivier, for welcoming me back to stay some days with you in Geneva, after many years in which I had not been there. And, thanks to all of you in Yansa, for providing a long term organizational backdrop against which a book like this actually makes some sense. It's an inspiring leap! To Javier, for being enthusiastic about the book, and for encouraging me to get involved in the strategic thinking of the initiative, even though I am still not currently formally involved in it in any way. To Diana, for using tried and tested indigenous Mexican methods of softening the blow of chili in my eyes, which, of course, happened right in the midst of discussions related to the initiative. Thanks to all of you for the inspiration, fun and friendship! We're in it for the long-haul together, and times are surely getting very interesting!

In the early stages of the book, I had hoped to make the book a collective editorial project with Massimo De Angelis, George Caffentzis, Sergio Oceransky, Ramón Fernández Duran, and myself. Although everyone gave valuable input into developing the proposal and outline for the book, each of these people has a very heavy work schedule, and unfortunately, it was not possible to work collectively on the book and I became sole editor. However, three of them, nevertheless, managed to have a perhaps larger presence, both visibly and behind the scenes, than most of the other contributors. I have used a large number of Sergio and George's already-existing texts, which fit well with the book and were kindly offered. And, while Massimo and I did not work directly on the book together, I am grateful to him for suggesting that we edit an issue of the web journal *The Commoner* together, on energy related struggles. Some of the articles that we commissioned for the journal were simultaneously commissioned for this book, and are being reproduced here, either in original or in modified form. We also coauthored an introduction for the journal that provided an important basis for the introduction of this book.

Jane Kruse and Preben Maegaard helped introduce me to the world of grassroots renewable energy perspectives and their successful implementation. As some of the chapters in this book show, the world of renewable energy is increasingly full of pirates! It's good to have some honest and sincere fellow travelers who appreciate the importance of democratizing energy. Preben kindly offered a number of the many texts he's written through the years, and also wrote two new chapters.

A number of people took the time to read through things I'd written and offered me feedback, which was very useful in the writing, rewriting, fine tuning, and above all, intellectual sharpening of the introduction and conclusion: Preben Maegaard, Sergio Oceransky, George Caffentzis, Silvia Federici, Massimo De Angelis, Ovidiu Tichindeleanu, Javier Ruíz, Ramón Fernández Durán, David Feickert, Dörte Fouquet, Rainer Hinrichs Wahl, Jane Kruse, Mia Wastuti, Greg Muttit, Brian Kohler, Yannis Margaris, Raj Patel, and Ewa Jasiewicz.

A big thank you to all the contributors for your extremely interesting texts, without which, to put it bluntly, this book would not exist, and for tolerating my insistent badgering before, at, after, and, in some cases long after, the deadline to send your texts. Above all, thank you for getting your second drafts in even in the midst of the economic-financial crisis, which, for many authors, meant a massive amount of urgent and difficult work in their own organizations and local struggles.

A number of people helped with the preparation of the book at the technical level, including translation. Thanks to Ann Stafford, Thomas Early, Nicolaj Stenkjaer, Nina Alsen, Claudia Roa and Adam Rankin, Diana Labajos, Craig Daniels, and Ben Pakuts for their invaluable help in this time-consuming, and, often thankless, task. Peter Polder was of great help in ensuring appropriate usage of technical terms related to oil, as well as helping with some of the more specialist information in the introduction. I am also grateful for the fact that he encouraged me to be more explicit about the need to think about how to manage scarcity in a collective and fair manner.

Several people were very helpful in suggesting and facilitating contacts, either to authors, potential authors, or others who had useful ideas to contribute to the discussion: Nicola Bullard, Sujin, Ramón Fernández Durán, Kevin Smith, Hermann Scheer, Esperanza Martinez, Louca Lerch, and Joe Drexler.

This book would have never happened in the first place, were it not for a chance encounter that happened at a talk I gave as part of a panel on "green capitalism" at the Renewing the Anarchist Tradition conference in Vermont in 2007, and the email exchange that followed. So, thanks Cindy Milstein, Carwil James, and others involved in putting that event together.

In Binghamton, New York, a small group of people were supportive at the beginning of the work on the book, helping with the intellectual concepts of the worldwide division of labour, and helping me feel confident in applying them to the world of energy: Dale Tomich, Richard Lee, Dennis O'Hearn, Yann Moulier Boutang, and others at the Fernand Braudel Center. These were the people who helped to keep me sane—at least figuratively speaking—in a wider institutional setting that clearly suffered from major insanities … And, above all, they did not try to convince me otherwise when I announced to them that I was leaving ahead of schedule.

I worked on the book a lot while living in Graz, Austria. There, a different group of people helped contribute to my well-being. In addition to being responsible for me learning about Graz in the first place, the people in the Rotor Association also found me a place to live, introduced me to their political artwork, gave me the necessary dosages of entertainment, and provided me with additional work right when I needed it. And thanks to the people at the Institute for Advanced Study of Science, Technology and Society (IAS-STS/IFZ) for giving me a welcoming, friendly, and intellectually-respectful environment to work autonomously in, and for the research fellowship that paid my bills.

In the last weeks before the book was completed, rather than working full time on it, I took a break, only returning to it ten days before the deadline. I spent three weeks in Colombia, participating in an international mission to document the destructive social and ecological implications of agrofuel production, as well as to have a number of meetings with grassroots organizations struggling in the field of energy. In this time, many people generously shared their time, energy, political and life experiences. Though the political content of the trip was decidedly heavy, and though its purpose was not directly related to work on the book, it nonetheless energized me enormously, enabling me to finish the book. It also inspired me to continue with the organizational work that will come once the book is published. Thanks to all the organizers and communities I met during this time, as well as to those in Europe who gave me useful suggestions of who to meet when I was there.

I would like to thank Zach Blue from AK Press for offering me the chance to have this book published; for his helpful and reliable communication as a publisher; for his useful comments; and for his continuing enthusiasm, positive feedback, and shared excitement all along the way. I also am grateful to Kate Khatib for publicity work and for others in the AK team who helped make this book happen.

Finally, some words of thanks and appreciation for my family: my parents Jack and Lenore, my sister Tanya, my brother Sasha, and my late granddad Chimen. Although none of you, except Sasha with his most recent work, are grappling with the questions discussed herein, you have always shown great interest in my work and provided me with loads of warmth, love, and emotional support. You have also, over the years, filled me with a desire to learn, to piece together the different parts of the puzzle—whatever the particular puzzle might be—and to approach questions with a strong respect for social justice. For all of this I am very grateful. Lenore's help with the last minute shortening of texts, at a time when I was under particular pressure from several unexpected non-book related issues, was also invaluable, as was her help with the back cover blurbs!

Without these people, the book would have surely taken shape in a very different, and probably much lesser form. Thank you all.

Introduction

RACING TO "SAVE" THE ECONOMY AND THE PLANET
Capitalist or Post-capitalist Transition to a Post-Petrol World?[1]

Kolya Abramsky

The Economy's a-Tanking and There's an Energy Crisis in the Air ...
Panic!

Either "peak oil" or climate change is to blame for our impending doom. And, to make matters worse, a whole lot of headless chickens are desperately trying to stabilize the world's stock markets and major corporations before the real chickens come home to roost ...
Panic!

One scenario tells us that oil production has just peaked (or is about to peak), and that coal, gas, and uranium production will also peak in the not so distant future. After the peak in production has passed, production will go down while demand continues to grow. Thus, energy prices and corporate profits will reach unprecedented levels, accelerating global inequalities: the already outrageously wealthy will become even more so, the middle class will quickly fall down the social ladder, the already dispossessed will become disposable, and the disposable will be starved to death. In fact, this is already happening ...
Panic!

---

1   This introduction draws heavily from the introduction to issue No 13 of *The Commoner*, "Energy crisis (among others) is in the air," which I coauthored with Massimo De Angelis (who is the editor of *The Commoner* webjournal and my coeditor on that issue. I worked on *The Commoner* and this book at the same time, and worked on both introductions in parallel. As such, it seemed appropriate to simply include large sections of text from *The Commoner* introduction, rather than attempting to quote or paraphrase the text. I proposed to Massimo that we could also coauthor the introduction to the book, but unfortunately his other work commitments did not allow for this. Despite drawing very extensively on the coauthored introduction, I have also made some important additions and modifications, and the two introductions are different texts. Consequently, the introduction to this book only bears my name. I am very grateful to Massimo for agreeing on this format, and also for his suggestion that we coedit the issue of *The Commoner* and coauthor its introduction. A number of texts in this book were commissioned for the book and *The Commoner* in parallel, though appearing first in *The Commoner*. In many cases the version included here is a substantially revised version than that which appeared in *The Commoner*, even if its title remains unchanged. In other cases, the same version has been used. The complete edition of the Issue 13 of *The Commoner* can be found at http://www.commoner.org.uk
The introduction also draws heavily on a number of different discussions I had with people from a range of different organizations and initiatives, as well as some unpublished documents relating to the grassroots mobilizations around the Copenhagen COP 15. Individual acknowledgements are included in the Acknowledgements section.

… Another scenario tells us that we are about to enter a new phase in the history of the planet, defined by what scientists call "non-linear effects" in the earth's climate—the process by which one change leads to another, which leads to another and so forth. We are about to reach several of these critical changes, known as "tipping points." When the first one is reached, there is no way back. The Earth's system then continues evolving, without us being able to do anything about it, until a new natural balance is reached. Nobody can predict what the chemical composition of the atmosphere or the average temperature of the earth will be in this new equilibrium. The process of change will be extremely violent, leading to the collapse of natural systems and social breakdown. It will happen very fast—it has already started, and we are witnessing its acceleration each year. The next decade is critical …

Panic!

And, should we choose to look, though very few do, we should see a third, and equally alarming story. This is the fact that the current global energy regime is characterized by immense inequalities and hierarchies. The average US citizen consumes approximately five times as much energy as the world average, ten times as much as a typical person in China, and over thirty times more than a resident of India. Peripheral zones of the world-economy have exported energy resources to core countries at a steady rate since the Second World War. For some oil-exporting countries this has been the basis of impressive economic growth (as well as social reforms). However, it has also greatly exacerbated long-standing global inequalities in levels of per capita energy consumption between inhabitants of core regions of the world and the rest of the world's population. Approximately 2 billion people throughout the world, particularly in rural areas in Southern countries, use traditional fuels (such as wood, charcoal, and dung) for cooking, a large proportion of these lack access to electricity in their homes. The lack of access to affordable energy services is a serious barrier to people's livelihoods and their possibility of a good life. And energy-poverty disproportionately affects women.

Panic!

★★★★★

And so, the urgency of "peak oil," and especially climate change, is ushering in a new scenario. The end of "the fossil fuels era" may be postponed, but it cannot be prevented. In all probability it cannot even be postponed much longer. A transition beyond petrol is not a question of ideological choice, but is increasingly becoming an imperative imposed by material constraints. Some kind of transition has become virtually compulsory and inevitable.

Changes within the energy sector are speeding up dramatically. A combination of ecological, political, economic, and financial factors are converging to ensure that energy production and consumption are set to become central to global political, economic, and financial dynamics. This is true of energy, in general, and of the globally-expanding renewable energy sector, in particular. The way that the world's energy system evolves in the years ahead will be intimately intertwined with different possible ways out of the world financial-economic crisis (which is also increasingly becoming a political crisis).

The crisis now wreaking havoc with the world-economy is resulting in the reduction of energy demand and emissions—at least temporarily. It has also resulted in slashed investments in the energy sector, both fossil fuels and renewables. This could, in fact, mean that a drop in the energy supply will catch up with the drop in demand. In other words, the economic crisis may well be accompanied by an energy supply crisis, meaning that the path back to economic growth that most governments seek may well be made impossible by a scarcity of supply of the fossil fuels necessary to make it happen.

The kind of massive and rapid reductions in CO_2 emissions required (and the corresponding changes in energy production and consumption that are necessary for this to occur) will not be possible without extensive changes in production and consumption relations at a more general level, involving fundamental change in how humans interact with nature.

The process of building a new energy system, based around a greatly expanded use of renewable energies, has the potential to make an important contribution to the construction of new relations of production, exchange, and livelihood that are based on solidarity, diversity, and autonomy, and are substantially more democratic and egalitarian than the current relations. Furthermore, the construction of such relations are likely to be necessary in order to avoid disastrous "solutions" to the multiple intersecting economic-financial and political crises.

The stark reality is that the only two recent periods that have seen a major reduction in global CO_2 emissions both occurred in periods of very sudden, rapid, socially disruptive, and painful periods of forced economic *degrowth*—namely the breakdown of the Soviet bloc and the current financial-economic crisis. Strikingly, in May 2009, the International Energy Agency reported that, for the first time since 1945, global demand for electricity was expected to fall.

Experience has shown that a lot of time and political energy have been virtually wasted on developing a highly-ineffective regulatory framework to tackle climate change. Years of COPs and MOPs—the international basis for regulatory efforts—have simply proven to be hot air. And, not surprisingly, hot air has resulted in global warming. Only *unintended* degrowth has had the effect that years of intentional regulations sought to achieve. Yet, the dominant approaches to climate change continue to focus on promoting regulatory reforms, rather than on more fundamental changes in social relations. This is true for governments, multilateral institutions, and also large sectors of so-called "civil society," especially the major national and international trade unions and their federations, and NGOs. And despite the patent inadequacy of this approach, regulatory efforts will certainly continue to be pursued. Furthermore, they may well contribute to shoring up legitimacy, at least in the short term, and in certain predominantly-northern countries where the effects of climate changes are less immediately visible and impact on people's lives less directly. Nonetheless, it is becoming increasingly clear that solutions will not be found at this level.

The problem has to do with production, not regulation. The current worldwide system of production is based on endless growth and expansion, which is simply incompatible with a long term reduction in emissions and energy consumption. Despite the fact that localized and punctual moments of reduction may well still occur, the overall energy consumption and emissions of the system as a whole can only increase. All the energy-efficient technologies in the world, though undoubtedly crucial to any long term solution, cannot, *on their own*, square the circle by reducing the total emissions of a system whose survival is based on continual expansion. This is not to say that developing appropriate regulation is not important—it is completely essential. However, the regulatory process is very unlikely to be the driving force behind the changes, but rather a necessary facilitation process that enables wider changes. Furthermore, regulation that is strong enough to be effective is only likely to come about once wider changes in production are already underway.

Energy generation and distribution plays a key role in shaping human relations. Every form of energy implies a particular organization of work and division of labor (both in general, and within the energy sector, in particular). The most significant social, economic, cultural, political, and technological transformations in history were associated with shifts in energy generation: from hunting and gathering to agriculture, from human and animal power for transport and production to wind and the steam engine, from coal to oil and nuclear fission as drivers of industry and war. All these transformations have led to increased concentration of power and wealth. And a very real possibility exists that the coming transformation in the world's energy system will result in similar shifts in power relations.

The combination of world economic crisis and the twin energy/climate crises have the potential to substantially increase the already brutal inequalities that exist today, hitting the world's most vulnerable people hardest. This will almost certainly produce economic and environmental refugees on an unprecedented scale. Some of these people will be able to migrate into the global centers where the planet's plundered wealth is accumulated, and will be exploited as cheap labor and used as scapegoats by racist politicians and societies. Most of them will have an even worse future. Already the buzz phrase "climate change" is being shouted to all corners of the wind as a justification for coercive policies that limit freedom of movement and association. And "peak oil" and "rising energy costs" are rapidly becoming an excuse for imposing austerity on both waged and unwaged workers and their communities throughout the world. In the energy sector itself, extraction efforts are being intensified on the backs of the several million workers in the existing, mainly fossil-fuel-based energy sector, as well as on populations that live in the vicinity of these fuel sources. Meanwhile, oil companies have been reaping record profits as a direct result of rising prices.

But we live in interesting times. The ecological and social carrying-capacity of our planet and existing social relations are overstretched, snapping in different places. This will trigger a major change in the next few decades, but nobody knows in which direction. Consequently, the most important single factor determining the outcome of this change will be the intensity, sophistication, and creativity of grass-roots social mobilization.

Although we are clearly only in the very early stages of these processes, it is already becoming increasing clear that people are not passively sitting back and allowing such scenarios to play out. The first half of 2008 saw fuel (and closely associated food) protests and riots spreading rapidly, in more than thirty countries throughout the world. These spontaneous social upheavals brought both urban and rural populations, and waged and unwaged workers into a process of common struggle. People everywhere, relying on energy to meet their basic subsistence needs, are beginning to question the "inevitability" of rising prices, insisting loudly and clearly that they should not be the ones to pay these rising costs. Struggling for cheap (or even free) and easy access to energy, they are claiming it as a human right—not a privilege.

Faced with the urgent task of collectively moving towards an equitable and ecologically-sensitive energy system as part of a wider process of collectively finding an emancipatory way out of the economic-financial crisis, we cannot afford to wait for the breakdown of the existing order in the hope that it will bring a happier future. On the one hand, there is the need for a far greater proportion of energy to be obtained from renewable energy sources than is currently the case. And on the other hand, we must develop new ways of cooperatively organizing our relations of production and consumption that do not require huge and ever increasing amounts of energy.

The idea that a massive introduction of "clean energy" or "renewable energy" on its own is enough to solve the problems at hand maintains the illusion that it will be possible to sustain current levels of energy consumption, levels that continue to expand unstoppably. Similarly, efforts centered around "energy efficiency" suggest that the solutions are technical, when in fact the question of necessary levels of energy demand is highly political. Rather than being inevitable, they depend on the way in which we collectively choose to organize ourselves.

ENERGY CRISIS AND TRANSITION: AN OPEN AND UNCERTAIN PROCESS

Today's energy system is an exceedingly complex process. It is tempting to reduce energy, and thus the energy crisis, to a single technical issue, however, technology alone is not going to solve the crisis, since what we are facing is an unprecedented political, economic, and social crisis, rather than a technological one. The terms "the energy sector" and "the energy system," though used throughout this book, are really very murky. The notion of a single, homogenous sector or system attempts to lump together many people and different interests in one boat. As such, an over-simplistic use of the terms risks masking structurally-existing material hierarchies and conflicts of interest.

There are no easy answers, and, alas, in case you were expecting an easy ride, this book is not a book of sound bites. Instead, the book seeks to unpack the seemingly innocent terms "energy sector" and "energy system." It does so by situating the current "energy crisis," "peak oil," and the "transition" to a post-petrol future within a historical understanding of the global, social, economic, political, financial, military, and ecological relations that energy and technology are parts of. It aims to probe the systemic relationships between energy production and consumption and the worldwide division of labor on which capitalism itself is based—its conflicts and hierarchies, its crisis and class struggle. A class analysis of energy helps to situate the contemporary evolution of the energy sector in general, and the expanding renewable energy sector in particular, within wider systemic dynamics. With this analysis, the book seeks to contribute towards anticipating and strategizing future scenarios in order to assess current options for collective action.

Today's energy patterns are the cumulative product of hundreds of years of historical development. The energy system is the outcome of many different social relationships through which human beings organize themselves in order to live, sustain, and reproduce themselves over time. The energy system is intimately intertwined with the expansion of the social economic and political relations of which it is a part. Crucially, it is not defined by individual nation states, even the "most important ones," but is a worldwide energy system, existing within the context of worldwide social relations and the worldwide division of labor that these relations are based upon.

Energy has twin and contradictory functions that exist simultaneously. On the one hand, energy is a highly profitable commodity for production and exchange in the world-market and an essential raw material in the production and circulation of other such commodities. And, on the other, it is fundamental to human life and subsistence. As such, energy is an important site of ongoing conflict and struggle, with one major aspect of these struggles being the ongoing tension between energy as a commodity for profitable sale and energy as a non-commercial means of subsistence. Struggles for control of energy (broadly along the lines of interstate, inter-firm and inter- (and intra-) class struggles) have had a crucial impact on the historical development of capitalism as a global set of social relations.

With the world's energy system on the verge of far-reaching change, it comes up for grabs; the struggle for who controls the sector, and for what purposes, is intensifying. It is becoming increasingly clear, to capitalist planners and those in anti-capitalist struggles alike, that some form of "green capitalism" is on the agenda. We are told from all sides that it is finally time to "save the planet" *in order* to "save the economy." However, what we are not told, with a deafening silence, is that, given energy's key role, this means that the transition process to a new energy system is, in effect, the next round of global class struggle over control of key means of production and subsistence.

Class struggle is inherently uncertain, however, and this is the main uncertainty of the transition process. Who will bring the transition about and for what purpose? Who will benefit, and at whose expense? Given that energy is relevant to class relations in general (since energy both replaces and enhances human labor), energy "crisis" and "transition" are also relevant to class struggles in general and not just those that exist within the energy sector itself.

It will take many years before it is clear whether capital can harness new combinations of energy that are capable of imposing and maintaining a certain stable (and profitable) organization of work in the way that fossil fuels have allowed, or whether we will find that a new energy system will not allow such possibilities, and perhaps even strengthens the material basis for anticapitalist struggles. We are in the early stages of what is likely to be a lengthy and complex struggle to determine whether capital will be successful in its efforts to force labor (i.e. people throughout the world, as well as the very environment itself which green capitalism proclaims to "save") to bear the costs of building a new energy system, or whether labor (i.e. social and ecological struggles throughout the world) is able to force capital to bear these costs. This struggle is already becoming central in shaping social relationships, and is likely to become ever more so in the coming years.

ENERGY AND CAPITALISM IN WORLD HISTORY

A discussion of energy cannot be separated from a discussion of capitalism, crisis and class struggle. Furthermore, the question of energy is also crucial to anti-capitalist resistance and the construction of non-capitalist alternatives.

Conflicts related to energy are becoming central in this process of global restructuring. The transition to a post-petrol energy system which is predominantly based on renewable energy must be understood in this context. For close to a century, the advent of coal, and later oil, meant that the widespread commercial use of renewable energy was largely abandoned, though it has always retained its non-commercial role and a small commercial role. However, the sector has been reactivated since the energy crises of the 1970s.

When considering the question of whether renewable energy might offer new possibilities for emancipation, or whether it will contribute to maintaining and strengthening existing forms of hierarchy and domination, it is crucial that we never lose sight of one simple fact above all others. *Capitalist relations arose during the era of renewable energies and their associated technologies.* Wind-powered sailboats conquered the world, windmills ground sugar cane on slave plantations, and land was drained by wind- and water-powered pumps. This was the energy basis of the Italian city states; British, French, Spanish, and Portuguese naval empires; and Dutch hegemony (Dutch hegemony also relied extensively on peat).

It was only *later* that the use of fossil fuels was to have a tremendous impact on capitalism's expansion. Artificial lighting played a crucial role in lengthening the working day. The coal powered steam engine developed hand-in-hand with the British-led industrial factory-based production system and the railway and steam ships. On the one hand, this enabled an unprecedented increase in the productivity of labor, thus greatly expanding output. And, on the other, it greatly expanded the geographical reach of markets for buying and selling raw materials, finished commodities, and labor. This allowed capitalism to become a truly world-reaching system of social relations.

The twentieth century shift towards petrol (combined with electrification) and the ability to harness atomic energy further intensified these processes. "Cheap" energy became an indispensable pillar of post-World War II economic growth in the United States and US hegemony globally. Increased energy inputs greatly expanded the capacity for transport, agricultural and industrial production. At the same time, further mechanization, automation, and robotization massively increased the productivity of labor, while the ability to provide cheap food, heating, transport, and consumer goods dramatically brought down the costs of reproducing the labor force. These latter factors, automation and lowering the price of reproducing labor, were both key to containing class struggle in the US. Elsewhere, energy-intensive agriculture and the "green revolution" were key to containing rural struggle throughout the world. All of these were essential cornerstones of the post-Second World War-Keynesian and developmentalist social pacts on which US hegemony was based. Above all, the ability to harness atomic energy gave certain states unprecedented military capacities. (As an aside, which cannot be explored in further detail here, it is also worth pointing out that at the same time, increased energy inputs also played a key role in the attempt to construct alternatives to capitalism. Lenin famously dubbed Communism: "Soviet power plus electrification.")

Summarizing, increasing energy inputs have played an important role in at least five key areas effecting worldwide class relations:

1) MECHANIZATION has enabled increased productivity of labor. In the context of capitalist relations means providing the basis for what Marx calls *relative surplus value strategies and wage hierarchy*.

2) ARTIFICIAL LIGHTING has lengthened the working day. In the context of capitalist relations this has provided a material basis for what Marx calls *absolute surplus value strategies*.

3) TRANSPORT has enabled an expanded geographical reach for markets in raw materials, labor and commodities, as well as reducing the circulation time of goods, money, and people etc.

4) COMMUNICATION technologies have made the working day more pervasive.

5) CHEAP FOOD, SHELTER, CLOTHING AND CONSUMER GOODS have lowered the cost of reproducing a planetary workforce, thus buffering reduction in wages, and intensifying differences within the wage hierarchies which exist throughout the world. For example, cheap food has largely been obtained through the agro-business model imposed on the world's farmers. This is a model that has increased food insecurity for many sections of world population who have been dispossessed of their the land to allow the land concentration necessary to the energy intensive agro-business model.

And, while it is true that energy has undeniably contributed to making certain tasks easier, paradoxically, in the midst of all the "labor saving" technology which energy inputs have enabled, no one really does any less work than they did before. The wage relation that shaped the factory has not been done away with, nor have the un-equal gender roles that shape so many households and kitchens been replaced. Rather than doing away with unequal and exploitative patterns of work, energy-intensive appliances, vehicles, machines, food, and materials have simply rearranged people's working patterns and structures. Alas, neither the smoothie maker nor the SUV have managed to abolish work. The diesel engine, originally designed to lighten the work load of poor urban workers, has proven to be the technological invention *par excellence* for decentralizing and expanding capitalist relations throughout the world.

The history of energy use is thus the history of the enhancement of the productive powers of cooperatively-organized human labor, on a global-scale. However, the form in which social cooperation is currently organized, capitalism, is one that reproduces and amplifies social injustice and environmental catastrophe.

TRANSITION AS A MATERIAL AND ORGANIZATIONAL PROCESS, NOT JUST AN ETHICAL ISSUE

Whether for pragmatic or ideological reasons, it is common to downplay the centrality of capitalist social relations and their role in climate change and energy production, trade, and consumption. Consequently, the conflicting nature of the transition process towards a new energy system is also downplayed.

An important result of all this is the widely-held belief that capital does not need to be expansive or at least that it doesn't have to be based on ever-expanding energy consumption. The liberal capitalists' discourse is based on a value judgment that says that continuous capitalist growth is desirable. That judgment is then naturalized, and becomes a tacit assumption that then forms the basis of pragmatic solutions to the material requirements of energy production and consumption in a given context of class relations. The closely-related "environmental" approach is based on a strong *ethical* desire for "change," but does not imagine challenging the fundamental value premises of capitalism or the *material* relations behind it.

Neither of these premises, nor the material requirements for their satisfaction, can be wished away for the sake of a pragmatic engagement. States and corporations will do anything in their power to maintain capitalist social relations as *the* fundamental form of reproducing our livelihoods. Furthermore, the experience of capitalist renewable energy regimes of the past stands as a reminder that social relations of production, based on enclosures and exploitation, are not exclusively associated with fossil fuels and nuclear energy. There is nothing automatically emancipatory about renewable energies.

Energy looks set to play a crucial role in the realignment of economic and social planning, following the deepening world financial-economic and, in all probability, a soon-to-follow political crisis. In order to re-launch a new cycle of accumulation, capital must tackle this energy crisis, and the world economic crisis creates a context in which to promote new attacks on the current composition of the waged and unwaged working class, on its forms of organization and resistance. A new wave of structural adjustments, expropriations, enclosures, market and state discipline will

most likely be attempted, together with new and creative forms of capitalist governance of social conflicts.

What is clear is that, when discussing solutions to the energy crisis, economic
liberal ideologues are quite open-minded. Rather than sticking to any one technology
to meet capitalism's ever-increasing energy need, which will never go away as long
as capitalist social relations continue, all possibilities are left open. These options
consist of a combination of oil, so-called "clean coal," natural gas, nuclear energy, and
a whole host of "renewable" technologies. Whether a new post-petrol regime crystal-
lize's in the face of different struggles is of course open—and what kind of regime and
at what pace it might take shape remains to be seen.

What happens will depend on how and to what extent capital is able to successfully restructure planetary relations and weaken and divide the worldwide circulation of struggles. The combination of financial-economic and energy-climate crises
gives capital great possibilities to justify its actions under the twin slogans "save the
planet" and "save the economy." Hence, the planners' coming pragmatism might help
capital to create a common ground with some sections of the environmental movement, a so called "green capitalism." Should this occur, it would, in all probability, be
the ruin of environmental and social justice causes. On the other hand, it might also
help emancipatory struggles throughout the world to further de-legitimize capital's
priorities in the management of these crises, especially if movements are able to recompose themselves across the global wage hierarchy and establish links furthering
models of social cooperation and production based on pursuits of values that are
alternative to capital's.

GLOBAL EVENTS IN THE WORLD OF ENERGY

Against this backdrop of world economic crisis, the "timeliness" of the issue can also
be seen in three separate institutional processes—each extremely important—that
are currently taking shape in relation to energy and climate change. The institutions
of the world-economy are already recognizing this new situation. In addition to the
recent Copenhagen debacle, the "timeliness" of these issues can be seen in terms of
two other important global institutional developments in the energy sector. In 2008,
the International Energy Agency World Energy Outlook anticipated an oil supply
crisis as soon as 2010 and called for an "Energy Revolution". This date is now already
upon us. And, in January last year, 75 countries from around the world met to establish the International Renewable Energy Agency, IRENA. The agency's membership
has expanded rapidly, and now boasts 143 countries.

In November 2008, the International Energy Agency (IEA), the energy watchdog
of all the oil-addicted western OECD governments, published its most noteworthy
report to date. Its now-yearly report, the 800 page "World Energy Outlook," seeks
to give a picture of the major issues the energy sector is facing, and to project what
would occur if existing energy policies were to remain unchanged until 2030. Many,
especially within the political and financial establishment, view the "World Energy

Outlook" (WEO) as a kind of "energy bible." Its results are seen as absolute truth, and its recommendations form the basis of all western energy policy.

In the Face of the Coming Energy Crisis, the International Energy Agency Calls for an "Energy Revolution"[2]

Until 2007 the WEO painted a picture of ever growing energy demand which would be met by correspondingly ever growing energy supplies. With today's energy mix, this means fossil fuels providing 80 percent of energy and 10 percent from nuclear. In 2007, the IEA issued a mild warning about the possibility that, in the near future, supply would no longer be capable of meeting demand. In 2008 it delivered the numbers. Surprisingly enough, until 2008, the IEA had never really carried out research on the supply side of oil, gas and coal. It always calculated demand and assumed supplies would automatically follow. After coming under increasing criticism for this by peak oil advocates, as well as some within the oil industry itself, the IEA undertook a major study into the ability of the world's 800 biggest oilfields to deliver. The results shocked many in the IEA. The average decline rate in these fields was not the moderate 3.7 percent the IEA had reported in 2006, but somewhere between 6.7 percent and 8.4 percent.

The projected 116 million barrels a day of oil production in 2030 which had been reported in 2006 were cut back to 106 million barrels a day in the new report. In November 2008, the credit crisis, long in the brewing, dramatically accelerated and intensified. The report's figures for energy demand was still showing an increasing demand. This would mean that by 2010 we would face a severe energy crisis and high energy prices for the foreseeable future. Furthermore, the report investigates the implications for the other side of the energy crisis, climate change, which would result from continuing to follow existing energy policies. Its answer is simple. Disaster. If the way we use energy is left unchanged, civilization will be swept away by 6 degrees of global warming. Thus, the report concludes that it is impossible to maintain today's energy course. In the press release announcing the report, the IEA predicted an 'energy crunch in 2010' and called for an 'energy revolution'. Essentially, the report is demanding that the old way of doing energy politics, as exemplified by its previous reports, must be scrapped and that governments should undertake a drastic change of course on energy in the coming years that involves breaking away from oil, gas and coal.

At the same time, the picture painted in the report nonetheless remains highly optimistic. The alternative scenario presented by the IEA for stabilizing the level of carbon in the atmosphere at 450ppm is seen by most climate scientists as a complete denial of the latest scientific findings that point to anything above 350ppm as being dangerous. Despite the strong language, the agency still underestimates the potentials of renewable energy, overestimates oil resources and advocates a strong presence of nuclear energy in the future energy make up. The report still denies the hard facts that oil production will start to crumble in the coming years because of underground, geological reasons. And, above all, it still calls for more investments in fossil fuel production and the opening up to the market of those countries that want to control their own energy resources. The economic crisis wreaking is cutting in to energy demand and emissions. Yet, it is possible that the path back to economic growth that most governments seek will be made impossible by a lack of the fossil fuels to make it happen. At the same time, this makes the conclusions of the report even more alarming. The economic crisis has also slashed investments in the energy sector, both fossil and renewables. This goes in the opposite direction of the IEA's call for bigger investments to be made in order to meet demand in the coming years. This could mean that a drop in the supply for

---

2   This section was written by Peter Polder, for which the author is greateful.

energy will catch up with the drop in demand. In other words, the economic crisis will
be compounded by an energy crisis.

Another important institutional development is at the level of "alternatives." After many years of preparation from grassroots renewable energy organizations, the
German government hosted the founding conference of the International Renewable Energy Agency (IRENA) in January 2009, in Bonn. IRENA is undoubtedly the
most progressive and far-reaching item on the international agenda of governments
and policy makers in relation to renewable energy, at least in terms of its original
conception; it now counts on more than 130 member states. However, the fact that it
is being established as a multilateral institution within the context of both capitalist
social relations and the nation-state-based system, as well as existing power relations
within the energy sector itself (in which the large fossil fuel and nuclear companies
dominate) raises important questions for grassroots struggles.

The following text gives the objectives of IRENA as well as the latest news of its
development, taken from the IRENA website.[3]

> *Many states already foster the production and use of renewable energy through*
> *different approaches on a political and economic level as they recognize the urgent*
> *need to change the current energy path. The current use of renewable energy, however,*
> *is still limited in spite of its vast potential—the obstacles are manifold...This is where*
> *IRENA—the International Renewable Energy Agency—comes in. Mandated by*
> *governments worldwide, IRENA aims at becoming the main driving force in promoting*
> *a rapid transition towards the widespread and sustainable use of renewable energy on*
> *a global scale.*
>
> *Acting as the global voice for renewable energies, IRENA will provide practical advice*
> *and support for both industrialised and developing countries, help them improve their*
> *regulatory frameworks and build capacity. The agency will facilitate access to all racing*
> *to "save" the economy and the planet relevant information including reliable data on*
> *the potential of renewable energy, best practices, effective financial mechanisms and*
> *state-of-the-art technological expertise. The International Renewable Energy Agency*
> *(IRENA) was officially established in Bonn on 26 January 2009. To Date 143 states and*
> *the European Union signed the Statute of the Agency; amongst them are 48 African, 37*
> *European, 34 Asian, 15 American and 9 Australia/Oceania States.*
>
> *IRENA's Preparatory Commission consists of IRENA's Signatory States and acts as*
> *the interim body during the founding period. The Commission will be dissolved after*
> *entry into force of the Statute, whichwill occur upon the 25th deposit of an instrument of*
> *ratification. The Agency will then consist of an Assembly, a Council, and a Secretariat.*
> *The Agency's interim headquarters are in Abu Dhabi, the capital of the United Arab*
> *Emirates. Bonn will host IRENA's centre of innovation and technology and Vienna will*
> *become the Agency's liaison office for cooperation with other organisations active in*
> *the field ofrenewable energy. Ms. Pelosse, from France, has been appointed as the first*
> *Interim Director-General of IRENA.*

And, the third important event is at the level of grassroots resistance to institutional "solutions." The UN COP 15 Climate summit took place in Copenhagen at the
end of 2009, its aim was to produce the protocol that will replace the Kyoto protocol.

---

3   See http://www.irena.org

The "Age of Climate Change Denial," with George W. as its chief global spokesman is over. Now, we hear a mantra shouted loudly, from all corners of the planet. It is time to "pull together to 'save the planet'". Indeed, one of the chief spokesmen of this rallying call is Bush's successor, the ever-so well spoken and intelligent President Obama (who, despite being renowned for being highly articulate, is, nonetheless, still a US president...as was revealed in no uncertain terms in Copenhagen).

A first international preparation meeting for grassroots mobilization was held in Copenhagen in September 2008, and an initial call to action was issued and translated into many languages. The original call is copied below.

A Call to Climate Action

We stand at a crossroads. The facts are clear. Global climate change, caused by human activities, is happening, threatening the lives and livelihoods of billions of people and the existence of millions of species. Social movements, environmental groups, and scientists from all over the world are calling for urgent and radical action on climate change.

On the 30th of November, 2009 the governments of the world will come to Copenhagen for the fifteenth UN Climate Conference (COP-15). This will be the biggest summit on climate change ever to have taken place. Yet, previous meetings have produced nothing more than business as usual.

There are alternatives to the current course that is emphasizing false solutions such as market-based approaches and agrofuels. If we put humanity before profit and solidarity above competition we can live amazing lives without destroying our planet. We need to leave fossil fuels in the ground. Instead we must invest in community-controlled renewable energy. We must stop over-production for over-consumption. All should have equal access to the global commons through community control and sovereignty over energy, forests, land and water. And of course we must acknowledge the historical responsibility of the global elite and rich Global North for causing this crisis. Equity between North and South is essential.

Climate change is already impacting people, particularly women, indigenous and forest-dependent peoples, small farmers, marginalized communities and impoverished neighbourhoods who are also calling for action on climate and social justice. This call was taken up by activists and organizations from 21 countries that came together in Copenhagen over the weekend of 13–14 September, 2008 to begin discussions for a mobilization in Copenhagen during the UN's 2009 climate conference.

The 30th of November, 2009 is also the tenth anniversary of the World Trade Organization (WTO) shutdown in Seattle, which shows the power of globally coordinated social movements.

We call on all peoples around the planet to mobilize and take action against the root causes of climate change and the key agents responsible both in Copenhagen and around the world. This mobilization begins now, until the COP-15 summit, and beyond. The mobilizations in Copenhagen and around the world are still in the planning stages. We have time to collectively decide what these mobilizations will look like, and to begin to visualize what our future can be. Get involved!

We encourage everyone to start mobilizing today in your own neighbourhoods and communities. It is time to take the power back. The power is in our hands. Hope is not just a feeling, it is also about taking action.

If there is one thing that the Copenhagen spectacle revealed with great clarity, it is that existing political institutions are completely unwilling to undertake the required changes on the scale and within the time frame necessary to solve the climate-energy crisis. Furthermore, the partially "green tinged solutions" that they are proposing are rapidly being dismissed by movements around the world as "false green capitalist solutions." Those few national governments that are in fact willing to push a more emancipatory vision of change are not capable of doing so, while those that are capable are not willing.

The failure of the Copenhagen talks, and the grassroots resistance that surrounded them, give explicit visibility to the structural conflicts at the heart of the climate-energy crisis, themselves part of a wider crisis of social relations. These conflicts, tensions and contradictions have been brewing for many years (there were international grassroots mobilizations around the COP process as early as 2000 in The Hague, growing much larger in Bali, 2007). In Copenhagen, they exploded into the open.

The conflicts exist, and cannot be wished away. Above all, Copenhagen shows the deceptiveness of the rhetoric that "we are all in the same boat and must pull together to solve the climate crisis." This is little more than a thinly veiled way of exhorting people throughout the world to pull together to shoulder the burden of a capitalist transition to a new energy system. A moment of structural conflict is not a moment for remaining neutral, but rather for making informed decisions and commitments about with whom to align and on what basis, in order to prepare for the long term and highly uncertain process of collective struggle that almost certainly lies ahead. The call by the Bolivian government for an alternative international climate conference in Cochabamba in April this year, and predominantly aimed at social movements, as well as more progressive governments, is an important development in this respect.

NAVIGATING THE CONFLICTS AHEAD: MAPPING THE WORLDWIDE ENERGY SECTOR IN ORDER TO OVERCOME DIVISIONS AND CREATE COMMONALITIES OF STRUGGLE

The challenge is to develop methods of collectively organizing that enable us to come through the current crisis in a way that puts an end to the system of organizing social life and production that is at the basis of both ecological disaster and social injustice. This raises the political question of how struggles can find ways of collectively organizing and acting together that do not pit one struggle against another, but instead give rise to a social force that is simultaneously able to set limits on capital and also create alternatives. This political recomposition is becoming increasingly urgent as the challenges posed by the socio-economic-environmental catastrophe are becoming ever more pressing. There is an urgent need to take informed decisions about with whom to align and on what basis.

Many different struggles related to energy already exist throughout the world, each with their different organizational forms and particular networks, though they frequently lack familiarity with one another and are working in isolation. In some instances, different struggles may even perceive each other with a certain degree of suspicion and distrust, or, worse still, as opponents to be fought against.

Of central importance is the need to create a common ground among people in struggle across the potentially dividing and contradictory lines of the issues of energy and climate change. It is vital that movements in the energy sector are able to develop a worldwide dialogue, common analyses, political perspectives, and long term collaboration processes. In particular, it will be necessary to find ways of building a long term process of overcoming and avoiding three important lines of hierarchy and division that already exist and have the potential to get much worse as the energy system undergoes changes in the coming years. These are: the relation between rural and urban communities and workers; the relation between workers in the "dirty" and "clean" energy sectors; and the relation between communities and workers in energy-producing regions and energy-consuming ones.

In particular, the choice of which technologies will play an important part in the energy system of the future is proving to be an incredibly conflictual issue. Another important issue here are diverging strategic choices and perspectives as to the best way of bringing about social and technological change, and the extent to which this can take place within existing power structures, or whether it requires a more confrontational approach towards these power structures and the construction of new social relations.

A clear example of opposing goals can be seen in the fact that many environmentalists are outright opposed to coal and nuclear energy, whereas worker organizations in these sectors are predominantly in favor of worker-led efforts at clean up. Away from the question of technology choice, important differences in strategies of how to relate to power can be seen in a number of areas. For instance, the dominant approach within many organizations in the renewable energy field is focused on lobbying global or multilateral institutions, such as the World Bank, International Monetary Fund, G-8 (and now also the G-20), European Union, or national governments. Similarly, the dominant strategy of workers' trade unions and other organizations, as well as the International Labour Organization, is to secure reforms within the context of a tripartite framework between capital, labor, and nation states (though strikes, occupations, and other forms of direct action still play an important role), and protect waged labor as the principle form of making a living. On the other hand, many in anti-capitalist struggles, including many rural and indigenous struggles, may use tactics that are more rooted in direct action, and seek to protect and promote non-wage-based livelihoods.

Another issue of particular importance in this regard is the fact that some of the most visible struggles today are about the ownership and control of hydrocarbon resources, not renewable energies themselves. The last decade has been characterized by intensive struggles in the existing petrol-based energy regime, such as in Bolivia, Venezuela and Iraq, as well as in Nigeria, Ecuador, and Colombia. Consequently, the sector has become increasingly difficult for neoliberal capital to control. This has major implications for wider global class relations and hierarchies in the existing division of labor, in terms of the relation between oil-producing and oil-consuming workers (waged and unwaged), and presents a serious threat to capitalism itself.

It goes without saying that hydrocarbon production, when inserted in capital's circuits, must follow the profit logic of capital and has very few other options. To shift away from boundless extraction of those fossil fuels requires a collective global process. Consequently, it does not make sense to blame people who happen to live in an area that has an abundance of hydrocarbons, since this is tantamount to a head-on attack on those people whose livelihoods and survival currently depend on these fuels. Rather, it is likely that some form of collective ownership of, and democratic and participatory decision-making process over these resources at a local or national level, offers a strong basis from which to contribute to the collective global process of a planned shift away from them.

Crucially, fossil fuel resources are geographically specific to only a few locations in the world. This means struggles in these areas are becoming increasingly strategic, whether they are interstate, inter-firm, or capital-labor struggles, and are likely to produce sharp local conflicts in the coming years. A collective and emancipatory transition process will not be possible if it is based on empty slogans. It is very likely that the next phase of emancipatory global struggles will be strongly rooted (though by no means exclusively) in the regions where there is a struggle over fossil fuel energy resources. It will be important that global networks of resistance are able to make themselves relevant and broad enough to include these struggles, where they are not already included.

However, the struggle over the ownership, control, and use of hydrocarbons (a major revenue source for social programs, land distribution, and grassroots community empowerment) is largely absent in current discussions between advocates of renewable energy and many of the more mainstream organizations that are active around climate change, including the different organizations mobilizing around the Copenhagen COP summit. Yet these struggles are fundamental means to generate and distribute wealth in those countries despite the fact that the use of these fuels undeniably contributes to carbon emission and climate change. The articulation between these struggles, the aspirations they posit, and the general issue of climate change and renewable energy is a problem that urgently needs to be tackled. Similarly, the comparative absence of movements from many of the oil and coal rich areas of the world (especially the Middle East, Caucasus, and China) within global anti-capitalist networks is a big obstacle that urgently needs overcoming.

PURPOSE, STRUCTURE, AND CONTENTS OF THIS BOOK

As the many chapters in this book show, a wide range of social struggles are emerging in relation to energy. An understanding of these struggles is important in order to assess both short term priorities for collective action, as well as longer term strategic orientation within struggles that may take several years to bear fruit, if indeed they ever do. The book aims to pose strategic questions as to how to open up spaces that can bring about and mobilize the kind of mass social and political force that is necessary for an accelerated transition to a decentralized, equitable, and ecologically-sensitive energy system, which contributes to a wider process of building emancipatory relations. In particular, an important aim of this book is to highlight the importance of ownership, labor, land, and livelihood in relation to a discussion of energy resources, their infrastructures, and technologies. The different chapters point to the fact that in order to get to the root of the problems, struggles in the North and South have to develop a collective global process to take decisions concerning energy.

Above all, the aim of this book is to contribute to a process of ensuring that any future transition to a new energy system is part of a wider movement to construct non-capitalist relations that are substantially more egalitarian, decentralized, and participatory than the current relations. It strives to offer long term perspectives in order to discern where axes of conflict and rupture lie, as well as where possibilities for common struggle in the short term might exist. In addition to the crucial question of which energy sources and technologies are the most suitable, there is also the question of how energy is used (or not used), in what quantities, and for what purposes.

If we make these decisions through capitalist markets, we end up stressed out, overworked, and murdered, divided and pitted against one another, while the planet goes to hell. If we make these decisions through the capitalist state, we end up repressed, silenced, and manipulated into believing that the sacrifices that are required of us to deal with this "emergency" and "crisis" are worth the suffering, since it will be the final crisis, and there will never be another "crisis" again, while in fact it will merely open up a new cycle of more of the same.

The book seeks to contribute to an appreciation of the open and political nature of the "energy crisis" and its "solutions," and to question the idea of "transition" as something fixed and predetermined. While technology is, and will surely continue to be, of great importance, the process of building an emancipatory post-petrol energy system will not be the inevitable result of technological fate. If such a system is to emerge, it will largely be the result of collective human activity and choices, intentional or otherwise. There is no single "transition" process waiting to unfold that already exists in the abstract. Multiple possible transition processes exist, and the actual outcome will be determined through a long and uncertain struggle. These struggles are already rapidly taking shape, and in all probability we are only in the very early phases of this process. This book seeks to help orientate people within these emerging conflicts so that they can actively anticipate, prepare for, and sharpen these struggles.

Many different actors and voices play their part in the energy sector, and the sector is criss-crossed by multiple conflicts and alliances. This book seeks to create a space where different voices from around the world, who come from different areas the energy sector, can share information and listen to one another. In doing so, the aim is to contribute towards the building of a critical common analysis, or rather map, of the current worldwide "energy crisis." It is hoped that this can help strengthen people's ability to act collectively in order to intentionally shape future developments in the energy sector in ways that contribute to a rapid and

smooth transition process, in the face of worldwide economic-financial and political crisis.

However, it is hoped that this book will go beyond information exchange and
the development of common analyses. By bringing organizational processes that are
frequently working in isolation into contact with one another, or at least making
them known to each other, it is anticipated that the book may be able to contribute
to concrete organizational processes, both in the short and longer term. As such, it is
intended to be a networking tool that can contribute to building the kind of collective
social force that is capable of bringing about an emancipatory "transition process."

Rather than appealing to politicians and "official decision makers," this book
especially seeks to reach self-organized grassroots organizations with similar ideas
and principles and from all continents, in order to contribute to the emancipatory
potential of renewable energy within the context of wider social change. It is hoped
that the book can make a significant contribution towards already existing networking processes between organizations, and the development of common communication tools to encourage increased exchange and knowledge of each other's work,
foster ongoing links and the creation of longer term collaborative initiatives. For this
reason, to ensure it has a maximum impact possible, *Sparking A Worldwide Energy
Revolution* is being published under a Creative Commons License. Translation into
other languages is encouraged.

It is hoped that this collective work might contribute to strengthening people's
collective capacity for exchange and support between different struggles in defense
of livelihoods, rights, and territories related to the global energy sector. This includes
several aspects: on the one hand, rural communities throughout the world, including
indigenous communities and communities of African descent, who are struggling
against the negative impacts of extraction, processing and transportation of energy
resources and the associated infrastructures. And on the other, workers in the existing energy sectors, as well as energy-intensive industries, and their communities
and dependants who are struggling to protect their livelihoods in the face of the
far-reaching structural changes that have begun and that are likely to intensify in the
years ahead.

Another aim is to encourage people's capacity for exchange and mutual support
of different struggles in defense of common/collective/cooperative or public ownership and control of energy resources, infrastructures, and technologies. This includes
fossil fuel resources and associated infrastructures (such as electricity generation and
distribution), which are being privatized due to bilateral, regional, or multilateral free
trade and investment agreements. And it also includes renewable energy resources,
infrastructures, and technologies, which are coming into the sights of investors. A
big challenge is to develop proposals and interventions collectively that allow these
vital resources to aid in the collective self-reliance of community organizations.

The book also seeks to create a conceptual framework for laying the foundations
for solidary, upward-leveling relationships between workers in different branches
of the energy sector, and the avoidance of downward-leveling competition between

them. A key question resulting from all this is: how can workers in the different areas of the sector avoid being pitted against one another in competition (which would almost certainly result in a downward-leveling relationship)? It will be important that workers across the different branches are able to build a process based in solidarity and mutual support, which aims at upward leveling between them.

This collection also seeks to create a framework for thinking about what kind of long term collaboration and cooperative projects and initiatives in non-commercial renewable energy technology transfer, open source technology research, education, training, and grassroots exchanges might be both useful and possible. This is especially important in relation to three broad social groupings: a) rural communities (communities and communities of African descent) whose territories contain abundant renewable energy resources; b) urban tenants and home owners, who could implement major changes in residential energy production and consumption patterns, c) energy sector workers in the fossil and nuclear industries, as well as workers in energy-intensive industries, whose livelihoods may be directly threatened by a transition to a new energy system.

Finally, the book also seeks to contribute to a long term strategic debate about how, and for what purposes, wealth is produced and distributed in society, and how people's subsistence needs are met, as part of a shift to a new energy system. The key means for generating society's wealth and human subsistence include: land, water, energy, factories, schools, etc. Especially important in this context are energy-intensive industries, such as transport, steel, automobiles, petrochemicals, mining, construction, the export sector in general, etc. The kind of far reaching change in the relations of production and exchange that are necessary for the scale and pace of the required energy shift, are difficult to imagine without these key means of generating wealth and subsistence being under some form of common, collective, participatory, and democratic control that is based around serving human needs rather than the profit needs of the (currently existing) world-market. However, following years of market-led reforms, and immense concentrations of wealth and power, we are very far from this reality. The dominant political strategy for achieving change is now, for the most part, rooted in a discussion of how to achieve minor regulatory reforms (at best including state ownership) rather than a more fundamental shift in control and ownership structure. This is true even in quite progressive and radical circles. Consequently, we urgently need to discuss what kind of short term interventions might help make such a political agenda more realistic to achieve in the near and medium term future.

The book is constructed in four sections, with fifteen parts and sixty chapters. The chapters combine analysis with stories of concrete developments and struggles. It starts by documenting the conflictive nature of the existing, predominantly-fossil-fuel-based energy sector, and then moves on to trace the emerging alliances, conflicts, and hierarchies that are starting to define the globally-expanding renewable energy sector. The final section of the book poses the question of whether a transition to a new energy system will take place within the framework of capitalism or as part of a process to create new social relations that seek to go beyond capitalism.

The book has been carefully structured to be read as a whole, from beginning to end. In this way, it seeks to build a collective map, based on the view as seen from some of the many different players within the sector. The chapters have been ordered in such a way as to trace relationships step-by-step in order to construct, from the bottom up, a view of the energy system. The result is an understanding of the worldwide energy system as a self-organizing, emerging whole that consists of many interrelated parts but which is larger than the sum of any of these individual parts. At the same time, it seeks to show that the future of this system is inherently uncertain and open. The focus of the different chapters moves back-and-forth between particular local dynamics within the energy sector to this wider systemic and global whole. Through this back-and-forth process a clearer understanding of the overall energy system is created, and is actually constructed through the very process of tracing the relations that exist between separate but interdependent parts that shape one another.

For this reason, readers are strongly encouraged to read the book in its entirety, from start to finish, but of course it is also possible to browse the book, as one would with an encyclopedia. Each chapter is a self-contained piece and can be read on its own and in whatever order the reader chooses. However, it is worth bearing in mind that reading it in this way will not give an overall sense of the world's energy system as a whole, so an important goal of the book will be lost.

Contributors include individuals, organizations or institutions, including:

- Those who struggle around the different aspects of climate change and the negative effects of market based "solutions."
- Those defending and promoting common/collective/cooperative or public ownership and democratic participatory control of energy resources, infrastructures and technologies, as well as cheap and easy access to energy, as a basic human right.
- Rural communities resisting the negative social and environmental affects of land-use conflicts due to energy extraction, infrastructure and transportation.
- Workers whose structural location means that they have a key role to play in any shift towards a new energy system, but whose livelihoods are potentially also at great risk from such a transformation.
- People with an expertise in renewable energy, and who are working to promote local, collective, and commonly-owned renewable energy and non-commercial technology transfer.
- Global anticapitalist and anti-war networks, especially in regard to wars over oil and other energy resources.
- Those researching the above themes.

These contributors have been chosen on the basis of their strategic location in regards to the worldwide division of labor, both in general and in the energy sector specifically. Their positioning clearly illustrates how the "local" and "global" dimensions of energy are interrelated and mutually shape each other. The authors who have contributed to the book are not intended to represent all the players in the sector, but rather one particular part of it, namely the one that points towards the possibility of bringing about a transition to a post-petrol future that is also part of a wider process of building emancipatory social relations.

Within the above framework, efforts have been made to ensure both gender and regional balance amongst contributors to the extent possible. Many of the chapters were written especially for the book. Other articles were previously published and, where necessary, have been updated.

A major challenge in putting together a book like this has been how to integrate so many broadly "common," yet nonetheless different perspectives, opinions, and viewpoints into a coherent common whole. In fact, there are remarkably few points of divergence, let alone points of major conflict or tension. Nonetheless, a few important ones do appear, both in terms of style and also in terms of perspectives. At the stylistic level, the bulk of the chapters focus more on social relations, while a smaller number contain technical descriptions and information. At the conceptual level, the texts which deal explicitly with capitalism have slightly different, and not always completely compatible, theoretical foundations through which to understand social change and history. The book does not seek to paper over these differences, but rather to create a space for debate that is broad enough to include these differences, and address them through dialogue. It is part of the process of slowly forming common positions, perspectives, and long-term goals.

Finally, it is important to end this section with a disclaimer. It is worth stating clearly that the views of any one author are not necessarily shared by any other. Each author speaks for him/herself and him/herself alone (either in a personal capacity or an organizational capacity if they have contributed in the name of an organization). Similarly, while the introduction and conclusion seek to tie together the different chapters in the book, and are based on considerable collective discussions with many of the different contributors and others active within the energy sector (for which I am very grateful), they are my sole responsibility, as the book's the editor, and do not necessarily reflect, or even attempt to reflect, the views of all the individual authors who have contributed to the book.

STRUGGLING FOR A TRANSITION BEYOND THE MARKET

This book is not intended to be neutral. Rather, it is intended to equip the reader with certain political perspectives which might be useful for sharpening the struggles ahead. Of crucial importance is the opposition to market based mechanisms and defense of some form of common ownership of society's key resources, sources of wealth production and sustenance. In particular, it is crucial that energy resources, their infrastructures and technologies are owned and controlled in such a way as to ensure that they remain a common good, at the service of human needs rather than private profit. According to the particularities of different local struggles, realities, and political traditions, this may include different forms of worker, community, co-operative, common, public, and in some cases state ownership. These are forms of control and ownership that, despite having important differences between them (especially in their degree of democratic participation), nonetheless share certain important considerations and aspirations.

Linked to this is the demand for access to cheap (or free) and reliable sources of efficient, safe, and clean energy as a fundamental human right, not a privilege or a service. Above all, energy must serve to satisfy human needs rather than exist as a commodity to buy and trade for profit in the world-market or to satisfy the needs of endless accumulation. It is also fundamental that the workers in the energy sector have decent working conditions and pay that allows them a dignified life. Furthermore, all of this is crucial regardless of what energy source is considered, and regardless of whether or not it is a high emitter of CO_2.

The fact that coal and oil are finite resources means that there is a long-term tendency in the direction of their phase-out, regardless of what intentional short-term interventions are carried out or not. Many proponents of renewable energy simply advocate leaving this phase-out process to the market. It is hoped that rising oil and coal prices will make these fuels increasingly less attractive. Efforts are focused on developing a renewable energy sector that is able to compete, rather than directly confronting, suppressing, and ultimately dismantling the coal and oil industries. However, leaving the phase-out of oil and coal to the market has at least three crucial implications.

First, such a phase-out is likely to actually prolong the use of fossil fuels. As long as these energy sources are profitable to extract and to use, they will be. Down to the last remaining drops of oil or lumps of coal. Although resources are finite, they are still relatively abundant. Even those analysts who give the most pessimistic (though realistic) perspectives on resource availability, such as those included in this book, do not predict a complete exhaustion of resources in the very near future. And, from the perspective of climate change, a prolongation of fossil fuel use is the exact opposite of what needs to happen, phase-out must be sped up, not prolonged.

Linked to this, the second consequence of a market-based phase-out of oil and coal will mean that the remaining oil and coal resources are frittered away for immediate profit rather than to build the infrastructure for a transition process. Given that building a new energy system will require massive amounts of energy inputs in a very concentrated period of time, this is a recipe for disaster.

The third important consequence is that leaving the transition process to the market is likely to be increasingly coercive and conflictive if competition is left to determine who controls the last of these resources and for what purposes they are used. This means competition between workers globally, competition between firms, and competition between states. This translates to massive inequalities, hierarchies, and austerity measures being imposed on labor (both in and outside the energy sector); massive bankruptcies of smaller firms and concentration and centralization of capital; and last, but not least, military conflicts between states.

Accepting a market-based phase out of oil and coal is accepting in advance that the rising price of energy and a transition away from coal and oil is paid by labor and not capital, when in actual fact the question of who pays still remains to be determined. The answer will only come through a process of collective global struggle, which occurs along class lines within the world-economy. It is important to correctly identify these lines of struggle at the outset, otherwise it will be a struggle lost before the fight even begins. Collectively planning energy use and fossil fuel phase-out is proving to be an enormously difficult social process, but it is likely to be far less socially regressive if based on cooperation, solidarity, and collectively-defined social needs, rather than if it is based around competition and profit.

On the other hand, as the renewable energy sector expands globally, it is becoming increasingly clear that the only possible basis for an emancipatory transition towards renewable energy is by ensuring that a significant proportion of the sector is held under common or public ownership for non-commercial use. This includes the relevant infrastructures, technologies, and knowledge. It is likely that, as the sector expands, so too will struggles over its ownership. Of particular importance here is the struggle for non-commercial technology transfer against the iron straitjacket of the international patent regimes.

Linked to this is the issue of workers in the emerging renewable energy sector. Predictions have to be made with caution. However, initial indicators suggest that, just as with other energy sources, renewable energy is slowly becoming a site of worker unrest. This is especially true if labor is also understood to include those whose land needs to be accessed for the production of renewable energy. As the sector expands, so too does the struggle over whether capital or labor should bear the costs. Most of the infrastructure for renewable energies (such as wind turbines, solar panels, and fuel stocks) simply do not yet exist on the necessary scale. The longer transition is postponed, the quicker it will have to occur when the existing energy regime loses its viability (either through gradual decline or sudden collapse, or some combination of both, according to location), as it almost certainly will in the very near future. It will be the workers in the new energy sectors who will have to deliver vast amounts of infrastructure at great speed and under great pressure.

The current period shows a system in crisis, characterized by increasing levels of systemic chaos, intensified social struggles (both within and outside of the energy sector), interstate rivalry, and a rapidly declining US hegemony. There are some fundamental similarities to past periods in which far reaching and rapid global energy shifts occurred, and there are good reasons for believing that such a rapid and far-reaching shift in the world's energy system may be possible again. In fact, current dynamics offer incredibly optimistic conditions for accelerating and collectively planning a rapid transition towards a renewable-energy-based regime.

However, such a transition process will not come about through persuasion alone. While ideas and communication are essential, they are not enough. It will be a long and frustrating wait if we are to expect the fossil fuel industries to simply dwindle into irrelevance as they miraculously become self-enlightened as to their destructive aspects. This process cannot merely be a battle of ideas, since a move away from these fuels entails major material conflicts of interests. Efforts do date show that such a process is almost certain not to happen voluntarily through a process of global consensus building.

The next ten years offer a unique window of opportunity. During this period, we are likely to face an acceleration of the system that has been constructed in order to run a crisis economy on the basis of growing inequality, oppression, racism, and war. It is not accidental that concepts such as "clash of civilizations," "permanent war on terror," "migration control" constitute the core of the discourse that has been fed into public consciousness by media corporations and most governments in recent years.

Barack Obama is undoubtedly a more benign, articulate, and generally sane person than our dear departed Mr. Bush, and his slogans far less crude. He will almost certainly offer some important reforms, and his approach to energy and climate, which revolves around the idea of "let's all pull together to save the planet AND the economy," is almost certainly the best possible one *within the neoliberal frame*. Yet, it would be profoundly unwise to lay all our problems—past, present, and future—at the door of an aberrational madman who has now been voted out of power. Obama's policies are caught between a number of different and conflicting interests, making his agenda for change far less radical than it might at first sight appear. Only a fundamental rethinking of the neoliberal model can generate change with the speed and on the scale that is needed to respond to the climate and energy crisis. As long as Obama and the Democrats on Capitol Hill are unwilling to challenge corporate power, and continue to operate within the same paradigm of corporate led-globalization, his policies are almost certain to give rise to inadequate half-measures and the solutions of capital, with all the social and ecological dislocation and brutality that this entails.

We can only avoid this if we take the initiative and build alternatives based on totally different values. The energy and climate crisis, and many other deeply-related crises, cannot be solved unless grassroots movements are able to abolish the current economic, political, and social order and build non-capitalist, egalitarian, and participatory societies. We cannot expect governments to do this. We need to get better at building infrastructure, at creating and multiplying working examples of positive futures. For this we need to organize ourselves substantially better, cooperate closely to expand the existing alternatives, and join our strengths to make our voices heard and inspire many more people into action. We can build our own energy systems—for the common good, not for private profit. We have the tools and the experience, we just need to get better at sharing them and putting them into practice. We also have to get hold of the means. It can be done. It depends on us.

Time is running. The clock is ticking.

Yet, there is much room for optimism.

A FINAL NOTE ON *SPARKING A WORLDWIDE ENERGY REVOLUTION*

There is a danger, especially with a book this size, that includes so much, that readers may feel it is claiming to paint "the complete and definitive" picture of the energy sector. Such a claim would be both arrogant and false. This book, while going to great lengths to be broad, extensive and coherent, in no way claims to paint a complete picture. It would be quite unprofessional not to include a mention of some of the key areas that have not been covered, or have only been included briefly. In some cases this was a deliberate choice, in others it is far from desirable and was the result of lack of space in the book, and my own limited time, knowledge, and contacts. Other topics were originally planned to be included, but the authors who were approached to contribute were unable to and alternative authors could not easily be found. Some authors who had agreed to write chapters were at the last minute not able to write their pieces due to unforeseen circumstances. This included important chapters from India, South Korea, Mali, and the US.

Three important areas were deliberately not included. Crucially, though climate change is an ever-present theme lurking in the background, this is not a book about climate change. It is a book about energy. There is already a lot of critical material on climate change, but far less on the issue of energy. Similarly, a lot of material about interstate competition for control of oil resources already exists, but not on the social dimensions of these conflicts. The third deliberate omission is the media and cultural issues surrounding energy. The book's intention is to trace some of the material processes and human relations on which the energy system is based. Importantly, it seeks to show that a transition to a new energy system requires a material process of building new social relations and not just a shift of ethical and cultural values (though this latter is of course crucially important as well).

The book's contributions cover a wide range of countries. However, there are some important omissions and weaknesses. Certain regions have been covered in greater depth than others, and some have hardly been covered at all. In particular, the coverage of Eastern Europe (including Russia) is non existent, and the coverage of the Middle East is limited to discussions about Iraqi oil. Arguably, the importance of Brazil as an emerging energy power could have warranted more in-depth treatment than it has been given. Another important aspect of the international dimension of energy that has not been explicitly tackled, though is touched upon briefly, is the restructuring of international organizations in the energy sector, such as OPEC or the International Energy Agency.

The social and ecological conflicts relating to certain technologies have been largely neglected, especially in relation to large scale hydro-electric dams. These struggles are extremely important. However, they are already quite well documented and widely known. Less deliberate was the relative omission of discussion on natural gas, solar, and hydrogen technologies due to lack of space, time, knowledge, and contacts. The book has also not attempted to provide an in depth analyses of specific sectors and how they use energy, for instance, industry, agriculture, cities, or transportation. Each of these topics would require a book in its own right. Similarly, a discussion of technology design, users, and ownership of knowledge has not been given the attention it arguably deserves. Similarly, there has been little discussion about the energy and raw materials that go into energy production, especially renewable energy, or on the effects that increasingly difficult conditions of extraction will have on workers in the oil industry itself—both of which are sure to become major issues in the future.

Finally, it is necessary to add a few words about timing. With a book this size, that has chapters written by many different contributors, from many countries, and in several languages, it is almost inevitable that, by the time the book comes out in the shops, there will be some out-datedness of individual chapters. A book of this complexity and scale cannot be produced from one day to the next. It has taken just under three years from when the author and the publisher first discussed the idea until the time that it is going to press, somewhat longer than originally anticipated. Work on the book first began in December 2007. A first draft was submitted in January 2009. A second, and almost final, version was submitted in early September 2009, after which time the copyediting and other preparations for publication occurred. It is going to press in June 2010. The chapters were updated by the different authors for the version submitted in September. However, it has not been possible to keep them more updated than this—there are simply too many chapters, involving too many authors, all of whom are extremely busy. The last years have seen important changes in the global landscape. Obama is no longer the new president of the US, but is now half way through his first term. The economic-financial crisis and people's responses to it has developed that much further, pushing ever more towards a worldwide political crisis. And, the Copenhagen climate change summit and the mobilizations around it have taken place, as has the Cochabamba conference that the Bolivian government called in response to the failed Copenhagen summit. This inevitably means that a fair amount of the surface detail described in individual chapters is outdated. This is far from desirable, but there was no way to avoid this problem. Everyone involved in the production of this book, from the individual chapter authors, to the editor, to the publishers, have done all they could to bring the book out as quickly as possible. The editor's introduction and final chapter have been updated in the weeks immediately before the book went to press, but, with approximately 50 chapters from about 20 countries, it was simply not possible to get the individual chapters updated. Having said this, this is not a major problem, since the book is referring to long term processes of change, and the broad issues that the book deals with, and the questions it raises, will be valid for many years to come. These are structural questions relating to the worldwide division of labor in the energy sector, and go beyond changes relating to particular individuals in power, or specific individual laws. The overall map that the book creates is completely unchanged by the fact that some of the surface detail has changed, and, in fact, many of the changes that have occurred were anticipated by the authors.

Setting the Scene:

This section seeks to pose the need for a transition towards a new energy system being part of a wider process of finding an emancipatory way out of the current economic-financial and, increasingly, political, crisis. The current "energy crisis" is, at least in part, one aspect of a wider crisis of social relations. The connections between how people organize their lives in terms of work, production, and exchange, on the one hand, and how energy is produced and consumed, on the other, are vital parts of this story. In particular, the chapters in this section seek to show the importance of energy production and consumption in relation to wider relations of production, exchange, and consumption in the capitalist world-economy. A two-way process is at work. Energy related conflicts are shaped by the world-economic (and political) contexts within which they are played out and are a part of. However, because of the centrality of energy to capitalist relations, struggles within the energy sector have, in turn, made an important impact in shaping these wider social relations. Importantly, the outcome of the coming period of transition and attempts at resolving the multiple crises is an open process. Nonetheless, while there are no inevitable outcomes, this does not mean that chance will be the deciding factor either. On the contrary, the outcome will be almost entirely shaped, directly and indirectly, by human action and struggle.

Chapter 1

PROMISSORY NOTES
From Crisis to Commons

Midnight Notes Collective and Friends

> "The bullet that pierced Alexis' heart was not a random bullet shot from a cop's gun to the body of an 'indocile' kid. It was the choice of the state to violently impose submission and order to the milieus and movements that resist its decisions. A choice that meant to threaten everybody who wants to resist the new arrangements made by the bosses in work, social security, public health, education, etc."
>
> —Translated from a flyer, "Nothing will ever be the same," written and distributed December 2008 in Greece.

CRISIS: WHAT IT IS, WHAT IT IS NOT

After 500 years of existence, capitalists are once again announcing to us that their system is in crisis. They are urging everyone to make sacrifices to save its life. We are told that if we do not make these sacrifices, we together face the prospect of a mutual shipwreck. Such threats should be taken seriously. Already, in every part of the planet, workers are paying the price of the crisis in retrenchment, mass unemployment, lost pensions, foreclosures, and death.

To make the threats more biting, there are daily reminders that we are in an era when our rights are everywhere under attack and the world's masters will spare no atrocity if the demanded sacrifices are refused. The bombs dropped on the defenseless population of Gaza have been exemplary in this regard. They fall on all of us, as they lower the bar of what is held to be a legitimate response in the face of resistance. They amplify, thousand-fold, the murderous intent behind the Athenian policeman's fatal bullet fired into the body of Alexis Grigoropoulos in early December of 2008 (described in the epigraph above).

On all sides there is a sense that we are living in apocalyptic times. How did this "end-of-times" crisis develop, and what does it signify for anti-capitalist/social justice movements seeking to understand possible paths out of capitalism? This pamphlet is a contribution to the debate that is growing ever more intense as the crisis deepens and the revolutionary possibilities of our time open up.[1] We write it in an attempt to penetrate the smokescreen now surrounding this crisis that makes it very difficult to devise responses and to anticipate the next moves capital will make. All too often, even within the Left, explanations of the crisis take us to the rarified stratosphere of financial circuits and dealings, or the tangled, intricate knots of hedge-funds/derivatives operations—that is, they take us to a world that is incomprehensible to most of

---

1   This issue is a slightly shortened version of a pamphlet written by members and friends of Midnight Notes. Originally published under Creative Commons License in 2009, the original, full version is available at http://www.midnightnotes.org/.

us, detached from any struggles people are making, so that it becomes impossible to even conceptualize any forms of resistance to it.

Our pamphlet has a different story to tell about the crisis because it starts with the struggles billions have made across the planet against capital's exploitation and its environmental degradation of their lives.

Crises in the twenty-first century cannot be looked at with the eyes of the nineteenth, which did not see class struggles as an important source of crises, but rather considered them to be automatic, inevitable products of the business cycle caused by the capitalist "anarchy of production." An intervening century of revolutions, reforms, and world wars has led to a revised view. First, a distinction between a real epochal crisis and a recession was recognized. The latter is a state of "disequilibrium" (i.e. part of the normal dynamic of the "ordinary run of things" periodically meant to discipline the working class). The former is an existential condition that puts the "social stability" and even the survival of the system into question. A second revision was the recognition that recessions and crises are not totally out of human control; they can be strategically provoked, precipitated, deferred, and deepened.

---

promissory |'pramə'sorē| adjective, chiefly *Law*
conveying or implying a promise: statements that are promissory in nature: promissory words.
archaic indicative of something to come; full of promise: "the glow of evening is promissory of the splendid days to come."
and:
promissory note, noun
a signed document containing a written promise to pay a stated sum to a specified person or the bearer at a specified date or on demand.

Capitalism's acclaimed automatic tendency to the full-employment of labor, capital, and land has long been disconfirmed by history. By the 1930s, even bourgeois economists saw that it might be necessary in real crises for the government to pull, kick, and stimulate the system when stuck far from full employment. But in devising tools to overcome the crisis of the Great Depression, they also realized that they could plan crises and recessions. Crises can never be eliminated, but they can be hastened and deferred by governmental action. Though dangerous, they can be used as opportunities to deliver coups in class confrontations to keep the system alive. They are the "limit experiences" of capitalism, when the mortality of the system is felt, and it is widely recognized that something essential must change—or else.

The last century has also shown the importance of class struggle in shaping crises, for workers (waged and unwaged, slave and free, rural and urban) have historically been able to precipitate capitalist crises by intensifying the contradictions and imbalances inherent in the system to the breaking point. This capacity makes it possible to understand workers' revolutionary potential: if they cannot put capitalism in crisis, how can they have the power to destroy capitalism in a revolutionary opening?

However, one thing remains true of genuine crises from the nineteenth century until now: they are the occasions of revolutionary ruptures. As Karl Marx insisted in 1848, crises' "periodic return put on trial, each time more threateningly, the existence of the entire bourgeois society." So for him, the approximately five to seven year business cycles end in crises when all of capitalism is put in question.

The word "crisis" gets meaning from its origin in medicine: "a point in the course of disease when the patient either descends to death or returns to health." In this case, the patient is capitalist society. That is why, for Marx and his comrades, the approach of a crisis was closely watched with much excitement, even glee, since it signaled to them the possibility of a revolution. They were confident that the system's ever-deeper crises would soon lead to the sounding of its death knell and the expropriation of the expropriators!

It is with this knowledge, from this perspective, and with a cautious joy that we approach the present crisis. Our discussion is in five sections:

(i) the long-term sources of the crisis;
(ii) its immediate causes and consequences;
(iii) the opportunities it affords to each class;
(iv) the constitution of commoning, i.e., the rules that we use to share the common resources of the planet and humanity; and
(v) the nature of revolutionary struggles arising out of the crisis.

1. CRISES PAST AND CRISIS PRESENT: FROM KEYNESIANISM TO NEOLIBERALISM AND GLOBALIZATION

A comparison is often made between the present crisis and the Great Depression, and, by extension, a capitalist "solution" is often sought after in a replica of the New Deal. However, the profound differences between the Great Depression and the present crisis prevent a return to New Deal policies.

Similarities between the two crises abound, of course. In both crises, the epicenter lay in speculative investments. Both crises can be seen as the results of capitalists' refusal to continue to invest in production in the face of diminishing returns. Most importantly, both crises can be read as products of over-production and under-consumption, resulting in gluts and a fallen rate of profit, all of which combine to freeze new investment and instigate a "credit crunch."

Many left analysts hypothesized that these common trends in capitalist society have led to "over-accumulation" or "stagnation"—in other words, to the inability of capitalists to find investment opportunities in commodity production that would provide an adequate rate of return. The argument is that, in a sense, capitalism was too successful in the 1980s and 1990s: it destroyed US workers' power to such a degree that they no longer struggled for wages high enough to buy the commodities produced, thus causing gluts, over-capacity, under-investment, etc. The emerging Leftist theory of our present crisis emphasizes the commercial failure of the system that led to a profits crisis. This is often called the "realization" problem, i.e., com-modities are over-produced and the working class' demand is restricted (to preserve profits), leading to under-consumption and difficulty investing in manufacturing industries at an acceptable rate of profit. The drive to make profits by attacking work-ers' wages undermines the very condition of profitability, since the commodities produced must be bought to make a profit!

The result, it is argued, is the "financialization" of the economic system, where, because investment in production is no longer profitable enough, more and more capital has been invested into making speculative loans and complex hedging bets. This financialization has benefited from and strengthened the effort to monetarize and marketize all actions within society, from eating dinner to planting seeds in a garden.

Indeed, it was the very objective of the dominant economic strategy of the last thirty years (often called "Neoliberalism") to bring the world-economy back to a pre-New Deal stage of "free market" capitalism—hence the similarities of the two crises. In this sense, today we can also say that capital is paying the price for its calculated disconnect between over-production and under-consumption. Ideally, over-accu-mulation can eventually be corrected by destroying and/or devaluing various forms of capital: unsold commodities, the means of production, and the wages of millions. FDR rejected this path (which had been the advice of the paleo-liberal economists who advised Herbert Hoover), because it seemed that revolution might result from the devastation wreaked by devaluation. Instead, FDR proposed the New Deal.

The New Deal solution—a combination of (1) the institutional integration of the working class through the official recognition of unions, (2) the stipulation of a productivity deal where increased wages would be exchanged for increases in pro-ductivity, and (3) the welfare state—is not in the cards today. The New Deal was struck in the context of an organized, rebellious workforce in the US, empowered by years of marches, by revolts against unemployment and evictions, and by thousands ready to march on Washington with their eyes turned to the Soviet Union.

We are in a very different world now. Although class struggle continues, in no way can today's waged and unwaged workers in the US match the political power and organizational level they achieved in the 1930s. The Keynesian policy (named after the economist and philosopher John Maynard Keynes) that inspired and theoreti-cally justified the New Deal was wiped out by the long cycle of waged and unwaged workers' struggles, which in the 1960s and 1970s attempted to "storm the heavens" and transcend the New Deal. These struggles circulated from the factories through the schools, the kitchens, and bedrooms, as well as the farms of both the metropoles and the colonies, from wildcat strikes, to welfare office sit-ins, to guerrilla wars. They challenged the sexual, racial, and international division of labor with its unequal exchanges and legacy of racism and sexism. In a word, Keynesianism was undone by the working class (waged and unwaged) in the 1970s.

Moreover, it was in response to these very struggles that, by the mid-1970s, capi-tal in turn declared "an end to Keynesianism" of its own and for a short time even adopted a program of "zero growth." This was just the prelude to the deepening of crisis in the early 1980s and to the broad reorganization that went on under the name of "neoliberal globalization" aimed at destroying the victories of the international working class: from the end of colonialism to the welfare state. Therefore, the crisis we are facing today is twice removed from that culminating in the Great Depression. It is problematic to use the 1930s as our guide for the next period, since the political composition of the working class in the US and internationally has changed so radically. It is more useful to consider the plan neoliberal globalization was intended to realize and to evaluate why only three decades later it has led to a new crisis.

Neoliberalism's overall solution to the crisis of Keynesianism was to devalue labor power, reconstitute wage hierarchies, and reduce workers to the status of apolitical commodities (as they were considered in the bourgeois economics of the nineteenth century). Neoliberalism took many forms in response to the different composition and intensity of workers' power: relocation of the means of production, deterritorialization of capital, increasing the competition among workers by expanding the labor market, dissipation of the welfare state, and land expropriation (see MN, 1997). It was a precise (and, at first, successful) attack on the three great "deals" of the post-WWII era, what we in the past (following P.M., 1985) have called the A-deal (the Keynesian productivity deal), B-deal (the socialist deal), and C-deal (the post-colonial deal).

- [A-deal] In the US and the UK, Reagan's defeat of the air traffic controllers' strike in 1981 and Thatcher's defeat of the miners' strike in 1985 were followed by an orgy of union-busting campaigns and continual threats to sabotage social security pensions and other guarantees (the "safety net").

- [B-deal] The ultimate triumph of Neoliberalism was the breakup of the Soviet Union, the collapse of the socialist states of Eastern Europe, and the Chinese Communist Party's decision to embark on the "capitalist road."

- [C-deal] In the "Third World," the debt crisis gave the World Bank and the IMF the ability to impose Structural Adjustment Programs (SAPs) that amounted to a process of recolonization.

In other words, with the arrival of Neoliberalism, all previous deals were off. Together, these developments ended the "mutual recognition" of working class and capital by fomenting worldwide workers' competition through the creation of a true global labor market. Capital could now sample workers like a bee in a field of clover.

The consequence of these combined developments was that, by the 1990s, the first sign of the inability of the system to digest the immense output disgorged by its multitudes of sweatshop workers worldwide appeared. According to this argument, the culmination of the 1997 Asian crisis was the stimulus for the full financialization of the system—the attempt to "make money from money" at the most abstract level of the system once making money from production no longer sufficed.

Capital's flight into financialization is one more move in the neoliberal effort to continually shift the power relation in its favor. Faced presumably with diminishing returns in the "real economy" and an inability to sell their goods, capitalists made two important moves: on one side, they leapt to the world of hedge funds and derivatives, and, on the other, intensified the availability of credit for the US working class, so that US workers would buy the goods that workers in China and other (mostly Asian) nations continued to produce at extremely low wages (compared to the US). The success of this game—whose eminent goal was deferring crisis—depended upon the high profits capitalists operating in China and in Third World nations could accrue because of the low wages, which were then invested in credit markets in the US, enabling growing financialization. This circuit came to an end only at the point in which the enormity of (both workers' and capitalists') debt sent its underwriters into a panic flight.

This account explains much, but it leaves out an important detail: though overproduction and under-consumption reduce the rate of profit, why is the resulting reduced rate of profit inadequate for capitalists to want to re-invest? Take an average capitalist: if s/he sold all the commodities produced in her/his firm, s/he would receive a 100 percent rate of profit; but with the "realization" problem, s/he only receives a 50 percent rate. Would that not be adequate? Even with a realization problem that required the destruction of half of what is produced, capitalists might still make a sizeable profit rate. This "inadequacy" is not inherent to capital in the abstract. Rather, it is based on capitalists' determination to make more, to demand a more rapid expansion of the system and of the profits of its owners. When capitalists deem a field of investment possibilities "inadequate," it means that the average rate of profit currently available is less than their expectation based on past experience. What, however, are the causes of an actual decline in the planetary rate of profit?

An actual fall is rooted in many factors, but there are two that are especially crucial for us: capital's inability (a) to increase the rate of exploitation by decreasing wages; and (b) to reduce the value of the constant capital (raw materials, especially) involved in the production of a commodity. The latter is especially due to the inability to pass along to workers the cost of the environmental damage caused by the extraction of the raw materials and the production of commodities. That is why the impacts of "economic" and "ecological" struggles on the average rate of profit are hard to distinguish in this crisis.

Let us consider the consequences of both (a) and (b).

(a) Globalization has helped to reduce wages in the last three decades in the US by bringing manufacturing production to the "periphery" (especially to China in the last decade), where prevailing wages are just a fraction of US workers'. If wages remained low there, the deal between US and Chinese capital would have been stable. Chinese workers would have provided super-profits for US capitalists and super-cheap commodities for cash-strapped US workers. However, though wages are relatively lower in China than the US, they have been rising rapidly. The Chinese average nominal wage has risen about 400 percent in the decade between 1996 and 2006, while the Chinese average real wage has risen by 300 percent between 1990 and 2005, with half of that increase between 2000 and 2005. This can have a profound effect on profitability long before wages in China become comparable to those in the US.

It would help to look at a simple hypothetical numerical example to appreciate this point: the wage of a Chinese worker might be a tenth of a US worker's wage, and the rate of profit for a factory with relatively little investment in machinery in China might be 100 percent. Though the doubling of the Chinese workers' wages would still make his/her wage one fifth of a worker in the US, other things being equal, the rate of profit would have fallen to 50 percent.

Thus, wage increases can cause a dramatic fall in the rate of profit without wages necessarily becoming equal in purchasing power to the wages of a Western European or North American worker. The first large-scale taste of this phenomenon in the neoliberal period was the workers' mobilizations in Korea and Indonesia that were the basis of the famous "Asian financial crisis" of 1997 we chronicled in "One No and Many Yeses" (Midnight Notes, 1997).

The lowering and stagnation of average wages in the US (but still at a relatively high level from a global perspective) has been accompanied by increases in Asian workers' wages that challenged the rate of profit long before they came close to being equivalent to wages in the US. Super-high levels of profitability can disappear well before suburbia, the car, and the Gucci handbags arrive en masse.

This problem of "realizing" the surplus value in the face of the actual or impending confrontation with workers struggling for higher wages and greater power at work led capitalists to turn to other avenues to earn the rates of return that they desired. But there is an inherent problem in this move as well: the ability to increase interest revenue though financialization is limited by the surplus value created in production and reproduction throughout the global capitalist system. The crisis in the financial sector arises from the confrontation with this limit. Since financial gains are—however indirectly—finally also extracted from real labor, one can readily understand that even a modest increase in Chinese wages could pull the rug from under the financial house of cards.

(b) The ecological/energy moment of the crisis appears most directly here. The reduction of the costs of constant capital can lead to an increase in the profit rate, but it crucially depends upon being able to "externalize" the harm it causes (i.e., to force those harmed by the pollution of raw material extraction, by the climate change caused by industrial production, or by genetic mutation produced by the spread of genetically modified (GM) organisms to quietly and continually submit to it without demanding that it cease). It is only when there is a mass refusal to allow this externalization to pass that ecological issues become "pressing" and an "emergency." Unless there is struggle against the harm and the tacit assumption of the costs, ecological damage is an aesthetic phenomenon like the smog in a Monet painting.

This struggle has now come out of the shadows and is threatening profitability throughout the system. There is a worldwide recognition that we aren't just in another round between workers and capitalists to see how to organize the economy; we are facing catastrophic climate change and generalized social and environmental breakdown in a world where "the civilization of oil" has placed a great part of humanity in cities and slums that were already reaching their breaking point before the crisis set in. It's frightening to see Mexico, for instance, with so many people barely surviving and the State and other oligopolists of violence already so intense, poised on the brink, with migrants returning from the USA … to what? One community recently came out with guns to cut off water to another that they considered was taking too much. What will happen when—as the scientists say is already determined—the average heat in these latitudes has increased three degrees, when every summer is as hot or hotter than the hottest on record?

There clearly cannot be any more profit-making business as usual. Indeed, in its disciplinary zeal, capitalism has so undermined the ecological conditions of so many people that a state of global ungovernability has developed, further forcing investors to escape into the mediated world of finance where they hope to make hefty returns without bodily confronting the people they need to exploit. But this exodus has merely deferred the crisis, since "ecological" struggles are being fought all over the planet and are forcing an inevitable increase in the cost of future constant capital.

So on both counts, with respect to wages and ecological reproduction, the struggles are leading to a crisis of the average rate of profit (and the rate of accumulation) and imposing a limit on the leap into financialization.

2. THE CRISIS OF NEOLIBERALISM: CAUSES AND CONSEQUENCES

Neoliberal globalization was an ambitious project. Had it succeeded, it would have changed the very definition of what it is to be human into "an animal that trucks and barters him/herself to the highest bidder" and would have returned labor power to its status in pre-Keynesian economics: a pure commodity receiving its value from the market. Why did neoliberal globalization fail?

To answer this question, we must turn to the struggles that people have made. Even though US workers may not display the level of militancy they had in the 1930s, broad movements have risen worldwide that in our view must be recognized as sources of the crisis. Certainly, these are not the only factors and possibly not the most immediate ones. Undoubtedly, for example, the lack of regulations on financial transactions was a factor in the non-linear complexity created by the meta-gambles in the derivatives trading that have destabilized the "markets."

Yet even the financial de-regulation that began under Carter and continued afterwards under Reagan, Bush, Clinton, and Bush Jr. was a moment of class struggle. De-regulation began in response to accelerating inflation that was due—in reality as well as in the minds of policy makers—to the power of US workers (on average) to raise money wages fast enough to prevent capitalist price increases (of food, energy, etc.) from cutting their real wage throughout the 1970s—a power that undermined the hoped-for conversion of OPEC into a financial intermediary and of petrodollars into vehicles for transferring value from workers' income into profit-earning investments.

The IMF's annual reports from that decade reveal that, by 1975, inflation was being identified as the number one economic problem in the world, and a key source of that inflation was identified as "structural rigidity in labor markets," IMF-speak for workers' power. By the time Carter and Volcker acted, accelerating inflation had driven many real interest rates below zero and threatened the viability of the whole financial sector. The strategy of deregulation included, among many things, the removal of anti-usury laws throughout the US that allowed interest rates to rise into the double-digits. It was a response to the power of workers to not only raise wages and other forms of income to the point of undermining profits—despite the capitalist recourse of basic good price manipulation and floating exchange rates—but also to block any recovery in the rate of growth in productivity at the point of production.

Many of the struggles in the 1970s in the US eventually were defeated, but since then there has been a new generation of struggles, both in the US and internationally, against neoliberal globalization that has proven decisive.

We focus on some of these struggles as conditions for the understanding of the political questions posed by the Crisis. Schematically, the sources of the Crisis include:

(1) the failure of neoliberal globalization's institutional changes;
(2) the failure to neoliberalize the structure of the oil/energy industry;
(3) the inability to control wage struggle (especially in China);
(4) the rise of land and resource reclamation movements (Bolivia, India, Niger Delta);
(5) the financialization of class struggle though the expanded use of credit in the US to supplement the fallen and stagnant real wage; and
(6) the inclusion of blacks, Latinos/Latinas, recent immigrants, and women into the "ownership society," undermining class hierarchy.

(1) Neoliberal globalization depends upon a framework of laws and rules that eliminate barriers to commodity trade and financial transactions, especially those transactions that emanate from the US, Japan, or Western Europe. The process of elimination began in the Keynesian era (with GATT), but took institutional shape with the formation of the World Trade Organization (WTO) in 1994. The WTO had an ambitious agenda of realizing the globalization of traditional trade and money transaction, but also services and intellectual property. It looked like nothing could stop this agenda from realization.

But it was stopped by a surprising convergence of:
(a) anti-structural adjustment riots and rebellions stretching from Zambia in the mid-1980s, through Caracas in 1989, to the Zapatistas in 1994;
(b) the anti-globalization movement in Western Europe and North America and its street demonstrations and blockades at the WTO, IMF, World Bank, and G-8 meetings; and
(c) the many Third World governments that refused to completely give away the last shreds of sovereignty (especially over their agricultural production) to organizations like the WTO, the IMF, and World Bank that were dominated by the US, Japan, and Western Europe. The reasons for this were not purely "patriotic;" they had much more to do with the power of the farmers' movements in their territory and the threat they posed to their own "sovereignty." The Doha Round at the WTO finally perished in particular because the Indian government officials just couldn't give away

any more on agriculture—although they would have loved to sacrifice their peasants for some high-tech stuff. The Indian movements have been mobilizing by tens and hundreds of thousands over the decade from 1998 to 2008 to stop the WTO (not to mention the Filipino, Korean, and Bangladeshi farmers).

Though often ignorant of each other's actions and intents, these rebellions, street demos, and "insider" resistances de-legitimized the "Earth is flat" globalization ideology and the attempt to enclose the world's remaining subsistence and local market farmers.

(2) The second moment of failure was the attempt to revive the flagging neoliberal globalization project after 1999 by war, especially in an effort to transform the oil and gas industries into ideal neoliberal operations through the invasion and occupation of Iraq (MN, 2002). This failure has been caused by an armed resistance that inflicted tens of thousands of casualties on US troops, but that, in turn, has suffered hundreds of thousands of deaths and injuries. It has had enormous consequences for neoliberal globalization. First, after six years of war in Iraq, the most basic of industries—the oil and gas industry—is still organized, both in Iraq and around the world, by two forms that are anathema to the neoliberal doctrine: the national oil company and the international cartel (OPEC) that tries to influence the market price for oil. Second, the leader of the neoliberal project, the US, has been severely weakened both militarily and financially by the effort. This became most evident when the US government declared victory (due to "The Surge"). It simultaneously was told by its own Iraqi "puppets" to leave the country by 2011, to dismantle its bases, and not to expect to see a neoliberal "Oil Law" soon! Surely the "puppets" spoke so harshly to their masters because they feared the violent reaction of the Iraqi people to the attempted giveaway of Allah's hydrocarbon gift.

(3) The neoliberal project of the "refusal of wages" has been quite successful in the US where the real wage has never regained its 1973 peak. That is why one cannot find a source of this crisis in the US wage struggle as one can for the crisis of the 1970s. All the typical indices of such struggle (e.g., strike activity) in the US have been depressed. There have been defensive struggles waged, with some success, to limit attacks on non-waged income, e.g., social security, Medicare, and food stamps. Moreover, there have been ongoing struggles against other attacks on the working class, e.g., on the terrain of women's rights, environmental protection, etc.

However, the neoliberal project depended on the ability to use competition in the international labor market not only in the US but throughout the world. This project has failed, especially with respect to Asian countries. We saw the failure of this control in Korea and Indonesia during the lead up to the Asian financial crash in 1997 (see MN, 1997). The major failure of this strategy since then has been in China, where the level of wage struggle has taken on historic dimensions, with often double-digit wage increases as well as thousands of strikes and other forms of work stoppages.

(4) The "New Enclosures" have operated through Structural Adjustment Programs and the fomenting of war that were meant to expropriate people throughout the Third World of their attachment to their communal land and its resources.

Certainly, they have driven millions of people from their land and communities in Africa and many parts of the Americas, if the increase in immigration rates and numbers of refugees is any indicator. But there has also been a powerful response to the attack on common lands and resources throughout Asia (especially in India and Bangladesh), in much of South America, and in parts of Africa. The Bolivian "water" and "gas" wars of the last decade have made it clear that the effort to privatize vital resources is a risky enterprise. Similar limits are being experienced in oil production in the Niger Delta, where there is now an ongoing war of appropriation waged by groups like the Movement for the Emancipation of the Niger Delta (MEND); such groups are demanding that the people of the Delta be recognized as communal owners of the petroleum beneath their soil, against the Nigerian government and the major oil companies. Indeed, there is a political limit being reached in oil exploration and extraction that Steven Colatrella has aptly called a "political Hubbert curve."

(5) The main function of the financialization of capital was to buffer accumulation from working class struggle by putting it beyond its reach and by providing a hedge against it by making it possible for capitalists to bet against the success of their own investments, hence providing insurance in any eventuality. What capitalist does not want to be able, for a small payment, to protect him/herself from a dramatic devaluation of the currency of the country they are investing in due to a spate of general strikes, or from the bankruptcy of a company that they are dealing with due to workers' wage demands?

Paradoxically, however, Neoliberalism has thrown open a new dimension of struggle between capital and the working class within the domain of credit. For a whole set of credit instruments and speculative investments were offered to US workers, from sub-prime mortgages, to student loans, to credit cards, to 401(K) pension management schemes. Workers used them because their inability to project their collective power on the job to achieve significant wage increases, guarantees for pensions, or health care forced them to try to expand into the financial realm. With the dismantling of the so-called welfare state, workers in the US had to pay a greater share of the cost of their own reproduction (from housing and health care to education) at the very moment when their real wages were falling. Workers demanded access to these requirements for reproduction through the credit system. Capital's "sharing" with workers of accumulated value through making credit available comes at a price: that workers' desires for access of the means of reproduction (home, auto, appliances, etc.) are aligned with capitalists' desires for accumulation. "Financialization" is not simply a capitalist plot; it too is a process and product of class struggle. True, there is an element of necessity in workers' response to the attack on their conditions of reproduction, but without necessity there is no agency either.

The entrance to the credit system is no workers' paradise, of course. Borrowing and the accompanying interest payments depress wages, sometimes quite substantially, and credit ties workers to the real estate and stock markets. However, it is an important achievement for workers to be able to "use someone else's money" in order to have a home without worrying about rent increases and paying the owners' mortgage and his/her taxes, to have the desire (real or fancied) evoked by a commodity satisfied today, to have access to education that might make for higher wages in the future, and to have an automobile that makes a wider range of jobs and social contacts possible in the lonely landscape that life in the US often presents. This dangerous working class strategy hovered between using the credit system to share in collective wealth and debt peonage!

In a way, though neither "consciously" nor in a coordinated manner (as so many things happen in capitalist society), many in the US working class have collectively attempted to turn the neoliberal vision of transforming everyone into "rational economic" agents against the system itself by taking the Bush Administration's "ownership society" rhetoric at its word. In so doing, they have brought the system into a crisis by implicitly threatening to refuse to pay their debt, i.e., to leave the key in the mailbox and walk out. As was pointed out long ago, if you owe the bank $1,000 and you can't pay, you are in trouble; but if you owe the bank $1,000,000,000 and you can't pay, the bank is in trouble. What is often not mentioned is that if 1,000,000 people each owe the bank $1,000 and can't pay, then the bank is still in trouble!

Financialization was meant to provide capital with a shield against the indeterminacies caused by class struggle, but it invited the working class into its very breast. This attempt by financial capital to play both sides of the equation (i.e., to have capital pay for protection against struggle and at the same time bring the presumably "tamed" agents of that struggle into the financial machine) is one basis of the contemporary crisis. True, though the working class' share of the total debt is sizeable, it is much smaller than US corporate or state debt. However, its quality is different. Corporate debt is intra-class, while national debt is omni-class, but working class debt is inter-class and potentially creates the greatest tension.

(6) This double character of financialization was intensified by the struggle of workers previously excluded from access to credit (blacks, Latinos/Latinas, recent immigrants, single women, and poor whites) to enter into the charmed circle of home mortgages, student loans, and credit cards. Financial capital significantly opened up to these new creditors in the twenty-first century, who previously could only borrow under the most onerous conditions from loan sharks and pawnshops. It answered their desire to be able to have legal claim to a house, car, desired commodities, and a better paying job, but with poison pills: sub-prime mortgages whose interest payments would balloon after three years, credit cards whose interest rates approached loan shark levels, student loans that would turn graduation into an entrance to wage slavery. These workers' pressure to be included into the neoliberal deal—i.e., one can have access to social wealth only on an individual basis and via non-wage income—was answered affirmatively by capital in the first years of the twenty-first century. It proved to be the initial point of destabilization of the credit system.

Does the deepening and widening of the circulation of credit into the working class mentioned in (5) and (6) deserve to be called a "struggle"? One might well question such a formulation, given the immediate denouement of the story—millions of foreclosures and bankruptcies, etc. But there is no doubt that there has been a struggle over conditions of payment and of bankruptcy (extending to workers), as well as struggles over legislation that would "rescue" homeowners from foreclosure. Many on the Right have taken this "credit revolution" as the cause of the crisis, since it let too many of the "unworthy" into the inner sanctum of credit. But this does not invalidate the actual struggle that had been launched by black workers, from the 1960s on, against "redlining" and other forms of credit discrimination. After all, debtors' struggles have traditionally been basic to the analysis of class history since ancient times. Why should these be excluded in the class analysis of the twenty-first century?

We do not attach a "price tag" to these six moments of struggle. Along with many other conjunctural factors, they combined to create a crisis of historic proportions in 2008. The failure of Neoliberalism's Wage and War doctrines, Globalization, New Enclosures, Financialization, and the Crisis of Inclusion together not only produced the economic "downturn," but the logical contradictions that infest them are transforming the present recession into a real crisis. It might be possible for there to be a "recovery" (as measured by increased GNP) in the near future, but if the contradictions are deepened and the failures intensified, capitalism could become "history."

3. CAPITAL'S IMMEDIATE RESPONSE TO THE CRISIS OPPORTUNITY

This crisis gives capital an opportunity in at least three aspects: (i) the reorganization of the power relation between financial capital and the rest of the system, (ii) the disciplining of the US working class' role as a debtor and player in the financial system, and (iii) the justification of environmental plunder, wage reduction, and land expropriation in the Third World through a revival of the "debt crisis." Let us take each one in turn:

(I) FINANCIAL CAPITAL'S AGONY OR ITS RENAISSANCE?

This crisis begins as a financial crisis (i.e., as the inability to pay back the principal and interest on debts or to pay for lost wagers made on a grand scale). Though most crises have a financial aspect, this clearly is one that poses fundamental challenges to the system's fate, for it makes a major transformation of the order and hierarchy within the sectors and phases of capital inevitable.

Will the crisis be the opportunity (in return for the enormous amount of capital that the financial sector is demanding of the state) to call for a complete halt or at least draconian regulation of many of the financial practices (especially Collateralized Debt or Mortgage Obligations, Structured Investment Vehicles, Credit Default Swaps, credit derivatives of all sorts, and maybe even of offshore banking—tremble little Switzerland!) whose collapse have put the everyday operations of industrial, commercial, and service companies large and small into jeopardy? Or will financial capital hold the rest of the system hostage by threatening to shut off lending and bring the credit system to a halt unless it gets its debts secured by the government on its terms?

We see an aspect of this conflict in the struggle over the "bail out" of the "Big Three" automakers versus the almost unanimous support on the highest level of government (from the Bush administration to the Obama administration) for the large multipurpose banks (Citigroup), insurance companies (AIG), and even invest-ment houses (Bear Stearns). The tremendous controversy—and now potentially fatal terms demanded of GM and Chrysler—over what is a relatively small sum compared to the swiftly granted billions for AIG is a sign that financial capital still has the upper hand in highest elements of the state.

But this is only the first round of a long drawn out battle that will lead, if capitalism survives, to a twenty-first century hybrid between two poles: (a) an intensely stringent regulatory regime imposed on financial innovations, with the capital released from the financial sector being directed to a new investment wave in "green energy" proj-ects (from wind turbines, to Clean Coal technologies, to nuclear power plants) and biotechnology; or (b) a victory of the financial sector, the final "de-industrialization" of the US, and a universal reconciliation with a regime of bubbles and crashes.

The first pole describes an outcome that is reminiscent of previous periods of recovery from intense "financialization" and speculation, from the "Bubble Act" of 1720 in Britain after the South Sea Bubble and the French bourgeoisie's retreat to gold in the aftermath of the 1720 Mississippi Bubble to the Glass-Steagall Act after the stock market crash of 1929. It is a return to Keynesianism, but with "green" char-acteristics and without nuclear-armed Communist states, whose existence was being used by workers in the US and Western Europe as a constant threat to capitalists.

The second alternative describes an outcome bitterly recognizing the unconscious anti-capitalist side of Margaret Thatcher's shibboleth, "There Is No Alternative," when applied to the hegemony of the financial sector in neoliberal capitalism with its hell-ish conclusion: the market is the best (since the only) way to allocate the resources of the planet, even though it leads to an ever shorter cycle of boom, bubble, bust, and depression. Can the US become, in the early twenty-first century, something of a late-twentieth century Britain, existing without a significant manufacturing or agri-cultural base (leaving this part of the division of labor to China and other continents of cheap labor)?

That is, the financial sector will be "nationalized" or the nation will be "financial-ized" (or some combination of both). Either alternative alone is equally improbable. Some chimera of a Keynesianism meant to revive the industrial base (with a large "green" sector) and another round of reformed Neoliberalism meant to re-legitimate financial capital's adventures will be constructed, unless there is another force in the field that can use the crisis to forge a way out of capitalism. In the short term, Keynes-ian and "green" policies will be pushed—perhaps aided by the fact that capital move-ments (with which sustained Keynesian policies are not viable) are low due to the current crisis context. Some regulation will be implemented, and definitively—after the depth of the crisis—some reconciliation with a regime of bubbles and crashes will be promoted.

(II) US WORKERS AS DEBTORS

Karl Marx, the great nineteenth century anti-capitalist analyst, saw financial capital as purely related only to capitalists. He pithily wrote in the 1860s: "Interest is a relationship between two capitalists, not between capitalist and worker." In other words, interest appears to be an income paid to a financial capitalist, based on the money loaned. How the loan is paid back with interest is irrelevant. Interest is logically autonomous of the production process (although for Marx it is vitally dependent on the exploitation of workers somewhere in the system). Most crucially for us, Marx writes as if workers never receive loans and pay interest. This is important, for the credit system is like a capitalist common, since it offers the capitalist (or the person who can pass as a capitalist) "an absolute command over the capital and property of others, within certain limits, and, through this, command over other people's labor." Value detached from its owners becomes a common pool resource that, though abstract, gives tremendous power to those who can access it. This power was not to be shared with workers, at least not in the nineteenth century.

Marx got many things right about the future of capitalism, but here he failed to see the absorption of the propertyless but waged working class into the financial system. When he looked at workers' debt, he saw only pawnshops. Since workers had almost no property that could be used as collateral to take out loans from financial institutions and they had almost no savings to be used as deposits in banks, they were never important direct players in the financial world. In fact, many mutual aid and credit union organizations sprang up in the nineteenth century because banks and other financial institutions considered themselves as having solely capitalists (large and small) as their customers, or workers were too suspicious to put their hard-earned savings into the hands of financial capitalists. This is no longer the case. Workers' pension funds are an enormous source of capital for the system, and their debts comprise a large share of total indebtedness in the US (household debt is about 30 percent of the total debt in the US). Consequently, when we speak of financial crisis in the twenty-first century, we must speak of inter-class conflict as well as conflict among capitalists.

As noted in the previous section, workers in the neoliberal deal have been using the credit system to enter into the realm of nonwage income, i.e., to get access to the value common that had previously been the sole privilege of the capitalists. In doing so, they have posed a collective threat and opportunity to capital. The question is: can capital operate in the twenty-first century without extensive working class participation in the credit system? Can capital return to the days before "life on the installment plan" and make credit the sole realm of capitalists again? There are many who are skeptical of either a definite "Yes" or a definite "No" to these questions for very good reasons, since the duplicitous character of financialization that we analyzed above cannot be easily "corrected." To block the working class completely (or even differentially) from access to the value of commodities, homes, and education via credit, without returning to the wage struggle, could be to invite an unacceptable level of class war; but to restart the machine with the working class having the same access to credit as it had before the Crisis could be to invite another repetition of the same cycle and struggle in short order. This is the capitalists' dilemma, of course, and they will have a devil of a time resolving it. But this process is not just simply a matter for capital to decide; much of the outcome lies in the actions of that sphinx, the global working class.

This dilemma intensifies the observation Marx made about the "dual character" of the credit system long ago: "on the one hand it develops the motive of capitalist production, enrichment by the exploitation of others' labor, into the purest and most colossal system of gambling and swindling, and restricts ever more the already small number of the exploiters of social wealth; on the other hand, however, it constitutes the form of transition towards a new mode of production." For the demand that the workers have increasingly made for access to the accumulated wealth their class has produced via the credit system also has the seeds of "a transition towards a new mode of production," even though it also is embedded in an equally colossal system of gambling and swindling.

(III) THE CRISIS OUTSIDE OF THE US AND WESTERN EUROPE: THE RETURN OF THE IMF AND WORLD BANK

The importance of debt as a weapon in the course of class struggle is not new. It was most clearly shown in the "debt crisis" of the early 1980s, when African peasants and South American factory workers were saddled with enormous debts because of variable interest rate loans negotiated by their countries' dictatorial governments behind their backs in the 1970s when real interest rates were low (and in some cases even negative). But in 1979 interest rates skyrocketed, leaving peasants and factory workers holding the bag for debts that were many multiples of their country's GNP.

This constituted the "debt crisis" of the early 1980s that made it possible to squeeze an enormous amount of surplus value from Africa, South America, and Asia by huge interest charges on old loans, and by new loans from the IMF and World Bank to pay back old loans on the condition that these governments adopt Structural Adjustment Programs (SAPs). SAPs made it possible to pry open previously closed economies; substantially weaken the target countries' working classes; and allow US, Western European, and Japanese capitalists to access workers, land, and raw materials at extremely low cost. They were the foundation of what became known as "globalization," and the IMF and the World Bank became globalization's central control agencies, opening up countries that threatened to refuse to play by the rules of "free trade." Up until the post-Asian Financial Crisis of 1997, the SAP-dominated countries of the former Third World provided much of the flow of capital to finance housing and stock market booms in the 1980s and 1990s. Afterwards, China almost alone would do this job.

All this happened in the face of a tremendous struggle from the mid-1980s to the early 2000s. There were literally hundreds of what became known as "IMF riots" throughout the planet as well as armed revolutions that continually pressured the IMF, the World Bank, and the governments of the US and Western European nations to renegotiate loans, change loan conditionalities, and even write loans off. The struggle against SAPs became an international one, stretching from the forests of Chiapas to the streets around the IMF and World Bank headquarters in Washington, DC. Moreover, beginning with the rise of oil and commodity prices in the twenty-first century, the IMF and World Bank were being shunned by their former "clients" (more accurately, former "debt peons"). This was especially true of many oil-producing countries like Algeria, Nigeria, and Indonesia that were able to pay off a substantial part of their old loans and/or attract loans outside of the SAP-framework of the IMF and World Bank, e.g., Argentina's loan from Venezuela. Although total external debt was not reduced (or even increased) for many countries, the monopoly role of the IMF and World Bank was shattered, making it possible for countries to ignore these agencies' draconian "recommendations."

The Crisis, however, can change the power relations once again by drying up the alternative sources of funding (e.g., the Venezuelan government will find it difficult to lend to a South American nation nearing bankruptcy in this situation). As a consequence, there will be the possibility of a revival of the power of the IMF and World Bank as the global lenders of last resort, with all the power that this role implies. For the external debt for many countries has far from vanished, and under the pressure of the crisis it will dramatically increase. Indeed, the G-20 governments have agreed to expand IMF reserves to $1 trillion, and the IMF has already imposed SAP-like conditionalities on several bankrupt East European nations. Going back to the vomit of SAPs would be a historic defeat and an invitation to a new wave of neo-colonialism.

One vehicle of return is global warming, which poses an ecological limit to the forced growth of capitalist regimes. Undaunted, the usual northern players (including the World Bank) are investing in a horrific series of "solutions" to global warming in the South, rather than reducing the causes of northern emissions. Agrofuels (Genetically Modified (GM) soya, African palm, sugar cane, jatropha, and all kinds of GM monstrosities in the near future) are menacing southern farmers with the greatest enclosures yet. Half of Argentina's arable land is already a "green desert" of GM soya, without speaking of Paraguay and Brazil, while the African palm has replaced a huge proportion of Indonesia's forests and is now being used to attack the Afro-descendant communities in Colombia. India is planning more than a million hectares of jatropha (which means expulsing about as many peasants). And Nigeria talks about industrial farming to counter struggles over oil and land in the Niger Delta.

The Crisis will put more power in the hands of the World Bank and IMF to open up the economies of the Third World to even more projects like these, while simultaneously (re)introducing the austerity programs that gutted already inadequate education, health, and social services. For example, carbon trading will allow the North to continue to pollute while financing dams and other "big" developments in the South. Through the IMF, SAPs, and "development," the "global south" will be made available to complement if not replace the Chinese workers that have been demanding higher wages. You have to hand it to those capitalists. They try to make a buck out of anything—even the end of the world![2]

---

2   As this section was written in April 2009, due to some datedness and space limitations in this book, Midnight Notes and Friends have agreed to deletion of the next section, entitled "3b Working Class Responses to the Crisis." It can be found, with the rest of the pamphlet, at http://www.midnightnotes.org.

4. THE CONSTITUTION OF THE COMMONS IN THE CRISIS:
EATING FROM A DISH WITH ONE SPOON

Struggles circulate, and open struggles against the consequences of the crisis will soon explode in the US. What apparently began as a financial crisis, which turned into an economic one, is soon to be called a "political crisis." The abject destruction that capitalists have created with their "management" of the two great commons of labor and the planet's eco-system will stop being considered a "tragedy of the common" (where no one in particular is responsible) and come to de-legitimate the capitalist class as a whole. These crises have been predicated on the presumption that labor and the planetary eco-system are common resources to be used and abused for the profit of anyone who has (or successfully pretends to have) the capital to appropriate them.

The capitalist class is unable to control the common pool of resources that make up our means of production and subsistence without creating terminal damage. Who can do better? Though many workers in the US might not rise to the challenge today and continue to look to their bosses for salvation, we still should say what the logic of the struggles indicates should be done. Let us be guided by the words of Thomas Paine in *Common Sense*, who, in a previous period of revolutionary crisis, noted that most everybody favored independence in the days before the Declaration of Independence was promulgated. The only issue was the timing: "We must find the right time," they said. Paine answered, "The time has found us!"

The Crisis has shown for all who have eyes to see that State and Market have certainly failed in their claim to provide a secure reproduction of our lives. Capitalists have conclusively shown (once more) that they cannot be trusted to provide the minimal means of security even in capital's heartland. But they hold hostage the wealth generations have produced. This pool of labor past and present is our common. We need to liberate, to re-appropriate that wealth—bringing together all those who were expropriated from it, starting with the people of the First American Nations and the descendants of the slaves, who are still waiting for their "forty acres and a mule" or its equivalent. We also need to construct collective forms of life and social cooperation, beyond the market and the profit system, both in the area of production and reproduction. And we need to regain the sense of the wholeness of our lives, the wholeness of what we do, so that we stop living in the state of systematic irresponsibility towards the consequences of our actions that capitalism fosters: throw away tons of garbage and then don't think twice, even if you suspect that it will end in some people's food, as smoke in somebody else's lungs, or as carbon dioxide in everyone's atmosphere.

This is the constitutional perspective we can bring to every struggle. By "constitutional" we do not mean a document describing the design for a state, but a constitution of a commons, i.e., the rules we use to decide how we share our common resources. As the indigenous Americans put it, in order to collectively eat from a dish with one spoon, we must decide on who gets the spoon and when. This is so with every commons, for a commons without a consciously-constituted community is unthinkable.

This means we have to craft a set of objectives that articulate a vision in any context of class struggle, turning the tables on capital at every turn. First, we need to establish what violates our rules as we are constituting the commons. What follows is a sample of such immediate taboos. We cannot live in a country:

- where 37 million people are hungry;
- where the cost of surgery kicks you out of your home;
- where going to school rots your mind and leaves you in debt peonage;
- where you freeze in the winter because you cannot pay the heating bill;
- where you return to work in your 70s because you have been cheated out of your pension;
- and where work that produces murder and murders its workers is sold as a path to "full employment."

These are very elementary taboos, but they have to be loudly pronounced. Though the system has shown itself to be bankrupt, many still listen to its siren songs.

The time has come for us in the anti-capitalist movement to propose a constitution of rules by which to share the commons of past labor and present natural resources and then concentrate on building political networks capable of realizing it. At revolutionary junctures in US history (like the Civil War, the Great Depression, the Civil Rights/Black Power Movement), a basic constitutional change within the working class is manifested in action (the years-long "general strike" of slaves in the South during the Civil War, the innumerable factory clashes, the "sit-ins," as well as many "hot" summer insurrections in city after city, respectively) and is "captured" by a law or even "a constitutional amendment" (like the 13th and 14th Amendments, the Wagner Act, the Voting Rights Act, respectively).

But US history is not alone in connecting crisis, revolutionary transition, and constitution. There has recently been a whirlwind of constitutional politics throughout the Americas south of the Rio Bravo in the last two decades. From the Zapatistas' call for a new Mexican constitution, to the many constitutional transformations in Venezuela, to the most recent Bolivian constitution that formally recognizes the commons, there has been a formal statement of *potencia* (or "power to") instead of *poder* (or "power over"). It is exactly this spirit that the Zapatistas, in "The Sixth Declaration of the Lacandon Jungle" (2005), have called for: "We are also going to go about raising a struggle in order to demand that we make a new Constitution, new laws which take into account the demands of the Mexican people, which are: housing, land, work, food, health, education, information, culture, independence, democracy, justice, liberty and peace. A new Constitution which recognizes the rights and liberties of the people, and which defends the weak in the face of the powerful."

We should formulate demands, objectives, programs of struggle around the main elements of our lives—housing, work, income—all in view of guaranteeing our livelihoods, building cooperation and solidarity, and creating alternatives to life in capitalism. We need to build a movement that puts on its agenda its own reproduction. We have to ensure that we not only confront capital at the time of the demonstration or the picket line, but that we confront it collectively at every moment of our lives. What is happening internationally proves that only when you have these forms of collective reproduction, when you have communities that reproduce themselves collectively, can struggles come into being that move in a very radical way against the established order. This is our constitutional politics. It is not a list of demands or grievances, but an expression of who we are becoming, i.e., our constituting our being.

For instance: Let's guarantee housing to each other. This means not only "No" to evictions, but the reoccupation of houses that have been abandoned, the distribution or occupation of the empty housing stock that lies all around us; the collectively decided self-reduction of rent of the kind that was carried out in Italy in the 1970s; the creation of new housing that would be organized collectively and built ecologically. Short of that we should build our version of "hobo jungles" on the steps of the White House, open soup kitchens there, show the world our empty pockets, our wounds, instead of agonizing in private.

For instance: Let our struggle over housing be a struggle for the reorganization of work reproductive of daily life on a collective basis. Enough of spending time in our solitary cages with trips to the mall as the climax of our sociality. It is time for us to join with those who are reviving our tradition of collective, cooperative living. This "year-zero" of reproduction that the capitalist crisis creates, as evinced by the mushrooming of tent cities from California to North Carolina, is a good time to start.

For instance: Let's struggle in such a way as to disable the mechanisms that perpetuate our exploitation and divisions. To ensure that our struggles are not used to divide people on the basis of differentially dished out rewards and punishments, we must continually raise the issue of reparations, i.e., the price paid and that continues to be paid for the racist, imperialist, sexist, ageist, chauvinist, ecologically-destructive deals US workers have accepted.

For instance: Let's call for a life where our survival does not depend on constant war on the people of the Earth and on our own youth. We must speak against war in Iraq and Afghanistan, and against the butchery in Palestine.

For instance: Let's speak against prisons, the politics of mass incarceration, and the obscenity of plumping employment and business profits by putting people in jail. We must call for the abolition of capital punishment … even for capitalists! And most importantly we need to redefine crime, exploding the logic that sees a horrendous crime when a proletarian robs a liquor store, but calls capitalists' crimes that lead to the death and destitution of thousands "accidents," "mistakes," or even "business as usual."

For instance: Let's also speak about male violence against women. What struggles for the constitution of the commons are we are going to make when every fifteen seconds a man beats a woman in the US? How much energy would be liberated for the struggle, if women did not have to fight men, often even to be able to fight the system?

For instance: Let's revive our social imagination after decades of defensive reactions to neoliberal enclosures and determine new constitutions of the commons. Of course, what our imaginations can suggest now is limited and only a preparation for attaining another level of power and capacity to envision. But even with this poverty, we can hear snatches of a medley of "musics from possible futures." Listen to two musicians in our midst:

"The future commons boils down to two elements: access to land (i.e., food and fuels); and access to knowledge (i.e., capacity to use and improve all means of production, material or immaterial). It's all about potatoes and computers."

"The wage system should be dismantled immediately. Given the existence of the internet, of 21st century accounting methods, and of direct deposit, it would be possible to immediately move toward a guaranteed income, at first in monetary terms, with everyone having access to an "account" upon birth, and with a responsibility to a minimum of socially necessary labor time—including housework of all kinds, art work, writing, etc., and political activity (participation in assemblies, sitting on juries, or whatever). This would create an incentive for cooperative living in that everyone that can reduce their housework hours through cooperation and can have more time available for other activities. This guaranteed income would replace the insurance, finance, welfare state agencies, and other sectors, freeing millions of people to participate in cooperative activities, reducing further everyone's socially necessary work time."

"The only feasible way of doing agriculture on this planet is intensive, mixed-crop, organic production. This form of agriculture is hopelessly unprofitable under current conditions—so a new type of cooperation between consumers and producers (in fact the abolition of this distinction) must be found, transforming agricultural work into a part of housework for everybody."

"The financial system should immediately be replaced by assemblies and community-based 'credit unions' that can decide where to put community resources, demystifying 'finance' as societal planning."

"If the livelihood of people is guaranteed by subsistence and general services on all levels, free sharing of intellectual production is possible without endangering the survival of its producers. The planet can become a sphere of free exchange of knowledge, know-how, and ideas. Additionally to this intellectual commons, a material commons must be instituted to establish a just distribution of resources."

For instance: …

5. CHARACTERISTICS OF REVOLUTIONARY STRUGGLES THAT MOVE BEYOND CAPITAL

The struggles that have brought on the crisis, especially those in Latin America, from Mexico to Argentina, have laid down the foundational experiences of contemporary struggle for the "constitution of the commons." We believe that these experiences are important for the US anti-capitalist movements, and we have tried to identify some characteristics of these struggles (especially those of the Zapatistas and other groups arising from indigenous Americans).

One of the most important distinctions to make (but most difficult to draw) is that between those that are on the "inside" (what we sometimes call "social democratic") and those that are "autonomous" or "outside." In a way, this distinction is a variant of one between "reform" and "revolution" in the anti-capitalist politics of the first part of the twentieth century when "reformist" social democratic parties were important institutions.

The "inside/outside" distinction, however, is not a spatial one, but one of political relation. "Inside" means demands on a (state/market) institution that is normally dedicated to reproducing the labor/capital relation, while "outside" means communal appropriation of de/non-commodified resources, perhaps in parallel with formal demands. Either can happen anywhere, just as commons can be maintained or created anywhere. The two aspects can be complementary or contradictory. For example, appropriation can be enhanced and/or undermined by demands made on an institution. Either can be means to build alliances and express needs beyond those making the demands. By analyzing inside/outside relationships and potentials in specific contexts, a movement can clarify its strategy.

The inside struggles are waged primarily within existing institutions and arenas, such as the state, corporations, the legal system, traditional civil society, or traditional cultural constructs, the goals of which are generally to increase working class income, commodity wealth, and power within the system, without directly challenging the capitalist organization of society or creating collective alternatives to the capitalist system. They typically take the form of demands on the system. However, they may at times be quite confrontational and push the bounds of capitalist legality and propriety. Such willingness to openly confront the system is very valuable, at least at this point in the US, since it has greater likelihood of transcending initial demands.

By contrast, "outside," autonomous struggles strive to create social spaces and relations that are as independent of and opposed to capitalist social relations as possible. They may directly confront or seek to take over and reorganize capitalist institutions (a factory, for example) or create new spaces outside those institutions (e.g., urban gardening or a housing cooperative) or access resources that should be common. They foster collective, non-commodified relations, processes, and products that function to some real degree outside of capitalist relations and give power to the working class in its efforts to create alternatives to capital. In the US many of these struggles appear as outside the formal economy. A number of MN friends have recently commented on these kinds of struggles. Massimo DeAngelis writes in a definitional spirit in *The Beginning of History*:

> When we reflect on the myriad of community struggles taking place around the world for water, electricity, land, access to social wealth, life and dignity, one cannot but feel that the relational and productive practices giving life and shape to these struggles give rise to values and modes of doing and relating in social coproduction (shortly, value practices). Not only that, but these value practices appear to be outside corresponding value practices and modes of doing and relating that belong to capital.... The "outside" with respect to the capitalist mode of production is a

problematic that we must confront with some urgency, if we want to push our debate on alternatives onto a plane that helps us to inform, decode, and intensify the web of connections of struggling practices. (DeAngelis, 2007: 227)

Chris Carlsson has mapped some part of this terrain in the US in his book *Nowtopia*, where he writes:

Community gardening, alternative fuels, and bicycling, on the other hand, all represent technological revolts that integrate a positive ecological vision with practical local behaviors ... Taken together, this constellation of practices is an elaborate, decentralized, uncoordinated, collective research and development effort exploring a potentially post-capitalist, post-petroleum future. (Carlsson, 2008: 45)

That is, the social democratic approach tries to use existing institutions to increase the power of the working class in its relation with capital, while the autonomous approach tries to move independent of existing institutions and to build a non-capitalist society.

This "outside"/"inside" distinction, however, is not easy to make. After all, just because you write on your banners in red and black that you are a Revolutionary Outsider, it doesn't follow that you are. "History" will have to judge, and often the answer is long in coming. Moreover, those who wish for a short answer should remember the warnings of our situationist friends who point out to us the difficulties in making this "inside/outside" distinction in a society that is dominated by the endless flow of images, metaphors, and dialectical hooks, where A is easily turned to not-A (and back again) in a flash, and the "outside" can easily be turned "inside out."

We believe, however, that working class struggles in the Americas are becoming increasingly autonomous, and this distinction between reformist and autonomous struggles is central to much of the political discussion that has been permeating Mexico, Venezuela, Bolivia, Brazil, Uruguay, Argentina, and Ecuador. It certainly has been central to the Zapatistas and the debate they initiated with their "Other Campaign" in 2005, when they offered a non-electoral alternative to the Obrador presidential campaign of the social democratic PRD (Partido Revolucionario Democratico). The "Other Campaign" was an extended, cross-Mexico conversation between the Zapatistas and local activists in dozens of communities, sharing experiences of struggle and asking how authentically-democratic politics might be constructed. We are learning from this rich discussion and are trying to walk in the direction it has pointed.

First, we must note the inevitability of many "inside" struggles. Indeed, most struggles against the destructive consequences of the crisis at this time in much of the world at least start from the "inside." But such struggles may escape the bounds of being "inside." Our intent is that the characteristics we identify below can help determine whether social democratic struggles create, or are likely to create, conditions that foster real alternatives to capital. That is, whether they foster or lead to "autonomous" struggles, rather than confine struggles to the systems' limits, perpetuate or recompose divisions within the class, or turn those involved off to any possibility of future revolutionary struggles.

Autonomous struggles, however, are far from free of the need for careful scrutiny and thoughtful evaluation. What are the characteristics of anti-capitalist "autonomous" struggles? After all, autonomous struggles may be co-opted or isolated, they may not generalize, they may privilege some class sectors over others, etc.

History has "many cunning passages," and not only may social democratic struggles develop in increasingly autonomous directions, but autonomous struggles can support, inspire, and guide struggles that emerge in an inside context. Some people might be involved in both forms. And in the real world, many struggles are likely to blur this schematic categorization, perhaps in their initial action, but also in their evolution (for example, the Greek battles sparked by the murder of Alexis Grigoropoulos in Athens). The following are a series of characteristics of revolutionary struggles that we have gleaned from this anti-capitalist experience, especially from the struggle against genocide and mass murder in the service of capital that has turned the tide in the last decade from Oaxaca and Chiapas to Tierra del Fuego.

1. The struggles subvert class hierarchy—between working class and capitalist class, within the working class, and within nations and internationally; between races; between women and men; between immigrants and citizens; and between diverse cultures. Their demands lead to greater equality if won (and perhaps even if not won) because of how the battle is fought. The needs of those "on the bottom" (the poorest economically, least powerful socially or politically) are to be put first in an explicit way that builds unity and sustainability.

Social democratic demands continue generally for access to wealth: wages and income, work time, job security, pensions, health care, housing, food (which may mean land in many cases), and education. (Some of these comprise the indirect wage—which is more apt to be in some ways socialized, a form of commons, even if within capitalism.) Do such struggles privilege the already relatively privileged/powerful, would "victory" lock into place greater inequalities? Similarly, do autonomous actions include or exclude the least powerful socially or economically?

2. The struggles increase class unity, bringing together different class sectors in positive, mutually strengthening relationships, overcoming divisions within the class. They go beyond single issues, connecting them, without diminishing the significance or value of those issues. This unity must become planetary. As another MN friend, Kolya Abramsky, writes in "Gathering Our Dignified Rage": do these struggles "expand and deepen global networks ... towards an accelerated process of building long-term autonomous and decentralized livelihoods based on collective relations of production, exchange and consumption that are based on dignified livelihoods" (Abramsky, 2008)? In an older terminology, these struggles increase the "political recomposition" of the working class, as defined by the editors of *Zerowork* in the mid-1970s: "the overthrow of capitalist divisions, the creation of new unities between different sectors of the class, and an expansion of the boundaries of what the 'working class' comes to include" (MN, 1992: 112).

3. The struggles build dignified inclusion in community. The walls of exclusion and apartheid come down in revolutionary struggles—including, in our time, the walls against immigrants, prisoners, gays and lesbians, and historically oppressed races and peoples. They respect the otherness *and* commonness of the other so as to be more aware of her/his needs, especially the less powerful at present. They aim to ensure that we all treat one another with dignity.

4. The struggles strengthen the commons and expand de-commodified relationships and spaces. The commons is a non-commodified space shared by the community. Social democratic versions include such things as health care, education, social security—however imperfectly realized. However, does the struggle also support bringing the bottom up, expanding inclusiveness and participatory control? On the other hand, are autonomous sectors able to avoid commodification (avoid being turned into business products or services for sale)? Even if they cannot do so completely, can they maintain a political stance and active behavior that pushes towards non-commodity forms? More generally, how can the working class on small or large scales create forms of exchange that are or tend toward being de-commodified? Create markets (forms of exchange) that do not rule lives and livelihoods? Reduce the reach of commodification and capitalist markets on people's life?

5. The struggles enhance local control and participatory control. "Local" is not a geographical term, it means that decisions are taken as close to those involved as possible; participatory means that all those affected have a real voice in the decisions. This puts on the table the issues of who makes decisions and how.

Much of what we know as autonomous action is local and almost definitionally includes "local control" of some sort. Social democracy historically does not. Indeed, one of its hallmarks is the reliance on a large, bureaucratic, intrusive, and hard to influence state apparatus. This state was the target of a widespread working-class attack in the 1960s, which, however, was turned against the working class and used by the right wing to promote Neoliberalism. Can the working class make social democratic demands/struggles that include the demand and fight for local and/or participatory control? (There were aspects of this in some early war on poverty programs, but these were eliminated or co-opted once the US state saw danger in its "miscalculation" on this.) More generally, do "inside" struggles help support "outside" struggles?

Are there ways to move social democratic struggles towards more autonomous action? Example: battles for government support of urban gardening may also push for control through local, participatory democratic bodies, rather than city or state government. Factory struggles may begin as "inside," but the participants may come to organize themselves in assemblies, etc., take over and control production cooperatively, and then set up cooperative support across factories and other sectors (as happened in Argentina after its economic collapse). Indeed, many union struggles (the quintessential "inside" struggle) reached a turning point that transformed them into outside struggles as an examination of "general strikes" will show. However, even in autonomous developments, participatory control is not guaranteed, either at the level of writing the rules or in ongoing practice. So in the various areas of reproduction (health care, food, education, housing) and production, what would participatory democratic control look like, and how can it be fought for in ways that win in the specific area and decrease divisions in the class?

6. The struggles lead toward more time outside of capitalist control. In particular, this means a shorter work-week for the waged and unwaged. It means recognizing "women's work" as productive, creating income for those doing this work as well as expanding who does it. How can we ensure that a shorter waged-work-week does not further empower men relative to women? Or some class sectors over other class sectors? That is, how can victories in the realm of time be egalitarian?

7. The struggles reduce the staggering wastefulness and destructiveness of capital, of lives, time, material wealth, health, and environment (air, land, and water), but these reductions happen in ways that do not penalize other workers. Example: in the US there is huge waste (as well as profiteering) in the medical insurance bureaucracy. Single payer proposals could eliminate lots of that—but also throw many people out of their jobs, intensifying inequality. What will have to be done so these folks are not economically destroyed? Of course, from a working-class perspective, things like the military and weapons production are destructive to the point of insanity, so should be eliminated. Reducing waste of some sorts may benefit some, while not benefiting others (for example, if it leads to reduction of waged work time, it may not help mothers with kids)—so inclusion must be considered when "capitalist wastefulness" is addressed.

8. The struggles protect and restore ecological health. Struggles facilitate a healthier, more holistic approach to the planet. For example, battles to save jobs in industries that foster ecological disaster need to be addressed; there are now and will be such battles.

Land, air, and water are of crucial importance. Agribusiness, global commodification, bioengineering, and war lead to pollution, erosion, dams, flooding, deforestation, global warming, diminishing diversity, and the death of land and oceanic ecosystems. In replacing agribusiness as the mode of food production, closer human relations to food production are to be fostered.

9. The struggles bring justice. Too often, exploiters and oppressors have acted with impunity. Thus the real criminals must be brought to justice for healing to occur. Revolutionary justice is bottom up, and new forms of enacting justice should be consistent with the other revolutionary characteristics, e.g., "No" to capital punishment even for capitalists.

Beyond capital. We have located these characteristics of revolutionary struggles from our knowledge of histories of struggles (especially in the Americas) and our own experiences. We do not claim they are definitive, but we do see them as interlinked. Our hope is that this necessarily incomplete list of characteristics of revolutionary struggles (since revolutions in their nature will create unforeseen realities and characteristics) can be remembered to protect our struggles from not being turned back against us, as has too often happened in the past, and can help create a world beyond capital.

CONCLUSION: CRISIS—WAR—REVOLUTION

Revolutionary struggles of the character we described above are undoubtedly being unleashed in the Crisis. However, there is a terrifying mediator between crisis and revolution—*War*—giving a somber edge to our joy.

It would be a pleasant denouement if capitalism simply stops existing after a long slow process of dissipation and another friendlier mode of production and subsistence takes its place without anyone noticing. Perhaps for a long time what we call capitalism might be replaced without the name of the prevailing mode being changed. After all, there is no logical necessity for huge, terrifying creatures to always have huge, terrifying endings. Might we not wake up one morning, long after a constant threatening drone has stopped, and say to our mates, "The drone has stopped," then go out to meet a new day? Couldn't our capitalist rulers depart as quietly as the Communist bureaucrats of the GDR in 1989?

This kind of ending is possible, but not probable. The system has many indices and self-sensors (e.g., the revenues derived as profits, interest, rent) with immediate consequences and alarms for its rulers. A fall in any of these revenues alerts its recipients that something is dramatically wrong, and they will demand action from the state to return their profits, interest, or rents to an "acceptable" level. Given the often unspoken but widely shared recognition that such a fall in these revenues is rooted in a reduced availability of surplus labor and the increased cost of non-human means of production (due to the ecological struggles), the hypothesis is that this reduction in the rate of profit needs to be "corrected" by increasing exploitation of workers and reducing the costs of production (especially of raw materials) by shifting the cost of ecological regeneration onto the working class.

The previous history of crises indicates that the preferred path to increasing exploitation and reducing costs directly passes through war, violence, and repression to terrorize workers and separate indigenous and agricultural people from their attachment to their land and its wealth. Certainly the possibility of an irenic capitalism was negated in the early 1990s with the initiation of the "fourth world war" (against people and states that refused the neoliberal New Enclosures) immediately after the end of the "third world war" (against communist states).

In this crisis too there will be conflicts in a still-to-be-envisioned "fifth world war," which will not just involve repetitions of neoliberal wars intended to discipline a recalcitrant subordinate state into "playing by the neoliberal rules" of world trade (like the invasion and occupation of Iraq). That is why we began and now will end this tract on crisis and revolution with the fatal bullet that pierced Alexis Grigoropoulos' youthful body. It eternally reminds us that capitalism in the final analysis is a cold, violent, and murderous system. Thus, the most important step in planetary "harm reduction," while we traverse the trajectory from crisis to revolution, is to disarm the state and capital as much and as soon as possible.

BIBLIOGRAPHY

Many Midnight Notes publications are available at http://www.midnightnotes.org.

Abramsky, Kolya. 2008. "Gathering Our Dignified Rage." Available at http://zapagringo.blogspot.com.

De Angelis, Massimo. 2007. *The Beginning of History: Value Struggles and Global Capital.*

Boal, Iain, et al. 2006. *Afflicted Powers: Capital and Spectacle in a New Age of War.*

Bonefeld, Werner (ed.). 2008. *Subverting the Present, Imagining the Future: Insurrection, Movement, Commons.*

Carlsson, Chris. 2008, *Nowtopia: How Pirate Programmers, Outlaw Bicyclists, and Vacant-lot Gardeners Are Inventing the Future Today!*

Cleaver, Harry. 2000. *Reading Capital Politically.* Second edition.

Holloway, John. 2002. *Change the World Without Taking Power: The Meaning of Revolution Today.*

Linebaugh, Peter. 2008. *The Magna Carta Manifesto: Liberty and Commons for All.*

Midnight Notes. 1992. *Midnight Oil: Work, Energy, War: 1973–1992.*

———. 1997. *One No, Many Yeses.* [Available at http://www.midnightnotes.org.]

———. 2001. *Auroras of the Zapatistas: Global and Local Struggles in the Fourth World War.*

———. 2002. "Respect Your Enemies—The First Rule of Peace: An Essay Addressed to the U. S. Anti-war Movement." [Available at http://www.midnightnotes.org.]

P.M. 1985. *Bolo-Bolo.*

Shukaitis, Stevphen, David Graeber, with Erika Biddle. 2008. *Constituent Imagination: Militant Investigations Collective Theorization.*

Zapatista Army of National Liberation (EZLN). 2005. *The Sixth Declaration of the Lacandon Jungle* ("Part VI: How We Are Going To Do It"). [Available at http://www.inmotionmagazine.com/auto/selva6.html#Anchor-14210.]

Chapter 2

A DISCOURSE ON PROPHETIC METHOD: OIL CRISES AND POLITICAL ECONOMY, PAST AND FUTURE[1]

George Caffentzis

> "So Foxy Loxy led Chicken Little, Henny Penny, Ducky Lucky, Goosey Loosey, and Turkey Lurkey across a field and through the woods. He led them straight to his den, and they never saw the king to tell him that the sky is falling."
> —The Story of Chicken Little

I. THE AGE OF CHICKEN LITTLE

There is definitely a sense of crisis in the air and many a Chicken Little is running down the road to tell the king that the sky is falling. There is a lot to tell! On the one hand, the Peak Oil zealots are pointing to the "end of the era of cheap oil" and the beginning of a permanent emergency for a capitalism addicted to an ever-diminishing supply of petroleum. On the other, the housing bubble has burst followed by the inevitable pain of millions of people whose homes have been foreclosed. Add to this the collapse of dozens of financial corporations and the efforts of thousands of jittery bankers trying to calm the even more jittery anxieties of millions of depositors and stockholders and you get the sense that Nature and Capital are joining forces to write in bold letters across the social skies: THE END IS NEAR.

People like myself, who have lived through a number of crises "real or fancied," are not so easily aroused by the apocalyptic pathos that accompanies the Littles' announcement. I think back with a superior smile at Marx's almost childish rejoicing over the financial crisis of 1857–58 that inspired him to write the glorious midnight notebooks we now call the *Grundrisse*. He often wrote until 4:00 AM in the winter of 1857–58, fortified by "mere lemonade on the one hand but an immense amount of tobacco on the other ... so that I at least get the outlines clear before the *deluge*" (quoted in Wheen 1999: 227). I treasure the notebooks, but I frown on Marx's expectation that a mere financial panic would bring a world-system like capitalism to the brink of catastrophe. The deluge Marx was expecting then did not come (at least not for more than a decade). After studying literally dozens of financial bubbles (and their bursting) and of commodity price explosions (and their crashes)—indeed, since the 1857–58 crisis also involved the price of gold, there was a meeting of commodity price and financial bubble then as well—I have become blasé over the prophets of doom (who were often hoping to make some profit on the side!).

---

1   This chapter was originally given as a talk at the Left Forum, Cooper Union. New York, NY on March 16, 2008. It is being reproduced here with permission from the author, in slightly modified and updated form.

The themes I have harped on in my writing are that (1) capitalism is not only crisis-prone but it is also crisis-creative (so whenever one sees a crisis one should not assume this is a problem for the capitalist class, even though it might be one for individual capitalists, for a crisis might end by putting the capitalist class as a whole in a more powerful position), as Naomi Klein has recently reminded us; and (2) the hope to find a short-cut to go beyond capitalism through Natural limits (whether it be "Peak Oil" or "Global Warming") is understandable, but it is misplaced—the only path for a positive "transition" from capitalism is through a political recomposition of the working class internationally (Klein 2007; Caffentzis 1992). The problem with the optimists of either variety is that they tend to disarm the anti-capitalist movement and can make us vulnerable to dangerous political assumptions. In other words, I am more concerned about Foxy Loxy's murderous intentions than Chicken Little's inferences from experience, even though, *eventually*, of course, Chicken Little will be right!

For all my insouciance, however, my comrades and I knew that a major crisis of global Neoliberalism was on the agenda long ago. The first sign was "the Asian financial crisis," which was ignited by a wage rebellion in the Eastern Asia (South Korea, Indonesia, Thailand) of the globalization era in 1996 (Midnight Notes 1997). The subsequent banking crisis echoes in Russia, Argentina, and Brazil and the "dot-com" equities crash in the US called for a new phase of globalization, often called the "war on terrorism." The second crisis was instigated by the military failures of the US invasions of Iraq and Afghanistan, since they bode ill for a world regime that required military dominance to back its financial and ideological dominance (with the dollar the "god of the market" and the universalization of commodification as the practical maxim). When the unity of the series dollar-market-gun collapsed, a situation similar to the period between World War I and World War II opened up … So you see, I too had my prophetic globe tucked somewhere in my pocket. I just did not see this awaited crisis around every corner and did not want to play the role of a gleeful "Chicken Little" that Marx played 150 years ago (Bologna 1973).

It is time, now, for me to take out my prophetic crystal. However, I will not join Henny Penny and the others on the road to the king. I make no prophesies in this presentation. I will instead set the stage for the methodological analysis of the many prophesies concerning the coming crises that will come. My main negative maxims in this effort are:

- the rejection of "oil and energy exceptionalism," i.e., the view that oil and energy are so important for the capitalist system that the "rules of the commodity" do not apply to them (basic commodities are *still* commodities); and

- the rejection of the fetishistic view of oil and energy production as being classless and workerless. One can read books and books about the magnates, shahs, and sheiks of the oil world, and books and books about oil geology but never learn that oil and energy is produced in a class society by workers (i.e., the oil-producing proletariat) who are involved in a class antagonism with capital at the well head, across the oil regions, along the pipelines, in the tankers, and in the cities of oil-producing countries. Their struggle is crucial for world history, but it is rarely mentioned in the history books. Petroleum fumes apparently produce strange abstractions. The avoidance of class struggle that would be impossible with coal (where the struggle of the miners is always front and center) is commonplace for oil!

In this article I will examine the impact of "Oil, Energy, and Environment" *on* Political Economy. I will further limit my efforts in "comparative crisisology" today to the impact of oil prices and the relations of production in the oil industry on the political economy of Keynesianism and global Neoliberalism. Finally, I will compare the commonalities of and differences between the crisis now developing and the main crisis of capitalism that I (and many others still breathing) lived through, i.e., the crisis of 1973–1983. In doing so, I will sketch out the role of oil prices and rents in the general situation of the coming crisis.

In fact, there are many aspects of the present that have an eerie resemblance to the "energy crisis" of the 1970s. First there is the volatile oil price: on March 4, 2008 "the highest trading price, $103.95 a barrel on the New York Mercantile Exchange, broke the record set in April 1980 during the second oil shock. That price, $39.50 a barrel, equals $103.76 today, when adjusted for inflation," while a year later it hovers around $50 a barrel (Mouawad 2008). Second is war: the US military defeat in Vietnam is echoed in the military quagmire of Iraq and Afghanistan. Third is the ideology of scarcity and apocalypse: the present anxiety expressed by the Peak Oil enthusiasts is reminiscent of the Club of Rome's widely heralded "Limits to Growth." Fourth is the monetary anxiety: the dollar's loss of its hegemonic role in world exchanges (especially oil exchanges) is similar to Nixon's cutting of the connection between the dollar and gold. This last change is further reflected in a golden mirror: the $750 per ounce peak in 1980 is matched (though not in real terms this time) by the return and surpassing of its nominal peak (gold would have to reach about $1850 per ounce to equal its 1980 price adjusted for inflation) in early 2008. I feel I'm in a situation now that is similar to the one in 1980 when I wrote "The Work/Energy Crisis and the Apocalypse," i.e., I knew that a new political economy was on the agenda, but I did not know yet all of its lineaments.

II. OIL AND THE CRISES OF TWO BOURGEOIS POLITICAL ECONOMIES: KEYNESIANISM AND GLOBAL NEOLIBERALISM

My general argument is that the oil industry played a crucial role in the crises of both the political economies of Keynesianism and global Neoliberalism. This should not be surprising, for oil and its energy substitutes are basic commodities that are essential in the production of all commodities (including labor power). Consequently, any specific form of capitalism in this era must be able to integrate the energy branches of industry, and the dominant political economy must conceptualize and strategize how this is to be done. Not any kind of integration will do. A particular energy regime must be compatible with and support the prevalent mode of the exploitation of labor. Once this integration breaks down and the ruling political economy confronts too many anomalies and bungles too many struggles, a crisis ensues both on the level of practice and theory. In this section I will sketch, first, how Keynesianism from the 1940s to the early 1970s was in sync with the international oil industry, and then how a revolution in the relations of property in the oil industry played such a central role in the overturning of Keynesianism. I do this because it can provide a reference point for our analysis of the present crisis and, hopefully, of how it can be resolved with greater power for the anti-capitalist forces of the planet.

1. KEYNESIANISM AND ENERGY

Keynesianism is many things, of course. Like Marxism, it is closely related to the life and thought of its "founder," John Maynard Keynes, and therefore to its founder's political and theoretical situation. This is not the place, however, to deal with these biographical and contextual matters. I will simply refer to a tradition of reading Keynesianism that emphasizes its class characteristics and therefore is most useful in analyzing the crisis of the 1970s (cf., Caffentzis 1999; Negri 1994; Cleaver 1979; De Angelis 2000). Let me present the key elements of this interpretation:

- Keynes (and his supporters) recognized that, since the Russian Revolution, the working class had become a crucial *independent variable* in the functioning of capitalism. It was both an antagonist and a motor of capitalist development. No longer could it be relegated to the status of "laboring species" (i.e., defined as a race that works) or a "factor of production," since it could step out of the system.

- For Keynes, the wage and therefore *the wage struggle* has become the center of capitalism, because it drives effective demand and must be kept in balance with increases in productivity. The state plays a vital role in this political economy, i.e., as a homeostatic mechanism interposed between classes to guarantee the productivity deal between the classes.

- Keynes also realized that "the enormous accumulation of fixed capital embodied in the assembly-line factories required a proportionate accumulation of capital in the working class ('human capital' as it was called later)" (Caffentzis 1992: 231).

This energetic conception of the working class and its reproduction is crucial to recognizing that the main power capital had over workers was in its ability to chart "technological paths of repression." It was crucial therefore for capital to have access to a cheap, dependable source of "counter-energy" that could power the machinery necessary for the production of what Marxists call "relative surplus value." What Renfrew Christie summarized long ago as a general condition of capital was even truer of Keynesianism, "It is only from capital's need for machines so that it can win the class struggle, and from energy's special relation with machines, that energy receives its particular importance [in capitalism]" (Christie 1980: 13).

The energy regime that was fashioned by the US, the UK, and the "Seven Sisters," the cartel of British and US transnational oil corporations, was typical of the

Keynesian period (roughly 1945–1973). The blatant collusion (later tempered into a "systems analysis" approach) among the major oil companies to set the price of oil both in the US and internationally was seen as simply the most extreme of these pricing arrangements found throughout the "monopolized" industries of the US and Europe at the time. The arrangements (which began as openly cartelistic and then became covert) made for a very predictable price (on average about $20 a barrel in real 2008 dollars according to my rough calculation) for a quarter of a century (cf. Blair 1976 on the "International Control Mechanism"). There were other, less contractual methods that were used to keep oil "cheap and predictable" in the face of anti-colonial struggles in the oil-producing regions of the planet. First, for most of this period, the US oil industry was the world's "swing" producer, and hence "uppity" countries like Iran in 1953 could be isolated and boycotted out of the market, if need be, with the US making up the difference in supply to support the international price. Second, if any oil-producing nation's working class and/or capitalists decided that they would take control of the oil production on their territory, then they would face a coup (as with Mossadeq's efforts in Iran in 1953) or a direct invasion (as in the case of Roosevelt's deal with King Saud in 1945 that committed the US to intervene militarily to defend the Saudi throne).

The Keynesian energy regime that brought together the "Seven Sisters" with the US and Britain military to organize the "stability" of the oil areas of the world, especially the Middle East was a crucial part of the larger Keynesian political economy. This regime—what Leonardo Maugeri calls "The Golden Age of Oil" (Maugeri 2006)—guaranteed a steady supply and low price of petroleum that made it possible to substitute machinery for labor at a rapid pace, with the added bonus of eliminating the centrality of obstreperous coal miners in the class struggle of Europe and the US. Maugeri, in the typical fetishized style of oil commentators, writes:

> Oil's success in fuelling modern economic development brought about the fastest process of energy source substitution in the history of humankind. As late as 1950, the chief energy source of the first industrial revolution, coal, still reigned over all rivals, supplying about 65 percent of world energy needs. But by the mid-1960s, oil had supplanted coal as energy king (Maugeri 2006: 77).

THE CRISIS OF KEYNESIANISM: 1973–1980.

The crisis of 1973–1980 was one of a whole political economy, it was not "just" an "energy crisis." It was a crisis of class strategy and theory as well as of unemployment, rust belts, and austerity budgets. My comrades and I at the time, in trying to express this point, called it a "work/energy crisis" (Midnight Notes 1979, Caffentzis 1992). What was at stake in the 1970s was a general relationship between classes that had been built up in the US from the New Deal in the 1930s. True, the dominant theme of the time was focused on oil and energy issues, especially questions of quantity (were the Club of Rome's claims correct?), form (was the nuclear-powered or the solar-powered economy going to be the alternative to oil?), and price (was there a tendency for the secular increase of oil prices?).

We argued at the time that the key issue was that workers internationally (in the US and Western Europe, as well as in the anti-colonial struggles in the so-called Third World) were both demanding a wage for the unwaged work they did and were imposing wage increases (beyond productivity increases) that put capital's accumulation strategy at risk. The crisis was first and foremost one of work and wages. Its "energy" aspect was due to capital's use of energy prices to overcome the struggles around and against work.

The relation of the "energy crisis" to the "crisis of Keynesianism" is the following: the class struggle in the US and Europe took the form of a direct wage struggle either at the factory proper or the "social factory" (by coalitions of waged and unwaged workers); while the class struggle in the oil-producing areas was an attempt to take control of the rents and transferred profits that were accruing to the "Seven Sisters" since the early twentieth century (by coalitions of national capital and the working class waged and unwaged).

These two simultaneous rebellions of the early 1970s struck at the heart of the Keynesian universe. The struggle in Europe and North America put into question the wages/productivity equation that was at the center of the accumulation process. The one in the oil-producing parts of the former colonialized world was demanding back its national resources (especially oil, a commodity that was being produced at a very high level of organic composition, *pace* Emmanuel!) that had been deliberately devalued and had been turned into a source of super-profits by the corporations of the imperialist powers, especially the US and UK. These two polar rebellions, taking place simultaneously, sabotaged the basic mechanism of Keynesianism, *viz.*, responding to workers' struggle in the factories of Detroit for "more money, less work," by automating the assembly line using cheap energy provided by a compliant oil-producing proletariat a world away.

These simultaneous struggles created the specter of stagnation, the stationary state, and "zero growth" for capital's theorists. Indeed, if there were political forces that could have created some kind of "political recomposition" at this time, world history would definitely have taken a different turn in the 1980s. Certainly, there was no "International" then that could have achieved (or even thought of) such a project.

Instead of *recomposition*, the crisis of Keynesianism brought *decomposition* for the working class internationally; the polarity of the very social forces and movements that triggered the crisis of Keynesianism was used against each other. Instead of creating a crisis *of* capital, capital turned the crisis against the working class internationally. The nationalization of the oil-producing companies in many countries took place in the early 1970s and the imposition of steeper oil rents returning to the national coffers led to the oil boycott of 1973. OPEC presented itself as the first commodity trading organization that would realize the dreams of the International Economic Order and reverse the injustices of centuries of colonialism and imperialism. This vision, however, was translated at the other pole of the Keynesian world as a wage nightmare. Unemployment, abandoned factories, austerity budgets, welfare cuts, the prison-industrial complex, began to take shape in the recessions of the middle and late 1970s. These signs of working class defeat were all laid at the door of the "Arabs" or "OPEC." The tools of vilification and the powers of racism were turned against workers at the other pole of the class struggle.

There was clear evidence that this stage of the crisis (when one crisis-provoking pole was used against the other) was planned, and the Yom Kipper War boycott met with the concealed approval of strategists of capital like Henry Kissinger (*the* Foxy Loxy *par excellence* of the time). As Mario Montano wrote long ago: "Behind the ritualistic position of diplomatic adversaries that the US and OPEC countries necessarily entertain during international bargaining sessions, stands their Holy Alliance" (Montano 1992: 127). This was the time when the Arab oil sheik was projected to be a thief of the US workers' future. Indeed, when the Iranian Revolution in 1979 led to another spike in the oil price, US workers expressed open hostility to Iranian immigrants and students in the streets and campuses of the US. What could have meant a major crisis for capitalism, however, became a pretext for cutting the wages of workers in Western Europe and North America, while creating an investment flow (then called "petrodollars") that was used to make loans to formerly-colonized countries (imposing a flexible interest rate that the "subprime mortgage" was to emulate in the early twenty-first century!) that in the 1980s forced them to near bankruptcy and then, under the pressure of the World Bank and IMF, to neoliberalize their economies. What a foxy trap!

2. GLOBAL NEOLIBERALISM AND OIL

This trap was successfully sprung and it immobilized worker struggles both in the First and Third Worlds. Keynesianism, however, had to be abandoned. The "Chicago Boys" and Neoliberalism took over theoretical and practical hegemony throughout the planet. This transformation was politically legitimated in the neoliberal regimes that took power at the end of the oil price crisis in 1979 and 1980, first with Thatcher in Britain, then Reagan in the US, and then through the "debt crisis" of 1982, the IMF/World Bank imposition of neoliberal structural adjustment programs (SAPs) throughout the Third World. These neoliberal regimes both in the "center" and in the "periphery" of the early and mid-1980s made it possible to set up the political arrangements that would make for a successful globalization of neoliberal capitalism on three counts: (1) the working classes of the neoliberalized world gave up on the Keynesian productivity deal in North America and Western Europe (wages would be correlated to increases in productivity) and the post-colonial developmentalist deal in the Third World (import substitution and the creation of a local market would generate employment); (2) the state was reduced as the place of surplus distribution (with tax cuts and austerity budgets); (3) the complete destruction of the "Chinese walls" against the free flow of capital in the form of money, equities, and physical equipment constructed during the long period from WWI to the end of import substitution regimes in the late 1970s.

Let me comment on each of them and determine their relation to the oil and energy industry.

In the Keynesian period, the state stopped being the exclusive club of collective capital and was interposed between the classes (and by a law of dialectics, it was divided against itself). In the neoliberal era the state abandoned this mediating role. It had to also abandon its role as the primary overseer of working class reproduction and regulator of capitalists' exchanges. The dictatorship of the market was to prevail. As Massimo De Angelis nicely put it, the state's job was to impose a practice of "good governance," i.e., "every problem raised by struggles can be addressed on condition that the mode of its addressing is through the market" (De Angelis 2007: 89). The "global" path to Neoliberalism is indicated by the fact that the formalization of neoliberal policies was the adoption of Structural Adjustment Programs (managed by the central agencies of global collective capital, the IMF and World Bank). Moreover, the rise of the World Trade Organization with its legal system that made it possible for corporations to sue sovereign states as standard procedure symbolized the triumph of this transformation in the 1990s.

The next feature characteristic of global Neoliberalism was the totalization of commodification and monetarization (what a Latinate sentence!). The previous barriers to commodification, especially those aspects of life involved in the reproduction of labor power, were to be battered down. Similarly, the barriers to the free flow of capital were to be annihilated, letting a tidal flow of hard currency—dollars, yen, pounds, marks (and eventually euros)—enter into previously unmonetarized parts of the world-economy. "Financialization," not industrialization, became the most obvious feature of global Neoliberalism, so that "hard currency (not labor) is the measure of all things."

The class nature of the global neoliberal deal is that the winners—those willing and able to "swim" in the seas of the free market—will receive substantial increases of income, *not wages*. (Indeed, wages were displaced as the primary class relation in the neoliberal economy by "ownership" income like equity in stocks or real estate.) Workers would be paid either far beyond (if you were neoliberally graced) or far below (for the majority) their "average individual productivity." The two "prices to pay" for this opportunity to "play in the field of dreams" is the loss of guarantees (since every worker was in competition with workers around the world) and the increasing division in the working class both nationally and internationally (since most workers were either unwilling or unable to "swim"). Inevitably, the neoliberal era brought about ever-widening wage divisions within the working class (with shining city centers surrounded by miles of slums), waves of immigrants, and the experience of "new enclosures," both in terms of the direct attack on communal land and other common resources.

For the oil and energy-producing proletariat a corollary of these axioms of a globalized neoliberal political economy is that the collective ownership (through the state or through communal rights) of the energy resources (especially oil and natural gas) of the national territory had to be abrogated. Thus the oil-producing proletariat's rent claims on international capitalism (mediated by the state) were to be declared null and void, i.e., the birthright of millions was to be sold for a bowl of spicy pottage. Under the dictate of the new political economy, all moments of the hydrocarbon energy cycle producing the most basic of commodities for contemporary capitalism—from ownership of the subterranean resource to extraction to refining to shipping—had to be commodified. The rules of the global market had to determine its oil price (especially since its price included a tremendous transfer of surplus value from the rest of the system in the form of profits and rents). Thus the oil and energy regime was to be determined by a commodity market similar to the emerging "spot" market. No longer could the global economy depend upon deals made on the basis of a price structure managed either by the Seven Sisters or by OPEC.

THE CRISIS OF GLOBAL NEOLIBERALISM, ITS ENERGY ASPECT

These were the dictates of global Neoliberalism. Though many of them were obeyed in dozens of Structural Adjustment Programs, those pertinent to the oil and gas industry were not, i.e., the attempt to undo the nationalizations of oil and energy that took place largely in the late 1960s and early 1970s, and to dismantle OPEC have failed, even though the spot market seemed to promise a "neoliberal" solution for the organization of oil and energy corresponding to the "globalization" of other commodities continues to operate. I read the failure to change the property relations in the oil and gas fields of Saudi Arabia (2001), of Russia (2004), of Venezuela (2002), of Iran (2007), and especially of Iraq (since 2003), along with many more "minor" setbacks, as crucial "events" in the larger failure of the neoliberal globalization model (Caffentzis 2004a and 2004b). For if energy commodities, the most basic of commodities, cannot be managed by neoliberal globalized means, this mode of accumulation is a dead letter in the long run.

We must remember that the nations listed above are the largest oil producers with the largest oil reserves on the planet. Consequently, the inability to shunt Iraq to a new neoliberal oil track, even when US troops have occupied it for five years, is a glaring testimony of the US government's impotence in "managing" the political terrain. Add to this gigantic failure the stalling of the neoliberalization of the Saudi gas industry after 9/11, the inability of the US government to protect Exxon from the Russian state, the failure of the US-supported coup against Chávez, the failure of the campaign to depose the Iranian state (disguised as an effort to stop the building of a nuclear weapon), the inability to gain concessions from OPEC on pricing, and one gets a dismal picture of the US government's capacity to play the rule enforcer of the neoliberal global order.

We must also remember that the so-called "minor" difficulties are not minor at all when added together. Some examples include:

• a long-standing and now armed rebellion of the local inhabitants demanding the rights to the petroleum under their feet in the Niger Delta;

• the "gas war" in Bolivia that pitted indigenous peoples against the expropriation of the hydrocarbons resources of the country;

- the Zapatista rebellion against the extraction of the oil reserves of the state of Chiapas, Mexico.

What we are seeing here are flash-points of the "fourth world war" that Sub-comandante Marcos has so eloquently spoken about. Capital is now driving exploration and extraction of oil to the "margins" of the world (where communalist ethics still prevail among indigenous people) and it is confronting a tremendous communalist resistance. In a hundred different spots of Africa, Latin America, and Asia, a "petroleum common" is being defended, often by force of arms. As Steven Colatrella has called it, there is a "political Hubbert's curve" that is taking shape under the pressure of a myriad of "micro-struggles" between the oil companies and the indigenous peoples who are imposing a major barrier to capitalist expansion of the oil industry. The "war of the flea" is so powerful partly because it is not categorized as a "war" at all!

Not accidentally, this crisis of the oil industry coincides and interacts with a crisis of the US proletariat, which is seeing its own future in the form of income "outside" the wage being devastated. The dream of wealth beyond work has been the proletariat's since its birth in the "Land of Cockaigne." With the inability to increase wages through collective struggle beginning in the mid-1970s and the increase in employment of women and children as the only way to maintain the family income, the US proletariat has been trying to find other ways to survive and prosper. These ways have been increasingly individualistic and parasitic on the market. In the 1990s many workers hoped to hit it big in the world of the stock market and in the stock options that were increasingly offered by companies in lieu of wage increases. In the boom, many became millionaires "on paper." When the "dot com" crash came in 2000–2001, the dream paper became worthless (and workers suffered more than capitalists). Almost immediately after the "dot com" crash, however, a housing price boom began to take off. This boom was also fueled by the neoliberal reorganization of the credit industry that made swift and unregulated movement of loans for real estate property possible. The previously-excluded sectors of the working class (blacks, immigrants, poor whites) demanded entrance to the so-called "American Dream" and they got it, but they had to swallow it with a poisoned pill (the sub-prime mortgage). This boom was pushed to its limits, and also has now crashed, this time with millions of workers (especially the late comers) homeless and pensionless.

The "class deal" Neoliberalism has offered to the "ambitious" and "energetic" part of the US working class is now dead. This constitutes a major crisis of neoliberal capitalism for the working class in the US, whereas the inability of imposing the neoliberal deal for the oil industry internationally is a crisis *for* capital. That is why one must be very careful in articulating what sense of "crisis" one is using at any moment. The political question of our day is whether capital will be able to turn the crisis from itself into a crisis of the working class internationally. The "war on terrorism" and the "surge in Iraq" have been military/ideological efforts to turn the US working class' catastrophe at home into the basis of a renewed effort to accomplish the goals of neoliberal capitalism abroad. Will capital again be able to do what it did in the previous crisis of 1973–1980? Certainly the Bolivarian movement in Venezuela has recognized the danger that such a possibility poses and has taken some steps to respond to it through an offer to sell, at a steeply-discounted price, oil to low-income communities in the US. This provides a model for class solidarity between the two poles of global Neoliberalism.

If capitalism is able to survive this period, one thing is now clear: *a larger state role will be decisive*. Inevitably, neoliberal political economy's main effort—to take state power out of the sphere of working class appropriation—will have to be compromised. The sovereign wealth funds that are now proliferating across the planet are signs that the state's role in investment will be crucial once again in the political economy of the coming period.

Will this huge planetary surplus (represented by the surplus value transferred into rents and profits that are being appropriated through oil prices by the states of oil-producing countries) be invested in a new "energy" regime not based upon the exploitation of work? Could the feared high price of oil become the lever for a transformation both of the energy and power problem of the planet? That will depend on whether this time around a relation of solidarity will be forged between the oil producing and the US proletariats.

This solidarity certainly will not emerge by simply calling for the US proletariat to stop being oil-consuming "hogs" and transform themselves into solar "angels." After all, the "down side" of Hubbert's Curve, in a sense, could be seen as a potential payback for a century of exploitation, forced displacements, and enclosures. It appears that the capitalist class is unwilling to pay reparations to the peoples in the oil-producing areas whose land and lives have been so ill-used. Capital's resistance to reparation is suggested by its horror, for example, of paying the Venezuelan state oil taxes and rents that will go into buying back land that had been expropriated from *campesinos* decades ago, and giving it to their *campesino* children or grandchildren. But Venezuela is just one country. After all, shouldn't reparations be paid to the people of the Middle East, Indonesia, Mexico, Nigeria, and countless other sites of petroleum extraction-based pollution over the last century? Capital wants to be able to control the vast transfer of surplus value that is being envisioned in the discussion of a new post-crisis energy regime, it does not want to see the surplus spent "unproductively," i.e., in a way that is not functional to accumulation … like paying reparations.

BIBLIOGRAPHY

Blair, John M. 1976. *The Control of Oil*. New York: Random House.

Bologna, Sergio. 1973. "Money and Crisis: Marx as Correspondent of the *New York Daily Tribune*, 1856–57." Translated and printed in *Common Sense*, Nos. 13–14.

Caffentzis, George. 1992. "The Work/Energy Crisis and the Apocalypse." In *Midnight Oil: Work, Energy, War 1973–1992*. Edited by the Midnight Notes Collective. Brooklyn, NY: Autonomedia.

———. 1999. "On the Notion of a Crisis of Social Reproduction: A Theoretical Review." In *Women, Development and the Labor of Reproduction*. Edited by Maria Rosa Dalla Costa and Giovanna Dalla Costa. Trenton, NJ: Africa World Press.

————. 2004a. "Oil, Globalization, and Islamic Fundamentalism." In *Globalize Liberation: How to Uproot the System and Build a Better World*. Edited by David Solnit. San Francisco: City Lights.

————. 2004b. "The Petroleum Common." In George Caffentzis, *No Blood For Oil!* accessed at http://www.radicalpolytics.org.

Christie, Renfrew. 1980. "Why Does Capital Need Energy?" In *Oil and Class Struggle*. Edited by Peter Nore and Terisa Turner. London: Zed Books.

Cleaver, Harry. 1979. *Reading Capital Politically*. Austin, TX: University of Texas Press.

De Angelis, Massimo. 2000. *Keynesianism, Social Conflict, and Political Economy*. London: Macmillan.

————. 2007. *The Beginning of History: Value Struggles and Global Capital*. London: Pluto Press.

Montano, Mario. 1992. "Notes on the International Crisis." In Midnight Notes Collective, *Midnight Oil: Work, Energy, War 1973-1992*. Brooklyn, NY: Autonomedia.

Maugeri, Leonardo. 2006. *The Age of Oil: The Mythology, History and Future of the World's Most Controversial Resource*. Guilford, CT: The Lyons Press.

Midnight Notes Collective. 1979. "Midnight Notes 2: No Future Notes." Available on http://www.midnightnotes.com.

————. 1997. *One No and Many Yeses*. Midnight Notes 11. Available on http://www.midnightnotes.com.

Mouawad, Jad. 2008. "Oil Tops Inflation-Adjusted Record in Set in 1980." *New York Times*, March 4.

Negri, Antonio. 1994. "Keynes and the Capitalist Theory of the State." In Michael Hardt and Antonio Negri, *Labor of Dionysus: A Critique of the State-Form*. Minneapolis: University of Minnesota Press.

Wheen, Francis. 1999. *Karl Marx: A Life*. New York: W. W. Norton & Co.

Chapter 3

BUILDING THE CLEAN ENERGY MOVEMENT: FUTURE POSSIBILITIES IN HISTORICAL PERSPECTIVE

Bruce Podobnik

As the first decade of the twenty-first century draws to a close, concern is growing around the world about the stability of the global energy system. People from all walks of life—including students, scientists, corporate executives, and government officials—are coming to recognize that serious threats are being fueled by conventional energy industries. Wars in centers of oil production in the Middle East are generating repeated political crises; energy price spikes are having economic reverberations; and ecosystem disruptions caused by climate heating are causing devastation all across the globe. Never before has the need for a clean energy revolution, capable of addressing many of these problems, been so apparent.

Efforts are now underway to draw concerned citizens into a massive, global movement dedicated to pushing this clean energy revolution forward. Even though the challenges facing this movement are significant, we can draw inspiration from far-reaching transformations that were achieved in earlier centuries. Indeed, the historical record shows that the world's energy industries have gone through periods of quite rapid and far-reaching change. This suggests that an even more fundamental change, toward a more sustainable energy system, can be achieved if a mass movement of people—from all walks of life and all regions of the world—can be mobilized.

In order to arrive at an understanding of the changes that can be achieved in the future, it is important to see what has been accomplished in the past. This chapter briefly describes major shifts that have occurred in the world's energy systems over the last two centuries (see my book *Global Energy Shifts* for a fuller discussion of these events). As will become clear, social struggles of various kinds have played key roles in these earlier shifts. We can learn from these earlier struggles, and help strengthen the global movement that is emerging to push for a clean energy revolution in the coming years.

GLOBAL ENERGY SHIFTS IN HISTORICAL PERSPECTIVE

The first modern energy system, based on coal, grew steadily in the nineteenth century and reached maturity in the twentieth century. Indeed, coal went from providing about 10 percent of the world's commercial energy in 1800 to over 60 percent in 1913. The second modern energy system, based on oil, expanded much more rapidly. In 1913, oil provided only around 5 percent of the world's commercial energy. By 1970, though, oil was supplying around 50 percent of the world's energy. Natural gas has also undergone a rapid process of growth, growing from 6 percent of the world's commercial energy in 1946 to 24 percent in 2000. What is perhaps most remarkable here is the speed with which huge volumes of energy can be incorporated into the world-economy. It is also important to remember that the exponential growth of oil oc-curred during a time of two world wars and a great depression. Clearly, massive shifts in global energy systems can take place even in very challenging circumstances.

Not all energy industries have undergone such rapid trajectories of growth, of course. Nuclear power experienced some expansion in the 1970s and 1980s, but then it plateaued at about 7 percent of the world's energy supply. Energy from hydro-electric facilities, meanwhile, underwent slow growth throughout the twentieth century—so that, by the year 2000, hydro-electricity was providing about 3 percent of the world's commercial energy. Meanwhile, all modern renewable energy systems (including wind, solar, geothermal, and modern biomass) provided only around one half of one percent of the world's commercial energy in the year 2000. This is, of course, a sobering statistic for anyone concerned with the environmental sustain-ability of modern societies.

It is important to point out that new energy systems have been superimposed on top of older systems, which themselves continue to expand. The shift toward in-creased dependence on oil and natural gas, for instance, has been layered on top of a still-growing coal system. At some point in this century, however, a deeper shift will have to be achieved, in which systems based on coal, oil, and natural gas are replaced by something else. Since coal and unconventional petroleum resources like oil sands are very plentiful, pressures are already building to shift toward greater reliance on these highly-polluting reserves. And some are trying to take advantage of climate heating concerns in order to promote a new generation of nuclear power stations. With the right social pressures, though, a shift toward cleaner, more environmental-ly-benign energy systems can be achieved in this century.

If we look at patterns of energy consumption, another important feature of the global energy system is revealed. Up until the end of WWII, nations in the global north were relatively self-sufficient in commercial energy terms. Since then, how-ever, countries in the global south have been exporting energy resources to the north at a growing rate. This energy trade has intensified long-standing global inequalities in levels of energy consumption. Currently, the average citizen in the United States consumes at least five times as much as the world average, ten times as much energy as a typical person in China, and over thirty times more than an average resident of India. Even in such major oil exporting nations as Venezuela and Iran, per capita consumption of energy is less than one-half and one-quarter of the US average, re-spectively. A starker illustration of these inequalities is reflected in the fact that about 40 percent of the world's population—over 2 billion people—still have no regular access to commercial energy products in their homes. One of the central challenges facing the world in this century will be to ease these patterns of inequality in the global energy system, which fuel resentment and climate policy paralysis.

The final aspect of the global energy system that must be highlighted is that its historical evolution has been strongly impacted by mass movements of people, who have struggled to change the energy trajectories of their communities and nations. Growth patterns in energy sectors are not dictated by the kinds of resources that are found in the ground. Instead, their evolution is driven by decisions made by governmental officials and corporate executives, and by resistance movements created by workers and citizens. As the next section shows, societies have been set on fundamentally new energy paths because of complex interactions between these different kinds of social conflict. We can learn from these earlier struggles, and increase our ability to create an effective mass movement on behalf of a clean revolution in the coming decades.

SOCIAL STRUGGLES AND GLOBAL ENERGY SHIFTS

Dynamics of social contestation have had far-reaching impacts on the evolution of global energy systems during the last two centuries. Specifically, three dynamics—those of geopolitical rivalry, commercial competition, and grassroots mobilizations—have interacted to produce energy shifts that have sometimes been quite rapid and far-reaching. Let me briefly describe how these social dynamics have interacted to produce changes in global energy systems over the last 200 years.

Throughout the history of the modern world, nation states have struggled against each other to win greater geopolitical and economic power. This geopolitical competition has often prompted political leaders to intervene in energy industries, since access to energy is closely linked to military and economic success. For instance, governments in Western Europe strongly promoted the expansion of coal mining after the Napoleonic Wars. As warfare became increasingly industrialized, this state intervention on behalf of coal intensified. What state agents helped set in motion was then greatly accelerated by private investments. Indeed, corporate competition to gain profits in coal-related mining and transportation sectors drove massive investment booms in new coal systems in the nineteenth century.

Starting in Western Europe, and then spreading to North America and beyond, public-private synergies fostered the growth of a global energy system based on coal. In fact, the tremendous expansion the world-economy went through in the nineteenth century was, in large part, made possible by the global diffusion of this coal system. However, social dynamics began shifting against coal in the late 1800s, thereby setting the stage for a rapid and far-reaching expansion of a new global energy system based on oil in the next century.

Just as oil was emerging as a distinct energy industry, coal mines were shaken by waves of labor militancy that disrupted operations and undermined confidence in that established energy system. From the 1880s to the beginning of the First World War, miners in Western Europe and North America were able to form national unions and carry out large strikes. During the Second World War and its aftermath, an even more dramatic wave of unrest swept through coal industries across the world. Coal miners distinguished themselves as the most militant of industrial workers during this era. They succeeded in improving wages and working conditions, and their struggles brought about a reduction in death rates in mines across the world. At the same time, this militancy had the unintended effect of pushing government authorities and private investors toward greater reliance on emerging oil industries. Overall, dynamics of social conflict clearly had the capacity to alter the trajectory of coal and open a window of opportunity for the expansion of oil.

Similar patterns played themselves out over the next century in the international oil system. The early consolidation of a petroleum-based industrial regime can be traced to the 1890s, when a naval arms race between the era's most advanced states began. By the onset of the First World War, most leading navies were in the process of conversion to oil. Government purchases of oil allowed private companies to invest increasingly large amounts of capital in new oil-related infrastructures. This public-private synergy accelerated during the inter-war period, and even more intensely after the Second World War, as military power and economic growth became ever-more reliant on oil-powered aircraft, vehicles, and ships. The expansion the world-economy experienced in the second half of the twentieth century was in large part made possible by the global diffusion of this oil-based energy system.

The combination of increasing social conflict in coal, and growing public-private support for oil, shifted the world solidly in favor of petroleum in the post-World War II period. In a time of massive new discoveries of crude oil reserves, shifting toward this energy resource seemed rational. However, by the 1970s it became clear that over-reliance on a single energy resource, which was itself increasingly produced by a relatively small number of nations, exposed the world-economy to a substantial level of peril. The first major disruption came during the 1970s, when the global oil system was fundamentally transformed by a wave of nationalizations. Governments in the most important production zones of the Middle East and Latin America seized ownership of a huge proportion of the world's oil reserves. Strikes by oil workers and mass demonstrations by citizens helped push political leaders forward on these nationalist campaigns.

Just as an earlier outbreak of labor militancy in coal created space for a shift toward oil, the nationalist shocks that swept through the international oil system in the 1970s created a temporary shift toward more efficient forms of energy consumption throughout the global north. Indeed, a variety of solar, wind, and other alternative energy systems went through a first phase of development and commercialization during this period of turmoil in the international oil system.

Though the threat posed by nationalizations was eventually contained, more intractable threats have emerged in centers of oil production in recent years. Repeated wars between western powers and Iraq have destabilized the Middle East, while rising tensions with Iran also pose significant challenges. Meanwhile, groups such as Al Qaeda are hoping to use new tactics to destroy energy infrastructure and sow unrest in this key oil region. Just as rising labor unrest in coal had the unintended effect of accelerating a shift towards oil in the twentieth century, growing social conflict in the Middle East could have the unintended effect of speeding a transition toward new energy technologies in the twenty-first century.

Environmental groups have also altered the trajectory of major energy industries in recent decades. This has been most evident in the case of civilian nuclear power. After having been heavily promoted by governments and utility companies during the 1950s and 1960s, a series of nuclear accidents helped spur the creation of massive social movements against this energy system in the 1980s. From the US to the UK, from West Germany to Sweden, and then in the Soviet Union and Japan, people marched in demonstrations and engaged in acts of civil disobedience against nuclear power plants and nuclear weapons. The ability of mass movements to contain nuclear power in many countries again demonstrates the capacity of grassroots mobilizations to alter the evolution of energy industries on a wide scale.

The historical record clearly demonstrates that social struggles of various kinds have had broad impacts on the evolution of global energy systems. Miners employed in the most dangerous, exploitative kinds of occupations organized themselves into unions and transformed coal industries across the world. Citizens and political leaders in oil-exporting nations mobilized against the wealthiest multi-national corporations in the world, and succeeded in nationalizing huge oil reserves. And environmentalists across the world created mass movements that restricted the expansion of nuclear power. In each case, mobilization strategies emphasized the creation of broad coalitions that drew in people from all walks of life, and from all political backgrounds. Similarly, moderate and radical activists each played important roles in coal unions, nationalist struggles, and environmental campaigns. These earlier campaigns can inform and inspire those who are now beginning a new, mass-based effort dedicated to refashioning the energy foundations of our world.

BUILDING THE CLEAN ENERGY MOVEMENT

The world stands at what is likely to be its last window of opportunity to shift toward a sustainable energy system, and avoid the full impact of the crises being fueled by conventional energy industries. There are some who argue that the trajectories of global energy systems are hard-wired, and that they cannot be fundamentally changed. But the historical record shows that social struggles have altered the course of large-scale energy industries in the past. Although mobilization strategies must be adapted to present circumstances, there are a few important lessons that can inform those who are now working to create a clean energy revolution.

The first important historical lesson is that movements that have succeeded in reshaping large-scale energy industries in the past have relied on the mobilization of large numbers of people. The coal system, for instance, witnessed the emergence of strong labor unions in virtually all important mining centers. Massive numbers of miners marched, went out on strike, and directly confronted company owners and government agents during their struggles to reform their industries. Similarly, the nationalist wave that swept through the oil system was propelled forward by huge demonstrations of citizens and oil workers across the Middle East and Latin America. And the containment of nuclear power was achieved by similarly large mobilizations of people in countries all across the global north.

There has been a tendency for those concerned about contemporary energy dangers to focus on getting government officials, corporate executives, and media celebrities to acknowledge the crisis. But the historical record shows very clearly that deep, enduring changes in energy industries require the mobilization of mass social movements. We cannot simply wait for visionary politicians to forge the way, though they will be an important part of the solution. We cannot rely on new energy entrepreneurs to resolve the crisis, though again they will be crucial allies. And we cannot be satisfied when a few media celebrities dramatically describe the dangers that are on our horizon, though their involvement is certainly helpful. Instead, history shows that we must draw large numbers of people, from all across the world, into a broad social movement that fights for fundamental change in the global energy system.

The second lesson that can be drawn from the history of social struggles in energy industries is that it is important to attract citizens from many different political and ideological backgrounds. In the case of coal, anarchists, socialists, and apolitical miners were pulled together into broad-based unions that drew strength from their ideological diversity. Similarly, a wide variety of motivations propelled citizens in oil-producing countries into mass movements that demanded the seizure of petroleum properties. Some were driven by patriotism and nationalism, while others were anti-western in their orientation, but they all agreed on the need to take control of the oil in their respective nations. And in the case of the anti-nuclear struggle, a remarkably diverse movement emerged that included housewives, green activists, scientists, and many other groups. In each of these earlier campaigns, the creation of broad, ideologically-diverse coalitions was essential to the rapid expansion and eventual success of each movement.

Fortunately, the emerging clean energy movement shows signs of following this ideologically-inclusive, coalition-building strategy. Efforts are underway to draw in environmentalists, indigenous rights advocates, community organizers, relocalization activists, and even religious evangelicals. This inclusivity is important, because individuals understand and respond to different kinds of messages about energy-related dangers. If the clean energy movement can build a diverse coalition of leaders, each of whom can speak effectively to constituencies from all across the political and ideological spectrum, it will more likely spread deep roots into societies throughout the world.

Just as there is a need to mobilize an ideologically-diverse group of people, the history of social movements demonstrates that a diversity of tactics must also be used if fundamental changes are to be attained. In the past, coal miners across the world made use of a whole range of tactics—including negotiations, boycotts, strikes, and occasionally violent uprisings—in their efforts to win improvements in their working conditions and wages. Similarly, the international oil system was rocked by movements that used political pressure, oil refinery occupations, and attacks on pipelines as part of the nationalization process. And the anti-nuclear movement used legal actions, media campaigns, mass marches, and civil disobedience campaigns to halt the construction of new power stations. Just as in all movements for social change, synergies that emerged between moderate and radical activists helped each of these energy-related campaigns significantly transform their industries.

Although it is still in an early phase of development, the clean energy movement is already demonstrating a willingness to employ a diversity of tactics in its efforts to transform large-scale energy industries. To a large extent, the tactics that have been used reflect the specific context of the struggles. In countries of the global north, the emphasis has been on mounting media campaigns, trying to change consumer behavior, marshaling voter pressure, and developing legislative and legal mechanisms for enforcing energy reform. In the global south, meanwhile, conflicts have tended to emerge around hydro-electric dam projects, oil industries, and mining projects—and they have often escalated from non-violent civil disobedience to violent confrontations between local residents and officials.

If the clean energy movement is to be strengthened as a global movement, then there is a need to develop moderate and radical tactics that can be used in all regions of the world. People in the global north need to move beyond their almost total reliance on moderate strategies, and make use of civil disobedience tactics, if they expect to contain the growth of coal, oil, and nuclear industries. Citizens of advanced industrial nations must put their bodies into the struggle, and their lifestyles on the line, if true change is to be achieved. Meanwhile, people in the global south need better access to legal mechanisms for containing ecological damage and directing development in appropriate ways. As the Indian Supreme Court has demonstrated, it is possible to forge legal statutes that begin to address these issues in the global south. These initiatives need to be expanded across the developing world, so that citizens have new ways to protect themselves from the impacts of conventional energy industries.

We live at a time when efforts by government officials to forestall catastrophic forms of climate heating are faltering. And we are witnessing the emergence of corporate-driven efforts to shift toward greater reliance on coal, unconventional petroleum resources like oil sands, and nuclear power. There are conscientious public officials and corporate executives around the world who understand the dangers posed by these projects, and who would like to move in a new direction. But these elites cannot enact fundamental energy reforms on their own. We need to build a mass movement that incorporates people from all across the world into a coalition that is firmly dedicated to transforming the global energy system. Large-scale mobilizations have succeeded in reforming energy systems in the past, and they can do so again in the coming years. We can draw inspiration from these earlier struggles, and build our own mass movement that can bring about a clean energy revolution in our lifetime.

This section explores energy's role in maintaining and reproducing class and gender relations, relations of production and reproduction.

Energy is a substitute and enhancer for human labor. This means that energy and human relations are intimately intertwined, with energy playing a fundamental role in the capitalist division of labor in general. However, the energy sector itself also has its own specific division of labor. Consequently, the energy system is far from homogenous. It is rife with inequality, hierarchy, and struggle. In particular, major struggles exist in relation to ownership, labor conditions, energy access and pricing, and land and ecological conflicts, and also in terms of shaping gender relations. Hierarchies also exist at the regional level, as different regions have different roles within the worldwide division of labor associated with the global energy system.

As the world's division of labor has undergone a profound restructuring in recent years, a process that is still ongoing, the world's energy sector, and division of labor associated with it, has also undergone restructuring. As US hegemony declines, a process massively accelerated by the current economic-financial crisis, an important process of interstate realignment and rivalry is getting underway within the interstate system. Energy is an important aspect of this process. Natural limits relating to peak oil and climate change are also becoming an increasingly important and unnegotiable physical reality. "Nature doesn't do bail-outs" is becoming an increasingly popular slogan in many countries. Thus, a combination of political, geological, and climatic factors is rapidly throwing the oil-based system into a major crisis, and points towards the urgent need to create a new energy system.

A major feature of the restructuring is the antiprivatization struggles that seek different forms of common, collective, cooperative or public state ownership of energy resources and infrastructures. These struggles are also frequently linked to struggles over access, including prices. Affected communities—workers and users of these energy resources—are at the forefront of such struggles, and often face harsh repression. Struggles are especially strong within the hydrocarbon and electricity sectors. Energy resources, infrastructure, and technologies are amongst the most important means of production for capitalism, as well as being one of its most profitable commodities. And, on the other hand, they also provide a fundamental basis for human life. As such, these struggles over the ownership of energy are part of the ongoing struggle to determine whether energy is used for satisfying the needs of producing for profit in the world market, or to satisfy human needs. Essentially it is a worldwide struggle over commodification of energy resources per se, and the degree to which they are commodified. Furthermore, struggles over the control of hydrocarbon resources can result in a strong collective political force advocating to not use these fuels in order to combat climate change, while at the same demanding reparations for providing a revenue base for moving towards building a new, renewable energy-based, system.

Chapter 4 | Part 1: Energy Makes the World Go Round and Work Makes the Energy Sector Go Round

MACHINERY AND MOTIVE POWER
Energy as a Substitute for and Enhancer of Human Labor

Tom Keefer

The past several decades have seen a wide-ranging debate over the question of the economic limits of growth, not only in regards to the scarcity of key natural resources, but also in terms of the stresses being put on the integrity of ecosystems integral to the continuation of life as we know it on our planet. Many of these concerns are grounded in concepts belonging to the school of "ecological economics" developed in the 1960s and 1970s by such thinkers as Nicolas Georgescu-Roegen and Herman Daly, who articulated a critique—grounded in thermodynamic principles—of both neo-classical economics and mainstream Marxism. Georgescu-Roegen and Daly brought attention to the inevitably entropic nature of industrial production and argued that any industrial economic system based on "drawing down" non-renewable low entropy sources of energy and raw materials would ultimately exhaust the resources it needed or would fall victim to the high entropy pollution and ecological disruption that it produced.[1]

The dependence of industrial capitalism on what Elmar Altvater has termed a "fossil fuel energy regime"[2] is a perfect example of the problems that Nicolas Georgescu-Roegen and Herman Daly outlined, as, in addition to the large amounts of carbon dioxide released into the atmosphere from the combustion of fossil fuels, increasing evidence suggests that on a global level, the extraction of conventional crude oil is reaching a point of peak production and that within the next decade it will begin an irreversible decline with grave consequences for the industrial order.[3] With non-

---

1   Entropy is a measure of disorder within a system. The stocks of fossil fuels and other minerals that are so essential for industrial society have been concentrated into low entropy deposits by the effects of millions of years of heat and pressure within the Earth's crust. Because industrial society feeds off the capture and use of these resources, which are then used up and dispersed in the course of production and consumption, industrialism is inherently entropic and its continued growth inherently undermines its long-term viability. See Nicholas Georgescu-Roegen, *The Entropy Law and the Economic Process*. Cambridge: Harvard University Press, 1971 and Herman E. Daly and Alvaro F. Umana, eds. *Energy, Economics, and the Environment: Conflicting Views of an Essential Interrelationship*. Boulder, Colorado: Westview Press, 1981 for a summary of these perspectives.

2   See Elmar Altvater, "The Social and Natural Environment of Fossil Capitalism" in the *Socialist Register 2007*. Halifax: Fernwood Publishing, 2007.

3   One of the clearest expositions of the peak oil thesis can be found in the recent "oil report" by the German Energy Watch Group. It is available at http://www.energywatchgroup.org/Oil-report.32+M5d637b1e38d.0.html. For one of the clearest arguments concerning peak oil from a thermodynamic perspective see Richard Heinberg, *The Party's Over: Oil, War and the Fate of Industrial Societies*. Gabriola Island: New Society Publishers, 2003. For an assessment of the role of oil scarcity in current geopolitical conflicts see Michael T. Klare, *Blood and Oil: The Dangers and Consequences of America's Growing Dependency on Imported Petroleum*. New York: Henry Holt & Co., 2005.

Middle Eastern supplies of natural gas facing their own peak and with alternative energies unlikely to meet the shortfall of oil and natural gas, global capitalism may find itself thrown into crisis as shortages of liquid fuels and high energy prices lead to skyrocketing price increases, disruption of the production and distribution of essential goods and services, and the sharpening of both global and local class struggles. Because of the ubiquitous use of oil and natural gas in generating electricity and heating, supplying fuel and fertilizer for industrial agriculture, and providing energy for transport, high energy prices will immediately be felt as significant cost of living increases for much of the world's population. Moreover, barring the discovery and widespread application of a new non-carbon-based energy system, the increasing cost and declining availability of oil and natural gas will encourage a widespread return to the use of coal and biomass—fuels that release greater amounts of carbon dioxide and toxic pollutants into the atmosphere than oil and natural gas.

The environmental problems associated with the use of fossil fuels have been the subject of numerous studies, international conferences, and well-meaning declarations, but to date there has been very little substantive analysis of what the root causes are of capitalism's addiction to fossil fuels and why capitalists are so unwilling to undertake the transition to a new energy regime. The failure to adequately grapple with this question stems from the fact that two of the most important schools of thought that hold important components of the analytical framework necessary for this undertaking—ecological economics and Marxism—miss crucial insights that the other brings to the debate. What is manifestly absent from most ecological economist thought is a critique of capitalism as a historically-specific economic system that is not only based on ever-increasing expansion, but is also compelled to substitute machinery (and the energy these machines require) for human labor in its quest to both achieve higher margins of profit and to undercut tendencies towards working-class self-organization and resistance. Moreover, in failing to recognize commodified, alienated, and exploited labor as lying at the root of the capitalist system, the ecological movement has largely been unable to see the intimate connections between preserving ecological diversity and replacing capitalism with an alternative economic and political order.

For its part, Marxism as a historical movement has paid little attention to the social, political, and ecological contradictions entailed by the inherently entropic nature of industrial production. With the notable exception of Marx and Engels, the Marxist movement has by and large failed to bring an adequate ecological analysis to bear on questions of capital accumulation and the ecological aspect of working-class resistance to capital.[4] In a time of potentially catastrophic climate change and the rapidly approaching exhaustion of easily accessible fossil fuel energy inputs, the old Marxist perspective that capitalism would develop the forces of production to a point at which they could be simply appropriated by the working class and used

---

4   John Bellamy Foster and Paul Burkett are two leading contemporary Marxists who have done much to remedy this weakness. See their book, *Marx's Ecology: Materialism and Nature*. New York: Monthly Review Press, 2000 and Burkett's book *Marxism and Ecological Economics: Toward a Red and Green Political Economy*. Leiden: Brill, 2006.

to construct socialism seems increasingly remote. Because of the historic failure of the international working class to overcome capitalism in the twentieth century, it is increasingly possible that by the time capitalism produces the kind of economic and ecological crisis that will delegitimize it on a global scale, it will have exhausted the world's easily available stocks of low entropy fuels and materials, leaving any alternative mode of production to be built on the ruins of today's industrial society.

In large part, the ecological blind spot within Marxism stems from the fact that the Soviet Union, the first society founded on Marxist principles, was forced by capitalist encirclement and threats of invasion to rapidly build up a fossil-fueled industrial base along the lines pioneered by developing capitalist economies. After the ebb of the wave of global revolutionary struggle that followed World War I and the resulting consolidation of Stalin's tyrannical regime, many of the progressive environmental and democratic perspectives within Marxism were disavowed. However, the fact remains that Marx displayed an ecological awareness far in advance of many of his contemporaries—and even of many of his critics today—and that many of his key ideas, especially his conception of the "metabolic" relationship between humans and nature must become a central part of the framework of contemporary environmentalism.

In this text I will argue that the analysis that Marx developed in *Capital* provides one of the most important starting points for understanding capitalism's addiction to fossil fuels and its existence as a global economic system responsible for today's ecological crisis. Following Marx's discussion of the role of machinery in the capitalist production process, I suggest that, in its transition from an agrarian form to an industrial one, capital came to rely on machinery as an indispensable tool to break workers' resistance, increase the productivity of the labor it commodified, and to aggressively spread the capitalist system across the world. Because modern machinery requires a cheap and reliable source of low entropy energy to keep its machines going, and because there are, at present, no ready alternatives to the fossil fuel energy regime, the capitalist system has always been dependent on finding and producing increasing amounts of fossil fuel resources. During the industrial revolution, fossil fuels provided the means to overcome both workers' resistance to dispossession and the very real natural limits of agrarian capitalism. Coal, oil, and natural gas became the lifeblood of the capitalist system—providing energies that, like labor power, must be kept coursing through the system lest fixed capital and processes of accumulation should come to a shuddering halt.

A Marxist analysis of the role of machinery in the development of capitalism, which is enriched by Georgescu-Roegen's and Daly's notions of the inevitably entropic nature of industrial production, provides a crucial framework within which to situate the problem of fossil fuel dependence, and the likely consequences for the capitalist system and the alternative modes of production that may follow it. Such an approach makes it possible to understand the peaking of world oil production and the beginning of the end of the age of fossil fuels as an epoch-making turning-point for contemporary class struggles—a perspective central to understanding and transcending global capitalism.

★★★★★

The "colossal productive forces" commanded by the bourgeoisie that Marx and Engels referred to in *Manifesto of the Communist Party* arose not only from capital's property relations and the "scientific" exploitation of human labor, but also from the way in which capitalism appropriated stocks of fossil fuel energy and channeled them in an ever-increasing flow into the production, consumption, and transportation processes crucial to capitalist accumulation.[5] At the root of industrial capitalism and its astonishing conquest and transformation of the world in the past 250 years is the fossil-fuel-powered machine. From steam-powered textile factories, locomotives, and steamships, to coal-fired foundries and electrical generating plants, to the automobile, dishwasher, vacuum cleaner, jet engine, and intercontinental ballistic missile, fossil-fueled machinery has transformed capitalism and the world we live in. Fossil fuels—coal, oil, and natural gas—are a rich source of stored up solar energy that contain huge amounts of readily accessible energy in a portable and accessible form. And in every year from the first commercial application of the steam engine in 1715 to the present day, the capitalist world-economy has incorporated an ever-increasing amount of this fossil energy in its economy.[6]

It is conventional to view the rise of capitalism and industrial society teleologically, as an inevitable consequence of scientific rationalism, the declining power of religion, the influence of the Protestant work ethic, or any number of other "inevitable" social and political processes. But this does not explain why an industrial capitalism based on fossil fuels first developed in eighteenth century England instead of in twelfth century China—whose manufacturers used rich coal deposits to produce more iron and steel than all of Europe did in 1800—or why, when the ancient Greeks invented steam-powered machinery, they did not apply it to increasing economic production.[7] To understand why the growth of industrial capitalism and the widespread use of fossil fuels to power machinery arose in eighteenth-century England and nowhere else in the world requires an understanding of the historical specificity of capitalist social relations and the economic laws of motion inherent to capitalism as an economic system.

Under capitalism, the machine is the predominant means by which human labor can be displaced from the production process and the best way to make the labor that remains more productive. In various pre-capitalist modes of production, ruling elites had little interest in displacing human labor from the productive process, as the societal surplus appropriated by ruling elites was acquired with the direct application of state backed coercive force and not through technological improvements to

---

5   Karl Marx and Frederick Engels "Manifesto of the Communist Party," in Karl Marx and Frederick Engels, *Collected Works*, Vol. 6. New York: International Publishers, 1976: pp. 477–517.

6   Valclav Smil, *Energy in World History.* Oxford: Westview Press, 1994: pp. 157–222.

7   Barbara Freese, *Coal: A Human History.* New York: Penguin Books, 2003.

production. In the feudal societies of Western Europe, strict written and customary laws determined all aspects of economic production, and innovations in the labor process were strictly regulated because it was feared that they could create dangerous social upheavals by displacing workers from the production process.[8]

The extraction of surplus under capitalism is fundamentally different than in pre-capitalist class societies, where the surplus was extracted from direct producers through the political power of the state, or through what Ellen Meiksins Wood calls "politically constituted property."[9] Under capitalism, the surplus is extracted through economic means, based on the wage labor/capital relationship and not through the direct coercion of the state (although the state clearly remains present to enforce capitalist property relations and to put down open revolt from the working class). Workers are denied free access to the means of production and must sell the only thing they have—their power to work—to capitalists in order to survive. Capitalists buy this commodity, what Marx called "labor power," and by setting it to work in the production processes that they control, they use it to produce commodities that are then sold on the market. The source of capitalists' profit is the fact that over the workday, the workers' labor power produces more than the costs of their subsistence (the wage that they receive) as it creates what Marx calls "surplus value," which is appropriated by capitalists as profit.

Because every capitalist is in competition with many other capitalists, and seeks ever-higher profits to reinvest in production, the key to continued accumulation lies in increasing the productivity of the labor power purchased from the worker. This growth in productivity may take place in what Marx called "absolute" terms—by lengthening the working day and by intensifying the pace of work—or in "relative" terms, by changing means and methods of production and thereby increasing the proportion of the worker's labor time that can be appropriated by the capitalist. Either way, Marx saw machinery as being fundamental to the increasing of both "absolute" and "relative" surplus value.[10]

While Marx identified a number of ways in which machinery could be used to increase the absolute rate of surplus extraction, because human beings can only be pushed to a certain level of exhaustion, increasing labor productivity through the substitution of newer and more sophisticated forms of machinery has been at the core of the continued development of capitalism and explains its dynamic growth and expansion. Conversely, the fact that the 12th-century Chinese economy, despite its technological complexity, did not operate along capitalist lines, explains its "failure" to take off along European lines.[11]

Marx outlined three key ways in which machinery was and continues to be central to the growth, expansion, and relative stability of labor-capital relations. Firstly,

---

8   See Ellen Meiksins Wood, *The Origin of Capitalism: A Longer View*. London: Verso, 2002.
9   Ibid.
10   Karl Marx, *Capital Vol. 1.*, p. 492.
11   See Robert Brenner and Christopher Isett, "England's Divergence from China's Yangzi Delta: Property Relations, Microeconomics, and Patterns of Development." *The Journal of Asian Studies* 61, no. 2 (May 2002): 609–662.

the introduction of machinery increases the productivity of the labor power that the capitalist has purchased and set to work, which means that, for the same wage and in the same amount of time, more goods can be produced per worker, thus leading to greater profits for the capitalist who first introduces this machinery. The increased availability of the cheaper commodities produced with this machinery has the effect of lowering the worker's costs of living and thus the overall cost of labor power on the market. Because the cost of the means of subsistence has decreased, capitalists can drive down wages in relative terms, as workers can buy more commodities for less money. The ultimate effect of this process is that a smaller proportion of the working day is spent by workers in laboring for their own reproduction, and thus a relatively greater proportion of time is spent working to produce surplus value for the capitalist. Secondly, the increase of productivity achieved with the introduction of machinery displaces workers from their jobs and creates what Marx called a "reserve army of the unemployed," which, by acting as a pool of desperate would-be wage labor, drives down the cost of wages to the subsistence minimum and provides a cheap labor force for new and emerging branches of industry.[12]

As well as the two macroeconomic processes described above, Marx argued that machinery was essential as a tool that could be used by capitalists to break working-class resistance at the point of production. Because the introduction of machinery reduced the need for muscular strength as a motive force in production, it allowed for the incorporation of women and children into the industrial workforce. As capitalism has expanded to new industrial zones across the planet, it has always relied upon the super-exploitation of women and children who are more easily disciplined and controlled than their male counterparts. In referring to the British example, Marx noted that the introduction of machinery and the replacement of male workers by women and children "at last breaks the resistance which the male workers had continued to oppose despotism of capital throughout the [earlier] period of manufacture."[13]

In addition to its effects in destroying working-class families and pitting workers against each other, capitalists also introduced machinery in specific circumstances in order to break strikes and overcome working class self-organization at the point of production. Conscious of this fact, workers often resisted exploitation by attacking specific types of machinery that were seen as having been introduced for the express purpose of breaking their class power. In the 1630s, a wind-driven sawmill near London was destroyed by a group of workers who feared the loss of their jobs, while in 1758 the first wool-shearing machine driven by water-power was burned down by some of the 100,000 people that it had thrown out of work. The Luddite rebellion in the early 1800s was perhaps the best example of this resistance towards the new fossil fueled machines introduced during the industrial revolution.[14]

Marx looked at the evidence provided by capitalists themselves in their own assessments of their production methods and argued that "the steam engine was

---

12   Karl Marx, *Capital Vol. 1.*, p. 532.
13   Ibid, p. 526.
14   Ibid, p. 554.

from the very first an antagonist of 'human power', an antagonist that enabled the capitalists to tread underfoot the growing demands of the workers, which threatened to drive the infant factory system into crisis."[15] Indeed, he added, "it would be possible to write a whole history of the inventions made since 1830 for the sole purpose of providing capital with weapons against working-class revolt."[16] Machinery was thus a crucial aspect of the process of primitive accumulation and dispossession as capitalists struggled to overcome and discipline a new industrial workforce against the old habits of communal solidarity and village living.

The point raised by Marx is an interesting one, for it describes a dialectical struggle between labor and capital in which class antagonisms play an active part in the technological development of capitalism. Just as state intervention in the form of the Factory Acts, which legislated maximum working hours in the factories ended up benefiting the richest capitalists who were able to invest in new machines to replace over-exploited workers, the resistance of workers at the point of production forced capitalists to invest in new machines to overcome the increasing organization and class consciousness of workers. As Marx and early capitalists were well aware, the development of capitalism was not a *fait accompli*—workers and the dispossessed were capable of pushing it into crisis through their struggles.

In addition to transforming work processes within established capitalist societies, the introduction of machinery was decisive in opening up the world to the dominance of the capitalist mode of production. Modes of production and specific branches of industry and non-capitalist countries that did not incorporate the use of machinery were easily overcome by industrial capitalism. Drawing a connection between economic warfare and the one-sided nature of colonial warfare then forcibly expanding the world-market, Marx argued that the result of competition between unequal processes of production "is as certain as is the result of an encounter between an army with breach-loading rifles and one with bows and arrows."[17] The widespread use of coal in the British economy led to greatly increased steel and iron output, and the use of these raw materials revolutionized warfare through the standardized production of modern weapons as well as forms of mass transport such as steamships and railways. This warfare was as much economic as military, since:

> The cheapness of the articles produced by machinery and the revolution in the means of transport and communication provide the weapons for the conquest of foreign markets. By ruining handicraft production of finished articles in other countries, machinery forcibly converts them into fields for the production of its raw material ... by constantly turning workers into "supernumeraries," large-scale industry, in all countries where it has taken root, spurs on rapid increases in emigration and the colonization of foreign lands, which are thereby converted into settlements for growing the raw material of the mother country, just as Australia, for example, was converted into a colony for growing wool.[18]

---

15 Ibid, p. 563.
16 Ibid.
17 Ibid, p. 578.
18 Ibid, p. 579.

It is, therefore, that the basis for the successes of Western imperialism and the domination of the capitalist mode of production throughout the world is fundamentally related to the expansion of machine production and its generalization through all branches of industry. The invention of fossil-fuel-powered machinery and its application to capitalist labor processes appears as a savior of the capitalist mode of production and the guarantor of its local and global domination. Without machinery to increase labor productivity, overcome working-class resistance at the point of production, and project economic and military might across the world, the question could be seriously posed as to how agrarian capitalism could have grown and expanded beyond the ecological limits that constrained it on the British Isles.

While stressing the importance of machine production to the capitalist system, Marx never suggested that it was technology itself that was the driving force of history or that changing class relations could be explained by recourse to technological determinism. As Marx argued, it was not the steam engine itself, but rather the use of machines under capitalism that had human labor as a motive force, that explained the development of machinery and the growth of industry under capitalism. Once production existed under a capitalist framework with machines powered by human labor power, the logic of capital meant that there would be significant economic rewards for any capitalist successful in replacing human labor with a cheaper alternative. In some cases that alternative was children's or women's labor, but ultimately it was far more profitable to completely drive out human labor as a motive force and replace it with fossil-fuel-driven machinery. This process occurred not because of the inevitable growth of more advanced forms of technology, but rather because capitalist social relations necessitated the constant improvement of labor productivity.

Marx's account thus differs from a technological determinist perspective, which sees technology itself as a driving force of history. Marx recognized the machine and, in particular, the steam engine as central to the industrial revolution, but he saw this revolution as having been put into motion by the laws of capitalist accumulation and its drive to increase the productivity of labor power. As Paul Burkett and John Bellamy Foster point out, the transformation of property relations that heralded the rise of agrarian capitalism was key to capital's control of the industrial labor process:

> After all, the ability of the capitalist to separate the tool from the worker and install it in the machine—and the subsequent application of science to the technical improvement of machinery on the capitalist's profit-making behalf—presumed that the worker had already been socially separated from control over the means of production.[19]

The introduction and application of machinery is thus fundamentally linked to class struggle, and to extracting surplus value from the working class. Under capitalism, technology is not some neutral force that spontaneously develops of its own accord, but a means by which individual capitals can out-compete their rivals, and a tool by which capital as a whole can collectively maintain its control over the working class. Certain technological innovations may result (as they did in Marx's time) in

---

19   Paul Burkett and John Bellamy Foster, "Metabolism, Energy, and Entropy in Marx's Critique of Political Economy: Beyond the Podolinsky Myth," presented at a Marxist Sociology session of the American Sociological Association Meetings, San Francisco, August 14–17, 2004, p. 18.

the disappearance of whole trades and industries, but the global process remains one of drawing ever-increasing numbers of workers into the capital-wage labor relationship as the technology of capitalist production advances relentlessly. However, there is an Achilles' heel to this process of advancement, and that is the finite amount of fossil fuels on the planet and the thermodynamic limits affecting all forms of energy appropriation required to power machines.

With the peaking of world oil production, capitalism will face a historic turning point. Its new short-term strategies of accumulation will be based upon securing the declining low entropy sources of energy, most of which remain within the Middle East, and striving to boost production of these resources to allow for continued economic growth. If it is to continue to grow, capitalism must shift to some alternative energy source in a manner every bit as transformative and revolutionary as the move from biotic energies to fossil fuel energy regime was, and end its dependence on fossil fuels. This source of non-carbon based energy must be cheap, nonpolluting, avoid contributing to global climate change, and be capable of integration within existing energy distribution infrastructures. Should capitalism not develop such a source of alternative energy, we can expect that the climate change feedback loop will be accelerated as coal and biomass are used to replace declining stores of oil and natural gas. At the same time, international competition for remaining stores of low entropy oil will be accelerated, and dramatic increases to the cost of living will lead to a global intensification of local, national, and international class struggles.

As industrial capitalism matures and its machines devour ever-increasing amounts of non-renewable fossil fuels, a point of crisis will be reached when capital will no longer be able to externalize its contradictions. This will provide a whole new set of opportunities for revolutionary forces seeking to transcend the capitalist economic system. However, it also poses grave dangers and requires a fundamental shift in how we view processes of economic production. With the depletion of easy to access fossil fuel reserves and the impacts of global climate change, humanity will be required to build an alternative to capitalism under conditions of declining labor productivity and under the solar energy constraints momentarily transcended by twentieth century industrial capitalism. Consequently, the implications for our theory and practice are significant, and deserve to be put at the center of any anti-capitalist revolutionary project.

BIBLIOGRAPHY

Elmar Altavater. "The Social and Natural Environment of Fossil Capitalism." *Socialist Register 2007*, Halifax: Fernwood Publishing, 2007.

Energy Watch Group. "Crude Oil—The Supply Outlook." Ludwig-Boelkow-Foundation, February 2008. http://www.energywatchgroup.org/Oil-report.32+M5d637b1e38d.0.html.

Brenner, Robert, and Christopher Isett. "England's Divergence from China's Yangzi Delta: Property Relations, Microeconomics, and Patterns of Development." *The Journal of Asian Studies* 61, no. 2 (May 2002): 609–662.

Foster, John Bellamy. *Marx's Ecology: Materialism and Nature*. New York: Monthly Review Press, 2000.

Freese, Barbara. *Coal: A Human History*. New York: Penguin Books, 2003.

Georgescu-Roegen, Nicholas. *The Entropy Law and the Economic Process*. Cambridge: Harvard University Press, 1971.

Heinberg, Richard. *The Party's Over: Oil, War and the Fate of Industrial Societies*. Gabriola Island: New Society Publishers, 2003.

Klare, Michael T. *Blood and Oil: The Dangers and Consequences of America's Growing Dependency on Imported Petroleum*. New York: Henry Holt & Co., 2005.

Marx, Karl. *Capital: A Critical Analysis of Capitalist Production*. Volumes 1–3. Moscow: Progress Publishers, 1970.

Wood, Ellen Meiksins. *The Origin of Capitalism: A Longer View*. London: Verso, 2002.

Chapter 5 ▌ Part 1

ENERGY, WORK, AND SOCIAL REPRODUCTION IN THE WORLD-ECONOMY

Kolya Abramsky

> *"Energy is the fundamental prerequisite of every life. The availability of energy is a fundamental and indivisible human right.... It is violated a billion-fold"*
> —WREA 2005

> *"From the capitalist perspective, energy is recognized as the fundamental* techno-logical tool for the international control of the working class. *First of all,* it is a replacement for labor. *Since World War II, capital has increasingly dealt with the working class on a daily basis by replacing labor with energy.... In its immediate ap-plication to the process of production, energy frees capital from labor. It follows that control over the availability and price of energy means control over the technological conditions of class struggle internationally and also control over economic develop-ment"* (emphasis in original).[1]
> —Midnight Notes

In order to understand the current so-called "energy crisis" and a possible future "transition to renewable energies and/or post-petrol future," it is crucial to consid-er the relations by which human beings produce wealth in the world-economy and how this labor force is reproduced and subverted over time. It is also important to consider the specific division of labor that exists within the energy sector, worldwide. There are two important tasks: a) mapping the worldwide division of labor within the energy sector, and b) tracing the relations that produce, reproduce, and shape this division of labor, and identifying how the different parts relate to one another, within a wider analysis of capitalist relations.

This chapter seeks to identify, and partially answer, three broad questions.

• How does energy relate to labor and its reproduction, at a general level?

• How does labor operate within the energy sector, specifically?

• How can an understanding of energy and labor contribute to under-standing current concepts such as "energy crisis" and "transition"?

A few brief words about both energy and labor.

Throughout history, different energy sources have been used at different times and places and in different combination with one another. There are various energy

---

1   Midnight Notes, *Midnight Oil: Work, Energy, War 1973–1992* (New York: Autonomedia, 1992)., p. 124.

sources or sectors, including whale fat, wood, peat, coal, oil, nuclear, wind, solar, natural gas, biofuels, hydro-electric, and cow dung. Each of these sectors has a specific division of labor associated with it, and each requires technology to transform the fuels for use as, for instance, motive force, heat, light, etc. (for example, petrol and the internal combustion engine, or coal and the thermo-electric power station). Finally, energy may be more or less commodified.

Labor is understood in the broadest sense of the word, including anyone whose labor (or land or other natural resources) needs to be harnessed and/or commodified in order to produce surplus value for capital. It does not prioritize industrial labor in the factory, nor urban labor over agricultural labor, nor waged labor over unwaged, nor "free" over "forced." Furthermore, it is based on the premise that real material hierarchies and conflicts of interest between workers exist. In order for production of goods and their sale for profit to occur in a continually expanding market, a worldwide pool of controllable labor must be replenished, reproduced, and expanded over time. This is known as social reproduction.

ENERGY AS A MEANS OF SUBSISTENCE

Energy is a crucial means of subsistence, due to its importance for food production and preparation, shelter, lighting, and heating especially. Without it, human life cannot exist and the generational reproduction process breaks down. If people lack access to energy, they have to have access to money in order to buy energy to survive.

As with land and other means of subsistence, the degree of separation between the energy producer and consumer is of great importance. The more that producers are separated from their basic means of subsistence, the more they are dependent on their own waged labor to buy the means of subsistence. Historically, the process of separating people from their means of subsistence has been necessary to create a pool of people with no other option than to work for wages, and thus provide the necessary labor force for capitalist production. Importantly, the degree of separation that may exist between people and their key means of subsistence, in this case energy, is neither permanent nor given, but is the subject of an ongoing process of struggle, conflict, and negotiation. This ongoing process is called primitive accumulation and dispossession.

This poses the question of ownership, control, and access to energy production and consumption. And, above all, which purposes does energy production serve? Crucially, is energy produced and consumed to serve the needs of capital accumulation (for which it is a crucial raw material and means of production) or does it serve subsistence needs for human survival? Fundamentally, these interests are diametrically and structurally opposed to one another, which gives rise to the struggle over commodification of energy, revolving around whether energy is a resource held in common outside of market relations, or whether it is commodified in order to sell for a profit in the world-market. And, to the extent that energy is already commodified, there is a struggle over the *degree* to which it has been commodified.

Currently, as is described in the other chapters of this book, common or public energy resources, from forests to oil fields, are facing increasing privatization through-out the world, especially through regional and multilateral free trade agreements, such as NAFTA, FTAA, or WTO. This is greatly affecting prices and people's ability to access reliable sources of energy, regardless of whether it is "clean" or "dirty."

Privatization and the enclosure of common or publicly-owned resources for profit elsewhere is reminiscent of the enclosure of commonly-owned and managed woodlands in Europe over the last several centuries, a process that was integral to the emergence of the European-centered capitalist world-economy. Importantly, it is forcing people to become increasingly dependent on money, and thus on waged labor, in order to satisfy their energy needs. As such, it is a crucial part of the process of expanding the world-market based on an availability of a worldwide labor pool. Expansion of the world-market means enclosure of commons—energy being one of the key commons. Once energy has been commodified, its pricing plays an important part in social reproduction, in relation to the magnitude of the price and the issue of *who* pays for it. Does capital pay for reproducing the labor force that it uses to extract profits, or is it able to shift these costs onto workers themselves, both waged and unwaged?

Finally, it is worth saying something about resistance. Next to struggles over control of land, there is perhaps no area in which such struggles for "commons" are more central than in relation to the expropriation of common energy resources and increased energy pricing. The Zapatista uprising in Mexico, ongoing since 1994, is partly in response to NAFTA's easing of the over seventy-year-old restrictions on for-eign ownership of Mexico's oil. Most recently, in Bolivia, Evo Morales has national-ized the country's gas fields. The last decade has also seen major struggles in relation to electricity privatization throughout the world, including in France, South Africa, South Korea, and Thailand. Privatization of forests is being resisted throughout the world, with women playing a leading role in the struggles. Many, if not most, of these battles have been internationally networked, with specific local struggles inspiring and informing one another, and with support offered and received through a range of global networks.

ENERGY RESOURCES EXIST ON LAND

Most energy resources exist in rural areas, and if they are to be harnessed by capital, that typically means the expropriation of land, or at least its control. Like energy, land is also a basic means to human survival. The current restructuring of the world-economy involves companies gaining expanded investment rights over an increasing geographical scope throughout the world, which undermines the territorial auton-omy of rural communities. As well, social and environmental constraints on invest-ment are being removed and ownership is being forcefully transferred from peasants to capital. So, in addition to a generalized expropriation, land that contains energy resources is particularly central.

Oil, gas, coal, and uranium exploration and extraction, as well as large-scale hydro-electric dams are having a major social and environmental impact on communities in the vicinity of these activities, which produce major social conflicts related to land rights, pollution, and (frequently violent) displacement. In relation to oil, there are struggles over displacement, pollution, and oil company-associated violence in Nigeria, Colombia, Ecuador, as well as several other countries. Particularly impacted are peasant, indigenous, communities of African descent (in Latin America), and fishing communities, many of which have communal land ownership structures intact.

In recent years, tactics used in resisting land appropriation or destruction have ranged from parliamentary struggles to autonomous community organizing, street protests, non-violent civil disobedience, and most recently, in Nigeria, armed struggle and kidnapping of oil company employees. In Colombia, the U'wa Community even threatened to commit mass suicide in the face of continued activities from OXY (Occidental) Petroleum. The construction of the world's biggest oil pipeline, the Baku-Tbilisi-Ceyhan (BTC), pipeline has also provoked protest from land rights and environmental activists, both within the affected countries and by their international supporters. In Venezuela, indigenous peoples are facing displacement from coal mining activities from a range of state owned and foreign multi-nationals. In the US, Navajo communities are being adversely affected in Black Mesa in Arizona, by the coal giant Peabody Coal. Millions have been displaced throughout the world by the construction of large hydro-electric dams, in India, China, Brazil, and Indonesia, amongst others. As the nuclear industry gears up for a renewed expansion, anti-nuclear struggles have also grown in strength, both in areas where power stations are to be sited, as well as in areas where uranium is mined, such as in Indigenous territories within the USA's Nevada/Arizona desert or in the uranium dumps and mines on aboriginal land in Australia. As with struggles over ownership of energy resources, these and many other struggles associated with energy-related conflicts over land-use have successfully sought international allies.

ENERGY AND WORK

In addition to energy providing means of subsistence, and that existing on land, it is also important for work in general.

- Mechanization has enabled increased productivity of labor—which, in the context of capitalist relations, means providing the basis for what Marx calls relative surplus value strategies and wage hierarchy.

- Artificial lighting has lengthened the working day (just as the more recent spread of information technologies have), which, in the context of capitalist relations, means providing a material basis for what Marx calls absolute surplus value strategies.

- Transport has enabled an expanded geographical reach for markets in raw materials, labor, and commodities, and has reduced the circulation time of goods, money, and people, etc.

• Cheap food, shelter, clothing, and consumer goods have lowered the cost of reproducing a planetary workforce, thus buffering reduction in wages, and intensifying differences in global wage hierarchies. For example, cheap food has largely been obtained through the agro-business model imposed on the world's farmers, causing increased food insecurity for many sections of the world's population whose land has been expropriated to allow the land concentration necessary for energy-intensive agro-business model. This has escalated the ecological crisis due to the fertilizer and pesticides used, and exposed increasingly large sections of the world's population to the swing of food prices in the world-market.

As such, energy has played an important role in shaping worldwide class relations *as a whole*, not just within the energy sector.

Mechanization is a particularly important process through which energy and human labor impact one another. The history of energy use is, for better or worse, a history of human (or animal) labor being replaced or supplemented by outside energy sources—wood, coal, gas, oil, nuclear power, windmills.

Paradoxically, in the midst of all this "labor saving" technology, no one really does any less work than they did before. The wage relation that shaped the factory has not been done away with, nor have the unequal gender roles that shape so many households been replaced, nor has unwaged labor disappeared. Rather than doing away with unequal and exploitative patterns of work, energy-intensive appliances, vehicles, and machines have simply rearranged people's working patterns and structures. In fact, the replacement of human beings by machines and robots has often created huge pools of deskilled and unemployed workers, and has frequently been met with resistance from workers.

However, it would be wrong to view the replacement of human labor as an unintended side effect of mechanization. Throughout the ages, mechanization has often been introduced *precisely* in order to replace and subvert human labor—that is, organized and rebellious human labor that threatens to escape the control of those who seek to control it, whether they are landlords, factory owners, or agricultural companies. The Luddites stand out famously here for smashing the looms that threatened their livelihoods.[2]

A more recent example of this can be seen in the South African gold mines. Facing strong resistance from miners in the post-World War II period, the mine owners invested heavily in mechanization in order to replace workers. This was seen as the most effective way of breaking class struggle. For every 10 kg of gold produced in 1950, ten men were employed and 99,000 KWh of electricity used. In 1975, five men were employed and 180,000 KWh of electricity were used for the same output.[3] This pattern has been an especially important element of class relations in the United States, and will be addressed in a later section.

---

2   Karl Marx, *Capital Vol. 1* (London: Penguin/New Left Review, 1976), p. 554.
3   Peter Norre and Terisa Turner, *Oil and Class Struggle* (London: Zed Books, 1980).

All of the above shows the importance of energy to the capital-labor relation *in general*, not just within the energy sector itself. Hence, a transition to a new energy system is of importance not just to labor within the energy sector but to *all workers throughout the world, both waged and unwaged.*

LABOR IN THE ENERGY SECTOR

Listen! We ought to be in a wood choppers union! Chop wood for breakfast! Chop wood, wash his clothes! Chop wood, heat the iron! Chop wood, scrub floors! Chop wood, cook his dinner![4]

This ship is a floating transporter of labor.... About 5 million emigrate to find work.... It's got 750 passengers.... You can tell by looking at faces and hands that many are farmers, country people.... The same poor sods who spent last night out on the sidewalk.... The same people who are pushed and shouted at.... Who wait in huddled groups, for some official to deign to notice their existence.... Their faces and their clothes are the color of the earth. Dark and Brown.[5]

The commercial energy sector has always involved the labor of many different people and geographical locations worldwide, relying on global commodity chains that operate within the wider context of capitalist relations, relations that are geographically uneven and hierarchical. Historically, energy sector workers (at least within the waged sector) and their unions have been well organized both within countries, and between countries. In May 2006, the International Federation of Chemical, Energy, Mine and General Workers' Unions (ICEM), represented approximately 20 million workers organized in 379 industrial trade unions in 123 countries.[6]

The fact that energy is a strategic raw material means that energy workers (as well as workers extracting and producing the raw materials associated with the sector) are strategically positioned. This has had contradictory effects.

On the one hand, there is a need to ensure high levels of output and to extract large amounts of surplus from them. This means that the energy sector has frequently involved highly coercive labor forms, especially in periods of intensified inter-firm and inter-state rivalry. Examples are numerous and include: coal mines using forced labor in the African colonies to fuel the rivalry between the European imperial powers,[7] and prisoner labor in the post-Reconstruction US South in order to provide for the US industrialization process.[8] The period prior to World War II witnessed a renewed wave of coercion in energy sectors, both in the US New Deal and in Stalin's rapid industrialization drive. Nazi Germany, lacking its own source

---

4   Miner's wife in the film, by Herbert Biberman, *Salt of the Earth* (Independent Productions/International Union of Mine, Mill and Smelter Workers, 1954).

5   Description of a ship transporting migrants for work in the oil industry in the Persian Gulf. Midnight Notes, *Midnight Oil: Work, Energy, War 1973–1992* (New York: Autonomedia, 1992), pp. 67–70. The similarity between this and classic descriptions of slave ships during the Atlantic slave trade is striking.

6   International Federation of Chemical, Energy, Mine and General Workers' Unions (ICEM): http://www.icem.org/.

7   George Padmore, *The Life and Struggles of Negro Toilers* (Hollywood: Sundance Press, 1931).

8   Alex Lichtenstein, *Twice the Work of Free Labor—The Political Economy of Convict Labor in the New South* (London/New York: Verso, 1996), pp. 105–126.

of oil, used a form of synthetic gasoline. Together with the industrial company IG Farben, the state set its armies of forced laborers to the horrendous task of producing this fuel from coal. In the events preceding the 1979 Iranian revolution, striking oil workers were literally pulled out of their houses at gunpoint and forced to resume production.[9] Contemporary examples include migrant labor in the Persian Gulf oil states. In Colombia, the country with the highest rate of murdered trade unionists in the world, oil workers have to survive in the face of paramilitary repression. As will be discussed more thoroughly in a later part of this book, labor conflicts are also emerging in the new energy sector. Brazilian sugar workers face conditions akin to slavery as they produce the raw material for US ethanol supplies.

On the other hand, the strategic positioning of energy sector workers has also given them a robust bargaining power in relation to their employers and governments (as well as other workers). Worker struggles in the energy sector have frequently resulted in improved conditions and wages, etc., and have also frequently had a chain reaction effect on the condition of workers in other sectors. Examples of this phenomenon are also numerous, and include the coal miners in the British general strike of 1926, and oil workers in the Iranian revolution of 1978–79.

Perhaps the contradictory positioning of energy workers is most visible in oil workers in OPEC countries. Oil workers' struggles played an important role in pushing the price of oil up in the 70s. The consequent high revenues from oil have, on the one hand, meant that many social reforms have been granted, such as education and health care (paid for by industrialization and "development"), but they have been combined with harsh repression.

UNWAGED LABOR IN THE NON-COMMERCIAL ENERGY SECTOR, THE PILLAR OF CHEAP REPRODUCTION OF LABOR

It is widely agreed that oil is the energetic bedrock of contemporary capitalism. In a sense this is completely true—it is certainly the main energy behind the production and consumption of commodities for the world-market, if we exclude the production of labor power, itself an important commodity in the world-market. However, it is precisely this exclusion of the production of labor that is problematic. Throughout much of the world, especially in rural areas, people do not satisfy their energy needs exclusively, or even predominantly, through the commercial use of energy, but rather through the non-commercial use of dung, wood, and other biomass that provide heat, lighting, and cooking fuel. More than one third of humanity, over 2 billion people, currently rely on these fuels for their daily energy needs. Collection of such fuels is most commonly done by women and children, as part of "domestic work," without access to wages and the (limited) protection that the so-called "formal economy" and its trade unions, or other organizations, may be able to offer.[10]

It is this "traditional biomass" energy, and not in fact oil, that makes a significant

---

9   Norre and Turner, op. cit., p. 299.
10   Hugh Warwick and Alison Doig, *Smoke—The Killer in the Kitchen: Indoor Air Pollution in Developing Countries* (London: Intermediate Technology Development Group, 2004).

contribution to maintaining the lives of approximately one-third of the planet's population, by meeting their needs for cooking, heat, and lighting. As such, these fuel sources are absolutely crucial for reproducing the worldwide labor force at extremely low cost.

The labeling of certain energy sources "modern" and others as "traditional" is based on the unspoken assumption that the current inequalities in the global energy system can actually be solved through a simple expansion of the existing system so that the number of losers ("traditional" energy users) is reduced, and the number of winners ("modern" energy users) increased. However, this is based on an assumption that those people without access to "modern" energy sources *can* actually catch up and access these sources. Yet, it appears as if "primitive" biomass fuels are not simply an anachronistic anomaly to the "modern world" but, rather, a fundamental part of its uneven nature, just as non-waged forms of labor are not a "pre-capitalist" anomaly, but rather a pillar on which waged labor can exist. "Modern" energy sources and technologies, such as oil, and "non-modern" ones are in fact related to one another. It seems that perhaps one is the underside of the other, and that oil cannot exist without the biomass. The complement and essential pillar of commercial energy in the world-market is non-commercial energy combined with non-waged labor.

THE USA—A COUNTRY OF "CHEAP ENERGY" AND EXPENSIVE LABOR

Let us turn to the USA, the biggest per-capita energy consumer in the world. The USA has utterly subordinated the rest of the world to its own energy—especially oil—needs. Two parallel pictures emerge: one of absolute selfishness and insensitivity to the energy needs of the rest of the world, and another of extreme vulnerability and dependence. Why has the US economy and population become so dependent on oil from around the world? And what are the effects of this dependency?

"Cheap" energy has been a fundamental pillar of post-World War II economic growth and hegemony in the USA. Access to abundant energy sources has been crucial to ensuring social peace within the USA, both within industrial and agricultural production, and in relation to the reproduction of basic subsistence for the country's workforce.

If labor is expensive and hard to control, one of the most successful strategies that landlords, corporations, and employers can adopt is to simply replace human beings with machines and robots, and subject workers to controlling and divisive discipline. Both tactics squeeze more labor out of workers in a shorter time period, thus intensifying their work. This was an important factor in the automation of the car factories in Detroit in the 1950s, a process that followed on the heels of a series of major strikes in the sector. Automation itself sparked numerous organized struggles by organizations such as the Dodge Revolutionary Union Movement (DRUM) and the League of Revolutionary Black Workers.[11] Black workers bore the brunt of these changes and disparagingly dubbed the process "niggermation." By 1970, the manufacturing sector of the US economy used 66 percent more energy, but only 35 percent more labor than it had in 1958.[12]

Cheap energy has also been essential to reducing the costs of living, in terms of food, shelter, clothing, and transportation. In other words, it has been essential for reducing the cost of reproducing the labor force, thus increasing capital's share of the surplus. Social unrest has been contained by facilitating high levels of consumerism that directly improve standards of living.

Consequently, in the US, capital's collective strategies to control labor, through the twin processes of mechanization and high levels of material consumption require abundant sources of cheap energy. Or, more accurately, they at least require the ability to control energy flows and prices. Energy prices, far from being inevitably decided by the so-called "invisible hand" of pure supply-and-demand, are in fact highly political.[13] Expensive energy can, at times, be useful for controlling the terms on which humans work. In the multiple and interconnected crises of the 1970s (political, economic, financial, energy, food, etc.), when social struggles were strong, a *direct* attack on labor (including wage cuts) would have been very difficult without provoking fierce resistance. A planned hike in energy (and food) prices was a highly effective *indirect* attack on wages in the US, as well as globally, since rising energy costs also meant a rise in the cost of living.

There are great problems, inequalities, conflicts, and vulnerabilities associated with the current US energy system, and in particular Big Oil. Yet it is merely a part of a bigger, and highly stratified, global energy system. These problems and inequalities are likely to become increasingly visible as global energy prices rise, and as new energy sources start to replace oil.

CONCLUSION—TRANSITION, CLASS STRUGGLE, AND UNCERTAIN OUTCOMES

The twentieth century, especially in the post-World War II period, saw "expensive labor" and "cheap energy" go hand in hand with one another. This has been an integral factor in preventing and containing class struggle throughout the world, especially in the US, where it was an essential component of US hegemony. Now, some kind of major global energy shift is certain to occur. The question is no longer *whether* a shift will occur, but rather what kind of shift it will be, based on which technologies? Crucially, on whose terms will the process be, and to what ends? And, above all, who will reap the benefits and who will pay the costs? What might the relationship between

11   Dan Georgakas and Marvin Surkin, *Detroit: I Do Mind Dying: A Study in Urban Revolution* (Boston: South End Press, 1975); Stewart Bird, Rene Lichtman, and Peter Gessner, in association with the League of Revolutionary Black Workers, *Finally Got the News* (Detroit: 1970); Charles Denby, *Workers Battle Automation* (Detroit: News and Letters Pamphlet, 1960); Charles Denby, *Indignant Heart: A Black Worker's Journal* (Boston: South End Press, 1989).

12   *Midnight Notes*, op. cit., p. 124.

13   An interesting discussion of the political nature of prices including energy prices, though unrelated to the USA, can be found in Bruno Ramirez, "The Working Class Struggle Against the Crisis: Self Reduction of Prices in Italy," *Zerowork*, 1, 1975.

workers in the renewable and non renewable energy sectors be? Who will be able to harness the labor necessary for production (as well as the knowledge, raw materials, and money)? How will changes in the energy sector change the relations between capital and labor, and between waged and unwaged labor forms?

As existing energy supplies becomes more expensive (in monetary, social, political, and ecological terms), there is likely to be a corresponding effort on the part of capital to cheapen labor (not just in terms of reducing wages, but also other costs of labor, especially shifting the cost of reproducing the world's labor force onto unwaged, and predominantly women's, work). And, if energy prices rise suddenly rather than gradually, we can also expect the assault on labor to be equally rapid and sudden. Given that cheap energy has been essential for reducing the costs of reproducing labor, who should pay the increased costs of reproduction? Will capital be able to shift the increasing costs of reproduction onto workers (especially unwaged domestic and agricultural labor, predominantly carried out by women) in various parts of the world? Or will workers refuse to accept this?

These conflicts are likely to be especially acute in the USA, where escalating labor costs have, at least partially, been kept at bay with cheap energy. The twin strategies outlined above have converted large (and dominant) sectors of the US working class into extremely big consumers of energy relative to the rest of the world. This has been an essential part of controlling worker struggle. Consequently, workers in the US are incredibly vulnerable to the massive changes that are currently underway in the world's energy system. Without preparation, it is likely that they will suffer an enormous and rapid assault, which could foreseeably result in a resurrection (albeit in new circumstances) of forms of labor that had been virtually abolished in the energy-rich countries of the global north—especially in the USA. This is especially likely if the US starts to "reindustrialize" in the wake of the world economic crisis, this time on the back of a battered work force. One has only to look to the streets, fields, and kitchens of India, to see the working (waged and unwaged) and living conditions that flourish when commercial energy is expensive and scarce, and labor is both plentiful and cheap.

On the other hand, there is the renewed worldwide class struggle within the worldwide division of labor as a whole, not just the energy sector. Given that cheap energy inputs have been so important for containing class struggle in the US, the rising cost of energy can make US capital vulnerable to renewed class struggle, rooted in the rising cost of living, which capital will attempt to push on workers, and the fact that it will become increasingly costly for capital to implement one of its most tried and tested mechanisms for containing class struggle, namely mechanization.

Not only is the question of class struggle in the US of crucial importance here, but also the issue of whether new global growth centers, such as China and India, will be able to harness energy (of whatever sort) in the same way as Britain and the US were able to do in order to control class struggle and become hegemonic powers in the world-system.

Considerations of the capital-labor conflict that are central to a discussion of energy add a considerable element of uncertainty to any discussion of energy crisis and transition. This invites cautious speculation about the extent to which renewable energy will provide a material basis for either the continued expanded reproduction of capitalist social relations or for the construction of non-capitalist social relations of production and reproduction, especially in the long term. There are no obvious or inevitable answers to these questions. They are not technical questions, but political ones. And, while there is plenty of room for more exploration of these questions, they are not fundamentally research questions. The answers lie with the concrete historical evolution of the energy sector, capitalist relations in the world-system, and the outcome of the intertwined struggles that shape these processes. This chapter has focused on the fossil fuel energy economy, and has not discussed the globally expanding renewable energy sector. However, as will be described in later chapters in this book, struggles are also shaping up in these other sectors. It is likely that we are only in the very early phases of a period of very intense energy-related struggles. There is an urgent need to appreciate the open nature of the "energy crisis" and its "solutions," in order to actively prepare for and participate in the struggles that these entail.

Chapter 6 ▌ Part 2: The World's Foremost Energy Sector in Terminal Crisis?

PEAK OIL
Past, Current, and Future Scenarios[1]

Energy Watch Group

Crude oil is the most important energy source in global terms. About 35 percent of the world's primary energy consumption is supplied by oil, followed by coal with 25 percent, and natural gas with 21 percent (WEO2006).[2] Transport relies on oil for well over 90 percent of its energy needs, be it transport on roads, by ships, or by aircraft. Therefore, the economy and the lifestyle of industrialized societies relies heavily on the sufficient supply of oil, moreover, probably also on the supply of *cheap* oil.

Economic growth in the past has been accompanied by rising levels of oil consumption. However, in recent years growth in the supply of oil has been slowing and production has now reached a plateau. This is happening despite historically high oil prices. It is very likely that the world has now practically reached peak oil production and that world oil production will soon start to decline, and the rate of decline is probably beginning to increase. The point in time when the maximum rate of global petroleum extraction is reached is known as "peak oil."

Because of the importance of oil as an energy source, and because of the difficulties of substituting oil with other fossil or renewable energy sources, peak oil will be a singular turning point. This will have consequences and repercussions for virtually every aspect of life in industrialized societies. Because the changes will be so fundamental, the whole topic is not popular. Colin Campbell put it this way: "Everybody hates this topic but the oil industry hates it more than anybody else."

However, as facts cannot be ignored indefinitely, public perception is also changing. Although the possibility of peak oil is now more frequently referenced in the media than it used to be, it is still regularly and ritually dismissed as being only a "theory." This is a signal that the conventional ways of explaining what is actually happening are obviously failing. The oil industry is now admitting to the fact that the "era of easy oil" has ended. And the International Energy Agency (the

---

1   This extract is from the Energy Watch Group report "Crude Oil: The Supply Outlook," EWG-Series No 3/2007, October 2007, authored by Dr. Werner Zittel, Ludwig-Bölkow-Systemtechnik GmbH and Jörg Schindler, Ludwig-Bölkow-Systemtechnik GmbH. The complete report is available for download at: http://energywatchgroup.org/fileadmin/global/pdf/EWG_Oilreport_10-2007.pdf.

The extract included here was prepared by Thomas Seltmann at the Energy Watch Group, for the purpose of this book, and they have kindly agreed to include it under the general Creative Commons License. However, the main report is protected by © Energy Watch Group/Ludwig-Boelkow-Foundation.

2   Our report was made in 2007, so quotes from the International Energy Agency's *World Energy Outlook* refer to the 2006 issue. Data from newer issues does not change the evidence of our reports or this chapter.

intergovernmental organization of the twenty-eight OECD nations and energy policy advisor to them), in stark contrast to its past messages, now warns of an imminent "oil crunch" occurring within a few years.

The purpose of our report is to give some background information for understanding the concepts and data relevant for the assessment of the future supply of oil. This is the basis for detailed projections of future world oil supply up to the year 2030.

Last, but not least, future developments will be affected by so many different factors, such as geology (frequently referred to as "below ground" factors) and economics and politics ("above ground factors"), that the setup of scenarios is as much an art as a science. However, it appears that "geology" is now dominating economics and politics, with geological limits now defining the upper limit of the future possible supply. Economic and political factors can only further constrain this boundary. The bandwidth of uncertainty is rapidly getting narrower.

Only oil that has already been found can be produced. Therefore, the peak of discoveries that took place a long time ago, in the 1960s, will some day have to be followed by a peak of production. After peak oil occurs, the global availability of oil will decline year after year. There are strong indications that world oil production is already near its peak.

1. KEY FINDINGS

"PEAK OIL IS NOW"

For quite some time, a hot debate has been going on regarding peak oil. Institutions close to the energy industry, like CERA (Cambridge Energy Research Associates), are engaging in a campaign that seeks to "debunk" the "peak oil theory." Our report is one of many by authors inside and outside ASPO (Association of Scientists for the Study of Peak Oil) showing that peak oil is anything but a "theory." It is real and we are witnessing it already. According to the scenario projections in this study, world oil production peaked in 2006. This study places peak oil a few years earlier than other authors (e.g. Campbell, ASPO, and Skrebowski) who are, nonetheless, also well aware of the imminent oil peak. One reason for the difference is a more pessimistic assessment of the potential of future additions to oil production, especially from offshore oil and from deep sea oil, which is due to the observed delays in announced field developments. Another reason is the earlier and greater declines that are projected for key producing regions, especially in the Middle East.

The most important finding is the steep decline of oil supply after peak occurs. This result—together with the timing of the peak—is obviously in sharp contrast to the projections the IEA made in their 2006 WEO reference scenario. However, the decline is also more pronounced compared with the more moderate projections made by ASPO. Yet, this result conforms very well with Robelius' recent findings in his doctoral thesis. This is all the more remarkable because a different methodology and different data sources have been used.

The global scenario for the future oil supply is shown in Figure 1 below.

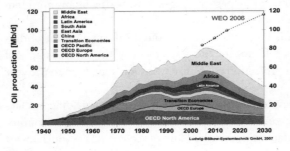

Figure 1: *Oil production world summary*

The projections for global oil supply are as follows:

- 2006: 81 Mb/d
- 2020: 58 Mb/d (IEA: 105 Mb/d)[3]
- 2030: 39 Mb/d (IEA: 116 Mb/d)

The difference between these projections and the IEA's projections could hardly be more dramatic.

A regional analysis shows that, apart from Africa, all other regions show a decline in production in 2020 relative to 2005. By 2030, all regions show significant declines relative to 2005.

Three examples of regional results for key producing regions are given below.[4]

OECD EUROPE

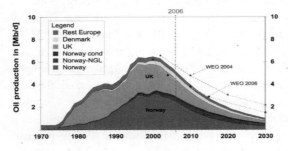

Figure 2: *Oil production in OECD Europe*

The projections for the oil supply in OECD Europe are as follows:

- 2006: 5.2 Mb/d
- 2020: 2 Mb/d (IEA: 3.3 Mb/d)[5]

---

3   Since IEA gives data only for 2015 and 2030, those for 2020 are interpolated; these data include processing gains.

4   Since IEA gives data only for 2015 and 2030, those for 2020 are interpolated.

5   For this comparison 2.3 Mb/d crude oil and 25 percent of OECD NGL are added.

- 2030: 1 Mb/d (IEA: 2.6 Mb/d)[6]

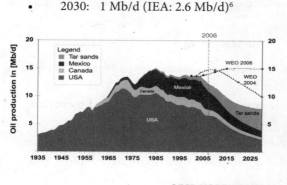

OECD NORTH AMERICA

Figure 3: Oil production in OECD North America

The projections for the oil supply in OECD North America are as follows:

- 2006: 13.2 Mb/d
- 2020: 9.3 Mb/d (IEA: 15.9 Mb/d)[7]
- 2030: 8.2 Mb/d (IEA: 15.9 Mb/d)[8]

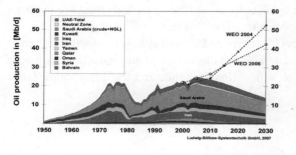

MIDDLE EAST

Figure 4: Oil production in the Middle East

The projections for the oil supply in the Middle East are as follows:

- 2006: 24.3 Mb/d
- 2020: 19 Mb/d (IEA: 32.3 Mb/d)[9]
- 2030: 13.8 Mb/d (IEA: 39.6 Mb/d)[10]

---

6   For this comparison 1.5 Mb/d crude oil and 25 percent of OECD NGL are added.

7   For this comparison 8.6 Mb/d crude oil, Canadian tar sand, and 75 percent of OECD NGL are added.

8   For this comparison, 7.8 Mb/d crude oil, Canadian tar sand, and 75 percent of OECD NGL are added.

9   28.3 Mb/d crude oil and 4 Mb/d NGL.

10   34.5 Mb/d crude oil and 5.1 Mb/d NGL.

This is the region where the assessment in this study deviates most from the projections made by the IEA.

2. FUNDAMENTALS

Observing oil production's history, which now extends over more than 150 years, we can identify some fundamental trends:

- Virtually all the world's largest oil fields were discovered more than fifty years ago.

- Since the 1960s, annual oil discoveries have tended to decrease.

- Since 1980, annual consumption has exceeded annual new discoveries.

- Until now, more than 47,500 oil fields have been found. However, over 75 percent of all the oil ever discovered is contained in just 400 of the largest oil fields (1 percent of all fields).

- The historical maximum of oil discoveries must be followed, at some point in the future, by a maximum level of oil production (the "peak").

3. UNDERSTANDING THE FUTURE OF OIL

In this section a few basic concepts are introduced in order to better understand the patterns that govern the future availability of oil. These considerations are the basis for the supply scenarios in subsequent parts of this text.

First, the concept of "reserves" is explained, as well as how it is used by different players. Then, the history of discoveries and the history of oil production are briefly described. Typical patterns of oil production over time and the influence of technology are also discussed.

3.1 RESERVES

DEFINITIONS:

Oil reserves are primarily a measure of geological and economic risk—of the probability of oil existing and being producible under current economic conditions and using current technology. The three categories of reserves generally used are proven, probable, and possible reserves.

Proven Reserves: defined as oil and gas that is "Reasonably Certain" to be producible using current technology at current prices, and with current commercial terms and government consent. This is also known in the industry as 1P. Some industry specialists refer to this as P90, i.e., having a 90 percent certainty of being produced. Proven reserves are further subdivided into "Proven Developed" (PD) and "Proven Undeveloped" (PUD). PD reserves are reserves that can be produced with existing wells and perforations, or from additional reservoirs where minimal additional investment (operating expenses) is required. PUD reserves require additional capital investment (drilling new wells, installing gas compression, etc.) to bring the oil and gas to the surface.

Probable Reserves: defined as oil and gas that it is "Reasonably Probable" will be produced using current or likely technology at current prices, and with current commercial terms and government consent. Some industry specialists refer to this as P50, i.e., having a 50 percent certainty of being produced. This is also known in the industry as 2P or Proven plus Probable.

Possible Reserves: defined as "having a chance of being developed under favorable circumstances." Some industry specialists refer to this as P10, i.e., having a 10 percent certainty of being produced. This is also known in the industry as 3P or Proven plus Probable plus Possible.

THE DIFFERENCE BETWEEN DISCOVERIES AND RE-EVALUATIONS

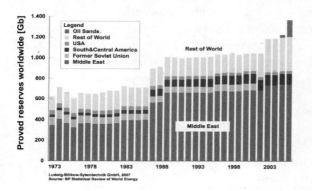

One of the prominent set of statistics existing in the public domain is the BP Statistical Review of World Energy (BP 2006). The oil reserve statistics refer to proven reserves and their development is shown in Figure 6 below.

Figure 5: Development of proved reserves of oil worldwide according to public domain statistics

Figure 5 shows an overall growth of proven reserves during the last decades (from 600 Gb in 1973 to about 1,400 Gb in 2006). Since consumption of oil also has increased considerably in this period, this is widely seen as a strong indication that a supply problem is not imminent.

The significant rise of proven reserves in the past occurred within a few years

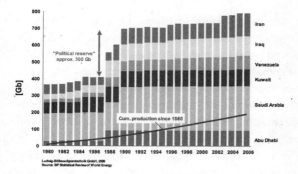

(1987–1989) and is confined to few countries. In this period, reserves increased by

40 percent—from 700 Gb to more than 1,000 Gb. This increase was entirely due to increases in OPEC countries. The latest increases, in 2006, of 163.5 Gb (sic!), account for Canadian tar sands. The details are shown below.

Figure 6: Development of proved reserves of oil in OPEC countries according to public domain statistics

All major OPEC oil-producing countries increased their reserves considerably, despite the fact that there were no corresponding new discoveries reported in this period. The reason given for the re-evaluation of reserves was that the reserve assessments in the past were too low. To a certain extent this may well be justified, since before the oil industry was nationalized in these countries, private companies may have had a tendency to underreport reserves for financial and political reasons.

However, there were also other reasons. OPEC production quotas are set according to reserves and also other factors. Therefore, each country had an incentive to defend their quota by keeping up with reserves. OPEC's real reserves are not transparent, especially since reserve estimates have not been adjusted since then, despite significant production levels. In this context, critical observers speak of "political reserves."

At any given point in time, reported reserves are the result of:

Reserves (as reported at the start of last period)

+ Re-evaluation of existing reserves (in last period)

+ New discoveries (in last period)

− Production (in last period)

_____

= Reserves (as of date)

In published statistics, the individual elements of the reserve calculation described above are, in most cases, not transparent. However, without this information, it is very difficult to assess the quality of reserve data.

It is frequently the case that revisions about field reserves are made due to earlier under-reporting. This guarantees that proven reserves reported increase year by year, thus hiding the real situation regarding new discoveries. This is a common practice used by private oil companies when reporting the size of reserves. During the lifetime of a producing field, the initially-estimated proven reserve is re-evaluated several times and is finally very close to the value that, at the beginning of the process, was internally known as the P50 reserve.

Also, with the help of these systematic upward revisions, it becomes possible to hide years of disappointing exploration results, and the quantities that have already been produced are smoothly replaced in the company statistics. This accounts for the fact that oil reserves have almost continuously increased for more than forty years, though each year large quantities were removed through production. The reserve figures used in financial contexts and shareholder meetings are thus completely different from those that address the question of how much oil has already been found and how much oil will still be found.

The main reason, however, for world reserves apparently remaining unchanged year after year is the reporting practice of state-owned companies. More than

seventy countries have reported unchanged reserves for many years, despite substantial production.

World oil reserves are estimated to amount to 1,255 Gb, according to the indus-

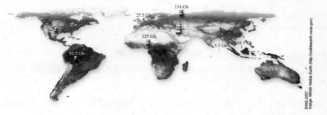

try database (IHS 2006). There are good reasons to modify these figures for some
regions and key countries. This leads to a corresponding EWG estimate of 854 Gb.
The greatest differences in estimations are for the reserve numbers for the Middle
East. According to IHS, the Middle East possesses 677 Gb of oil reserves, whereas
the EWG estimate is 362 Gb.

Figure 7: World oil reserves (EWG assessment)

Proven and probable reserves of crude oil are an important factor in determining future production possibilities (whereas looking solely at proved reserves will
always be misleading). However, proven and probable reserves are but one factor,
and other determinants are equally important. Many assessments that rely solely on
reserve data tend to overlook relevant facts. Apart from that, reserve data for many
major oil producing regions are not very reliable.

3.2 DISCOVERIES

When trying to assess the amount of oil that can still be expected to be discovered
in the future ("yet to find"), the statistics on proven and probable reserves discussed
above are obviously not very helpful. The same is true for the assessment of their future
production potentials. For these purposes, an analysis of past discoveries (measured
as proven + probable reserves) and related production profiles is far better suited.

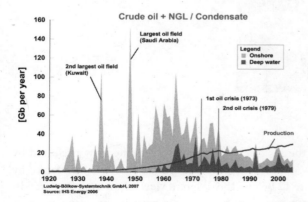

Figure 8 below shows the oil discoveries annually since 1920 and also the annual production rates (IHS Energy 2006). Past discoveries are stated according to best current knowledge (and not as the reserve assessments at the time of discovery)—a method described as "backdating of reserves." Therefore, the graph shows what "really" was found at the time and not what people thought they had found at the time.

Figure 8: History of oil discoveries (proven + probable) and production

Since about 1980, annual production exceeds annual new discoveries. This is obviously not sustainable. The peak of discoveries must eventually be followed by a peak of production.

Figure 9 shows the long-term trend in discoveries: the big oil fields were found quite early on—in 1938 the world's second largest field, Burgan (32–75 Gb), was found in Kuwait; in 1948 the world's largest field with 66–150 Gb, Ghawar, was discovered in Saudi Arabia (Robelius 2007). Today, more than 47,000 oil fields are known. However, these two largest fields, between them, contain about 8 percent of all the oil found to date. Subsequently, thanks to better exploration technology, many more fields have been discovered in many parts of the world, the high point of discoveries occurring in the 1960s. The average size of new discoveries, however, has been declining over the years. Even higher oil prices in the wake of the oil price crises in the 1970s were unable to reverse this trend. One important lesson can be learned: contrary to the assumptions of many economists, no empirical relation exists between the price of oil and the rate of discoveries.

At the end of the 1990s, there was a new increase in discoveries. This was due to exploration successes in the deep offshore regions in the Gulf of Mexico, off Brazil and off Angola, as well as the discovery of the 6–10 Gb Kashagan field in the Caspian Sea. Meanwhile, deep sea exploration seems to have already peaked and discoveries are once again declining.

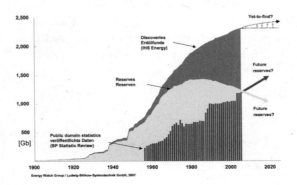

The difference between the history of proven reserves (the view preferred by "economists") and the history of proven + probable reserves (the view preferred by "geologists") is shown in Figure 12 below. The different views show opposing trends: proven reserves look as if they can stay constant or even grow in the future, whereas proven + probable reserves are steadily approaching a limit, with the possibility of perhaps eventually finding 200–300 Gb at some point in the future.

Figure 9: History of proven reserves, proven + probable reserves, production, and remaining proven + probable reserves

A possible criticism of the cumulative curve showing proven + probable reserves is the fact that re-evaluations of past discoveries are included, but possible future re-evaluations are not accounted for. Therefore, future reserve assessments might lead to an upward shift of the curve. Although this criticism is a valid one, it will not affect the estimate of the amount of yet-to-find oil, nor will it affect possible future production profiles much.

When subtracting the cumulative production from the cumulative proven + probable reserves, one gets the history of remaining reserves. Remaining reserves (proven + probable) have been decreasing since about 1980. Even when assuming constant consumption levels in the future, remaining reserves will decrease faster in the future because of declining new discoveries.

Discrepancies between public domain statistics (e.g. BP) that report only proven reserves as assessed for the previous year, and industry databases (e.g. IHS Energy) that report proven and probable reserves and backdate reassessments, are a major reason for the differences between conventional forecasts (e.g. by IEA) of the assessment of future oil discoveries and production and the approach presented in our report. Of relevance for production forecasts is the fact that reserve reassessments are usually made for producing fields. However, these reassessments do not influence the field's production pattern and, especially when production has already started to decline, the decline is not affected by upward revisions of reserves.

For the most part, future growth in production can only result from the development of yet-undeveloped discoveries. Therefore, the distinction between reassessments of reserves and new discoveries is of the utmost importance.

3.3 PRODUCTION PATTERNS

THE GENERAL PATTERN

The different phases of oil production can be described schematically by the following pattern: in the early phase of the search for oil, the easily accessible oil fields were found and developed. With increasing experience, the locations of new oil fields were detected in a more systematic way. This led to a boom in which more and more new fields were developed, initially in the primary regions, later on all over the world. Those regions that are more difficult to access were explored and developed only when sufficient new oil could not be found anymore in the easily accessible regions. As nobody will look for oil without also wanting to produce it, in general, the development of the most promising fields has followed shortly after their discovery.

In every oil province the big fields are developed first, and only then the smaller ones. As soon as the first big fields of a region have passed their production peak, an increasing number of new, and generally smaller, fields have to be developed in order to compensate the decline of the production base. From there on, it becomes increasingly difficult to sustain the rate of production growth. A race begins that can be described as follows: more and more large oil fields show declining production rates, the resulting gap has to be filled by bringing a larger number of smaller fields into production. But once the rate of discoveries falls, this is no longer possible at a

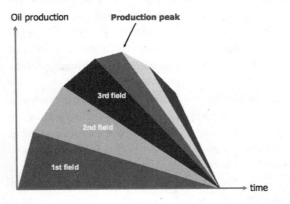

sufficient rate. Eventually, these smaller fields reach their peak much faster and then contribute to the overall decline in production. As a consequence, the region's production profile, which results from the aggregation of the production profiles of the individual fields, becomes more and more "skewed" and the aggregate decline of the producing fields becomes steeper and steeper. This decline has to be compensated for by the ever faster connection of more and more ever smaller fields, see Figure 14.

Figure 10: Typical production pattern for an oil region

So, the production pattern of an oil province over time can be characterized as follows: increasing the supply of oil will become more and more difficult, the growth rate will slow down, and costs will increase until the point is reached where the industry is no longer able to bring a sufficient number of new fields into production fast enough. At that point, production will stagnate temporarily and then eventually start to decline.

This pattern can be observed very clearly in many oil provinces. However, in some regions this general pattern was not prevalent, either because the timely development of a "favorable" region was not possible due to political reasons, or because production was held back for longer periods of time owing to the existence of huge surplus capacities (this was the case in many OPEC countries). However, the more existing surplus capacities are reduced, the more closely the production profile follows the described pattern.

3.4 PEAK OIL IS NOW

This chapter has discussed indications that a peak is imminent. However, let it be said that the question of the exact timing of peak oil is less important than many people think. There is sufficient certainty that world oil production is not going to rise significantly anymore and that world oil production will definitely start to decline soon.

On a global level, the development of different oil regions took place at different times and at varying speeds. Therefore, today we are able to identify that different production regions are in different stages of maturity. With this empirical evidence we can validate, using many examples, the simple considerations that have been described above.

PRODUCTION IN COUNTRIES OUTSIDE OPEC AND THE FORMER SOVIET UNION (FSU)

It is observable that total production in the countries outside of the former Soviet Union and OPEC increased until about the year 2000, but since then total production has been declining. A detailed analysis of the individual countries within this group shows that most of them have already reached their production peaks and that only a very limited number of countries will still be able to expand production, particularly Brazil and Angola.

The stagnation of oil production in this group of countries is attributable to the peaking of oil production in the North Sea, which occurred in 2000 (1999 in Great Britain, 2001 in Norway). Global onshore oil production had reached a plateau much earlier and has been declining since the mid-1990s. This decline could be balanced by the fast development of offshore fields, which now account for almost 50 percent of the production of all countries in this group. The North Sea alone has a share of almost 40 percent of the total offshore production within this group. The peaking of the North Sea was decisive because the production decline could no longer be compensated by the timely connection of new fields in the remaining regions, and it was only possible to maintain the plateau for a few years.

Furthermore, a steady degradation of the quality of the oil produced can be observed in almost all regions that have passed their peak. This poses an additional challenge for the existing downstream infrastructures: refineries have to operate with oil of decreasing quality. The proportion of inferior oil is steadily increasing and this further drives up the price for the remaining higher grade oil.

WORLD'S BIGGEST FIELDS IN DECLINE

Crucial for further developments is the peaking of production of Cantarell in Mexico, the world's biggest offshore field and one of the four top producing fields in the world. This field, discovered in 1978, even today contributes one half of all Mexican oil production. However, it reached a plateau some years ago and started to decline in 2005. The field then declined dramatically from 2 Mb/d in January 2006 to 1.5 Mb/d in December 2006. Double-digit yearly decline rates are expected in the coming years.

With Cantarell, three of the four biggest producing fields are now in decline: the others being Daquin in China and Burgan in Kuwait. The status of Ghawar in Saudi Arabia is not known for sure, but the field is very likely now also in decline.

Once production in the largest fields declines, it becomes more and more difficult to maintain overall production levels, as has been pointed out already.

CONCLUSIONS

The major conclusion drawn from this analysis is that world oil production peaked in 2006. Production will start to decline at a rate of several percent per year. By 2020, and even more by 2030, global oil supply will be dramatically lower. This will create a supply gap that will be difficult to close, within the time frame, with increased contributions from other fossil, nuclear, or alternative energy sources.

The world is at the beginning of a structural change in its economic system. This change will be triggered by declining fossil fuel supplies and will influence almost all aspects of our daily life.

The transition period now underway probably has its own rules, rules which will only be valid during this phase. Things might happen that we have never experienced before and that we may never experience again once this transition period has ended. Our way of dealing with energy issues will probably have to change fundamentally.

Chapter 7 ▌ Part 2

A SHORTAGE OF OIL TO SAVE OUR CLIMATE?
On the Permanent Oil Crisis, Climate Change,
and the Interaction Between the Two

Peter Polder

O ver the next ten years, our daily lives will change drastically. It is increasingly clear that the way we deal with energy is unsustainable in the most literal sense. It is impossible to continue in this way. Or, to quote Fatih Birol, chief economist of the International Energy Agency, "the wheels will fall off our energy system." And, this is not a warning for something in the distant future, but for 2010. And, that's problematic, because most of us are unable to imagine anything except unlimited, abundant, cheap energy. No longer able to travel where and when we want and use as many electrical appliances as we want? Eat imported fruit and meat or drink bottled water?

The two main drivers behind the radical change in our lives will be peak oil and climate change. Both are widely underestimated, and the interaction between them is still barely explored. Only a handful of people are seeking to understand what the combined results of the two processes will be and how the interplay with one another will develop. They are both extremely complex processes, each with their own dynamics, and the interaction between them is full of surprises.

PEAK OIL, THE END OF THE CHEAP OIL ERA

Peak oil is the most unknown of the two. It is often misunderstood to mean that oil is running out. This is not the case. On the contrary, peak oil is the moment when geological factors cause oil production to decline. Oil may still be produced, but a little less every day.

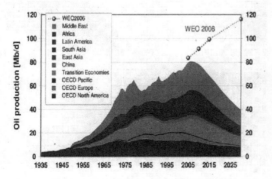

Figure 1: ASPO peak oil forecast and IEA 2006 consumption forecast.

Peak oil is seen in a single oil field, but also in an oil-producing country or the entire world production. More than sixty oil-producing countries have already passed their production peaks and are now seeing their oil production decline each year. This is despite an increase in the amount of drilling, technology, and investment. It is very likely that within a few years (somewhere between 2012 and 2018) the moment will occur when the entire world production peaks. From that moment on, less oil becomes available each year. However, besides scarcity, another process is at work. The remaining oil is decreasing in quality, becoming more difficult to exploit, more expensive to drill for and more polluting. And that is a major problem in a world where no less than 35 percent of its energy supply and 95 percent of its transport depends on oil.

The clearest indication that we are approaching peak oil is the continued decline in the quantity of oil from new discoveries. In the 1950s and 60s gigantic new fields were found, but the discovery trend has, since that time, been a declining one. Since 1980, less new oil has been found than has been consumed, and oil geologists agree that no new giant oil fields will be found, except under the polar caps and the deep sea. A reversal of the declining discovery trend is therefore unlikely.

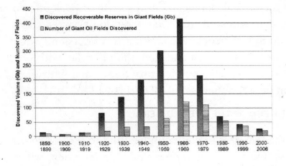

Figure 2: Conventional oil discoveries per decade (source, Robelius, 2008)

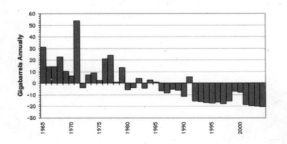

Figure 3: Reserves / Consumption ratio. (source; based on IEA data)

For natural gas and coal the same picture can be made, albeit with other production dynamics, because of their nature. Natural gas and coal data also show that reserves will reach a peak much earlier than previously thought, and there are already the first visible signs of scarcity.

Furthermore, there are a whole host of above-ground factors such as investment cycles, a shortage of technical staff, (geo-)political tensions, and greatly increased demand for steel and fossil fuels from China, India, and the Middle East. Therefore, the quantity of oil accessible to the West will drop more sharply than if it was just a question of geological decline in production. It is possible that around 2020, 20 percent less oil will be available for Europe.

THE PRICE OF OIL AND THE CREDIT CRISIS

The price of oil steadily increased between 2003 and 2008, from about $30 to a peak of $147 per barrel in July 2008. Many people expected that prices would steadily rise to a level above $200. However, this was not what peak oil experts predicted, nor has it been how prices have developed since the summer of 2008. The key word is volatility. Now that the supply and demand balance is extremely tight, the price is extremely sensitive to rumors, distortions, speculation, and other imbalances in the market. The price will bounce up to great heights, but also to unexpected lows. Now that the credit crisis is slashing economic growth numbers to zero, both the demand for oil and the price of oil is going down. On top of that, financial problems forced most traders in the oil markets to dump their contracts.

To this must be added the close interdependence of our money system with the trade in oil. International trade is still dominated by the dollar. In the 1944 Bretton Woods Agreement, the dollar was backed by the gold in Fort Knox. When that commitment was dropped in order to finance the Vietnam War, the trade in oil took over the role that gold had previously played. Because all oil is traded in dollars, there has been a continuous, and growing, need for dollars. This allowed the system of US federal banks to keep printing money, and thereby allowed the US government to continue spending money. The flow of cheap energy also provided for continued economic growth, and so consumers and businesses could carry on borrowing money. The chance that the loans would be paid back with interest was quite high.

Between 2003 and 2008, oil prices continued to rise and, at first, the economy seemed immune to high oil prices. However, the long duration of the increasingly high prices had an impact. One of the effects was the rising cost of gasoline in the US, which undermined the household budget, especially for poorer American households. These households were encouraged by the US government, as well as extremely low interest rates, to take on cheap mortgages. When the price of gasoline, and numerous other products, began to rise, people couldn't pay back their debts and thus the first domino in the credit crisis toppled.

One of the consequences of the credit crisis was a sharp drop in oil prices, from a peak of $147, to a valley of around $90, following which the price increased again to $130, only to then crash to a level below $50. The falling oil price makes many people think that the oil problem is solved. The economy goes into a depression, and so there is less demand for oil and the price collapses. Peak oil? No problem! Nothing is less true. First, a low-price oil means demand will rise again. This might give rise to an ongoing cycle in which the price and demand for oil oscillate up and down. Secondly, as peak oil approaches, oil production is getting more expensive and difficult. More and more projects have oil production costs of around \$70–80 per barrel. This means that the sharply declining prices in October 2008 have resulted in numerous projects being abandoned or postponed. On top of this, many projects and companies also face difficulties due to their credit flows drying up.

Alternative energy projects are facing the same problems of credit and falling energy prices. And, in a declining economy, consumers and businesses, but also governments, are tempted to invest less money in sustainable energy transition. Depending on how far the economy collapses and the demand for oil declines this could sharpen an energy crisis.

THE INTERACTION BETWEEN PEAK OIL AND THE CLIMATE CRISIS

The main questions that arise about the interaction between climate change and peak oil are:

- Are the assumptions about fossil fuels on which the IPCC climate models are based correct?
- What will be the potential CO_2 production from fossil fuels if you put "peak oil" into the models?
- How does climate change affect production of fossil fuels and other energy sources?

IPCC CLIMATE MODELS AND FOSSIL FUELS DATA

Before I start, let me say that questioning the climate models of the Intergovernmental Panel on Climate Change is seen by some as heretic. However, the criticism that I am placing on the models has nothing to do with what climate skeptics do. I believe that there is a good scientific basis for the claim and do not doubt the seriousness of climate change and the science behind it. The human impact on climate change is undeniable. Nevertheless, I want to say that the data on fossil fuels under the current climate models are leading to a gross distortion. There is both an over-estimation of the reserves that exist and an absence of a realistic understanding of production dynamics in the translation of that data into climate models. This means that all current IPCC climate models are misleading.

Several researchers have inserted other assumptions about fossil fuels into the climate models. In the forty climate models used by the IPCC, there are four base scenarios with assumptions about the future, which are reflected on technological, economic, and policy developments. Another influential factor is the sensitivity of climate to CO_2. In other words, how much CO_2 in the atmosphere is needed for a number of feedback mechanisms to accelerate and cause runaway climate change? For too long, the IPCC proclaimed that global warming would show a gradual trend. The climate is a chaotic system, and if it is disturbed too much it makes a turnaround and seeks a new balance. Feedback mechanisms such as the release of methane from the thawing permafrost, the collapse of the warm Gulf Stream, or the souring of the oceans by an excess CO_2 can be abrupt and cause rapid disruption of the climate. The scientific debate is not yet over, and so a range of assumptions has to be made. These range from a very low sensitivity (cover with 5 degrees of warming above pre-industrial levels) to a very high sensitivity (cover at 1.5 degrees of warming above pre-industrial level). The scientific consensus is moving increasingly towards acknowledging a turning point with 2 degrees of warming at 350 ppm of carbon in the air. Humankind has already caused half a degree of warming and boosted the level of carbon to 390 ppm. To make an assessment of the future on the basis of the above assumptions gives a lot of different possible outcomes. Depending on which assumptions are made about economic and political developments and the sensitivity of the climate system, less fossil fuel does not automatically mean a less disastrous outcome for the climate. We might just have enough fossil fuel left to fry our climate.

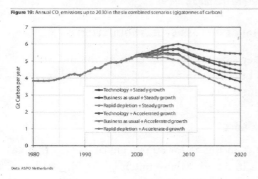

Figure 19: Annual CO₂ emissions up to 2030 in the six combined scenarios (gigatonnes of carbon)

Data ASPO Netherlands

Figure 4: global carbon emissions and peak oil, (source: Less oil, more Co₂? ASPO Netherlands, April 2009)

Although oil production declined, the amount of CO_2 per barrel of oil increased. This is mainly caused by the increasing share of unconventional oil. The best-known form of these unconventional oils are the tar sands in Canada, but in Venezuela and Russia similar sources of oil can also be found. The tar sands in Canada are loaded onto trucks and transported to factories where huge quantities of natural gas and water are used to boil out the oil and process it in order to make syncrude, artificial oil.

Another unconventional "oil" source is made available by transforming natural gas into diesel (GTL), as is done in Qatar and other Gulf states, or by converting coal into diesel, which is already happening in South Africa and China. In the US and Russia, several investors are planning CTL factories. All these variants on oil provide significantly higher CO_2 emissions. The increase in the share of unconventional oil is expected to rise to about 7 million barrels a day by 2030. The big question is how quickly the production of these unconventional oils will be increased and whether or not climate policies will affect this development.

If climate policies put a brake on this trend, then, with the peak in oil production, total CO_2 emissions would decrease. On the other hand, should a permanent oil crisis stimulate this development, pushing climate considerations to the side, this would mean less oil and more CO_2. At this moment, it is the latter scenario that is unfolding.

Table 4: Carbon emission factors of various grades of oil

	Emissions (gram carbon-equivalent per MJ of refined product)		
	Production emissions	Combustion emissions	Total emissions
Petrol	5.6	20.1	25.7
Diesel	4.4	21.1	25.5
Tar sands (low emission limit)	9.3	20.1	29.4
Tar sands (high emission limit)	15.8	20.1	35.9
Tar sands (average used here)	12.55	20.1	32.65
Extra heavy oil (low emission limit)	9.3	20.1	29.4
Extra heavy oil (high emission limit)	15.8	20.1	35.9
Extra heavy oil (average used here)	12.55	20.1	32.65
Oil shale (low emission limit)	13.00	20.10	33.00
Oil shale (high emission limit)	50.00	20.10	70.00
Oil shale (average used here)	31.5	20.1	51.5
Gas-to-liquids (low emission limit)	7.10	20.20	27.30
Gas-to-liquids (high emission limit)	9.50	20.20	29.70
Gas-to-liquids (average used here)	8.3	20.2	28.5
Coal-to-liquids (low emission limit)	20.70	21.10	41.80
Coal-to-liquids (high emission limit)	28.60	20.10	48.70
Coal-to-liquids (average used here)	24.65	20.6	45.25

Data: Brandt & Farrell (2006)

Figure 5: emissions by type of fuel, (source ASPO Netherlands, 2009)

THE EFFECT OF CLIMATE CHANGE ON FOSSIL FUEL PRODUCTION

Little attention is paid still to the other side of the coin, both by the industry and among climate activists. This is the fact that, in some areas, climate change may seriously hamper the extraction of fossil fuels.

Three examples:

CLIMATE CHANGE DISRUPTS GAS FLOW TO EUROPE

By 2020, Europe will be dependent on Russia for 80 percent of its gas. That gas comes mainly from the frozen tundra in Siberia where much of the infrastructure is built on permafrost. From pipelines and drilling platforms, to the villages and cities of the workers, maintenance roads and airstrips, everything has a foundation on permanent frozen ground. However, more and more permafrost in the arctic region is beginning to melt. The number of destabilized houses, roads, and pipelines is growing. Russian scientists estimate that eventually half of the pipeline network will almost certainly need to be rebuilt. However the Russian gas companies have a reputation for under-investing in maintenance and infrastructure. If the necessary investments do not take place, the risk of higher production outage grows. The development of new fields will be more expensive, and infrastructure will have to be built on floating structures.

STRONGER HURRICANES DISRUPT OIL PRODUCTION IN THE US

In the Gulf of Mexico, one quarter of all US oil is extracted. Also, the third largest field in the world, the Mexican Cantrell, lies in this area. There is strong evidence suggesting that warmer seawater means that more and stronger hurricanes are formed. When Hurricanes Katrina (category 5) and Rita (category 3) ploughed through the heart of the US oil industry, large-scale damage and prolonged loss of production resulted. The region has never recovered its pre-Katrina production levels. In 2008, the same area was hit by Gustav and Ike. The damage on production platforms was light, but the damage to the refineries on the coast, especially in the Galveston, Texas area, was huge, leading to diesel and petrol shortages in parts of the US that lasted for months.

The chances of next year's hurricane season again resulting in large scale damages are huge. Consequently, insurance companies and local mining authorities are demanding higher building and safety standards, making oil more expensive. In addition, the insurance premiums in the sector have gone up. Given that the region is reaching peak production, this will make the extraction of oil less attractive.

HEAVY RAINS DISRUPT COAL MARKET

In January 2008, coal prices exploded. In just one month, the price quadrupled. The reason lies in a series of extreme weather events, caused by global warming, that severely hit three of the six main coal exporters in the world: extreme snowfall in the coal regions of China, extreme rainfall in Australia and Indonesia, and on top of that, a series of blackouts in South Africa that crippled the mining and shipping of coal. The heavy snowfall in China disrupted rail traffic between the coal regions of the Chinese inland from the coast, where most coal-fired power stations are located. Furthermore, the coldest winter in years caused an extra demand for coal. To prevent shortages and blackouts, China suspended all its exports and began greatly increasing its imports of the fuel. The heavy rainfall in Australia and Indonesia flooded a number of major coal mines. In some cases, they were put out of business for months. While it is true that January 2008 saw a bizarre combination of incidents, the probability of heavy rainfall again disrupting coal exports is nonetheless real. And, although coal has the image of being abundant, the reality is that only six coal exporting countries provide 80 percent of the market.

OIL UNDER THE ARCTIC

There is, of course, another side to this story; climate change has caused an accelerated melting of ice around the Arctic Ocean. This is the last area of the world that has large oil and gas reserves that are not controlled by national oil companies. Around 2015, the Northern Ice Sea will, at least in the summer, be ice free.

Media reports that 90 billion barrels of oil can be found in this region are based on investigation of the always optimistic USGS (United States Geological Service). As a result, geopolitical tensions in the area are growing. The western media mainly focused on the Russian flag being placed on the seabed. Canada's move to build two military bases to provide additional sea patrols in the region attracted much less attention.

However, the USGS report has been heavily criticized. First of all, the report is not so much about oil, but oil equivalent. In the arctic, about 80 percent of the "oil" is in fact natural gas, the largest portion of which is already known and is on the Siberian coast. The USGS recalculated the natural gas to its equivalent in barrels of oil, and then the media only talked about oil. More importantly, other researchers, including the renowned commercial research firm Wood MacKenzie, estimate that less oil and natural gas can be found than was in fact reported by the USGS and the media. Under the most optimistic scenario, it is projected that production from the

Arctic will contribute some 4.6 million barrels oil eq. per day of oil and 9.7 million barrels oil eq. per day of gas at peak. Whereas oil can be shipped to markets relatively inexpensively, gas is a much more complex story. And does the melting of the Arctic Sea really make oil accessible? With the melting of the ice, the number of icebergs, and thus the danger to shipping and expensive drilling platforms, increases. The area remains a hostile and extreme climate. In the winter, it is dark all day and the sea is frozen. Royal Dutch Shell expects that oil production around the North Pole is only attractive at a price of $200 per barrel. The question is whether there is enough demand for oil at this price.

NON FOSSIL ENERGY ALSO AFFECTED

It is not only the fossil fuel industry that suffers from climate change. Hydro-power dams are increasingly affected by droughts. More and more dams temporarily shut down when there is insufficient water. In dry regions such as the American South-west, which largely depend on dams, this is resulting in an increased risk of black-outs, which occur most frequently during the time when demand for electricity to power air conditioning is at its highest.

Another striking effect is the changing energy demand under the influence of a changing climate. In recent years, North America and Europe had remarkably mild winters and an increase in extremely hot days. This means less gas for heating build-ings and more electricity for air conditioning and cooling.

Finally, there are the increased investments in renewable, non fossil energy and energy efficiency. Both the rising prices of fossil fuels and also climate change have caused a huge wave of investment in renewable energy sources and nuclear energy. For both, however, current growth trends are insufficient to fill the growing gap be-tween demand and supply of energy or decrease the emission of CO_2.

Quantifying the impact of climate change on energy production is an impos-sibility. What is clear is that the likelihood of large-scale production disruptions will rise, and that the investment needed to prevent these disruptions are mind-boggling. Climate change thus changes the way we use energy.

WHAT THIS MEANS FOR DAILY LIFE

First and foremost, our way of dealing with energy will change drastically in the com-ing ten years. Saving energy, reducing demand, and renewables are not a form of ide-alism anymore, but a matter of survival. Economic growth will be extremely difficult in the coming ten to twenty years. Almost all our daily activities—shopping, going to work, the work itself, visiting friends, vacation—will change. New technologies will enter our lives, such as driving on bio-gas or electricity, plastics from plant material. Some things will disappear—cheap tickets to sunny destinations, for example. Those of us with the ability to adapt early and quickly, will gain from that change. Those people that don't will face poverty. Energy will therefore be at the center off political struggles. Our daily life twenty years from now will be drastically different.

Chapter 8 ▌ Part 2

NO BLOOD FOR OIL
A Retrospective on the Political Economy of Bush's War on Iraq[1]

George Caffentzis

After the gigantic worldwide show of popular will on Saturday, February 15, 2003, the anti-war movement was able to claim to have put a new player in the field besides the miserable protagonists of the Iraq/US war: Bush and Hussein. This figure was the refuser of a war with a banner on which was written: "No Blood for Oil." Who was this person? What did the banner mean? What challenges did it pose?

In this discussion I want to make some elementary retrospective reflections on this slogan and see what future the protester was pointing to. I will do this by reading the slogan on four different levels, each more general than the previous one.

LEVEL 1. NO BLOOD FOR OIL, LITERALLY

We should neither be reductive nor jump to conclusions, but there is plenty of evidence to show that the Bush Administration planned the war as a way to plunder the oil fields of Iraq.

It is widely known that Iraq's presently-known oil reserve of more than a 100 billion barrels is the second largest on the planet and that "the undiscovered oil in the Middle East [including Iraq] is very likely the largest untapped supply in the world." As a retired petroleum geologist unequivocally answered when asked about whether Iraq or Iran had more untapped oil: "It's Iraq. We plugged and abandoned any well that wouldn't make 5,000 barrels a day. Threw 'em back in the water." Iraq's oil reserve was worth potentially more than $3 trillion at the time of the invasion. Moreover, Iraqi oil is very inexpensive to produce and is one of the world's "sweetest," i.e., it produces fewer pollutants on combustion.

At the time of the invasion, however, even though the US government and corporations imported 2.3 percent of their total oil from Iraq, US-based oil companies were unable to directly profit from oil production there. In fact, the Saddam Hussein regime had made a number of important agreements with French, Russian, and Chinese oil firms assuring them of very attractive deals in oil production once the sanctions were ended. The British and US firms, however, were given clear notice

---

1   This article was originally published in 2003, as part of the ebook *No Blood For Oil! Energy, Class Struggle, and War, 1998–2004*. It is being reproduced here with permission from the author, and with minor modifications and updatings. The original version, published under Creative Commons License, is available here: http://radicalpolytics.org/caffentzis/06-no_blood_for_oil.pdf, under the title "No Blood for Oil! The Political Economy of the War on Iraq." While some details have changed, the main political argument is still completely valid. As this presents a perspective that is quite different from other perspectives on the war on Iraq, it is worthwhile to include this chapter here. As both Bush and Neoliberalism go out the back door together, it is both a fitting requiem and also a stern reminder that the demise of both do not in any way signify an end to the blood, oil, and war mix that this chapter describes.

that they would not be welcome in a post-sanctions era, *if* Saddam Hussein and/or the Ba'ath Party were to remain in power.

Therefore, the only way for the US (and British) oil companies to gain profitable direct access to Iraqi oil was through a war that would violently and irrevocably end the Hussein/Ba'ath Party rule and bring in a new government that would cancel the deals with the French, Russian, and Chinese companies. That is why the first objective of the US military was to secure the oil fields in the invasion of Iraq. Further, the US government *assumed* that its troops would occupy the country for many years and would have a general as a military governor, in the style of Douglas MacArthur in post-WWII Japan. It was also assumed that the occupation would be paid for with the sales of Iraqi oil.

Anyone familiar with the oil industry-connected backgrounds of key figures in the Bush Administration, starting with George W. Bush himself, should not have been surprised by this plan of plunder that the "No War for Oil" slogan revealed and protested. The US oil-related corporations (including Haliburton, VP Cheney's former company) were definitely poised to find opportunities in the "rebuilding" an Iraqi oil industry destroyed by US bombs and/or Hussein's "scorched oil" tactics.

Such a blatant plan of theft and plunder could only be accomplished by military means. The consequences for the Iraqi people were to prove devastating, even if the invasion was relatively swift. The subsequent struggles among Iraqis and against the US occupiers would inevitably be bloody indeed.

The slogan "No Blood for Oil" on this level rejected the obvious gangster behavior of the Bush Administration (and the Blair echo) with brevity and justice. *S/he who affirmed the slogan wanted to stop this act of brigandage pure and simple and treated Bush's and Blair's "high-minded" (and poorly crafted) rationalizations for invasion as crude, shameful parodies of justice.* Surely, s/he branded any oil company that profited from such an adventure as a criminal, calling for the boycott of it and its tainted products.

LEVEL 2: NO BLOOD FOR PRIVATIZATION OF OIL RESOURCES

Though plunder was definitely part of the Bush Administration's plan, there were other more global issues suggested by the slogan, for the US has been the leader in imposing neoliberal/globalization policies around the planet. Thousands of nationalized companies and agencies have been privatized due to structural adjustment programs imposed by the World Bank and IMF while many forms of "restraints to trade" (including "price fixing" cartels) have been abolished by the international trade agreements now coordinated by the WTO. The US government, not surprisingly, is the dominant partner in the World Bank, the IMF, and the WTO.

Though one commodity after another had been "neoliberalized" by 2003, oil had escaped this fate. Most of the nationalizations of oil companies took place between 1969 and 1973, but it had been almost impossible for these companies to be privatized, even though the national telecoms and airlines were put on the auction block in many of these same countries (e.g., Nigeria).

Similarly, though there had been an attempt to destroy international price fixing cartels in most commodities via treaties like the one that created the WTO, oil and OPEC had been exempted from the rules of the neoliberal global regime. This was unusual since oil is the commodity that is both most basic (i.e., being involved in the production of most other commodities) and the most traded (i.e., the highest value of international sales) while OPEC is the most blatant "cartel" in the world.

This exemption of oil and OPEC from neoliberal standards was at the heart of the Republican Party's critique of Clinton's energy policies. Thus in the waning days of the Clinton era, there was a Congressional Hearing on "OPEC's Policies: A Threat to the US Economy," chaired by Benjamin Gilman (R-NY) who charged that Clinton remained "remarkably passive in the face of OPEC's continued assault on our free market system and our antitrust norms."

With the Bush Administration's rise to power, OPEC was increasingly seen as a hostile entity—especially after 9/11—that had to be subverted and either replaced or abolished.

This hostility was intensified by the recognition that the main political figures in OPEC at the time (aside from Iraq's Ba'ath regime) were either politically hostile to or unable to impose neoliberal policies. In Iran, there were the desperate Islamic clerics, in Saudi Arabia there was a ruling class that was divided between globalization and Islamic fundamentalism, in Venezuela there was the populist government of Chávez, in Ecuador there was a government that was nearly seized in a rebellion by the indigenous, in Libya there was Gaddafi (need more be said?), in Algeria there was a government that just narrowly repressed (and collaborated with) an Islamicist revolutionary movement, in Nigeria and Indonesia there were "democratic" governments with questionable legitimacy that could have collapsed at any moment. There was simply too much class struggle in an area of high-tech production (oil production) that these leaders and governments were not able to control.

This list of OPEC leaders constituted a "rogues" gallery from the point of view of the thousands of capitalists who were sending a tremendous portion of "their" surplus value to OPEC governments via their purchases of oil and gas. With such a composition, OPEC was hardly an institution to energize a neoliberal world.

Of course, OPEC was not always a political or economic problem. In the 1960s and in the early 1970s, OPEC was a relatively pliable organization, while nationalization and monopolistic pricing were still acceptable elements of the Keynesian political economy of the day. Iran was under the Shah, the Ba'athists had just lost their Nasserite zeal, Ghaddafi's fate was still undeveloped, Venezuela was a tame neocolony, Indonesia was ruled by the communist-killer Suharto, Nigeria was under the control of General Gowan, and the Saudi Arabian monarchy's Islamic fundamentalism was considered a quaint facade under which the movement of billions of "petrodollars" could be reliably recycled back into the US-European economies.

But that was then. From the Bush Administration's viewpoint, OPEC needed to be either destroyed or transformed in order to lay the foundation of a neoliberal world that would be able to truly control of the energy resources of the planet. The

Bush Administration put as much pressure as possible on OPEC's members. In April of 2002 there was a US-supported *coup d'etat* in Venezuela against the Chávez government, the leading price hawk in OPEC. It failed. In August 2002, it was Saudi Arabia's turn. The RAND corporation issued a report claiming that the Saudi Arabian monarchy was the "real enemy" in the Middle East and should be threatened with invasion if it did not stop supporting anti-US and anti-Israeli groups. But that verbal threat was nullified by the Bush Administration in the controversy that followed.

All in all, the Iraqi government was clearly the weak link in OPEC. It had lost two wars it recently instigated, it was legally in thrall to a harsh reparations regime; it could not control its own air space, and it could not even import freely, but it must have UN accountants approve of every item it wanted to buy on the open market. Ideologically and economically it lay prostrate.

A US-sponsored Iraqi government committed to neoliberal policies would definitely be in a position to undermine OPEC from within or, if it departed, from without.

Such a transformation would have made it possible to begin a massive investment in the energy industry, which was seen as a possible alternative to the spectacular failure of the high-tech sector that had just dissolved trillions of dollars into nothing in the dot com crash of 2000/2001.

Given the exceptional political-economic character of the oil commodity, it is not surprising that this gift of hundreds of millions of years of the meeting of organic life and the heat of the earth's core should generate so much violence in a capitalist world. The protester's sign now appeared to be saying: *no blood was to be spilt to preserve the energy system envisioned by Bush and Co.* S/he was calling for the system to be scrapped before we all became bloodied for oil. Some new way of distributing the earthly commons needed to be devised, since the present and future pricing/profit system that would lead to one war after another could not be allowed to continue.

LEVEL 3: NO BLOOD FOR NEOLIBERALISM

One of the Bush Administration's main diplomatic failures was to give the impression that this new "world domination" strategy was a product of a spontaneous Nietzschean will to power. Similarly, their claim that the urgency of the Iraq invasion and take-over was due to some imminent threat to national security posed by Hussein's weapons of mass destruction had been rejected even by many of their most loyal defenders. The Bush Administration was responding to an emergency, but it was not a military one ... it was a political-economic one.

The neoliberal system of capitalist accumulation (what we in the US call "globalization") that replaced the Keynesian system in the late 1970s had been in deep crisis since 1997, and the Bush Administration needed to respond to this crisis or it too would be thrown out by its masters (if not by its subjects!). I need not inform you of the story that now conventionally begins in Thailand in July 1997 with the collapse of the "bhat." This was not the first financial crisis of the neoliberal model (there was the Mexican crisis of 1995 we should remember), but the Thai crisis began a series of events that directly led to the crisis situation that Bush faced in 2003. Nor need I trace this series for you through the dramatic collapse of the stock market bubbles throughout the planet leading to the destruction of trillions of dollars of values (paper though they were), the stagnation in Europe and Japan, and even the decline of profitability in US capitalism. This constituted the first major crisis of Neoliberalism.

The Bush Administration's answer to this crisis was war. How could this be? What did war have to do with this political-economic crisis? Of course, there are many reasons for such a correlation in the past that are not to be slighted. For example, war is a classic device of ideological and juridical control of a population dissatisfied with an unrelenting economic crisis (after all, the late Chief Justice Rehnquist reminded us that "in war the laws are silent"). As another example, there is "war Keynesianism," i.e., the use of war expenditures to stimulate demand for capital and consumer goods in order to jolt the system out of a far-from-full-employment equilibrium. These could have been reasons for the Bush Administration's answer to the crisis, but they do not deal with the fact that the crisis of Neoliberalism was global, and that the US government was now "responsible" for the survival of neoliberal globalization as a whole.

The main problem facing neoliberal globalization was that, for it to "work" at the level of the system as a whole, the participant nations and corporations must follow the rules of trade even when they are going against their immediate self-interest. In a time of crisis, however, there is a great temptation for many participants to drop out of or bend the rules of the game, especially if they perceive themselves to be chronic losers. What country would keep the recalcitrants (both old—those who refused to be part of the game—and new—those who recently dropped out) from proliferating? Up until the post-1997 crisis, most of the heavy work of control was done by the IMF and World Bank through the power of money and the threat of being kept out of the global credit market, but since then it had become clear that there were countries that would not be controlled by structural adjustment programs.

The most obvious case was Argentina, but there were other, quieter dropouts in Africa and South America. The most illustrious recalcitrants were the Bush-baptized "axis of evil" nations—Iraq (one of the last of the national socialist states), Iran (one of the last fundamentalist states after the demise of the Taliban), and North Korea (one of the last of the Communist-Party-ruled states)—but there were many other Islamic, national socialist, and communist governments that had not transformed their economies into neoliberal form. By 2003, it was clear that this list would undoubtedly grow unless there was a check, in the form of a world police force, that would increase the costs of an exit.

At that moment, in order for Neoliberalism to function properly, there needed to be the equivalent of the role Britain played for the liberal capitalist system of the nineteenth century. Clinton and his colleagues believed that the US government could eventually use the UN as such a force. The Bush Administration disagreed. According to Bush, the US needed to act in its own name to enforce the rules of the neoliberal order (even though many of its adherents were unwilling to do so) and that, at times, action had to be military. In the end, it was only with the construction of a terrifying Leviathan that the crisis of Neoliberalism would be overcome and the regime of free trade and total commodification could finally be established for its Millennium.

The invasion of Iraq (the "oil" of the slogan) was a step in this construction process that was seen by Bush and Co. as a sacrifice of US human and capital resources for the greater capitalist good. The internal debate in the UN was part of a complex negotiation process that ultimately was meant to determine the conditions of US interventions, not their elimination. That is why the protester's sign did not say, "No Blood for Oil ... unless the UN says so!"

The Bush project of "saving Neoliberalism" might have been possible if there had promised to be but a few recalcitrants to and migrants from the neoliberal order. However, the antiglobalization movement proved to be right in doubting the likelihood of this. For Neoliberalism has proven unable to deliver on the "sustained growth" that rises all ships, even in its halcyon days. On the contrary, it had not even raised the 20 percent of the population it had claimed to do in its inception. This means that many ruling classes and even more working classes around the planet are going shopping at Porto Alegre to look for another system.

There will be wars aplenty in the years to come if the US wishes to play the British Empire of the twenty-first century. For what started out in the nineteenth century as a tragedy, will be repeated, not as farce, but as catastrophe in the twenty-first. Thus the slogan, "No Blood for Oil," was *a rejection of the series of wars that were being planned by the Bush Administration in its "war on terror" for the years ahead, aimed at terrorizing the recalcitrants of the neoliberal order into cooperation.*

LEVEL 4: NO BLOOD FOR CAPITALISM, PERIOD

The protester's sign's slogan has been interpreted on three different levels so far: first, as a refusal to spill blood for the plunder of Iraq's oil resources; second, as a refusal to spill blood in order to impose privatization and "free market" practices on the oil industry internationally; third, as a refusal to spill blood to preserve the rules of the neoliberal global regime. On the final level, I want to think about "No Blood for Oil" as a revolutionary slogan similar perhaps to the "Land, Peace, and Bread" of the Russian Revolution, i.e., a concrete demand that at first sight seems quite moderate and practical, but, due to the context, it becomes revolutionary. After all, the world is complex, and having "revolution" painted in red on one's banner does not make the bearer revolutionary!

The slogan itself is neither anticapitalist nor against war. It commits one to be against a war for oil, but not necessarily against war for other things. Nor is it absolutely anticapitalist, for the sign is conditional. It seems to be saying, "I reject the spilling of blood in order to continue with the profit-making from 'oil' (or indeed any other vital stuff). Human blood transcends the value of any commodity, and a system that can only run on the exchange of blood for oil is a corrupt and obnoxious Molloch." The slogan seems to be offering a reformist alternative: if "oil" can be commodified and sold at a profit without the expenditure of blood, then let it continue.

A tame, non-aggressive capitalism was apparently an acceptable one to the bearer of the banner who gave the impression of challenging the "world leaders" at the UN to come up with such a non-violent capitalism.

However, capitalism in any of its forms—neoliberal, Keynesian, liberal, or mercantile—cannot meet the challenge of the slogan. It must produce war and blood, since it cannot satisfy the minimal demands of the human race as a whole, much less of its terrestrial environment. We have 500 years of experience, in general, and 150 years of oil production and commerce, in particular, to support that claim. Since non-violent capitalism (especially in the oil sector) cannot exist, and the slogan's advocates will not part with their own or others' blood to preserve it, period, then the slogan was calling for revolutionary refusal of capitalism, however reformist the slogan sounds.

The end of the Bush Administration, the intensification of the crisis of Neoliberalism, and Barack Obama's ascension to presidential power since 2008, do not change the logic of this conclusion, though they might change the atmospherics. For Obama, just like every other president before him, is committed to the satisfaction of capitalism's energetic requirements. Thus it should not be surprising that even before his promised draw down of occupying troops in Iraq begins, a "surge" of new troops are being deployed into Afghanistan, making it clear to the protestor with the sign that Obama too adheres to the basic equation of his predecessors: blood=oil.

Originally presented in New York, Feb. 2003, and revised in Feb. 2009.

CLIMATE CHANGE, ENERGY, AND CHINA—TECHNOLOGY, MARKET, AND BEYOND[1]

Dale Jiajun Wen (on behalf of Focus on the Global South)

Climate change looms as the biggest threat to human civilization. In order to prevent climate calamity, no one can continue business as usual: developed countries have to cut emissions drastically to prevent climate disaster, and developing countries have to be engaged as well.

China has already overtaken the US as the world's largest CO_2 emitter. How do we combine the need and right to development with the right to a viable climate future? This chapter will discuss energy and emission trends in China, the already-felt impact of climate change there, the ongoing government efforts to address the challenge, the diverse perspectives of various sectors on the topic, and some current analysis on controversial issues like border tax adjustment and technology transfer (these two issues are often discussed when people talk about China and climate change). It will also raise questions regarding the current proposals like the various market and techno-fix approaches.

SECTION 1: ENERGY AND EMISSION TRENDS IN CHINA: CHINA AS THE PERPETRATOR AND VICTIM AT THE SAME TIME

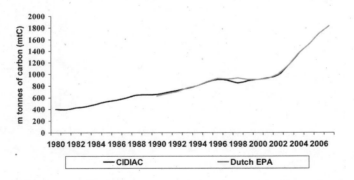

Figure 1 (above): China's emission growth since 1980.[2] Black line: data from Carbon Dioxide Information Analysis Center, gray line: data from Dutch EPA.

---

1   This chapter was commissioned for the book by Focus on the Global South. They had already commissioned the writing of it as an Occasional Paper Report. It was published as Occasional Paper 6, in February 2009, with the title: "Climate Change, Energy and China—Technology, Market and Beyond," under Creative Commons Attribution. The version reproduced here has been slightly shortened by the editor, owing to space limitations for the book. The original, full version is available at http://focusweb. org/pdf/occasionalpaper6.pdf.

2   Figure from "China's Carbon Emissions: Theirs or ours?" by Jim Watson and Tao Wang at http://www.wilsoncenter.org/events/docs/jim_watson_presentation.pdf.

China has enjoyed spectacular economic growth in the last quarter century—the average 9 percent annual growth rate is unparalleled in modern history. Despite the improvement in energy efficiency, the energy demand of the country has grown considerably. Especially since 2000, the energy sector in China has been growing faster than the country's GDP. The leaps in the annual energy use are frequently exceeding the expectations of even the Chinese government and planning agencies. This results in rolling blackouts, which have become a normal condition in some parts of the country due to supply shortage.

In 2007, China overtook the US as the world's top CO_2 emitter, several years earlier than previously projected by IEA.

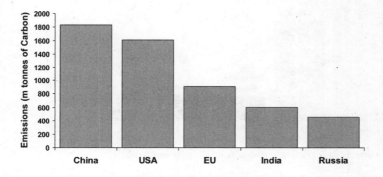

Figure 2: National carbon emissions in 2007 (Estimates by Dutch EPA).[3]

In terms of cumulative emissions, from 1904 to 2004, carbon dioxide emissions from fossil fuel burning in China made up only 8 percent of the world's total over the same period, and its cumulative emissions per capita only ranked 92nd in the world. It must be pointed out that even with the huge increase of emissions of China, its per capita emission is just one quarter of the US, and 60 percent of the EU levels.

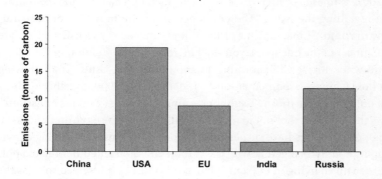

Figure 3: Per capita carbon emissions in 2007 (Estimates by Dutch EPA).[4]

---

3   Ibid.
4   Ibid.

One big reason for China's fast growth of carbon emissions is that it has become the "world's factory," or more precisely, the "factory owned by the world." Many companies, including some of the most environmentally toxic ones, are subcontractors or direct sub-units of multi-national corporations from the US, Europe, and Japan. They are churning out more and more cheap consumer goods for western consumers, while most of the profits are amassed by multi-national corporations that control the brands and distribution channels. In essence, China is the kitchen, while the west is the dining room.

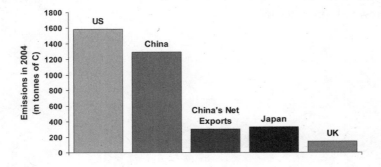

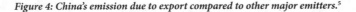

Figure 4: China's emission due to export compared to other major emitters.[5]

According to estimates by Tao Wang of the Tyndall Center for Climate Change Research of the University of Sussex, the emissions from exports from China in 2004 accounted for 1,490 million tons of CO_2, while emissions avoided due to imports was 381 million tons of CO_2. This shows that 23 percent of China's emissions were due to net exports. This estimate is lower than some estimates made by government officials and researchers, who claim that one third of China's emissions are due to exports.[6]

In June 2007, Chinese Foreign Ministry Spokesman Qin Gang made the following comment regarding the issue: "The developed countries moved a lot of manufacturing industry into China.... A lot of the things you wear, you use, you eat are produced in China. On the one hand, you shall increase the production in China, on the other hand you criticize China on the emission reduction issue." The following figure shows how China's carbon emission has soared since 2000, together with its export. It not only raises the thorny issue concerning "who owns China's emissions," but also shows the failure of the "not in my backyard" type of elite environmentalism. Indeed, developed countries have successfully exported their manufacturing activities to developing countries together with the carbon emission and other related pollution. As we are still all living in the same planet, this must be addressed soon as the greenhouse gases cannot be outsourced to the moon.

5   Ibid.
6   "Inequality, trust and opportunity" by Olivia Bina and Viriato Soromenho-Marques. http://www.chinadialogue.net/article/show/single/en/2535-Inequality-trust-and-opportunity.

SECTION 2: IMPACT OF CLIMATE CHANGE ON CHINA AND THE WORLD

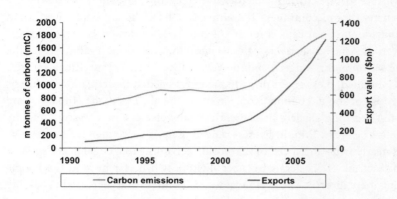

Figure 5: carbon emissions and exports grow hand in hand.[7]

There is great geographical disparity between carbon emitters and those impacted by climate change. Emissions of carbon dioxide greatly vary between places, due to differences in the level of development, technological capacity, and affluence. In 2000, 28 percent of global carbon emissions came from North American territories, and only 0.09 percent came from Central African territories. Yet, Central Africa is where global warming will cause the greatest human suffering. In fact, it has already started—as estimated by the World Health Organization—there were between 40 and 120 extra deaths per million inhabitants in 2000.

In China, the impact of climate change is already obvious in certain areas. Take the Qinghai-Tibetan Plateau as an example: many locals notice that "glaciers are melting, the temperatures are rising and rainy seasons have become unpredictable." While some urban dwellers there may welcome the warmer, more comfortable weather, the changing climate is foreshadowing doom for the local ecology and economy. Mado County in Qinghai Province (where the Yellow River originates) used to have more than 1,000 lakes, now there are less than 300. The top reason for the disappearance of lakes, according to a Tibetan environmentalist that I recently interviewed there, is climate change. According to him, "many of these lakes are seasonal and shifting. They come and go depending on the local snowfall and rainfall. From season to season, year to year, some may disappear while others appear in new places. That is the normal process. So it takes us a while to realize that we are having much fewer lakes today compared to thirty years ago. It is warmer nowadays and there is much less snowfall." The disappearance of highland wetlands and the degradation of grassland have already cost the livelihood of many nomadic herders. In Mado County, it is estimated that around one fourth of the herders have become ecological refugees—they have been relocated and are totally dependent on government welfare now.

---

7   Figure from "China's Carbon Emissions: Theirs or ours?" by Jim Watson and Tao Wang at http://www.wilsoncenter.org/events/docs/jim_watson_presentation.pdf.

The Qinghai-Tibetan Plateau has a small population itself, thus the government can afford welfare for the current ecological refugees. But this will no longer be the case if the current trend continues. The Himalayas have the largest concentration of glaciers outside the polar caps. They are literally the "ice-tower" or "water tower" of Asia. Seven of Asia's great rivers—the Ganges, the Indus, the Brahmaputra, the Salween, the Mekong, the Yangtze, and the Yellow River—are fed by Himalayan glacier melts. Combined, these rivers provide the water needed for irrigation, industry, and the daily use of about 3 billion people in Asia. The glaciers of the Himalayas are also the fastest receding glaciers in the world. Many glaciers are retreating rapidly at 15–25 meters per year. "Mount Everest is heating up at twice the speed of China's average and nearly triple the speed of the world," according to a Greenpeace spokesperson.[8] The victims of accelerated glacier melting will be far beyond the people who are living there directly. The decline of water resources and increased variability of water will negatively impact almost half the human population. For China, this would further exacerbate its already serious water deficiency.

China is facing one of the world's worst water shortages. Per capita, it only has 35 percent of the world's average fresh water resources. The water distribution is also highly uneven. The country is divided into two regions: the "dry North," referring to all areas north of Yangtze basin, and the "humid South," which includes the Yangtze River basin and everything south of it. The north has two-thirds of the country's cropland and one-fifth of the water. The South has one-third of the cropland and four-fifths of the water. Climate change may further this imbalance. Climate models predict that global warming would cause less rainfall in northern China and more rainfall in southern China. This is consistent with observations in recent years. The Huayuankou station of Yellow River has showed a decreased flow of 5.7 percent per decade. There has been a continuous drought in the North China Plain since the 1980s, while flooding disasters have happened more frequently in southern China. This trend has been especially enhanced since the 1990s.

Besides water crisis, climate change may threaten China's food security. Global warming could—if the worst predictions of scientists come true—lead to a drop of between 20 and 37 percent in China's yield of rice, wheat, and maize over the next twenty to eighty years, according to a report published in September 2004 by the Chinese and British governments.[9] In a more recent report commissioned by Greenpeace,[10] scientists from the Chinese Academy of Agricultural Sciences have warned that temperature rise, water scarcity, and loss of arable land could cut China's overall food production by 14 to 23 percent by 2050.

In 2008, a series of winter storm events affected large portions of southern and central China. Heavy snows, ice, and cold temperatures caused extensive damage. It

---

8   VOA May 30, 2007 news, "Greenpeace Says Global Warming Melting Himalayan Glaciers, Threatening Millions," available at http://www.voanews.com/english/archive/2007-05/2007-05-30-voa13.cfm?CFID=156738277&CFTOKEN=48752938&jsessionid=6630181bb7ae86034078753b5f372e2a6925.
9   "Investigating the impacts of climate change on Chinese agriculture," http://www.china-climate-adapt.org/.
10   http://act.greenpeace.org.cn/event/olympic/climate-agriculture.pdf

was China's worst winter in half a century. In early 2009, a severe drought in northern China—considered the country's breadbasket—hit almost 43 percent of the country's winter wheat crop. The expectation of withered harvest has already driven up world wheat prices. All these events are consistent with the trend of global warming: more extreme weather conditions, more droughts in the dry north. They foreshadow a turbulent climate future.

SECTION 3: CHINESE GOVERNMENT POSITION AND ACTIONS ON CLIMATE CHANGE

Fully realizing the ongoing impacts and the predicted vulnerability of China to climate change, the Chinese government is taking the issue very seriously. In June 2007, the National Development and Reform Commission (NDRC) issued "China's National Climate Change Program," the country's first global warming policy initiative. In it, the government outlined measures ranging from laws, economy, administration, and technology, which aim to reduce greenhouse gas emissions and prepare the country for both mitigation and adaptation. In October 2008, the government released a white paper on climate change, which summarizes China's ongoing effort to combat climate change, as well as clarifies China's position in international climate negotiations.

As China and the US are the world's biggest producers of greenhouse gases, the US has often used China as an excuse for inaction. But contrary to common awareness, China is already implementing a comprehensive and aggressive energy policy that tackles greenhouse gas emissions. The following is an incomplete summary of China's current goals and actions in addressing climate change.

TO REDUCE NATIONAL ENERGY INTENSITY (UNIT ENERGY PER GDP) BY 20 PERCENT IN 2010 COMPARED TO THAT OF 2005

As industry is the biggest energy consumer and greenhouse gas emitter so far, Chinese policies are now focused most strongly on improvement of industrial efficiency to reduce emissions. China's leaders' view is that energy conservation and efficiency should be addressed before searching for new fossil fuel sources.

The "Thousand Enterprises Program" identified 1,008 top energy consumption enterprises (33 percent of the country's energy consumption), and incentives have been applied in order to improve their energy efficiency. The program's goal is to reduce China's coal consumption by 100 million metric tons, approximately 5 percent of annual CO_2 emissions for China. The program is essentially a contract between the government and industry, or negotiated targets with commitments and time schedules on the part of all participating parties. A number of government departments and entities are involved in the top-1,000 enterprise program, including the Department of Resource Conservation and Environmental Protection of NDRC (which promotes energy saving in China), the National Bureau of Statistics (which collects and manages statistical information of enterprises in China), the state-owned Assets Supervision and Administration Commission (which manages major state-owned enterprises in China), the Office of National Energy Leading Group, and the General

Administration of Quality Supervision, Inspection, and Quarantine.[11]

China is replacing old inefficient power plants and factories with state-of-the-art new units. In 2007, the government announced a timetable for thirteen industries in different areas to close down backward production facilities as part of the latest Five-Year Plan period. In 2007, small thermo-power generating units, which produce 14.38 million kilowatts of energy, were stopped. At the same time there were reductions of about 46.59 million tons of iron-smelting capacity, 37.47 million tons of steel-making capacity, and 52 million tons of cement production capacity. More than 2,000 heavily polluting paper-making plants, chemical plants, and printing and dyeing mills were ordered to close down, as were 11,200 small coal mines.

The government has recently reformed the passenger vehicle excise tax to encourage the production and purchase of smaller-engine vehicles, and to eliminate the preferential tax rate that applied to sport utility vehicles (SUVs). The fuel efficiency standard for motor vehicles is increasingly stringent. While the Corporate Average Fuel Economy (CAFE) Standards in China are lagging behind that of Japan and Europe (who are world leaders in this aspect), they are far above the US.

The government is setting goals and taking actions for energy-efficient lighting. With subsidies from the government, 50 million energy-saving bulbs are now being distributed to households all over the country, and within the coming three years more than 150 million energy-saving bulbs will be distributed.

Green building initiatives are underway. By October 2007, 97 percent of all new urban construction across the country conformed to energy saving standards for the design stage, and 71 percent for the construction stage, a respective increase of 1 and 17 percentage points over 2006. Energy-saving renovations to existing buildings are also carried out—tasks have been assigned to different regions to install measured heating equipment and complete energy-saving renovation to a total of 150 million m² of floor space.

TO RAISE THE PROPORTION OF RENEWABLE ENERGY (INCLUDING LARGE-SCALE HYDROPOWER) IN THE PRIMARY ENERGY SUPPLY BY UP TO 10 PERCENT BY 2010, AND 15 PERCENT BY 2020

In 2005, China set two wind power goals—5 GW by 2010 and 30 GW by 2020—but it has consistently outpaced them. While 500 MW of new wind capacity was installed in 2005, the pace of installation accelerated considerably in 2006, with 1.3 GW installed—an amount equal to the total over the previous two decades. By 2007, it had already reached 5 GW, and it raised its 2020 target to 100 GW. China is now the fifth largest wind energy producer in the world. China's solar industry is also growing rapidly, having produced 35 percent of the global supply of solar photovoltaic in 2007 (up from 20 percent in 2006), most of which is exported to other markets. China already accounts for 70 percent of global production and use of solar hot water heating systems.

China is already the world leader in renewable energy capacity (with 42 GW in

---

11   More detailed info about the program can be found at http://ies.lbl.gov/iespubs/2007aceee.pdf.

2005, excluding large hydro projects). In 2005, China tied with Germany for the largest national investment in renewable energy, excluding large hydro-power, which amounts to $7 billion. This was primarily directed to small hydro and solar hot water projects.

Other policy goals include a 20 percent increase in forest coverage by 2010, and an increase of annual volume of carbon dioxide in carbon sinks by 50 million tons by 2010 compared to that of 2005.

China is not the only developing country that is taking concrete actions to combat climate change. Often unknown to western readers, the unilateral measures by developing countries including China, when implemented, are expected to significantly reduce emissions even if compared to the commitments by Annex 1 countries in the Kyoto Protocol.

The Center for Clean Air Policy's 2006 report titled "Greenhouse Gas Mitigation in China, Brazil, and Mexico: Recent Efforts and Implications" said:

> Unilateral policies and programs adopted by China and Brazil between 2000 and the end of 2005, if fully implemented, were projected to be greater in 2010 than those to be achieved by the United States' voluntary carbon intensity reduction goal and approximately 40 percent of the domestic reductions to be achieved in the 15 EU countries under their Kyoto Protocol target. As discussed above, a number of additional measures have been adopted since the end of 2005 in these countries which are expected to further reduce emissions. These reductions are significant when compared with the reductions in developed countries under various commitments or proposals.[12]

The Report further states that:

> With full implementation, combining the measures identified in our earlier report with these new measures yields total annual GHG emissions reductions in China, Brazil, and Mexico that are greater than the annual reductions under the Kyoto Protocol (without the US), EU's reduction commitments in 2020, and the reductions estimated in the early years of the main US legislative proposals with a total reduction of 2,100 MMTCO$_2$e (2,100 Million Metric Tons of CO$_2$ Equivalent).

SECTION 4: EMISSION REDUCTION, BINDING COMMITMENT OR NOT?— COMMON BUT DIFFERENTIATED RESPONSIBILITIES AND GEOPOLITICS

"Common but differentiated responsibilities" as outlined in UNFCCC is one of the guiding principles of the Chinese government's position on international climate negotiations. To cite the October 2008 government white paper, China sticks to the following principles to address climate change:

> To uphold the principle of "common but differentiated responsibilities," which is a core principle of the UNFCCC. Both developed and developing countries are obligated to adopt measures to decelerate and adapt to climate change. But the level

---

12   http://www.ccap.org/docs/resources/64/Developing_Country_Unilateral_Actions_2007_Update.pdf

of their historical responsibilities, level and stage of development, and capabilities and ways of contribution vary. Developed countries should be responsible for their accumulative emissions and current high per-capita emissions, and take the lead in reducing emissions, in addition to providing financial support and transferring technologies to developing countries. The developing countries, while developing their economies and fighting poverty, should actively adopt adaptation measures, reduce their emissions to the lowest degree and fulfill their duties in addressing climate change.

But how do we interpret "common but differentiated responsibility" on a practical level? At the December 2008 Poznan climate negotiations, Chinese representatives argued for a "per capita accumulative emission convergence" as representing the equity principle. China, along with the rest of the Group of 77 (G-77), stressed that developed countries have continued to fail to fulfill their financial commitments as well as drag their feet in technology transfer, and that some significant progress must be made on these fronts. They pressed the developed countries to implement their finance and technology transfer commitments as already outlined in UNFCCC as a condition for serious discussion on some other issues that developed countries are pushing for.

However, given China's status as one of the biggest emitters, and citing its impressive economic growth in the last two decades, there are growing pressures from countries in the west that China should unilaterally commit itself to binding emission reductions without pre-conditions. What do the Chinese think about the issue?

Hu Angang, a public policy professor at Tsinghua University in Beijing thinks that China should bind itself to international goals to slash greenhouse gas emission without conditionality. As reported by Reuters in September 2008, Hu's suggestions to China's leaders, as well as a recent essay, argues that China could emerge as an economic and diplomatic winner if it vows to cut gases from industry, farms, and transport that are trapping increasingly dangerous levels of solar heat in the atmosphere. "It's in China's own interest to accept greenhouse gas emissions goals, not just in the international interest," he suggested. According to his recent paper published in *Contemporary Asia-Pacific Studies*, China's greenhouse gas pollution would continue rising until around 2020. The country would then "dramatically" curtail emissions, cutting them by 2030 to the level they were in 1990 and then half of that by 2050.[13]

Hu's position is a minority view in China, which he himself has acknowledged. Among Chinese scholars and NGO activists who are working on climate issues, I have yet to meet anyone who agrees with his notion, even though most of them agree that China should try its best to cut emissions and explore a low carbon development pathway as soon as possible. The difference is mostly due to different understandings of international politics. In the same interview with Reuters, Hu revealed another reason for his advocacy: "Like joining the WTO, this should be used as international pressure to spur our own transformation." While he undoubtedly thinks that China's

---

13   "China government adviser urges greenhouse gas cuts," by Reuters News on 08 September 2008, by Chris Buckley, available at http://www.reuters.com/article/reutersEdge/idUSPEK19898020080908.

WTO accession is a great success, not everyone agrees.

China has made huge concessions during the WTO accession in certain sectors. For example, once the stipulated tariff reductions were fully implemented, China's agricultural sector would be more open than that of Japan and South Korea. Between 2000 and 2002 (China joined WTO in 2001), the income of 42 percent of rural households decreased in absolute terms. Largely due to the rural exodus caused by social economic factors including the WTO, it is estimated that China has to keep its economic growth rate at 8 percent minimum to keep unemployment at bay.

Given these facts, there is no wonder that there are ongoing debates about China's WTO accession. Internal debates aside, the international impact of China's WTO accession cannot be ignored as well. According to a third-world trade activist who preferred to remain anonymous, Europe, the US, and Japan have often used China's example in recent WTO talks to pressure other developing countries to give more concessions. The common argument is, if China has agreed to this and that, why can't you?

Given such domestic and international background, many scholars and activists think that it is important for China to avoid the same mistakes similar to the WTO accession in international climate talks, instead of repeating the "success" as perceived by Prof. Hu. This is why the official position of the government has lots of traction among Chinese climate researchers and activists. Domestically, they agree with Hu that China would be one of the biggest victims of global warming if the crisis were not abated. Thus they ardently support the ongoing measures by the government to reduce emissions, and many are pushing for even more drastic actions. Internationally, they think that as a leader of developing countries, China should take a strong stand for the advocacy of development rights and equity principles to preserve the policy space for developing countries in general. Furthermore, it should also use its power to push developed countries for implementation of existing commitments and further commitments. After all, the developed countries have contributed to 75 percent of accumulative greenhouse gas (GHG) emissions with only 20 percent of the global population. As the biggest accumulative emitter and per-capita emitter, the US has withdrawn from any climate agreement so far. And the emissions by Europe and Japan have continued to climb despite the binding commitments in Kyoto Protocol. If this trend is not reversed, the climate future would be doomed even if developing countries disappeared completely (thus reducing their share of GHG emissions to zero).

SECTION 5: PUBLIC OPINIONS AND VOICES FROM THE "CIVIL SOCIETY"

Since the early to mid-1990s, the Chinese government has allowed environmental NGOs to proliferate. Presumably, it hopes that these NGOs can fill in a gap in public education and help to address the country's pressing environmental problems. Environmental NGOs have rapidly moved into the newly opened political space. Right now, environmental groups are probably the fastest growing non-governmental organizations in China. Many international environmental NGOs, like the Nature

Conservancy, Conservation International, World Wildlife Fund, and Greenpeace, have established offices in China as well.

Environmental NGOs are very active in the campaign for energy efficiency. For instance, in July 2007, forty NGOs jointly launched the "20 percent Energy Saving Citizen Actions," in response to the government target of improving energy efficiency by 20 percent by 2020. In March 2007, eight NGOs, including the Friends of Nature, Oxfam Hong Kong, Greenpeace, Action Aid China (AAC), Global Village Beijing, Worldwide Fund China (WWF), Green Earth Volunteers, and the Institute of Public and Environmental Affairs came together to initiate the "Chinese Civil Society's Response to Climate Change: Consensus and Strategies" project. The aim of the project was to raise the level of awareness and concern about climate change within Chinese civil society, to seek common positions and strategies based on Chinese realities, and to call for common actions to combat climate change. Over 200 NGOs joined a survey, and dozens of NGOs participated in several rounds of consultations and workshops.

The project produced two reports: the first report, "The Feasibility Study on Chinese Civil Society's Response to Climate Change," summarizes the perspectives and positions of various governments and civil society groups around the world in the international climate negotiation, and aims to help Chinese civil society form positions and strategies on climate change based on Chinese conditions and realities; the second report, "Climate Change Impacts on China: Thoughts and Actions for Chinese Civil Society," attempts to establish a common perspective for Chinese civil society on the topic. In the latter report, the consensus positions on global warming of the participating NGOs are presented as follows:[14]

Positions of Chinese Civil Society

In order to avoid the worst impacts of climate change, countries around the world should take immediate actions. Chinese civil society hence calls for:

Position One: The governments of the world to set a common goal to tackle climate change under the auspices of the United Nations Framework Convention on Climate Change.

Position Two: To differentiate responsibilities between developed countries and developing countries in tackling climate change.

The developed countries to take the lead to drastically cut their GHG emissions and to provide assistance to the developing countries in areas such as technology transfer and funding through effective mechanisms.

Developed countries and developing countries should explore low carbon sustainable development together.

Position Three: The Chinese government should participate more proactively in international efforts to tackle climate change, taking responsibilities of global climate protection while securing the right to social and economic development.

---

14   See in "A Warming China: Thoughts and Actions for the Chinese Civil Society" at http://www.greenpeace.org/raw/content/china/zh/reports2/social-action.pdf.

The Chinese government should reform its economic development model and its energy structure to implement its energy efficiency target and to promote faster development of renewable energy, therefore controlling its GHG emissions.

Position Four: To apply the principle of social equity in drafting and implementing the adaptation and mitigation policies; to raise the capacities and conditions of the vulnerable groups and regions on adaptation; to prevent and reduce negative effects of policies, technologies and market mechanisms on the local environment when mitigating climate change.

Position Five: The Chinese government to encourage and ensure the participation of civil society in the climate change policy-making process and implementation and monitoring processes.

While such actions by these environmental and development organizations should be praised and encouraged, one should also realize the ambiguous position they occupy in the public sphere. On the one hand, environmental conscience is increasing, and green NGOs are growing rapidly. On the other hand, they are increasingly being accused of acting like foreign agents who are trying to stop China's development, especially when they are engaged in public debate. While such accusations bear little or no truth at all, the heavy dependence on international funding makes it difficult for many environmentalists to defend themselves. Such accusations, when coming from some sections of the public, also serve as a sober reminder that nonprofits are only part of the civil society, instead of representatives of the civil society.

Terms like "NGO" (non-governmental organization) and "civil society" are in many circumstances used interchangeably, and it is often assumed that non-profit organizations represent NGOs.[15] Another often-held assumption is that a growing middle class would foster more accountability and more open civil society, thus leading towards a liberal democracy. Unfortunately, these assumptions are not necessarily true. A Chinese professor once commented wittily, "not all organizations from civil society are good or progressive. To give an extreme example, the mafia is also one form of civil society."[16]

Instances of citizens' self-organizing are indeed growing rapidly in China. While the above example of a joint statement on climate represents the better part of civil society and is encouraging, there are opposite examples of middle class organizing. One recent case involves the ongoing debate about gas price. With the recent crash of the oil price, there are talks to finally implement the long discussed fuel tax. This has caused lots of resentment and organized opposition among the rising middle class—many think that it is their given right to imitate the US lifestyle, just as then-President George Bush declared at the Earth Summit in 1992, "the American way of life is non-negotiable." In November 2008, organized by a Beijing law firm, 1,773 private car owners submitted a letter to the government, complaining that the current

15   One such example is the above statement: a group of non-profit organizations came together and worked out a joint announcement, and called it "positions of Chinese civil society."

16   Sadly, in certain areas of rural China, mafia is indeed the fastest growing type among all the NGOs. He Xuefeng, a leading expert on China's rural development, has documented such cases.

gas price was not as low as that of America and lobbying against the planned fuel tax. They demanded that the oil price should also "get on track with the world"—a catch phrase often used in the reform era, stipulating that China should copy the rules of the west. In most circumstances this phrase has been used, "the west equals the world," a very problematic bias indeed. These car owners went one step further: the US equals the world. They did not compare the gas price to that of Japan or Europe, where the high population density and other resource constraints are more comparable. (As of December 30, 2008, gas price in Beijing was around 5.15 Yuan/litre [0.54 Euro/litre or $2.86/gallon], while similar grade gas costs around 1.10 Euro/litre in Germany, making it twice as expensive). They also did not complain earlier in 2008, when the gas price in China was much cheaper than in the US. When raw oil price skyrocketed from $70–140 US per barrel, the gas price at the pump only increased by 20 percent, which was made possible through a combination of direct government subsidies and the loss-making operations by the state-owned oil companies, because the government took these active measures to dampen the shock.

This group of 1,773 car owners is only the tip of the iceberg—they are organized enough to lobby the government. While on the other side, as far as I know, only ten professors and a handful of energy experts have come out in support of the fuel tax, and no environmental group has taken a position, probably for fear of offending the car-driving middle class—or more precisely, the elite class, which comprise less than five percent of the population. Exactly because of this elite status, car owners are the most organized and vocal part of the "public." With many media professionals part of the car driving elites or expecting to join soon, they are the most dominant "public" voice in the ongoing fuel tax debate. In Chinese newspapers, these 1,773 car owners are often being portrayed as heroes in defense of "public" interest against the "evil" government and "evil" state-owned oil companies. There are lots of opinions on the internet criticizing automobile-based growth—for example, some Chinese bloggers went as far as proposing a 100 percent car purchase tax and suggesting that the money be used to subsidize public transportation, but one seldom reads such ideas in the printed press.

Given all these, it is not surprising that, in the latter part of 2008, the best-selling book related to the subject of global warming was titled *Global Warming: Unreasonable Scare*. It is the Chinese translation of a book by two American authors, Dennis T. Avery and S. Fred Singer, titled *Unstoppable Global Warming: Every 1,500 Years*. The authors claim that global temperatures have been rising primarily—or entirely— because of a natural cycle. It's not very dangerous, and humans can't stop it anyway. The middle class are happy to read what they would like to hear, instead of the reality they need to know. Similar to many urban elites in other parts of the world, China's middle class are largely sheltered from the negative impact of climate change: it is at most an inconvenience, if not outright conspiracy.

The ignorance of the consuming elites is especially depressing when one realizes how many Chinese are already negatively impacted by climate change. As mentioned above, a significant number of herders in Qinghai-Tibetan Plateau have had to abandon their previous livelihood and become welfare recipients. In Northwest China, hundreds of thousands of people are being driven from place to place because of droughts and the encroaching desert. Farmers in many places are reporting shifting weather patterns and more unpredictable rainfalls that are hurting agricultural production. Unfortunately, many of these people do not necessarily link their "local" problems with global issues like climate change (at least not yet), let alone articulate it. And they are largely voiceless. During the last quarter century of market-oriented reform, herders and farmers in China had been increasingly marginalized. In most cases they are not seen as a constituency of the environmental movement either. So far, most environmental NGOs, especially those based in Beijing, have focused their efforts on educating and converting the more conscientious part of the urban elites. If they can reach areas away from the comfort zone of their middle-class enclave and reach the real grassroots who are suffering the consequences of environmental degradation, they will gain a much larger support base, and improve their own understanding of environmental challenges, including global warming.

SECTION 6: BORDER TAX ADJUSTMENT

Influenced by the US green-labor alliance, one key demand of the American climate community is the right to unilaterally implement border tax adjustment (BTA) to protect jobs. The claim that American workers are losing manufacturing jobs to China is often used as an argument. Let us first examine this premise. Is China really stealing jobs from the US and other parts of the world? Yes, huge amount of manufacturing has been relocated to China. As explained earlier, one major reason for the rapid increase of China's GHG emissions is that it has become the industrial platform of the world. But, contrary to what many think, China's export-oriented growth has not created a net increase in China's manufacturing jobs. On the contrary, China experienced massive job losses. From 1995 to 2002, manufacturing jobs decreased by 15 percent—from 98 million to 83 million.[17]

This seemingly paradoxical phenomena was caused by machines replacing labor. China used to have a machine tool industry built for a populous country. For example, compared to the western machines, Chinese textile machines employed ten times more workers, but required much less initial capital investment (and were likely to be less energy intensive as well). But in the relentless pursuit of efficiency and profit during the reform era, foreign machines (mostly imported from Germany and Japan) became increasingly favored. In 1997, former Prime Minister Zhu Rongji ordered the destruction of massive numbers of locally-made machines. As a result of such transformation, large numbers of textile workers have been laid off, even though Chinese textiles gained a bigger market share around the world. In former state-owned enterprises (SOEs), an eight-hour work day was the norm, and workers got one day off every week. With the massive privatization of SOEs, sweatshops became more widespread, twelve-hour work days became the norm in many coastal factories, and now workers are lucky to get one day off per month.

---

17   China Statistics Yearbook, 2002.

Between 1996 and 2002, manufacturing jobs decreased by 22 million globally. Thus China's job loss of 15 million in the same period accounted for two-thirds of the global shrinkage. Besides the massive net job loss, China's transformation into a global industrial platform has created more wealth for transnational corporations instead of its own citizens; although much manufacturing happens in China, it is the western companies that capture the lion's share of the profits. Again, take China's "highly competitive" textile industry as an example: Chinese producers receive less than 10 percent of the profit, while more than 90 percent of the profits go to western companies that control the brands and distribution channels. Rather than blaming China for stealing jobs from the world, we should instead understand global restructuring according to neoliberal rules and how it destroys jobs around the world. In this light, China may not be seen as the culprit but rather a participant of the current development model; a small minority of Chinese have joined the global elites in the process,[18] while the working class are being marginalized just as elsewhere.

Popular media in the US often blames China for the manufacturing job losses. However, American ruling elites are perfectly aware of the facts. In a congressional testimony in May 2005, William H. Overholt, Chair in Asia Policy Research from the conservative think tank, RAND Corporation, acknowledged that "rapid Chinese globalization has required stressful adjustments. State enterprise employment has declined by 44 million. China has lost 25 million manufacturing jobs."[19]

His numbers were even bigger than the Chinese government numbers cited above, as he was referring to a longer time frame.

It is really sad that instead of looking into these facts and analyzing what is wrong with the system, the US unions are often buying the misguided narrative that blames other workers who are supposedly "stealing" their jobs. A Chinese labor activist once commented on this tragic reality of global labor movements, "it seems to me that it is the big capitalists who have learned the most from Marx: they have unity through institutions like the WTO and IMF while the working class in different countries are often being pitched against each other." Viewed from such an angle, the border tax adjustment advocated by US unions is another knee-jerk response, instead of a well-thought-out policy option resulting from careful examination.

If the purpose of border tax adjustment is to prevent employment leakage,[20] it is questionable how effective such protectionist measures can be without addressing the deeper structural problems outlined above. Also, there are better ways to protect jobs. For example, one possibility is for American workers to support Chinese proposals to reduce and eliminate preferential treatments of transnational corporations.

---

18   Such people are often called the "comprador class" in China, meaning Chinese representatives of foreign (often western) interests.

19   "China and Globalization," William H. Overholt, Testimony presented to the US-China Economic and Security Review Commission on May 19, 2005, available at http://www.rand.org/pubs/testimonies/2005/RAND_CT244.pdf.

20   In a November 2008 climate change conference in Washington DC, a labor leader from AFL-CIO gave a twenty minute presentation about how jobs are being lost to China, and why BTA is needed to protect American jobs.

In order to attract foreign direct investment, the Chinese government has implemented many favorable measures like the lower tax rate enjoyed by foreign corporations compared to domestic ones.

There are growing calls now to reduce and eliminate such super-citizenship treatments of multi-national corporations from many sectors in China. US unions can support such efforts, as it can be a truly win-win situation for workers on both sides of the Pacific. As there will be less tax incentive to relocate to China, US workers can better protect their jobs. For Chinese workers, a bigger percentage of the corporations' profit will stay within their community, instead of being siphoned off. The key is for the global working class to explore ways to work together to make capital more accountable and rooted, instead of being pitched against each other.

If the purpose of BTA is to prevent carbon leakage, there are also many problems on this front as well. First, how is leakage defined? Empirical data hint that almost all new energy-intensive installations in developing countries, such as those for steel, cement, chemicals, etc. are more efficient than existing ones in developed nations. So the baseline emissions can be lower in developing countries' new installations than in developed ones. Second, BTA undermines the principle of common but differentiated responsibility, and can be perceived by many developing countries as a back door maneuver to force them to take on similar levels of mitigation. This is counterproductive to confidence-building. There are much more clever and sustainable ways to get carbon/energy-intensive industries from developing countries into a global deal.

In 2007, realizing the resource pressure created by the rapid export increase of energy intensive products including steel and cement, the Chinese government first reduced tax rebates, then further imposed an export duty on such products. The voluntary "border tax adjustment" measure was taken up by taxation authorities with advice from the State Environment Protection Agency, and it significantly lowered the exports of the targeted products (40 percent for certain categories of products). Now the State Environment Protection Agency is researching the feasibility for a full range of green taxes. Developing countries should be strongly encouraged to take such measures. On the one hand, it addresses the competitive concern of *developed* countries to a certain extent, while on the other, it may serve *developing* countries in the long run. After all, most developing countries are poorer in resources than developed ones on a per capita basis,[21] so large volume export of resource and energy intensive products is probably not for the long-term benefit of the country, even if the production is more efficient in a narrow economic sense.

However, these measures should remain voluntary instead of mandatory for a certain time frame, as developing countries need the policy space to decide for themselves instead of being forced to take a similar level of mitigation responsibility prematurely. Border tax adjustment by the importing countries should only take

---

21   According to WWF 2008 Living Planet Report, the per capita biocapacity is 3.7 global hectare for high-income countries, 2.2 global hectare for middle-income countries, and 0.9 global hectare for low-income countries.

place as the last measure of penalty, say, against the US if it continues to refuse their responsibility as Annex 1 countries, or against certain sectors of a developing country if it refuses to take the voluntary measure after a certain agreed-upon grace period. Instead of unilateral measures as currently proposed, it would be more efficient and more equitable if the system was implemented under UNFCCC. The border tax collected should go into a general fund, where the money can be used for mitigation and adaptation measures in developing countries.

Unfortunately, border tax adjustments as proposed or practiced by the western countries are going exactly the opposite direction. In June 2009, EU and US made a WTO complaint regarding China's export tariff on a series of energy intensive goods including charcoal, citing unfair competitive advantage for Chinese firms. And it was only March 2008 that EU decided to charge a five-year anti-dumping tariff on Chinese charcoal. It seems the real logic is, "When we don't need so much charcoal, you are dumping; when we do need more charcoal, your export tariff is unfair. When things are not perfectly aligned with our interests, you must be doing something wrong." It is especially ironic that this WTO complaint was made at the same time the west is drumming up support for border tax adjustment against developing countries. If they really care about carbon emissions, why do they want to prevent developing countries using export tariff to reduce energy intensive exports? Or is border tax adjustment just another excuse for trade protection, as it has always been suspected by many developing country observers?

SECTION 7: WHERE IS THE OPEN SOURCE MOVEMENT FOR THE CLIMATE?—THE ISSUE OF INTELLECTUAL PROPERTY AND TECHNOLOGY TRANSFER

Advocates of intellectual property rights from the west often claim that it will provide a stimulus of innovation and catalyst for the deployment of environment-friendly technologies. But in reality, there are plenty of examples to the contrary. One such case can be found in the Montreal Protocol, allegedly one of the more successful international environmental agreements. Corporations have patented refrigerants that do not destroy the ozone. Instead of stipulating measures like compulsory licensing to facilitate the rapid adoption of such technology around the world, corporations are allowed to continue to charge high monopoly prices that many developing countries cannot afford, while compromises are being made to postpone the phase-out period. For example, in the case of hydrochlorofluorocarbons or HCFCs, Article 5 countries (developing countries) only have to freeze production on January 1, 2016, then eliminate it on January 1, 2040, in exchange for the unconditional protection of corporate patents. Usage of certain types of HCFCs like HCFC-141b, HCFC-142b, HCFC-22 has been in sharp increase in recent years, mostly due to increasing refrigeration in China and India. As a result, 2006 saw the worst depletion of the ozone layer in history (UNEP. 2006, "2006 Antarctic ozone hole largest on record"). These HCFCs are also powerful global warming gases, often tens of thousand times more potent than CO_2. In a strict economic sense, this arrangement in the Montreal Protocol can even be argued as a win-win compromise: the western corporations continue to enjoy the benefits of monopoly patents, and the developing countries continue to enjoy the low cost of HCFCs until 2016. The loser is the environment and our shared planet.

It is not only the developing countries that suffer from the obstacles created by the current intellectual property system. One revealing example is the case of Enercon, one of the most innovative wind energy companies in the world. Enercon is the third-largest wind turbine manufacturer in the world and has been the market leader in Germany for several years. One of its key innovations is the gearless (direct drive) wind turbine in combination with an annular generator. As gearbox problems are responsible for most down time in conventional wind turbines, this new design significantly improves efficiency and reduces maintenance needs. However, Enercon has been prohibited from exporting its wind turbines to the US until 2010 according to a WTO ruling, allegedly due to infringement of US patent 5083039 held by Kenetech. Enercon claims their intellectual property was stolen by Kenetech and patented in the US before they could do so. Kenetech made similar claims against Enercon. During an investigation by the European Parliament, a US National Security Agency employee revealed that detailed information concerning Enercon was passed on to Kenetech via ECHELON.[22] In early 2008, Enercon reached a cross-patent agreement with its competitor General Electric (which holds US Patent 7397143, a later patent partly based on US patent 5083039). During this long drama of international espionage and legal battles, neither Kenetech (which went bankrupt in 1997) nor General Electric have built or installed any direct-drive wind turbines based on the disputed technology. In short, in this particular case, all that the WTO rules and IP rules have achieved is to prevent the deployment of this climate-friendly technology in the US until now. Once again, the environment loses.

One beauty of knowledge and ideas is that they are non-competitive and non-exclusive, unlike most material goods. If you have an apple and I have a pear, and we make an exchange, then I only have an apple and you only have a pear. If you have an idea and I have another idea, and we make an exchange, then both of us will end up with two ideas. My use of a certain technology does not prevent you from using the same technology. But the current intellectual property system treats knowledge as a rival and exclusive resource: if I patent an idea, nobody else can use it unless they can pay the monopoly price. There are better ways to stimulate innovations and deploy technologies than commodifying and monopolizing knowledge in this way. One successful example is the vibrant open-source and free software movement in the IT industry. The "free software" and "open-source" movement has millions of followers who contribute their time freely. It has produced impressive technologies including Linux and OpenOffice. These products are great low cost or even zero cost

---

22   ECHELON is a name used in global media and in popular culture to describe a signals intelligence (SIGINT) collection and analysis network operated on behalf of the five signatory states to the UK-USA Security Agreement (Australia, Canada, New Zealand, the United Kingdom, and the United States). The above case regarding ECHELON is documented in a EU Parliament investigation, and its report available at http://www.europarl.europa.eu/sides/getDoc.do?pubRef=-//EP//TEXT+REPORT+A5-2001-0264+0+NOT+XML+V0//EN&language=EN.

alternatives for consumers around the world, and viable substitutes to software from industrial monopolies like Microsoft. Instead of conventional copyright or intellectual property, free software often uses the following principles of "copyleft," which means:

1. the freedom to use and study the work,
2. the freedom to copy and share the work with others,
3. the freedom to modify the work,
4. the freedom to distribute modified and therefore derivative works,
5. all derived work should be distributed under the same or equivalent "copyleft" license.

It promotes free sharing and further development of ideas and knowledge, instead of validating the monopoly of knowledge.

I have spent a fair amount of time trying to convince my Chinese friends that climate change is a real threat instead of another conspiracy by the rich countries to stop the economic growth of the developing countries. Oftentimes it is frustrating, but it has its reward as well; sometimes one is being asked sharp and thought-provoking questions. One such question comes from a friend working in the IT industry. He gave me quite a powerful argument, as paraphrased below:

> If global warming is really a serious threat to human civilization as you are telling me, then where is the open-source movement for the climate? I am an active participant of the free software movement. Every week I spend more than ten hours of my free time on it, like millions of other tech guys around the world. We all understand that the free software we help to create and distribute probably hurts the profit margin of the whole IT industry. But there are more important things in life than making money at all costs. So this is what we do to make the world a bit better and fairer. Unless I see a comparable movement for the climate, I will always suspect that you guys are just another interest group, and the whole climate change thing might be some hype to sell certain kind of proprietary technology of the west.

I was at a loss to argue against his suspicion: he and the movement he is in have walked the walk, while the climate community has only largely talked the talk. The technology transfer mechanism under UNFCCC has yet to transfer one single piece of equipment or technology to developing countries. Then there is the World Business Council for Sustainable Development (WBCSD), a CEO-led global association of some 200 companies dealing exclusively with business and sustainable development. WBCSD did establish an Eco-Patent Commons project in early 2008, where companies can pledge eco-friendly patents to the public domain. Companies can choose which patents they want to put into the "pool"—one patent is enough to get in and claim the badge of honor. So far, seven companies (IBM, Nokia, Bosch, Xerox, Dupont, Pitney Bowes, Sony) have joined it, but what they have donated are hardly breakthrough or potentially big sales technologies. During the December 2008 Poznan talk, WBCSD representatives called it "completely unacceptable for industry" that a UN climate agreement would include compulsory licensing of patents. They want technology transfer only to take place through projects that require the participation of multi-nationals. All these make the earlier Eco-Patent Commons initiative look like a greenwash exercise—or even worse, a typical cynical attempt to head-off compulsory licensing.

"Where is the open source movement for the climate?" This question from someone outside the environment movement could be a challenge for everyone who works on climate-related issues, whether in the government, business, or non-profit sector. Until we produce a Linus Torvalds or Richard Stallman[23] of climate-related technology, until some significant eco-friendly technologies are put into the public domain, the suspicion that the climate community is just another interest group will always linger in many people's mind. We have to walk the walk to prove otherwise. Global warming is one huge crisis of the commons, and we need collective efforts and ingenuity to rebuild the commons. Ideas of reciprocity as embodied in the "copyleft" principles are better suited for this purpose, instead of further commodification as promoted by the current IP regime.

Besides the hurdles presented by the IP regime, another block to talking constructively about climate and technology is that so many people assume that the ideas to be shared in a "climate commons" will come mainly from TNCs, or high-tech professionals (people like Linus Torvalds or my IT-industry friend) who are altruistic enough to devote time and energy to open-source. In fact, the ideas and technologies that need to be shared are not necessarily "high-tech" and will also come from communities across the world: Indian river valley farmers refining their non-carbon customary irrigation systems, Brazilian farmers seeking to restore and promote mixed agriculture, Chinese peasants using biogas digesters to turn wastes into fuel and green fertilizer, British Transition Towns, and so forth. The problem now is that what is referred to as "technology transfer" at the international level (in the UN, etc.) means the elimination and erasure of such technologies in favor of the purchase or the negotiation of the transfer of technologies that the western TNCs would like to sell to the rest of the world. The Indian, Chinese, or Brazilian villagers, of course, have no patents on their technologies and so they are freely available already—but they are being squashed (and often by the international climate apparatus itself, including the Clean Development Mechanism (CDM), foreign investment, etc.) instead of being exchanged with the rest of the world.[24] What is the best way to make such community-based knowledge and technology benefit more people? A parallel can be drawn with indigenous knowledge on medicinal plants. Attempts to co-opt such knowledge into the existing intellectual property regime often results in biopiracy and even deprivation of access. The monopoly of intellectual property has to be questioned if we want to prevent a similar fate for community-based eco-technologies.

---

23   Linus Torvalds is a Finnish software engineer who initiated the development of Linux Kernel. Richard Stallman is a US software engineer who pioneered the General Public License and started the free software movement.

24   Documented cases can be found in Larry Lohmann's "Carbon Trading: a critical conversation on climate change, privatisation and power" (Uppsala, Sweden: Dag Hammarskjold Foundation, 2006).

SECTION 8: BEYOND TECHNO FIX: IS THE AMERICAN DREAM STILL POSSIBLE OR DESIRABLE?—EXPLORING THE REAL POSSIBILITY OF A LOW CARBON ECONOMY

In comparison to many other countries (especially the US), China is taking more concrete actions on the ground for fostering clean energies, efficiency, and so on. While such efforts are laudable and one can only hope that the US will follow suit, we still have to ask: will such techno-fixes be enough for the big challenge? Let's examine some facts.

Global warming is just one aspect of the global environmental crisis, thus it has to be addressed in the context of global governance and sustainable development. China's strong focus on energy efficiency and technology fixes has its ideological roots in ecological modernization theory,[25] an idea coming out of Scandinavia. It is an optimistic, reform-oriented environmental discourse. It puts its confidence in modernization and technological innovation—by improving energy and resource efficiency, technology advancement can solve the environmental crisis and promote economic growth at the same time, thus a "win-win" scenario.

Given this theory, one would expect that developed countries are better models of sustainable development. Unfortunately, this is far from the reality and the US obviously does not follow the Scandinavian model. According to data from the Living Planet Report 2006[26] by the World Wildlife Fund, one can calculate that if Chinese people copy the American lifestyle with the current US technology level, we would need more than one planet. We need five planets if everyone consumes at US levels.

At the Poznan climate talks in December 2008, China said that development itself is the great contribution to addressing climate change. Thus, the development space and rights of developing countries should be guaranteed. But one thing missing from the mainstream discussion of development—whether by China or any other country—is the crucial question—what kind of development? Take the biofuel debate as an example. Even the language and options of the current biofuel discussion expose a distinctive northern bias. Regarding the possibilities of biofuel, all we hear about are industrial scale bioethanol or biomass-generated electricity. Why? Because people in the north have taken it for granted that electricity is a necessity instead of an improvement after other more basic needs are fulfilled, and ethanol is needed to drive the automobiles. In contrast, there is hardly any mention of other modes of utilizing bioenergy, such as direct burning of biomass, or biogas digesters. More than 300 million families in the world (or about 20 percent of humanity) still depend on the direct burning of biomass (mostly wood) for cooking. Most of them use open fire or simple three-stone pits which are highly inefficient. The resulting smoke and toxic emissions cause 1.6 million deaths a year. In many places (for

---

25   There are exceptions to this generalization. For example, China's Environmental Protection Agency has pioneered green GDP accounting, and some scientists from Chinese Academy of Agricultural Sciences are advocating organic farming as both mitigation and adaptation measures of global warming. Both have deviated from the standard ecological modernization theory. One should realize that there are different school of thoughts in the Chinese government, just as in most western governments.

26   http://assets.panda.org/downloads/living_planet_report.pdf.

example Haiti), the quest for fuel wood is also a driving force of deforestation and the consequent emission increase. Yet the technology for rapid improvement already exists. Properly-designed stoves built with local material and local labor can reduce fuel consumption up to 80 percent, as well as significantly cut down emissions of smoke and organic volatiles. When we talk about development of bioenergy, the first priority should be adapting the design of efficient stoves to conditions of each locality, and rolling out the technology using local resources so that the 20 percent of the poorest of humanity can take better care of their environment as well as fulfill their development needs at the same time. However, when people think about development and technological advances, few would ever think of fuel-efficient woodstoves or other appropriate technologies. Instead, the usual images include more electronic appliances, consumer goods, and cars.

On the issue of cars, it is especially sad that in blind worship of the US lifestyles, China has abandoned its previous focus on public transportation and bicycles, encouraging, instead, an automobile-oriented lifestyle. In stark contrast, Cuba imported millions of bicycles and bicycle production lines from China in the 1990s (partly in response to the energy crisis generated by the collapse of the former Soviet Union), while China imported millions of cars and multiple automobile production lines from the west. In 2004, China became the world's fourth-largest producer and third-largest consumer of automobiles. The number of car owners is growing at 19 percent annually.

Apart from increased dependency on imported oil and growing emissions, the massive explosion of private automobiles is harming the well-being of many Chinese, especially the poor. Public buses are getting slower and slower because of traffic jams. For example, the average bus speed in Beijing was 10 miles per hour in the 1980s; it decreased to 5 miles/hour in the 1990s. Nowadays, it is further reduced to a crawling 2.5 miles/hour. More and more roads are closed to bicycles to make room for cars, highways and urban sprawl are swallowing huge swathes of land, which is creating many landless peasants. The estimated number of landless peasants today ranges between 40 million and 70 million, while there were none thirty years ago. Even if we suddenly had a magic technology to make all cars infinitely more efficient (zero fossil fuel demand, zero emissions), there is another resource constraint: the urban sprawl generated by an automobile-centered infrastructure could eat up so much arable land, that it would threaten China's food security. If only 50 percent of the Chinese population drive a car, would the remaining 50 percent have places to walk and bike or even have enough land to grow food?

While technological fixes (for example, improving energy efficiency and reducing emissions per car) are important, one also has to ask other more fundamental questions as well: How do we want to organize our lives? What kind of urban and rural landscape do we want to have? What kind of transportation system should we have? There is a limit to technology fixes without paradigm shifts. After all, the fuel efficiency of automobiles cannot compete with that of bicycles, no matter what the level of technology.

The following photo was taken in summer 2006, near the city center of Amsterdam, the Netherlands. Since then I have used it in many talks in China, asking the audience to guess when and where it was. No one even came close. The two most frequent guesses are some Chinese city twenty years ago or some Southeast Asian city today. Even though I mostly talked to progressive audiences who care about social justice and sustainability, they were all deeply brainwashed in this respect: modern cities should be a land of automobiles, while a land of bicycles is a sign of backwardness. It is intriguing that so many Chinese audiences think that a photo of today's Amsterdam is of some Chinese city twenty years ago. In a sense, they are not wrong. Just like today's Amsterdam, back then, cities were designed for people and bicycles—in most city roads, bike lanes were as wide as or even wider than auto lanes. This was by no means achieved by chance. Some westerners may assume it was simply because China was too poor to afford automobiles, but low per capita GDP did not prevent Manila or Bangkok from becoming auto-traffic hell decades ago.

In a 1970 interview with American progressive William Hinton,[27] China's first

Prime Minister mentioned the air pollution problem caused by automobiles in a certain Japanese city, and said that China would not imitate automobile-oriented urban growth. He probably knew nothing about peak oil or climate change, but he had enough information to realize that given China's large population and resource constraints, private automobiles would be an unaffordable luxury for the majority of the people. So the government decided to focus on bicycles and public transportation to serve the masses. In a related observation, William Hinton noted how little material difference there was between Beijing, the capital city, and Zhang Zhuang (a rural village he frequented), which is another manifestation of the "serving the people" instead of "serving the elites" policy orientation at the time. Unfortunately, the wisdom

27   Appendix 1: interview with Prime Minister Zhou Enlai, in "Shen Fan," William Hinton, 2008 Chinese edition.

that the late Prime Minister had thirty-eight years ago is being forgotten by Chinese leadership and many of its people today. Today's China is marked by a rapidly growing gap between the rich and poor, and cities are increasingly transformed for cars. So, is China's recent auto frenzy good development? Aren't we just blindly copying the worst mistakes of the west? The same question should be asked about China's rising middle class and their newly-found obsession with consumerism.

SECTION 9: THE CURRENT ECONOMIC CRISIS: GREEN HOPES OR BLACK FEARS?

The financial crisis that originated from the US has created huge job losses in China. Due to decreased demand in the US, there have been massive factory closures in the coastal export region, and there will be more. In many cases, factory owners simply disappeared in the middle of the night, leaving hundreds of workers without their due salary. It is estimated that 10 million migrant workers have returned to their rural villages, with another 20 million lingering in the cities searching for jobs. To combat the economic slowdown, China has announced a RMB 4 trillion ($586 billion US) economic stimulus package with many new investment projects. Local governments have followed suit with their own plans, which in total may reach a gigantic RMB 10 trillion. Most of them are infrastructure projects.

The word "crisis" for the Chinese means danger and opportunity at the same time. The ongoing economic crisis, as bad as it is, could offer an opportunity for China to re-examine its export-oriented and resource-intensive growth model. So far, the signals from the Chinese government are mixed. For example, there is a lot of talk about using the opportunity of lower oil prices to implement a fuel tax, which will help to curb oil consumption and encourage a move to clean energy in the long run. On the other hand, some government officials are encouraging consumers to buy more cars, in order to stimulate the economy. Such confusion is to be expected. After all, many advocates and practitioners of the market-oriented reform in the last quarter century have held the unspoken conviction that the eventual purpose is to copy the US system. Now with the storm originating from the US, the center of laissez faire capitalism, many people are struggling to understand and cope.

Many of the infrastructure projects announced in the stimulus package will be energy and resource intensive, repeating the process by which China spent its way out of the 1997 Asia financial crisis. There is nothing wrong with infrastructure building itself. The global South needs development to pull itself out of poverty and environmental destruction, just as the poorest 20 percent of humanity (many of whom are in China), who still cook with open fires, desperately need more efficient stoves and biogas digesters. The question is: what kind of infrastructure? Solar panels, wind turbines, and improved power grids require one-time intensive input, but may lay the groundwork for a future low-carbon economy. On the other hand, more highways and cars will soon become a liability for the future.

For rural China, where the majority of Chinese people still live, there are many possible projects (not all of them resource intensive) that can bring long-term environmental, economic, and social benefits. Many irrigation canals and water works are in serious disrepair and deterioration. Restoration and new development of water works can greatly improve resilience of rural economy to droughts and floods, so they can be better prepared for the changing climate. The same thing can be said about re-planting of windbreaks, networks of trees to protect arable lands from soil erosion, etc.

The massive overuse of chemical fertilizers and pesticides has caused serious soil degradation as well as undermined food safety. Now, with millions of migrant workers going back to their home villages, it is a golden opportunity to promote the more labor-intensive, but socially/environmentally-friendly, organic agriculture—as many experts point out, organic agriculture is an effective mitigation and adaptation measure against global warming. The list can go on and on, if one can open up the imagination and think out of the existing development paradigm. The material benefits of many such projects will take some time to realize, thus local governments and people may be reluctant to take on such projects, as we have all been so-entrenched in the culture of "instant rewards and short-term gain" in the last few decades. However, doesn't the ongoing economic crisis offer the perfect reason for us to question such a culture?

As pointed out by Lord Stern in the famous Stern report: "Climate change is the biggest market failure." In fact, it is a bigger market failure compared to the more obvious financial market failure of the ongoing economic crisis. We are in both crises because there is something fundamentally wrong with our way of organizing our society. At this junction of global environmental, social, and economic crisis, we urgently need to ask: What kind of world do we want to live in? What kind of development do we really need?

Chapter 10 ▌ Part 3

FOR DEMOCRATIC, NATIONAL DEVELOPMENT OF NORTH AMERICA'S ENERGY RESOURCES

Energy Workers Unions[1]

Energy workers from Mexico, the United States, Canada and Quebec together with our social partners in civil society and hemispheric solidarity movements, declare to our respective members and citizens in each country our commitment to democratic, national development of our energy industries.

We are meeting at the time of the Montebello summit of the Security and Prosperity Partnership (SPP) that links our countries in a new political and economic framework for continental integration based on the security agenda of the George Bush presidency. This agenda has the complicity of President Calderon and Prime Minister Harper, but has no democratic mandate from the people of Mexico, Canada, or the United States.

We share the concern of civil society movements that the SPP is a new and powerful instrument created by government and corporate elites to shape the destinies of our nations without democratic participation or oversight. We reject the security agenda of the SPP, which links NAFTA and trade to the limiting of civil liberties, mass surveillance, racial profiling, and the failed and disastrous military and foreign policies of George W. Bush. We challenge the neoliberal assumptions of prosperity that have led to increasing disparities of wealth and power in each of our countries.

---

1   This statement was issued in Montreal, August 18, 2007, on the occasion of the Montebello summit of the Security and Prosperity Partnership (SPP). As it is a statement, original spelling has been used, including non-US spelling standards. It was signed by the following organizations: Unión Nacional deTrabajadores de Confianza de la Industria Petrolera (UNTCIP); Sindicato Mexicano de Electricistas (SME); Alianza Nacional Democrática de los Trabajadores Petroleros (ANDTP); Sindicato Único de Trabajadores de Industria Nuclear (SUTIN); Comisión Nacional de la Energía; Frente Auténtico del Trabajo (FAT); United Steelworkers (USW); Syndicat des employé-es de techniques professionnelles et de bureau d'Hydro-Québec—section locale 2000 SCFP; Syndicat des spécialistes et professionnels d'Hydro-Québec—section locale 4250 SCFP; Syndicat des employé-e-s de métiers d'Hydro-Québec—section locale 1500 SCFP; Syndicat des technologues d'Hydro-Quebec—section locale 1500 SCFP; Canadian Union of Public Employees (CUPE); Communications, Energy and Paperworkers Union of Canada (CEP); SCEP Section Locale 121—Montreal Shell Refinery; Fédération des travailleurs et travailleuses du Québec (FTQ); Centrale des syndicats démocratiques (CSD); Confédération des syndicats nationaux (CSN); Conseil central du Montréal métropolitain (CSN); International Federation of Chemical Energy Mines and General Workers' Unions (ICEM); Mexican Action Network on Free Trade (RMALC); Réseau Québécois sur l'intégration continental (RQIC); Fédération des femmes du Québec; Association droit a l'énergie—SOS Futur; Coalition of Québec—Vert—Kyoto et Association Québécoise de lutte contre le pollution atmosphérique (AQLPA); Common Frontiers Canada; North South Institute; KAIROS; and Council of Canadians.

Permission to reprint it here was obtained from the Communication, Energy and Paper workers union, CEP. The original can be found here: http://www.cep.ca/cep_on_line/spp/spp_statement_e.pdf.

However, as energy workers we are compelled first of all to respond to the SPP energy agenda. Through the SPP and the North American Energy Working Group, the governments of Mexico, United States, and Canada have formed an unprecedented collaboration with energy corporations to promote the continental integration of our energy industries and infrastructures. Nine working groups have been working intensively to integrate oil, natural gas, electricity, nuclear power, hydrocarbons, science and technology and regulatory agencies. While these working groups bring together government, regulators, and corporations at the highest level, they have excluded labour, environmentalists, and civil society movements, and circumvented the oversight of our elected legislatures.

The SPP-corporate agenda of substituting continental corporate rule at the expense of national and local plans of development includes:

- The complete integration of electricity grids between our countries and the continuing deregulation of electricity in each country to promote electricity generation for export.

- The promotion of a continental integrated natural gas system and imports of liquefied natural gas to meet a continental shortage of natural gas, which is expected within a short period of time.

- The "streamlining" of regulatory processes and deregulation in each country for cross-border oil pipelines, including a five-fold increase in Canadian tar sands production, and continuing privatization of energy industries.

- The direct intervention of the US to guarantee the security of energy installations.

These and other elements of the SPP-corporate energy agenda are unsustainable and sacrifice the needs of workers and communities in each country to the profits of energy corporations. This is an agenda that fails to address the need for each country to reduce greenhouse gas emissions, including a new round of far-reaching goals after 2012. Nor does this corporate-continental model of energy development respond to the needs of national economic development or recognize the primary role of energy industries for community economic development.

We share a concern that the promotion of biofuels and ethanol puts at risk agricultural economic stability and food sovereignty in North America. North American farmers and consumers must not be sacrificed to facilitate unsustainable, speculative investments in new biofuel industries.

Energy workers in each of our countries have fundamental and urgent concerns over the misguided energy policies that are being pursued in the context of the SPP.

UNITED STATES

- Bush/Exxon opposition to world efforts to combat global climate change.

- Deregulation of electricity resulting in Enron corporate fiascos.

- Rising energy costs for working families and industry.
- Closure of fifty oil refineries in last twelve years.
- Growing dependence on foreign oil.

MEXICO

- Unconstitutional privatization of Mexico's constitutionally-protected energy industries.
- Threat to privatize PEMEX.
- Oil industry operating at 80 percent capacity and petrochemical industry at 50 percent capacity.
- United States prohibitions on development of Mexico's nuclear sector.
- Neoliberal economic policies.
- Trade union freedoms for energy workers.
- The weakening of the guiding role of the state with respect to energy and development.

CANADA

- Failure to meet Kyoto targets.
- Canadian energy security needs.
- Tar Sands development based on bitumen exports.
- Natural gas exports and loss of petrochemical industry jobs.
- Electricity deregulation and market failures.

The energy industries in each of our countries must be guided by the common principles of democracy and sustainability.

We affirm the responsibility and the right of democratically elected governments to establish national and local energy policies, to defend and promote public ownership of energy production and distribution, and to regulate the activities of private sector energy corporations within the context of national and local policy. Access to energy resources for basic human needs is a right of citizenship and must not be denied by unfair markets and corporate greed. Energy resources in each of our countries are publicly owned and must be democratically managed in the public interest.

Electricity grids, home heating and transportation fuels, and energy sources for industry are necessary and strategic factors in national and local economic development. These industries provide good jobs that are family and community sustaining. We reject the model of energy development that sacrifices local generation and supply systems to be replaced by continental corporate grids and never ceases its obsession with eliminating labour. We support the right of local communities to demand that energy resources are processed locally to achieve the highest possible value.

Energy workers understand the historic transformations that are necessary to achieve global energy sustainability. The petroleum, gas, coal, and other carbon-based industries will be impacted by measures to address global climate change and dramatically reduce greenhouse gas emissions. Large scale hydroelectricity and nuclear power are also faced with many formidable environmental challenges. Energy workers understand the necessity for conservation and energy efficiency, and new renewable energy industries, as well as for new policies in each country that may impact our employment security. We are ready to be part of the solution on the basis of Just Transition that ensures that workers and communities do not unfairly shoulder the burden of social and environmental change.

Sustainability and national and local development of energy resources cannot take place without democratic involvement of workers and communities. Energy policy will not achieve these goals without the voices of energy worker unions and communities.

We condemn the policies of union avoidance by many energy corporations and the failure of our respective governments to assure the right of workers to freely organize in independent and democratic trade union structures.

We commit to forge a new hemispheric worker to worker solidarity to ensure the growth of our unions and the negotiation of strong collective agreements with employers. Through the ICEM, the UIS-TEMQPIA (Unión Internacional de Sindicatos de Trabajadores de la Energía, el Metal, la Química, el Petróleo e Industrias Afines) and other international trade union bodies, we will establish strong networks and respond to calls for solidarity when our membership engages in trade union and community struggles.

We commit ourselves to establish co-ordination between this forum and the Energy Workers Forum of Latin America and the Caribbean to share experiences and joint actions with respect to energy integration plans.

We will continue to work with our social partners in the hemispheric solidarity movements to bring workers of each country together and to jointly challenge the harmful consequences of unfair trade agreements and neoliberal globalization policies.

Energy policies will shape our world in the twenty-first century. These policies will lead either to democratic, sustainable development or to global environmental disaster and new wars of aggression. Energy workers, their unions, and social partners in Mexico, Canada, and the United States will act together for democratic, sustainable national development of our energy resources.

Chapter 11 ▌ Part 3

EUROPEAN ENERGY POLICY ON THE BRINK OF DISASTER
A Critique of the European Union's New Energy and Climate Package[1]

Sergio Oceransky

European energy policy is on the brink of disaster. The European Commission published, on January 23, 2008, a policy package on energy and climate that has been heavily influenced by the nuclear and fossil-fuel industries, as well as by large power utilities. This is a coherent and well-designed strategy to ensure the continued centralization of the energy system, and of the political and economic power associated with it.

FROM LONDON TO BRUSSELS

The European Commission has just presented its proposal for an energy and climate policy package that includes the Directive on the Promotion of the Use of Renewable Energy Sources (RES), the revision of the EU Emissions Trading Scheme (EU ETS), and a new directive on Carbon Capture and Storage (CCS).

Great efforts have been made, not least by the UK government, to ensure that the legislative package is based on instruments that have proven totally unsuccessful in terms of promoting renewable energy (RE) deployment and reducing greenhouse gas (GHG) emissions, and that will displace existing successful policies and alternatives. It also includes an outright destructive agrofuels policy, which, if implemented, will strengthen the existing trend to transform large-scale centralized RE into a source of social conflict, remove RE's potential contribution to the common good, and in some cases, even turn them into a further (and potentially very powerful) contributor to environmental destruction.

---

1   This is a selection from a previously published piece by the same author, entitled "Confronting the Nuclear Resurgence: British Government's Manoeuvres, EU Policy, and the Nuclear-Fossil Collusion." It was published as a special issue of the *Nuclear Monitor*, on January 28th, 2008. No. 665. *Nuclear Monitor* is the regular publication of the World Information Service on Energy (WISE) and the Nuclear Information & Resource Service (NIRS). It is reproduced here with permission from both the author and WISE. The article has been divided in two pieces for this book, and another selection is included as Chapter 31 "Confronting the Nuclear Resurgence: British Government's Manoeuvers, EU Policy, and the Nuclear-Fossil Collusion." It has also been shortened considerably, owing to space limitations in this book.

This original article was written when the new energy and climate policy framework of the European Union was taking shape, as a contribution to the heated debate around it. The debate is over and the EU policy has been passed but the contents of the text are still relevant to discussions on energy and climate policy issues. While there were some important changes in the EU package that was in fact passed, many of the issues discussed here were included in the final package. The complete text of the original article can be downloaded at http://www10.antenna.nl/wise/665/Special/665_Special.pdf.

For updated analysis of the modified EU climate and energy package that actually was approved, as well as the EU's commitment to renewable energy in the face of the economic-financial crisis, see a range of articles the European Renewable Energy Federation's website, at http://www.eref-europe.org/.

On a superficial reading, the policies outlined in the package seem to be based on a positive approach. They are based on the decision taken in March 2007 by the European Council (the Heads of State of all EU countries), which endorsed:

- a minimum unilateral reduction of 20 percent in GHG emissions for the EU (to be extended to 30 percent if other dirty countries reduce their emissions),
- an indicative target of 20 percent reduction of the EU's energy consumption compared to projections for 2020 (to be obtained through energy efficiency),
- a 20 percent minimum mandatory target for the share of renewable energies in overall EU energy consumption, and
- a 10 percent minimum mandatory target for the share of sustainably-produced biofuels in transport petrol and diesel consumption.

These targets, to be achieved by 2020, seem to reflect a sincere concern for sustainability and a strong political will to promote renewable energies through mandatory action. The policy package defines specific rules to implement these guidelines.

A closer look reveals a large number of destructive policies hidden behind a convenient green façade. The directives are plagued with problems in their own right, but their combination makes the policy package far more damaging than the sum of its components.

Given the complexity of the policies involved, all the components will be briefly described before examining them and their interconnections in more detail.

The Commission proposes to introduce a European market for renewable energy certificates, which is incompatible with the only successful RE policy (known as "feed-in tariffs"). This proposal is highly lucrative for large power utilities, which will make immense windfall profits and regain complete control over the power sector by pushing out independent power producers. It denies a fair opportunity for public supply of RE at the local level, keeping it firmly in the hands of large energy companies. In addition, this virtual market will make RE more expensive and therefore less competitive in comparison with fossil and nuclear energy, delaying the necessary transition to a 100 percent renewable energy system. It will have a devastating effect on promising technologies (such as photovoltaic solar or thermo-electric power, wave energy, etc.), condemning them to irrelevance instead of giving them the opportunity to reach the leading role that they should and can play in our energy supply.

Even more serious will be the immediate consequences of the EU's fixation on biofuels as the way to solve the myriad of problems that plague transport policy. Due to the limits of our planet, it is simply impossible to produce 10 percent of the increasing amounts of fuel that we consume from organic matter in a sustainable manner, even if more stringent criteria than the inadequate set proposed by the European Commission would be adopted. Production of biofuels on such a scale will have immensely destructive indirect consequences along the complex web of relations that interconnect the global food system, natural and plantation forests, and biofuel-production networks. Our environment and social relations will suffer immensely, since biofuels will link the price of oil with the price of all the basic components of life-sustaining production (including food, land, and water). However, it will also allow the established oil sector to maintain its power, and the car industry to continue making profits from inefficient technologies. In the face of growing eco-system destruction, scarcity of land, water, food, and fuel, and rising social tensions, it is imperative to phase out fuels (both fossil fuels and biofuels) and to base our need for mobility on electricity derived from renewable sources such as the sun, the wind, and the waves.

Regarding the directive regulating the second phase of the Emissions Trading Scheme (ETS), the Commission proposes to auction emission rights, instead of distributing them for free amongst the largest polluters as they did in the first phase. This could be viewed as a positive move. The free distribution of emission rights to the dirtiest industries during the first ETS phase was an outrageous example of nega-tive redistribution, a clear contradiction of the "polluter pays principle" (turning it into "polluter gets paid principle"), and generally an insult to intelligence. However, the auctioning of polluting rights is certainly not the answer. This so-called "cap and trade system" not only amounts to a privatization of the atmosphere: it also puts into the hands of large corporations one more powerful instrument to manipulate pro-duction costs; bring smaller competitors to bankruptcy; and concentrates economic, political, and physical power. It cements an emerging market where enormous speculative profit margins (the best basis for economic concentration) are only pos-sible if a continuous demand for carbon credits is maintained. This produces a very strong incentive to keep an active carbon economy alive and kicking, and therefore contributes to the marginalization of RE.

However, public opinion makes it difficult to keep RE on the backburner in order to sustain the carbon economy, without offering any alternative. Cosmetic measures are required in order to save face, in a context where climate change and other environmental concerns play an increasingly important political (and elec-toral) role. This is the main reason for the scientifically- and economically-absurd push for Carbon Capture and Storage (CCS). CCS offers the opportunity for the power sector to claim that they are working on supposedly "clean" fossil energy—further delaying the urgently needed (and perfectly feasible) quick transition to a 100 percent RE-based decentralized energy system. CCS also increases the amount of energy required to produce energy, offering a perfect vehicle to increase profits on behalf of the environment. This leads, in the case of several capture technologies, to higher levels of other pollutants being emitted into the atmosphere. But the main problem with CCS is that even if the capture technologies would work perfectly, there is simply no space to store all the carbon emitted by fossil fuel-based power plants, and no certainty that the carbon that can be stored will remain where we put it—actually, for all we know, it is far more likely that it won't. The only supposedly-reliable and economically-viable "solution" is pumping liquefied carbon back into oil and gas fields (or into saline underground water). This makes it technically easier to extract the last remaining reserves out of those fields. Therefore, public funds that should be used to foster REs will instead bolster the already astronomical profits of oil corporations.

In addition, the carbon market artificially created by the ETS (and, in the UK, guaranteed by government intervention) provides the conditions on which the nuclear industry can present credible business plans, thus overcoming the most important obstacle to its grand renaissance. This is a brilliantly concealed way to make taxpayers foot the bill for the revival of the nuclear industry. Direct subsidies would be a political liability, since public opinion would not accept transparent payments to maintain a source of energy characterized by such economic, political, technical, environmental, and security problems. However, state intervention to maintain the price of carbon can be sold to the public as environmental policy (despite the complete absurdity of such a claim), since almost nobody understands the obscure technicalities of this speculative market.

The policy package proposed by the European Commission keeps renewable energies in the corner, strengthens the artificial market for carbon, and presents CCS, nuclear, and biofuels as the only viable alternatives to confront climate change. It therefore contains all the ingredients necessary to increase the economic and political power of the fossil and nuclear sectors and of the power utilities. But the package has also been designed to foster power concentration at the national/geostrategic/military level. The industries and politicians behind the package, and the bureaucrats at their service, have no hesitation in sacrificing public interest in pursuit of their interrelated and mutually-reinforcing interests.

This is the general picture, now some of the specific mechanisms of this policy package will be explained in more detail.

CERTIFICATE TRADING:
MAKING RENEWABLE ENERGY IRRELEVANT AGAIN

Every single study about the promotion of renewable energy (RE) reaches the same unambiguous conclusion: the only policy that has proven effective in Europe in achieving large-scale, fast, and cheap RE deployment is the so-called "feed-in tariff." In countries with feed-in laws, power utilities are forced by law to buy renewable energy from all producers who meet the required quality standards, and to pay prices fixed by the law on a long-term basis. The prices are different for each technology (for instance, for solar photovoltaic energy they are higher than for wind energy) and sometimes they are also different depending on the local conditions (for instance, in Germany wind energy producers in locations near the coast get less per kWh than producers in the interior, where there is less wind). The objective is to make it possible for everyone to invest in renewable energy equipment, since the law guarantees a modest but worthwhile profit on a long-term investment—and therefore provides access to loans for such investments. The tariff is revised every few years, generally getting reduced for new projects as the price of RE equipment goes down. The price of RE equipment goes down due to the experience and the economies of scale produced by the proliferation of independent power producers (IPPs). The combination of all these positive effects has enabled the take-off of the RE sector in the countries that apply well-designed feed-in laws on a consistent basis. All studies on this matter (including those done by the European Commission) unanimously conclude that the feed-in law has proven to be the only successful RE policy in Europe. More importantly, in countries with successful feed-in laws, small and medium-sized IPPs are rapidly growing, taking some 1.5 percent of the incumbent industry's market share each year.

The feed-in tariff does not create a fixed market share for RE, instead it provides the conditions in which investment in RE can happen successfully. All countries that adopted effective feed-in tariffs (in particular, Denmark until 2001, Germany, and Spain) have witnessed an exponential growth in the sector and the emergence of a new and dynamic RE industry. They also produce the cheapest renewable electricity and have the largest share of IPPs. The extra price paid by the power utilities is diluted in the electricity bill of all electricity users, making no impact on state finances, and is hardly noticeable for the consumers. For instance, the feed-in tariff in Germany has added an average of EUR 1.5 to the monthly electricity bill of households, and in exchange it has avoided the emission of 97 million tons of CO_2 in 2006, produced a €21.6 billion turnover (also in 2006), and created around 320,000 jobs (out of a total of around 600,000 RE jobs in the whole world). The German industry has certainly not suffered a loss of competitivity due to the feed-in tariff—in contrast, it has developed a very promising (and rapidly growing) new area of activity, export, and expertise.

The other major policy used for the promotion of RE, only used by the UK and four other countries (Belgium, Italy, Poland, and Sweden), fixes a minimum target of renewable energy to be achieved by energy utilities and creates a market for RE certificates to be traded towards the fulfillment of this target. Sometimes there are specific targets for specific technologies (in order to avoid that, all investment goes to the cheapest technologies). This system normally fosters the creation of "green electricity" markets at the consumer level too: consumers are offered the option to pay more for renewable energy (although in fact they receive the same electricity as everyone else, since they are connected to the same network), and that extra money is devoted to RE projects. The price paid for renewable electricity is normally higher in certificate-based RE schemes than in countries with feed-in laws, which makes RE unnecessarily expensive and uncompetitive. However, IPPs cannot participate in this market, since they cannot get loans for the initial investments. The reason is that the price of the electricity that they generate is uncertain, hindering the long-term planning required to finance RE projects. Therefore, this system has an extremely poor record: the targets are hardly ever reached, the renewable electricity is more expensive, and most RE projects remain in the hands of power utilities and other large corporations. Obviously, power utilities prefer this system, since they have almost complete control and make large profits from the few renewable energy projects that come into being, while in a feed-in system they simply pass on the cost to the consumers but do not (or should not) make any profit.

The results of both systems are clearly illustrated by the situation in Germany and the UK. In Germany, on-shore wind energy receives 8.36¢ per kWh, and the country installed more than 20 GW of capacity between 1999 and 2006. In the UK, the same electricity receives between 13 and 14¢ per kWh, but less than 2 GW of capacity were installed between 1999 and 2006. In Germany, with one of the worst RE potentials in the world (not much wind, not much sun, nothing much of any other RE source), the share of RE in electricity production is 12.5 percent, up from 4.7 percent in 1998. Germany reached its indicative RE target for the year 2010 already in 2007, and the sector continues growing vigorously, three and a half times faster than in the UK as far as wind energy is concerned. In contrast, only 2 percent of electricity production and 1.3 percent of the final consumption of energy in the UK is renewable (the lowest percentage of any major European country), although the UK has one of the best renewable energy potentials in Europe (including the best wind, wave, and tidal potential), and could therefore produce the cheapest RE electricity. The UK policy choices make it impossible to reach the indicative target for 2010, and much less the mandatory 2020 target: the UK Government's 2007 Energy white paper admitted that present policies will only deliver a 5 percent contribution from renewables to the UK's energy by 2020. However, the Cabinet refuses to change its policy.

Despite (or due to?) the appalling record of certificate trade, last summer the UK government pushed its way through in the European Commission in order to extend its certificate-based system to the complete European Union. It proposed creating, through this Directive, a EU-wide market for tradable "Guarantees of Origin" (GO or GoO, another name for certificates) for renewable energy. This responded to the demands made by large electricity corporations. The UK government has positioned British bureaucrats in key positions in the process of drafting energy policy; therefore, last summer's operation (taking place while most people's attention was elsewhere) was successful: the first draft of the new RE Directive presented certificate trade as a *fait accompli*.

The justification for introducing this measure is that member countries need a flexibility mechanism for the fulfillment of their share in the 20 percent target on renewable energy, since the Council decision of March 2007 made this target mandatory. This means that all countries will need to contribute to its fulfillment— the countries that have a larger share of RE will increase their share more than the countries with little RE, but all will have to do something. And since some countries have more RE resources (such as wind, sun, etc.) than others, the Commission considers that they should be allowed "to support renewable energy produced and consumed in another member state instead of deploying more expensive domestic resource."[2]

The problem with this argument is that certificate trading is incompatible with

---

2   Letter sent by Commissioner Piebalgs to the European Renewable Energy Federation (EREF), dated 14th December 2007.

feed-in laws. Their coexistence is politically and economically unfeasible: you cannot ask energy consumers in one country to collectively pay the marginal extra costs of energy that will "count" towards the RE target of another country. Certificate trading undermines the basic tenants on which the feed-in tariff is based, and leads to its disappearance. This in turn has a large number of negative consequences, which ultimately render RE irrelevant.

In order to understand why this is so, one has to grasp the difference between markets characterized by scarcity and speculation, compared with markets based on plenty and security. Certificate trading is of the first kind, while feed-in laws establish the second kind of markets. This point is well explained in the following text from Tomas Kåberger, of the International Institute for Industrial Environmental Economics, Lund University, Sweden:

> The artificial market of certificates is small. The demand is non-elastic, there is strong demand to reach the compulsory quota, but then no more—not at all. As a result large suppliers can control the certificate price depending on their supply of certificates to the market. Thus the large producers can create price changes. At the same time the large companies have no problems surviving such price fluctuation on a minor part of their total market.
>
> They will be able to hold back their investments and supply of certificates, and let the certificate price rise so as to make investors build windmills, etc. Then the large power industries would start investing and sell off certificates to lower the price, wait a few months and then start buying the capacity from other investors who face cash-flow problems. That is what I would like to do if I was director of Vattenfall. With competition in the Swedish market, overcapacity led to low prices. Large, often state-backed, companies bought almost all smaller competitors. Then prices increased.
>
> Later, again, with emission trading. Power companies got a surplus of emission rights for free. They held back their emission rights from the market to increase the price. Then increased the price of electricity as if they had to buy certificates on the margin, making billions of euros. And then there was a chicken-race until someone started selling off the surplus emission rights and the market collapsed. They are not stupid—and they get rich.

These concerns do not exist in a country with feed-in laws. The prices are guaranteed, everyone is free to invest on the basis of a decent and secured return, and there is no space for speculation. In contrast, an EU-wide certificate market would lead to a downward competition between countries towards lower support for REs, since no country wants to pay for a good support scheme if foreign companies can benefit from it and speculate with prices at their will.

Another consequence of certificate trading will be that only the currently cheapest RE technologies will be developed, which is particularly bad for photovoltaic solar power (PV) and emerging technologies (such as solar thermo-electric, geothermal, wave, tidal, and even off-shore wind). Due to a fundamentally-flawed accountancy that disregards externalities and long-term impacts, PV is valued by the market to be about four times more expensive than conventional energy sources. However, increasing production of PV panels in the last years has brought down the price at an amazing pace, on the basis of good feed-in tariffs introduced by a handful of countries. The price in 2005 was half of that in 1995, and at the current cost-reduction speed, it is likely to take less than twenty years for PV electricity to be cheaper than that produced from fossil fuels (including coal) or nuclear reactors, even in countries with meager solar resources. From a total figure of 1.246 new MW installed in the EU in 2006, Germany installed 1.153 MW, despite not being the sunniest country in Europe. But if feed-in tariffs are replaced by certificates, the development of the sector will be brought to a standstill.

This is not only a problem for specific technologies: it represents a grave hindrance to the transition to an energy system that is 100 percent based on REs. This transition requires all RE sources and technologies (not only the presumably cheapest), since otherwise it is not possible to secure a balanced and stable energy system. For this reason, all emerging technologies (including a range of energy storage technologies) must be promoted to mature technically and reach economies of scale. The trade in RE certificates undermines this process.

Certificate trading also creates new transaction costs to producers of renewable energy. The RE sector is forced to cover the costs of a mandatory system that demands the annual production and tracing of RE certificates (the so-called "Guarantees of Origin"), while the producers of nuclear and fossil-fuel-fired power plants don't cover this expense. According to Dr. Dörte Fouquet, Director of the European Renewable Energy Federation (EREF), "the German Government estimates additional costs of such a scheme would be €100 billion until 2020 for the consumers in the EU-27. For Germany alone, it is estimated, that the costs for renewable electricity compared to the present feed-in costs will almost double."

Managing such a system, and making the best use of opportunities for speculation, is a comparatively smaller burden for large corporations than for independent power producers (IPPs). This contributes even more to market concentration in the hands of oligopolies.

In addition, RE certificate trade concentrates all RE investments in the regions with the best potential. A high density of RE projects owned by distant corporations provides powerful nourishment for local opposition. It is only logical for local communities to reject projects that endow them primarily with the impact of wind turbines and solar panels, while the profits go elsewhere. The impact goes well beyond the landscape, affecting also social relations. For instance, wind turbine proliferation often affects negatively the price of nearby property with less wind resources, while the places with good resources get good rents from project developers. This leads to tensions and divisions in communities. In contrast, policies oriented towards local collective ownership and a fair distribution of benefits, combined with feed-in tariffs, have resulted in strong local support for RE projects, and have the best track record in terms of speed and positive engagement in RE deployment.

According to Dr. Dörte Fouquet, the UK government's decision to push for the introduction of an EU-wide certificate market was based on a paper prepared by its Industry Department, which claimed that the electricity prices for industry would triple if they had to fulfill the EU 20 percent binding target on RE. This report is based on bizarre and baseless arguments, but it shaped UK policy since it fit very well the long-term strategic interests of the Government and the nuclear and fossil lobby. On 23 October 2007, *The Guardian* published internal documents which described the British government's plans to undermine REs and to press for the inclusion of nuclear power in the 20 percent target:

> Leaked documents seen by *The Guardian* show that Gordon Brown will be advised today that the target Tony Blair signed up to this year for 20 percent of all European energy to come from renewable sources by 2020 is expensive and faces "severe practical difficulties." John Hutton, the secretary of state for business, will tell Mr. Brown that Britain should work with Poland and other governments skeptical about climate change to "help persuade" German chancellor Angela Merkel and others to set lower renewable targets, before binding commitments are framed. Ministers are planning a U-turn on Britain's pledges to combat climate change that "effectively abolishes" its targets to rapidly expand the use of renewable energy sources such as wind and solar power.

However, the British proposal to establish a EU-wide market for RE certificates was rejected by countries that had already developed a sizable RE industry such as Germany and Spain, but also Slovenia, Latvia, and other countries that are part of the European Feed-In Alliance. The proposal was also received as a war declaration by the small and medium Independent Power Producers (IPPs), organized in the European Renewable Energy Federation (EREF); by the RE industry, organized in the European Renewable Energy Council (EREC); and by a diversity of NGOs. The Commission insisted, but saw itself eventually forced to water down its plans to create a mandatory EU certificate trade market. They brought in a provision that would have allowed countries to request permission to withdraw a part (and only a part) of their RE production from the EU certificate market. Such requests would be decided upon by the Commission on an annual basis, and would be valid for only one year; new permissions would have to be requested each year, not less than six months in advance. The British bureaucrats at the commission thus came up with an outlandish method to invest themselves with the power to decide about European renewable energy policy. They were extraordinarily obstinate: they were reportedly acting even against the will of their own Commissioner, who seemed not to feel able to do anything about it. However, they saw themselves forced to water down their proposal even more when the Legal Service of the European Commission declared in unambiguous terms the illegality of this regulatory framework for certificate trade.

The draft directive presented on 23rd January 2008, therefore, changed the terms of the certificate trade. The draft foresees that only the countries that are up-to-date in the annual evolution of their RE target can "export" certificates, and that all countries can create "a system of prior authorization" for certificate trade in order to protect their RE policy. However, all countries are still forced to create certification agencies, and to issue certificates for each MWh of renewable energy produced. Therefore, the added costs are imposed also on countries that have no interest in certificate trade.

In addition, the Directive foresees that the Commission will evaluate the situation with regards to certificate trade and may submit further proposals to the European Parliament and to the Council.[3]

The press release of the European Commission clearly indicates their intention to continue pushing for an EU-wide RE certificate market: "As long as the EU's overall target is met, Member States will be allowed to make their contribution by supporting Europe's overall renewables effort, and not necessarily inside their own borders."[4] Despite the good progress achieved by the organizations and governments that defend the feed-in tariff, the Commission will keep up the pressure in favor of a mandatory EU certificate market. The current draft Directive already forces all countries to put into place the costly bureaucratic structure necessary for an EU-wide mandatory market. They might have to wait some time, but a mechanism to introduce mandatory trade at a later point has already been built into the draft Directive: before the end of 2014, the Commission "shall assess the implementation of the provisions of this Directive for the transfer of guarantees of origin between Member States and the costs and benefits of this."

The Commission's press release already hinted at the direction of their future assessment. It claims that certificate trading "would shift investment to where renewables can be produced most efficiently, which could cut EUR 1.8 billion from the price tag for meeting the target," even though all serious studies on the matter (including the Commission's) contradict this view. The British neoliberal fundamentalists that produced this claim use a simplistic method to reach these conclusions: they calculate how much investment would be needed to meet the target if it was concentrated in optimal locations and using the currently cheapest technologies, and compare it with the investment needed to meet the target if it is spread all over Europe using a technology mix. They consciously leave everything else out of the picture. They know that certificates will wipe out independent power producers and concentrate the whole market in the hands of utilities, who will only do a minimum investment on RE in order to keep certificate prices as high as possible (making RE much more expensive). But their oversimplistic calculation allows them to present certificates as cost-saving policy.

It would not be surprising if in the next few years, the corporations and governments that support a EU-wide mandatory market would arrange a small amount of certificate trading in such a way that it does result in lower prices than if it had not taken place. This would provide a solid argument to the Commission to push for mandatory trade, at least for the percentage of RE produced beyond the minimum target (which is supposed to increase from year to year until it reaches an average 20 percent). If feed-in countries refuse, then countries such as the UK are most likely to refuse to fulfill their share of the target.

The current draft Directive suggests that the next step that the Commission is

---

3   European Commission press release "Boosting growth and jobs by meeting our climate change commitments," 23 January 2008, reference IP/08/80.

4   Ibid.

likely to take (possibly long before 2015), is making certificate trade mandatory for "excessive" RE. This is a serious (though obscure and technical) issue with important repercussions for our future energy mix, so it is worth exploring in detail.

The 20 percent EU target has been divided into different country-specific targets. Countries that already have a large share of RE have to contribute more than countries with less RE (according to a bizarre rule that rewards anti-RE countries such as the UK), and richer countries have to contribute more than poorer countries. The target for each country also differs from year to year: it increases until it reaches that country's target in 2020. If all countries reach their country-specific targets in time, their combination produces a 20 percent share of RE in the EU as a whole.

According to the current draft, there is a limit to the protection that member countries can offer their RE producers from the EU-wide RE certificate market created by the Directive. That limit is each country's minimum share of the target for a given year. This is what is implied in Article 9(2) of the current draft:

> Member States may provide for a system of prior authorisation for the transfer of guarantees of origin to persons [including juridical persons, i.e. companies] in other Member States if in the absence of such a system, the transfer of guarantees of origin is likely to impair their ability to comply with Article 3(1) [i.e. to fulfill their contribution to the 20 percent target in 2020] or to ensure that the share of energy from renewable sources equals or exceeds the indicative trajectory in Part B of Annex I [i.e. to ensure that their contribution to the target progresses according to the calendar set by the Directive].
>
> The system of prior authorization shall not constitute a means of arbitrary discrimination.

This innocent-looking piece of bureaucratic jargon has wide-ranging consequences. The Commission (and the UK government) will make good use of it, and of other provisions that make the judgment about certificate trade a matter of economic performance. On the basis of a few initial and well-managed "successful" examples of cost-cutting due to certificate trade, they are likely to push for mandatory trade for RE produced beyond the minimum targets. They will get the active support of countries that are not interested in RE and would prefer to buy certificates of RE generated in other countries.

The introduction of mandatory certificate trading beyond the minimum targets will provide a very strong incentive to downgrade RE promotion policies. From a government's perspective, there is no point in promoting RE that will count toward other countries' target requirements. The policy downward spiral will take place, in a less dramatic form than if certificate trade was mandatory for all RE, but it will take place nonetheless. No country will see the sort of robust and healthy growth of RE, beyond official targets, that was witnessed in Denmark (before 2001), Germany, or Spain.

Another likely consequence of this is the disappearance of the feed-in tariff entirely. In feed-in countries renewable electricity prices are lower than average, and certainly lower than the prices for certificates. Producers in those countries (especially the large utilities) will challenge in court a policy that only gives access to the higher certificate price to RE produced in "excess" of the annual target. They will argue that this limitation is a silent appropriation of the profits that they would be able to make if all the RE produced would have access to the EU-wide certificate market. This would be, de facto, a legal challenge to the feed-in system in place, which will be used by the utilities in order to create insecurity and discourage investment by independent producers, regardless of the final outcome reached (many years later) by the court.

The price of RE certificates (and therefore of RE as such) will be in the hands of large energy corporations, which will manipulate them to get oligopoly-based windfall profits. They will drop the price to bankrupt independent producers from time to time, in order to minimize investment in RE. The rest of the time they will keep the price of certificates high enough to ensure that RE remains a set of niche technology in a niche market controlled by them, and therefore providing them with exorbitant profits. Renewable energies will once more be confined to marginality in the midst of a nuclear revival, but the few existing wind farms and solar installations will surely be displayed in every single advertisement of energy corporations.

It is to be hoped that several governments will oppose this move, but they might sell out. The UK government has two powerful cards under its sleeve that might result in the inclusion of nuclear power in the RE target and a slow but sure introduction of a mandatory EU-wide certificate market. The two cards, which can be particularly effective at weakening the position of the German government, are transport emission reduction policy and carbon quota allocation. Both are described in the following sections.

GERMAN CARS: THE WEIGHT OF TRADITION

The current draft directive gives a privileged and exclusive treatment to biofuels. It is the only RE source for which a binding minimum target is set for all EU countries. The 20 percent RE target refers to all forms of energy (electricity, heat, and transport), and each country is in principle free to choose where to concentrate their efforts. But that freedom is relative: all of them have to use at least 10 percent of biofuels in transport.

This privileged treatment is not accidental: it reflects the power of oil corporations and car manufacturers. The very existence of oil corporations would be threatened if we move towards a fuel-free economy. The car industry also has a lot to lose if the highly inefficient combustion engine is replaced by electricity-driven engines, since they would lose most of the post-sale business that they make by selling unreliable nineteenth century technology that requires regular check-ups and recurrent replacement of components. Car manufacturers, the oil industry, agribusiness, and biotechnology companies are working together to ensure that agrofuels represent the backbone of transport emission reduction policies.

In contrast, the current draft directive on RE has a mandatory 10 percent target for agrofuels, which implies sanctions against the countries that do not reach it. The most important specific reason for this discrimination in favor of agrofuels is directly linked with the support of the German government to its car industry.

In early 2007, the Commission wanted to impose a mandatory efficiency standard for cars in order to reduce CO_2 emissions to an average of 120 gr. of CO_2 per km. In response, car manufacturers launched a heavy campaign, which was most articulate and aggressive in Germany. Many "prestige" car manufacturers claim that they won't be able to survive in the market unless they sell large and heavy vehicles that can reach very high speeds in a very short time, and therefore refuse limitations on the amount of fuel to be wasted in their engines. The German car industry was particularly aggressive, publishing full-page ads with direct threats to close down their factories in Germany and lay off their (still substantial) workforce if mandatory efficiency standards would be imposed. As a result, Angela Merkel's administration (with her own personal involvement) became the governmental speaker of "prestige" car manufacturers, and bargained efficiency down to an average of 130 gr/km. The deal was signed on the understanding that the 10 gr/km difference would be made up for with the mandatory use of agrofuels. For this reason, the European Council chaired by Angela Merkel in Spring 2007 included a specific mandatory 10 percent minimum target for agrofuels as part of the guidelines for the EU energy and climate policy package.

Therefore, the childish obsession with size and speed felt by affluent (mainly male) car buyers is one of the key reasons (although not the only one) behind the strengthening of one of the most disastrous policies ever devised.

CARBON EMISSIONS TRADE: COMPETITION ON UNEQUAL BASIS

There have always been many good reasons to oppose the trade in emission "rights" of carbon and other greenhouse gases. Now there are two more reasons. First, this market will be used to subsidize the nuclear revival (at least in the UK, and probably in many other countries too). Second, the UK is most likely to use the negotiations around the EU Emission Trading Scheme (EU ETS) to get nuclear power accepted as contributor to the RE targets, as well as to dismantle the only effective policies for the promotion of RE.

EU ETS has so far completely failed to deliver greenhouse gas reductions. Last year, European governments agreed that avoiding dangerous climate change means keeping the eventual temperature rise below 2°C. Since we have already seen a rise of just over 0.7°C and cannot now prevent another 0.7°C rise, there is not much room to manoeuver: drastic reductions need to happen within the next decade. However, the main instrument to achieve a reduction of emissions in the EU is a lousy system that only produces profit for large polluters.

Under the EU ETS, large emitters of greenhouse gases must annually report their CO_2 emissions, and they are obliged every year to give an amount of emission allowances to the government that is equivalent to their CO_2 emissions in that year. The first phase of EU ETS (2005–2007) involved about 12,000 polluters, representing approximately 40 percent of EU CO_2 emissions. These large industries got emission allowances free from their governments, who were supposed to give them less than they would emit under a business-as-usual (BAU) projection, in order to force them to innovate to reduce their emissions, or to buy allowances from others.

According to the Climate Action Network, during the first phase of EU ETS only two of the twenty-five EU states (UK and Germany) asked the participating industry sectors to reduce emissions compared to historic levels. In the fifteen old EU member states as a whole, allocations were 4.3 percent higher than the base year, and more than 90 percent of the polluters emitted less than their quota of free credits. In May 2006, when it was clear that too many allowances had been given away, trading prices crashed from about EUR 30/ton to EUR 10/ton. After an initial slight recovery, the price declined further to EUR 4 in January 2007 and below EUR 1 in February 2007, reaching an all time low of EUR 0.03 at the beginning of December 2007. Therefore, the system did not result in any reduction of emissions whatsoever. Instead, it produced amazing profits for the polluters, particularly the energy utilities, many of whom added the cost of the allowances in the energy price, even though they got the allowances for free. Several high-profile court cases have found them guilty of fraud, and imposed heavy fines on them for making profits based on their oligopolist position. Interestingly, there seems to have been a well-coordinated EU-wide strategy to withdraw allowances from the market in order to maximize the price of allowances, and therefore the illegitimate price rose, before the race to sell began.

The prospects for the second phase of EU ETS look just as grim as for the first phase. The National Allocation Plans include a reduction of 7 percent of greenhouse gases (now all GHGs are included, not only CO_2) under the official business-as-usual (BAU) projections. But according to independent estimations, in fact all the National Allocation Plans except for Portugal, Spain, and UK result in higher emissions than the independently estimated BAU. Therefore, the second phase will also create further speculation and nothing else. In addition, it has been suggested that it will be possible to buy credits for emission reductions outside of the EU. The EU ETS is therefore likely to result in a major overall increase in EU emissions. Partly in response to this, the Commission cut eleven of the first twelve Phase II plans it reviewed, accepting only the UK plan without revision.

The second phase of EU ETS also introduces the auctioning off of a great part of the allowances, although heavy polluters have obtained opt-outs and a delay in the date of entry of auctions until 2012. The exact terms of the opt-outs and delays are still under negotiation.

EU countries have highly asymmetrical positions in this negotiation. The UK, which produced the idea of EU ETS and pushed for its imposition, is the strongest player. It hardly manufactures anything and it plans to build up its nuclear industry thanks to the competitive advantage that nuclear power will obtain from high carbon prices. The UK also has the strongest and most dynamic financial markets and hosts almost the entire European carbon market, thanks to the experience gained by the UK Emissions Trading Scheme in advance of the introduction of EU ETS. Carbon markets operate on the basis of the same speculative tools (such as futures and options) as financial markets. Therefore, the more money that goes through carbon markets, the more revenue for UK-based speculators. As a result, the UK is interested in reducing the distribution of allowances as much as possible: high carbon prices do not affect negatively its competitive position (in fact they are beneficial, since they affect negatively countries with a large manufacturing base), and they bring high dividends to city-based carbon brokers.

This is in stark contrast to countries like Germany, where many energy-intensive industries are located. These industries claim that their ability to compete with imports from non-EU countries will be affected by high carbon prices, due to the extra costs added with buying emission allowances. This argument is highly questionable, since there is a large potential to innovate and save energy or reduce emissions. But the fact remains that these industries already warned of a risk of relocation outside of the EU, and several governments (including Germany's) take this threat very seriously.

The decision to publish and negotiate the draft directives on renewable energies and the EU as a package makes it much easier for the UK to use EU ETS as a strong bargaining tool. Being the only country whose EU ETS National Allocation Plan has been accepted by the Commission without comments, it does not need to negotiate its own share of excessive allocation. It can therefore demand other political concessions in exchange for accepting other countries' excessive allocations.

This explains very well why the UK energy bill announced on the 10th of January 2008, and currently pending approval by the Parliament, includes a commitment to develop "national" and publicly-funded mechanisms to keep the price of carbon high if the EU ETS price is too low. It gives the nuclear industry reassurance that they will continue being competitive with regards to fossil energy in the UK, regardless of whether other EU countries over-allocate emission "rights" to their polluters. It therefore provides the UK Cabinet freedom to use the carbon emission bargaining tool in the negotiation of the energy and climate package, and therefore obtain the acceptance of nuclear power as contributor to the RE target and/or the destruction of feed-in tariffs across the continent. And at the same time, it allows the government to project an image of environmental concern. One must admit that it is a brilliant example of political manipulation.

Carbon Capture and Storage is the third element of the energy and climate package. It is a concession to the oil and coal industries, in order to compensate for the losses that it will suffer due to EU ETS and to ensure that the fossil-nuclear mix remains the backbone of the energy supply.

The energy package also plans another mechanism to compensate power utilities for the auctioning of emission rights: certificate trading. The German Association of Industrial Energy Users and Self-Generators calculates that auctioning of allowances from 2012 onwards means that the utilities will lose roughly €5 billion per year in unjustified windfall profits. A recent study by Fraunhofer Institute ("Increased auctioning in the EU ETS and trade in guarantees of origin for renewables: A comparison of the impact on power sector producer rents") concludes that they will basically gain the exact amount of new windfall profits through introduction of RES certificate trading. It is a strange coincidence that both amounts are roughly equivalent.[5]

MOVING FORWARD—RENEWABLE ENERGY FOR THE COMMON GOOD

The debate about this policy package offers a unique opportunity to collectively shape a good set of policies that will accelerate the transition to a 100 percent renewable energy system. A future-oriented Directive must also be socially-oriented and give communities a leading role in this transition, allowing them to utilize decentralized RE resources in order to generate local collective prosperity, in order to avoid opposition to new projects and speed up the adoption of renewable energies as a public service. But in order to generate the social and political energy required to win this battle, we must be able to provide convincing positive alternatives, and to organize alliances around them. It is important to create awareness about the connection between all aspects of energy policy and their deep interconnection with the evolution of our societies, and to produce an inspiring political platform that encompasses all these aspects.

---

5   http://www.isi.fhg.de/e/working percent20papers/WP_ETS-auctioning.pdf

Chapter 12 ∎ Part 3

ENERGY SECURITY IN AFRICA WITH RENEWABLE ENERGY[1]

Preben Maegaard

More than any other continent, Africa needs an energy revolution and independence from the international fossil fuel economy; a change to renewable energy and energy autonomy is paramount for survival. Africa depends largely on the import of fossil fuels to meet a significant and growing part of its modern energy needs, which has had perverse effects on the economy and lives of Africans. Renewable energy is the only viable alternative with the potential, when properly managed, to improve quality of life on a national and continental scale.

The current sky-high oil prices are disastrous for the fragile economies in most African countries. With the much lower oil prices of the past, several African countries were already spending half of their foreign trade expenditure for the import of oil. With oil prices doubling and $200US per barrel in the foreseeable future, the misery we already see in Darfur and other regions will spread, and people will suffer and continue to bleed.

ENERGY RICH AFRICA

The end of the fossil oil era has the potential to foster energy innovation based on Africa's tremendous renewable energy resources. The continent has an abundance of wind resources, biomass, and not least, solar energy, all in sufficient quantities for satisfying future energy needs. Africa has all of the renewable resources. What it needs is access to know-how and practical technological solutions.

Within this context I ask why the Africa Energy Forum, gathering July 2 to 4, 2008 in Nice, France, will focus almost entirely on the conventional energy system, rather than on renewable energy. The Africa Energy Forum brings together senior government officials and private-sector executives to discuss opportunities to expanding public and private power. I received an invitation but was unable to attend. I asked for a better representation of the renewable energies as most of the presentations of the Forum focus on conventional energy options that will not be affordable for the masses of the African continent.

NO RENEWABLE ENERGY AT AFRICA ENERGY FORUM!

The prompt answer from the director of the conference revealed that the

---

1   Originally the chapter on regional perspectives in Africa was meant to be written by Ibrahim Togola, Director of the Mali Nyetta Folkecenter for Renewable Energy. However, due to last minute time constraints it was not possible. Ibrahim suggested this chapter, written by his close collaborator and colleague, as an alternative. This text has been previously published online at http://www.folkecenter.net/gb/news/fc/re_africa/. Reproduced here with the permission of the author.

priorities of the conference were already decided and did not intend to include renewable energy. The conference manager, Rod Cargill, e-mailed to me:

> One thing is certain, conventional power is pivotal to Africa's economic growth. To
> claim that Africa's problems of poverty would be alleviated by relying on renewable
> energy is folly. The number of failed renewable energy projects in Africa over the
> last 20 years is unacceptable, and verging on the irresponsible. These failed projects
> have setback development by raising aspirations and then failing to deliver, thus
> curtailing self-help in Africans.... The aim of the Africa Energy Forum is ultimately
> poverty alleviation in Africa. We are well aware of the difficulty of bringing power
> to rural communities and the consequences of untrammelled power expansion
> on climate change. But we believe that cooperation between all power providers
> is the only way to achieve our objective. We find a strong reaction in Africa to the
> moralizing of western countries, particularly when they are the ones selling the
> renewable technology.

Well-known suppliers of renewable energy solutions like Sharp, Kyocera, Vestas, Solar World, Enercon, and many other world brands within wind and solar
power are not the sponsors, yet the conventional fossil fuel energy sector will be well
represented.

One might get the impression that the solar and wind industries, despite a large
annual turnover of €60 billion, are still not considered a professional sector. Some
might say the sector should be considered, as a whole, responsible for a " ... *number
of failed renewable energy projects in Africa over the last 20 years as they failed to
deliver* ... " and thus prevented energy change in Africa.

DISASTROUS SOLAR PROJECT BY ESKOM AND SHELL

Even though renewable energy has seen tremendous technological achievements,
there will of course be failures, as happens in any other innovative sector. In Africa
the renewables sector undoubtedly suffered their most severe setback ever when oil
giant Shell tried to pave the way for solar power but failed with their widely adver-
tised solar initiative in rural South Africa.

In 1998, Shell Renewables and Eskom, South Africa's national electricity supplier,
embarked on a joint venture to supply homes in the remote and rural communities
of South Africa with a unique solar home system. This project was the largest commercial, rural solar electrification venture ever undertaken. The aim was to bring
illumination to 50,000 rural homes in South Africa.

After some years, their token payment system failed, and systems were either
not functioning or panels were stolen. As the project did not succeed organizationally, technically, and commercially, Shell was forced to withdraw. What was planned
as a model for the 2 billion people globally that have no grid electricity, ended up as
a disaster for the reputation of solar power.

But fortunately we find in Africa numerous successful renewable energy solutions that obviously did not get the same negative response as the disastrous Shell
top-down project. In Namibia, small solar shops charge mobile phones, and many

towns and villages get their electricity from PV. In Kenya, thousands of solar systems give light to homes after sunset. Egypt and Morocco already have wind farms and are planning many large-scale uses of wind and solar energy. They have better resources than most industrialized nations and have begun to mobilize their own industrial capacity to collect in full the fruits of their natural resources.

DECENTRALIZED SOLAR TECHNOLOGY CAN IMPROVE THE CONDITIONS OF LIFE

I have personal experience from solar projects in Uganda and Mali, which are in no way comparable with the Shell experience. At the 2004 project inauguration, every solar panel installed was serving the rural population with electricity for schools, clinics, and other basic institutions. Some of the installations had been producing electricity for years. With extremely modest financial resources, dozens of villages are now demonstrating that modern technology can improve the conditions of life among the poorest in rural areas—93 percent of the population does not have access to electricity.

The nearest power line may be 100 km away and will never find its way out to the thousands of villages where you find the majority of the population. In the last ten years, solar cells fortunately have become more efficient and reliable and can deliver electricity to schools and clinics, improve the supply of water, all for the common good. Meanwhile, the residents with sufficient income have started to buy their own solar installations. This energy revolution, admittedly still at its very beginning, has been made possible by a small dedicated team at the Mali Folkecenter (www.malifolkecenter.org) that has implemented other pioneering projects in some of the poorest countries of the world.

ENERGY SUPPLY AND REVITALIZATION OF LOCAL ECOSYSTEMS

In 2006, the rural commune of Garalo, in the south of Mali, celebrated the implementation of a biofuel project based on jatropha oil. The facility will help bring biofuel-generated electricity (245 KW) to approximately 8,000 residents of the Garalo commune and possibly later to the rest of the people in the surrounding villages. For the 70 percent of Malians who live in rural communities, this project shows that living rurally does not have to mean a cash-crop dependent economy with no running water, or that the only alternative for electricity is petroleum generators.

The Sahel environment is fragile and arid, yet jatropha is resilient and can grow under these harsh conditions. Jatropha can thrive in the region's difficult land and restore eroded areas, effectively generating environmentally-friendly energy, helping reduce CO_2 emissions, and helping to revitalize local ecosystems. Such projects will also stimulate the economy and create disposable income, which, in turn, can be used to develop healthcare, education, small-business needs, living conditions, and much more. The project will be closely monitored and documented, so others interested in similar initiatives can learn from this experience. Jatropha is expected to transform Garalo, offering residents greater opportunities, stable energy prices and a chance for sustainability.

Chapter 13 ▎ Part 3

MULTI-NATIONAL COMPANIES AND THE ENERGY CRISIS IN LATIN AMERICA[1]

Erika González, Kristina Sáez, and Pedro Ramiro on behalf of Observatorio de Multinacionales en América Latina (OMAL) Asociación Paz con Dignidad

In most Southern countries, the energy model has been articulated on the basis of the rules established by western governments and international financial institutions. This is no surprise given that, since the 1980s, with the development of neoliberal policies, these countries have been left with little choice but to globalize economically, including the globalization of their energy industries.

In Latin America, Africa, and Asia, energy consumption rates are rising dramatically,[2] but these values refer to rates within urban centers and in industrial zones. When it comes to hydrocarbons, the fact that they are considered to be an exploitable resource has provoked a whole host of tensions and armed conflicts in Southern countries.[3] Moreover, in spite of being energy-exporting countries, many of these countries are obliged to import refined oil products for their local consumption. As a consequence, they are strongly dependent on international crude oil prices.[4]

The global demand for electricity has increased substantially in the last decades and it has been forecasted to double by 2030. A quarter of the world's population does not have access to electricity, and most of these 1.6 billion people are located in South Asia and Sub-Saharan Africa.[5] As such, energy resources are still being denied to these impoverished populations, who do not benefit from the quality of life that access to those resources would bring them.

Southern regions are very attractive for foreign corporations given that they are rich in terms of natural resources and biodiversity, and because the widespread poverty among their populations ensures an inexpensive workforce. They are also attractive for the energy sector due to their development of several policies: on a global scale, these Southern countries supply energy resources to those countries that have a higher demand for energy. At a local scale, they provide Transnational Companies

---

1   This chapter was originally written, for the purposes of this book, in Spanish. It was translated into English by Diana Labajos, Craig Daniels, and Ben Pakuts.

2   According to the International Energy Agency, between the years 2004 and 2030 the global demand for primary energy will increase a 60 percent and Southern countries will experience the highest increase. Hydrocarbons will constitute 85 percent of the energy to be consumed.

3   Twenty-four countries out of the world's forty-nine main oil producers are plagued by tension and armed conflicts (sixteen out of those twenty-four are Southern countries that do not belong to OPEC). In thirty-eight of the forty-nine, or three-quarters of them, there are substantial violations of human rights and fundamental liberties.

4   Ramón Fernández Durán, *El crepúsculo de la era trágica del petróleo: Pico del oro negro y colapso financiero (y ecológico) mundial*, Virus and Ecologistas en Acción, 2008.

5   International Energy Agency, *World Energy Outlook 2004*, Paris, OECD, 2005.

(TNCs) unrestricted access to their energy sector. As a result, far from fostering economic and social strength, this take over of the Southern countries' markets by TNCs has led to harmful impacts on both their populations and their environments.

In this context, Latin America becomes a model case of the possible consequences of having foreign countries' governments and TNCs struggling for control of the regions' resources, in order to exploit their potential as a new market. In Latin America, as in other Southern countries, natural resources play a strategic role,[6] which means that TNCs and foreign countries use their political, judicial, legal, and economic power to control them, thereby dragging these rich but impoverished countries into major conflicts.

TRANSNATIONAL COMPANIES AND THE CONTROL OF ENERGY IN LATIN AMERICA

Latin America has 9.7 percent of the world's oil reserves and 4 percent of the world's gas reserves; 13.5 percent of the annual world crude oil production is supplied by the region.[7] Although Latin America is not one of the largest hydrocarbon producing areas on the planet, it plays a key role in international geopolitics, given that foreign oil companies have easy market access and because of their role as energy suppliers for the US—the largest consumer of hydrocarbons in the world.

The description of the energy situation in Latin America is completed with a mention of the abundant renewable energy resources that exist there. The region's water, air, agrofuels, and, to a lesser extent, sun are all attracting foreign direct investment. Of the primary energy consumed in the region, 18 percent originates from biomass and renewables.[8] The multi-nationals see the large hydroelectric, wind, and agrofuels projects that are currently being developed in Chile, Mexico, and Brazil, among other countries, as an abundant source profits.

Venezuela has the largest hydrocarbon reserves in Latin America, with 69 percent of all oil reserves and 60 percent of all gas reserves. Brazil, if the new untapped offshore reserves are confirmed, has the second largest crude oil reserve in the area, while Bolivia has the second largest gas reserves. It is no surprise then to know that Venezuela and Bolivia have become priority targets for energy TNCs—69 percent of the foreign investment in Bolivia in 2005 went to the extractive industry.[9] Nor is it surprising that the governmental reaction to this has been to assure and protect public management of the oil in order to assert control over one of the country's key axes of development.

Although national oil companies are still players in the Latin American hydrocarbon sector—as is the case with PDVSA in Venezuela, PEMEX in Mexico, Petrobras in Brazil, and ECOPETROL in Colombia—privately-owned multi-national

---

6   Erika González, Kristina Sáez and Jorge Lago, *Atlas de la energía en América Latina y Caribe. Las inversiones de las multinacionales españolas y sus impactos económicos, sociales y ambientales*, OMAL—Paz con Dignidad, 2008.

7   BP, *BP Statistical Review of World Energy 2006*, London, 2007.

8   International Energy Agency,. *Energy Balances for Latin America*, Washington, 2005.

9   UNCTAD, *World Investment Report 2007. Transnational corporations, extractive industries and development*, New York and Geneva, United Nations, 2007.

companies are pushing to increase their control over these resources (see table 1). Repsol YPF, with headquarters in Spain, has been the leading multi-national hydrocarbon company in Latin America since 1999, the year the company acquired the Argentinean state-owned firm, Yacimientos Petrolíferos Fiscales (YPF). Repsol YPF owns 95 percent of the Latin American hydrocarbons reserves, and it has obtained 85 percent of its exploitation results. Although Repsol YPF has the second highest income of all Spanish TNCs, its activities have negative impacts on human rights, labor relations, indigenous peoples, and the environment. For instance, Repsol has been publicly sued for having its operations located in seventeen indigenous reserves in Bolivia. They also have been sued for polluting the Mapuche territory in Argentina.

Company	Sales (in millions)	Country of Origin
Repsol YPF	16.900	Spain
Royal Dutch Shell	9.757	United Kingdom / Netherlands
Exxon Mobil	8.208	United States
Chevron Texaco	7.532	United States
Petrobras	4.437	Brazil
BP	2.782	United Kingdom

Table 1. List of major oil companies in Latin America and the Caribbean by sales. Source: CEPAL, La inversión extranjera en América Latina y el Caribe. Santiago de Chile, United Nations, 2007.

In the last few years, some Latin American governments have increased state control over hydrocarbons. For example, in Venezuela, the constitution in force prevents the state-owned firm PDVSA from being privatized. The price the government had to pay was a *coup d'état* in April 2002. In Bolivia, massive popular protest against the energy corporations, owing to the fact that the companies had looted the national energy resources, led to the gas war in October 2003, which was the starting point for the later hydrocarbon nationalization process.

In the electricity sector, the main energy sources used for generating electricity are water (68 percent) and natural gas (12 percent).[10] At a global level, the state of affairs is somewhat different, as coal accounts for 40 percent of global electricity generation. Access rates to electricity mirror the inequality conditions present in the region: industrial consumption and well-off citizens are supplied by huge electrical power stations, while around 46 million people have no access to this service.[11]

The interest multi-nationals have in the electricity sector lies not only in the control of generation, but in managing the whole chain of supply (from electricity production to marketing). This way they can secure the profits derived from the continuous growth in electricity consumption. Also of great importance are the energy integration projects—such as the Central American Electrical Interconnection

---

10   Energy Information Administration (EIA), International Energy Annual 2006, Washington, DC, US Government, 2008.

11   International Energy Agency, *World Energy Outlook 2004*, Paris, OECD, 2005.

System, or the Initiative for the Integration of the Regional Infrastructure of South America—because they contribute to the expansion of giant electric infrastructures, which facilitate the connection between the areas where electricity generation takes place and the end-point consumers.

Nowadays, the electricity sector in Latin America and the Caribbean is strongly controlled by the leading multi-national electricity companies, mainly European. Among others are the French EDF; the French-Belgian Suez-Tractebel; the Italian Enel; and the Spanish Endesa, Iberdrola, and Unión Fenosa. All of these companies have been fortified through successive mergers, and have benefited from the privatization processes that resulted from the neoliberal reforms. Therefore, in practice, there is a clear conflict between the people's needs and rights to access electricity and the financial windfall that electricity generation and distribution represents for the firms.

Endesa is the leading private electricity company operating in Latin America. Its purchase of Chilean energy giant Enersis led the way for its expansion throughout Latin America. The company is already well known worldwide due to the effects resulting from its megaproject at Ralco's hydroelectric dam, located in Mapuche territory, Chile. The construction of this dam affected the territory at the environmental, social, and cultural levels.

As far as Iberdrola is concerned, the company has received strong opposition to its activities at the wind farm, La Venta, located in Oaxaca State, Mexico. There, lands have been expropriated and rural communities pushed to the wayside. Spanish transnational Unión Fenosa has been called into question because of its deplorable management of the electricity services in Colombia, Nicaragua, and Guatemala.[12]

In just a decade and a half, Spanish TNCs have become established leaders in the banking sector, in the telecommunications sector, and also in the energy sector. As these corporations' profits have increased year by year, their activities have impacted the environmental, social, and cultural health of the region; all of which greatly affect the lives of people living there.

In this sense, the Latin American population is highly critical of the impact that the energy multi-nationals have had as they came into the region. According to the Latinobarómetro 2007, 77 percent of the population believe that oil should be managed by the state and 76 percent think that electricity should be in public hands.[13]

THE UNFULFILLED PROMISES OF NEOLIBERAL REFORMS

The dominance of TNCs over Latin America's energy sector is the result of the drastic reform packages promoted in the Washington Consensus for southern economies. These reform packages, which began in the 80s and peaked in the 90s, were aimed at reducing the states' intervention in the economy, privatizing state-owned enterprises and liberalizing the markets. These measures promoted the opening of regional markets to foreign companies. Upon entering southern markets, TNCs

---

12   Pedro Ramiro, Erika González, and Alejandro Pulido, *La energía que apaga Colombia. Los impactos de las inversiones de Repsol y Unión Fenosa*, Barcelona, Icaria—Paz con Dignidad, 2007.

13   Corporación *Latinobarómetro, Informe Latinobarómetro 2007*, Santiago de Chile, 2007.

have heavily invested in the energy sector, with the aim of gaining control of natural resources, including oil, coal, and gas. These companies have taken renewable resources, like water and land, for the construction of large hydroelectric dams, the erection of wind turbines, as well as the exploitation of wide swathes of land for the production of agrofuels. Although, in principle, these are renewable energy projects, their enormous scale is resulting in the privatization of large territories and the displacement of the population who live there. Yet, with a different process, these are the very people who could be central to a sustainable management of the energy system.

As a consequence, all the countries holding substantial hydrocarbon reserves in Latin America and the Caribbean, except for Mexico and Cuba, undertook strong sectoral reforms during the 90s. Consequently, many states sold part of their oil firms to TNCs, either by granting concessions for the exploitation of oil fields, or by the selling off of entire public corporations. In Argentina, YPF was sold under Menem's government and the same happened in Bolivia under President Sanchez de Lozada's rule, when Yacimientos Petrolíferos Fiscales Bolivianos (YPFB) started being managed by private hands.[14] In Brazil, foreign entry in Petrobras was negotiated during Cardoso's government. As for Venezuela, Caldera's administration boosted the so-called *apertura petrolera* process, which consisted of making exploitive contracts with multi-national corporations. In Colombia, the privatization process started relatively recently when, in 2006, the state-owned company, Ecopetrol, was privatized.

The appropriation of hydrocarbons by the multi-national enterprises has affected all of these countries in a similar manner. In relation to oil, operations such as free disposition and export/import have often been done without paying any tariffs. For instance, taxes have only risen to 18 percent in Bolivia. In addition, by directing oil production towards the export market, TNCs have created an energy shortage in these countries; there are a variety of reasons for the scarcity of energy in countries such as Bolivia, Argentina, and Ecuador. On the one hand, the governments only earmark a small proportion of the resources (e.g. petrol and gas) for domestic consumption, despite the fact that the new governments are attempting to increase domestic supply. An example of this is the tense negotiation process between Bolivia, an exporter of gas, and Brazil and Argentina, both of which are importers. On the other hand, the cost of refined products consumed by the population has risen. This has meant that supply has become inaccessible, especially for the most impoverished rural population that still suffers today. Rather than being due to lack of resources, the scarcity of energy affecting these countries is due to neoliberal policies.

So what energy situation awaits these countries once their sources begin to run out? The outlook is very pessimistic: The consuming regions will apply pressure to maintain their quota of energy, while, in turn, the inhabitants of the producer countries will assert their right to the resources in their territory. Historically, such a situation has resulted in conflicts in which the military arm of the state has attempted

---

14   Marco Gandarillas, Marwan Tahbub, and Gustavo Rodríguez, *Nacionalización de los hidrocarburos en Bolivia. La lucha de un pueblo por sus recursos naturales*, Barcelona, Icaria—Paz con Dignidad, 2008.

to repress the protests of the majorities in order to serve the needs of the economic and political elites.

Simultaneously, TNC delegates have accessed State-owned company boards of directors, thus taking control over decision-making in the energy sector away from the state. This is not to mention the results of the multi-nationals' strategy of petrol and gas exploration and extraction on other levels. In terms of labor, it has given rise to enormous cuts in staff and the deterioration of working conditions. In relation to the environment, a loss of biodiversity, the contamination of rivers and soils, and the deforestation of national parks and biosphere reserves are all becoming a reality, among the many other problems. Petrol countries, such as Bolivia and Ecuador,[15] have responded to the reality of serious environmental impacts by enshrining the right to a healthy environment in their new constitutions. On many occasions, the activity of the petrol transnationals has forced the indigenous population to be displaced from their ancestral territories, has meant the loss of their means of production, of natural resources, culture, worldview, etc.—effects that are felt even more severely by women. This also results in the infringement of human rights, especially those of social movement and trade union leaders who are critical of the activity of the multi-nationals.

The reform of the electrical system has taken place simultaneously with the privatization of the oil sector. Historically, states monopolized the management of electricity. However, this model was harshly criticized by neoliberals, and specifically targeted by the structural reforms boosted by international financial institutions such as the IMF. Measures taken in those reforms were to split the electrical system into the processes involved—generation, transportation, distribution, and commercialization—then to create companies for the exploitation of the different processes, followed by their sale to foreign partners. The reforms were justified as being required in order to solve the electricity supply crisis and to improve the electrical supply infrastructure for the Latin American population. In the 1970s, states held a monopoly control over electricity, and electrical services were financed from the public coffers and from multilateral organizations. The latter played an especially important role throughout the 80s, as a result of the explosion of the debt crisis. In the case of the electrical sector, this economic crisis manifested itself in a scarcity of investment in electricity generation and distribution projects, as these projects were generally very costly. The lack of an adequate infrastructure, coupled with an increase in demand and the deterioration of the transportation networks, meant that episodes of supply cuts were frequent in the 90s. This resulted in a lack of energy, affecting the majority of Latin America's population, especially in rural zones. According to their promoters, reforms would bring the spread of the electrical supply to other consuming areas, the end of corruption in state companies, and more financial resources.[16]

However, after twenty years of private electricity management, the reality is considerably different: there has been little investment in infrastructure because the

15   Political Constitution of Ecuador 2008, Chapter 5 on Collective Rights. Articles 86–91. Political Constitution of Bolivia 2009, Title III. Environment, Natural Resources, Land and Territory.

16   Mª José Paz, Soraya González and Antonio Sanabria, *Centroamérica encendida*, Barcelona, Icaria—Paz con Dignidad, 2005.

companies have given preference to short-term benefits. As a consequence, electrical quality and supply have not really improved. To give an example, in Nicaragua, where Unión Fenosa holds the monopoly for electrical distribution, electrical coverage in the rural area has only marginally increased in this period, rising from 40 percent to 43 percent.[17] Companies have also reduced running-costs so as to maximize profitability. Accordingly, power cuts have continued and the power grid has not reached areas that are "cost-ineffective" for the companies. In addition, foreign investments have been focused on buying privatized companies rather than in building up local production capacity. Private electricity management, far from benefiting governments, has resulted in a double-edged sword: governments do not take in any profits from the electricity sector's exploitation—as it is in private hands—and because electricity is a basic service, governments have to subsidize the sector when there are financial difficulties.

TENSION AROUND THE CONTROL OF ENERGY

The rivalries for control of energy between foreign TNCs, states, communities, collectives, and workers, etc. have caused strong social conflict, political tension, and lawsuits in international courts of justice. The competition between states and TNCs is not based on equal terms; on the contrary, the population fights to keep the sovereignty of their energy resources while having to face the corporations' influence over the mass media, the economy, politics, and legislation. Furthermore, there is not only tension between the populace and the TNCs, but also political and social conflicts that arise from the network of energy dependence between countries.

A large fraction of Latin American hydrocarbon reserves are located in the Andean community. In order to intensify the exploitation of these reserves, the TNCs operating there have used all the power at their disposal, including military force and the police. In Ecuador, oil corporations have had an agreement with the army, since 2001, to secure their facilities. In Peru, the case is very similar: the armed forces have political and military control over certain strategic areas.[18] Military presence has been denounced in both countries, as social actions have been violently repressed and indigenous peoples have been stopped from passing into their territories. Accordingly, indigenous communities have been the most affected by oil extraction activities, and are the ones who have put up a resistance most directly orientated towards the defense of their territorial wealth and natural resources. Their struggles, which seek to put an end to the intrusion of TNCs into their territories, have taken the form of demonstrations, legal denunciations, and political pressure, etc. Other indigenous peoples that have been affected include the Waoranis in Ecuador; the Machiguengas, Yine, and Asháninka in Peru; and the U'was and Guahibos in Colombia. In this country, the armed conflict that has continued for the last forty years has made the militarization of oil extraction areas even graver than in Peru. In Colombia

---

17   CEPAL, 2009. Anuario estadístico de América Latina y el Caribe, 2008.
18   Alfredo Seguel, "La invasión de las empresas petroleras en la selva amazónica de Perú," *Blog de la Red de Comunicacion Ucayali*, October 2008.

the corporations' control over the natural resources has also worsened the violation of human rights. In fact, some oil corporations are co-responsible for inhumane crimes, as they have financed mercenary intelligence companies, supported the US army, and funded military units with proven human rights violations.[19]

If we look at the energy scenario of the Southern Cone, Bolivia stands clearly as the main gas exporting country. The main recipient countries of Bolivia's gas are its neighbors, Argentina and Brazil, though it plans to expand its markets by exporting gas to Uruguay and Paraguay. According to Bolivia's Minister of Planning and Development, Carlos Villegas, the country exported 42 million cubic meters of gas in 2008, 24 million cubic meters of which went to Brazil. For its part, Argentina receives 7 million cubic meters every day. Hydrocarbon reserves of this country are decreasing because of the entry of the TNCs and the lack of investment in the country.

In Bolivia, two TNCs control the hydrocarbon sector: the Spanish Repsol YPF and the Brazilian Petrobras. The latter exploits the biggest gas fields and the export pipelines going to Brazil. Therefore, they have great influence and can exert pressure over the Bolivian government. This enterprise's pressure on the government reached its peak in 2006, when the government, headed by Evo Morales, enacted the decree to nationalize hydrocarbons. What happened was that the Brazilian government and the Petrobras executives put a lot of pressure on the Bolivian government so as to protect their interests. They threatened to take them to the International Centre for Settlement of Investment Disputes (ICSID)—an arbitration court of the World Bank—stop investing, or even leave the country. The Spanish government also used diplomacy, the mass media, and their political power to defend Repsol YPF's business in Bolivia.

According to Bolivian law,[20] it is the state company, Yacimientos Petrolíferos Fiscales Bolivianos (YPFB), that defines policies related to petrol.[21] In practice, however, the refounded YPFB finds itself subject to multiple constraints when it comes to exercising control over the hydrocarbons. These include barriers such as external and internal political interference, the conditions of the global market, and the international price of gas, etc. Nonetheless, Bolivia is making advances. The state takeover of the country's most important refineries, which were previously controlled by Petrobras, is an example of this.

As part of this analysis, it is also important to mention the process of energy integration that is occurring within the framework of the Bolivarian Alternative for the Americas (ALBA, for its initials in Spanish), a process driven by Venezuela, and the Peoples' Trade Treaty (TCP), which is being promoted by Bolivia. The objective of ALBA and the TCP is to put an end to the economic dependence that the exportation of raw materials gives rise to. These integration programs seek to stimulate endogenous development and sovereignty in the realm of production, as

19   Pedro Ramiro, Erika González, and Alejandro Pulido, *La energía que apaga Colombia. Los impactos de las inversiones de Repsol y Unión Fenosa*, Barcelona, Icaria—Paz con Dignidad, 2007.

20   Supreme Decree DS 28701

21   In theory, it defines the conditions, volume, and prices, both for the internal market, as well as for export and the country's own industrialization processes.

well as introduce lower energy prices so that the poorest sector of the population can have access.

Chile depends on Argentinean gas imports, while Argentinean gas comes from Bolivia. This fact is important when looking at the regional conflicts related to gas. In 2003, the gas Argentina imported from Bolivia was used for producing more than half the energy consumed in Chile.

In Central America, tensions around energy come from the fact that these countries strongly depend on the hydrocarbons they import, which means that their disposition is subject to the international energy market prices. Moreover, one of the biggest electrical infrastructures, the Central American Electrical Interconnection System, is being built in the region, with the intention of connecting electrical energy-producing areas, coming mainly from large hydroelectric dams, with the consumers. In order to do so, the power grids from the different countries will be connected, from Mexico to Colombia. This mega-system will obviously have numerous impacts on the region's environment and society.

Lastly, crude oil production in Mexico has been managed by the state-owned company Pemex. However, lobbying by TNCs has been successful, and in October 2008, the Mexican senate approved an energy reform by which private capital can enter into Pemex.[22] Foreign company stake-holding in Pemex is currently being contested through civil resistance.

THE ROLE OF NATION STATES IN THE ENERGY CRISIS

For the nation states, keeping or regaining control over the hydrocarbon and electricity sector management is an asset, especially because they can increase private company taxes. In Bolivia, for instance, the most productive oil fields taxes can reach 82 percent of the value of oil production. As such, the state's increased income leads to less dependence on international and multilateral funding, allowing it to implement wider social reforms with the oil revenues. Another positive outcome of the states' control over the sectors is the control of private companies' activities in relation to corruption and other offenses. As for the population, it is a step forward in regaining their energy sovereignty, as they can push governments to hear them when making decisions in regard to hydrocarbons or electricity.

The global market energy production has strong negative effects. Large hydroelectric plants and oil or gas extraction lead to deforestation, alteration of waterways, pollution, and displacement of indigenous and rural populations. Nonetheless, social pressure on governments may force them to create regulations more effective in guarding people's and environmental rights. We should take into account that state and regional administrations are often to blame for the unclear management of the profits made from energy production. The partnership between the state administrations and foreign corporations hardly ever leads to investment in the population's needs but rather to the benefit of the corporations.

---

22   Erika González and Kristina Sáez, "La maldición de los recursos energéticos en América Latina," *Pueblos*, no. 32, June 2008.

All in all, it is important to have a state-owned company administering the whole hydrocarbon production chain. This would be the first step towards establishing limitations to private activities, with the aim of regaining peoples' sovereignty over their territories, their natural resources, and their economic future. State management allows for energy production and demand to be coordinated throughout the whole country, and also offers an accompaniment and means to defend those communities who wish to undertake local alternative energy projects under the management of the collective itself. For that purpose, while hydrocarbon nationalization is not really enough, it is nonetheless still the *sine qua non* of conditions for achieving peoples' sovereignty. The ability of the population to decide on its energy resources is premised on the possibility of realizing effective political and legal control over those who manage these resources. It thus follows that the managing body must open channels of popular participation in order to allow debate about a plan for managing these resources in a way that prioritizes a socially and environmentally sustainable use of energy.

Undoubtedly, in the future, measures such as those mentioned above should go hand in hand with others aimed at facing the end of fossil fuels and climate change— including solutions that favor the transition towards the use of renewable energy sources. An example of this is the type of energy policies that Venezuela is initiating through programs such as the Operative Plan in Renewable Energies and the Energetic Revolution Mission, both of which seek to use renewable resources to supply isolated communities with energy. Nevertheless, the very first and fundamental step towards favoring people power over TNCs is to let the social majorities of the planet manage their natural resources.

A debate around the most suitable economic models for social justice and environmental care is needed. Until now countries rich in energy resources such as Venezuela and Bolivia have followed an exploitation model that only benefits a privileged minority. In this sense, social necessity has defined the model, as the need of states to supply the population's basic needs has oriented state management towards the exportation of energy resources. By regaining sovereignty over their resources and deciding what to do with them, while remaining producers, these countries place themselves in a fairer place within the global energy system. However, unfortunately, this does not secure their energy autonomy, as Brazil imports gas from Bolivia, and Mexico imports Peruvian gas, for example.

Environmental sustainability of oil production, hydroelectric dam construction, and biofuel production sets the limits for the current energy model in Latin America and other southern regions. It also brings to light the need to change the paradigm: the energy matrix will need to be adjusted toward an economic system where consumption is equitably distributed among the people, not dependent on foreign resources, and where energy resource exploitation meets social and environmental sustainability.

Chapter 14 ▌ Part 4: Community and Worker Struggles Over Ownership and Control in A Transition to A Post-Petrol World

STRUGGLES AGAINST PRIVATIZATION OF ELECTRICITY WORLDWIDE

David Hall on behalf of Public Services International Research Unit

This chapter looks at campaigns against electricity privatization: their extent, what were the issues, who was involved, their use of court cases and electoral campaigns, and the development of alternative policies.[1]

1.1 SCALE OF CAMPAIGNS

There has been widespread opposition, throughout the world, to all forms of electricity privatization—the sale or privatization of distribution networks; the sale or privatization of existing power plants; the creation of new, private power stations through IPPs (Independent Power Producers); and management contracts or leases to operate networks or power stations. This opposition has come from a wide range of civil society groups, including trade unions, environmentalists, consumer organizations, community organizations, peasant and indigenous groups, and political parties. In some cases governments have also adopted a position of passive resistance to World Bank proposals.

The table lists campaigns that could be described as at least partly successful in terms of their own objectives.

CAMPAIGNS AGAINST ELECTRICITY PRIVATIZATION

Key to active groups: U=unions; C=consumers; B=(local) business; N=other NGOs; E=environmentalists; P=political parties

Source: PSIRU database

Country	Location	Year	Result	Decision mechanism	Active groups
Australia	NSW	1999, 2008	State utility corporatized, not privatized	Election (regional)	U,P
Brazil	All	Ongoing	Limited privatization of utilities, generators	Elections (national and state)	U,P,N,C
Canada	Ontario	2002	Halted privatization of utility Ontario Hydro	Court ruling	U,C, E, N
Colombia	Cali, all	1997-date	Opposed privatization of municipal utility Emcali	-	U,N,E
Dominican Republic	all	2003	Renationalized electricity distributors	Government decision	C,U

---

1   This chapter is based on a number of PSIRU reports published over the last ten years, all of which are accessible at www.psiru.org.

Ecuador	National	2004	Rejected privatization of state company	Parliament, court ruling	
France	National	Ongoing	Delayed privatization of state company EdF	-	U,P
Germany	Leipzig, Dusseldorf etc	Ongoing	Campaigns against privatization of stadtwerke	Referenda	U,C,P
Hungary	National	Ongoing	Oppose privatization of state electricity co MVM		U,P
India	Maharashtra	2004	Nationalization of Dabhol IPP (Enron)	Government decision	N,E,U
India	Karnataka	2000	Rejection of Cogentrix IPP plan	Court ruling	E,N
Indonesia	National	Ongoing	Rejection of privatization of electricity utility PLN	Court case	U, N
Kenya	National	Ongoing	Opposition to privatization of distribution and generation companies		
Mexico	National	Ongoing	Defer privatization of electrical utilities	Parliament, court ruling	U,P,N
Nigeria	National	Ongoing	Delay unbundling and privatization of state company		U, C
Pakistan	National	Ongoing	Opposition to privatization of WAPDA		U,
Philippines	National	Ongoing	Opposition to privatization of state electricity co NAPOCOR		U,N
S. Korea	National	Ongoing	Opposition to privatization of state electricity co KEPCO		U,N
Senegal	National	2002	Collapse of privatization plans	Government decision	C
South Africa	National	Ongoing	Eskom remains public utility	-	U,C,N
South Korea	National	2004	Withdrawal of plans to privatize and liberalize Kepco	Government decision	U,N
Taiwan	National	Ongoing	Opposition to privatization of state electricity co		U, N
Tanzania	National	Ongoing	Opposition to privatization of distribution and generation companies		U,N
Thailand	National	2004	Withdrawal of planned sale of shares in state electricity co.	Government decision/uprising	U, E,
Uganda	National	Ongoing	Delay of privatization of distribution company and creation of IPP.		U,C
Zambia	National	Ongoing	Partial success in preventing privatization of distribution and generation companies		U,C,P

It is a striking feature of the table that these campaigns of resistance have happened in all regions of the world, and in the high income OECD as well as in developing countries. This resistance to electricity privatization has been part of a more general public resistance to all forms of privatization in public services. Despite initial public support for the idea in the 1990s, experience of privatized utilities, in particular, rapidly generated opposition. Opinion surveys in Latin America, for example, showed a dramatic change by 2000, and hostility continued to grow. By 2003, a speaker from management consultant, Deloitte, told a World Bank meeting that there was hardly a country left where politicians dared to support privatization of electricity:

"Growing political opposition to privatization in emerging markets due to widespread perception that it does not serve the interests of the population at large," which they attributed to a number of features of privatization: "Pressures to increase tariffs and cut off non-payers; loss of jobs of vocal union members that will be hard to retrain for the new economy; the perception that only special interests are served—privatization is seen as serving oligarchic domestic and foreign interests that profit at the expense of the country ..."[2] From then on, the pressures have been reduced, at least in developing countries, because most multi-national companies have taken strategic decisions to withdraw. But continuing pressures for privatization in Europe, and general ideological motives, continue to produce privatization initiatives.

The campaigns have thus tended to be long continuous struggles, because the proposals for privatization have been continuous. A typical example is the case of Ecuador, where government attempts to privatize electricity assets have repeatedly encountered organized resistance including unions, provincial and local governments, indigenous organizations, and others. In 2002, these campaigns forced the abandonment of proposals to sell electricity distributors, after Ecuador's Congress passed a resolution rejecting the privatization, and a Constitutional Court ruling that the sales were unconstitutional. A further attempt at privatization was abandoned in February 2004 when there was not a single tender for any of the companies.[3]

OPPOSITION TO PRIVATIZATION IN LATIN AMERICA

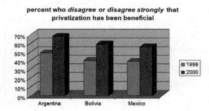

percent who *disagree* or *disagree strongly* that privatization has been beneficial

Source: Hall et al (2005)[4]

The greatest issue has been the level of prices, but campaigns have also focused on job losses, failure to invest, unreliability, inefficiency, environmental impact, policy, loss of public accountability and/or national control, and corruption. The campaigns have been reinforced by a growing body of empirical research critical of experiences with privatized electricity.

1.2 ISSUES

In some cases these concerned a single power station or local facility—e.g. the Co-gentrix campaign in southern India—or a single city's utility, such as the Emcali

2   "The Declining Role of Foreign Private Investment," Matthew Buresch, Deloitte Emerging Markets World Bank Energy Forum 2003. http://www.worldbank.org/energy/week2003/Presentations/EnergyForum1/BureschWBForumpresentation.pdf

3   *World-markets Analysis* April 13, 2004: "Energy Minister Replaced in Ecuador."

4   David Hall, Emanuele Lobina, and Robin de la Motte, 2005. "Public resistance to privatisation in water and energy." *Development in Practice*, Volume 15, Numbers 3 & 4, June 2005.

campaign in Colombia; in other cases they covered a whole country, such as the campaigns in Mexico, Thailand, and South Korea. They include cases where existing systems have been successfully defended so far, while developing or maintaining services, such as in South Africa and USA (California). And they include cases where privatizations have failed to take place or been rolled back, such as Senegal and the Dominican Republic. There are other cases, not listed, where privatization has been terminated as a result of an exit decision by the company concerned, such as Orissa, in India, where AES abandoned a generation and a distribution company.

Resistance to large price increases was central in a number of campaigns that succeeded in rejecting privatizations. In Senegal, for example, the government refused to meet the demands for price rises of three successive multi-nationals—Hydro-Quebec, Vivendi, and AES—as a result of which even the World Bank abandoned the plan for privatization of the electricity utility.

Other campaigns have been based on defending community interests, as well as resistance to implied price rises, such as the campaign against Enron's private power plant at Dabhol (in Maharashtra, India), which was based on a long-term power purchase agreement. The campaign was supported by energy NGOs opposed to the project on social, economic, and environmental grounds, and by the local communities around the plant whose livelihoods were seriously damaged by it. Demonstrations by the local communities were brutally suppressed (leading to the unique case of an Amnesty International report on Enron). The power station was finally nationalized by the Indian government. The long-running campaign against the proposed private power station at Bujagali falls, in Uganda, was also based on the impact on the environment and on local communities. The project was abandoned by the company originally involved, AES, but has since been revived.

Resistance to, and conflicts over, electricity privatization have often formed part of greater political processes and struggles. In Thailand, for example, a series of demonstrations and strikes were organised by the Thai electricity workers union from 2004 onwards, highlighting the dangers of privatization in terms of higher prices, the risk of corrupt allocation of shares to cronies, and the risk of foreign control developing through buying of shares. In March 2004, the government backed down and announced the cancellation of the EGAT privatization plans. Following an election, the government revived the plans; further strikes and demonstrations then formed part of a movement that culminated in the overthrow of the government, and its replacement by military rule, followed by new elections, new privatization proposals, and further action in 2008. In Pakistan, the introduction of IPPs in the 1990s with excessively generous power purchase agreements resulted in the distribution company, WAPDA, becoming unviable, as a result of having to buy power at prices higher than it could charge consumers. The privatization proposals were strongly resisted by the union, and attempts were made to prosecute the companies involved in IPPs for corruption, but these prosecutions were dropped at the insistence of the World Bank, and instead WAPDA was taken over by the military—a precursor of the subsequent military takeover of the country—and the union was banned. In Venezuela, the distribution company was privatized to the US multi-national AES before Hugo Chávez became president. AES, supported by the US government, strongly resisted Chávez's proposal to renationalize the company—this wish to defend AES' investment was one factor in the failed coup attempt against Chávez in 2002. By 2007, however, AES itself wanted to withdraw, and was content when the company was finally renationalized in 2007.

1.3 UNIONS, COMMUNITIES, ENVIRONMENTALISTS AND POLITICAL PARTIES

Most of the campaigns have been led by trade unions. This is based on the clear economic interest of workers whose jobs and working conditions are threatened, but the unions have generally campaigned on wider issues of public interest, including prices and accountability. Environmentalists have been involved in many campaigns, and, in some cases, have taken a leading role. Community groups, especially where a private power plant threatens the local environment, and consumer groups, who became increasingly wary of the price rises associated with privatization, have also been widely involved.

The electricity campaigns show a variety of relations to political parties. In Australia, unions have used their specific relations with the Labour Party to obtain a policy position from Labour against privatization, and then campaign in elections for the Labour Party on this issue. For example, a union-led campaign succeeded in influencing the results of elections in New South Wales, so that the electors rejected the Conservative party, which was proposing privatization of electricity, in favor of a Labour party policy of public sector, corporatized Electricity Companies. This followed similar election results in Tasmania, where the Labour party defeated Conservatives proposing electricity privatization; and electricity privatization has also been rejected in South Australia and Queensland, leaving Victoria as the only state that has privatized power.

South Korean unions, by contrast, have waged a long campaign against privatization of electricity, gas, and other utilities, without relying on any one party for support. Their campaign included parliamentary pressures, general strikes, and research, and, more recently, collaboration with environmental groups and others. The privatization of the electricity utility Kepco has still (early 2004) not taken place.

1.4 COURTS AND REFERENDA

In some cases, campaigns have succeeded after winning court cases. Examples of such successful court actions can be seen in India (ruling against the legality of a proposed power station on environmental grounds), Canada (reversal of proposed Ontario electricity privatization), and Indonesia (ruling that proposed privatization of power system contravened the constitution).

In Canada, the government of the province of Ontario proposed to privatize the transmission grid in 2001. A campaign against this privatization was led by the union, CUPE, but included broad support from environmental and community organizations. A court case was brought, arguing that the government had no explicit power to sell the shares, and therefore the privatization could not take place: the court ruled in favor of the union, and the privatization was abandoned after a change of government following the next election.[5] The court also explicitly ruled that the union did have status to bring a public interest case:

> It has long since been recognized that unions have an interest in matters which transcend the "realm of contract negotiation and administration".... To borrow [from a case of the Supreme Court of Canada] "the interests of labor do not end at some artificial boundary between the economic and political." Inherent in this proposition is the notion that interests of labor are expansive and are meant to include more than "mere economic gain for workers."[6]

The background in Indonesia was similar to Pakistan: a series of IPPs were established in the 1990s through corrupt agreements with the Suharto dictatorship, with power purchase agreements setting prices so high that the national distribution company, PLN, became commercially unviable, because it was forced to buy electricity at prices higher than it was charging its customers. Attempts to prosecute the IPP companies for corruption were again resisted by donor companies, and the government subsequently proposed the break-up and privatization of PLN itself. The electricity workers union and others opposed this strongly, and brought a case to the constitutional court, winning a ruling that the privatization was in breach of the constitution.

In Europe, a number of campaigns in Germany and other central European countries have made use of laws that enable campaigners to force referenda if they can acquire sufficient signatures. In Leipzig, for example, a referendum was triggered in 2008 by a campaign against proposals to sell the municipal works company (Stadtwerke) responsible for electricity distribution, as well as water supply and other services. In a 40 percent turnout, an overwhelmingly majority voted to halt the privatization. Similar campaigns have won referenda against privatization of municipal electricity and other services in Hamburg, Leipzig, Dusseldorf (although the privatization went ahead despite a majority vote against), and in Switzerland.

2. DEFENSIVE DEMANDS AND PUBLIC SECTOR ALTERNATIVES

Some of the campaigns have adopted limited defensive positions, protecting the status quo against a proposed privatization, without advocating or supporting policies for reforming an existing public sector system without privatization. Campaigns against the development of specific IPPs, for example, may not advance any alternative method for increasing generating capacity. Some union-led campaigns have been solely concerned with preventing the loss of jobs that usually accompanies privatization, without acknowledging problems with the existing system that might require some alternative reforms for the sake of the public interest.

---

5   Jamie Swift and Keith Stewart, 2005. "Union Power: The Charged Politics Of Electricity in Ontario." *Just Labour* vol. 5 (Winter 2005). http://www.justlabour.yorku.ca/Swift_Stewart.pdf.

6   *Payne v. Ontario* (Ministry of Energy, Science and Technology). http://www.sgmlaw.com/en/about/Paynev.OntarioMinistryofEnergyScienceandTechnology.cfm

An outstanding example of this is the union campaign in Thailand, which has successfully prevented the privatization of the state electricity company, EGAT, by direct action in the form of strikes, demonstrations, and political action against the governments that have proposed privatization. Their opposition to privatization was supported by other organizations, including a number of environmental groups that developed quite detailed proposals for alternative reforms. The unions, however, neither advanced any alternatives of their own, nor offered support for the environmentalist alternative proposals.

These features are not surprising nor are they necessarily limitations of the campaigns. Defending the interests of an existing community, or an existing workforce, against a threat from far more powerful forces is a central and legitimate function of action to control the impact of powerful political and economic forces. Most campaigns depend on organizational power and mass mobilization for success, especially when confronted with initiatives from governmental or international institutions that show no interest in consulting or acknowledging the legitimacy of the interests of people concerned. The interests of different groups can sometimes be effectively pursued through separate campaigns.

However, campaigns provide an organizational base and create a political opportunity to develop policy proposals for the sector. A key element of this is identifying reforms that are in the public interest, rather than principles derived from market ideologies. The World Resources Institute offered a general approach to structuring alternative reforms around clearly-identified and agreed-upon public interest objectives. In a report that examines the varying experiences with electricity reform in the 1990s in six countries—Argentina, Bulgaria, Ghana, India, Indonesia, and South Africa, major problems with the goals and processes of electricity reform in nearly all the countries studied are identified.

> By focusing on financial health, reforms in the electricity sector have excluded a range of broader concerns also relevant to the public interest. In this study, we have examined the social and environmental concerns at stake in these reforms. We have found that not only are they inadequately addressed, but that socially and environmentally undesirable trajectories can be locked-in through technological, institutional, and financial decisions that constrain future choices.

The report put forward four clear recommendations for what it calls "a progressive politics of electricity sector reform," including:

- Frame reforms around the goals to be achieved in the sector. A narrow focus on institutional restructuring driven by financial concerns is too restrictive to accommodate a public benefits agenda....

- Structure finance around reform goals, rather than reform goals around finance....

- Support reform processes with a system of sound governance. An open-ended framing of reforms will reflect public concerns only if it is supported by a robust process of debate and discussion.

- Build political strategies to support attention to a public benefits agenda.[7]

Examples of alternatives can be categorized according to three major issues: the need to extend systems to ensure universal connections, the need for transparency, and the need for efficiency.

The first example concerns the most successful extension of electricity services in sub-Saharan Africa, in South Africa. The ending of apartheid, following a massive liberation struggle, created an almost revolutionary situation open to political initiatives, including a program for electrification of cities and the countryside through the state electrical utility, ESKOM: "a period of political change and policy disruption were essential to the program's initiation, and the critical role played by organizations and individuals outside of national government in helping shape new electrification policies and strategies."[8] There was no formal role at all for international institutions such as the World Bank. Instead—unlike nearly all other programs in Africa—a central role was played by organizations representing citizens. A key body on the whole process was a public multi-stakeholder planning institution, the National Electrification Forum (NELF), "a broad-based stakeholder body with participants from Eskom, municipalities, the DMEA, unions and others [supported by] ... university-based electricity researchers ... and the energy policy analysts/activists in the ANC." NELF formed an arena where stakeholders could negotiate the shape of an electrification program, which would be both politically legitimate and practically implementable, based on a political acknowledgement of the social function of electrification and its funding from [public finance]: the end result was the "transition of electrification from a socially desirable (but economically limited) activity to an imperative, brought about broadly by a powerful democratic drive and commitment to service delivery (including the electoral significance of achieving targets)."

Secondly, improving transparency and public accountability is a significant issue because in almost all countries, the pre-existing public sector organizations have become unpopular because of a lack of responsiveness to public concerns both on an individual and a collective basis. Improving transparency also creates a more favorable political environment for campaigns. The outstanding example of development of this kind of alternative is the work of the Indian energy group, Prayas (www.prayaspune.org). Prayas recognizes the achievements of the existing Indian electricity model, based on state ownership, self-sufficiency, and cross-subsidy to agriculture and households. In fifty years, capacity has increased fifty-five fold, with 78

7   Dubash N. (ed.) 2002. "Power politics: Equity and environment in electricity reform." World Resources Institute. August 2002. http://www.wri.org/governance/pubs_description.cfm?pid=3159

8   Bekker B., Eberhard A., Gaunt T., and Marquard A. (2008) "South Africa's rapid electrification programme: Policy, institutional, planning, financing and technical innovations" doi:10.1016/j.enpol.2008.04.014

million customers, and half a million villages connected. There are, however, limits
to these achievements, and real problems in the sector: half the population is still
unconnected, and there are power shortages, weak accounting and metering, and
huge financial losses. Prayas advocates the application of three principles of transparency, accountability, and participation (TAP): "all the governance functions and
governance agencies are made amenable, on mandatory basis, to full transparency
to the public, direct accountability to the public, and meaningful participation of
the public.... The three major governance agencies—the state, the utilities, and the
regulatory commissions—could be TAPed in a variety of ways. However, the space
and capabilities of civil society institutions will be the important determinants of
successful TAPing of these agencies" (Prayas 2001). Prayas showed the potential of
these principles by demanding public consultation on price-setting by the distribution authority in Maharashtra, India, and then successfully advocating operational
changes which enabled significantly lower tariffs. It has published a series of booklets
on this kind of approach.[9]

Thirdly, diagnosis of the problems may identify efficiency as a key issue for
popular reform. In 2008, the National Union of Electricity Employees (NUEE) in
Nigeria, where 70 percent still lack electricity connections, argues that privatization
will lead to the continuing denial of electricity connections to those who currently
do not have access to electricity. The union argues instead that: "the crisis has more to
do with corruption and problems relating to the management of the sector.... Some
of the alternatives proffered by NUEE include the efficient use of Thermal stations
and gas, accurate billing and payment for electricity consumption." [10] Echoing the
success of the South African multi-stakeholder organization, the NUEE has called
for a summit, "targeted at all stakeholders in the Sector ... to assist governments
and the various stakeholders in developing joint strategies and actions to extend and
improve the efficiency of the electricity services and also develop alternative sources
of generating power." [11]

3. CONCLUSION

The scale of these campaigns shows a far stronger public aversion to privatization
than was expected in the 1990s. Their political impact is remarkable, and sufficient
in itself to explain their successes. The range of groups and interest involved also
show that these have rarely been narrow, single interest group issues, but based on
broadly shared concerns about privatization. An economic analysis of why electricity
privatization is generally dysfunctional still needs to be developed. The campaigns
have also generated some interesting approaches to alternatives, addressing issues
of transparency, accountability, and efficiency, but only rarely issues of renewable
energy sources.

---

9   See http://prayaspune.org/peg/energy_home.php.
10   NUEE, 2008. "Electricity in Nigeria: Challenges And Way Forward."
11   NUEE, 2008. "Electricity in Nigeria: Challenges And Way Forward."

Chapter 15 ▌ Part 4

COMMUNITY RESISTANCE TO ENERGY PRIVATIZATION IN SOUTH AFRICA[1]

Patrick Bond and Trevor Ngwane

In spite of South Africa's alleged "economic boom,"[2] the harsh socio-economic realities of daily life actually worsened for most when racial apartheid was replaced by class apartheid in 1994. That process occurred in the context of a general shift to global neoliberal power, instead of prior Keynesian eras in which middle-income countries like South Africa were permitted to build an industrial base and balance their economies through inward oriented strategies.

South Africa suffered enormously from neoliberal policies that increased income inequality (with the Gini coefficient soaring from below 0.6 in 1994 to 0.72 by 2006)[3] and doubled the official unemployment rate (from 16 percent in 1994 to around 32 percent by the early 2000s), as ecological problems became far worse, according to the government's 2006 "Environmental Outlook" research report, which noted "a general decline in the state of the environment."[4] Social unrest and the rise of social movements reflect the discontent: there were 5,813 protests in 2004–05, and subsequently, an average of 8,000 per annum.[5] Until China overtook in early 2009, this was probably the highest per capita rate of social protest in the world during the late 2000s.

Matters will not improve, in part because of macroeconomic trends. The most severe problem is the vulnerability that South Africa faces in hostile global financial markets, given the 2008 current account deficit of 9 percent of GDP, one of the world's worst. It is also highly likely that investment and economic activity will be deterred by ongoing electricity shortages, given that it will take a generation for sufficient capacity to be added, and that the government confirmed its desire in early 2008 to continue offering a few large smelters and mines the cheapest electricity in the world, instead of redistributing to low-income people.

The electricity generation shortfalls of January–March 2008, which led to consistent surprise "load shedding"—entire metropolitan areas taken off the electricity grid—were due partly to a lack of new capacity built by national power generator

---

1   The authors—based at the University of KwaZulu-Natal Centre for Civil Society (http://www.ukzn.ac.za/ccs)—presented this paper to the Gyeongsang University Institute for Social Studies (supported by the Korea Research Foundation's grant KRF-2007-411-J04602). Thanks are also due to numerous collaborators in other institutions and justice movements.

2   Russell, A. (2007), "Post-apartheid phase two: Zuma's leadership of the ANC needs to prove skeptics wrong," *Financial Times* 19 December; MacNamara, W., A. Russell and W. Wallis (2007), "Post-apartheid phase two," *Financial Times*, 20 December.

3   Joffe, H. (2008), "Growth has helped richest and poorest," *Business Day*, 5 March.

4   http://www.info.gov.za/speeches/2007/07062911151001.htm

5   Nqakula, C. (2007), "Reply to Question 1834, National Assembly, 36/1/4/1/200700232," Cape Town, 22 November.

Eskom since the early 1990s (when excess capacity had risen to more than 30 percent), the running down of coal supplies, and rain damage to incoming coal. But the main reason was the increased electricity consumption of metals smelters due to the 2002–08 speculative uptick in commodity prices. Indeed, even earlier, the economy's five-fold increase in CO_2 emissions since 1950, and 20 percent increase during the 1990s, can largely be blamed upon supply of the world's cheapest electricity by Eskom to mining houses and metals smelters.

Emitting twenty times the carbon tonnage per unit of economic output per person than even the United States, the SA energy sector's reliance upon fossil fuels is scandalous. Not only are vast carbon-based profits fleeing to the mining houses' offshore financial headquarters but, despite consuming huge amounts of electricity, the smelters create very few jobs. Instead of cutting back on these sorts of projects, and turning the subsidies to renewables, the government decided to augment coal-fired generation with dangerous, outmoded Pebble Bed technology (rejected by German nuclear producers some years ago). Renewable sources like wind, solar, wave, tidal, and biomass are the suggested way forward for this century's energy system, but still get only a tiny pittance of government support.

Behind this gluttonous and reckless consumption of electricity in South Africa is a long history of cheap energy for big capital that was made possible by the availability of large amounts of poor quality coal and an incestuous relationship between the coal mines and Eskom, the government-owned electricity company. A history of state intervention in securing the energy needs of the mines, agriculture, and industry established the principle of keeping electricity as cheap as possible for the benefit of big capital.[6] The ANC government has not changed this arrangement. But grassroots organizations have challenged these policies through policy advocacy, public conscientisation, international alliance-building, and the court system.

POWER TO THE PEOPLE

The ordinary Sowetan working-class electricity consumer is a good case study, because of extraordinary political mobilizations that have occurred in the Johannesburg "South Western Townships" (Soweto), including the student uprising of 1976. In the same spirit, using the same rhetoric and songs, a new movement against extreme electricity price increases arose in 2000, the Soweto Electricity Crisis Committee.

Sowetans experienced high price increases due to a huge reduction in central-local state subsidies. As a result, an estimated 10 million people were victims of electricity disconnections. According to the government, 60 percent of the disconnections were not resolved within six weeks. This confirmed that the blame lay with genuine poverty, not the oft-alleged "culture of non-payment" as a hangover of anti-apartheid activism. Likewise, of 13 million given access to a fixed telephone line for the first time, 10 million were disconnected due to unaffordability. The bulk of suffering caused by the rescinding of vital state services was felt most by women, the elderly, and children.

---

6   Fine, B. and Z.Rustomjee (1996), *The Political Economy of South Africa: From Minerals-Energy Complex to Industrialisation*, London, Christopher Hirst and Johannesburg, Wits Press.

Ultimately these problems are the outcome of neoliberal capitalism. The state's post-apartheid urban policies tended to amplify rather than counteract the underlying dynamics of accumulation and class division, despite electricity having been central in the anti-apartheid struggle. The first acts of sabotage by a then recently-banned ANC in the early 1960s were to bomb electricity pylons. The choice of target was symbolic given the economic importance of electricity and the fact that black working class areas were deliberately not electrified by the apartheid regime at the time. In the 1980s, when townships like Soweto were granted electricity, the residents launched a municipal services payment boycott that included electricity as part of their struggle against apartheid. This campaign was later adopted by the ANC, and its aim was to underline the illegitimacy of apartheid (local) government authorities and to make South Africa "ungovernable."

The slogan "electricity for all!" resonated with and moved the masses during apartheid days, in part because black households were denied electricity until the early 1980s as a matter of public policy (World Bank loans to Eskom during the 1950s–60s accepted this as a matter of course, though surplus value raised from black SA workers repaid those very loans). Hence one of the most popular African National Congress military tactics was the limpet mining of electricity pylons.

But the late apartheid regime and the capitalist class established their own agenda and kick-started the process of electricity commodification in a 1986 white paper on Energy Policy which called for the "highest measure of freedom for the operation of market forces," the involvement of the private sector; a shift to a market-oriented system with a minimum of state control and involvement; and deregulation of pricing, marketing, and production. After apartheid was replaced in 1994, similar language was found in the Urban Development Strategy (1995), the Municipal Infrastructure Investment Framework (1997 and 2001), and the Energy white paper (1998). The latter called for "cost-reflective" electricity tariffs so as to limit any potential subsidy from industry to consumers.

Asked why cross-subsidization of electricity prices to benefit the poor was not being considered, the state's leading infrastructure-services official explained, "If we increase the price of electricity to users like Alusaf [a major aluminum exporter owned by BHP Billiton], their products will become uncompetitive and that will affect our balance of payments."[7] (Alusaf pays approximately one tenth the price that retail consumers do, without factoring in the ecological price of cheap power at the site of production and in the coal-gathering and burning process.)

Rising electricity prices across South African townships had a negative impact during the late 1990s, evident in declining use of electricity despite an increase in the number of connections.[8] Most poor South Africans still rely for a large part of their lighting, cooking, and heating energy needs upon paraffin (with its burn-related health risks), coal (with high levels of domestic household and township-wide air pollution),

---

7   *Mail and Guardian*, 22 November 1996.
8   Statistics South Africa (2001), *South Africa in Transition: Selected Findings from the October Household Survey of 1999 and Changes that have Occurred between 1995 and 1999*, Pretoria, pp.78–90.

and wood (with dire consequences for deforestation). The use of dirty sources of energy has negative consequences, especially for women's health, leading to respiratory diseases and eye problems. Women are traditionally responsible for managing the home. They are more affected by the high cost of electricity, and spend greater time and resources searching for alternative energy. Ecologically-sensitive energy sources, such as solar, wind, and tidal, have barely begun to be explored, notwithstanding the enormous damage done by SA's addiction to fossil-fuel consumption.

The 1994 Reconstruction and Development Programme (RDP) mandated higher subsidies, but far stronger continuities from apartheid to post-apartheid emerged thanks to neoliberal pricing principles and the consequent policy of mass disconnections, preventing the widespread redistribution required to make Eskom's mass electrification feasible. As protests began in earnest from 1997, and the African National Congress witnessed rising apathy before the 2000 municipal elections, the ruling party introduced a "Free Basic Services" monthly package of 50 kWh of electricity per household, but it proved far too little.

Eskom continued to be a target of criticism, especially from environmentalists who complain that coal-burning plants lack sufficient sulfur-scrubbing equipment and that alternative renewable energy investments have been negligible. Moreover, labor opposition mounted. Having fired more than 40,000 of its 85,000 employees during the early 1990s, thanks to mechanization and overcapacity, the utility tried to outsource and corporatize several key operations, resulting in periodic national anti-privatization strikes by the trade union federation.

But it was in Soweto that the resistance became world famous and internationally networked. In 2001, domestic consumers paid an average price to Eskom of US 3¢ per kWh, while the manufacturing and mining sectors paid only half that amount. Two years earlier, in 1999, Soweto residents had experienced three increases—amounting to 47 percent—in a short period, as Eskom brought tariffs in line with other areas.[9] This reflected the move towards "cost reflectivity" and away from regulated price increases, in order to reduce and eventually eliminate subsidies, so as to achieve "market-related returns sufficient to attract new investors into the industry," said Eskom.[10]

When prices became unaffordable and payment arrears began to mount, Eskom's first strategy was disconnection and repression. Eskom decided in 2001 to disconnect households whose arrears were more than $800, with payment more than 120 days overdue. An anticipated 131,000 households in Soweto were to be cut off due to non-payment, according to Eskom—even though the company had only 126,000 recorded consumers in the township.[11] Johannesburg Metro authorities decided, in an act of solidarity, to cut off water, and began evictions, selling off residents' houses in order to recoup the debts owed, in an attempt to pressure people to pay Eskom arrears.[12] A survey of Soweto residents found that 61 percent of households had experienced electricity disconnections, of whom 45 percent had been cut off for more

9   Star, 15 July 1999.
10   Eskom (2001), Annual Report 2001, Megawatt Park, Johannesburg.
11   Eskom (2001), "Eskom Targets Defaulters," Press statement, Megawatt Park, 27 February.
12   Saturday Star, 10 March 2001; Star, 17 May 2001.

than one month. A random, stratified national survey conducted by the Municipal Services Project and Human Sciences Research Council found that 10 million people across South Africa had experienced electricity cutoffs.[13]

The impact of disconnections can be fatal. One indication of the health implications of electricity denial and of supply cuts was the upsurge in TB rates, as respiratory illnesses are carried by particulates associated with smoke from wood, coal, and paraffin. Because of climate and congestion, respiratory diseases are particularly common in Soweto. In a 1998 survey, two in five Sowetans reportedly suffered from respiratory problems.[14]

Survey respondents reported many fires in the neighborhood, often caused by paraffin stoves, many of which were harmful to children. Eskom's disconnection procedures often resulted in electricity cables lying loose in the streets.[15] Residents were unhappy not only about the high reconnection fees charged, but the fact that Eskom used outsourced companies that earn $10 per household disconnection. No notification was given that supply would be cut off, and residents were not given time to rectify payments problems. Eskom can disconnect entire blocks at a time by removing circuit breakers, penalizing those who do pay their bills along with those who don't. All these grievances provided the raw material from which the Soweto Electricity Crisis Committee (SECC) and its Operation Khanyisa emerged.

SOCIAL RESISTANCE TO COMMODIFIED ELECTRICITY

The SECC was formed in June 2000, through a series of workshops on the energy crisis, followed by mass meetings in the township. Operation Khanyisa ("light up") allowed for mass reconnections by trained informal electricians. Within six months, over 3,000 households had been put back on the grid. The SECC turned what was a criminal deed from the point of view of Eskom into an act of defiance, and also went to city councilors' houses to cut off their electricity, to give them a taste of their own medicine, and to the mayor's office in Soweto. SECC were soon targeted for arrest, but 500 Sowetans marched to Moroka Police Station to present themselves for mass arrest; the police were overwhelmed. By October 2001 Eskom retreated, announcing a moratorium on cut-offs, and the SECC announced "a temporary victory over Eskom, but our other demands remain outstanding."

- commitment to halting and reversing privatization and commercialization;
- the scrapping of arrears;

---

13   McDonald, D. (2002), "The Bell Tolls for Thee: Cost Recovery, Cutoffs and the Affordability of Municipal Services in South Africa," Municipal Services Project Special Report, http://qsilver.queensu.ca/~mspadmin/pages/Project_Publications/Reports/bell.htm.

14   Morris, A.; B. Bozzoli; J. Cock; O. Crankshaw; L. Gilbert; L. Lehutso-Phooko; D. Posel; Z. Tshandu; and E. van Huysteen (1999), "Change and Continuity: A Survey of Soweto in the late 1990s," Department of Sociology, University of the Witwatersrand, pp. 34–35, 41.

15   In a shack settlement outside Cato Manor in Durban, this problem caused the death of eleven children in 2001 (Mail & Guardian, 16–22 March 2001).

- the implementation of free electricity promised to us in municipal elections a year ago;

- ending the skewed rates that do not sufficiently subsidize low-income black people;

- additional special provisions for vulnerable groups—disabled people, pensioners, people who are HIV-positive; and

- expansion of electrification to all, especially impoverished people in urban slums and rural villages, the vast majority of whom do not have the power that we in Soweto celebrate (SECC 2001).

The *Washington Post* took up the story in a front-page article in November 2001:

> SOWETO, South Africa—When she could no longer bear the darkness or the cold that settles into her arthritic knees or the thought of sacrificing another piece of furniture for firewood, Agnes Mohapi cursed the powers that had cut off her electricity. Then she summoned a neighborhood service to illegally reconnect it.
>
> Soon, bootleg technicians from the Soweto Electricity Crisis Committee (SECC) arrived in pairs at the intersection of Maseka and Moema streets. Asking for nothing in return, they used pliers, a penknife and a snip here and a splice there to return light to the dusty, treeless corner.
>
> "'We shouldn't have to resort to this," Mohapi, 58, said as she stood cross-armed and remorseless in front of her home as the repairmen hot-wired her electricity. Nothing, she said, could compare to life under apartheid, the system of racial separation that herded blacks into poor townships such as Soweto. But for all its wretchedness, apartheid never did this: It did not lay her off from her job, jack up her utility bill, then disconnect her service when she inevitably could not pay.
>
> "Privatization did that," she said, her cadence quickening in disgust. "And all of this globalization garbage our new black government has forced upon us has done nothing but make things worse ..: But we will unite and we will fight this government with the same fury that we fought the whites in their day." [16]

A few weeks later, ANC Public Enterprises Minister Jeff Radebe visited Soweto to offer a partial amnesty on arrears, which the SECC declined as inadequate. The focus then moved to fighting prepayment meters. From the SECC and similar campaigns emerged an umbrella group, the Anti-Privatization Forum.

How serious a threat was the SECC at this stage? The ruling party's main intellectual journal, *Umrabulo*, carried a 2003 article by Tankiso Fafuli (later to become ANC councilor for Pimville), that gives a flavor of the challenge:

> On the 24th September 2001 the Soweto Electricity Crisis Committee [SECC] convened a rally at Tswelopele hall in Pimville zone 7. A wave of agitation permeated through the gathering, which influenced the attendants to march to councillor George Ndlovu's house in ward 22. Councillor Ndlovu with his family was held at ransom and the electricity box of his house was ransacked.... The incident prompted the branches

---

16   Jeter, J. (2001), "For South Africa's poor, a new power struggle," *Washington Post*, 6 November.

of the ANC in both wards to convene a special joint forum in the evening wherein a vigorous debate ensued on the political challenges posed by SECC ... [which] has successfully earned the respect from the community and thus the ANC could no longer tread willy-nilly in every territory....

In the initial stages of community mobilization, the key message from these forces was that the ANC in power has not only abandoned its historical constituency (i.e. the working class and poor), but has begun to unleash terror against it. This terror—they argue—is in the form of electricity and water cuts conducted against the weak and poor. Electricity cuts that intensified during the winter of 2002 were presented as naked savagery unleashed by a liberation movement against its people who are largely destitute ... these struggles have resulted into an open confrontation like the shooting between employees of Eskom and residents of Dlamini in Soweto in the year 2001. Such readiness and agitation for extreme action is encapsulated in Duduzile Mphenyeke's (SECC secretary) statement when proclaiming that "In every struggle there are casualties." In explaining Operation Khanyisa the SECC has stated in some of its public forums that people must chase away Eskom "agents" tasked to cut electricity cables with whatever means necessary and that "councillors must be made to taste their own medicines"....

The Pimville rally mandated the SECC to expand its scope of demands beyond electricity cuts and to begin to include a demand for houses, a stop to eviction/ relocation, and access to free basic water among other issues. This is essentially a call to develop a broad united front that goes beyond SECC and the electricity issue.... [The Anti-Privatization] Forum also creates the imperative link between the shop floor struggles against right-sizing (retrenchments), casualization of labour, and the struggles waged against water and electricity cuts in the townships. As a result, the APF synchronizes the struggles waged by SECC, Dobsonville Civic Association (DCA) against electricity and water cuts in Soweto with those fought by among others SAMWU [South African Municipal Workers' Union] on the shop-floor against retrenchments, as a result of privatization ... trade unionists have played key roles in some of the APF campaigns and marches. It is this ability to link these cuts of services and electricity to privatization that creates a strong and broader appeal— not only to ordinary residents but trade unionists, intellectuals, and development activists—and the capacity to make inroads within the frontiers of the Tripartite Alliance.[17]

This is an extraordinary admission of the SECC's community popularity as well as the sophisticated way the new movement expanded its organizing reach and agenda. Subsequent years were spent in issue linkage. The APF and SECC adopted socialism as their "official" vision. The World Summit on Sustainable Development (WSSD) in August 2002 also helped raise the SECC's profile. A memorable *Mail & Guardian* front page on 16 August framed elderly SECC stalwart Florence Nkwashu in front of riot police with the headline "We'll take Sandton!" Two weeks later, the SECC was central to the memorable 30,0000-strong march from Alexandra to Sandton, the largest post-1994 protest in South Africa aside from trade union mobilizations. The "Big March" was roughly ten times larger than one aimed at supporting the WSSD (by the ANC, trade unions and churches) held along the same route later that day.

---

17   Fafuli, T. (2003), "Beyond dreadlocks and demagogy," *Umrabulo*, 18 June.

To the outside observer, that 2002 demonstration was the peak for many of the "New Social Movements" that emerged since the late 1990s. For the SECC, there were several years ahead in which attention shifted to water rights, culminating in the victories against prepayment meters and inadequate free supplies in 2008. In its journey it has faced many challenges including organizational crises due to internal political differences. It has set itself new challenges including running candidates in the 2006 local government elections where it won one seat in the Johannesburg City Council, which it uses to amplify its campaigns to a broader audience. Recently it helped form an electoral front of community and left organizations to run candidates in the national elections on a red-green platform, but lacked the finances required to formally register.[18]

CLIMATE PRIVATIZATION

Meanwhile, the SA government's own stumbling attempts to address electricity shortages and the worsening climate crisis provided further opportunities for communities to link energy access and CO_2 emissions campaigning. The government appeared co-opted by the Minerals Energy Complex—the phrase that captures the fusion of state, mining houses, and heavy industry—especially in beneficiating metallic and mineral products through smelting. As Ben Fine and Zav Rustomjee showed, throughout the twentieth century, mining, petro-chemicals, metals, and related activities that have historically accounted for around a quarter of the GDP typically consumed 40 percent of all electricity, at the world's cheapest rates. David McDonald updates and regionalizes the concept a decade later in his edited book, *Electric Capitalism*, finding an "MEC-plus": "South Africa's appetite for electricity has created something of a 'scramble' for the continent's electricity resources, with the transmission lines of today comparable to the colonial railway lines of the late 1800s and early 1900s, physically and symbolically."[19]

Eskom fostered a debilitating dependence on the (declining) mining industry, causing a "Dutch disease," in memory of the damage done to Holland's economic balance by its cheap North Sea oil, which, in South Africa's case, is cheap but very dirty coal. As one study found, South Africa is "the most vulnerable fossil fuel exporting country in the world." if the Kyoto Protocol is fully extended (because of the need to make deep cuts).[20]

Eskom is amongst the worst emitters of CO_2 in the world when corrected for income and population size, putting South Africa's emissions far higher than even the energy sector of the United States—*by a factor of twenty*.[21] To deal with this legacy, the government adopted a *Long-Term Mitigation Scenario* in mid-2008, to great fan-

---

18   The Socialist Green Coalition's platform is available at http://www.sgc.org.za/

19   McDonald, D. (Ed) (2008), *Electric Capitalism*, Cape Town, Human Sciences Research Council Press.

20   Spalding-Fecher, A. (2000), "The Sustainable Energy Watch Indicators 2001," Energy for Development Research Centre, University of Cape Town, Cape Town.

21   International Energy Agency (2000)a, "CO_2 Emissions from Fuel Combustion, 1971–1998," Paris; International Energy Agency (2000b), "Key World Energy Statistics from the IEA," Paris.

fare, calling for cuts in CO_2, but beginning in 2050. Meanwhile, the rollout of at least a $100 billion worth of new coal-fired plants ensued. Moreover, the 2004 National Climate Change Response Strategy endorsed carbon trading, specifically the Kyoto Protocol's Clean Development Mechanism (CDM), declaring "up-front that CDM primarily presents a range of commercial opportunities, both big and small." The carbon trading gimmick allows Northern firms to buy World Bank Prototype Carbon Fund investment allowances in CDM projects so they can continue emitting at species-threatening rates, instead of cutting emissions.

The October 2004 "Durban Declaration on Carbon Trading"[22] rejected the claim that this strategy will halt the climate crisis, insisting that the crisis is caused by the mining of fossil fuels and the release of their carbon to the oceans, air, soil, and living things, and must be stopped at source. By August 2005, inspiring citizen activism in Durban's Clare Estate community forced the municipality to withdraw an application to the World Bank for carbon trading finance to include methane extraction from the vast Bisasar Road landfill, which community activists insisted should instead be closed. The leading advocate, long-time resident Sajida Khan, died two years later, but her struggle to halt the "privatization of the air," as carbon trading is known, lives on. The only way forward on genuine climate change mitigation is to leave fossil fuels in the earth.

Hence "Keep the oil in the soil" and "Leave the coal in the hole" are regular slogans of African energy activists ranging from the South Durban critics of deadly petrol refining in residential communities to the Niger Delta critics of deadly petrol extraction from residential communities. The hard work of winning more civil society organizations to this position, especially organized labor, continues. A Nigerian journalist explains:

> Human rights activists from across the African continent that converged in Durban, South Africa recently for a conference which was convened by Oilwatch Africa and GroundWork South Africa have warned that Africa is facing another round of colonisation that threatens livelihoods and ecology. The thrust of the conference was the renewed focus on Africa as one of the fastest growing sources of oil and gas for the global markets amidst tightening oil supplies, spike in oil prices, low sulphur content of the oil found in Africa and an equally growing appetite for fuel by emerging global economic powers like China, India and Korea.... Nnimmo Bassey, executive director Environmental Rights Action and Friends of the Earth Nigeria included in his presentation entitled "The Future of Crude Oil is Already History" a profile of the environmental degradation in the Niger Delta in the last 50 years, stressing that fallouts of oil exploration include socio-economic displacement of the locals, pollution-induced sicknesses and violent conflicts in the region.... Ivonne Yanez, co-ordinator of Oilwatch South America, explained that an initiative on keeping the oil underground, was taking placing in Yasuni Forest Reserve ... in Ecuador. Calling on Oilwatch Africa member countries to emulate the Yasuni struggle since the human and environmental costs of fossil fuel extraction far outweighs any gain that

---

22   The Durban Declaration on Carbon Trading was adopted by civil society organizations that met in Durban in October 2004, with the specific aim of halting the carbon trade as a "false solution" to the climate crisis.

accrues from it. Activists from countries such as Ghana, Eritrea, Ethiopia, Mauritius, among others also took time to share their ugly experiences. All were unanimous that oil extraction activities as shown in the cases of the Niger Delta or Angola, South America and several other places have been a curse rather than blessing to the indigenous people under whose soil oil is being tapped.[23]

In addition to campaigning against fossil fuel extraction, South African environmental activists insist on higher renewable energy subsidies to kick start the solar, tidal, and other methods of harnessing the country's vast potential resources. However, less than 10 percent of state R&D spending on energy went to renewables since 1994 (compared to 90 percent for nuclear).

CONCLUSION

Reviewing this complex terrain of energy and social activism leaves us with several conclusions about the prospects for decommodifying electricity for poor people and shifting the generation to renewable production in a red-green synthesis:

- South Africa became more unequal during the late 1990s, as a million jobs were lost due largely to the stagnant economy, the flood of imports and capital/energy-intensive investment that displaced workers (especially in the strike-rich manufacturing sector)—and these trends had enormously negative implications for the ability of low-income citizens to afford electricity;

- billions of Rands in state subsidies are spent on capital-intensive energy-guzzling smelters, where profit and dividend outflows continue to adversely affect the currency;

- the price of electricity charged to mining and smelter operations is the lowest in the world;

- little is being spent on renewable energy research and development, especially compared to a dubious nuclear program;

- greenhouse gas emissions per person, corrected for income, are amongst the most damaging anywhere, and have grown worse since liberation;

- electricity coverage is uneven, and, despite expansion of coverage, millions of people have had their electricity supplies cut due to commercialization and privatization.

All of these problems are being countered by critiques from civil society. However, most challenging is the paucity of constructive *collective* work carried out between the three major activist networks that have challenged government policy and corporate practices: environmentalists, community groups, and trade unions. This is partly due to serious political setbacks suffered by progressive forces, including

---

23   Chimeziri, U. (2008), "Activists demand end to oil exploration in Africa," *Financial Standard News*, 5 October.

internecine divisions and material differences in class interests.

Overcoming these will require a highly-enhanced politics that must be able to reconcile differences of interest between the various sectors of civil society. What unites is the certainty that if the capitalist destruction of the environment is allowed to continue all are sunk. There is a need to challenge the power of capital because, while the rule of profit dominates the world, all solutions tend to fall flat. Humanity needs to stop digging out the coal and re-employ coal miners in socially-useful activities. The truth is that this will be next to impossible to implement unless power shifts to the hands of ordinary people and away from the monied elite.

In South Africa, the ANC's pro-capitalist policy means that wasteful white elephant projects continue: the Coega industrial complex; the expansion of the Lesotho Highlands Water Project mega-dams; huge new soccer stadiums for the 2010 World Soccer Cup; the corruption-ridden R43 billion arms deal; and the R20 billion+ Gautrain elite fast rail network. To these we can add the multi-billion rand nuclear and coal power stations that Eskom plans to build.

In contrast, activists will have to intensify their work, to get any of the spending the society requires redirected into providing a sufficient minimum free basic supply of electricity, into rolling out the power grid to unserved rural areas as well as to Southern African societies who have long contributed cheap labor to South African mines, and to cutting back CO_2 emissions via major state investments in renewables. But if the apparent impossibility of acquiring AIDS medicines from 2000–03 or reversing water privatization in 2006–08 are useful examples, these are the kinds of challenges that compel South African activists to rise up and shout, "Amandla!" (Power!)—"Awethu!" (To the People!)

Chapter 16 | Part 4

RECUPERATING THE GAS:
Bolivia in its Labyrinth[1]

Marc Gavaldà

In the last decade, conflicts ignited by Bolivia's takeover of its hydrocarbon resources has turned the country into an example for countries willing to defy global capitalist power. In a see-saw-like motion, the Bolivian state handed over all its companies and resources to global corporations, only to fight to recuperate them just a few years later. From the depths of the neoliberal abyss, a profound rejection has emerged, expressed in a popular yearning to recover that which had been lost. The radical nature of the massive protests demanding that those in power should retake control over the resources meant that several governments had to be toppled before this goal could be achieved. Yet, despite having come a long way, the popular movements still have a long fight ahead of them.

PETROL AND GAS—THE STORY OF BOLIVIA'S ILLUSORY WEALTH

Bolivia's history as an extraction-based economy has meant that the country has always been enmeshed in conflict. Be it silver or tin, wood or rubber, disputes over ownership and use of Bolivia's resources have been a constant throughout the country's history. Over the course of five centuries, the country's natural resources have been extracted for the enrichment of foreigners, leaving the majority of the population with no other option than to perceive exportation as simply a source of illusory wealth. And today the story revolves around hydrocarbons. Half of Bolivia's territory has hydrocarbon potential and concessions have been granted over 2,811,157 hectares. There are currently forty-four contracts in operation, spanning six departments.

Furthermore, the petrol industry has an added ingredient. It is as polluting as mining, but with an even wider territorial footprint. Owing to the fact that the state has granted concessions to the petrol companies so that they can explore entire blocks of territory, the petrol frontier is expanding. Exploration now spans millions of hectares of forested areas, areas that, until now, had been subject to very little intervention. However, the territorial invasion and the enormous environmental destruction is such that people are very unwilling to believe that they will ever even get any of the illusory and derisory wealth resulting from petrol operations.

Proof exists that Bolivia's indigenous peoples were aware of the advantages of using petrol centuries ago. The Chiriguanos, a subgroup of the Guarani people, called it *itami* and used it for torches and flamed arrows. The priest Álvaro Alonso Barba, from the parish of Tarabuco, made reference to Chiriguanos carrying bitumen in

---

1   This chapter was oringally written in Spanish, and was translated by Kolya Abramsky.

jugs in his 1647 work, *El Arte de los metales* [The Art of Metals] (Royuela, 1996). The ordinances of Aranjuez, dictated exclusively for the administration of the New World Colonies, by Carlos III between 1780 and 1783, contain references to the Earth's so-called "bitumens and juices" (Mariaca, 1966).

Since then, and especially throughout the last century, the situation with regard to ownership and use of the resource has changed. As such, the conflict surrounding ownership and control of petrol and gas goes back as far as their exploitation.

THE HISTORY OF CONFLICTS OVER HYDROCARBONS

During the first period, which spanned close to a century, struggles emerged that gave voice to the conflict over ownership and use of hydrocarbon resources, though it was still premature for them to speak in terms of environmental conflicts. However, the fact that communities were constantly being run roughshod over was enough to prepare society for the desired reassertion of state control over natural resources.

As early as February 1867, before the world had woken up to the petrol age, General Mariano Melgarejo awarded a ten year concession to the Germans Merkest and Hansen (Mariaca, 1966). Over the course of the next century, the pendulum swung away from the private control of these resources towards their nationalization, in accordance with the political and military conjuncture of the period.

Between 1932 and 1936, Bolivia and Paraguay were pitted against one another in the Chaco War. This war, which was cheered on by Standard Oil and Royal Dutch Shell, the two business giants of the period, resulted in a painful loss of life and territory. Consequently, by the time the war was over, a strong national sentiment over ownership of hydrocarbon resources had developed. The condtions were ripe for founding Bolivia's state-owned petrol company, YPFB, in 1937. The new company was founded on the seizure of the assets of the North American company, Standard, which received $1.7 million in compensation.

The state company continued discovering and operating new sources, and by 1954, it had succeeded in meeting the country's entire demand for petrol. The following year, however, President Víctor Paz Estensoro opened the doors to foreign investments, through the Davenport Code [Código Davenport], a piece of legislation that was drafted in the United States in order to benefit its private companies (Orgaz, 2005). The new favourable conditions—such as the royalties falling to just 20 percent—attracted ten foreign companies, with the Gulf Oil Company being the one that produced the most petrol reserves.

The gradual disinvestment in YPFB, combined with the wider policies of the US-aligned dictator René Barrientos, favored Gulf Oil's dominance. This was the company that had been given the responsibility for the nascent business of exporting gas to Argentina. In 1968, reinvestment of its profits into new wells enabled Gulf Oil to control 80 percent of Bolivia's petrol and 90 percent of its gas (Royuela, 1996).

In September 1969, following the death of General Barrientos, the military junta presided over by General Ovando Candía nullified the Davenport Code. One month later, after declaring a "Day of National Dignity" [*Día de la Dignidad*

Nacional], Marcelo Quiroga Santa Cruz, minister of Hydrocarbons and Mines (who later disappeared during the dictatorship of General Banzer), nationalized the Gulf Oil Company's assets (Quiroga, 1997). The North American pressure resulting from the expulsion of Gulf Oil forced the state to become indebted to the tune of $78 million.

In 1972, under the dictatorship of Hugo Banzer, foreign companies once again entered the country, this time under Shared Risk contracts based on a 50-50 share of profits. In 1985, with the collapse of the price of tin, inflation, and massive unemployment, YPFB was able to maintain its position as the main state company, generating $3.57 billion in profits between 1985 and 1995.

THE CORRUPT 90S: SELLING OFF THE STATE FOR A SONG

In 1995, against the backdrop of the "New Political Economy" and IMF-imposed Structural Adjustment Programmes, an annual tax of 65 percent on YPFB's gross earnings—including its profits—was decreed, and the company's investment projects were liquidated. Thus began the dismantlement of the state, a process that culminated in 1996 with the Capitalization Law (read as "privatization"). This was expressed in the Hydrocarbon Law 1689, a piece of legislation which granted concessionaries the right to freely trade in hydrocarbons, both domestically and on the world-market. Furthermore, the concessions also included granting property rights over the hydrocarbons extracted at the well mouth, a crucial element since it contravened Article 139 of the State's Political Consitution, which stipulated that hydrocarbons are national goods of the state, are inalienable, and may not be subjected to external authority. Hydrocarbons were constitutionally defined as inviolable public property.

The Capitalization Law paved the way for selling off the state petrol company YPFB at a very low price. Before its privatization, YPFB had been in charge of the entire production process, including both the upstream phases (exploration and production) and downstream (refining, industrialization, transport, storage, and export), but under the new legisalation, the process was divided up. Thus, the new "capitalized" (semi-private) companies, Chaco (Amoco) and Andina (Repsol YPF), took control of exploration and production, while Transredes (Enron and Shell) took over pipeline transportation, and Petrobras took on the refining. The companies were faced with a genuine bargain. The petrol fields and the pipeline networks were acquired solely on a promise to invest $834 million, while the refining complex together with the trade networks were handed over for $122 million. Apparently, the fact that the total value of the reserves was, at that moment, $13 billion was ignored. Furthermore, the valuable geographical information, which YPFB had generated throughout the course of its existence, was not taken into consideration when calculating the price, and was simply handed over for free to the petrol companies.

The "new legal framework" underlined the changes underway with regard to control of hydrocarbons. It established some norms, including equal conditions for foreign and national companies, with contracts guaranteed by the MIGA and ICSID (bodies in international investment law). Several decrees were enacted which prohibited the publication of information concerning the privatization process as well as about the operations of petrol companies in Bolivia (4th August 1997). Decree 26259 even went as far as proclaiming petrol companies' right to have their voluntary honorariums, per diems, transport, and their own consultancy teams, comprised of public civil servants paid for by the Bolivian authorities (Cedib, 2006).

Arguments that demonstrate *exactly how* Bolivia benefits from the capitalization and the Shared Risk Contracts established in partnership with the transnational companies are hard to come by. Effectively, the new legislative framework established one of the lowest taxation levels—just 18 percent—found anywhere in international petroleum law. Worse still, it was decided that hydrocarbons "existing" before capitalization would be taxed at 50 percent while "new" ones would be taxed at 18 percent. However, 65 percent of the fields that had already been discovered by YPFB were actually classified as "new" fields, including some of the most important reserves, like San Alberto, which was discovered in 1992, and inexplicably categorized as "new" in 1998. Bolivia's economic loss engendered by this reduction of state income in favour of transnational companies is $3.15 billion, a figure not too dissimilar to Bolivia's external debt (more than $5 billion). Even the Bolivian Confederation of Private Businesses went as far as declaring that privatization of the hydrocarbon sector has been one of the major causes of Bolivia's fiscal deficit (Mariaca, 2004).

Ironically, the exploitation of gas by the companies does not allow for energy sovereignty. For example, only one city (Tarija) has a network with which to supply houses with gas, and the large majority of Bolivia's population is forced to use wood or to pay world-market prices for Bolivian gas that has been processed in Brazil.

It also must be stressed that the new legisation did not establish any means of control or supervision over field operations, leaving the companies to decide whether or not to declare their committed investments and output levels. Sectoral organizations, financed by the companies themselves, were given a supervisory role in order for the companies to, effectively, regulate themselves. Thus, for example, in the absence of any other verifing body, the companies have minimized their declared losses or profits. This is in contrast to the situation prior to privatization when YPFB used to obtain average annual profits of $220 million, while registering an average profitability of 23 percent (Intermon-Oxfam, 2004).

With magnate Gonzalo Sánchez de Losada at its helm, the Bolivian state defended the capitalization process. Supposedly, the fact that the Administrators of Pension Funds (AFPs) owned a 34 percent stake in the capitalization meant that every citizen would benefit from it, acquiring a part of the shares in the new companies. However, representatives of these administrators are appointed without any civic participation whatsoever, and instead are appointed in a process predominantly controlled by the Spanish bank, BBVA. Furthermore, the state also promised the capitalized companies would generate $134 million in profits per year—or $420 per citizen of retirement age—though, in 1997, the profit was just $45 million.

THE EMERGENCE OF A SOCIAL FORCE FOR REPOSSESSION

In the few years since the country's resources were handed over for international financial capital, the dazzling promises have lost their allure, and a sense of reality has returned. The country was in a state of collapse and was being directed by corporate sharks. It was beginning to suffer an unbearable hangover from the corrupt neoliberal partying that took place in the 90s.

Disilllusionment had set in. A number of reasons for this stand out, including the construction of the mega-infrastructure necessary for exporting to the detriment of domestic consumption, the weakness of a state that lacked income, and increasingly visible environmental outrages. Above all, people began to see the transnational companies as having usurped the country. Social discontent began to heat up, as debate about the laws and the economic impacts of the capitalization became more widespread and people began to understand what was going on. This was the driving force behind the social mobilization that crystallized in the Gas War of October 2003.

Furthermore, it was during this period that socio-environmental and territorial conflicts first became visible. This was due to the fact that populations were so directly affected by the advance of the oil frontier towards the Amazon region, the construction of gas pipelines and incidents of extreme pollution such as the fire at the Madrejones well, and the spillage of petrol in the Desaguadero river.

It is widely agreed (Crespo and Fernández, 2003; Crespo, 2006; Ceceña, 2004) that the Water War in Cochabamba served as an example of transnational privatization, and nurtured an awareness that mobilizing to take back stolen resources was a possibility. The social organizations gained strength and assumed a leading role in shaping the country's history (Rivero, 2003; Linera, 2008). From that moment, different episodes of intense social struggles achieved great successes thanks to the massive popular support vested in these struggles. In the water wars in Cochabamba and El Alto, the coca war, the confrontations of the Achacachi, and black February (where protests against the *Tarifazo* scheme for taxation of salaries, designed by the IMF, resulted in dozens of deaths in February 2003), the social movements made the state give way and they won gains in their respective mobilizations.

Because of this level of organization, the people were in a position to rise up with great strength when they found out that Gonzalo Sánchez de Losada was negotiating the export of gas to the United States, without public knowledge, during his second term in office. Once again, the country was swept with national aspirations to own the resources, affecting the hearts of rural and urban populations alike. The abstract hope for a rational use of gas generated the biggest movement in the country's history. The popular desire to appropriate the gas resources was able to become such a groundswell because of the existence of some widely disseminated publications (CEDIB, 2003; Iriarte, 2003), fiery debates that took place in the city squares, and also the connections that Bolivian social movements had recently made with global resistance networks.

In February 2003, protests which sought to stop the Sánchez de Losada government from implementing a *Tarifazo* on the miserable salaries of workers resulted in dozens of people being killed in the streets of La Paz. In the wake of this, in mid-2003, leaders of the social and indigenous organizations came to know about the plans to sell natural gas to the United States (via Chilean ports). Information about the sale, together with the contents of the Supreme Decree 24.806, which awarded contracting companies ownership of hydrocarbon extraction for the next forty years, began to spread like wildfire. Similary, awareness of the destructive consequences of YPFB's capitalization and the new hydrocarbons law [*Ley de Hidrocarburos*] also became widespread. Slowly but surely, this information led to a radicalization of a confrontation between the government and broad social sectors.

The agreement to export gas was what lit the fuse for the Gas War. Behind the backs of the Bolivian people, and under heavy pressure from the US ambassador, a deal was struck to sell gas to Brazil at half of what it would otherwise have been worth. The project involved the consortium Pacific LNG, which is comprised of Repsol YPF, British Gas, BP, Total, and the US company Sempra. It sought to export 22 million cubic metres every day to the United States, using a gasline to Chile where it would be liquified for transport, by sea, to Mexico and California. According to one estimate, despite projections of corporate profits of almost $1.9 billion annually, Bolivia will only receive $190 million (Gómez, 2004).

In September, fueled by a series of social and worker demands, social upheaval spread to Altiplano and several other cities in Bolivia, and in the face of the blockades that spread through much of the country, the government responded heavy-handedly. On the 20th September, in an attempt to clear a road where some tourists were trapped, a military contingent did not hesitate to open fire on the Aymara population of Warisata, killing five and wounding twenty-nine. The government's repression only made the protests grow in size and strength. A general strike was declared in the city of El Alto and hundreds of miners began to march on the government buildings. Within a few days, the country's major roads had been blocked, and La Paz was besieged and surrounded. Soon after, people found themselves unable to satisfy their basic needs, as crucial supplies to the city were cut off. However, the government, in its stubborn insistence on defending a contract that only benefitted the transnationals, did not think twice about escalating the scale of military repression in order to regain control of the streets. Without a doubt, this bloody conflict culminated when the military escorted some fuel trucks that were trapped in El Alto. The rich neighborhoods of La Paz were already starting to feel the scarcity of gasoline, and the president, whose own security was at stake, ordered the route to be opened, despite the fact that the road was littered with corpses.

The events of the Gas War have been interpreted in a few different ways. Edgar Ramos gives an extensive chronicle of each day of the conflict, analyzing in depth how the socio-cultural practices from El Alto motivated the mobilization. He charts the development of seven parallel battlegrounds: military, police, mediatic, psychological, politico-trade union, economic-financial, and medical (Ramos, 2004). Juan Perelman interprets the Gas War as a revolutionary insurrection, which, being a movement without leaders, was spurred forward and led by the self-determination of a multitude unwilling to be governed (Perelman, 2004). Testimonial memory of the Gas War has also been collected, including interesting maps of the different episodes and deaths that occurred during the conflict, as well as a study of the socio-economic and family backgrounds of those affected (Auza, 2004). Yet another perspective is offered by Gretchen Gordon and Aaron Luoma: they interpret the Gas War as a rejection of a political system that welcomed with open arms an economic policy written elsewhere for the benefit of others (Gordon and Luoma 2008). The author of the present article describes elsewhere how the politicians, prisoners to the fury of the masses, paved the way for a constitutional solution to the crisis that enabled Carlos Mesa to take over the reins while Goni and his collaborators fled in a chartered helicopter to the United States, a refuge for criminals seeking immunity (Gavaldà, 2007).

Edgar Ramos presented the relatives of the dead with a vigorous testimonial about the human consequences of the Gas War and brought individual political-administrative, military, and police leaders before a political and penal court (Ramos, 2004). Gretchen Gordon and Aaron Luoma document the composition of the Association of Families of the Fallen in the Defence of Gas and its work towards securing the extradition of Gonzalo Sánchez de Losada and two of his ministers (Gordon and Luoma, 2008). Currently, there is a broad campaign to bring to trial and punish those responsible for the October massacre.

OIL-RIGGING THE REFERENDUM

The flames of the Gas War could not be extinguished until the president resigned and a promise was secured from the incoming president, Carlos Mesa, that he would undertake four initiatives. These initiatives, known as the October Agenda, include holding a national referendum over the sale of gas, modifying the Hydrocarbon Law, revising the capitalization process, and convening the Constituent Assembly.

A referendum was held on the 18th July 2004, but was boycotted by many organizations on the grounds that the questions simply sought to perpetuate the existing conditions (Cedib, 2004; Pasoc, 2004). Not only were the questions intentionally formulated in such a way as to ensure a positive response, but they were so ambiguous that "voting yes did not necessarily mean either changing the policies or continuing with them" (Cedla, 2004; Quiroga/Arce, 2004). In the words of President Mesa, "It is not possible to modify the contracts with the petrol companies since this would be a declaration of war against the world." Thus, when the time came to translate the referendum into a new law, Congress refused to approve the proposal. Thus, the terrain around the new legislation became increasingly treacherous, and the major mobilizations of May 2005, known as the Second Gas War, resulted in an erosion of power that plunged the country into a period of ungovernability.

A NEGOTIATED NATIONALIZATION

The power vacuum was filled by an interim government that held power until new elections took place later that year. All of the candidates presented gas as the focalpoint of their policies, speaking in varying degrees about regaining control of it. At one extreme, the right wing offered to increase exports, while the other, Evo Morales' party, offered "nationalization and processing of the gas without confiscation."

And that is what Morales did. Once he gained power with an absolute majority, Morales went to the country's richest gas wells and announced Nationalization Decree No 28701 (known as "Heroes of Chaco"), a measure that shocked the transnationals, provoking them to rapidly mobilize their diplomatic corps and threaten disinvestment and international arbitration.

However, the move was not so bad for the companies afterall (Cedla, 2005); six months later, none of them had abandoned the country, and revised exploration and production contracts were already in place. And, although its income from royalties has increased, the state's control over hydrocarbon operations is taking shape under the shadow of Neoliberalism (Cedib, 2007; Magnhild, 2007; Poveda/Rodríguez, 2006). The contracts—forty-four of which were negotiated and signed in 2006—evaluate the feasibility of operations from a stricty commercial, rather than environmental standpoint, and the costs of environmental repair are shifted onto the state. They were drawn up hastily and erroneously. Furthermore, the new law was controversially approved in congress, without discussing its appendices which secretly awarded enormous benefits to the petrol companies, such as recovering any investment. All of this sparked popular discontent, this time directed against Evo Morales and his "nationalization without confiscation."

OBSTACLES ALONG THE ROAD TO NATIONALIZATION

The reconstruction of YPFB is advancing at a snail's pace. Three years after the nationalization decree the company itself has already seen several different presidents. In July 2008, the state aquired its first drill, yet prior to the privatisation, YPFB counted on sixteen such drills. In order to break out of its inactivity, the state petrol company has invited other companies to become partners in "the forty fields where YPFB has preferential access *viz á viz* foreign companies." State companies from other countries began to show interest in the gas, including PDVSA from Venezuelan, Argentine Enarsa, and Russia's Gazprom. Repsol YPF and Total, both private companies, have also come into the fold. Some social sectors are suspicious of these moves, fearing that once these companies have gained a foot in the territory, should a conflict of interest arise between them, the YPFB might end up siding with the transnational companies against the affected populations.

Taking back control of the capitalized companies is proving to be a long and drawn out process. Morales' government needs to show the people that the nationalization is advancing, but, in some cases, recovering control has simply meant buying a minimum percentage of shares. On the 1st May 2008, the president announced the results of recovering Andina. Flag in hand, he addressed the large crowd of people who filled the Plaza Murillo and championed the nationalization. He proceeded to tell the crowd how Repsol had demonstrated "great responsibility" in conceding to sell 1 percent of its shares in Andina, allowing YPFB to become the majority stakeholder. The presidential discourse ended with Morales hugging the president of Repsol, Antoni Brufau, who was unable to hide his satisfaction as he awoke from the Bolivian nightmare. Two years after the Nationalization Decree, "Heroes del Chaco," Repsol has not lost a single contract in the country. With the transfer of just a handful of shares it was able to ensure the continuity of a company that had been aquired during the corrupt Goni years.

Later, on the 2nd June 2008, Morales' government issued Supreme Decree No. 29586, which nationalized the shares of the monopoly tranport company, Transredes, a company that is part of the consortium TR Holdings S.A, which is comprised of Shell and Ashmore. The latter company had initiated an international arbitrage case before the International Centre for the Settlement of Investor Disputes (ICSID), in which it sought to sue the state for $500 million. At the last moment, the crisis was resolved by drawing up a transactional contract "of Recognition of the General Rights and Liberation and Reciprocity of Obligations."

In April 2009, the government also took control of Chaco S.A, marking a new step in the road towards nationalization. At that time, the company, which was privatized in 1996, was in the hands of Pan American Energy, with 60 percent controlled by the BP group. Since April, YPFB has gained 98.97 percent of the shares of the Chaco petrol company, the balance belonging to ex-workers in the petrol industry.

THE ROAD AHEAD

Against this landscape, it is only natural that the social organizations feel their expectations have not been fulfilled. In fact, some local outbreaks of rage have flared up once again in petrol communities such as Camiri. March and April 2008 saw mobilizations for an indefinite general strike demanding a "genuine" [verdadera] nationalization of the surrounding petrol fields. The Bolivian government held out for 8 days before using the military to repress the strike. In order to finally expropriate Andina (a subsidiary of Repsol YPF) and to effectively take back control over Camiri and Guairuy's oil fields, 2,000 soldiers were mobilized.

On the other hand, despite the "nationalization," the refounding of YPFB, the purchasing of refineries, the recovery of the major stakes in shares of some of the capitalized companies, and the establishment of joint ventures with companies such as PDVSA, Bolivia has still not managed to meet its own fuel needs. This has become the transnational companies' last remaining trump card. Having failed to dethrone a power hostile to their interests through the electoral process, they are nevertheless able to destabilize the government, or at minimum bend it to their interests, by cutting off the fuel supply. The most recent episodes of the bloody regional violence have ushered in a growth of autonomist activities, and the petrol companies have not been able to remain outside these developments. Links have even been uncovered between the fierce struggle of the Civic Committee of Santa Cruz [Comité Cívico de Santa Cruz], with its self-declared quasifascist discourse, and the Bolivian Hydrocarbons Chamber, the body that oversees the transnational petrol companies that operate in the country.

To concentrate attention solely on the conflict over ownership and control of the hydrocarbon resources would be hugely short-sighted on a number of different counts. And explaining this in purely monetary terms means that a whole constellation of socio-environmental problems are ignored. The hydrocarbon sector has an ecological impact and a territorial footprint, both of which are causing irreversible environmental damage, which has repercussions on local populations, and will continue to affect future generations. A whole mosaic of indigenous peoples exist whose social and cultural survival is under threat from the millions of barrels of oil and gas that are trapped beneath their feet. PetroAndina SAM, a mixed company formed by PDVSA and YPFB, recently announced that they are set to begin operations north of La Paz and Beni, a clear indicaton of the dangers ahead. Just because Bolivia is rich in petrol and gas, the country should not become complacent about the problems associated with the source of its wealth. In addition to the peoples' struggle to reclaim control of the hydrocarbons, they will also need to mobilize in order to diversify the energy model towards sources of clean energy and the protection of cultures and ecosystems. Such mobilizations are necessary in order to prevent half of the country's territory, the bulk of which is covered by forest, from being turned into a dark well.

BIBLIOGRAPHY

Auza, Verónica. *Memoria Tesitomial de la "Guerra del Gas,"* CEPAS-Cáritas. El Alto, 2004

CEDIB (Centro de Documentación e Información Bolivia). *El Referéndum del 18 de Julio. Evitemos que Mesa legalice lo ilegal.* Cochabamba: CEDIB. June 2004.

———. "Informe de caso: intervención de Repsol YPF S.A. en Bolivia," Sesión del Tribunal Permanente de los Puebos, Enlazando Alternativas 2, http://www.alternativas.at

———. *Nacionalización de Hidrocarburos en Bolivia. Dossier hemerográfico Mayo 2006–Abril 2007.* La Paz: Azul Editores, 2007.

CEDLA (Centro de Estudios para el Desarrollo Laboral y Agrario). *Ley de Hidrocarburos 3058. ¿Recuperación real de los hidrocarburos?.* Serie Para que no nos mientan. n. 2. La Paz: CEDLA. November 2005.

FOBOMADE (Foro Boliviano sobre Medio Ambiente y Desarrollo). *Geopolítica de los Recursos Naturales y Acuerdos Comerciales en Sudamérica.* La Paz: FOBOMADE, April 2005.

FOBOMADE, *Gritos, Voces y Sonidos.* Fobomade, La Paz, 1999.

Gavaldà, Marc, *La Recolonización. Repsol en América Latina, invasión y resistencias.* Icaria: Buenos Aires, 2004.

Gavaldà, Marc, *Las Manchas del Petróleo Boliviano.* Fobomade-Olca: La Paz, 1999.

Gavaldà, Marc, *Viaje a Repsolandia. Pozo a pozo por la Patagonia y Bolivia.* Amigos de la Tierra-Tutuma: Buenos Aires, 2007.

Geretchen Gordon and Aaron Luoma, "Petróleo y gas: riqueza ilusoria debajo de sus pies," *Desafiando la globalización: Historias de la experiencia boliviana,* Ed. Jim Shultz and Melisa Crane. Plural, 2008.

Gómez, Luis, *El Alto de pie. Una rebelión aymara en Bolivia.* Comuna-Indymedia-HdP: La Paz, 2004.

INTERMÓN OXFAM, "Repsol en Bolivia: una isla de prosperidad en medio de la pobreza," 2004, http://www.intermonoxfam.org.

Magnhild, Julie. *Why was the gas not nationalized?. A case study of Bolivia's 2006 "nationalization".* Universitet i Tromso, 2007.

Mamani, Walter, Suárez, Nelly, García Claudia, *Contaminación del Agua e impactos por actividad hidrocarburífera en Aguaragüe.* PIEB: La Paz, 2003.

Mariaca, Enrique, *Mito y Realidad del Petróleo Boliviano.* Amigos del Libro: La Paz, 1966.

———. "Historia de los descubrimientos de gas y los contratos de exportación como marco de la propuesta de una nueva ley de hidrocarburos," *Relaciones energéticas Bolivia-Brasil,* ed. Fobomade, La Paz, 2004.

Molina, Patricia, "Petrobras en Bolivia: petróleo, gas y medio ambiente," *Petrobras: ¿integración o explotación?*. FASE: Río Janeiro, 2005.

Montoya, Juan Carlos, *Efectos ambientales y socioeconómicos por el derrame de petróleo en el Río Desaguadero*. PIEB:La Paz, 2002.

Orgáz, Mirko. *La nacionalización del gas. Economía, política de la tercera nacionalización de los hidrocarburos en Bolivia*. La Paz: C&C Editores, April 2005.

PASOC. Pastoral Social Cáritas. *Análisis de los resultados del Referéndum sobre hidrocarburos*. Documento. n. 11. Santa Cruz, PASOC. 2004.

Poveda Ávila, Pablo, Rodríguez, Alvaro. *El gas de los monopolios. Análisis de la política de hidrocarburos en Bolivia*. La Paz, CEDLA. August 2006.

Quiroga, Marcelo, *Oleocracia o Patria*, Obras Completas Vol. 5. Plural/CID: La Paz, 1971.

Ramos, Edgar, *Agonía y Rebelión Social*. Plataforma Interamericana de Derechos Humanos: La Paz, 2004.

Royuela Comboni, Carlos. *Cien años de hidrocarburos en Bolivia (1896–1996)*. Los Amigos del Libro: La Paz, 1996.

IRAQI OIL WORKERS' MOVEMENTS:
Spaces of Transformation and Transition[1]

Ewa Jasiewicz

Five years into the war and occupation of Iraq, the US and UK administrations, international oil companies, and occupation-installed Iraqi elites are laboring hard to open up Iraq's massive oil reserves to their own long-term investment and control.

Possessing 115 billion barrels of proven reserves, with possibly twice that amount un-discovered, Iraq has the second largest reserves on the planet—approximately 10–20 percent of the global total. What makes Iraq's oil potential more important is that Iraqi oil is amongst the cheapest to extract ($1.50 per barrel compared to approximately $30 per barrel of tar-sands extracted hydrocarbons). It has a reserves-to-production ratio triple that of neighboring Saudi Arabia—a staggering 173 years. The ratio is calculated at current levels of productivity and demand and the unextracted potential of current producing and discovered fields. The quality of Basra Sweet Light Crude is also of a high purity, meaning a less capital- and energy-intensive refining process.

Geo-politically, Saudi Arabia as a key ally of the United States has become increasingly volatile. When Al Qaeda attacked Saudi's Abqaiq oil processing facility in 2006, the price of oil leapt by $2 per barrel. The US pulled out most of its troops and military infrastructure in 2003.

Oil is also more than a strategic commodity in its "crude" use-value sense. Traded in dollars, it also secures the value of the US Dollar and keeps the US economy financially lubricated, under-writing the currency with each transaction, compelling national treasuries to stash reserves of dollars to pay for it—if US and allied governments and companies control oil supplies that is. If these alliances break down, as in the case of Iran, which has diversified all of its external reserves away from the dollar and is trading with oil-dependent (90 percent of energy supplies) Japan in Yen, it is the US economy that could be made to "scream." Securing Iraqi reserves for US companies and allies to ensure their trade in dollars has security implications for US currency and the US economy. How much would a state invest to secure the future of its' currency? How do you value currency? Worth trillions?

US and UK authorities expected post-invasion Iraq to represent a more stable and acquiescent petro-state, given the removal of Saddam Hussein and the establishment of neoliberal free-market and authoritarian legislation, which began with 100 orders passed by the first pro-consul Paul Bremer in 2003.

---

1   This article also appeared in Issue No. 13 of *The Commoner*, a special issued devoted to energy and social struggles, edited by Kolya Abramsky and Massimo De Angelis.

LOCKING-IN NEOLIBERALISM

Bremer's hundredth order locked in and re-legitimized the previously passed 99 orders. The Iraqi Constitution, which was written in a matter of weeks under conditions of duress, according to some Iraqi law-makers, and under the heavy influence of US Ambassador Zallamy Khalilizad, who circulated US-drafted copies of a model constitution, also enshrines free-market policies for liberalizing the energy sector.

Article 110, frequently quoted by oil executives keen for privatization deals, decrees: "the federal government and the governments of the producing regions and provinces together will draw up the necessary strategic policies to develop oil and gas wealth to bring the greatest benefit for the Iraqi people, *relying on the most modern techniques of market principles and encouraging investment*" (my italics)— opening the door to the liberalization of the oil sector in the interests of foreign investors.

Still off the law books, however, is legislation allowing oil companies to effectively own Iraqi reserves and secure long-term investments—the absolute key to raising IOC share price, growing core-business, and gaining competitive advantage in energy markets. Through their allied oil companies, the British and US governments would be able to leverage political and economic influence over competing economies such as India and China, but also to mitigate the risk by having the potential to restrain the developmental capacity of a potentially non-aligned Iraqi government that could be hostile to Israel—the most important strategic ally of the US in the region.

HISTORY REPEATING ITSELF

This tactic of stunting economic capacity was deployed during the life-span of the Iraqi Petroleum Company, the consortium of Shell, BP, Total, and Exxon Mobil which originally signed a concession with the British-installed monarchy of King Faisal. At the time, Iraq was occupied under the British mandate, an occupation that became "Iraqified" with a paid off ruling monarchy and elite, enticed and maintained by oil revenue rents. Meanwhile a restive population mounted insurrection after insurrection until the monarchy was deposed by the coup of Abdel Karim Qasm in 1958.

Under Faisal, the IPC deliberately left fields undeveloped in order to fulfill its own quotas and market agendas and render the Iraqi government relatively weak. These companies had their seventy-five year concessions axed and were eventually booted out of the country under the nationalizations of the 1970s.

The past thirty years have seen a succession of nationalizations by governments laying claim to common energy sources, meaning the international oil companies now own approximately 4 percent of global oil reserves. For the likes of Shell and BP, Iraq represents a pendulum swing back in their favor after thirty years of declining influence and reserves.

The key to transferring ownership of these resources from state control to international oil company control is the ratification of the Iraqi Oil Law.

THE IRAQI OIL LAW—BREAKING AND ENTERING

A document of seismic political and economic power, its signing would have global implications for the growth of the global oil industry—corporate and state—and pave the way for the break-up of Iraq and an economic empowerment of an already politically- and militarily-empowered Iraqi ruling class.

The Oil Law currently on the table was influenced by nine multi-national oil companies, the IMF, and the UK and US governments, all of which saw copies of the original draft within weeks of its completion. The law has over-run more than five US administrations and IMF deadlines in the past few years, and was the top priority for the Bush administration to pass before Bush and oil-industry partner Dick Cheney left office.

The law, if passed in its current form, would create new realities on the ground by allowing regions to create their own oil industries, signaling the dismemberment of the Iraqi National Oil Company and potentially the creation of a host of new, regionalized oil and gas companies—private, and part state-and privately-owned.

The law establishes an entity known as the Federal Oil and Gas Council—a fifteen member, politically-appointed body made up of sectarian regional representatives— which would have ultimate decision-making power over what contracts are signed, with which companies, on what terms, and for how long.

The sectarian conflict fostered by the US and UK occupation has already produced new facts on the ground—namely the movement of millions of internal refugees fleeing sectarian violence and and the creation of new communities, divided along sectarian lines. Baghdad is currently divided up into sectarian cantons, sealed by concrete walls.

The US' "Awakening Councils"—known as the Sawa movement—is a network of paid-off tribal militias working in the service of US interests in Iraq. The Sawa councils, located mainly in Anbar province, are being groomed for local government under long-term US occupation. Incentivization for separation has been dressed up in the language of economic and political empowerment, namely the creation of a separate central so-called "Sunni" state with authority over the development of its oil and gas reserves, of which there is estimated to be a considerable amount in the Western desert where the Akkas Gas Field lies, only a few miles from the Syrian border and currently targeted for control by Shell.

WAR ZONE, CARBON COMFORT ZONE

The privatization of Iraqi energy by both the international oil companies and regional, occupation-supporting and -supported elites represents a win-win situation for the US and UK occupation authorities: guaranteed security of supply and stability of contract, enshrined with treaty status through the Oil Law and protected on the ground by Iraqi militias, paid by oil revenues, and private military security companies—US and British, yet employing local staff, all backed up by permanent US military bases.

The result could be a triple-lockdown preventing local resistance against these "facts on the ground," fracturing a potential resistance that could force a change in government and provoke a possible abdication from contractual responsibilities (known as the "obsolescing bargain"—a state claiming decisive power over the use of resources exercised recently by Venezuela and Bolivia). In this context, Iraq's oil industry would become highly militarized, as it has in Nigeria, Colombia, and Saudi Arabia, protected by concentric circles of concrete, and aerial and land surveillance.

The financial gains to be made through the development of oil and gas reserves risk an entrenched dependency on fossil fuels for the accumulation of capital and growth at the expense of alternative energy sources and development. Its a common process known in the industry as "Dutch Disease," a form of "putting all of one's eggs in one basket," which puts the economy at high risk of external market shocks or shifts in the energy market.

The entry and establishment of IOCs on Iraqi terrain, owning reserves for three decades, would not just entrench sectarian divisions, conflict, repression of the population and peoples' movements, but also the military occupation.

As well as the occupation, the economic occupation of fossil fuel resources by corporations would enforce the reliance on fossil fuels, on the refining infrastructure needed, and on the structures and industries they fuel, as well as on the related market structures, commodities, and systems it supports.

In short, Iraq can be seen as a major refueling zone for free-market corporate capitalism. A war zone, but a carbon-comfort zone for the dwindling IOCs that seek "energy security" for their own reserve tallies and energy fiefdoms.

IRAQI OIL WORKERS—A NEW SOCIAL MOVEMENT

Iraq's oil industry was the only industry that kept going during the wars, through the sanctions and uprisings. The United Nations Security Council's sanctions regime remained in place for thirteen years. Barely any spare parts, fertilizers, or materials could be imported into the country. While many private sector companies slowly went bust, and key public sector businesses began to mechanically fail and become decrepit, the oil sector, despite also being worn down and partially damaged due to the Iran-Iraq war and subsequent gulf wars, remained onstream and ongoing.

That consistency meant that oil workers in the huge Oil and Gas sector, kept coming to work and socializing and working with purpose, while many other public sector workers found themselves still paid and going to work, but without any meaningful work to engage in, no industrial power or sense of personal fulfillment and usefulness.

Collective bargaining and strikes to resist oppressive employers or the government were unavailable tools. The oil sector was probably one of the most repressed and highly-surveilled industries in the country. Workers talk of union officials carrying guns and issuing threats against them; your union official could have you killed, and your boss really was most probably a fascist. Both in cahoots with one another, the reality of "workplace organization" was state unions acting as a second line of regime-defense and surveillance, behind the existing lines of security forces and secret agents.

But repression in the workplace did not impede workers' sense of purpose, their feeling of power and responsibility. Oil was, and still is, the backbone of the Iraqi economy, and oil revenues under the "Oil-for-Food Programme" were literally putting food on Iraqi tables up and down the country. Oil workers were (and still are) incredibly conscious of their own power and value to the economy. This power was underscored by—to paraphrase Iraqi Federation of Oil Unions president Hassan Jumaa Awad—"heroic" and "mujahedeen"-like (resistance-fighter-like) grassroots reconstruction efforts by workers themselves.

Workers threw out KBR-subcontracted workers and banned military contractors from worksites in the summer of 2003. They knew the company represented "Dick Cheney" and "The American Occupation," and they wanted to retain control of their workplaces and do the necessary reconstruction themselves.

In the Iraqi Drilling Company alone, twelve drilling rigs were reconstructed using black market and cannibalized parts, repairing what had been damaged and looted following the 2003 war. Celebrations would be held after each autonomous reconstruction was completed. Ingenuity, invention, and tenacity flourished under the sanctions.

Management and worker relationships in some sectors of the industry became cooperative and mutually respectful, with workers themselves—senior technicians and engineers—managing maintenance and reconstruction processes through and in spite of the wars and sanctions, in a "collective war-effort" approach.

The shared experience of Iraqi oil workers, particularly in the South, where the bulk of the industry lies and where a major uprising took place in 1991, has been formative for creating the conditions for a social movement.

The Kurdish uprising in '91 had some success, in terms of an autonomous zone being created, free from Ba'ath dictatorship repression yet under the control of the US authorities and the two main Kurdish ruling class parties—the Kurdish Democratic Party and the Patriotic Union of Kurdistan. The South, on the other hand, suffered a brutal crackdown, and those who fought had to keep their heads down and carry on under ever-more precarious, surveillanced, and grief-heavy conditions.

SHARED RESISTANCE

But the shared experience of resistance, repression, and economic responsibility/power created undercurrents of organized resistance, unspoken and intuitive relationships of a depth that was sensual in the most intuitive, mentally, and spiritually intimate sense. Compounded by religious faith, these unspoken, evident truths of collective experience created the conditions for trust, self-organization, and a unity of purpose, and conviction that has resulted in powerful union organization that goes beyond workplace issues of wages, health and safety, compensation, and managerial repression and into the realms of a spiritual quest to guard Iraq's resources from tyranny, be it corporate, neoliberal capitalist, or dictatorship capitalist.

Nationalism is a major facet of this resistance identity, in the sense of a "national good," and unsectarian agenda. Mature political forces, present since the union's inception but more pronounced and better armed now, are trying to steer, hijack, and co-opt the union.

Even so, the union has rejected calls for localized compensation for pollution caused by the oil industry for fear of coming across as sectarian—it was Iraqi exile activists who urged union leaders to cover this in their demands as a pre-requisite for the improvements of conditions.

PRIVATISATION IN ISLAM

The IFOU has a mixed political leadership including communists and Muslims. The membership is overwhelmingly Muslim, and the community of the Mosque is an essential relationship of support for the union and a part of the members' community and collective, as well as individual, consciousnesses and conscience.

One of the areas of agreement between the two ideological strands of belief is a definition of privatization and capitalism as inherently anti-human and exploitative. One union leader—who has recently been ordered out of Basra by the Iraqi Oil Minister and into a different oil company in Baghdad—explained the following to a group of workers some years ago, as an Islamic interpretation of privatization: "In any production process of work, you have the following: the human being, energy, the means of production, and capital. In capitalism or privatization, the pinnacle principle, the most important goal is Capital, in second place of importance the means of production, thirdly energy, and in the very last place—the human being. In Islam, as we know, it is the human being that has the most value and is at the top of all priorities."

To value meaningful work or education as a means of self-betterment, as a means to evolve and become a better human being, is in line with some interpretations of Islamic or spiritual principles, but is not exclusive to Islam. The right to this evolution was cited in a statement of demands against the Oil Law that was signed by all of Iraq's unions in 2006, and also forms a central tenet of the IFOU's organizing principles:

> Since work is the qualitative activity that sets apart the human experience, and it is the source of all production, wealth, and civilization, and the worker is the biggest asset to the means of production (we honour humanity) we demand that this law includes an explicit reference emphasizing the role of all workers in matters of oil wealth and investment, to protect them and build their technical capacity, both in and outside Iraq.

Environmental protection is rooted in Islam. The Quran states that humanity is to act as "caliph" to the rest of nature, co-existing with it rather than dominating it, and working to preserve and maintain global ecology. It states that humanity should make gardens instead of working to satisfy greed.

This is not to say that the IFOU has an environmental policy or that there have been discussions about or an understanding of the contribution that the oil industry makes to global warming and the science behind it. Far from it. By and large, oil in Iraq is seen as liberation, an asset that, if managed properly, for the collective good, can free Iraqis from poverty, lift up the working class, educate, house, clothe, feed, and progress generations ahead toward better lives than they ever have had—if the revenues are steered into the public sector and finally out of the hands of dictatorship and private capitalist gain.

Oil and the industry is a source of pride, identity, and advancement. So how can an ecological critique of capitalism and the oil industry evolve under these conditions of consciousness and a culture of dependency and intertwined identity with oil? There may be a social movement dedicated to keeping oil out of the hands of the multi-nationals, but what if it simply wants to keep it pumping and selling and fuelling catastrophic climate change—only under workers' control? Even under the most egalitarian, and ideally horizontal conditions, this reliance on oil can appear as a brick wall and a death sentence for ecology—under different terms and conditions than those usually found, but the ecological and ultimately-capitalist facts are the same.

Or are they?

JOINING THE DOTS AFTER SHOCK

Do we dismiss social movements in this critical sector because their interests seemingly do not cohere fundamentally with our own? I would argue that there is a coherence and the space—crucially, a potential for the creation of a space—for an eventual coherence and co-operation of sorts.

Who are the "we?" "We" are the ecological justice and anti-capitalist movement. It's a movement that, at times, appears to be together in its critique of climate change as a consequence of industrialized capitalist expansion and economic growth, but in some ways avoids it publicly or does not "join the dots," in the sense of global production, consumption, energy ownership.

Focusing on local, domestic carbon emissions is not a bad thing and is essential for motivating the personal sense of responsibility necessary for engagement and involvement in social movements. But de-carbonization in the UK necessitates a de-carbonization of UK oil companies, still in the top five of the FTSE 100 and responsible, in the case of BP, for twice the annual carbon emissions of the UK domestic energy use.

"The Carbon Web" of oil companies' interdependent relationships with banks, consultancies, law firms, educational and cultural institutions, and unions spins out further than the UK. It is global, and unraveling the web and its monopolization of energy commons, means responding to it where it is strongest, at its front lines, and its point of re-enforcement, as well as where it is the weakest and being challenged and contested.

Discourses on climate change have veered at times into changing individual behaviors (aviation, personal responsibility for flying), which is positive in itself, but can fall short of expanding into an enunciated public articulation of the role of avia-tion in economic growth ideology. The war on Iraq opened the oil-control motive to the public imagination. As with the enduring image of the gouged out Canadian tar sands, the war opened up, with mine-like exposure, the possibility for challenging government and IOC ideologies of "energy security" and a fossil-fueled free-market growth for the next thirty years in this country, and opened debates of resource sov-ereignty, oil grab, and US imperialism in Iraq.

The moment of war was mined by numerous groups for political advantage pre-cisely because of the psychological shock it dealt to the public imagination and the possibility for new ways of seeing that came with it. The shock may be wearing off here, but militarized energy security policies and their neoliberal context are still shocking Iraq and need re-exposure and integration into the climate-change narra-tive. We cannot talk about ecological justice/climate justice/just transition without including oil producers—state and grassroots—in energy consumption, ownership, and movement.

The ecological movement has steered well clear of the struggle of oil workers in Iraq. What self-respecting climate change activist wants to throw in their lot with those busy pumping the black stuff out of the ground? "Oil Workers," the last work-ers' taboo, along with "miners" if we see a resurgence of the industry in the UK as planned by government. How can one support those who want to speed up climate change and are at the physical frontier of the raw perpetuation of it? These are some of the questions and contradictions at play when Iraq and oil come together. Why? Because these people are some of the most powerful in the world. As oil is a strategic commodity, those in a position of physically producing are also in a position to influ-ence a change and a shift in its production.

The Iraqi Federation of Oil Unions is one movement in this strategic position and has proved itself a force that the likes of Shell, BP, Exxon, Indian and Chinese oil companies, and oil-addicted governments of the world cannot ignore.

ALIENATING ALLIES?

To ignore the potential in the oil workers' movement as a space to combat the growth of the oil industry from its grassroots is to lose hope. It is to lose one of the most vis-ceral and paradoxically-organic relationships in the production of the industry and its power, and to close the door on some of the most important people that ecological liberation and anti-capitalist movements need to be engaging with.

Narratives of a just transition, debates on climate change, and introductions of the concepts of ecological debt, of keeping oil in the ground in return for compen-sation, though problematic alone, are unlikely to be even be uttered in Iraq with any impact, if international oil companies gain control of the country's oil for the next thirty years. I am not arguing that these debates will happen if big oil and the Iraqi ruling class don't come to control Iraqi oil, nor am I arguing that revolutionary workers' control of Iraq's oil is even likely, but our movement is about revolutionary potential and the creation of space and possibilities, and about solidarity.

TABOO TODAY, TURBULENCE TOMORROW

Despite a close personal relationship with leaders of the Iraqi Federation of Oil Unions, I myself have never had a debate about climate change with them. The subject of fossil-fuel energy and climate change and the contribution of oil to it, is a taboo. Those seeking to tarnish international solidarity and critiques of the oil-grab agenda have labeled activists working on the issue as cynical and self-interested environmentalists who want to keep Iraq's oil in the ground, who have no interest in supporting Iraqis' development. Raising these issues now risks feeding into this narrative.

My own support work with the union was, and still is, based on reinforcing their strategic position as a grassroots resistance force to the occupation and US imperialism and the refueling of capitalism. I didn't suddenly shed my ecological beliefs, and I still believe that there is hope and a necessity to be able to speak about climate change with workers' movements at the crucial point of the production, but that this potential and power can only develop if those workers and related popular movements have control of energy. Keeping these spaces open demands solidarity and support.

THE FIRE SOMETIME ...

The fire in Iraq is the ongoing military occupation and the corporate and state struggle for control of Iraqi oil. Maybe if there were no counter-force at the grassroots fighting this fire, we would have no space and human relationships to engage with and support, but there is.

Iraq is a tipping point in terms of the control and supply of energy to imperial powers and imperialistic oil companies fading and ascending, vying for power through strategic control of supply and the power to re-produce and perpetuate that power.

As we read, this struggle over the last bastion of easy oil on the planet is ongoing, and the outcome undecided. If the major IOCs and their governmental ruling-class partners succeed, the space for movements to challenge these interests will be severely restricted and their opposition and organizing on the ground in Iraq, severely repressed. There is still everything to fight for, and it is a fight, not for "more oil" or "an oil industry in workers' hands but still for the oil industry," it is a fight with a *long-term* view, and it is a fight in defense of this strategic space of resistance, energy, and *alliance* for an ultimately different world beyond capitalism and a shared, sustainable energy commons. A world where a narrative and practice of ecological co-existence and a non-exploitative energy commons evolves as a popular narrative of liberation.

Chapter 18 ∎ Part 4

WAY OUT FOR NIGERIA: NO MORE OIL BLOCKS!
Let's Leave the Oil Under the Ground[1]

Nnimmo Bassey (on behalf of Environmental Rights Action)

As the Niger Delta boils and Nigeria looks towards a bleak future with diminished oil revenues, the oil corporations operating there continue to garner obscene profits. This happens because the corporations are not paying the environmental costs of their operations and because ecological debts go unattended to. Local communities have shouldered the burdens while the corporations laugh all the way to the banks, secured by their opaque Joint Venture agreements.

The trend of profits made by oil companies over the past couple of years is very telling. These companies reap profits in the face of whatever woes the world is confronted with.

In 2007, Shell's net profit rose to $11.56 billion from $8.67 billion the year before.[2] According to reports, Exxon, the world's largest privately-held oil company, reported a 14 percent rise in profit, for a record $11.68 billion, which was adjudged to be the largest ever for a US corporation. In the first quarter of 2008, Exxon made nearly $90,000 of profit a minute![3]

Today, we expect Shell to declare another big profit, underscoring the fact that the Niger Delta environment is still not receiving the attention it deserves. Spills remain unattended to at Ikarama in Bayelsa State, Ikot Ada Udoh in Akwa Ibom State, Uzere and Iwerekhan in Delta State. Today we demand that they use their "profit" to clean up their mess in the Niger Delta.

The convulsions currently gripping the global system have directly impacted the economic outlook of Nigeria. Banks and other money-gobbling corporations have begun to bob belly-side up, and citizens of the world have been forced to bear the brunt of their profligacy. What we are witnessing may be on a new scale, but certainly it is not a novel thing. We do well to note that crises of capital would always heap the burden on the producer and consumer while the middlemen constrict both and live off their blood.

The major challenge of the Nigerian State is related to the collapse of crude oil

---

1   This was originally published as an ERA briefing, on 29th January, 2009, in Lagos, Nigeria. It is being reprinted with permission from the author. The original is available at: http://www.oilwatch.org/index.php?option=com_content&task=view&id=610&Itemid=224&lang=fr.

2   http://www.iht.com/articles/2008/07/31/business/oil.php.

3   http://wsws.org/articles/2008/auG-2008/oil-a06.shtml. This report indicated that "The major US oil companies appear headed for a combined $160 billion in profits for 2008. That compares to $123 billion in 2007. Exxon and other oil companies have rewarded their CEOs with multi-billion dollar payouts. Last year Exxon CEO, Rex Tillerson, cashed in $16.1 million in stock options in addition to his $1.75 million salary. He also received a $3.36 million bonus. Conoco Chairman James Mulva received $31.3 million last year."

revenue from an unprecedented height of about \$150/barrel to below \$40/barrel. This crash revealed that behind the cheap piles of petrodollars we can see the active fingers behind the forces that shape the market. We quickly note at this point that the so-called market forces are not as free as international financial institutions would want the world to believe.

Some Nigerians are equally worried that even the cheap oil we depend on for revenue may soon be set aside due to the real possibility that the world will move on to new alternative energy sources. If that happens and crude oil attracts less attention, what will be the consequences for the Nigerian economy?

While these are legitimate concerns, they also present us with a great opportunity to transform our environment and, by extension, our economy. And this is why we are making this proposal.

Cheap petrodollars drove us into believing that the only problem with money was how to spend it. They drove us into debt and debased our sense of nationhood. Cheap petrodollars turned Nigerian politics into a struggle for the control of the national purse and led to a regime that converted public funds and properties into private control. That has been the visible outcome of privatization in our nation. Cheap petrodollars invited the jackboots into Dodan Barracks and into Aso Rock,[4] and rocked and overturned every sense of common good and collective ownership in our dear nation.

The drive to maintain the flow of foreign exchange into the national coffers made it impossible for the government to see that a safe environment is a basic requirement for citizens to be productive. The government overlooked the fact that in a largely subsistence economic system, where the vast proportion of the citizens thrive outside of the formal economy, the first thing that must be secured for national health and productivity is an environment that supports the people's efforts in family farming and livelihoods. The grave inability to grasp this truth allowed oil companies (national and transnational) to operate with impunity in the oil fields and to pollute, destroy, and dislocate the very basis of survival of the people in the region. This inevitably spread to the entire nation since we run a quirky unitary federalism.

We have a clear proposal of how to turn the crises into a real opportunity to break from an ignoble system and move on to a sustainable path. As they say, it will require sacrifice, especially the jettisoning of our firmly-held prejudices.

QUENCH THE FLARES

The issue of gas flaring is a key one that must be addressed once and for all. Worldwide, an estimated 168 billion cubic meters of natural gas is flared yearly, and 13 percent of it is in Nigeria (at about 23 billion cubic meters per year). After years of paying lip service, the Nigerian state must wake up to its responsibilities to protect the lives of Nigerians. The many health impacts of gas flaring are well documented and include leukaemia, bronchitis, asthma, cancers, and other diseases.

---

4   Dodan Barracks, in Lagos, Nigeria, was the seat of government in the years of military dictatorship. The current state house in Abuja is known as Aso Rock.

In economic terms, Nigeria sends over $2.5 billion worth of gas up in smoke annually, using 2005 estimates. If we assume that this rate held good for the ten years, we are talking of $25 billion wasted. For each additional year that the government refuses to act on this, the amount wasted continues to grow, as does the number of dead due to the poisonous nature of the gases.

We are worried that, at a time when the world is seeking ways to combat global warming, we are busy cooking the skies through gas flaring. From pronouncements on climate change emanating from government agencies it is obvious that the government cannot plead ignorance of the massive contributions of gas flaring to global warming. This places every citizen of this country, and indeed the world, at risk. There can be no excuse for this unhealthy and uneconomic act.

At this point we want to quote a 1963 confidential communication from the British Trade Commissioner to the UK Foreign Office:[5]

> Shell/BP's need to continue, probably indefinitely, to flare off a very large proportion of the associated gas they produce will no doubt give rise to a certain amount of difficulty with Nigerian politicians, who will probably be among the last people in the world to realise that it is sometimes desirable not to exploit a country's natural resources and who, being unable to avoid seeing the many gas flares around the oilfields, will tend to accuse Shell/BP of conspicuous waste of Nigeria's "wealth." It will be interesting to see the extent to which the oil companies feel it necessary to meet these criticisms by spending money on uneconomic methods of using gas.
>
> In the longer run, Shell/BP is going to have to consider very carefully how it should explain publicly the large outflow of capital that is likely to take place towards the end of the decade … it will no doubt come as something of a shock to Nigerians when they find that the company is remitting large sums of money to Europe. The company will have to counter the criticisms which will very probably be made to the effect that the company is "exploiting" Nigeria by stressing the very large contribution it is making to Nigeria's export earnings.

From the above quote, it is clear that the oil corporations have been engaged in this action for at least half a century now. The fifty-year-old script of pacification-by-underhanded-play requires urgent critical political, environmental, and socio-economic examination and replacement.

It was not until the 1979 Associated Gas Reinjection Act that routine gas flaring was finally outlawed in Nigeria. Section 3 of the Act set 1984 as the deadline after which companies could only flare gas if they have field(s)-specific, lawfully issued, ministerial certificates. There are over a hundred flare sites still emitting a toxic mix of chemicals into the atmosphere in the Niger Delta. Through this obnoxious act the country lost about $72 billion in revenues between 1970 and 2006, or about $2.5 billion annually.[6]

The Gas Flares Prohibition Bill before the Senate proposes that the penalty for gas flaring be the market price of the gas being flared. It is a good effort, but the

---

5   Quoted in ERA/CJP, "Gas Flaring in Nigeria: A Human Rights, Environmental and Economic Monstrosity," Amsterdam, June 2005. This booklet can be found at both http://www.climatelaw.org and at http://www.eraction.org.

6   ERA Fact Sheet on Gas Flaring, December 2008.

government must *order* the immediate stoppage of gas flaring even if it means shutting down the offending oil wells.

DETOXIFY THE LAND

Stopping gas flaring will mark a major step towards detoxifying the Niger Delta environment. The other steps needed are two-fold: first is the immediate auditing of all oil spills, drilling mud and cuttings discharges, production-water handling, and other related polluting incidents in the entire Niger Delta. Second is the immediate and thorough clean up of the environment to international standards, such as those set by the World Health Organization (WHO) for safe drinking water and air quality.

These steps will make it possible for the people to farm and fish with reasonable hope of achieving living incomes from these activities. Life expectancy would also increase beyond the current forty-one years, as the environment would once again become people-friendly.

NO MORE OIL BLOCKS

ERA proposes that Nigeria should learn that there is no future in crude oil as the major revenue earner. We propose that, as a starting point, Nigeria should not make any new oil block concessions. We agree that existing fields should continue to be exploited, but at internationally-acceptable standards. Halting the giving out of new oil blocks would not mean a major loss in revenue. To start with, the current lowering of oil prices is also leading to production cuts, which means that the current fields can meet Nigeria's quota for quite some time. Leaving the oil underground does not translate to losses, but savings, and we must learn to conserve. The oil under the ground is still our oil. We must not exploit every resource simply because we have it. This is simple wisdom. Nigeria must step back and think!

Generally, it is believed that the world will soon witness a peak in oil production and this will coincide with the world having used more than half of all currently proven reserves.[7] It is estimated that Nigeria reached her own peak oil level a couple of years ago. The official figure of Nigeria's production stands at 2 million barrels/day. The plan to increase this production level to 5.2 million barrels/day by 2030 is a thinking that fits our profligate pattern. At this time, the country should be working to stop the daily theft of crude oil from the fields. That amount, which estimates place at between 200,000 and 1,000,000 barrels/day, would serve either to boost production or to increase/sustain reserves.

ECONOMIC CONSIDERATIONS

Let us assume that Nigeria would have probably been in a position to increase her crude oil production from 2015 by, say, 2 million barrels/day from new oil blocks that

---

7   *Multi-national Monitor*, "The End of Oil" (Washington: January/February 2007 edition), p. 6. This issue of the *Multi-national Monitor* illustrates, among other things, that the "Corporate control of energy policy and energy resources, especially in the United States, the country that consumes more energy than any other, is the single greatest obstacle to slow and hopefully reverse the world's headlong rush to disaster."

we are demanding should not be given out to the bidders. By this simple act, Nigeria would have kept the equivalent tonnes of greenhouse gases out of the atmosphere—curbing global warming through an infallible technology of carbon sequestration. This is a foolproof step that requires no technology transfer and does not require any international treaty or partnership.

If Nigeria were to trade that amount of carbon using any of the available market mechanisms for tackling climate change, such as the so-called Clean Development Mechanism, the country would surely earn some good income from keeping the oil under the ground. But we do not support the use of market mechanisms for this purpose. We would rather suggest the halting of the massive capital flight from Nigeria to boost the economy and offset whatever may be seen as "loss" of projected revenue from crude.

But let us do some calculations here, assuming crude oil prices stabilize at $30/barrel over the next several years. In that case, 2 million barrels/day would mean daily revenue of $60 million or an annual income of $21.9 billion. Now, assuming our population stands at 140 million, this means that the amount due to each citizen would be $156.43/year. If we factor in production costs (including staff salaries, payment of the military, etc.) and company profits, we can safely say that the amount would be less.

ERA proposes that, rather than exploit new oil fields with the attendant pollutions, human rights abuses, and malformed political system, we keep the oil under the ground and require every Nigerian to pay $156/year as a crude oil solidarity fund (for want of a better name). This will bring additional revenues to whatever the country makes from current oil fields, including the corked ones.

ERA recognizes that not every Nigerian can afford to pay $156/year into the national coffers. We can reasonably expect about 100 million Nigerians to enthusiastically make this payment if the benefits are carefully made public. Those who can pay multiples of the minimum amount would make up for the remaining 40 million Nigerians who could not pay. International aid agencies, philanthropists, as well as other countries can be approached to symbolically buy some barrels and the entire budgeted income would be met.

Moreover, by 2015 there would be more Nigerians and the burden would thus be less.[8] We also consider that the Naira would regain strength as corruption goes down and as governance becomes more transparent. If that happens, the Naira equivalent of the amount to be contributed by each Nigerian would further decrease. Note that these payments would not need to commence until 2015 and this will give us sufficient time to take caravans around the nation to explain the beauty of this economic move.

It is our considered opinion that the best foot forward for Nigeria is to halt new oil field developments and to leave the oil under the ground.

---

8   At a growth rate of 2.025 percent. See https://www.cia.gov/library/publications/the-world-factbook/print/ni.html.

9-POINT BENEFITS OF NO MORE OIL BLOCKS:

1. Carbon capture and storage, thereby tackling climate change.
2. No oil spills and gas flares from new oil fields.
3. No destruction of communities or high sea environments.
4. No socio-economic ills related to oil field activities.
5. Nigerians would have a direct stake in how national revenues are spent. There would be greater accountability and transparency. Moreover, hawks would no longer gather for so-called "excess crude cash."
6. Halt to the corrupt nature seen in the oil blocks allocation exercises.
7. No bunkering since the oil will be left in the ground.
8. Safe and clean environment.
9. Reduction, and ultimately elimination, of violent conflicts in the Niger Delta.

OIL IN A DEAD END

Decades of oil extraction in Nigeria have translated into billions of dollars that have spelled nothing but misery for the masses. It is time for Nigeria to step back and review the situation into which she has been plunged. The preservation of our environment, the restoration of polluted streams and lands, the recovery of our dignity will only come about when we stand away from the pull of the barrel of crude oil and understand that the soil is more important to our people than oil.

Oil blocks licensing has become a bazaar in Nigeria.[9] Huge signing fees are exchanged as though the players in the game were soccer or music stars. This signals the fact that there is something fundamentally faulty about the entire enterprise. This is the time for all Nigerians to demand that no more oil block should be given out for exploration or exploitation. Nigeria was richer through her great agricultural produce before the ascendancy of crude oil as the major foreign exchange earner for the nation. Crude oil brought about crude actions in every realm of national life. ERA is making a modest contribution to give Nigeria a better future by urging the nation to look away from oil and, at the same time, keep a stable economic platform from which to leap unto greater heights—through agriculture with supporting governmental structures. We must end our decades-old dependence on oil rents that has damaged our national psyche and sense of commitment to nation building.

Let every Nigerian contribute to the national purse. This will make it clear to politicians that when they misappropriate public funds they are indeed stealing from the suffering people. Our life and our future are in our hands.

---

9   New reports, showing this, abound. See, for example, Obinna Ezeobi, "FG suspends oil bid rounds," *The Punch*, Saturday, 23 August 2008 at http://www.punchontheweb.com/Articl.aspx?theartic=A rt200808231593070.

Chapter 19 ∎ Part 4

LEAVE THE OIL IN THE SOIL
The Yasuní Model[1]

Eperanza Martinez

Yasuní is a national park and biosphere reserve located on Ecuador's eastern border. It is the long-standing traditional territory of the Huaorani people, and is currently the hunting ground of Peoples in Voluntary Isolation.

Ecuador's most important petrol reserves have been located within this park. The park contains two petrol blocks, ITT and Block 31, both of which are located deep inside the park. They are estimated to contain 969 million barrels of probable reserves.[2]

Ecuador presented a proposal concerning these oil reserves to the rest of the world, demanding that the international community should pay $350 million in compensation over a ten-year period, in exchange for these reserves remaining unexploited. This amount is equivalent to half of what the country would profit were it to exploit the ITT oilfield. At the time the proposal was made, Block 31 had already been handed over to the company Petrobras. However, the company has subsequently left the country and the Block has once again come under state control.

The proposal was presented in Ecuador in 2007, at a moment when the world's most remote, vulnerable, and fragile areas were being opened up to the oil industry. Not only is this a process that jeopardizes the planet, it also threatens to provoke an unprecedented climate crisis and is bringing about extreme impoverishment for the majority of the world's economies.

The proposal questions the currently-existing global model and proposes a new one. The existing model is a destructive one, based on extracting and burning resources that have been formed over the course of millions of years, and using the wealth of third-world countries to subsidize the industrialized ones. Importantly, the proposed new model seeks to initiate a transition to a post-petrol Ecuador.

The proposal consists of:

- Not extracting the crude oil from the subsoil.

- Chanelling international resources in the form of compensation, donations, and symbolic sale of the crude oil that will remain un-exploited.

- Creating a capitalization fund whose interest could provide a

---

1   The proposal to keep the crude in the subsoil is spearheaded by the Camapaña Amazonía por la Vida [Amazon Campaign for Life]. This article is a synthesis of proposals that have been made within the context of the campaign for Yasuní in Ecuador. It has been translated from the original Spanish by Kolya Abramsky. He is grateful to Peter Polder for assistance with ensuring accuracy in translating technical terminology associated with the oil industry.

2   Probable reserves are based on median estimates, and claim a 50 percent confidence level of recovery. The French Petroleum Institute set the figure at 846 million barrels. However, the company Nacional insists on using 960 million barrels as the reference figure.

permanent source of income.

• Using these funds to embark on a model of self sufficiency with regard to food production and energy supply, in order to work towards constructing a post-petrol Ecuador.

YASUNÍ PROFILE:

Location: In the provinces of Orellana and Pastaza betweeen the Napo and Curaray Rivers.

Area: 9,820 km².

Important dates: National Park status received in 1979, Biosphere Reserve status in 1989, and intangible zone status in 1999.

Estimated population: 3,000 Huaorani, Tagaeri, Taromenane, Oñamenane

Discovery of petrol resources: Shell drilled the Tiputini-1 well in 1948. Petroecuador drilled 3 exploratory wells in Ishpingo, Tambococha, and Tiputini in 1992.

Proven reserves (1P, or possible): 412 million barrels.

Probable reserves (2P, or probable): 920 million barrels.

Possible Reserves (3P, or possibly probable): 1,531 million barrels.

Biodiversity: The highest recorded anywhere in the world.

Characteristics of the crude oil: from 14.5 to 15 Degrees API, once it has been converted into heavy crude.

INITIAL STEPS TOWARDS A POST-PETROL ECUADOR

The concept of transitioning to a post-petrol world has come about through a lengthy process of collective struggles. These have taken the form of struggles against war, against pesticides, against plastics, against consumerism, and, above all, against the impacts resulting from the operations of the petroleum sector itself.

Throughout the last quarter of a century, petrol has constituted the fundamental backbone of Ecuador's economy. Its role is no less central today. However, it is also an area in which the state has had to face major conflicts. Such conflicts are over contractual forms that are damaging to the state, and abuses directed against communities by oil companies. The sector has also spawned severe environmental conflicts.

Ecuador is a rural country endowed with an enormous biodiversity—agricultural, and animal and plant wildlife. It has an abundance of fresh water, sun all year round in most regions, and does not suffer from extreme climatic conditions. The memory of having contributed to the domestication of crops—potato, cocoa, cassava, maize, beans, tomatoes, fruits—that feed the world is still vivid. Ecuador has the best conditions imaginable for having a well-fed and working population, however, the country has been converted into a petrol country, which has led to impoverishment, loss of sovereignty, and contamination.

Ecuador's development may be described as centralized, based on exclusion and privatization, and it has continually renounced its sovereignty. Consequently, a point has been reached that makes a transition away from petrol an inevitable necessity, rather than an ideological choice.

Despite the fact that the process of charting a new course aimed at bringing about authentic changes is still in its early stages, the new constitution nonetheless contains some important first steps. This constitution incorporates many elements that offer, at least potentially, a legal basis to keep the crude oil in the sub-soil. This includes the following:

• Prioritizing support for the development of new sources of clean, low impact, and and decentralized energy (Art. 413).

• Food sovereignty as the backbone of the country's agricultural production model (Art. 281).

• State control of non-renewable natural resources (Art. 313/317).

• An international agenda responsible for global environmental issues (Art. 416 lit. 13).

• Protection of territory belonging to Peoples in Voluntary Isolation (Art. 57).

• Prohibiting the petroleum sector from operating in protected areas (Art. 407).

• Energy sovereignty shall not affect food sovereignty, nor water supply (Art. 15/Art. 413).

• Recognition of nature's right to exist and preserve its vital cycles and structures intact (Art. 72).

• Recognition of the rights of persons and peoples to demand the fulfilment of the above-mentioned rights belonging to nature (Art. 72).

• Recognition of the right to comprehensive restoration of damage done to the environment, both as part of the rights accorded to nature (Art. 73), as well as the rights of persons to the environment. (Art. 397).

• Recognition of the Precautionary Principle based in the affirmation that the state must apply precautionary and restrictive measures on activities that may lead to the extinction of species, destruction of ecosystems, or permanent alteration of natural cycles (Art. 73).

• Recognition of the Prevalence Principle, a principle that stipulates that in case of doubt concerning the scope and applicability of environmental laws, these laws will be applied in the manner that offers the strongest possibility for protecting nature (Art. 395).[3]

STEPS ALONG THE WAY

30th March 2007: The President of the Republic declares that the first choice with regard to the ITT field is the option of receiving international compensation in

---

3   The new Constitution recognizes nature as a subject with rights, prohibits petrol operations in national parks, recognizes the territories of Peoples in Voluntary Isolation, and encourages a new vision of development.

exchange for leaving the oil unexploited.

5th June 2007: It is officially announced that the compensatory funds should be raised within a time period of one year.

16th January 2008: Decree 847 issued, authorizing the establishment of a Trust Fund.

9th February 2008: Decree 882 creates the Technical Secretariat for the Yasuní Initiative.

12th August 2008: Decree 1227 extends the time period until 31st December 2008, and German parliamentarians take up the proposal.

17th September 2008: A double bidding is announced for the ITT.

November 2008: Director of Petroecuador proposes to initiate early exploitation of the Tiputini field. In other words, he proposed to knock one of the Ts from the original ITT project, Ishpingo Tambococha Tiputini.

7th January 2009: The Ministry of Petrol and Mines announces the decision to open the Block up for bidding.

16th January 2009: Ecuador's Chancellor, Fander Falconí, announces a further six-month extension.

5th February 2009: Decree 1572 issued, extending the time period for implementing the initiative to not exploit Yasuní/ ITT's oil indefinitely.

THE PROS AND CONS OF OIL OPERATIONS: THE CASE OF YASUNÍ

One of the problems that must be dealt with when taking on these petrol operations is the question of petrol reserves themselves. These are managed capriciously. At times estimates are inflated in order to push up the price of company shares, while at other times they are deflated in order to negotiate better terms of trade with other states.

Various figures are in use concerning the ITT reserves, though, in general, the petrol company puts the figure at 900 million barrels of proven reserves, and Block 31s are cited at 60 million.[4] However, the reserves of both Block 31 and the ITT project will rapidly decline; it is anticipated that, after five years in operation, they could only supply an estimated 112,000 barrels daily, and then only for the first fifteen years.

Ecuador's state petrol company has calculated that these resources could bring in a total income of $700 million, however, what has not been calculated, at least not by the company itself, are the environmental costs that these operations would incur. Given that numbers usually have more power of persuasion than testimonies or verbal arguments, let us look at some key statistics that illustrate the likely scale of the environmental impact.

THE IMPACTS OF DRILLING

According to the petroleum industry, 500 m³ of solid and between 2,500–3,000 m³ of liquid wastes are generated for every vertical well that is drilled. The figures are

---

4   The petrol industry claims there are 920 million barrels of proven reserves, whereas the French Petroleum Institute puts the figure at 846 million. We have taken 960 million, which includes Block 31, as the basis of our calculations for the amount of waste generated.

20–30 percent higher for directional (i.e non-vertical) wells.

The plan is to drill 130 wells in the ITT field. This would mean 65,000 m³ of solid wastes (equivalent to 13,000 dumper truck loads of 5 m³ each) and between 325,000 and 390,000 m³ of toxic liquids (equivalent to more than 65,000 truckloads). The companies propose leaving this waste under the drilling rig, which would result in the waste being spread far and wide with the first rains. And, if horizontal drilling were to be used,[5] the figure could be as high as 78,000 m³ of solids (equivalent to 15,600 truckloads) and between 420,000–504,000 m³ of liquids (84,000–100,000 truckloads). And, of course, if the figures are in fact double, as Sinopec's proposal outlines, the volume of waste generated would also double.

Another factor that must be taken into account is the lifespan of the wells. In the case of heavy crude, the wells collapse rapidly, making it necessary to open new wells in order to continue extraction.

IMPACTS CAUSED BY THE WATER FROM PRODUCTION[6]

Water from production is a type of sedimentary water that is the product of 150 million years of natural processing. It contains very high levels of chlorides and heavy metals. Concentrations of sodium chloride and other solids may reach as high as 100,000 ppm (miligrams of solids per liter of water).[7]

This excess of salts is crucial as it improves the solubility of other elements, including the radioactive element radium. Additionally, the temperature of the water reaches 80°C.[8] This water also contains particles of soluble hydrocarbons, as well as the chemicals used to separate them from the petrol, and to protect the installations. Amongst others, these particles include anti-emulsions, antiparaffin, and biocidal chemicals.

Based on the assumption that the combined reserves of ITT+ Block 31 are 960 million barrels, this would mean sending 8,649 million barrels (1,375,052,616 m³) of production water into the environment.[9]

However, reinjecting such a large quantity of water presents great difficulties, and may in fact be impossible, owing to the extremely high cost of the production water it would require. In the production, water would be released into the Yasuní environment, though one proposal is to send it to Shushufindi, a city that already shows signs of over-saturation due to the discharge of water from production, and

---

5   Horizontal drilling is used to get more of the oil out of the ground. Heavy oil is too sticky and slow-flowing for a vertical well to be able to extract all of the oil.

6   When oil is drilled, other substances also come to the surface. Most of this is water, known as "water from production" or "production water."

7   In comparison, the maximum concentration in sea water is only 35,000 ppm.

8   The temperature of this water normally closely follows the Earth's average thermal gradient. This is a measure of temperature increases at increasing depths below the Earth's surface. Temperatures rise between 25–30°C for every 3–6 km of depth, the depths at which petrol exploration occurs. Beneath a certain temperature oil doesn't form, and beyond a certain temperature it dissolves.

9   This calculation is based on an average of seventy-five barrels of water being generated by every twenty-five barrels of petrol, the figures generally used for heavy crude. They have been applied to Block 16, to the Eden Yuturi field or the crude from AGIP, all of which have a geological make-up similar to that of ITT.

that has the most serious pollution indicators in the country. However, even inject-ing the amount that can be reinjected will pollute the subterranean waters of the Tiyayacu Formation.[10]

Because of its composition, both in terms of the chemicals it contains and its temperature, the production water is extremely toxic for the environment once it has been brought to the surface. Most freshwater organisms are unable to tolerate its high levels of salinity, and die.

It is calculated that more than 2,000 species of fish live in the Amazonic rivers, many of which are still to be identified. The rivers are also home to a range of organ-isms that enable species at the top of the aquatic food chain to exist. The productivity of the rivers are highest in the flood areas where developed food chains exist and where the majority of the Amazonic fish lay their eggs. Thus, toxins enter the food chain, making their way right up to the final consumers in the chain: humans.

The creatures that live in the Amazon region, wild and domesticated alike, are deficient in salt—and especially severely lacking are the mammals. The salty water that is cast into the environment attracts peccaries (a type of wild boar), deer, and other animals, and as they drink, they also ingest toxic susbstances. And, on a seper-ate, but related note, soil pollution is likely to result in roots being suffocated, sucking vitality from the vegetation, and in many cases killing it.

Waste from the petrol industry contains bioaccumulative substances that in-crease in concentration in an organism or in the food chain over time. They contain carcinogens, as well as teratogens and mutagens, and so are directly responsible for numerous illnesses.

ATMOSPHERIC POLLUTION

Changes in local and global climate result from the destruction of the region's mature forests as well as from the burning of fossil fuels.

Forests, water, and climate are intimately connected to one another. Mature forests capture water, maintaining the ecosystem and the local temperature in equi-librium. Tropical forests absorb large quantities of solar radiation. Consequently, large scale deforestation results in an increase in the "shininess" of the planet's sur-face, which results in increased amounts of solar energy being reflected into outer space—a phenomenon known as the "Albedo effect." This is a fundamental effect in the control of climate warming.

Deforestation occurs for major roads, for encampments, for heliports along the route of the pipelines and alongside all the other infrastructure that the industry requires. However, the most serious deforestation is associated with the construction of roads built in order to maintain infrastructure, and the consequent settlement that such work requires.

Significantly, the petroleum industry itself is one of the largest consumers of fossil fuels in the area, meaning that its operations are a major source of atmospheric

---

10   The Tiyayacu Formation is known to be one of the most important fresh water reserves in the world.

pollution. In 2005, for every ten barrels extracted one was burnt at the point of extraction. The most serious polluter is the heaviest crude oil; burning 960 million barrels of heavy petrol would generate 460 million metric tonnes of CO_2.[11]

DESTRUCTION OF BIODIVERSITY

According to a report carried out by various scientists in 2004,[12] the Yasuní National Park[13] protects the highest level of biodiversity on the planet. This region's diversity in many different taxonomical (animal and plant) groups is outstanding, both at the local and global level. Scientists from the World Wildlife Fund, have declared the Humid Napo Forest (Bosque Húmedo del Napo) 1 of the 200 most important areas in the world to protect. The Yasuní conserves one of the largest proportions of Amazonic wildlife and has been identified as one of the twenty-four priority areas for the world's wildlife. It is calculated that it contains 165 species of mammals, 110 species of amphibians, 72 species of reptiles, 630 species of birds, 1130 species of trees, and 280 species of lianas.[14] Just one hectare of these forests contains almost as many species of trees and bushes as can be found in the entire USA and Canadian territory combined.

This enormous biodiversity owes its existence to the fact that the area, which is now the park, was a refuge for life during the Pleistocene era. During this epoch, glacier formation was such that the majority of the Amazon region was untouched and remained grassland.[15] These refuges later became the life source from which jungles could become repopulated, enabling new species to evolve.

IMPACTS ON THE HUAORANI PEOPLE

Both ITT and Block 31 are territory belonging to the Huaorani. They are also the hunting grounds for Peoples in Voluntary Isolation. As these peoples are hunter-gatherers, they move around deep within the park, roaming as far as the two petrol blocks.

This fact further increases risk. The zone in question is part of territory that belongs to the Tagaeri, Taromenane, and Oñamenane peoples who have decided to avoid all contact with the external world, and have repelled all attempts to contact them or occupy their territory. These authentic warriors are the last free beings in Ecuador. Living in so-called "societies of abundance," they produce the minimum necessary to satisfy their needs.

At the time when contracts for Block 16 were signed, the risks involved for the

---

11   Official calculations put the figures at 846 million barrels and 407 million metric tonnes of CO_2. The difference in the two sets of figures is due to the fact that it has still not been possible to incorporate Block 31 into the calculations.

12   Scientists Concerned for Yasuní National Park. 2004. Technical advisory report *Yasuní National Park's Biodiversity, the Significance of Conserving it, the Impacts of Roads and Our Position Statement.*

13   Yasuní National Park was created in 1979; ten years later, it was declared a World Biosphere Reserve by UNESCO.

14   A liana is any of various long-stemmed, usually woody vines that are rooted in the soil at ground level and use trees, as well as other means of vertical support, to climb up to the canopy in order to get access to well-lit areas of the forest. Lianas are especially characteristic of tropical, moist, deciduous forests and rainforests.

15   The Pleistocene period lasted from 2 million years ago until just 10,000 years ago.

Huaorani people were already spelled out loud and clear. Though it was proposed that great care should be taken to avoid the anticipated impacts, the results have been nevertheless been dramatic. Illnesses, impoverishment, conflicts.

It is reported that massacres were carried out in May 2003 and possibly also in May 2006, which has underscored to both society and the state alike, the risks inherent in intervening in these territories.

On 10 May 2006, the Inter-American Commission on Human Rights resolved to adopt precautionary measures on behalf of the Taromenani and Tagaeri peoples.[16] These measures entail interventions aimed at guaranteeing their rights and protecting their lives.

On the 18th of April 2007, President Rafael Correa announced that the government was adopting a policy aimed at safeguarding the survival of these peoples. Assuming responsibility for protecting their fundamental rights, he committed himself to eliminating the threat of extermination, and to defending the human rights (both collective and individual) of these Peoples in Voluntary Isolation.

RISING INSECURITY ON ECUADOR'S TRIPLE BORDER WITH COLOMBIA AND PERÚ

Comparing petrol zones with non-petrol zones, together with a consideration of where coca is produced, allows us to trace the interconnections that exist between these resources. It is a well known fact that petrol exploration requires nearly all the same ingredients as those required for processing coca leaves into coca paste and finally into cocaine. Several of the substances used by the petrol industry also make up the chemical basis for the cocaine industry. These include white gas, sulphuric acid, hydrochloric acid, nitric acid, sodium hydroxide, and potassium permanganate.

Developing the ITT fields will be directly responsible for the opening up of roads and illegal settlement, as well as other activities including cutting down forests and biopiracy. And, of course the production of illegal crops for narco trafficking. The simple reality is that the trinational border is a high risk zone.

Furthermore, this situation culminates in environmental disaster, social pressure, and extremes of violence, all of which turn the border zone into a national security problem for Ecuador. To these national security-related conflicts must be added the internal conflicts that are arising due to the state's inability to satisfy the demands of local populations.

TOWARDS A MODEL OF DIFFERENTIATED RESPONSIBITIES

While Ecuador is obviously pursuing these policies out of its own interest, it also recognizes the need for "differentiated responsibility" when it comes to taking action against global warming. In this sense, the bulk of responsibility lies not with Ecuador but with the major fossil fuel-consuming countries.

The 1992 Convention on Climate Change recognized that responsibility for global warming lay with the North, not the South. It was recognized, at the international level, that the existing economic model, based on boundless economic growth,

---

16   http://www.cidh.org/annualrep/2006sp/cap3.1.2006.sp.htm.

is unsustainable; however, by the time the Kyoto Protocol was signed in 1997, the business community had managed to distort everything. A new market, this time in carbon, was opened up.

The Convention on Climate Change established the principle of "common but differentiated responsibilities." Several key elements of this principle need to be borne in mind.

- Recognition of common responsibility for protecting the environment. In other words, the environment is understood to be a common space that is shared by a diversity of nations, peoples, and also species. In this sense, its protection requires everyone's participation.

- Recognition of the common character of the atmosphere and the fact that, until now, it has been used in an unequal manner. Recogntion of the reality that different users have different histories in terms of their specific contribution to atmospheric pollution, and that they also have varying abilities to intervene, provide support, or offer funds.

- Taking into account the specific circumstances of each country, both in terms of their contribution to atmospheric pollution and also in relation to the role they should take in tackling the problems. Recognition that a history of colonization, occupation, and appropriation of wealth has given rise to a world in which the North has capitalized itself on the back of the decapitalization of the South. Allowance is made for apportioning different degrees of culpability, and assigning different roles in order to operate within the existing unequal framework.

- Departing from the traditional understanding that only the developed countries have a global understanding, based on a recognition that the southern countries are less unsustainable then the developed ones, and an understanding that unsustainability can be measured in terms of levels of entropy.

- Posing a concept of environmental rights and obligations that lies outside the Market. However, despite this, in the wake of the signing of the Convention, intervention was scaled down to simply attempting to create a carbon market, and providing nothing but market based solutions to the problem of climate change.

The Yasuní model is based on a recognition of the above principles of common and differentiated responsibilities and they form the basis of its demand for international compensation.

THE COMPENSATION MODEL ENVISAGED IN THE YASUNÍ PROPOSAL

Various mathematical models have been deployed to calculate compensation levels. Using only industry data on the impacts and the costs of reducing CO_2 levels, calculations have been made about the rhythm of extraction and possible costs.

In order to calculate the compensation, the useful lifetime of the ITT petrol project has been calculated and brought up to present values with the aim of estimating with greater precision the compensation that should be demanded. Estimations of the present value are based on a 6 percent annual level of discount.

The value of the petrol is approximately $80 per barrel. Of this, a high percentage is production costs. Petroecuador calculates that, on this basis, the state could receive $700 million net. The initial figure demanded by the President of the Republic was $350 million over ten years.

Compensation is to be linked to the savings that would result from the reduction of CO_2, calculated as being between $5.9 and $8.5 billion, based on the present value of between $2.5 and $3.6 billion. From these figures, the president would determine an appropriate equivalent amount of global compensation.

Other referencial statistics that have been used are:

• According to British Petroleum, the marginal cost of extracting a barrel of petrol is between $2 and $7.

• According to the World Bank, it costs $20 to remove a tonne of carbon from the atmosphere. The European Emissions Trading Scheme puts the cost at $14.

• It is estimated that a barrel in the subsoil is valued at $5.

GUARANTEEING THE PROPOSAL

Many different mechanisms were explored during the first year of this initiative. Not all of them were critical of the traditional neoclassical focus, nor even the neoliberal model. For example, an attempt was made to deploy an external debt-based mechanism in order to guarantee that the oil would not be exploited, in combination with a carbon bond within the framework of emissions trading. Just as excuses are sought in the international arena, so too in Ecuador.

Finally, the Yasuní Guarantee Certificate (CGY) was designed. While attempts have been made to distance these certificates from carbon bonds, the argument that it will be impossible to get the money outside of the market, nonetheless threatens to turn them into a new market mechanism.

Yet, the Yasuní ITT proposal was originally conceived as a political proposal that sought to go beyond the market. Such a proposal needs to be replicable if it is to become an effective strategy for putting a halt to climate change.

According to the official proposal, this guarantee does not accrue any interest, maintaining its nominal value at the lowest level possible. However, should the real prices of petrol and/or carbon certificates rise in the future, the value of the guarantee will be indexed in accordance with a weighted average between the relative prices of West Texas Intermediate[17] petrol and the carbon certificates in the European Emis-

---

17   West Texas Intermediate, also know as Texas Light Sweet, is a type of oil that is used as a benchmark. It is the oil that is traded on the NYMEX. The price for this oil is what you will see in the

sions Trading Scheme market. This indexed value will only come into effect in the event that the Ecuadorean state fails to fulfill its promise to leave the petrol in the ITT block unexploited. In other words, the real guarantee that backs the plan is the sum total of the ITT field's unexploited petrol reserves.

This mechanism functions because the countries with obligations to lower their emissions recognize "no emissions" as part of their efforts to confront climate change. By investing its certificates, Ecuador aspires to compete both spatially and for resources with the carbon bonds. In order to achieve this, the Ecuadorean government has initiated a series of close relations with governments and parliaments in other countries

One of the key elements in the original proposal was the creation of a capitalization fund that could provide a permanent source of income, i.e., not for only ten or twenty years. These funds would be used to finance activities that contribute to breaking the country's dependence on petrol in order to set it on the path towards the new objectives laid forth in the constitution. Growth and development are to be replaced by *sumak kausai* (good living).

Differences and even tensions exist. There are those who, critical of the neoliberal agenda, want to ensure that this process remains outside of the market, and those who, for more pragmatic reasons, advocate opening the model up to mechanisms that are achievable within the framework of already-existing mechanisms.

These tensions are visible in the discussions that are underway about prioritizing investments in reforestation and forest conservation. However, this attempt to recycle "non-emitted" carbon, and thus make a double profit, is once again based in the old neoliberal model of emissions tradings.

It is clear that for the proposal to move forward and for the crude to remain in the sub-soil, new paths will have to be charted, at both the national and international levels. The process cannot be separated from the historical struggles of the peoples who inhabit petrol zones throughout the world. That is the cry of those who struggle to put a halt to the destruction of the atmosphere and who dare to question petrol's role at the center of the capitalist model. And, in this sense, it is these people in struggle who will define the road that lies ahead.

The initiative is part of an agenda aimed at transforming Ecuador into a postpetrol country. Such a transformation presupposes an agriculture that based around food sovereignty and the avoidance of agrotoxins, activities that involve low levels of energy consumption, deconcentration of the cities, and overcoming the cults of plastic, the automobile, and motorways.

When all is said and done, it is an agenda for *sumak kausai*.

---

newspapers as "the price of oil," although in Europe, Brent (coming form the Brent field in the North Sea) is also used as a benchmark.

Section 2

From Petrol to Renewable Energies: Socially Progressive Efforts at Transition
Within the Context of Existing Global Political and Economic Relations

The renewable energy sector has made some impressive achievements in a number of places throughout the world. Although small compared to other branches of the energy sector, the sector is now standing on its feet, in terms of both technical and organizational capacity, and is ready for a massive global expansion.

It has required a great effort to bring the sector to this level. Several major obstacles have hindered a rapid global expansion of the sector. There has been immense opposition, coming both from governments, the existing energy sector and also at times some sectors of organized labor. The renewable energy sector has been grossly discriminated against in terms of the unequal subsidies that the fossil fuels and nuclear sectors receive.

However, despite these formidable obstacles, certain experiences in renewable energy use have simultaneously resulted in a high level of renewable energy capacity and use, and also shown a path of community empowerment, autonomy and energy sovereignty at least on a local level. These are indeed "best practices", as they are frequently referred to in the renewable energy sector itself. Such "best practices" have occurred in a number of different renewable energy technologies, including wind, biogas, solar, small scale hydro and other energy sources. These experiences have emerged in both high wage and low wage countries, core and periphery. And, in parallel to efforts to develop "best practices" there has also been an effort to diffuse these practices around the world. It is essential that knowledge about these experiences is disseminated as broadly available.

This section documents these positive experiences in some detail, in order to pave the way for subsequent chapters in the book which attempt to develop a more sophisticated analysis of the conflicts that are emerging as the renewable energy sector expands globally. These conflicts are making it increasingly apparent that there is in fact no such thing as a "national model" or easily replicable "best practices", but rather a complex process of global commodity chains. However, it is crucial that we nonetheless understand these "success stories", within the context that they emerged, in order to defend their gains (which are increasingly coming under threat) and also to ensure that the experience gained from these stories is successfully harnessed by an emancipatory transition process, rather than having them squashed in their infancy and co-opted for other ends.

THE EMERGENCE OF THE WIND ECONOMY IN GERMANY
Land as a Reusable Resource[1]

Klaus Rave

A lot of activity in German wind energy began in Schleswig-Holstein, and as such, much can be discovered and learned from this land between the seas. Today, Schleswig-Holstein still describes itself as the leading wind province in Germany, even if other federal states have built up a greater capacity. The beginnings go back to the 1980s, and three factors contributed substantially to this success: the given geographical conditions, economic developments, and political decisions. Nowadays, the use of wind power for electricity generation is no longer a marginalized sector—a wind economy has emerged.

STARTING FROM THE EXISTING ECONOMIC CONDITIONS

Industry in the north was concentrated on maritime technology, until the 1973 oil price crisis caused contraction in the shipyards. The shipyards epitomized an "old" and heavily-subsidized industry with few future prospects, though a strong connection has always existed between shipbuilding and shipping, and wind and waves. Traditionally, many service providers and suppliers have developed around the seafaring business, including financiers, classifiers, and insurers—long-established local services with a global business perspective. There were not only shipwrights, but also farmers, and rural agriculture, with a large number of full employment businesses that characterized the other, rural side of the region.

The continuing conflict with nature was, and remains, a common feature of both activities: the force of wind and water. Technical solutions developed to deal with natural resources were part of everyday working life, and had to be in order to access, develop, or use natural energy sources. Neighborhood and historical traditions played a further positive reinforcing role, like in Denmark and Friesland in the Netherlands, where there are thousands of centuries-old windmills for milling grain. And from a very practical viewpoint, maritime operating engineers, who usually worked alone, were in great demand ashore due to their ability to improvise.

Many farms close to shore had long been dependent on energy generation plants of their own. They were frequently pioneers both in agricultural areas as well as in

---

1   This article was previously published in a book edited by Hermann Scheer and Franz Alt, *Wind des Wandels*, published by Ponte Press Verlags GmbH, Bochum. It is being reproduced here with permission from Hermann Scheer and the author. It has been translated from the original German by Ann Stafford.

energy supply, and they lived on good land that produced high yields, allowing them to finance larger investments, and get larger loans from banks. Their heavy soil was known as "*klei*" (clay) in the local dialect, and those who had "much *klei* on their feet" were considered, in the North, to have a lot of money. Thus, capital was an important factor in dealing with nature and technology. The soil was a reusable resource: for cattle, for cabbage, for grain—and, as for windmills, they offered security for the bank.

CONSEQUENCES OF LOCAL POLITICS

There is a need to remain realistic about windmills in local affairs. Schleswig-Holstein has many villages, including a surprisingly large number of small, politically independent municipalities. Altogether, approximately 780 of the approximately 1100 municipalities have less than 1,000 inhabitants. These municipalities are managed in a voluntary capacity. Full-time administrative services are provided for them by ministerial or district offices. They possess depots for the fire brigade and are members of village school administration boards, and the mayor has a house with a coat of arms. These municipalities have financial reserves and are free from debt. Voter communities determine the municipal elections, and political parties are rare. That politics provide local identity and there is less real or ideological competition adds another favorable factor.

THE ENERGY DEBATE

For decades, there have been extremely intense controversies over energy policy in Schleswig-Holstein. These controversies were triggered in the 1970s by disputes about the use of nuclear energy. The Danish "*Nej tak*" ("nuclear power, no thanks") was widespread in the SPD (Social Democratic Party) in Schleswig-Holstein, and led to disputes with its own SPD Federal Government under Helmut Schmidt's "Stop Bonn's nuclear program!" The intense energy policy debate at the regional level also had repercussions at the federal level.

THE BEGINNINGS

The first large wind park—big by the standards of the time—was a result of the regional energy debates. It was built by Schleswag and HEW and inaugurated by then-prime minister of Schleswig-Holstein, Uwe Barschel, in September 1987. A piece of energy history, HEW was a proud and independent Hamburg power supply company. The Preussenelektra AG had a majority holding in Schleswag, a regional provider, and the province and the district administrations also had holdings. The latter had actually founded the company as a pure network operator. Today, the companies are Vattenfall and E.ON Hanse. These two shareholders, the legal successors of the past companies, illustrate the rapid change. The West Coast wind energy park constructed thirty plants of 25, 30, and 55 kW capacity, totaling 1 MW. Later, the field was expanded and altered. An information pavilion was built and continues to operate. An important beginning was made despite, or because of, controversial debates on energy.

Such controversies also occurred in the immediate neighborhood where the GROWIAN—a large wind plant with wing diameters of 100 meters and pendular hubs—was set up. Though an over-ambitious research project, it nonetheless provided many important findings, and it made the Kaiser Wilhelm-Koog the most surveyed piece of land on the wind maps of the world. Thus, an important prerequisite was created for setting up a testing field and the Kaiser Wilhelm-Koog Windtest company was established. Its now more than fifty engineers have made an essential contribution to the development of wind energy use. Today, the company also has branches in Spain and the USA.

Thus, science and technology came to the rural areas in the form of cooperation with the Windtest, which was very important for the successes of renewable energy. A division between the rural population and development engineers was successfully prevented. The corporation itself is the greatest employer in their communities, and a successful and popular host for delegations from all over the world—Chinese and Russian guests were amongst the early visitors, and provincial and federal prime ministers have always enjoyed going there. Another factor in their success is that they have repeatedly directed attention towards new developments, and promoted acceptance of the competence and potential of wind power.

A SPECIAL KIND OF TECHNOLOGY TRANSFER

A further positive aspect of the "failure of GROWIAN" has to be mentioned because it still affects widespread promotion: initially, large mechanical engineering companies were successful in securing subsidies, as is frequently the case in Germany, and their development engineers had much scope for experimentation. Metaphorically speaking, while tractors were manufactured in Denmark for generating electricity, Porsches and Goggomobiles (microcars), with numerous cylinders and diesel or petrol engines, were simultaneously built in Germany. Another innovative product, and a new design that was more important than its direct economic success, indicates that ENERCON from Aurich is a great exception. Run by Aloys Wobben, the company's founder and owner, it consistently and continuously develops new products.

When the German subsidies ran out, electricity generation could not match Danish competition, causing the companies MAN and MBB to terminate their production. Their engineers began to seek new jobs in this exciting field with a promising future. Thus, excellent engineers also came to the not so industrialized Schleswig-Holstein, to companies and sites that they would have otherwise considered as too remote. For example, in Husum and Rendsburg, medium-sized companies such as the Husumer shipyard and the Jacobs Energy company seized this opportunity. Engineering companies like Aerodyn have become world-renowned specialists. Consequently, research funding had a sustainable unintentional secondary effect, which is another significant element of success for the development of wind economy in the region and the rural areas.

There is a question hanging over medium-sized companies that are embedded in the rural areas: will such perspectives remain viable given the consolidation of the manufacturers of wind power plants that is taking place at present? This consolidation is driven by increasing project size and growing and diversifying international markets. Production sites will be relocated. However, construction, development, and, especially service, will remain local. The latter is particularly labor intensive, providing for 1–2 permanent employees per installed MW. Within the framework of a new global division of labor between construction, production, maintenance, and accompanying services, a success factor of the future could especially consist of skill enhancement and advanced vocational training, pilot applications (in relation to the overall value added chain), as well as continuing education contributing to competitiveness.

ISLAND SOLUTIONS

Special pilot applications exist—and always have—on the region's islands. Except for a small amount required to cover local needs, the islands Halligen and Sylt are indeed free of wind generated electricity. Special projects were developed on three islands.

Fehmarn is a favourable location for agricultural businesses with high yields, community-owned wind parks, and the largest re-powering project carried out till now. There is also plenty of sunshine. Clouds signify wind, and sunshine means no wind: a frequently observed natural phenomenon. The first hybrid project in the world was initiated on this island, which, like Freiburg in Southern Germany, is very sunny. This is a combination of wind with photovoltaic cells and gas from sewage.

The innovative large wind plant project on the island of Helgoland was less successful, and lightning found this first offshore high tower off the mainland extremely attractive; after the second lightning strike, the plant could no longer be insured under economically-acceptable conditions.

However, a quite individual path was and continues to be taken by the island of Pellworm, a marshy island near Föhr and Amrum that is characterized by agricultural farms. In the 1980s, a few small wind power plants were being tested there, as well as the first solar panel field. The potential of renewable energies was continuously explored and was embedded in the form of future agricultural farms, as well as sustainable tourism. Today, the island is not only self-sufficient, but it can even export surplus electricity, and generate income. A large, modern community-owned wind park was set up; the solar panel field was modernized by E.ON Hanse; a new biogas plant was put into operation; and thermal energy even heats the public swimming pool—to illustrate the sustainable energy supply and diversity in Schleswig-Holstein.

EFFECTIVE SUPPORT

All of this would not have been possible without funding. Then newly-elected prime minister of Schleswig-Holstein, Björn Engholm, and the minister for energy, Jansen, who supported a clear abandonment of atomic power, established an investment program. Wind power plants could be subsidized with up to nearly 25 percent of the investment amount. The Federal 100 MW program, later 250 MW program, subsidized the price per kilowatt hour—first at eight, and later at six pfennigs in a broad test. Additionally, in 1992, the provincial government formulated a clear target for its energy strategy, a first for Germany. 1,200 MW was to be produced by 2,000 plants creating approximately 25 percent of the electricity supply by 2010. The provincial plans extended their target to 1,400 MW, and like the target for the mainland wind energy sector, this objective was not only met, but exceeded long before 2010.

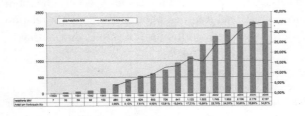

Figure 1: installed MW Proportion of consumption (percentage). Sources: Statistics Agency for Hamburg and Schleswig-Holstein. Ministry for Science, Economy and Transport of Schleswig-Holstein, German Wind Energy Institute.

NEW CONFLICTS

Two important components were promoted by specific funding policies. The quality of the sites was systematically improved, then a fixed amount of DM 800 could be acquired, and in exchange for funding, and the qualified applicants committed to provide data to specific institutes.

Schleswig-Holstein developed the first real wind map. The data of all registered wind plants and sites is published each year. More transparency is hardly possible; calm periods, good and bad years, the chronology does not lie. Creation and availability of this information is also a characteristic of the region and its intention to use wind power efficiently—they've became known for electricity generation, not tax evasion.

A quite central aspect of the funding policy needs to be emphasized. When it came to concrete planning, local councils were also skeptical at first. Who should have how many wind plants in which areas? Decision making was often endangered in the smaller communities because too many landowners wanted to build a mill and therefore were biased in regard to their voting. And when it wasn't just the landowner who was biased, it was the father, sister, or son and daughter. As a result, plans could be made but not carried out. Conflicts were public. Sometimes these were between local inhabitants and owners of second homes, sometimes, of course, between nature conservationists, in the narrower sense, and environmentalists with a greater perspective—i.e., between micro and macro ecologists.

People in the best-located villages, and on the Baltic island of Fehmarn—where building land could be expected to become wind-turbine-land—asked themselves, "Am I sitting in the first or second row?" We all know the expression "Taking the wind out of the sails." If the wind energy plant was prevented, no wind could move the propellers, and economic success would be threatened. The answer came from the discussion on the best use of subsidies. Was it possible to save money if several plants were linked together to a wind park and the infrastructure costs were shared? Would it make more sense to concentrate on the best locations and optimally configure the wind energy plants? Could the federal and provincial subsidies be combined as well as suitable plants? Farmers who knew their fields well also learned more, for example the degree of field effectiveness.

COMMUNITY-OWNED WIND PARKS

And so, the community-owned wind park concept was born. One's own land, one's own value creation, one's own business tax: individual choice aligned with community value. Symptomatically, this development started in the landfill areas, in soil that, for the most part, was not even 100 years old. It was not only farmers and landowners who joined together there to invest in a wind park; all inhabitants were invited to participate in an association. Five, eight, ten wind energy plants, investment amounts of several millions were planned and carried out.

Of course, the association had to place their tax location in the community and generally became the largest business tax payer. Community identity was strengthened. Something new had been ventured. New economic ways of thinking and behaving were triggered by the discussions concerning these investment plans. New employment possibilities arose: maintenance work is always local, even though the wind energy business has now become global. The considerable resignation that existed with regard to the future of agriculture in general, and of the milk farmers in particular, was at least partially reversed, and transformed into a positive perspective. The ongoing flight from the land was at least partially stopped.

According to agricultural consultants, and also local banks and savings banks, those regions and the farmers who invested in wind energy at a very early stage are today expanding their farms and making large-scale investments for their future. Agricultural farms are developing positively in these regions, with re-powering already playing a role to some extent, and in the very favorable locations—Fehmarn and the polders, as well as other small areas—utilization needs are optimized. These areas have experienced sharply rising investments: four times 2 MW, instead of six times 300 kW, for example, or even eight times 3 MW, instead of twelve times 600 kW. Rural liquidity is available, locational quality is assured, profitability has been improved, and the range of participants has diversified.

The Schleswig-Holstein map has changed, considering the economic dynamics of rural areas. Many citizens are convinced that wind energy can contribute to interrupting or even reversing flight from rural areas. Even those who were at first skeptical now admit that the rural areas would have been depopulated without wind energy, and that the credit supply to the rural population by a dense network of savings banks, peoples' banks, and Raiffeisen banks would also have been threatened. In view of the precarious agricultural future resulting from Brussels, Berlin, and Kiel, new business opportunities have only been provided from wind power, and now also biogas. The model of community-owned wind parks has made an essential contribution to strengthening acceptance of wind power.

"COMMUNITY OWNERSHIP"

Schleswig-Holstein has developed a model that is also generating increased interest at the EU level, and "Community Ownership" is now a household name. Certainly, a defining factor of the success, on the local and global scale, is that further developments of this concept, which take place in Schleswig-Holstein, are created by potential operators, agricultural consultants, and committed sponsors of the provincial government.

COMMUNITY-OWNED WIND PARKS AT SEA TOO?

As already shown, community-owned wind parks have a long and positive tradition in Schleswig-Holstein. Can this model also be transferred to the sea? Or does the era of rural participation end with offshore plants? Are offshore plants even a threat to the necessary change in energy policy for which wind power expansion is a strong catalyst? Nine committed west coast windmill owners with considerable experience in the development of community-owned wind parks pondered this issue and initiated the Butendiek project. The objectives were clearly defined:

- Construction and operation of an offshore community-owned wind park in the North Sea, approx. thirty-four km west of the island of Sylt;
- Eighty wind energy plants of at least 3 MW each, for a total of at least 240 MW effective output;
- Annual energy yield of approximately 800 million kWh (with 3 MW plants);
- 200,000 private households to be supplied by Butendiek with "green" electricity. (This represents almost all of the households in the provincial districts of Nordfriesland, Dithmarschen, and Schleswig-Flensburg.)

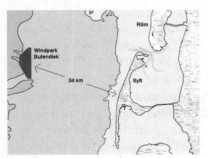

Figure 2: The distance from the island of Sylt is considerable: no infringement of the protected area of the national park.

The main idea was and remains: to achieve high acceptance of the wind park by the integration and participation of as many citizens as possible from the very beginning. As mentioned above, the nine founding initiators from the Nordfriesland (North Frisia) region all had long years of experience in the organization and management of wind parks. Consequently, they received the necessary trust to be able to raise the required venture capital for the intricate planning phase. A further 8,412 shareholders participated, and of these, over 4,600 were from Nordfriesland. They subscribed DM 500 shares with the associated option of a participation in the actual project phase.

Commission fees and other soft costs were strictly avoided so that the added value would remain, to a large extent, in the hands of all shareholders and thus remain predominantly in the region. The contractors themselves directly solicited funds at numerous events in all regions of the province, but primarily on the west coast and in the polder region. Nobody would have expected that the people from Dithmarschen and North Frisia would have been able to raise thousands in venture capital. It was a double mark of confidence: confidence in wind power as a growing economic sector, as well as in the group of nine people themselves. It is also confirmation of the theory that engaging with the soil and the sea leads to common positive effects: agriculture and maritime economy as the core of a new dimension in wind power economy.

The following is a brief account on the special geographical and geological features:

- Water depth: approx. 17–20 m.
- Soil structure: sand and gravel.
- Deep foundations with mono-piles (large steel tubes) that are fixed in the sea bed at a depth of 30 m.
- Exact test results from six drilling operations and thirty pressure probes are available.
- Wind speed: approx. 9.7 m/sec at 80 m of hub heigh.t

A number of specific ecological questions had to be assessed. The starting point was that the use of wind power means active climate protection, which is a macro-ecological statement. But it is also necessary to provide the answers to micro-ecological questions: an environmental impact assessment of sea bed creatures (Benthos), fish, migrating birds and non-migrating birds, harbor porpoises, and seals was carried out from 2000–2003, as well as numerous airplane and ship counts in the planned wind park area since the winter of 2000. The results: no substantial impairment of these protected subjects by Butendiek; compatibility with the targets of Natura 2000; discussions with nature conservation organizations beginning in summer 2000; extensive monitoring during the construction phase; several years of monitoring during the operation of the wind park.

An intensive examination was also carried out concerning the problems of shipping and fishing, and an extensive risk analysis by German Lloyds was commissioned. The result was that risk of collision and damage was classified as very low for ships and the wind park, since there is little traffic in this area. For the wind park itself, extensive safeguard measures were defined (radar, lights, AIS [automatic identification systems], etc.), and fishing is not permitted in the wind park area (losses are countered by fish stock recuperation).

Grid connection was a special challenge from the beginning, in view of costs as well as planning and realization. The German law on the acceleration of infrastructural planning provided the decisive breakthrough: the law stipulates that a network operator must invest and configure adequately. The planning goes on, and is coordinated with other offshore plants nearby. The technical feasibility of feeding into the 380 KV grid has already been confirmed by the network operator E.ON. At present, negotiations on commercial and technical modalities are taking place.

The power supply of the island of Sylt could be provided with "green" electricity—surely not a bad sign for an island threatened with continuous land loss, and increasing storms of greater ferocity as a result of climate change. As well, a new 155 KV cable improves the security of supply for the island.

In December 2002, the planning and building permission was granted by the BSH—the Federal Maritime and Hydrographic Agency, since the planning affects a foreign trade zone—and the negotiations with the companies involved were largely completed by the beginning of 2006. However, price increases for steel, raw materials, wind energy equipment, and more sophisticated demands for collaterals and guarantees by the banks have impeded a profitable realisation. Thus, a strategic partnership with Airtricity was formed in spring 2007, to which the Butendiek shareholders agreed by a 97 percent vote in June 2007. Airtricity, an Irish project developer for green electricity, assumes the responsibility for contract negotiations, financing, construction, and launch of the wind park. Butendiek has the possibility of participating up to 50 percent in the wind park. The objective remains: joint operation of the wind park, with each partner owning a 50 percent share.

This is the current state of a very special maritime story. On land, as well as at sea, the use of wind power leads to a new exemplary challenge with regard to planning requirements. It advances land recovery from the seas, with an ecological, as well as an economic, perspective. Offshore wind power plays a major role in maritime policy, especially since a green paper is just being redrafted at the European level.

MULTIFACETED FINANCING

Sponsoring institutions, including the development bank and the investors' banking house made constructive contributions. Some of them applied innovative approaches, as, for example, the Commerzbank in Brunsbüttel, whose then-branch office manager, Hugo Denker, had already used favorable interest rate financing from the KfW (Credit Institution for Reconstruction) at the beginning of the 1990s. This credit was intended for air pollution control measures. Surely wind power also contributes to clean air, Denker said, and so secured the participation of the branch, making Commerzbank the largest financing body of wind energy, with a portfolio of over €3.5 billion (with the support of the Hamburg branch office, which had some different ideas).

Thus the idea of a community-owned wind park was born very early in Schleswig-Holstein, both on the west coast and on the Baltic Sea island of Fehmarn. Along with the construction of smaller private wind parks, the public could also participate in wind parks that were brought to market only regionally. In some areas this has changed the conditions of generating income to such an extent that wind energy is now the main source of income.

The first big wind park of the Husum shipyard was merchandised nationwide in 1988. Uwe Niemann, the managing director of the shipyard, really knew more about financing a ship than a wind park, and consequently, he applied that financing model to the wind park. The resulting success of this plan made it possible to fund the biggest wind park in Europe, and also reduced people's reservations about the economics of wind power. For the shipyard, it was a kind of conversion measure. This was momentous for the further development of the wind converters as the serial production within the framework of semi-industrial manufacture was a significant signal. They got approval for the "wind fence," which stretches along the North Frisian coast up to the Hindenburgdamm linking the mainland with the island of Sylt. Fifty-two of the new class of 250 kW plants have been constructed there, and stretch out like a string of pearls. This pleased some, but also annoyed many visitors who travel with their not very small cars on the train to Sylt. Some journalists from Hamburg have already made this journey and processed their often very personal impressions on wind power in not very knowledgeable, not to mention prejudiced, articles on this way of generating electricity.

PLANNING TO AVOID CONFLICTS

The struggle of many civil society initiatives against wind energy should not be concealed; many operators have ignored the legitimate wishes of the population. For example, the noise emitted from a wind energy plant on one's own property may be music to the ears of the operator—he is likely to even wake up if the plant noise stops at night—but the neighbors are considerably disturbed if certain rules are not observed, which means these rules are absolutely necessary. Significant doubts were also expressed by conservationists and bird protectors, whose claims, however, proved-to-be mostly wrong. Nevertheless, the doubts could only be dispelled after extensive testing, which did occur. And some plants were built in locations that would never be permitted today.

The potential for conflict was taken very seriously and—for the first time in Germany—designated areas for wind parks and individual plants were drawn up. The company WINDTEST Kaiser-Wilhelm-Koog coordinated the work of some districts. Together with the responsible authorities at various levels, as well as stakeholders and affected persons, it was possible to designate areas that avoided potential conflict.

Chapter 21 ▌ Part 5

AN AUTHENTIC STORY ABOUT HOW A LOCAL COMMUNITY BECAME SELF-SUFFICIENT IN POLLUTION-FREE ENERGY FROM THE WIND AND CREATED A SOURCE OF INCOME FOR THE CITIZENS[1]

Jane Kruse and Preben Maegaard

SYDTHY MUNICIPALITY

The municipality of Sydthy is a district of lovely landscapes. The 322 square kilometers between the North Sea and the Fjord offer an unusual variety of landscape, characterized by tracts of blown sands in the west, and lush, rolling hills in the east with a large number of tumuli that bear witness that this is an area where people have been cultivating the land for thousands of years. About 11,800 people live here.

Sydthy offers more space than most other places. The average population density is as low as 37 persons per square kilometer, compared to 122 in Denmark as a whole, but the households are larger here than in other places, averaging 2.4 persons.

It is a real rural community. Only half the population live in towns or villages, and farming is still of central importance, which means that a culture of independence dominates the lives even of those who work for wages.

The average income for those with employment is €26,300. In Denmark as a whole, the average is €28,600. This part of the country has always been frugal, but perhaps the quality of life is above average. That can hardly be measured, but the frequency of theft and violence is, certainly, significantly lower than in most other areas in Denmark.

WIND POWER IN SYDTHY—A SUCCESS STORY

The 145 windmills that are harvesting energy out of the almost permanently blowing winds place Sydthy in a class of its own when speaking about energy policy. The majority of the wind energy comes from 200–300 kW units but some of the newest windmills belong to the 600 kW class. By 2002, megawatt-size windmills had not yet been installed in Sydthy, but they will appear in the coming years as part of a repowering program, during which small-size windmills, initially up to 150 kW will be replaced by large megawatt windmills leading to a significant increase in the energy production from the wind.

A large majority of Sydthy's windmills are scattered throughout the agricultural

---

1   This article was originally written in August, 2002. It is being reproduced with the authors' permission. What has happened since then is discussed in Chapter 43, "Denmark: Wind Leader in Stand-By Politically-Induced Paralysis in Wind Power's Homeland and Industrial Hub."

landscape at sites that are well known as good wind resources. Out of the 145 units only 20 were installed in regular wind farms in geometric patterns. This is the preferred solution among the central landscape authorities, but it is generally criticized by the local residents because of the windmills' distance from the owners, and their dominance in the landscape compared to the existing, more dispersed ones that compliment the contours of the landscape and the location of the farm buildings.

Before installation, the wind potential is carefully investigated by "the wind atlas method." Guaranteed electricity production by the windmill supplier is often achieved within 5 percent of the predicted annual production. This in itself provides high confidence in the investment from the side of the owners and the financial institutions.

There are hardly any areas in the world that can show such massive utilization of the power of the winds. The windmills produce more than 100 percent of the power consumed in the area, a feat that has only taken a few years. It is no more than 20 years since the first modern windmills were built by experimenting master-smiths.

The following account offers an overview and explanation of this revolutionizing development, which not many people would have imagined possible.

Sydthy, situated between the sea and the fjord, is one of the most windswept areas in the country. However, you could easily point out other areas equally favored by the winds, where the exploitation of the energy is much less intensive. Other, and more complex, explanations are needed. In order to evaluate them, it is necessary to move beyond the horizons of Sydthy. Sydthy can be seen as a positive example of wind energy where the energy policy of the government and Folketing (parliament) is combined with an unusually high degree of popular activity.

One has to investigate to what degree NIVE (a local energy organization) and the Nordic Folkecenter for Renewable Energy have played a part as initiators and mediators, as well as the role that the local power utility has played as partner and opponent. The local and regional planning authorities became decisive agents, not least during the 1990s when the windmills' capacity and size developed rapidly.

One might see Sydthy, which utilizes a high share of wind energy by exploiting the prevailing natural energy resource, as the future laboratory for wind power. The introduction of windmills has not caused local conflicts and them being rejected as it has in many other local communities. In those cases, residents have protested strongly against this new form of energy technology and blocked the progress from atomic power and fossil fuels, and towards the clean renewable energy solutions of the future.

In contrast, a 1996 opinion survey, based on interviews with almost 1,000 residents and representatives of the local population, clearly demonstrated massive good-will toward wind energy. Of those polled, 80 percent expressed a positive attitude to the local windmills. Especially surprising was that people living closest to the windmills were the most positive. The negative minority primarily consisted of senior and retired citizens in the towns.

The conclusions of the investigation were quite clear: ownership and direct economic participation in the windmills create a tolerance to their visual impact in the neighbourhood, which is significant.

Because the sympathy increases the closer you live to the windmills, we can observe a clear indication that, in order to obtain a high share of wind energy, involvement by joint ownership paves the way for maximum utilization and a transition to renewable energy without causing local conflicts.

However, by the turn of the century the region was fighting against a number of new problems that other areas will also experience when it comes to realizing the national targets for wind power capacity. One question in particular becomes urgent: how do you resolve the conflict between the aesthetic impact on the landscape and the demand for a continued growth in renewable energy sources when you demolish relatively-small, community-owned windmills and replace them with megawatt machines with predominantly single or non-local ownership? This is a process that clearly distorts the previous well-balanced economical and ecological structure in the neighborhood.

THE DANISH WINDMILL TRADITION

As I rode my bicycle about in Northern Jutland on my lecturing tours before and during the last war, it was impossible to avoid noticing the many windmills on the farms. The farms were self-sufficient in electricity. At that time I did not know that this state of things originated in an idea issuing from the folk high school of Askov, and that it was not only a technical issue, but that a far-reaching social idea behind it: Giant business corporations must never be allowed to monopolize the power production. It should be taken care of in small local communities and on the individual farms.

This is how folk high school professor, Richard Andersen, saw the landscape of Jutland a little more than half a century ago (in the preface to H.C. Hansen's *Forsøgsmøllen i Askov*, 1981). A statistical handbook from the beginning of the twentieth century tells us that 35,000 "wind engines" were registered on Danish farms, to which number should be added 2,000 grain mills. The classical Danish landscape was very much characterized by mills.

The special Danish windmill tradition originated with scientist and Askov folk high school professor, Poul la Cour. From 1891 on, he conducted extensive research and product development in the field of practical utilization of wind energy. The first experimental mill was built, with subsidies from the state, at Askov in 1891, and as early as in 1895, Askov was illuminated by means of wind energy—certainly a breakthrough of worldwide dimensions. In 1897, a new and bigger experimental mill was built—still in the "Dutch" style, like the old one.

From here the movement began to disseminate, and from the beginning of the twentieth century almost all larger farms in the inhabited Danish landscape were equipped with a *klapsejler*, or wind engine (a windmill a system of adjustable, narrow, horizontal slabs made from wood for the blades).

The windmills delivered mechanical energy for grinding, threshing, pumping water, and also for the production of electricity for lighting and radios. This resulted in an enormous improvement of rural life. The windmills were able to provide nearly all the conveniences that otherwise could only be satisfied in the cities.

To meet the needs of installation and maintenance of the new energy source, Poul la Cour organized the education of rural electricians who became very valuable in the ongoing modernization of Danish agriculture. For some decades, prosperity and welfare improved, making the rural lifestyle attractive compared to neighboring countries that did not offer similar opportunities for the rural population. Importantly, the windmill was a key factor in this development.

After a fire and re-construction in 1929, the Askov mill worked on until 1968, the year when so many old things were discarded. It was also in the 1960s that the farmers effectively stopped maintaining the iron windmill constructions and pulled them down.

The history of the *klapsejlers* has a special Sydthy angle, as the foundry in the village of Hurup (Hurup Jernstøberi) was one of the country's major producers of these windmills. It produced no less than 1,000 of these proud farm mills.

In 1929, writer Poul Henningsen, wrote a goodbye poem to the "wind engine:" "No one can avoid the evening of life, the times are changing for the motor power. Everything has its chance, and you have had it." The power station produced electric power, the petrol engine was triumphant, and few people thought that wind power had any future.

Among the few people who went against the spirit of the times after World War II was J. Juul. In 1951, this engineer started full-scale experiments, first with a double-bladed 11 kW windmill, then in 1953, with a three-bladed, 45 kW asynchronic generator for alternating current—the Bogø. By 1957, his research and innovative ideas resulted in an experimental and extremely-successful 200 kW windmill in Gedser. Demonstrating high reliability and efficiency, it was in continuous operation until 1968.

At the time, nobody realized that this was building a bridge to the future. However, his epoch-making principles of construction are in fact the experimental point of departure for the pioneering work on windmills that began in the 1970s.

THE BIG ENERGY CRISIS

The historical turning point was the 1973 energy crisis, which caused something like a shock to the Danes who had grown used to a life of affluence. At the same time, the debate on utilizing nuclear power in Denmark worked as a forceful stimulant for bringing alternative energy sources onto the agenda. The slogan "sun and wind" made it possible for the many people who were active in the movement against nuclear power to say not only "No," but also "Yes" to an alternative.

The energy crisis caused two scenarios to emerge. One was the movement "from above," originating from government and legislator initiatives, seconded by research at the atomic power experimental station, Risø, and mastered by the big central power stations.

But at the same time a movement "from below" arose. This was rooted in a new popular awareness of energy and environment. Experiments were made and experience was eagerly exchanged during the latter half of the 1970s, especially in Central and Western Jutland.

While the media especially singled out Tvind School's giant, and still operational, 2 MW windmill in Ulfborg for their praise and attention, many other initiatives were also in the offing. The first steps towards commercial production were taken around 1978, and in the following years a quite new, dependable, and distinctly "Danish design," concept emerged. During the 1980s, the mills returned to the inhabited Danish landscape.

Wind power utilization reached a popular level that went far beyond the planners' calculations. In 2002, wind power represented a total capacity of almost 3,000 MW, including off-shore wind energy, which will gain increased importance. The goal of the energy plan was more than fulfilled by that point, as the national target was originally 1,500 MW, by 2005.

Around 90 percent of the windmills in Jutland were built by private customers for energy distribution. So, by 2002, nearly 20 percent of the country's electricity consumption was coming from wind energy, a much higher proportion of which was west of the Great Belt, which divides Denmark into two separate electricity systems without connecting cables. In the western part of the country independent power producers representing cogeneration, wind energy, biogas, and so on deliver 60 percent of the electricity needs, replacing coal power from central utilities. The bulk of this share has been achieved in less than ten years and is of historical significance.

An important cause of this growth, which had hardly been anticipated at the end of the 1970s, was the guaranteed minimum price system of pollution-free energy. In the original legislation, the leading principle had been that windmills should be owned by people living in the mill's neighborhood, and that private individuals could only own shares in them corresponding to their household's private consumption. Farmers were allowed to install one windmill on their property. The intention was to create broad popular involvement and local ownership in the development of Danish wind energy.

Today, this perspective may be less striking. The 1993 tax reform favored mills owned by individuals and gave less favorable conditions to those owned by a community. Furthermore, it became possible to buy a tiny piece of land suitable for windmill installation and add it to one's own property, resulting in the loosening of the rule saying that you should live near the mill.

The maximum size of shares has been raised from 9,000 kWh per family to 30,000 kWh per household member, aged eighteen and over. As of 2001, there is no regulation of ownership. Anyone, including investors from abroad may own windmills in Denmark in accordance to globalization and liberalization policies. All this has led to windmills becoming strictly investment projects.

WIND POWER AND COMMUNITY POWER

In a process running parallel with the government and power-utility-based initiatives, grassroots, do-it-yourself people and master smiths joined in a common task after 1973. The job, both idealistic and business-conscious, was to develop mills. By and by, this joint effort came to form the foundation of the present, globally-oriented windmill industry.

Seeing the standardized and elegant windmill concept that we have now become used to, it may be difficult to imagine the diversity and insecurity that reigned in the mid-70s. A long series of technical options had to be tried out, and many disappointments experienced.

A broad exchange of experiences and open access to information were decisive conditions for the gradual shift toward functional and efficient mills. Engineer Juul's experimental work during the 1950s contributed decisively to turning the development in the direction of what later became the special Danish concept. It was, however, necessary to learn about his experiments from the United Nation's renewable energy conference reports from 1960.

During the bi-annual windmill sessions, initially arranged by the Organization for Renewable Energy (OVE), lively discussions and comparisons took place, contacts were made, and strategies and initiatives were decided. It was possible to share the experiences of many experimenting windmill builders, inventors, and other creative people who were contributing to the development of the emerging wind industry.

Here we find the incubator that overcame research and developmental challenges that large professional laboratories and corporations did not have the practical and economic tools to solve. The early sessions, which were to be of decisive importance in the course of the technological development, were coordinated by Preben Maegaard, chairman of OVE and later director of the Folkecenter, and his workmate in OVE, Lars Albertsen.

A key question was how to get a real and professional equipment manufacturer going. The Tvind School people had stipulated that their findings, however important, were not to be utilized for profit. But NIVE (the local development organization), represented by Preben Maegaard and Ian Jordan, was eager to find ways to make industrial production of windmills possible. Their aim was to stimulate a regular serial production by involving the mechanical industry and organizing consortiums with the required production skills in already-existing, small- and medium-sized companies that were especially motivated to enter into the emerging renewable energy sector. Instead of building on a total concept (e.g. the Riisager Mill, produced from 1976): NIVE saw that it might be possible to produce industrially by going in the opposite direction, by seeing the mill as a number of components coming from a variety of industries like tower building, fiberglass, electronic controls, machinery, etc.

The Danish Blacksmiths' Association showed an especially serious understanding of this manufacturing concept and successfully transfered technical knowledge and know-how within its membership of 2,000 independent companies. Twenty-five years later, the sector is still benefiting strongly from their role as supplier to the windmill industry.

THE COOPERATIVE WINDMILL—A CASE STUDY

During the mid-80s, people began to form mill cooperatives (guilds) on a shareholder basis. The Helligsø windmill cooperative, Simonshøj, may be seen as an example of this bottom-up movement.

The cooperative was formed in March 1988, of an initiative that came from a local teacher, Bjarne Ubbesen. At that time there were only two major mills in the area, but Ubbesen was inspired to start his work by taking part in meetings of people who took an interest in windmills.

The driving force was not a dream of economic gain—the enterprise was quite insecure actually. What they wanted was to produce pollution-free energy. According to calculations, a windmill could be called "pollution free" when it had been in operation for one year, in the sense that the energy production had by then made up for the consumption of resources necessary to build the mill.

Bjarne Ubbesen gathered people who were interested in the project, concentrating on the local area and limiting himself to families living within a radius of about 5 km. It was very important to him that the families living nearest to the site join. Only one refused, saying he was against mills on principle, but, his sons joined the project.

The most important reason for hesitation, overall, was the size of the investment in relation to the (in)security of the profit. The guild was formed on March 3, by fifty-one member-owners of the 200 kW windmill.

At that time, it was possible to own eight shares at 1,000 kWh per family. The return from eight shares was approximately €700 per annum, making an additional income of €270–450 per annum after payment of installments and interest.

The greatest challenge for the windmill co-op was their cooperation with the local power utility, Thy Højspændingsværk. The ruling principle was that the windmill guild would have to pay the actual costs of connecting the mill to the power utility and making the necessary grid reinforcements.

The cost was €45,000, and the guild had good reason to be dissatisfied because costs varied very much from one place to another. Several other cooperatives paid only €3,000 for being connected. Despite much attention from the national television and writings in the press, the windmill guild did not succeed in getting these conditions fixed.

Bjarne Ubbesen was of the opinion that the power utility's attitude was "political," in the sense that the station profited from the connection with the windmill, enabling it to generally renew its power lines.

The guild had an annual general assembly attended by between forty and fifty members. After the year's results had been presented, accounts approved, and the coming year's budget decided, a social dinner was enjoyed by all.

In the early phases, when the project was being built and while it was still new, this gathering was the space for many good talks among neighbors. Everybody was eager and curious. During the first years many members visited the windmill regularly to keep an eye on the energy production meter.

The windmill guild has strengthened the local community and thus counter-acted the tendency towards the closing down certain aspects of village culture.

The 200 kW windmill turned out to be a far better business enterprise than anyone had dared hope. The price of the mill was €160,000, of which €3,000 went to buy land, €2,000 for the specialists, and, finally, the unfortunate €45,000 for being connected to the power station. The windmill runs with great stability, with annual maintenance costs of between €700–1,400—primarily the costs of regular servicing.

Economically, the mill has been a success. The guild's original expectation had been a return rate of 12–13 percent interest, but the actual rate has been more than 25 percent per annum.

TOWARDS AN ECOLOGICAL COMMUNITY

We are convinced that the change to an ecological community is less a question of money, subsidies, timetables, and diagrams than of talent, cooperation, past experi-ence, and perseverance.

In this chapter, we have given a brief sample of what happened when, from early on, as Thy engaged in the development of wind power and was victorious. The local pioneers contained an extremely active and creative environment for development, involving engineers and enterprises all over the country.

Regrettably, this has not resulted in the emergence of a widespread, local wind-mill industry, which must be put down to mischance. Nonetheless, Thy, and in par-ticular Sydthy, do have unusually many windmills, which contribute strongly to the local economy.

In the late 1990s, more than all electric power consumed in Sydthy was produced by privately-owned local mills, bringing the citizens an income of €7–8 million per an-num through the sale of electricity. This replaced power that would otherwise be pro-duced from coal brought from Australia and South Africa. The change from fossil fu-els to the energy sources of the future is not exclusively a question of technology and planning, but also of new ways to organize and cooperate in the local community.

Renewable energy is, by nature, decentralized, and in Thy it has been possible to organize things in a way that makes new technology a part of ordinary people's everyday life. Not only has this served local development and the environment, but it is also a manifest instance of how individuals and households may play an active part in changing the social system and create a model reaching out far beyond the borders of the local area and the country.

As almost all the mills are owned by people living in the area, this has meant an extra average income of between €1,500–1,800 per household. That income did not exist before the windmills.

This aspect is of great interest today because it means that the windmills are regarded in the same manner as other human activities, while at the same time pro-ducing power that holds no future threat to the climate nor of international conflict in order to secure energy requirements. Seen in the long-range perspective, a very great change has begun.

Chapter 22 ▎ Part 5

THE ROLE OF RENEWABLE ENERGY SOURCES IN THE DEVELOPMENT OF CUBAN SOCIETY
The Lessons to be Learned[1]

Conrado Moreno Figueredo and Alejandro Montesinos Larrosa

The disintegration of the Soviet Union that began in 1989, together with the intensification of the US-imposed economic blockade, precipitated the collapse of the Cuban economy. One result of this was a national energy crisis. However, this created the conditions for a greater use of renewable energy sources.

The energy and economic crises in Cuba, which began in the early 1990s, imposed an accelerated rhythm on research and other activities related to the use of renewable energy sources, as well as a strengthening of the education and training of energy specialists.

Cuba is an example of a country that suffered an energy crisis from one day to the next. Yet, due to certain deliberate and coordinated responses which were driven with support from the Cuban Government, it survived. Given that an energy supply crisis is likely to happen in many countries in the very near future, many global lessons can be drawn from the Cuban experience. In this chapter, we seek to introduce readers to the situation in Cuba over the last fifteen years and the way in which the country faced its crisis.

RENEWABLE ENERGY SOURCES UP TO 1990

Throughout the first half of the twentieth century, Cuba's energy matrix was typical of an underdeveloped capitalist country. In 1959, the date of the triumph of Cuba's Revolution, scarcely 56 percent of the population was served with electricity. Fidel Castro described this situation in his speech, known as "History will absolve me" [La historia me absolverá]: "2,800,000 of our rural and suburban population lack electricity," and "the utilities monopoly is no better; they extend lines as far as it is profitable and beyond that point they don't care if people have to live in darkness for the rest of their lives."[2]

SCIENTIFIC AND TECHNOLOGICAL TAKE-OFF

The two decades following the triumph of the Revolution (1959–1980) were characterized by both structural and conceptual changes. This was true for socio-economic aspects, in general, and also for planning the country's energy sector, specifically. All areas of Cuban society underwent a scientific-technical revolution: access to universities was democratized and the curricula for study programs at the three existing

---

1   This article was written in Spanish for the book and translated by Kolya Abramsky.
2   Translator's note: I have used the English translation of this quote, which can be found at http://www.marxists.org/history/cuba/archive/castro/1953/10/16.htm.

universities (Havana, Oriente, and Central de Las Villas) were changed, resulting in the training of professionals capable of putting the country's new development plans into practice.

In 1973, Cuba's national grid, known as the National Electroenergetic System, was established and put into operation. In 1975 the Cuban Academy of Sciences [Academia de Ciencias de Cuba] created the Solar Energy Group. This was the first time that Cuban researchers dedicated their work exclusively to researching renewable energy sources with a view toward introducing them into the country.

The Cuban Communist Party [Partido Comunista de Cuba] held its First Congress in 1976, where it approved the government program, "Investigations into the use of solar energy in Cuba." This resulted in the development of compact solar heaters, solar stills, and solar driers, as well as technologies for employing solar energy in the cultivation of micro algae.

At the end of the 1970s, the Advisory Group on Energy, a branch of the Energy Saving Working Group, was set up under the Ministry of Basic Industry (MINBAS).

ORGANIZATION AND RESEARCH

The Cuban Academy of Sciences represented Cuba in the Solar Energy Program established by the COMECON (Council for Mutual Economic Assistance) in 1981.[3] The following year, with support from this program, a prototype 1 kW photovoltaic installation, made from mono-crystalline silicon, was built for supplying electricity to a rural house. In addition to this, the Physics Faculty at the University of Havana began work on preliminary research into manufacturing solar cells using gallium arsenide.

An important milestone in Cuba's energy development was the creation of the National Commission on Energy in 1983. In addition to stimulating the rational use of energy, and researching national energy sources, this body also embarked on a national program for mini-, micro-, and small-scale hydroelectric projects.

In 1984, the State Organs of Central Administration went about establishing and promoting research and development groups in nearly all provinces in the country. These were dedicated to generalizing the use of renewable energy sources, especially hydropower, biogas, biomass, solar thermal, and wind. The First Forum on Spare Parts and Advanced Technologies—later renamed the Forum for Science and Technology—was held in 1987.

During the 1980s, several research and development centers were established, of which the following stand out: the Centre for Research in Solar Energy, established in 1984; the Institute for Materials and Reagents (IMRE), established in 1985; and the Institute for Investigations on Telecommunications (IIDT), established in 1986.

MAIN UNDERTAKINGS

Throughout the first half of the twentieth century, Cuba imported all the fossil fuel that it needed, and hydropower resources were exploited only to a very limited

---

3   COMECON was founded in 1949 as an intergovernmental entity for assisting and coordinating the economic development of the socialist countries.

degree. Biomass, from the sugar industry, and wood, were the most widely-used energy carriers, especially in the sugar industry. The use of wind energy was limited to a small number of windmills, which were used to pump water, and the use of solar energy was virtually unknown. Between 1959 and 1989, electrical generation capacity grew eight-fold, and petrol refineries increased their capacity by a factor of nearly three. Electrification reached 95 percent of the population.

This period was characterized by the extensive use of fossil fuels and technologies that came from the Soviet Union and the socialist countries of Eastern Europe. The advantageous terms of trade within COMECON meant that these were the fuels on which Cuba's development was based and sustained. In 1989, the year the Berlin Wall came down, imports of fossil fuels from the Soviet bloc countries were at 12 million tons. Given these favorable conditions, there was no urgent need for developing the country's own energy sources. Nevertheless, several programs related to energy development were in fact carried out, and human resources were built up in a wide range of specialties.

The greatest advances were made in relation to biomass—in particular, sugar cane. New bagasse-powered steam generators and driers were developed, and the use of cogeneration techniques was also increased.[4] Other sources of wood-based biomass continued to be used, but were based on a more coherent policy with regard to the use and reforestation of affected regions.

Despite being used inefficiently for the most part, bagasse nonetheless met 30 percent of the country's energy needs during this period. Other locally-existing biomass fuels—rice husks and, to a lesser extent, sawdust and wood chips, the waste from coffee, coconut shells, and other sources—had varying potential. Although there was a good potential to produce biogas from organic materials, this technology was not considered. Biomass energy was used mainly for producing electricity, supplying heat to industrial processes, and cooking food.

No great leaps were made in relation to hydropower, and the amount of electricity generated using this technology did not increase substantially. However, important studies and plans for its application came to fruition in this period. Of major significance was the plan to create a large network of dams throughout the country and to install tens of mini-, micro-, and small-scale hydroelectric plants. In 1989, the installed power was 55 MW, out of an estimated potential of 600 MW.

The production of flat solar collectors and solar thermo-tanks also began, and involved cooperation between several different national institutions. This resulted in the installation of more than 350 solar heaters in certain buildings of particular social importance, such as hospitals, nursery schools, and old people's homes, as well as those for other social objectives. By giving specific social priorities to the application of renewable energy, Cuba's approach to this problem differs from that of most other countries—which fits in with the wider priorities of the Cuban revolution, more generally.

---

4   Bagasse is the solid waste that remains after the juice has been extracted from sugar cane.

CUBA'S ENERGY CRISIS

FEATURES OF THE CRISIS

Between 1989 and 1993, the island's Gross Domestic Product (GDP) was cut in half, falling from $20 billion to $10 billion. Unable to exchange sugar for petrol with Moscow, the country had the majority of its petrol supplies cut off in a very brief period. Cuba's total volume of imports, in all sectors, fell by 75 percent. This especially affected food, spare parts, agrochemicals, and industrial equipment.

Lacking petrol, industrial production fell, factories closed, public transportation was reduced drastically, blackouts became frequent and agriculture and food production were paralyzed. The years following 1989, when the State embarked on a dramatic restructuring of the economy, are known as the "special period" (*período especial*). This required the reduction of the country's dependence on fossil fuels and a shift toward a greater use of renewable energy sources.

Use of biogas, biomass, solar thermal, solar photovoltaic, wind, and hydro power became both more extensive and more intensive. The waste from sugar cane, Cuba's main export crop, was used to fuel steam boilers in the country's 156 sugar factories. Excess electricity was then fed into the grid. During that period, the country counted on more than 220 micro-hydro systems, which supplied electricity to 30,000 Cubans. In addition to the more than 9,000 wind pumps that had been installed over the years, the island was also able to benefit from the commissioning of a 0.45 MW wind park, which was connected to the national grid.

It must be highlighted that, by 1992, petrol imports had fallen 40.6 percent in relation to 1989, causing an extremely serious deterioration in electricity provisions. A crisis that lasted for several years.

The early 1990s also saw the discontinuation of work on the nuclear power station at Juraguá. While the underlying reason for this was a lack of funds, the potential risk of an installation of this type also played a part in the decision. Similarly, the construction of the large Toa-Duaba hydroelectric complex was also paralyzed, for a number of reasons, including, especially, the ecological damage it threatened in what is considered to be Cuba's most important biosphere zone. Faced with these circumstances, the Cuban State took the decision to intensify the exploitation of the country's own petrol sources, and, by 1997, extraction had risen to 1.5 million tons (meeting approximately 15 percent of the country's petrol requirements). Due to the high sulfur content of this petrol, however, it was only used as fuel in certain industrial installations, including specific electric power stations. This was despite the knowledge that its use would cause further deterioration of the equipment, especially the steam generators.

By 1997, the frequency and duration of the blackouts had been gradually reduced. This was true both in relation to those blackouts that had been deliberately scheduled to save fuel, and those that occurred due to unforeseen faults. The problem remained serious, but not as serious as it had been between 1992 and 1994. This relative improvement was one of the first successes of the Cuban government's emergency measures, which sought to adapt Cuba's economy to the difficult conditions encountered from its insertion into the newly-emerging, uni-polar world, in such a way as to avoid irreversibly compromising the most important social goals, such as education, health, and social security, which had been achieved since 1959.

The costly dependence on imported petrol, even in these moments, forced the government to expand the extraction of Cuban crude oil, as well as its use, to the maximum level possible. In order to achieve this, profit-sharing agreements were drawn up with companies from different European countries, and they were contracted to carry out risky prospecting operations. As well, several generating plants were repaired and modernized in order to make them suitable for a more intensive use of local oil, while simultaneously reducing their fuel requirements.

In addition to these measures, it was also decided to make use of the accompanying gas. So, in 1997, a joint venture—with a Canadian company—to construct a 220 MW electric power station using gas-fueled combined cycle generators was embarked upon.

Naturally, a country facing the kind of fossil fuel limitations that Cuba faced has to seriously consider exploiting other energy sources that exist there. Thus, in 1993, the Cuban government took the decision to establish a programatic platform that would plot a new course of action to tackle the energy crisis that the country was going through.

PROGRAM FOR THE DEVELOPMENT OF NATIONAL ENERGY SOURCES

Under the direction of the Cuban government, the Program for the Development of National Energy Sources was proposed in May 1993. It was approved by the Council of Ministers within the month, and the National Assembly of Popular Power (the Cuban Parliament) analyzed and approved it in June of the same year.

The program's objectives were to gradually reduce the imports of fuel, optimize the use of domestic resources, and improve the efficiency of energy consumption. The main measures outlined in the program were:

- Substitute imported oil with an increased use of Cuban crude and accompanying gas in electricity generation.
- Achieve a greater efficiency in the use of bagasse and other agricultural wastes from the sugar industry, so that the industry would be able to become self-sufficient in meeting its own energy requirements, and allow for greater amounts of electricity to be fed into the national grid.
- Expand the use of hydropower, solar thermal, solar photovoltaic, wind, biogas, and industrial wastes (both agricultural and urban).

In order to facilitate implementation, the program was divided into three stages, but, due to the country's difficult economic situation when the program was drawn up, it was not possible to create an exact schedule for implementation of the different stages.

Successful implementation of the program is illustrated by the six-fold increase in the national production of crude oil between 1991 and 2001, as well as the more than seventeen-fold increase in relation to accompanying gas.

The program also enabled greater precision in prioritizing the best use of the resources available at any given moment, which was crucial in accelerating interventions that solely depended on Cuban efforts, and did not require any external investments. In this way, the intelligence, creativity, and labor of the Cuban people could be most effectively harnessed.

INTERVENTIONS AND RESULTS

Several bodies set up during this period were mandated as the driving force behind a process of increasing renewable energy use, building up human resources, and achieving sustainable development. The following bodies are of the utmost importance: the Centre for the Study of Renewable Energy Technologies (CETER), in 1992; the Centre for Research in Sugar Thermal Energy (CETA), in 1992; the Villa Clara Biogas Group, in 1993; the Area of Research and Development of Hydropower, in 1994; the Commercial Division EcoSol, belonging to Corporación COPEXTEL S.A., in 1994; the Integrated Centre for Appropriate Technology (CITA), in 1995; the Centre for Energy Efficiency Research (CEEFE), in 1996; the Solar Energy Technology Applications Group (GATES), in 1997; the Centre for the Management of Energy-Related Information and Development (CUBAENERGÍA), in 2001; and the Renewable Energy Front (FER), in 2003.

The work carried out by the Cuban Society for the Promotion of Renewable Energy Sources and Environmental Respect (CUBASOLAR), which was founded in 1994, stands out.

In the technological arena, the Solar Heater Factory in Morón is worth highlighting, as is the photovoltaic panel assembly workshop that is part of the Ernesto Guevara Electronics Factory Complex in Pinar del Río.

The following interventions are noteworthy with regard to the use of renewable energy sources in these years:

- Installation of Cuba's first wind park in the Isla de Turiguanó, in 1999. This benefited from solidarity from various Spanish and German organizations and bodies.

- Design and construction of 10 kW wind turbines in Bayamo, with support from the Folkecenter for Renewable Energy in Denmark, under the guidance of specialists from CETER. This wind turbine was connected to the grid in 1999, in Cabo Cruz, Granma.

- Drawing up of the first national wind map and evaluation of the wind potential, based on measurements taken at twenty sites throughout the country.

- Electrification of more than 4,000 schools, doctors' surgeries, hospitals, and rural houses, as well as other facilities serving social and economic objectives, using photovoltaics. One major achievement resulting from this, among others, is that all Cuban schools, without exception, have access to electricity.

• Installation of dozens of gravity aqueducts, windmills, hydraulic ram pumps, and other technologies for supplying the population and small scale agriculture and livestock rearing with water.

• Installation of hundreds of solar heaters and driers, as well as efficient wood and other biomass stoves, mainly for cooking in rural schools.

• Publication, since 1997, of *Energía y tú* [*Energy and You*], a magazine that popularizes scientific knowledge, and, since 2002, the scientific magazine, *Eco Solar*.

• Installation of dozens of mini-, micro-, and small-scale hydroelectric plants.

From 1993, the country's electrical generation capacity began to recover, resulting in increased consumption levels. In 1997, the Cuban government approved a three-pronged approach aimed at increasing the efficiency of electricity provision, and gradually eliminating blackouts:

• Modernization of thermoelectric plants and accelerated incorporation of crude-oil use.

• Construction of new facilities and harnessing of new capacities for using accompanying gas.

• Development of an Electricity Saving Program in Cuba (PAEC), under the Ministry of Basic Industry, and the Program for Saving Energy (PAEME), under the Ministry of Education.

By the end of 2002, the installed capacity in the national grid had increased to 3,300 MW and the availability of the electric power plants reached 71 percent—that is, they operated 71 percent of the year without stoppages.

Despite these achievements, the country's energy situation remained very critical in the years 2003–2005. This was for a number of reasons:

• The generation was based on large and inefficient thermoelectric plants, averaging twenty-five years in operation. Blackouts were frequent, especially during periods of peak demand.

• Between 2002 and 2005, the availability of thermoelectric plants declined from 71 percent to 60 percent, owing to frequent problems. Furthermore, these plants were high consumers in their own right.

• Large losses in the electrical transmission and distribution networks.

- Residential electricity fees that discouraged savings.
- A large quantity of inefficient household appliances in use in Cuban homes.
- 85 percent of the population used kerosene to cook with, and there were great difficulties in ensuring its availability for every family.

These difficulties had major repercussions on Cuban society, and required serious solutions, capable of overcoming the crisis.

CUBA'S ENERGY REVOLUTION

In 2005, Fidel Castro announced the start of an Energy Revolution, based on the following principles:

- Rational use of energy, aimed at maximizing savings in end use and using high-efficiency technologies.
- Prospecting, knowledge-building, exploitation, and rational use of Cuba's own energy sources, both renewable and non-renewable.
- Generation of electricity near the place of consumption, in conjunction with a gradual improvement of the transmission and distribution networks.
- Adoption of electricity as the preferential energy carrier for cooking food, with the exception of those who are already served with manufactured and natural gas.
- Development of renewable energy technologies for generalized use and with an increased importance in the country's energy balance.
- Proliferation of an energy culture that is oriented towards achieving an independent, secure, and sustainable development, which is in defense of the environment and based on the participation of the people as a whole.

Based on these principles, the following programs were orchestrated and put into practice.

ENERGY SAVING AND EFFICIENCY-SAVING AT THE POINT OF END USE

In recent years, Cuba has successfully designed and carried out an energy saving policy that has enabled it to have an increasingly rational and efficient use of hydrocarbons. Simultaneously, energy saving projects have been implemented by the population on a national scale. There has been a massive substitution program for domestic appliances, including, among other things, distribution of more than 2.5 million refrigerators (replacing 95 percent), replacement of nearly 10 million high-energy incandescent lightbulbs with compact florescent bulbs (100 percent), replacing more than 300,000 air conditioning units (81 percent), more than 1 million ventilation systems (100 percent), more than 200,000 televisions (22 percent), and *all* the electric hydraulic pumps. In addition to all of these improvements, high-efficiency stoves have also been widely distributed.

INCREASING ELECTRICAL COVERAGE

Two lines of action were taken to implement this program: decentralized generation, and ongoing improvement of the transmission and distribution lines. An addition of more than 3,000 MW was installed in combustion engine electric generators over the last two years. These units, which are more efficient than centralized power stations, have been installed in more than 200 locations and are connected to the national grid.

All together, more than 6,000 emergency combustion engine electric generators have been installed in places of social and economic importance, allowing them to operate independently of the national grid. Amongst others, this includes health centers, food preparation centers, stations for pumping and purifying water, schools, and hotels.

The main benefits of this distributed generation are the following:

•   Low values of own plant consumptions and low indices of fuel consumption: 200–220 g/kWh generated.

•   Availability is now higher than 90 percent.

•   Should a fault occur in any single unit, there is no impact on the national grid.

•   Generators are able to operate at full capacity within a short time period.

•   Reduction of losses in transmission and distribution.

•   Generation from stand-alone micro-systems for disasters, or defense needs.

USE OF RENEWABLE ENERGY SOURCES

The program's main achievements are listed below.

WIND ENERGY

•   Accelerated prospecting of the wind resource in the country's windiest locations. There is an estimated potential of 4,500 MW.

•   Installation of 100 modern wind measurement stations in 32 zones, covering 11 of the country's 14 provinces.

•   Putting in motion two trial wind parks between 2007 and 2008. The park in Gibara is 5.1 MW and the other, on the Isla de la Juventud, is 1.65 MW. These are in addition to the demonstration wind park in Turiguanó, already in operation for close to ten years.

•   Production of the first Cuban Wind Map.

HYDRO-POWER

The total estimated potential for the country is 552 MW (this figure does not include the potential of the large hydro power station, Toa-Duaba, which was mentioned at the beginning of this article, and has been discontinued). There are 180 units currently in operation, with an installed capacity of 62 MW. This includes small-scale, mini-, and micro-stations, thirty-one of which are synchronized to the national grid. The intention is to have reached 70 MW of installed capacity by 2008, and over 100 MW by 2010.

SOLAR THERMAL ENERGY

More than eight thousand solar heaters are in operation. Work is underway to expand the Solar Heater Factory in Morón for low cost production of units adapted to Cuban conditions. Massive experiments with vacuum tube solar heaters are also underway.

SOLAR PHOTOVOLTAIC ENERGY

More than 8,000 stand alone photovoltaic systems are currently in operation in schools, doctor's surgeries, communal TV rooms, and rural housing.

A pilot project in underway for a 60 kW solar power station that will be synchronized to the national grid.

INCREASED PROSPECTING AND PRODUCTION OF PETROL AND GAS

The last three years have seen an increase in the number of wells drilled and areas that have been explored—both for petrol and gas.

ENERGY SOLIDARITY VERSUS THE BLOCKADE

Since its beginnings, the Cuban revolution has shown solidarity to other peoples. Cuban medical brigades have offered their services in the African continent since the 1970s. Thousands of doctors, teachers, and other Cuban solidarity aid workers are currently working around the planet—predominantly in Third World countries.[5]

Such an approach is ideologically rooted in the doctrines of the Cuban National Hero, José Martí, as well as in the teachings of Ernesto Che Guevara and Fidel Castro.

In contrast to this, successive US administrations have spent nearly fifty years maintaining an iron economic, commercial, and financial blockade against the Cuban people. Since this embargo was first imposed, every sector of the economy and services has been a priority target of US aggression. The energy sector has not been exempt.

This blockade is the main obstacle that prevents a more efficient use of energy resources and a diversification of energy sources, including renewables, which would allow the country to somehow alleviate the negative impact of high petrol prices. The threats faced by third world businessmen who are interested in doing business with

---

5   Translator's note: the Spanish original is *cooperantes*. I have translated this as "solidarity aid workers," as distinct from from the phrase "development aid workers" that is the predominant basis of international capitalist cooperation.

Cuba in the field of energy, are but one example of how the Bush Administration is prioritizing a strategy that seeks to slow down Cuba's sustainable energy development and make it more costly.

However, the Latin American and Caribbean countries are undertaking their own strategies of development and regional integration, such as the Bolivarian Alternative for the Americas [*Alternativa Bolivariana para las Américas*, or ALBA] and Petrocaribe.

THE BOLIVARIAN ALTERNATIVE OF THE AMERICAS AND PETROCARIBE

The Bolivarian Alternative of the Americas is a proposal for an integration with a difference; while ALCA (the Free Trade Area of the Americas) responds to the interests of transnational capital and pursues absolute liberalization of trade in goods, services, and investments, ALBA places its emphasis in the struggle against poverty, and social exclusion. As such, it expresses the interests of the peoples of Latin America.

ALBA establishes mechanisms for creating cooperative advantages between nations, in order to allow for compensating the asymmetries that exist between the hemisphere's countries. It is based in the cooperative use of compensatory funds that attempt to rectify the disparities that result when weaker countries are put in a disadvantageous position in relation to the major powers. Hence, the ALBA proposal grants priority to Latin American integration and the negotation of sub-regional blocs. It opens up new consultation spaces in order to acquire a greater understanding of the positions of the Latin American and Caribbean peoples and to identify spaces of common interest that enable the construction of strategic alliances and present common positions in negotiations processes. The challenge is to weaken the effects of dispersion within negotiations, in order to avoid sister nations remaining divided or getting absorbed into the whirlwind, pressured to sign a rapid agreement—as was the case with ALCA.

The ALBA is a proposal that attempts to rethink development agreements, based on achieving consensus around the goal of an endogenous national and regional development that eradicates poverty, corrects social inequalities, and ensures a growing quality of life for the region's peoples. ALBA is the pinnacle of the fresh consciousness that is being expressed in a new generation of political, economic, social, and military leadership in Latin America and the Caribbean. Today, more than ever, there is a need to relaunch Latin American and Caribbean unity. The ALBA, as a Bolivarian and Venezuelan proposal, adds to the struggle of the movements, organizations, and national campaigns against Neoliberalism—struggles that are currently multiplying and being articulated among themselves through the length and breadth of the continent. In no uncertain terms, this is a demonstration of the historic decision, by the progressive forces of Venezuela and Cuba, to demonstrate that Another America is Possible.

According to the president of the Republic of Cuba, Raul Castro, the ALBA is "a superior form of association between countries. It is an instrument to confront Neoliberalism and the financial crisis, and opens the door to a transition towards more just and equitable societies." As of 2009, ALBA's members are Bolivia, Cuba, Dominica, Honduras, Nicaragua, and Venezuela.

★★★★★

Petrocaribe is a new type of regional energy project. It is founded on principles of interaction, solidarity, and complementarity. Through the use of preferential payments aimed at benefiting society, it is a decisive step along the road to continue developing the forces of regional integration and cooperation. Petrocaribe's main objectives are to contribute to energy security, social-economic development, and the integration of the Caribbean countries, by way of the sovereign use of the region's energy resources. It is based in ALBA's principles of regional integration, as described above. In other words, it seeks to contribute to the transformation of Latin American and Caribbean societies, in order to make them just, cultured, participatory, and solidary. As such, the process is conceived as a fundamental part of the elimination of social inequalities, improved quality of life, and an effective participation of the peoples in the shaping of their own destinies.

In this context, as of November 2007, thirty-one projects to replace incandescent bulbs with energy-saving ones were in progress in thirteen member countries of Petrocaribe. The result has been less expenses incurred from increasing energy plants capacity and purchasing fuel. Cuba also collaborates in the electrification of schools, housing, and other installations in Venezuela, Bolivia, Haití, and other countries in the region.

In parallel to these developments, Caribbean countries are cooperating in the construction, assemblage, and the putting into operation of more than 1,000 MW of new generating capacity, using combustion engine electric generators; repairing electrical grids; training technicians and specialists; operating and maintaining power stations; and evaluating renewable energy use and potential; among other areas.

Cuba will continue offering, in a modest and disinterested way, whatever contribution is within its reach. Cubans firmly believe in solidarity, cooperation, and people's collective ability to develop in a way that is beneficial for all.

EPILOGUE

The threat of climate change, the exhaustion of fossil fuels, and the foreseeable use of energy technologies such as agrofuels, are all subtle reminders of the nuclear bombs dropped on Hiroshima and Nagasaki.

A country's development should be measured by its level of social development and never by its levels of consumption or the amount of resources it squanders. Cuba's social development has reached a level such that its understanding of basic human rights includes not only the right to life, to independence, and to liberty, food, health, education, housing, work, and social security, but also the right to a general and well-rounded culture.

There is a need to raise the level of culture and consciousness associated with energy, so as to break the neoliberal schemes and practices that are being imposed or copied, as well as eliminate the waste and increase the energy efficiency and savings in end use. Such a cultural and consciousness shift are indispensable conditions for any efforts to improve the relation between energy supply and demand, and caring for the environment.

Furthermore, it is impossible to achieve a sustainable energy system, let alone sustainable development, within a neoliberal capitalist system. A humanist, solidary, revolutionary, and socialist ethic is also necessary.

Chapter 23 ▋ Part 5

DEVELOPMENT, PROMOTION, DISSEMINATION, AND DIFFUSION OF HOUSEHOLD BIOGAS TECHNOLOGY IN RURAL INDIA

Raymond Myles, on behalf of INSEDA

DEVELOPMENT OF LOW-COST BIOGAS TECHNOLOGY IN INDIA

DESIGN AND DEVELOPMENT OF HOUSEHOLD BIOGAS PLANT MODELS

Biogas technology is not new to India. The Matunga Leper Asylum in Bombay produced biogas using anaerobic digestion of human waste for lighting purpose as early as 1897. Biogas from cattle manure (dung) using the principle of anaerobic digestion was first used at the Indian Agricultural Research Institute (IARI) in 1939.[1] The technology has evolved and improved over the years, involving a series of prototype models known as Gram Laxmi (meaning "goddess of village wealth"). These were developed by the Khadi & Village Industries Commission (KVIC), an autonomous body within the Ministry of Industries that promotes rural and cottage industries including handloom products made by village/rural artisans. This low-cost, easy-to-build technology, is suitable for rural areas and small villages, owing to the fact that it uses an abundant local material, namely animal dung (especially bovine), which allows for local control of energy production and use. This has led to the popular adoption of the technology. In 1956, a sanitary latrine (toilet) linked biogas plant was developed in Maharashta state, based on a steel gasholder floating inside a water jacket on top of the well shaped digester. In 1960, Research and Design (R&D) in biogas was initiated at various other institutes in India. However, the KVIC model dominated the Indian scene for about two decades.

In the late 1970s, following visits to China (a pioneer in biogas technologies), the Ministry of Agriculture paved the way for a few R&D institutions to develop fixed dome biogas technology. The first workable prototype of this biogas plant design was the GGRS, Ajitmal (in the Etawah district of Uttar Pradesh).[2] This fixed dome plant was named the Janata model (meaning "People's" model), and was 30 percent cheaper than the earlier KVIC model. In 1978, the Ministry of Agriculture released the Janata model in four different sizes (2, 3, 4, and 6 m^3), with a view to popularization and widespread diffusion. A group of grassroots development NGOs, supported by a national level technical service NGO called Action For Food Production (AFPRO), decided to popularize the Janata biogas plant (JBP) through technology transfer efforts begun in 1979. This group of NGOs initially operated as an informal network, growing from

---

1   In India, the term "cattle" commonly refers to all the bovine population, i.e. both cattle and buffalo; therefore, unless specifically mentioned, both bovine and cattle refer to cattle and/or buffalo.

2   GGRS is Gobar Gas Research Station. *Gobar* is the Hindi word for bovine dung.

ten NGOs in 1980, to seventy NGOs, with ninety Biogas Extension Centers (BECs) by the end of 1995. The organization played a crucial role in the promotion, transfer, and diffusion of this technology throughout the country. After fifteen years in the making, the network was registered as an independent national association in 1995. The new organization, Integrated Sustainable Energy and Ecological Development Association (INSEDA), was headquartered in New Delhi. The founding Director of INSEDA was the author of this chapter. Some of the group's original NGO members had first come together as early as 1980 in order to build demonstration plants and to organize trainings on building such plants. This was done with a view to undertaking systematic expansion of the model, and the organizations involved gradually became a strong network of grassroots NGOs. The network also worked to promote and disseminate, through technology transfer, other appropriate sustainable energy technologies (SETs) for widespread use, especially in rural India. This new network of autonomous NGOs coming together around a common technological theme was a great benefit, resulting in the organizations sharing and learning from each other's rich, practical grassroots experiences in the implementation of biogas development program. Furthermore, the important data and useful information that they produced has had an impact at the regional, national, and international levels, influencing national government and overseas donor agencies' policies in favor of biogas technology and other sustainable energy technologies.

The NGO network gradually realized that, in order to reach the wider rural population, it would be necessary to bring down the cost of the biogas plant even further, as well as to simplify the construction techniques in such a way that would not compromise either the plant's strength or its quality. With backing from the NGO network, the technical specialists and engineers at AFPRO undertook research in the early 1980s to develop a better and cheaper model. This process benefited from regular feedback and suggestions from the network's approximately fifty member organizations. In this way, a new improved model of biogas plant was developed. It was about 20 percent cheaper than earlier models. In 1984, the new model was christened "Deenbandhu," meaning "friend of the poor." With financial assistance from the Ministry of Rural Development's Council for Advancement of Rural Technology (CART), more than 100 Deenbandhu plants (DBPs) were constructed.[3] Throughout 1984–85, these DBPs were used to both evaluate and demonstrate how the technology functioned under different agro-climatic conditions. Based on feedback from NGO groups and farmers, the designs of five family- (household-) sized DBPs were finalized in 1985–86. Subsequently, a centrally sponsored scheme, the National Project on Biogas Development (NPBD), was set up by the Department of Non-Conventional Energy Sources (DNES) for promoting and diffusing the Deenbandhu biogas model.[4]

The three most popular Indian designs of biogas plants are the KVIC, Janata, and Deenbandhu. Some other models were also approved in the mid-1980s and early

3   CART is now known as the Council of Advancement of People's Action and Rural Technology (CAPART).

4   DNES was later renamed twice, first becoming the Ministry of Non-Conventional Energy Sources (MNES), and then the Ministry of New and Renewable Energy (MNRE).

1990s for extension work, though they did not take-off widely or gain much acceptance, so they are not discussed in this chapter.

Based on feedback from grassroots NGOs and the practical experience gained from being involved in the design and development of biogas plants and the technology transfer and extension program, the Director of INSEDA undertook the development of a new design of household biogas plant. After a few years of experimentation, he was successful in developing a new low-cost household biogas technology, mainly for application in rural India. It was designed in such a way as to reduce surface area, with a view to cutting down the amount of building materials required, as well as reducing labor time and overall building costs. However, its major breakthrough was the fact that it completely replaced bricks with locally available and environmental-friendly biomass materials (such as bamboo). The fact that it can be built almost entirely from locally-available building materials means that this new technology offers opportunities for self-employment and income generation in rural areas, especially to poorer unskilled, semi-skilled, and skilled laborers. Another consideration taken into account in the design of the new model was that rural women should be provided with appropriate training so that it could generate self-employment for them in their own villages, mainly in their spare time and in the lean agricultural season. Approximately 40–45 percent of the cost of building the plant goes in the form of wages to the local rural people (especially rural women). Based on this, the designer christened the new model the "Grameen Bandhu" (GBP), meaning "friend of the rural people." The designer built the first demonstration plants for three farmers in the mid-1990s and they are still working well.

RURAL HOUSEHOLD (RHH) BIOGAS PLANT

GENERAL DEFINITION

The term household (Hh) digester or plant is commonly used to describe a simple family-sized biogas (FSBG) plant that operates under ambient temperature. It does not require any complicated or expensive external devices or mechanisms to control its internal temperature. The capacity of a Hh digester unit or family-sized biogas plant normally allows it to produce enough gas to meet all of the cooking needs and about two to four hours of the household's lighting needs. There is an enormous potential for installing such units in rural villages throughout most of the developing countries, since peasants have easy access to the raw materials. Rural models are commonly known as rural household (RHh) biogas plants.

INDIAN BIOGAS PLANTS

In India, the capacity of RHh or FSBG plants is defined as the quantity of biogas that is produced over a twenty-four hour period. It is measured in cubic meters (m^3), liters (lt), or cubic feet (cu ft or ft^3). Thus, a 1 m^3 biogas plant refers to the rated capacity of that particular unit that has been designed to produce 1 m^3 (or 1,000 lt or 35 ft^3) per day (twenty-four hours) under optimum conditions.

The smallest size units, 1 m^3 capacity, are only able to meet the cooking and lighting needs of a small and comparatively poor rural family of three or four members. Athough rarely used in Indian villages, the biggest family size biogas plant, $6m^3$ capacity, are able to fully meet the domestic cooking and lighting needs of a comparatively affluent and large joint rural family of fifteen to twenty members. In addition to meeting all their cooking and lighting needs, it also supplies enough energy to boil water, feed their domestic farm animals, and operate pumps for irrigation, chaff cutting, and feed grinders.

A COMPARISON BETWEEN INDIAN AND CHINESE RURAL HOUSEHOLD BIOGAS PLANTS

Both India and China started developing and implementing household biogas plants for large-scale rural applications in the late 1950s and early 1960s. In both countries these plants have been built in the millions, on a far greater scale than in any other countries in the world. It is useful to compare the broad similarities and differences between the Indian and Chinese biogas models.

In both countries the models used are hydraulic digesters/plants. These are plants in which a liquid (fresh slurry or influent) enters the digester from an inlet side and subsequently leaves the digester's outlet side in the form of digested slurry or effluent. Common to both Chinese and Indian designs is that the total solid (TS) content of substrate or feedstock (input material) is always less than 20 percent. Once the initial loading is done to fill the digester up to the required level, regular feeding is done with manure slurry. Each day, between 60–80 percent of the daily input materials leaves the digester in the form of effluent.

However, the Indian biogas models can be best described as semi-continuous hydraulic digesters, since almost all the material that is fed in initially as fresh manure at the time of loading eventually comes out as digested slurry after it has fermented inside the digester. In contrast, the majority of Chinese biogas models are what are known as semi-batch hydraulic digesters. This means that the initial loading is done in batches using seasonal crop wastes and residues. After loading they are sealed airtight. Additionally, the majority of Chinese digesters are also connected with latrines and pigsties from which they receive daily organic wastes in liquid and semi-liquid form. These wastes are flushed directly into these digesters. Accordingly, about 60–80 percent of the effluent is periodically removed using buckets, and applied in the fields or used as top dressing in the standing crops. The rest of the material in the form of digested or semi-digested sludge is retained inside the digester. The Chinese models are emptied every few months, cleaned and then batch loaded again with fresh seasonal crop waste. Normally, this process takes place twice a year. On the other hand, the semi-continuous flow design used in India, only requires emptying and cleaning very infrequently, at intervals of between five and ten years, if operated correctly. This makes it a very popular design. Furthermore, if properly constructed by a well-trained master mason, and in accordance with the construction manuals, none of the three Indian biogas models run the risk of gas leakage, and they save unnecessary work as they do not require maintaining a manhole and cover.

THE ROLE OF NGOS IN THE GROWTH AND EVOLUTION OF BIOGAS TECHNOLOGY DIFFUSION IN INDIA

The KVIC played the pioneering role in promoting the family-sized (rural household) biogas model, by adopting the Gram Laxmi-III for popularization and dissemination throughout India in 1961. By 1974, the KVIC and its local agents had built 7,000 units. Over the next seven years—1975 to 1981—another 100,000 plants were built. A comprehensive National Project on Biogas Development was launched by the Indian government during 1981–82. This gave a big impetus to India's biogas program and accelerated the pace of implementation several-fold—about half a million plants were set up in the next four years. By 1987, the number of biogas plants had increased more than 8-fold, a quantum leap which took the total to 836,198. According to the 1981 census, India's bovine population was 240 million (180 million cattle plus 60 million buffalo). Based on this, planners estimated that 14–16 million family-sized plants, with an average capacity of 3–4 m^3, could be fueled from bovine manure alone.

While the period until the end of the 1960s was critical for indigenous development of biogas design and limited demonstration projects, the 1970s was the period when planned R&D efforts and major efforts to popularize biogas technology really took off. The successful development of a much cheaper fixed dome design, the Janata BGP, during this period made it easier for the government to commit to and plan an ambitious target, to which it devoted financial, physical, and manpower resources. After 1975, a mass popularization program, involving demonstration and training programs, was launched, and R&D in biogas-related areas got underway at several different institutions.

Up until the 1970s, biogas in India had been initiated and implemented in a relatively centralized manner, operating through normal government bureaucracy, as well as the KVIC. The promotion and administration of the biogas program during the 1970s was largely a top-down process in which relevant government departments played an important role.[5] However, from 1980 onwards the Ministry of Agriculture (MOA) and later on the DNES (now MNES) took the lead in developing innovative implementation strategies. The centrally-sponsored National Project on Biogas Development scheme was the first targeted attempt at reaching the wider population through a network of both official and unofficial agencies, and for promoting a comprehensive and systematic transfer and extension of family-sized biogas plants, mainly to rural areas. This approach still continues today, with only minor modifications.

THE ROLE OF NGOS IN THE DISSEMINATION AND DIFFUSION OF HOUSEHOLD PLANTS FOR RURAL USE, IN COMBINATION WITH OTHER SUSTAINABLE ENERGY SOLUTIONS

FIRST PHASE (1979–1982)

a) In 1979, NGOs decided to promote the fixed dome Janata (JBP) model plant.

---

5   In the 1970s, this was either the KVIC; Ministry of Agriculture (MOA); and the Department of Science and Technology (DST)—Ministry of Science & Technology (MOS & T).

This was 30 percent cheaper and sturdier than the existing (and popular floating gash-older (KVIC) model.

b) One national technical development NGO took the lead in promoting the Janata plant. The three main objectives were:

- The transfer of construction skills to rural masons—through brick-by-brick training in farmers' fields, selected by grassroots NGOs.

- The transfer of knowledge to the functionaries of grassroots NGOs.

- To establish a biogas network of grassroots organizations, with a view to (1) undertake a systematic and decentralized process of promotion and extension work for low cost household biogas plants in rural areas, and (2) create a momentum to influence national policy in favor of biogas technology.

SECOND PHASE (1982–1984)

a) Popularization and extension of JBP and development of a new low-cost fixed dome household BGP model by the NGO network.

- Infrastructure development to establish an informal network of grassroots NGOs operating in different regions of India.

- Strengthening the network through capacity-building, information dissemination, increasing the number of NGOs, and starting "Biogas Extension Centers" (BECs).

b) Network members launch the biogas program in mid-1983, throughout the country, with financial support from an overseas funding agency.

c) Planning and organizing different types of training programs and systemic extension work in rural areas, aimed at different levels of NGO functionaries as well as for the biogas "End Users."

d) Systematic practical training of over 5,000 masons in fixed dome BGP construction.

e) The network develops, in a participatory way, a new low cost fixed dome Hh plant, "friend of the poor" Deenbanhu model, which is available in five different sizes.

THIRD PHASE (1985–1989)

Field evaluation, demonstration, promotion, transfer, popularization, dissemination, training and extension of Deenbandhu model by NGO network. This included:

a) Testing and comparative performance evaluation of 5 different capacities Hh Deenbandhu BGP, resulting in finalized design in 1985.

b) Preparation of field manual of DB model and its approval by the Ministry of New Energy Sources, between 1986–87, approved for extension under NPBD.

c) The implementation of household biogas plants in rural India was undertaken by the NGO network over a period of twelve years, with the assistance of an overseas funding agency. It was carried out in two stages, each stage lasting six years.

d) Systematic promotion and extension of DB plant was taken up by this network, supported by an overseas funding agency, from 1983–89:

- The NGO network built a total of 42,000 household BGPs in the first stage of the project. 30–50 percent of the cost of BGP building was subsidized under the NPBD; the balance was met by plant owners, either with their own resources and/or taking bank loans.

In some states, plant owners were also provided additional subsidies by the State Governments.

- The average ratio of overseas funding to local resources generated by the NGO network was about 1:5, thus making it one of the most successful programs.

- Network members undertook experiments for biogas generation using other biomass and alternative building materials for DBP construction.

e) Some of the members of the NGO network were also involved in the transfer of Deenbandhu biogas plants to other developing countries.

FOURTH PHASE (1990–1995)

Implementation of second stage of biogas program by the NGO network.

a) The stage II (October 1990–September 1995) program of the NGO network was taken up for the further extension of biogas, as were other activities related to biogas technology.

b) Some of the important achievements of the second stage, initiated in 1990, were:

- the construction of over 35,000 Hh plants (mainly DBP model);

- the implementation of appropriate capacity-building activities aimed at strengthening the existing and new NGO members in the network as well as providing post-plant maintenance services to existing plant owners;

- R&D aimed at improving plant design and scientific utilization of BGPs based on digested manure for crop production;

- decentralization of the network by promoting Regional Consultative Groups (RCGs) for solving regional level problems;

- enlarging the network to include new NGOs.

c) Experimentation showed that BG digested manure would provide a cost effective solution for crop production in rural India.

d) Transferring the fixed dome (Deenbandhu model) biogas plant from India to other developing countries by some of the NGO members of the network.

e) By early 1994, increased demand for decentralization had enabled the network to grow to include seventy grassroots NGOs, operating ninety BECs.

f) One of the regional groups became an autonomous body, the Sustainable Development Agency (SDA), allowing it to receive direct funding from MNES under NBPD.

g) Further technical work on improving the existing fixed dome BGP and developing a new model to bring down costs further.

h) The 1992 Annual Biogas Workshop of the NGO network took the decision to register the network as a national autonomous body with the following aims:

- to formalize the informal biogas network;

- to respond to new and emerging challenges in the field of renewable energy as well as systematic promotion of new Renewable Energy Technologies (RETs);

- to systematically promote sustainable energy based ecological and environmental development programs.

By the end of September 1995, the NGO network had built a total of 85,000 household BGP. Following consultation meetings, the constitution of the new organization of the biogas network was finalized. The national organization, Integrated Sustainable Energy and Ecological Development Association (INSEDA), was registered in December 1995, with its national office in Delhi.

FIFTH STAGE (OCTOBER 1996-2004)

The conversion of INSEDA into a broad-based network to implement a sustainable energy and ecological and environmental development program:

a) By the end of 1997, INSEDA had over fifty members, who had built 100,000 Hh plants throughout rural India since the early 1980s.

b) By the end of 2004, the total number of household plants under the NPBD had risen to over 150,000.

c) Important achievements of members through socio-technical interventions of INSEDA, from 1996 to 2004 are summarized below:

- development and installation of 20,000 ferro-cement DB model by one of the member NGOs;

- fixed dome model fabricated and built using Bamboo Reinforced Cement Mortar (BRCM);

- INSEDA, in collaboration with the Foundation for Alternative Energy (FAE) (Slovakia) and financial support from the International Network of Sustainable Energy (INFORSE)/Forum for Energy Development (FED) (Denmark), prepared "Training Material" for Distant Education in Renewable Energy Technology (DIERET).

SIXTH PHASE (2004-2008)

Work has focused on the development and promotion of sustainable rural energy and a renewable energy-based eco-village development (EVD) project and program with active participation of target groups to promote people-centered, eco-friendly, and environmentally-sound sustainable human development (SHD).Work, together with INFORSE, has promoted sustainable energy solutions for poverty reduction in Bangladesh, India, Nepal, and Sri Lanka. This has included the preparation of training manuals and CDs on RETs, and finance and accounting for RET implementation for capacity building and information dissemination. There have also been efforts to develop proposals to implement biogas plants in rural India by mobilizing funds through carbon credit generation using the Clean Development Mechanism (CDM). Thus, by giving incentives to different stakeholders, biogas programs can sustain themselves without depending only on government subsidies.

PROPOSED FUTURE STRATEGY FOR PROMOTION AND IMPLEMENTATION OF HOUSEHOLD BIOGAS PLANTS

ANALYSIS OF PROBLEMS IN THE PRESENT IMPLEMENTATION STRATEGY

The bovine population in India was 240 million (180 million cattle + 60 million buffalo) in 1981, and, based on this, the total potential for the 3–4 m³ family size (household) plants was estimated to be between 14 and 16 million biogas units. However, according to estimates in 1995–96, the bovine population in India had gone up to 340 million (a rise of over 100 million over 15 years). Meanwhile, over a decade of experience in implementing a national biogas extension program throughout the country has shown that the average quantity of fresh manure produced per animal would be 8–12 kg, of which an average of about 75 percent (6–9 kg/animal) is comparatively clean and can easily be collected by rural families for their individual biogas plants. Hence, it can be estimated that the average feasible size of rural household biogas plants would be of 1, 2, and 3 m³ capacities. This differs slightly from the earlier estimate of household plants with a capacity of 3 and 4 m³. Under field conditions, these three sizes of plants (1, 2, and 3 m³ capacity) would also operate more efficiently and require less space and daily maintenance. Taking all this into account, the present average potential of biogas plants in India can be safely revised upwards to something in the range of 25–30 million rural household units. Taking this revised potential, even if the rate of plant construction is doubled from the present target of 150,000 family size units annually to 300,000 biogas plants annually, it will still take between 75 and 100 years to install 25 to 30 million plants. This is based on the assumption that the present bovine population remains static, though, in the decade since 1996, the bovine population increased substantially, and therefore the number of household biogas could be revised further upward.

One of the main reasons for lower annual biogas targets has been that the NPBD is perceived as a government-owned program, and the number of plants built are restricted by the provision of subsidies under the MNRE's NPBD. This has meant that the biogas-implementing agencies and individuals see themselves merely as contractors or turn-key agents who are working for a single, centralized agency. The people (the plant owners or the prospective-plant owners) are understood to be mere beneficiaries of the program, rather than stakeholders. Under the present approach, the biogas end users do not participate in any effective, meaningful, or intelligent way. A healthy environment that enables them to fully evaluate the pros and cons of biogas technology does not exist. Biogas is viewed as merely a technology that will solve their energy or manure problems, rather than as a tool for their holistic and sustainable development. The "Ongoing Strategy" for biogas implementation does not hold much scope to ensure involvement of rural people, especially peasant women. They neither see themselves as the "primary stakeholders" nor understand the socio-economic and environmental and ecological implications of investing in the biogas plants. Often, external agencies push them, through subsidies, to accept biogas technology, even when they are not fully convinced of its benefits. In order for the biogas program to be accepted by the rural people at large, there is a need to ensure their full participation at all levels. The one-time subsidy only provides an incentive to build a plant, not to maintain and operate it efficiently once built. Thus, there is an urgent need to review the entire approach of biogas implementation.

The first planned program for the promotion of household biogas plants in India (1961–80) aimed to popularize it as an energy device. From 1980, following involvement from NGOs/Voluntary Organizations (VOs) whose focus was comprehensive rural and agricultural development, serious attempts were made to integrate biogas into their respective livestock, agriculture, water, sanitation, and women's development programs. This was a major departure from the earlier approach of treating biogas plants almost exclusively as energy-producing units. The NGOs/VOs started promoting biogas as "Biogas and manure plants," rather than just a device to produce methane for the purpose of energy. As a result of this concerted effort and the demonstrations in farmers' fields organized by NGOs/VOs, by the early 1990s, farmers were beginning to see the relevance of an effective biogas-manure plant. Over this period, they had actually seen the results of using digested slurry from the BGPs on crop production. However, in spite of these efforts, biogas technology has not yet been fully internalized by the villagers, owing to the way in which it is being promoted. Its diffusion is far from the level that is necessary to reach the desired goal of 25 million or more household biogas plants. Experience has shown that it is not enough merely to guarantee the success of a given technology—on the strength that it is tested, mature, and sound technology—by subsidizing its cost. No proper analysis exists, based on a systematic study of authentic field data, to ascertain how much of the subsidy has gone (and goes) to the plant owners themselves and how much ends up with other vested interests.

Due to substantial increases in the cost of building materials, wages, and accessories in recent years, the overall cost of all the presently-approved Indian models has gone up substantially. Now even building the least expensive model, such as a biogas-manure plant, is generally unaffordable, though some villagers are able to get one.

Unfortunately, farmers have not been adequately educated about the socio-technico-economic benefits. Not surprisingly, they do not see the biogas plants as a sound socio-economically viable and technically-feasible investment that will be profitable in a foreseeable time frame. Decision making in rural India is vested with male family members, and as they lack a critical awareness of the problems associated with cooking, they don't see the drudgery of their women-folks who collect the firewood and cook in smoke-filled, polluted kitchens using the traditional cooking stoves. Women, adolescent girls, and young children inhale the equivalent of about twenty packs of cigarettes each day from burning wood, biomass, and dung cakes for cooking.

In 1982, the Indian government launched the NPBD. By the end of 2008, it had reached its target of building over 4.5 million household biogas plants in rural areas. However, the total achievements and impact of this program have nonetheless left much to be desired at the grassroots level, despite the fact that, overall, it has been successful. At the present rate of implementation, based on government subsidies, it will take 100 years to build 150,000 Hh, let alone accomplish the massive potential of building 25 million family-sized units.

Despite the availability of trained manpower, and organizations ready to work on the program throughout the country, there was a low level of implementation. A large number of plants failed due to following reasons:

a) Defects and Failures due to incorrect implementation, including
- using wrong designs,
- using flawed construction techniques,
- using low-quality building materials,
- construction of plants with the wrong capacity,
- construction by untrained or improperly trained masons and technicians.

b) Problems due to improper operation and maintenance, including a
- lack of knowledge about feeding digester and utilization of biogas and digested slurry from the plant,
- lack of knowledge about the importance of daily care and skills and knowledge about general maintenance.

c) Lack of appropriate technical skills and training at the grassroots levels in
- identification and testing plants for defects,
- repairing failing plants.

d) During the NPBD initial stages, a large number of plants failed due to too much emphasis being placed on achieving targets. However, after the first five years, during which a large number of technicians, supervisors, and engineers had been trained, things started to improve. The NGOs mainly used a development-oriented extension approach backed by regular and effective follow-up, and post-installation maintenance services. This was possible with financial support from a number of overseas donor agencies. By the end of the program's first decade, these efforts by national and state-level governments, NGOs, donors, and others, together with proper monitoring and physical verification of plants, brought the failure rate down to about 10 percent. Ongoing learning and the accumulation of experience means that improvements are made each year, and the number of defective plants constructed continues to decline.

e) Whenever possible, the NGO/VO network has tried to integrate implementation of biogas plants with their new and ongoing developmental programs. However, even the NGO/VO's biogas program suffered from its need to achieve targets. The time-consuming nature of a target-based approach meant that parallel aspects such as building critical awareness, motivation, education, and technical literacy to plant owners and potential plant owners were neglected. Lack of funding to employ appropriate staff was also a major problem, which led to rural people, the district, and the block level government functionaries becoming too dependent on NGOs (especially in the regions where there were a large number of defective plants). The status of NGO/VOs was reduced to that of any other turn-key agents for a government sponsored scheme. Being a central government scheme, the NPBD had the major drawback of being overly structured, and consequently, was too inflexible to encourage innovations at the grassroots level, despite being a well-conceived program based on good intentions and high levels of commitment from the staff involved. Implementation problems gradually declined, nevertheless non-functioning of plants due to operational and maintenance difficulties began to increase. This was due to the fact that people assumed that the biogas technology had reached a stage of maturity and that, therefore, there was very little need for follow-up after installation. At the same time, as coverage spread throughout a wider geographical area, the costs involved meant that it became increasingly difficult, if not impossible, for NGO/ VOs to do effective follow-up. Periodic field surveys of the plants constructed within the NGO network have found that 90 percent of the non- or poorly-functioning biogas plants suffer from their owners either operating them incorrectly, or caring for and maintaining them badly, or both. Only about 10 percent are actually affected by faulty implementation. The author visited a number of plants each year and found that the operational efficiency of even the best-operated and regularly-maintained plants was continuously decreasing. The blame usually lies in the fact that the owners have not received adequate training and information about the technology. The decentralized implementation of a rural-oriented technology requires adequate preparation and a critical understanding of local situations. Without this, there are bound to be problems.

f) The successes of the NPBD's implementation strategy was based on three main principles:

- the technology should be good (near fool-proof from the engineering point of view, and well-constructed by technical and skilled persons),

- implementation should be subsidized to partly offset the high initial investment cost and to generate the interest amongst farmers and others,

- post-plant installation services, backed by appropriate guarantees against faulty construction, lasting for a minimum of three years, should be provided by the implementing individuals and agencies.

g) Inadequate funding was earmarked for user training camps to educate the potential rural users of the technology.

The situation at present is that a culture of dependency has been created. NGOs/ VOs have become heavily dependent on government, and the rural population has, in turn, become dependent on the respective NGOs and VOs for providing "A to Z" services for construction of their plants. In such a situation, if the subsidies should be withdrawn or if the NGOs/VOs withdraw from biogas implementation, the entire program would be in danger of collapsing. The major expectation of NGO/VO involvement is a greater emphasis on building critical awareness and practical training amongst the rural masses about this technology. This, in due course, should have enabled the rural people themselves to reach a level of understanding to accept and take-up this program on their own, making their own innovation with the help and guidance of the respective NGOs and VOs operating in their regions. Beyond this point, the NGOs and VOs were only expected to play inspirational, facilitating, and supportive roles. But, things have gone wrong. If this trend continues, even if the the number of technically-sound plants goes up each year, the number of inefficient (poorly operated) and operationally defunct plants will, nonetheless, carry on rising.

SUGGESTIONS FOR FUTURE IMPLEMENTATION STRATEGIES

A new and innovative strategy is required to reverse the present trend of total dependency on external support and resources. The new approach should be process-oriented, striving for interdependency. It should seek to foster end users' participation, treating them as the primary stakeholders in the biogas development program. This is only possible if the key role of functional education and technical literacy is fully recognized and accepted as an integral part of the RET programs, in general, and biogas program, in particular. This is an area in which NGOs/VOs can play an important role, owing to their accumulated experience in the field.

Implementation should be shown to be linked with livelihood, generation of employment, self-employment, and as providing a source of additional income to landless, peasant women and unemployed youth in rural areas. The NGOs/VOs have a high level of commitment, motivation, credibility, training, knowledge, and awareness about the local socio-political situation, which positions them well to motivate and involve the poor, weaker, and marginal sections of the rural communities at village or panchayat level, in the construction, marketing, and post-plant installation services.

Biogas plants should also be seen as environmentally-friendly village-level enterprises that generate micro-level employment in rural areas and can enhance rural people's income, as well as providing regular off-farm jobs to the poor through self-employment and wage earning in the local fabrication/building and marketing of the low cost biogas plants.

Providing short term employment should be seen as an intermediate objective. The long term goal should be the promotion of a semi-market approach, through the development of rural entrepreneurs for the implementation of biogas and other RETs as a micro-enterprise, on a sustainable basis. The micro-enterprise development should be planned and implemented in a systematic manner at the village level, in combination with a semi-market-oriented strategy for promoting and implementing the technologies. This would involve, amongst other things, appropriate technical skills development of villagers (especially aimed at women, youth, and artisans) for users and potential users of renewable energy and environmental technologies.

The new approach and strategy would also require systematic development of simple skills in marketing and management, and gaining enough knowledge to realistically ascertain the viability of an enterprise and calculate the economic returns from it, for a given specific micro-level local situation. This would enable individuals and groups to implement renewable energy technologies in clusters of 15–30 villages, over a 10–20 km radius. In other words, the future thrust of implementation for the biogas program should employ a combination of extension and marketing approaches in a cluster of contiguous villages by local groups like Mahila Mandals (MMs), Community Based Organisations (CBOs), and Self-help Groups (ShGs). Such groups, in turn, can involve unemployed rural youths and local artisans (master masons and bamboo weavers, etc.). Key to the success of a large-scale biogas-implementation program would be an adult education and technical literacy program about biogas technology and other low-cost, appropriate, rural-oriented renewable energy, ecological, and environmentally-sound technology. This would have to be appropriately supported by establishing village level infrastructures and bases to set a faster pace for realizing the entire potential, but gigantic task, of building the desired 25–30 million biogas plants. If such a program were launched now, the results would start becoming noticeable in about five years, and rural households would feel the impact within ten, even if there was a phased withdrawal of subsidies at some point in the future.

There is also a need for promoting biogas plants using a cluster approach through a massive dissemination of skills and knowledge, followed by a semi-market-based extension approach for popularization and commercialization of R Hh biogas units. The best way of carrying out this would be to involve Non-Government Voluntary Organizations (NGO/VOs) as an integral part of their developmental programs. Such a program could be supported by government, financial institutions, and overseas funding agencies, involving both multi-lateral and bilateral funding.

SUMMARY

This review and analysis of biogas programs in India has highlighted the need for following an appropriate strategy that uses a comprehensive approach for a rural technology diffusion program. It has to be based on lessons learned from the past four decades of biogas popularization and dissemination, which has identified all the key stakeholders, including the people, and identifying their strengths and weaknesses.

Those involved in the biogas development program should recognize rural people as the primary stakeholders, so that a new program and strategy can be planned and developed around them. Implementation should be carried out in phases, based around a people-oriented education and technical-literacy program.

A rural-based program targeting diverse socio-cultural groups within the village community would have a longer gestation period than a more centralized program. Therefore, until it is able to sustain itself, NGOs/VOs would require total financial support in the form of 100 percent grants from funding agencies. Support would also be needed to further develop capacity-building of grassroots level NGOs/VOs.

It would be necessary to support the creation of a cadre of grassroots trainers in participatory training methodologies, a process requiring external support and resource agencies that are able to equip NGOs/VOs to impart appropriate functional education and socio-technical, marketing, and management skills and knowledge to rural people.

The development of rural entrepreneurs and creation of micro-enterprises should be the key to decentralized and sustainable implementation of biogas program. Participatory development of promotional and campaign materials, posters, small and simple pictorial booklets, and simple training materials and teaching aids will need to be developed in local languages, and using pictures and songs. The NGOs/VOs have the commitment, motivation, experience, and the will to successfully implement such programs.

The new strategy should also recognize the need for participatory development of simple socio-technical indicators to measure change, ensure transparency. They should have built-in mechanisms for regular participatory monitoring and evaluation for mid-course corrections and improvements, and the program should be flexible. Any reasonable mistake would have to be seen as a learning experience and to be thoroughly analyzed with other mistakes, before making any modifications in the strategy.

There is a need to build in some "market-oriented approach" to ensure greater accountability, from the top to the lowest level, including achievable targets set by all the stakeholders, and monitoring of money spent and return on spending. Evaluating the reasons for successes and failures and how to ensure better performance and achievements should become an integral part of the future strategy of decentralized implementation of the rural household biogas program.

The biogas network of grassroots NGOs has amply demonstrated that they can play a very crucial and positive role in realizing the above goals, by following a participatory strategy and using an educational and development-oriented socio-technical implementation approach, rather than the purely target-oriented approach followed until now.

TRANSITION TO AN ENERGY-EFFICIENT SUPPLY OF HEAT AND POWER IN DENMARK[1]

Preben Maegaard

In Denmark, combined production of heat and power supplies almost 60 percent of the electricity and 80 percent of the demand for heat. The change to combined heat and power, from centralized and decentralized CHP has created a heat and power structure that can be gradually transitioned entirely to renewable energy.

The establishment of local consumer-owned and municipality-owned Combined Heat and Power plants (CHPs) since 1990 has shifted ownership of a significant share of power production gradually from conventional low-efficiency, centralized power production to local, independent, not-for-profit energy supply. This transition to decentralized CHP happened in parallel with the building of 3,000 MWel of new wind power, with 85 percent owned by community power cooperatives and farmers referred to as Independent Power Producers (IPPs). This transition represented the single most important initiative to reduce CO_2 emissions in Denmark.

By 2001, a total of 45 percent of the 36 TWh of power used in Denmark was being produced by IPPs. Of that 45 percent, wind power accounted for 20 percent and local CHP 25 percent. As a consequence, the central power utilities (now owned by Vattenfall, DONG Energy, and E.ON) had their share of the power market reduced to around half of the domestic demand for electricity. Thus, it took only ten years to dramatically shift almost 50 percent of the power production from inefficient, centralized, fossil fuel power supply to local, municipal, or consumer-owned companies.

Coincidently this is the amount of time it takes to build one atomic power plant—or roughly 1200 MWel. It should be mentioned that Denmark has not and is not planning to build any atomic power plants; this source of supply was ultimately withdrawn from the energy plans in 1985.

In order to understand how other communities can benefit by following the lead of the Danish CHP and district heating model it is important to understand the history and framework that was developed and the subsequent advantages that they provided to the people of Denmark. This Danish model shows that a decentralized heat and power system owned by the consumers can provide a sustainable energy future.

In 2007, the municipality of Thisted in Northwestern Denmark received the European Solar Prize for its outstanding achievement of providing nearly 100 percent of the demand for collective heating and electricity from wind, straw, wood, geothermal, organic waste, and solar sources. The energy prices are some of the

---

1   This chapter was previously published in the magazine *Cogeneration* in February 2008. The original is accessible here: http://www.folkecenter.net/mediafiles/folkecenter/pdf/CHP_in_Denmark__1990_-_2001.pdf. It is being reproduced here with permission from the author.

lowest in Denmark.

Such examples can be multiplied, and reinforce the need for a European moratorium on central power production from coal. It also demonstrates that conventional fossil-fuel-based power production can be phased out with improved safety of supply.

HISTORY OF DISTRICT HEATING IN DENMARK

Between 1955 and 1974, fuel oil made up nearly 100 percent of the Danish heating supply for individual use—for district heating, as well as the production of electricity. Typical residential homes employed fossil fuel burners to provide space heat and domestic hot water. This form of heat generation was problematic, as it was expensive, dirty, and required maintenance on a regular basis.

In the 1950s, the idea of district heating systems that provided a cost effective, efficient solution for communities to get heat without the maintenance and at a reduced cost, started to come up. The majority of district heating loops in Denmark were installed between 1960 and 1998, and they were predominantly owned by the members of the community that they were supplying. This gave control to the people and ensured that energy was distributed to the communities at fair prices. In addition, the savings due to the increase in efficiency could be reinvested in the community or given back to the energy consumers in the form of lower heating costs.

District heating from big CHP, using fuel oil, started in cities including Odense, Aarhus, and Aalborg, but after 1978, the plants gradually changed to coal and natural gas. Steam was produced for power and hot water (up to temperatures of 80–90 degrees Celsius); from the condensers, the hot water was supplied to the district heating loops thus increasing the total efficiency of the power generation dramatically.

In 1986/1987, the Danish Energy Agency and The Steering Group for Renewable Energy, within the Danish Board of Technology, implemented programs to encourage the use of decentralized CHP district heating distribution in decentralized community-owned networks for towns and villages that were getting their heat from existing district heating boilers or individual fossil fuel heating systems.

The programs started with a few demonstration plants, ranging in size from 100 to 3,000 kWel. In 1990, the triple tariff system was introduced with tariffs for peak, medium, and low-load operation. The power was fed into the national grid. To encourage the building of local, consumer-owned CHP, a premium per kWh of power production of DKK 0.10 (€0.013) was introduced.

These polices paved the way for towns of 500–40,000 people to implement CHP district heating systems using gas turbines, gas engines, solid municipal waste, and biomass. Smaller towns and plants typically used CHP gas engines and small biomass combustors, while the larger towns employed gas turbines, or a combination of all of the technologies. Systems were designed based on the fuel available, the geography, and the needs of the cities and towns.

TECHNOLOGY

Since 1990, the favored technology has predominantly been natural-gas-powered

CHP engines coming from a basket of European and North American manufacturers. Stationary natural gas engines used in combined heat and power applications boast a factor four reduction in CO_2 emission, compared with conventionally-generated thermal coal power for the same produced power and individual supply of heat.

This is because:

- The heat can be used if the system is placed in the community increasing the total efficiency of the system to over 85 percent compared with the best thermal coal power plant at 44 percent electrical efficiency.

- Natural gas has a tenth of the SOx, half of the NOx, and a third of the CO_2 produced from combustion of coal. The cost of removal of these pollutants in a coal generation plant is significant.

- The cost to install a gas engine is 30 percent lower per kW installed than a coal plant, and even cheaper if removal of the emissions from coal is factored in. Additionally, natural gas-fired engines can be installed in six months, as opposed to the five years needed for a thermal coal plant.

- Gas engines are manufactured in big numbers and are cheap, while central power plants are one-of-a-kind technology. Shipping of the gas engine and the rest of the plant is done by trucks and trains. The gas engines can be installed in existing or new buildings without any noise impact for the neighborhood.

With these benefits it became possible for local district heating companies, owned by municipalities or consumers, to build their own CHPs and offer cheaper heat to the households. This was one of the driving forces that encouraged a rapid change to local CHP.

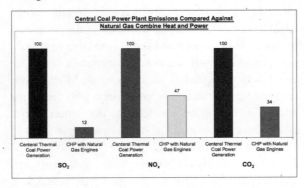

Figure 1: Comparison of emissions from central coal power production to natural gas combined heat and power for the same quantity of power produced.

IS CHP RENEWABLE?

CHP using natural gas is not renewable, but local CHP creates the basis for a decentralized energy structure that can later be changed to local renewable energy sources.

Stationary gas engines can run on a variety of fuels that can be tailored to local fuel availability.

These fuels can include local:

- Biodiesel
- Plant oil
- Biogas
- Gasified biomass
- Landfill gas

If these alternate fuel sources are not available, natural gas can be used as transitional fuel while the community determines what fuel can be utilized in the future. In essence, district heating with CHP provides the initial framework of a renewable energy power and heat system.

ADVANTAGES

Advantages of community based CHP units are vast, but the main benefits are:

- Reliability: Natural Gas engines are extremely reliable as they are used in the harshest of conditions where this stability is of the utmost importance. Typically, these engines are installed in transcontinental gas compressor stations, drilling rigs, offshore oil platforms, and villages not served by the national power grid.

- Community Autonomy: Having a power and heat producer in the community provides the locals with autonomy—giving the "power to the people." This gives the community the ability to ensure that the power is developed in an appropriate manner.

- The Ability to Incorporate Renewable Energy in the Future: Having CHP with district heating opens opportunities to incorporate large fractions of renewable energy in the form of biogas, solar thermal heating, wind for heat, biomass gasification, plant-oil-based fuels, and combustion of locally-based biomass.

- Scalability and Flexibility: Local CHP are scalable and flexible to operate. This makes it easy to increase capacity in the future and matches well with the incorporation of wind and solar power in the supply system.

- High Efficiency: Stationary Natural Gas CHP units boast an electrical efficiency of 42 percent and, with heat recovery of the jacket water, exhaust, lube oil, and turbo charger, can achieve an overall efficiency of over 85 percent (power 42 percent plus heat 43 percent).

- Cost Effective Heat and Power: With high total efficiencies, the cost of power and heat can be reduced. For example, according to Eurostat, Denmark had the third lowest European power prices (without taxes) in 2001, with only Sweden and Finland lower due to their high hydro contributions. In contrast, France, with 80 percent of its power coming from atomic energy, had a higher power price than Denmark with its thousands of independent power producers.

INDIVIDUAL USE OF CHP

Single users of CHP can include apartment buildings, supermarkets, railway stations, hospitals, hotels, schools, commercial buildings, and industrial businesses.

In addition to community power systems, single users could also benefit from the use of CHP, as is the Reichstag, the German parliament building, which is using its plant oil CHP unit as the main power and heat supply, and using the grid as the power back up. If the infrastructure is already in place then there is little to no capital cost to switch the operating philosophy—it is simply a management decision.

A small individual CHP (up to 10 kWel) is simply an energy box that makes the family house autonomous with its own supply of heat and electricity. The energy box can be combined with solar energy and other renewable energy. Fuel for the energy box may be biogas, other biomass gases, plant oil, ethanol, or solid biomass for external combustion in stirling-type engines.

If they were to be mass produced, like automobiles, such energy boxes could become very cheap compared to conventional power plants and CHP. A 100 kW car costs around €100 per kW, 90–95 percent less than a conventional power plant. Besides the drive train, included in the price of a car are five wheels, seats, stereo, windows, and many accessories that, in principle, make it much costlier to manufacture than an energy box of similar capacity. Therefore an energy revolution based on mass-produced CHP solutions has tremendous environmental, resource, and cost perspectives. The solution may be a derivative of hybrid car technology and pulse-operation.

COAL MORATORIUM IN DENMARK, 1990 AND 1997

Parallel to the implementation of CHP in Denmark, the first moratorium on building new coal-fired power plants was established in 1990. It did not require special legislation, as building of new power capacity must have the permission of the energy minister, and it was not given with reference to the need for improved efficiency in the production of power and environmental concerns. Only CHP could meet such requirements.

It was also decided by the Minister of Energy, as part of the Energy 2000 plan, that permission would not be given to build conventional power plants without utilization of the heat produced. However, a dispensation was issued for two 450 MWel CHP units, on the condition that similar coal-based, conventional-power capacity was decommissioned and scrapped.

In 1997, the government presented the Energy 21 plan, which had even more focus on energy efficiency and renewable energy. The 1990 moratorium on new coal-powered plants was maintained. As a consequence, new central CHP were built that had straw and wood as the primary fuels, in addition to 10–20 percent natural gas for superheating (in order to obtain sufficient power-generation efficiencies).

Liberalization and discontinuation of the renewable energy programs by the government in 2001 did not lead to the building of new coal-fired CHP or conventional power plants. The central power producers requested that the fuel mix of some of the coal-free power stations be changed, however without a substantial increase in the use of coal, for which permission cannot be expected. Thus the coal stop has been maintained.

Change of the tariff structure and market prices for power meant that most of the decentralized CHP owned by IPPs since 2004 produce less power but attain the same revenue as they had previously. These CHP function as a reserve capacity within the overall power supply, and will, in general, not be in operation when wind power or central coal power CHP cause low prices of electricity. Government policies in favor of the central utilities and liberalization has increased the use of coal power and reduced overall efficiency in the energy system, as more of the demand is now met by boilers and not local CHP. Also, the sale of electricity to neighboring countries has resulted in increased use of coal and, in 2006, the first increase in CO_2 emission for two decades.

As another consequence of the present government policies, no new wind power has been installed in Denmark since 2003. The state owned DONG Energy has become an international operator that builds big offshore wind farms in UK, and has built a 1,600 MW coal power plant in Lubmin, Germany. DONG reports that investments in new power production in Germany and UK are much more profitable than in Denmark, causing heavy protests from the Danish renewable energy community, who insist that best Danish practice within consumer-owner CHP should be transferred to Germany.

THE 1997 COAL-STOP RESULTED IN CONCRETE POLITICAL INITIATIVES

In order to achieve the CO_2 targets, the following initiatives have been accepted and implemented by the energy sector:

- The biomass agreement of the parliament from 1993, with a supplementary agreement from 1997, secures that at least 1.4 million tons of biomass (straw, wood chips, and willow) will be used in Danish CHPs. In addition an increased used of biogas and landfill gas was planned.

- In 1998, the energy minister instructed the Danish power stations to build offshore wind turbines with a capacity of 750 MW within ten years.

- Central and decentralized CHP plants were changed from using coal to natural gas, household waste, and biomass.

- In March 1997, the Danish parliament passed a coal-stop, which implies that permission to build power stations that use coal will no longer be issued. Basically, all of the more than 400 district heating stations have changed from using coal and oil to environmental-friendly CHP production based on natural gas and waste—or to using biomass and waste. This reorganization was launched in 1990 according to the heat supply law and was finally implemented in 1998.

- The development of industrial CHP based on natural gas was subsidized.

- With the passing of the Green Taxes Law in 1995 came a tax on sulphur at 20 DKK (€2.8) per kilo emission of sulphur (10 DKK per kilo sulphur dioxide). Until January 1, 2000, fuel for the production of electricity was exempt from the tax, as that tax was replaced by one on the consumption of electricity at 0.013 DKK pr. kWh.

- Maximum annual quotas for sulphur dioxide and nitrogen dioxides released from the big power stations.

- A CO_2 tax was imposed on the trades and industries. The revenue from the tax is sent back to the sector to subsidize investments in energy efficiency.

- Building codes were tightened to secure a lower consumption of energy for heating new buildings.

- Information campaigns were carried out and counselling given for households and companies urging them to reduce their energy consumption.

- Denmark should work persistently to pass common and coordinated initiatives to reduce the energy consumption and the greenhouse gas emission in the EU.

DANISH POWER SECTOR RESTRUCTURED IN 2004

In 2004, the organizational structure of the Danish power supply was dramatically restructured—the consumer-owned power companies were commercialized as part of a political compromise. Distribution, transmission, and production became independent sectors, each with their own framework.

- Distribution is the responsibility of local not-for-profit cooperatives, municipalities, or companies with a concession.

- Power transmission (over 60 kV) is the responsibility of Energinet.dk, a new, wholly state-owned company.

- Production of power comes from:
  1. central power plants owned by DONG Energy (owned by the Danish state), Vattenfall (owned by the Swedish state), and E.ON (German)
  2. local consumer-owned CHP, and
  3. wind power, with 85 percent owned by IPPs and the rest of the central power companies.

This section explores the controversial question of appropriate choices concerning energy sources and technologies, and the struggles that this is provoking.

A major obstacle preventing change in the energy system is the existing highly centralized energy sector (predominantly fossil fuel and nuclear), which has spent the last fifty years actively waging war on the renewable energy sector. The renewables sector has responded timidly, claiming neutrality and hoping to win a battle of ideas. However, it has shied away from the material and organizational conflict required in order to be able to confront these industries head-on and to defend itself against a roll-back and dismantlement of the sector's gains.

Governments and industry are engaged in efforts to expand nuclear and coal's role. However, grassroots struggles are emerging against these efforts. Regardless of whether nuclear and coal are socially and ecologically desirable or not, neither option is actually viable in the long term. Both are non-renewable resources that, like oil, are subject to "peaking." The existing energy industries also seek to prolong the use of oil, through the use of unconventional oils, such as tar sands and agrofuels. Both of these energy sources have a major social and ecological impact and are also provoking strong resistance, both in producer and in consumer countries.

The low cost of often highly exploited and repressed labor in the non-renewable energy sector has served as a hidden subsidy that, until recently, has contributed to ensuring that fossil fuels remain more competitive. The expansion of the renewable energy sector, to incorporate low wage areas of the world-economy, provides an important material basis for the sector to be able to compete with the fossil and nuclear sectors. It will be important for workers in the "clean" and "dirty" branches of the energy sector to ensure that growth in the renewables sector does not expand on the basis of downward levelling between them. In this regard, differing perspectives regarding the desirability or not of coal and nuclear perhaps constitute the single biggest obstacle to a worker-ecologist alliance taking shape.[1]

---

1   It was hoped to include a global over-view of the issues facing coal miners, written by a miner's trade union. However, due to time pressures, the person approached was unable to write this important chapter. Readers are invited to explore an excellent recent overview of the global coal sector, entitled: "Coal Mining and Trade Unions—Overview of Coal Industry, Problems and Challenges." and available at http://www.icem.org/ files/PDF/Events_pdfs/2007CoalConfINDIA.pdf. This report was presented at the International Coal Conference of Trade Unions, in Kolkata, India, December 14–16, 2007, organized by ICEM.

This section shows that the choice of energy technology (and source) implies a structural conflict between different, and competing, industrial sectors. Complex relationships link the different branches of the energy sector to one another, as well as workers within these branches. There is a need to identify lines of conflict and possibilities for alliances.

THE TECHNO-FIX APPROACH TO CLIMATE CHANGE AND THE ENERGY CRISIS
Issues and Alternatives[1]

Claire Fauset on behalf of Corporate Watch

As peak oil and the climate crisis loom, choices about solutions are ever more important. However, the debate on what direction to take to solve our energy crisis is surrounded by hype and vested interests. This chapter investigates the large-scale technologies that corporations and governments are putting forward as solutions to the twin climate/energy crises (including carbon capture and storage, hydrogen, agrofuels, and geoengineering), explains why they are unlikely to prevent the emerging catastrophes, and goes in search of more realistic and socially just solutions.

Making the right decisions about technology is vital, but many of the technologies being put forward as solutions simply won't work, will worsen the situation, cause significant environmental destruction, or are not going to be available within a short enough timeframe to avoid dangerous climate change. Even combined, they would fail to address the whole problem. For example, there can be no big technofix for deforestation, which currently causes around a fifth of all greenhouse gas emissions.

Technofixes are very appealing. They appeal to leaders who want huge projects to put their name to. They appeal to governments in short electoral cycles who don't want to have to face the hard choice of changing the direction of development from economic growth to social change. They appeal to corporations that expect to capture new markets with intellectual property rights and emissions trading. They appeal to advertising-led media obsessed with the next big thing, but too shallow to follow the science. They appeal to the rich world population trained as consumers of hi-tech gadgets. They appeal to (carbon) accountants. Technofixes appeal, in short, to the powerful, because they offer an opportunity to maintain power and privilege.

ASKING THE RIGHT QUESTIONS

Proposed technological solutions often fail to address the complexities of the problems at hand because they fail to ask the right questions. Agrofuels are indeed the solution to the transport problem if one asks the very limited question "how can people run their cars without oil?," rather than the more complex question "how can people get where they need to go without contributing to climate change?" Answers to the latter question might include limiting the need for travel by relocalizing jobs and

---

1   This chapter is a very summarized version of a full length report written by Claire Fauset, "Technofixes: A Critical Guide to Climate Change Technologies," Corporate Watch, 2008. http://www.corporatewatch.org.uk/?lid=3126

services, or investment in low-carbon public transport. Asking the right questions in a time of necessary change can lead to solutions which, far from being merely poor substitutes for old ways of doing things, are in fact better alternatives with real social benefits. Emancipatory social change can happen in a crisis. But social change is about much more than technology—a systemic framework is needed to assess the proposed technologies.

QUESTIONS FOR ASSESSING JUST AND EFFECTIVE CLIMATE CHANGE MITIGATION AND ENERGY-PRODUCING TECHNOLOGIES

Who owns the technology?

Not just the hardware (power stations, pipelines) but the patents and other intellectual property? Some technologies in particular—second-generation agrofuels, hydrogen, nano-solar—are likely to be dominated by a few companies owning fundamental patents and charging royalties for their use.

Who controls the technology?

This is a question of control—and of democracy. If supplies are short, who gets them—those in need or those who can pay? Who should decide what the solutions to climate change are and which technologies represent the best way forward? How can these decisions be made democratically with participation from the people who will be most affected?

Who gains from the technology? Who loses?

Is the balance of winners and losers just or equitable? For example, agrofuels may enable people to keep driving their cars, but will push up food prices and cause land conflicts. New technologies can also improve social justice: for example, deployment of small-scale hydroelectric systems can make reliable, cheap, controllable electricity supplies available to people in areas without a centralized grid.

In most discourse on climate mitigation, economic efficiency is prized above social justice, but promoting new technologies that do not help social justice will entrench and exacerbate existing problems, making them all the harder to deal with in the future. Preferring those new technologies that intrinsically promote equality, democratic control, and accessibility has wider benefits than the simple reduction of greenhouse gas emissions.

Inter-generational justice must also be considered—does a technology impose costs on future generations without conferring any benefits? This is a particularly important consideration for both nuclear energy and the application of carbon capture and storage technologies.

How sustainable is the technology?

Greenhouse gas emission reductions alone are not sufficient evidence of a technology's benefits. Does the technology deplete other resources, for example by consumption of rare minerals or through its impact on natural ecosystems and biodiversity? Does it have other pollution impacts, such as hazardous waste? Does it encourage or rely on other damaging activities? For example, carbon capture and storage relies on coal mining and encourages greater oil extraction when used for "enhanced oil recovery." Can the technology continue to be used in the long term without increasing negative impacts?

When will it be available?

Climate science shows that emissions need to start falling within the next few years, and fall massively in twenty to thirty years. Technologies that are unlikely to be available at an effective scale within that timeframe are not helpful. Resources should be diverted from these to more immediately available systems, and to ones that can be proven to work.

The focus of governments and corporations on emissions targets for 2050 can also be viewed as part of a distraction strategy. 2050 is conveniently distant—a target for 2050 allows time to continue business-as-usual in the short term in the expectation of future technological breakthroughs. Tough targets for 2050 are not tough at all. Where are the techno-fix plans for a peak in global emissions by 2015? If carbon capture and storage cannot be widely deployed within the next fifteen years, it is effectively of no use and a distraction from efforts to reduce emissions now.

OVERARCHING PROBLEMS WITH TECHNO-FIXATION: IGNORING THE SCALE AND SOURCE OF THE PROBLEM

Focusing on technological solutions ignores how the problem of climate change is caused, why it continues to worsen, and how much needs to be done to stop it. Even the IPCC now suggests that 85 percent cuts in global greenhouse gas emissions are needed by 2050.[2] Technology simply cannot deliver these levels of reduction without accompanying changes to demand, which will require economic and social transformation.

Technologies that encourage consumers to maintain high energy use and fossil fuel dependency—such as carbon capture and storage—fail to address unsustainable consumption levels that are the basis of rich-country economies and the cause of both climate change and other critical sustainability crises such as peak oil, declining soil fertility, and fresh water supplies.

The central problem is consumption—or more appropriately over-consumption—of fossil fuels and of forest and land resources—and the key motivation behind this over-consumption is corporate profit: in a word, capitalism. Technological improvements will not tackle over-consumption or growth in demand; this requires radical changes to economic systems. Without such changes, any technology-based emissions reductions will eventually be eaten up by continued rising demand for energy and consumer goods—efficiency gains will be converted into greater consumption, not reduced emissions in the long-term. This is due to the simple fact that capitalism has to have an increased energy base in order to continually expand production and consumption. It cannot use less energy.

Techno-fixation has masked the incompatibility of climate change solutions and unlimited economic (and energy) growth. A rational approach to a certain problem

---

2   IPCC, *Fourth Assessment Report. Climate Change 2007: Synthesis Report. Summary for Policymakers*, Table SPM.6, 2007. http://www.ipcc.ch/pdf/assessment-report/ar4/syr/ar4_syr_spm.pdf.

and a set of uncertain solutions might be to say that consumption should be limited to sustainable levels from now, with the possibility of increasing in future, *when* new technologies come on stream. Instead, the approach taken has been to continue consuming at the same destructive levels, with the expectation that new technologies *will* come on stream. A rational solution is impossible because our economic system forces us into irrational short-termist decisions. To make rational decisions, a new framework of social relations would need to be built.

The persistent claim that a solution is just around the corner has allowed politicians and corporations to cling to the mantra that tackling climate change will not impact economic growth. In 2005, in his address to the World Economic Forum, Tony Blair said, "If we put forward, as a solution to climate change, something that would impact on economic growth, *it matters not how justified it is*, it will simply not be agreed to [emphasis added]."[3] While this view may be slowly changing, it has delayed real action for years.

NEITHER THE CLIMATE CRISIS NOR THE ENERGY CRISIS CAN BE VIEWED IN ISOLATION

At the G-20 summit in April 2009, world leaders acted in a seemingly schizophrenic manner. On the one hand, they were quick to use lofty rhetoric about "green new deals" that would "save the planet" *in order to* save the economy. Yet, on the other, they continued as though the economic crisis could be dealt with in isolation from the energy and the climate crises, agreeing on a package of fiscal stimulus intended to bail out energy-intensive dinosaur industries and economic systems of the past, rather than building for the future. But the triple whammy of climate change, peak oil, and economic meltdown exposes the total failure of our dominant capitalist system.

Climate change is not the only crisis currently facing the planet. Peak oil (the point at which demand for oil outstrips available supply) is likely to become a major issue within the coming decade; while competition for land and water, deforestation, destruction of ecosystems, soil fertility depletion, and collapse of fisheries are already posing increasing problems for food supply and survival in many parts of the world. That's on top of the perpetual issues of class and colonialist exploitation and capital accumulation placing power and resources in the hands of the elite at the expense of the majority of people in the world.

Technological solutions to climate change generally fail to address most of these issues, except where they may reduce oil use. Yet even without climate change, this systemic environmental and social crisis threatens society, and demands deeper solutions than new technology alone can provide.

SCARCITY OF INVESTMENT

Governments spend a limited amount of money on mitigating climate change. Investment in energy R&D (research and development) increased massively in the 1970s as a result of the 1973 OPEC oil embargo, but in the last thirty years, R&D

---

3   "Blair bid for backing on climate," BBC News, 26 January 2005. http://news.bbc.co.uk/1/hi/sci/tech/4210503.stm viewed 2/2/08.

investment as a proportion of GDP has continually declined to the point where it is roughly comparable to pre-1973 levels.[4] Where this investment goes is a major issue. While it makes sense to research many options for mitigating climate change, time and resources are limited.

Some proposed technologies rely on things that simply don't yet exist; synthetic microbes that "eat" carbon dioxide and excrete hydrocarbons, a safe and efficient system for distributing and using hydrogen vehicle fuel, nuclear fusion power. This is not, in itself, an argument against any investment in these technological possibilities, but it is an argument against reliance on such future technological breakthroughs. Claims that something that doesn't already exist can solve a known problem, and that it should take most of the available resources, should be viewed simply as a stalling tactic on the part of vested interests.

Other technologies exist, but are benefiting from ongoing improvement—the efficiency and cost-effectiveness of photovoltaic solar panels; devices for exploiting wave and tidal power, energy-efficient electrical appliances, for example. These areas can be relied on to improve, though the timescale may be unpredictable. This is where technological investment needs to focus.

At present, it is the technologies that allow business-as-usual that are receiving the lion's share of investment, regardless of either potential benefit or feasibility. Investment in agrofuels or carbon capture and storage means less investment in wave power, in decentralized energy or in economic and social changes to limit the need for high energy consumption.

In 2008, the US government invested $179 million (£89 million) in agrofuels.[5] €10 billion (£7.9 billion) is being spent on an international, experimental nuclear fusion reactor in France.[6] Diverting this money away from more immediately practical solutions makes the target of peaking greenhouse gas emissions by 2015 less achievable. It both delays the transition to a low-carbon economy and endangers the future by making devastating climate change more likely.

TRANSITION

Transition—the period of change between the high-emitting societies of today and a distant sustainable future—is a hot topic. But while this change must come, the "transition" discourse coming from governments and corporations is frequently a cover for arguments that would permit short term use of technologies that are known to be unjustifiable in the long term—geoengineering, first generation agrofuels, "carbon-capture ready" coal-fired power stations are argued to be necessary now. But why?

4   JA Edmonds, MA Wise, JJ Dooley, SH Kim, SJ Smith, PJ Runci, LE Clarke, EL Malone, GM Stokes, "Global Energy Technology Strategy: Addressing Climate Change: Phase 2 Findings From an International Public-Private Sponsored Research Program," May 2007, Battelle Memorial Institute.

5   President George Bush, "State of the Union Address," 23/01/07. Full transcript available at http://www.america.gov/st/texttrans-english/2007/January/20070123210844abretnuh0.9462549.html, viewed 2/2/08.

6   "France gets nuclear fusion plant," BBC News, 29 June 2005. http://news.bbc.co.uk/1/hi/sci/tech/4629239.stm, viewed 2/2/08.

Largely to prevent serious change to the rich world's over-consuming lifestyles and to ensure new fields of growth for powerful energy companies.

The discourse of transition delays real action. When is the real transition to a low-emission, more equitable society even going to start? How long is it going to last? Will it ever start at all?

SOME OF THE MAJOR TECHNOLOGIES

Which technologies are we taking about? And what are the specific issues with each? This section gives a summary.

Agrofuels, carbon capture and storage/"clean" fossil fuels, and nuclear energy are all technologies that are at the absolute forefront of governmental and corporate energy strategies. These are also crucial technofixes, and can be critiqued accordingly. These technologies are dealt with extensively in other chapters in the book, with whole sections devoted to discussing the social and ecological problems associated with their use. However, because of this, and owing to space limitations, they are not dealt with any further in this chapter.

In addition to these energy and fuel sources, there are a number of other specific technologies, including:

Hydrogen: Hydrogen is a carrier of energy not a source in its own right. A primary energy source—coal, gas or electricity generated from these or other sources—is required to produce it. Using hydrogen as a vehicle fuel (the main application being considered) would be colossally expensive to introduce, would probably mean a commitment to long-term fossil fuel consumption and, most importantly, producing the hydrogen and compressing or liquefying it to use as a vehicle fuel could have a worse impact on the climate than using petrol.[7]

For hydrogen to be viable as a vehicle fuel it needs numerous technological breakthroughs in all major areas including production, distribution, and storage.[8] For hydrogen to reduce greenhouse gas emissions would require a glut of renewable electricity, or universal carbon capture and storage within a decade.[9] The likelihood of

---

7   Manufacturing hydrogen from natural gas emits 9.1 kg carbon dioxide per kilogram of hydrogen. Its climate impact is at least as bad as petrol, even before taking into account the substantial extra emissions from liquefying or compressing it, then transporting it. Powering BMW's new hydrogen car with electrolysis hydrogen using electricity from the UK grid would create around four times the emissions of its petrol equivalent. 59989-1033g/CO_2 per km for liquefied hydrogen produced via electrolysis or 364g/CO_2 per km for hydrogen production + 240-288g/CO_2 per km for liquefaction = 604-652g/CO_2 per km for liquefied hydrogen produced from natural gas. For comparison: the petrol car the H7 is based on, the BMW 750, emits 271g/km.102. A Toyota Prius emits 104g/km, a Renault Megane emits 117g/km, a vicious gas guzzler like the Porsche Cayenne emits 310g/km.

8   Production: It is impossible to produce hydrogen for vehicles whilst reducing carbon emissions unless carbon dioxide is captured and stored, a technology that is not currently unavailable, or unless we had a massive surplus of renewable energy. Distribution and storage: Hydrogen is highly reactive and corrosive therefore difficult to store. It also has to be liquefied and stored at -253 degrees centigrade, or compressed, making it grossly inefficient as an energy carrier. For more information, see Joseph Romm, *The Hype About Hydrogen*, Island Press, 2003,

9   Replacing the UK's vehicle fuels with hydrogen produced via electrolysis would take more than the country's entire present electricity consumption.[Tyndall Centre for Climate Change Research, *Decarbonising the UK—Energy for a Climate Conscious Future*, 2005, p. 74. http://www.tyndall.ac.uk/media/

the infrastructure it would need being put in place within four or five decades is slim. The cost of infrastructure to supply just 40 percent of the USA's light-duty vehicles with hydrogen has been estimated at over $500 billion (£250 billion).[10] So, the technical and economic issues mean this is an extremely unlikely to be a climate change solution.

Hydroelectricity and tidal barrages: Hydroelectric dams and tidal barrages have devastating impacts on local ecosystems,[11] and in the case of large dams, methane emissions have a major climate impact.[12] Other water power technologies such as wave power, tidal stream turbines, or tidal lagoons are less developed but potentially more sustainable.

Biomass: Burning biomass is humanity's oldest energy technology—the majority of biomass fuel used globally is still made up of traditional heating and cooking fires and stoves.[13] The main use for biomass in large-scale electricity production is in co-firing biomass with coal in power stations. Co-firing with biomass faces major problems of scalability. For example, producing 10 percent of the UK's electricity using willow as biomass would require an area of land one-quarter the size of the UK.[14]

Biomass does have a role to play where it can be harvested and used sustainably on a smaller scale. The production of gas fuel from agricultural waste including manure (biogas) also shows potential for sustainable expansion, though only to meet a small proportion of total energy demand.

Geoengineering: The term "geoengineering" refers to the large scale manipulation of the environment to bring about specific environmental change, particularly to counteract the undesirable side effects of other human activities. Technologies proposed include blasting the stratosphere with sulfates; mirrors in space; covering the

---

news/tyndall_decarbonising_the_uk.pdf]. Renewable energy currently counts for only 7 percent of UK electricity supply. Covering current needs and an increase of over 100 percent would require growth in renewables beyond any current projections.

10   Marianne Mintz et al, *Cost of Some Hydrogen Fuel Infrastructure Options*, Argonne National Laboratory Transportation Technology R&D Center, January 2002. http://www.transportation.anl.gov/pdfs/AF/224.pdf

11   See for more information: Claire Fauset, *Technofixes: A Critical Guide to Climate Change Technologies*, Corporate Watch, 2008. http://www.corporatewatch.org.uk/?lid=3126

12   At best, a dam gives one-tenth of the greenhouse effect of generating the same power from fossil fuels. (*Dams and Development: A New Framework for Decision Making*, World Commission on Dams, 2000, p. 75)

13   Greenpeace International, European Renewable Energy Council (EREC), *Energy [R]evolution: A sustainable world energy outlook*, January 2007, p. 7

14   Drax power station aims to produce 10 percent of its output from biomass, which would require 450,000 hectares of willow on 3-year rotation coppice. Drax burns around 10 million tonnes of coal a year (*Annual Report and Accounts 2006*, Drax Group plc, March 2007, p. 21. http://www.draxgroup.plc.uk/annual2006/files/page/5183/complete.pdf). Drax estimate that it takes 1.5 times the amount of biomass to replace a given weight of coal ("Alstom to build £50m biomass plant for Drax," *The Guardian*, 20 May 2008), so 1.5 million tones of biomass is required. The willow grown for Drax's trial yielded just under 10 tonnes per hectare ("Drax Goes Green with Willow," *The Guardian*, 19 March 2004), meaning 150,000 hectares would be needed. Grown on 3-year rotation (the fastest possible) means 450,000 hectares would be required to supply Drax with 10 percent biomass. On this basis, an area one-quarter the size of the UK would need to be planted with willow in order to produce 10 percent of the UK's electricity from co-fired biomass.

deserts in reflective plastic; planting or burying trees; planting shiny crops; painting surfaces white to increase reflection of solar energy; and dumping iron fertilizer in the oceans.

Reflecting the sun's energy: Some scientists are proposing to increase the amount of solar energy that is reflected back into space. Once any of these schemes is embarked upon it must be maintained for as long as the carbon dioxide emissions that it aimed to counteract remain in the atmosphere (centuries to millennia), regardless of any negative impact the scheme is found to have.

Sulfates in the stratosphere: When volcanoes erupt they release sulfates which are known to have a cooling effect on global temperatures by reflecting solar energy back into space. Some scientists are proposing increased levels of sulfates to simulate this effect. However, the sulfates will have unknown impact on ecosystems, including ozone depletion[15] and localized climatic impacts potentially causing major droughts.[16] Nobel prize winner Paul Crutzen, who advocated research into sulfate aerosols as a last ditch solution to global warming, predicted around half a million deaths as a result of particulate pollution.[17]

Ocean fertilization: One set of schemes for carbon dioxide capture centers on encouraging the growth of phytoplankton in the oceans, which take up carbon dioxide as they photosynthesize. In theory, some of this carbon dioxide might not return immediately to the carbon cycle. Exactly how much carbon dioxide is sequestered, and for how long, has not been quantified.

Ocean scientists, including the IPCC, have warned that this technology is potentially dangerous to ocean ecosystems, unlikely to sequester much carbon dioxide, and has the potential to increase levels of other dangerous greenhouse gases such as nitrous oxide and methane, to increase ocean acidification in deep ocean waters, and deplete nutrient loading in surface waters potentially leading to "dead zones."[18]

Overarching issues with geoengineering: Geoengineering rests on the assumption that humans are masters of the universe and the natural world, and can control and engineer its systems. Climate change has shown that humans do not and probably never

---

15   Brandon Keim, "Geoengineering Quick-Fix Would Wreak Ozone Havoc," *Wired*, April 24, 2008.

16   IM Held, TL Delworth, J Lu, KL Findell, and TR Knutson, "Simulation of Sahel drought in the twentieth and twenty-first centuries." *Proceedings of the National Academy of Sciences, USA* vol. 102, no. 50, 13 December 2005, p. 17891–17896. DOI: 10.1073/pnas.0509057102. Also M Biasutti and A Giannini, "Robust Sahel drying in response to late twentieth century forcings." *Geophysical Research Letters*, vol. 33, no. 11. DOI: 10.1029/2006GL026067, 8 June 2006.

17   Paul J. Crutzen, "Albedo Enhancement by Stratospheric Sulfur Injections: A Contribution to Resolve a Policy Dilemma?" *Climatic Change* 77, 2006, 211–219.

18   T Barker, I Bashmakov, A Alharthi, M Amann, L Cifuentes, J Drexhage, M Duan, O Edenhofer, B Flannery, M Grubb, M Hoogwijk, FI Ibitoye, CJ Jepma, WA Pizer, K Yamaji, 2007: "Mitigation from a cross-sectoral perspective." *Climate Change 2007: Mitigation. Contribution of Working Group III to the Fourth Assessment Report of the Intergovernmental Panel on Climate Change* [B Metz, OR Davidson, PR Bosch, R Dave, LA Meyer (eds)], Cambridge University Press, Cambridge, UK and New York, NY, USA. Michelle Allsopp, David Santillo, and Paul Johnston, *A scientific critique of oceanic iron fertilization as a climate change mitigation strategy*, Greenpeace Research Laboratories Technical Note, September 2007. http://www.greenpeace.to/publications/iron_fertilisation_critique.pdf. Ken O. Buesseler et. al., "Ocean Iron Fertilization—Moving Forward in a Sea of Uncertainty," *Science* Vol 319 11, January 2008.

will understand the planet's systems well enough to try to artificially engineer a re-balancing of the scales that over-consumption has tipped. Additionally, none of these schemes could be developed, tested, proven to be safe, and scaled up for decades to come.

THE OTHER PARTS OF THE SOLUTION—ALTERNATIVES TO TECHNO-FIXATION

Technological change is part of the solution. But only part. It is useful only as long as it is compatible with—and preferably supports—other changes to the way society works.

ECONOMIC CHANGE

Western government approaches to climate change consist largely of expecting the market to deliver emissions reductions. But the market doesn't want to deliver emissions reductions—it wants to deliver profits. The green capitalist approach is asking the wrong question. Instead of asking how to continue to grow the economy while living on the limited resources left on this planet, it should be asking: Why is economic growth seen as more important than survival?

The current global economic system is based on the assumption of indefinite growth. Growth of the whole global economy means consumption of an ever-increasing amount of goods, using an ever-increasing quantity of energy, mineral, agricultural, and forest resources. Even if energy intensity per unit of economic activity can be reduced, ongoing growth eats up the improvement, and overall energy consumption still rises. Renewable energy alone cannot decouple consumption from climate change; just because energy sources are called "renewable" does not mean there is an infinite amount available that can be accessed sustainably. Demand for energy is a political issue. Our energy priority should be to satisfy human survival needs, not to keep a worldwide division of labour, which is based on profit, in place.

Economic growth itself is not a measure of human well-being—it only measures things with an assessed monetary value. It values wants at the same level as needs, and, through its tendency to concentrate profit in fewer and fewer hands, leaves billions without the necessities of a decent life. Replacing the idea of growth as the main objective of the economy would require not just regulation and reform, but fundamental changes to financial and social systems, to the operation of large corporations (also based on the assumption of unlimited growth), and to people's own expectations of progress and success. Building a new paradigm of economic democracy based on meeting human needs equitably and sustainably is at least as big a challenge as climate change itself, but if human society as we know it is to succeed, the two are inseparable.

POLITICAL ACTION

Politics is about decision making. Effective and just solutions to climate change need decision making that involves everyone who is affected by the results of the decision—not just deals between those who stand to profit. As well as being in thrall to economic growth, current political systems are not equipped to deal with long-term issues.

Five-year election cycles and a party system based on petty point-scoring make it almost impossible for politicians to co-ordinate a decades-long process of change. Effort needs to go into long-term planning and real, hard decisions. If this is to happen, co-operation and maturity will be needed, along with a re-engagement of the population in real politics. The hold of corporate interests over political decisions must be broken—privileging profit over sustainability and equity defies democracy and leads to exactly those wrong answers current governments are pursuing.

Political change isn't simply a matter of making the right arguments to leaders. The global political elite has had access to all the science and information they need to demonstrate the imperative of radical and urgent action. No further evidence is needed of their inability to act for the good of humanity. Rather than political action from above, it is struggle involving a twin process of confrontation and construction from below that will change things. Policy makers will only implement these changes from above to the degree that they are forced to from below. Is it possible to organize a collective social force that is strong enough to confront power and build alternatives in a way that imposes change rather than just wishes for change? It has to be possible. The vast majority of people in the world will not commit social and ecological suicide to enable the minority to preserve their privileges.

SOCIAL CHANGE

The social change approach means taking systems that are unsustainable and finding ways of meeting people's needs within the limits of the planet's resources through co-operation, lifestyle change and appropriate technology.

We need to construct sustainable ways of managing our food systems, transport systems, housing, land use, and economic activity. These sustainable solutions will, for the large part, be small-scale and localized, with solutions meeting the needs of local populations. Achieving this means co-operation at a community level.

Technologies are a useful part of the solution, but techno-fixation isn't. There are other changes already available that can be quicker to deploy, more effective, cheaper, more equitable, and have a greater guarantee of success.

The beginning of the path towards a sustainable solution to climate change could look something like where we are now. The science is uncontroversial. There is a groundswell of public opinion. Politicians and corporations are giving lip service to the solutions, but the public does not trust them. Mainstream politicians are even starting to question the logic of perpetual economic growth. The seeds for change are being sown. The seeds of this movement are already here.

There's a huge amount of work to be done to make these seeds germinate and flourish, but it can be done. There is still time.

Chapter 26 ▌ Part 6

DEVELOPMENT OF ICELAND'S GEOTHERMAL ENERGY POTENTIAL FOR ALUMINUM PRODUCTION— A CRITICAL ANALYSIS

Jaap Krater and Miriam Rose on behalf of Saving Iceland

Iceland is known for its geysers, glaciers, geology, and Björk, for its relatively success-ful fisheries' management and its rather unsuccessful financial management. But this northern country also harbors the largest remaining wilderness in Europe, an end-less landscape of volcanoes, glaciers, powerful rivers in grand canyons, lava fields, swamps, and wetlands teeming with birds in summer, and plains of tundra covered with bright colored mosses and dwarf willow.

In 2006, 57 km² of one of the most magnificent areas of the country, the wild highland plateau north-east of the large Vatnajökull glacier, was inundated for Eu-rope's largest hydro complex, the 690 MW Kárahnjúkar dams. The energy from the dams went to a single new aluminum smelter built by the American transnational corporation Alcoa. On the day of the flooding, 15,000 people (out of a population of 320,000) demonstrated against the project. The protests against the Kárahnjúkar dams launched a wider movement aimed at protecting Iceland's wilderness from heavy industry.

Icelanders, who had been divided over the perceived costs and benefits, were shocked by the devastation wrought by the project. Since the flooding, strong winds in the highlands have eroded silt from the rising and falling water table, and dust storms are affecting an area much vaster than the reservoir. Mud rains fall in the eastern fjords where many local industries closed after the smelter was built. Seal colonies in the delta of the dammed rivers are diminished, and some of the most important breeding grounds of vast colonies of rare skua, geese, and duck species are gone. 3 percent of the Iceland's landmass is affected by the Kárahnjúkar project.[1]

Impact of large dams on climate has been found to be higher than previously assumed due to methane emissions from reservoirs,[2] and it has recently become clear that this is also significant for high latitude reservoirs such as Kárahnjúkar.[3] Dam-ming Iceland's glacial rivers also prevents the flow of mineral-rich silt (containing calcium and magnesium) to the sea. These nutrients feed marine phytoplankton, the beginning of most marine food chains. The damming of Iceland's glacial rivers

---

1   Icelandic Society for the Protection of Birds. 2008. Environmental facts and figures of the Kárahnjúkar project [online]. http://www.fuglavernd.is/enska/Kárahnjúkar/statistics.html [Accessed 3-12-2008]

2   Krater, J. 2006. "Elke stuwdam is een ramp." *Trouw*, 1-20-2008

3   Duchemin, E., Lucotte, M., Canuel, R., Soumis, N., 2006. "First assessment of methane and carbon dioxide emissions from shallow and deep zones of boreal reservoirs upon ice break-up." *Lakes & Reservoirs: Research and Management* 11, 9–19.

not only decreases food supply for fish stocks in the North Atlantic, but also impacts oceanic carbon absorption, and therefore the global climate.[4]

The promise of environmentally-friendly hydropower turned out to be a false one for the dams in east Iceland. Now, similar promises are being made for geothermal energy as a clean power source. In this chapter we review the development of geothermal energy in particular and examine its sustainability, environmental impact, and some of the associated social and economic issues related to recent industrialization in Iceland.

CHEAP ENERGY, MINIMUM RED TAPE

Iceland, with its vast possibilities of hydroelectric and geothermal energy, became an appealing target for heavy industry corporations such as Alcoa, Rio Tinto Alcan, and Century Aluminum. In a world increasingly concerned about carbon emissions, the clean image of hydroelectric and geothermal energy is appealing. Though heavy industry processes have an implicitly-high environmental impact, they can be made to appear greener by using "renewable" energy. To this end Iceland was granted an exemption for "green-powered" industrial emissions under Kyoto, and pollution control schemes are lenient, encouraging industrial investment.[5]

The wholesale of Iceland's energy resources began in 1995 when the Ministry of Industry and Landsvirkjun, the national power company, published a brochure entitled "Lowest energy prices!"[6] The brochure glorified the country as having the cheapest, most hard working and healthiest labor force in the world, the cleanest air and purest water—as well as the cheapest energy and "a minimum of environmental red tape."

For ten years, former Prime Minister David Oddsson (who would become the central bank director largely blamed for the collapse of the Icelandic economy) led the campaign to attract energy-intensive, and therefore often highly-polluting industries. In 1998, Century Aluminum constructed their first smelter in Iceland at Hvalfjörður, to be expanded eight years later. Three to five new aluminum smelters were planned. The existing Alcan (now Rio Tinto) smelter and a steel factory were to be expanded and an anode factory erected. An energy master plan was drawn up to harness the 30 Twh of electricity needed; dozens of dams would be built in every major glacial river, and nearly all geothermal areas would be exploited.

Not everyone agreed with the projects; in 2004, at the third European Social Forum in London, Icelandic environmentalists made an international call for help. That year, the international campaign Saving Iceland, was formed to oppose the masterplan.[7] Summer action camps were held in four consecutive years. A number of

---

4   Gislason, S.R., Oelkers, E.H., Snorrason, A., 2006. "Role of river-suspended material in the global carbon cycle." *Geology* 34, 49–52.

5   For example, RT-Alcan's smelter at Straumsvik is allowed to dispose of it's highly toxic spent potlining in an adjacent landfill site that is exposed to regular sea flooding, ten miles south of Reykjavik. Rio Tinto Alcan. 2008. Pot linings [online]. http://www.riotintoalcan.is/?PageID=111 [Accessed 12-12-2008]

6   Icelandic Marketing Agency (MIL) (1995). "Lowest energy prices in Europe for new contracts; your springboard into Europe. MIL, Reykjavik.

7   Saving Iceland: http://www.savingiceland.org.

years of direct action, and mainstream protests by celebrities, including Sigur Rós and Björk, as well as Icelandic intellectuals, have seen the cancellation of some of the most damaging projects. Still, construction of a number of new dams in the Thjorsá and Tungnaá rivers are planned to start in 2009, and are intended to provide power for expansion at Rio Tinto Alcan's existing smelter, a data center, and a number of silicon refining plants by corporations whose names are kept hidden by Landsvirkjun.

CHEAP IMPORTED LABOR

Large dam projects in the majority world have been associated with mass displacements and "cultural genocide" on an enormous scale.[8] Comparatively, the social impact of the developments in Iceland is small. Nonetheless, cheap energy and labor is just as important to corporations operating in Iceland as elsewhere. Special arrangements are made by governments for subsidized borrowing and tax cuts, and loans for expensive dams and geothermal projects are taken by the state-owned power company at the taxpayers' risk, while the price paid for energy is kept secret, and depends on world price of aluminum. Thus the taxpayer directly subsidizes every ton of aluminum when its market price drops.

Imported cheap labor and low workers' rights standards are routinely used on construction sites. More than a dozen Chinese and other foreign workers died in the construction of Kárahnjúkar, and, more recently, two Romanian workers suffocated in geothermal drill pipes on the site of a work camp near Reykjavík, where workers sometimes toil up to seventy-two hours a week, sometimes in seventeen hour shifts.[9] Workers are effectively confined to the camps for their three to five month work periods, going out to the capital once a month.

"KUWAIT OF THE NORTH"

Now that Icelanders have realized the full impact of Kárahnjúkar, public opinion is less favorable to large dams, and power companies have shifted their focus to geothermal exploitation. Currently the Hengill area east of Reykjavik is being developed on a large scale for the recently-completed expansion of the Century Aluminum smelter in Hvalfjörður. Test drilling is taking place in four fields—Krafla, Bjarnarflag, Theistareykir, and Gjástykki—in the north of the country for a new Alcoa smelter near Húsavík. Brennisteinsfjöll, Krísuvík, and Reykjanes fields, southwest of Reykjavík, are planned to be developed for a new Century smelter. The national power company plans to triple geothermal power capacity to 1,500 MW, on top of the 575 MW currently generated by geothermal, of which a large proportion already goes to the two existing smelters in the Reykjavík area. Also, a new public-private consortium has been formed to develop deeper drilling of geothermal fields, which would amplify the scale of geothermal production and power generation potential.[10] Ulti-

---

8   McCully, P., 2001. *Silenced rivers: The Ecology and Politics of Large Dams*. Blackwell Publishing, New York.

9   Personal communication with a number of anonymous workers at Hellisheiði.

10   At the time of writing, investments in most projects were put on hold because of Alcoa and Century ceasing capital injection due to economic uncertainty and the slump of aluminium demand.

mately, it is proposed that all of the economically-feasible hot spring areas in Iceland will be exploited for industrial use, including a number of sites located in Iceland's central highlands, the beautiful heart of Iceland's undisturbed wilderness.[11] Landsvirkjun, without any irony, has termed Iceland "the Kuwait of the North."[12]

GEOTHERMAL PROMISES

Geothermal potential with current technology is found at hotspots on the earth's surface, where magma intrudes into the rock bed and heats porous rock to high temperatures.[13] Electricity is generated by drilling into these reservoirs and powering turbines with high-pressure steam emitted from boreholes. The original geothermal power stations and boreholes supplying domestic needs in Reykjavik are small-scale installations that efficiently provide electricity, hot water, and heat from sources in close proximity to the city, and these are fairly sustainable.

As with any form of energy generation, there are environmental issues with geothermal exploitation that should be taken into account. These impacts are exacerbated significantly by the greater scale and intensity of production that energy-intensive industries require. But the quick-to-embrace enthusiasm for any technological solutions that promise to be a way out of our fossil-fuel addiction have tended to gloss over the downsides of geothermal exploitation and promote its intensive commercial use. Geothermal energy has the image of being sustainable, carbon neutral, and of low environmental impact. How does this image compare to reality?

RENEWABLE

Geothermal reservoirs have a sustainable production level if the surface release of heat is balanced by heat and fluid recharge within the underground reservoir.[14] This happens naturally in undisturbed hot springs, which have remained at more or less constant temperature over hundreds of years, but these recharge rates are generally not sufficient for exploiting economically.[15] The geyser hot springs at Calistoga, California experienced a 150 percent decrease in production over ten years, due to rapid exploitation to meet economic requirements, and there have been many similar cases.[16]

Also, opposition has surged as the link between borrowing for previous heavy industry projects and Iceland's severe economic depression has become evident (Krater, J. 2008. "More power plants may cause more economic instability." Morgunblaðið, 26-10-2008.)

11   Pálsson, B. 2007. "Iceland deep drilling: a project at risk." Presentation produced by NORD-NET for Landsvirkjun [online]. http://www.vsf.is/files/691972290Innovation percent201.pdf [Accessed 13/12/2008].

12   Landsvirkjun. 2004. "Now to tame the waterfalls of Iceland." *Living Science*, 8, 50–55.

13   For an overview of global geothermal potential in a sustainability context, see MacKay, D.J.C. 2008. *Without the Hot Air*. UIT, Cambridge.

14   Rybach. L. and Mongillo, M. 2006. "Geothermal Sustainability: A Review with Identified Research Needs." *GRC Transactions*, 30, 1083–1090.

15   Rybach, L., 2003. "Geothermal energy: Sustainability and the Environment." *Geothermics* 32, 463–470.

16   Sanyal, S.K., Butler, S.J., Brown, P.J., Goyal, K., Box, T., 2000. "An investigation of productivity and pressure decline trends in geothermal steam reservoirs." *Proceedings World Geothermal Congress*, Japan, 5, 873–877.

Extracting super-heated steam and fluids eventually causes a drop in the pressure and temperature of the reservoir. Re-injection of fluids maintains pressure but has a cooling effect, and best available technology cannot fully re-inject all extracted fluids, as significant amounts of steam and wastewater are released into the environment.[17]

Boreholes are usually modeled for only thirty years of production.[18] Recovery of reservoirs used for commercial energy generation takes 100–250 years before being viable for exploitation again, while in shallow, decentralized heat pump systems used for home-heating recovery time roughly equals production time.[19] Another problem is that geothermal hotspots like Iceland are seismically active zones. In Iceland, it has occurred that two-thirds of boreholes in a field were destroyed by quakes.[20]

Compared to the geological time scale of oil regeneration, geothermal energy *is* relatively renewable. However geothermal energy cannot *truly* be called a renewable energy source, and boreholes need to be decommissioned after a few decades.

CARBON-NEUTRAL

Geothermal gases are rich in various elements and chemical compounds (such as sulfur). Carbon dioxide is present in quantities reflecting this chemical make up, which is distinct to each area. In Krafla (North Iceland), CO_2 makes up 90–98 percent, the rest being hydrogen sulfide.[21]

Calculations based on the national power company, Landsvirkjun's site study for current North Icelandic geothermal developments reveal that the 400 MW of boreholes planned for a single Alcoa smelter in Húsavík will release 1,300 tons CO_2 per MW.[22] An average gas powered plant would produce only slightly more—1,595 tons

17   Þórleifsdóttir. Á. 2007. "Geothermal Exploitation in the Reykjanes Peninsula Area." *Saving Iceland Winter Conference*, 01-12-2007. Reykjavík.

18   E.g. VGK (2005), Environmental Impact Assesment for Helisheidarvirkjun [online]. http://www.vgk.is/hs/Skjol/UES/SH_matsskyrsla.pdf [Accessed August 15, 2007].

19   Rybach, L., 2003. "Geothermal energy: sustainability and the environment." *Geothermics* 32, 463–470.

20   Sæmundsson, K. (2006). "Assessing Volcanic Risk in North Iceland." ISOR—Icelandic Geosurvey. http://www.hrv.is/media/files/Volcanic percent20risk_web.pdf [Accessed August 4th, 2008].

21   Landsvirkjun (2008). Krafla key figures and specifications [online]. http://www.lv.is/EN/article.asp?catID=277&ArtId=306 [Accessed 13-12-2008].

22   Sigurðardóttir, R. Unpublished. "Energy good and green." *Bæ bæ Ísland* (Bye Bye Iceland), to be published by the University of Akureyri and Akureyri Art Museum.
The data in this study is arrived at by calculation of the figures in site surveys for the Krafla, Bjarnarflag and Þeistareykir geothermal plants. Sigurðardóttir has experienced threats and harassment by Landsvirkjun, the national power company, since 2000. In that year, she concluded the formal environmental impact assessment for a proposed large dam, Þjórsárver, a Ramsar treaty area, by stating there were significant, irreversible environmental impacts. The national power company did not pay her and refused to publish the report. Since then, Sigurðardóttir has been refused all Icelandic government commissions. Since then, practically all EIAs for geothermal and hydro plants and smelters have been commissioned to the companies HRV and VGK, construction engineers rather than ecological consultancies and "the leading project management and consulting engineering companies within the primary aluminum production sector" (HRV. 2008. Primary aluminium production [online]. http://www.hrv.is/hrv/Info/PrimaryAluminumProduction/ [Accessed 13-12-2008]).

per MW.[23] The total of 520,000 tons CO_2 for these fields alone is almost equivalent to all road transport in Iceland.[24]

In Iceland, a single site emitting over 30,000 tons requires an emissions permit. Conveniently, figures for current geothermal power stations hover just under that figure. Either way, Icelandic authorities do not consider emissions from geothermal plants anthropogenic and do not include them in greenhouse gas inventories, although currently operating plants emit 8–16 percent of the country's total emissions.[25]

MINIMAL ENVIRONMENTAL IMPACT

Geothermal fluids contain high concentrations of heavy metals and other toxic elements, including radon, arsenic, mercury, ammonia, and boron, which are damaging to the freshwater systems into which they are released as wastewater. Arsenic concentrations of 0.5 to 4.6 ppm are found in wastewater released from geothermal power plants; the WHO recommends a maximum 0.01 ppm in drinking water.[26] Hydrogen-sulfide (H_2S) is a main component of geothermal steam and is responsible for the rotten egg smell of geothermal areas. It is corrosive and classified as very toxic.[27] H_2S is a heavy gas and can linger in valleys, polluting local populations,[28] it forms sulfur-dioxide (SO_2) in the atmosphere causing acid rain. Geothermal power accounts for 79 percent of Iceland's H_2S and SO_2 emissions.[29]

In 2004, sulfur pollution in Reykjavík reached levels regarded as "dangerous."[30] In 2008, sulfur pollution from the Hellisheiði power station, thirty km away, was reported to be turning lamp-posts and jewelry in Reykjavík black. A record number of objections were filed to two more large geothermal plants in the same area, which would have produced more sulfur and carbon emissions than the planned smelter they were supposed to power, and plans were put on hold.

In the North the town of Reykahlið will become exposed to 32,000 tons of H_2S per year if the geothermal power plants (for which feasibility studies are now complete) are built.[31] High levels of sulfur pollution are associated with increased mortality from respiratory diseases.[32]

23   US Govt. Energy Information Administration. 2008. Voluntary reporting of greenhouse gases program. [online]. http://www.eia.doe.gov/oiaf/1605/coefficients.html [Accessed 13-12-2008].

24   Ministry of the Environment, Iceland (2006). Iceland's Fourth National Communication on Climate Change. http://unfccc.int/resource/docs/natc/islnc4.pdf [Accessed August 15, 2007].

25   Armannsson, H., Fridriksson, T., Kristjansson, B.R., 2005. "CO_2 emissions from geothermal power plants and natural geothermal activity in Iceland." Geothermics 34, 286–296.

26   Kristmannsdottir, H., Armannsson, H., 2003. "Environmental aspects of geothermal energy utilization." Geothermics 32, 451–461.

27   European Economic Community. 1967. Council directive 67/548/EEC on the approximation of laws, regulations and administrative provisions relating to the classification, packaging, and labelling of dangerous substances. Brussels, Belgium.

28   Ibid, 23.

29   Statistics Iceland. 2007. Emission of sulphur dioxides (SO_2) by source 1990–2006 [online]. http://www.statice.is/Statistics/Geography-and-environment/Gas-emission [Accessed 12/12/2008]

30   Benediktsson, O. 2004. "Open letter to the minister for the environment regarding operating licenses for an anode factory at Katanes in Hvalfjordur." University of Iceland, Reykjavík.

31   Ibid, 20.

32   Shwela, D. 2000. "Air pollution and health in urban areas." Review of Environmental Health. 15, 13–42.

Landscape impact is another significant factor. Each geothermal borehole drilled only produces a few megawatts of power, and may be located across a large area, and connected to the main power station with pipes and roads. Numerous test holes are drilled for every borehole that goes into production. A currently ongoing project, the proposed expansion of Hellisheiði, demands more than 100 boreholes in a stunning area of wilderness, and provides 160 MW—less than half of what is needed by the smelter it will power.[33]

Areas like Hellisheiði are globally rare, very beautiful, and scientifically interesting. Icelandic geothermal areas are characterized by strikingly colorful landscapes, hot springs, lavas and glaciers, and are biologically and geologically endemic to the country. In the extreme conditions of heat and salt found at each hot spring or cave, extremophiles, unique mosses, and bacteria develop, such as *Hveraburst*, a heat tolerant moss found only in Iceland's Hveragerði hot spring area. Research into these primeval species is in its infancy, and already has led to greater understanding of the formation of life on earth and the possibilities of evolution of extra-planetary life. Irreversible disturbance to these wild areas for power plants includes roads, power lines, heavy lorries, and loud drilling equipment. It has also been suggested that depletion of one geothermal reservoir can result in the drying of surrounding hot spring areas.[34] Thus the direct environmental impact of geothermal extraction may be much larger than previously thought, and landscape is a key consideration.

100 PERCENT RENEWABLE, DOUBLE THE EMISSIONS

In conclusion, the impacts from geothermal energy on a large scale such as is currently happening in Iceland, are greater than generally assumed. As regards climate issues, Iceland may end up in an extraordinary position. The Icelandic ministry of environment has calculated that if only some of the planned industrial projects continue,[35] greenhouse gas emissions in 2020 will be 63 percent higher than in 1990 (assuming that emissions from geothermal and hydro plants are nil).[36] If all projects continue and emissions are taken into account, Iceland's climate footprint, powered by 100 percent "green" energy, could double (again, this figure excludes emissions from geothermal or hydro plants).

This is made possible because the country was not just granted a generous 10 percent increase under Annex 1 of the Kyoto Protocol, but also took advantage of a specific exemption for emissions of heavy industry powered by "renewables."

33   VGK (2005), Environmental impact assesment for Helisheidarvirkjun [online]. http://www. vgk.is/hs/Skjol/UES/SH_matsskyrsla.pdf [Accessed August 15, 2007].

34   Ibid, 24.

35   Enlargement of RT-Alcan and Century smelters, of the Icelandic Alloys/Elkem steel factory and construction of an Anode plant. This does not include the new Century Aluminum (Helguvik) and Alcoa (Husavik) smelters. Century has recently received an emissions permit for the new smelter, but Alcoa hasn't. RTA is not expanding production at its smelter by as much as originally planned, and the status of the Elkem expansion and Anode plant is currently unclear.

36   Ministry of the Environment. 2006. Iceland's Fourth National Communication on Climate Change. Ministry of the Environment, Reykjavik, Iceland.

Iceland has also been mentioned in proposals for a European (or even global) green energy super grid.[37] The calculations brought forward here suggest that it is not worthwhile to replace gas-powered plants by Icelandic geothermal. If that electricity is to be used for growth of heavy industry, it is quite arbitrary for the climate whether that would be in Iceland or mainland Europe. The aluminum industry is set to increase its emissions by a fifth by 2020 (See Appendix 1: The aluminum industry, climate, and green energy) and this includes its embrace of non-fossil energy.

As an alternative, Landsvirkjun has taken to lobbying data center corporations, silicon refineries, and other energy-intensive industries with better public images than Rio Tinto to come to Iceland. If such plans go ahead, Iceland would become a large hard disk for the global Internet. Again, moving gas-powered servers from Europe to geothermal-powered servers in Iceland does not significantly decrease emissions.

And there is another reason not to embrace these projects. Wilderness areas are becoming rare globally, with over 83 percent of the earth's landmass directly affected by humans, and the Icelandic wilderness is one of the largest left in Europe.[38] It provides important regulating ecosystem services and has aesthetic, scientific, medical, cultural, and spiritual significance for humans. However, we believe all landscapes, ecological systems, and forms of life have their own intrinsic value and right to develop for themselves, rather than for the sole benefit of mankind. We believe the dominant world-view that sees the natural world as a collection of "resources" has greatly contributed to severe ecological and social crises. To recover from the consumption paradigm, we must redefine our environmental ethic and what it means to be human to include a profound sense of the fragile and beautiful interconnection of life on earth.

Proponents of heavy industry in Iceland have stated that it is the country's "ethical obligation" to sacrifice the country's wild areas for the sake of the environment.[39] While this is, more likely than not, moral opportunism on the side of those who are to benefit from the projects, the technological or pragmatic environmentalism in favor of super grids and mega data centers comes down to a proposal to sacrifice unique ecological areas for the greater good of living a resource-intensive lifestyle, "sustainably." In contrast, for anyone who identifies with a natural area, it is easy to understand why it has a value of its own. Given the rarity of wild lands in this context, the value can be seen as far greater than that of any of our possessions; it is in a sense, invaluable.

What can perhaps be concluded from this Icelandic green energy case study

---

37   E.g: Monbiot, G. 2008. Build a Europe-wide "super grid" [online]. http://e-day.org.uk/solutions/charities/14536/george-monbiot-build-a-europewide-super-grid.thtml [Accessed 13-12-2008].

38   Columbia University and Wildlife Conservation Society. 2008. "Last of the wild database and human footprint atlas." Center for International Earth Science Information Network, Columbia University. http://www.ciesin.columbia.edu/wild_areas [Accessed 13-12-2008].

39   The Economist. 2008. "Testing metal—when thinking globally requires unpleasant action locally," Economist.com, Green.view, 29-9-2008. http://www.economist.com/world/international/display-story.cfm?story_id=12323257 [Accessed 12-12-2008].

is that application of a technology that has been thought of as renewable, climate-friendly, and low-impact on the large scale that is associated with fossil fuels, makes it a lot like the technology it was supposed to replace. It has certainly been argued that technological systems tend to reproduce themselves independent of the specific technologies.[40][41] Simply applying a different technology to address issues that are not entirely technological is not addressing the problem of our over-consumptive life-styles. But it can end the existence of a place that is not like any other, irrevocably.

---

40   E.g. Mander, J. 1992. *In the Absence of the Sacred*. Sierra Club, San Francisco, CA.
41   Krater, J. 2007. "Duurzame technologie, een contradictie?" *Buiten de Orde*, Summer 2007.

APPENDIX 1. THE ALUMINUM INDUSTRY, CLIMATE, AND GREEN ENERGY

The aluminum industry is the world's most energy–intensive industry, and also one of the most polluting.[1] Aluminum is derived from bauxite soils, mainly found in the tropics and subtropics. Five tons of bauxite is strip-mined to produce one ton of aluminum. Large-scale deforestation of tropical forests, caused by shallow open cast mining, creates soil erosion and water pollution, and has displaced and destroyed the livelihood of numerous indigenous peoples in Australia, India, Brazil and elsewhere, a process which continues to this day.[2] Bauxite is refined to produce alumina and leave red mud, a caustic mixture of heavy metals and radionuclides, which is known to cause silicosis, cancer, and other diseases associated with radiation.[3]

Alumina is smelted using carbon anodes and aluminum fluoride to remove the strongly-bonded oxygen. This part of the process is the most energy intensive and produces inorganic fluorides, SO_2, CO_2, and perfluorocarbons (very strong greenhouse agents) in the airborne waste, as well as solid spent pot linings containing cyanides and fluorides. Approximately 30 percent of aluminum is used for arms production and defense, the remainder is used for cars, planes, and construction, packaging, and disposables.[4]

---

1   Switkes, G. 2005. *Foiling the Aluminum Industry: A Toolkit for Communities, Activists, Consumers, and Workers*. International Rivers, Berkeley, CA.
2   Das, S. and Padel, F. 2006. "Double Death: Aluminium's Links with Genocide," *Social Scientist* 34 (3/4), 55–81. For example, the Dongria Kondh in Orissa, Eastern India are under threat of being forcefully removed from their land to allow mining of Niamgiri mountain, a rich bauxite reserve, by Vedanta, a UK-based mining corporation. Pressured by Vedanta, the Indian supreme court removed the Dongria's constitutional right as tribal people to decide on development of their land. (*Survival International*. 2008. Dongria Kondh [online]. http://www.survival-international.org/tribes/dongria [Accessed 13-12-2008]).
3   Cooke, K. and Gould, M.H. 1991. "The Health Effects of Aluminium, A Review." *Journal of the Royal Society for the Promotion of Health*. 111, 163–8.
4   Das, S. and Padel, F. 2010. *Out of This Earth: East India Adivasis and the Aluminum Cartel*. Orient Black Swan, New Delhi.

CRADLE TO GRAVE

Metal giants have not enjoyed a particularly good reputation. Rio Tinto was described by motion in the British parliament in 1997, as "the most uncaring and ruthless company in the world" for human rights, anti-unionizing, and total disregard for indigenous people,[5] and was pulled up again in 2000 for war crimes, environmental destruction, and racism.[6] Recently, the corporation was thrown out of the Norwegian Government pension fund for similar reasons.[7]

Century Aluminum's Icelandic smelter has been accused of forcing injured workers back to work[8] and of producing illegal amounts of fluorine pollution causing health problems.[9] The company is working with the Sassou government of Congo-Brazzaville, a single-party regime that came to power in fraudulent elections in 2002, to develop large-scale, open-cast bauxite mining.[10] Its bauxite mining and refining[11] in Jamaica has been responsible for large-scale rainforest destruction and water pollution.[12]

Alcoa has been convicted numerous times for toxic waste dumping in the US,[13] old-growth and rainforest destruction and displacement of indigenous people in countries such as Brazil,[14] Suriname,[15] and Australia.[16] Alcoa has lost popularity in Iceland for its intimate association with the US military, which is categorically de-

---

5   Clapham, M., UK Parliament, House of Commons. 1998. Rio Tinto Corporation. Early day motion 1194. HMSO, London.

6   UK Parliament, House of Commons. 2000. Weekly Information Bulletin, 16-12-2000. HMSO, London.

7   *Survival International.* 2008. "Norway sells shares of unethical Rio Tinto" [online]. http://www.survival-international.org/news/3700 [Accessed 16-12-2008].

8   Morgunblaðið. 2008. Injured Century and Elkem workers forced back to work [online]. http://www.mbl.is/mm/frettir/innlent/2008/08/11/thryst_a_ad_ovinnufaerir_starfsmenn_snui_aftur_til_/ [Accessed 14-12-2008].

9   Iceland Review. 2008. Pollution from smelter damages teeth in sheep [online]. http://www.icelandreview.com/icelandreview/daily_news/?cat_id=16539&ew_0_a_id=309548 [Accessed 14-12-2008].

10   AZ Materials News. 2007. Century Aluminium to Build Aluminium Smelter in Republic of Congo. http://www.azom.com/News.asp?NewsID=7734 [Accessed 20-6-08].

11   Zadie Neufville. April 6, 2001, "Bauxite Mining Blamed for Deforestation" [online] http://forests.org/archive/samerica/bauxmini.htm [Accessed 20-6-08]. Mines and Communities report, "Bauxite Mine Fight Looms in Jamaica's Cockpit Country," 24 October 2006. http://www.minesandcommunities.org/article.php?a=6513 [Accessed 20-6-08]. Al Jazeera (2008). Environmental damage from mining in Jamaica, June 11, 2008 News. Available through http://www.youtube.com/watch?v=vJa2ftQwfNY&eurl=http://savingiceland.puscii.nl/?p=2192&language=en [Accessed 20-6-08]

12   Al Jazeera (2008). Environmental damage from mining in Jamaica, June 11, 2008 News. Available through http://www.youtube.com/watch?v=vJa2ftQwfNY&eurl=http://savingiceland.puscii.nl/?p=2192&language=en [Accessed 20-6-08]

13   Fernandes, S. 2006. "Smelter struggle: Trinidad fishing community fights aluminum project." *CorpWatch.*

14   Lynas, M. 2004. "Dammed Nation." *The Ecologist* 33 (10).

15   Gaspar, R. 2007. "Prosecutor states that impacts caused by Alcoa in Pará are serious." *Amigos da Terra Amazônia Brasileira.* http://www.amazonia.org.br/english/noticias/noticia.cfm?id=242981 [Accessed 14-12-2008].

16   Western Australia Forest Alliance. 2008. Alcoa clearing Jarrah forest [online]. http://www.wafa.org.au/articles/alcoa/index.html [Accessed 14-12-2008].

nied by Alcoa Iceland (although it has a website dedicated to its military products).[17] In Honduras, an Alcoa car parts factory was accused of treating workers worse than sweatshops do. The basic pay of 74¢ an hour covered 37 percent of an average family's most essential needs, and in the last three years, wages fell by 13 percent. Workers would be forced to urinate and defecate in their clothes after being repeatedly denied to use the bathroom and women would have to take off clothes to prove they were menstruating. Protests by workers in 2007 led to 90 percent of the trade union leaders being fired.[18]

Nonetheless, Alcoa claims to be one of the world's most ethical and sustainable companies, according to a host of international awards listed by the company.[19] Their website (subtitled "Eco-Alcoa"—"Click here to see how Alcoa is part of the solution") is dominated by articles on community projects and energy saving initiatives, and with former Greenpeace and WWF directors at the helm, they are doing well to promote a green image. In a recent presentation, Alcoa state they are on the cutting edge of green corporate thinking, embracing recycling and green energy, and even claiming they will be carbon-neutral, as an industry, by 2020.[20] Are these promises coming true?

RECYCLING

Recyclability of aluminum is probably the most important selling point for the industry: "It's more like reincarnation than recycling."[21] Recycling aluminum is indeed 95 percent more efficient than primary production; still, it takes the same amount of energy as producing new steel.[22] Alcoa sources only 20 percent of its aluminum from recycling. Overall recycling rates are 33 percent and, according to US Aluminum Association figures, going down.[23]

RENEWABLE ENERGIES

The aluminum industry has long been closely tied to the hydro-industry,[24] and over

---

17   Magnason, A.S., 2008. *Dreamland*. Citizen Press, London.

18   National Labor Committee and COMUN. 2007. The Wal-Martization of Alcoa: Alcoa's high-tech auto parts sweatshops in Honduras rocked by corruption and human rights scandal; a major challenge to CAFTA [online]. http://www.nlcnet.org/article.php?id=447 [Accessed 14-12-2008].

19   Alcoa. 2008. External awards [online]. http://www.alcoa.com/global/en/about_alcoa/sustainability/home_external_awards.asp [Accessed 14-12-2008].

20   Overbey, R. 2005. Sustainability, what more should companies do?, In Alcoa Conference Board Session on Sustainability. Alcoa. http://www.alcoa.com/global/en/news/pdf/conference_board.pdf [Accessed 12-12-2008].

21   Ibid., 63.

22   Das, S. and Padel, F. (unpublished). *Out of this earth: East India Adivasis and the Aluminium cartel*.

23   Container Recycling Institute. 2004. Aluminum can waste reaches the one trillion mark—recycling rates drop to lowest point in 25 years [online]. http://www.container-recycling.org/assets/pdfs/trillionthcan/UBC2004CRIPressRel.pdf [Accessed 12-11-2007]. Institute, C.R. 2006. Aluminum can sales and recycling in the US 1996-2006 [online]. http://www.container-recycling.org/images/alum/graphs/recsale-tons-96-06.gif [Accessed 12-12-2008].

24   McCully, P., 2001. *Silenced Rivers: The Ecology and Politics of Large Dams*. Blackwell Publishing, New York.

half of smelting is hydro-powered.[25] Due to the low economic return per energy unit, smelting is increasingly geared towards countries with low energy and labor costs,[26] whether hydro (e.g. Brazil, Congo, Iceland, Greenland), natural gas (Trinidad, Congo-Brazzaville), or coal (South Africa, India). Indirect greenhouse gas production from dams and geothermal power stations are not included in the industry's audits.

REDUCING GREENHOUSE GAS EMISSIONS

Aluminum production accounts for ca. 1 percent of global greenhouse gas emissions, producing 13.1 tons of CO_2 equivalent per ton of aluminum.[27] Technological advances have led to 20–25 percent emissions savings in the smelting process in recent decades, but overall emissions are increasing and there is no concrete intention to reduce them. In fact, Alcoa predicts a 20 percent increase of CO_2 emitted per year from ca. 335 million tonnes of CO_2e in 2000 to ca. 400 million tonnes in 2020 (see Figure 1).[28]

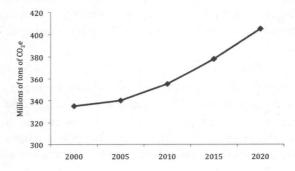

Figure 1: Projection of greenhouse gas production by the aluminum industry (Based on Overbey/Alcoa 2005)[29]

CARBON NEUTRAL

However, Alcoa states that around that time, cars will contain more aluminum, be lighter and thus save fuel. This saves carbon emissions, and in 2017, the amount saved will be roughly the same as the increase in emissions by the aluminum industry. Thus, the industry can be carbon neutral whilst producing 20 percent more greenhouse gases. The fallacy of this reasoning is easy to see: imagine we would drive even more and in larger vehicles than Alcoa is projecting. In that case, the industry

25   Harnisch, J., Wing, I.S., Jacoby, H.D., Prinn, R.G., 1999. Primary aluminum production: Climate policy, emissions and costs. Epd Congress 1999, 797-815.

26   Switkes, G. 2005. *Foiling the Aluminum Industry: A Toolkit for Communities, Activists, Consumers, and Workers.* International Rivers, Berkeley, CA, p. 69.

27   Das, S. and Padel, F. 2006. "Double Death—Aluminium's Links with Genocide." *Social Scientist*, 34 (3/4), 55–81.

28   Overbey, R. 2005. Sustainability, what more should companies do?, In Alcoa Conference Board Session on Sustainability. Alcoa. URL http://www.alcoa.com/global/en/news/pdf/conference_board.pdf [Accessed 12-12-2008].

29   Ibid., 73.

would be carbon neutral even earlier: if I buy an aluminum Hummer, I save more than when I buy an aluminum Fiesta. Even if crediting would work that way, Alcoa assumes the aluminum industry gets all the credits, not the car manufacturer or consumer.

The aluminum industry, like all mining industries, has a severe environmental impact and a consistent record of human rights violations. Because the industry is, in all aspects, "part of the problem," it is vitally important for corporations such as Alcoa to join the green bandwagon and proclaim that "it is part of the solution." However, ecologically-responsible primary aluminum production is not a reality. If Iceland is the model for green-heavy industry, one must question whether that is possible at all.

APPENDIX 2: HOW POWER PLANTS CONTRIBUTED TO ECONOMIC INSTABILITY

BY JAAP KRATER, IN *MORGUNBLAÐIÐ* AND *ICELAND REVIEW*, 22-8-2008

In times of economic crisis, it is tempting to embrace new megaprojects such as new power plants and aluminum smelters. But will this realistically improve Iceland's economic prospects?

Prime minister Geir Haarde recently explained on the talkshow, *Mannamál,* that one of the main reasons for the fall of the Krona was the execution of heavy industry projects, like the construction of Kárahnjúkar and Alcoa's smelter in Reyðarfjörður. Haarde's comments were not surprising. Before construction of Kárahnjúkar many economists predicted the negative impact on inflation, foreign debt, and the exchange rate of the ISK. If more large projects are undertaken, what will the cost be for the Icelandic taxpayer?

Of course there are some economic benefits from new smelters, but that "is probably outweighed by the developments' indirect impact on demand, inflation, interest rates and the ISK exchange rate," stated a report by Glitnir in 2006 on the impact of aluminum expansion in Iceland. The report expected an increase in inflation and a depreciation of the ISK.

"Kárahnjúkar will never make a profit, and the Icelandic taxpayer may well end up subsidising Alcoa," said the eminent economist Thorsteinn Siglaugsson, after publishing another report on the profitability of the Alcoa dam in East Iceland before construction commenced.

How did the Fjardaál smelter contribute to Iceland's economic crisis? The state had to borrow the $2 billion for the construction of the country's largest dam. That led to a more than significant increase in the current deficit, which is now felt in increased inflation and depreciation of the currency. The economic cost now needs to be coughed up.

Note that any schemes that demand new power plants associated with a significant amount of borrowed capital will have this effect, whether it is an expensive dam or a power plant meant for aluminum, a silicon refinery, a data center, or some other similar facility.

It is quite simple: If you borrow money, you will have to pay back in one way or the other.

Of course, once they are built, smelters bring in some amount of income to the country and, so it is argued, there are local economic benefits from a new smelter. They provide jobs. What has hardly been researched in Iceland, though, is how much these new jobs displace jobs in existing local industries.

Industries around Reyðarfjörður have had to shut down as a consequence of employment competition from the smelter. Many of the new houses that were built are empty. Between 2002–2008, an average of seventy-three more people moved each year from the Eastfjords to the southwest than the other way round. The smelter still depends on many foreign workers. Local communities where large projects such as Fjardaál get constructed become completely dependent on foreign investment, an undesired and unsustainable condition that destroys local resilience.

There is another reason not to construct more smelters in Iceland. The price that the aluminum giants pay for energy to Landsvirkjun is linked to the world price of aluminum. If supply is increased this will lower the price of aluminum, decreasing revenue for Iceland. One might think that a few hundred-thousand tons of aluminum will not impact the global market. The reality is that it is not the sum of production that determines the price but rather the friction between supply and demand. A small amount of difference can have a significant effect in terms of pricing. Demand for aluminum is already slumping in the US and Europe. It will in China too when growth slows down there, which is likely to happen before Alcoa's and Century's planned new smelters could come online, considering the world economic outlook.

The metal corporations compete amongst themselves. Because of this, it is not just the global price that determines their profitability. The bottom line is eventually determined by how cheaply they can produce. For aluminum, profitability is fundamentally determined by one thing: energy costs. In Iceland, energy prices are rock bottom—the lowest in the world. It is not a coincidence that as Alcoa's Fjardaál smelter went online, 400 workers in Rockdale, Texas were laid off as smelter operations there closed down. In the US, Alcoa pays much more for power.

This is why Alcoa, Century, Rio Tinto, and Norsk Hydro all want new smelters in Iceland and in third world countries with cheap energy such as Trinidad and the Congo. When demand slumps, expensive plants can then be shut down in favor of cheap ones, such as the proposed smelters at Husavik and Bakki. As inflation stays high and energy revenues low, the Icelandic taxpayer pays the price.

Construction of new power plants, smelters or other large scale projects will have some short term economic benefit as funds are infused into the economy, but, as Geir Haarde recently confirmed, after execution comes the economic backlash. These megaprojects in a small economy have been compared to a "heroin addiction."

Short-term "shots" lead to a long-term collapse. The choice is between a short-term infusion or long-term sustainable economic development.

The "shot" of Fjardaál overheated the Icelandic economy. What was called the "Kárahnjúkar problem" led to an all-time high in the value of the Krona, hurting export and the fish industry in particular. With the all-powerful currency, banks overplayed their hand and went into a spending spree. Drugs make you lose sight of reality.

There has been a lot of critique of the proposed plans to develop Iceland's unique energy resources. Those in favor of it have generally argued that it is good for the economy. Anyone who gives it a moment of thought can conclude that that is a myth. Supposed economic benefits from new power plants and industrial plants need to be assessed and discussed critically and realistically. Iceland is coming down from a high. Will it have another shot, or go cold turkey?

FACT SHEET
"Clean Coal" Power Plants[1]

Nancy LaPlaca

COAL-FIRED POWER GENERATION

More than half of the electric power generated in the US comes from coal-fired power plants, which are also the largest single source of greenhouse gases. Coal-fired power plants emit:

- 66 percent of sulfur dioxides (SOx, or acid rain)
- 40 percent of carbon dioxide (CO_2)
- 33 percent of mercury
- 22 percent of nitrogen oxides (NOx).[2]

Coal is the most CO_2-intensive fossil fuel, emitting about three pounds of CO_2 for every pound of coal burned. The US burns over 1 billion tons of coal every year. There are 492 coal-fired power plants in the US, with an average size of 667 megawatts (MW) and an average age of 40 years.[3] A 500 MW coal-fired power plant produces about 3 million tons/year of CO_2, adding a total of approximately 1.5 billion tons/year of CO_2 to the atmosphere. If 60 percent of the CO_2 from all these plants were captured and compressed to a liquid for geologic sequestration, its volume would equal the US oil consumption of 20 million barrels/day.[4] A large coal-fired power plant emits the equivalent CO_2 of 1 million SUVs. Coal-fired electrical generation has been the largest single source of pollution in the US (and the world) for over thirty years.

WHAT IS IGCC?

IGCC (Integrated Gasification Combined Cycle) is a type of power plant that gasifies coal into synthetic gas (syngas) to power a gas turbine. The heat from the gas turbine exhaust then generates steam to run a steam turbine. None of the basic

---

1   Originally published on the Energy Justice website at: http://www.energyjustice.net/coal/igcc/. Republished here with permission from the author. This text was written during the Bush period, and unfortunately, the author did not have time to update the text. However, the basic information and criticisms directed against "clean coal" as a solution to the current energy/climate crisis still remain valid. The version reproduced here has been slightly shortened owing to space limitations in the book.

2   Ilan Levin and Eric Schaeffer, *Dirty Kilowatts: America's Most Polluting Power Plants*, Environmental Integrity Project, May 2005. http://dirtykilowatts.org/Dirty_Kilowatts.pdf.

3   "Form EIA-860 Database, Annual Electric Generator Report," US Department of Energy's Energy Information Administration, 2005 data set. http://www.eia.doe.gov/cneaf/electricity/page/eia860.html.

4   Massachusetts Institute of Technology, *The Future of Coal: Options for a Carbon-Constrained World*, 2007, Executive Summary, p. ix. http://web.mit.edu/coal/.

technologies—coal gasification, gas turbines, and steam turbines—are new. It is the *integration* of these into electric power plants that is new, and presents engineering challenges.

There are 160–250 proposed new coal-fired power plants in the US; 32 proposed to be IGCC.[5] A September 2004 study, commissioned by the US Department of Energy (DOE), found that, despite a long history of gasification, only two gasified coal plants whose primary output is for electrical generation have been built.[6]

Although IGCC is promoted as being capture "ready," the key word is "ready"— no IGCC plants are actually capturing and storing CO_2 in commercial quantities.

TWO CURRENTLY OPERATING IGCC PLANTS IN THE US

The two IGCC plants currently operating in the US are the Polk plant in Tampa, Florida and Wabash River in Indiana. Although many petroleum and chemical plants employ gasification, the Polk and Wabash River plants use coal to generate electrical power with combined cycle turbines. Very little research has been done on using low rank sub-bituminous coal, such as Powder River Basin (PRB) coal. Existing plants use bituminous coal.[7]

According to Xcel Energy, the fourth largest electrical utility in the US, it costs more to use western coals, such as Powder River Basin, as IGCC feedstock.[8] Western sub-bituminous coal decreases plant performance due to its higher moisture content and lower heat value compared to eastern bituminous coal.

IGCC plants burn either coal or "petroleum coke," an oil refinery residue. All contain high levels of toxins, and "pet coke" contains high levels of sulfur.

Generally, conventional pulverized coal (PC) plants operate at 32–38 percent efficiency, while IGCC plants operate at 36-39 percent efficiency.[9] However, capturing CO_2 increases costs significantly, and has only been demonstrated at a handful of sites, in amounts that are a small fraction of total CO_2 emissions.

IGCC FEASIBILITY

Bush administration policies ramped up the push for "clean" coal, and Obama's policies continue in a similar vein.[10] A number of studies have looked at "market barri-

5   NETL (National Energy Technology Laboratory), Department of Energy, *Tracking New Coal-Fired Power Plants: Coal's Resurgence in Electric Power Generation*, January 24, 2007, p 24.

6   Booz Allen Hamilton, *Coal-Based Integrated Gasification Combined Cycle (IGCC): Market Penetration Recommendations and Strategies*, study for the Department of Energy's National Energy Technology Laboratory, September 2004, p. ES-1.

7   EPA Final Report, Environmental Footprints and Costs of Coal-Based Integrated Gasification Combined Cycle and Pulverized Coal Technologies, July 2006, EPA-430/R-06/006, p. ES-1. http://www.epa.gov/air/caaac/coaltech/2007_01_epaigcc.pdf.

8   Xcel Energy PowerPoint presentation, *Colorado IGCC Demonstration Project, An Overview of Project Concepts and Objectives, Prepared by Xcel Energy*, February 2006, slide 7 of 16.

9   Minnesota Pollution Control Agency, Comparison of Nitrogen Oxides, Sulfur Dioxide, Particulate Matter, Mercury and Carbon Dioxide Emissions for IGCC and Other Electricity Generation, p. 7, Docket E-6472/M-05-1993.

10   P.L. 109-58: The Energy Policy Act of 2005. http://legalectric.org/f/2007/01/ago_docs-_1696085-v1 excelsior_energy_final_emission_comparison_anne_jackson.DOC.

ers" to widespread IGCC implementation. IGCC "uncertainties" include lack of standard plant design, lack of a market, performance guarantees, and high capital costs.[11] These uncertainties question whether the technology is commercially viable.

IGCC veteran Stephen D. Jenkins testified in January 2007 that IGCC technology won't be ready for 6–8 years, has limited performance and emissions guarantees, and that commercial-scale CO_2 capture and storage has not been demonstrated.[12]

HIGH COSTS

Capital costs for IGCC plants are estimated to be 20–47 percent higher than traditional coal plants.[13][14] In 2004, Indeck Energy Services testified before the Illinois State EPA that IGCC's "capital costs are 30 percent higher."[15] On top of this, construction costs in general (including concrete, steel, and labor) have risen from 100–300 percent in recent years, driving up the costs of all sorts of power plants.[16] The Department of Energy reports that IGCC is seen as too risky for private investors, and requires large subsidies from the federal, state, and local governments.[17]

In 2006, the EPA estimated that capturing 90 percent of CO_2 emissions from IGCC plants would increase capital costs 47 percent, and the total cost of electricity 38 percent.[18] "Capture" does not include transportation of gas or storage. According to the DOE, IGCC is seen as too risky for private investors, and requires enormous subsidies from the federal, state, and sometimes local government.[19] Extensive research is required before a commercial-scale IGCC plant could capture, transport, and store its CO_2.[20]

---

11   Booz Allen Hamilton, p. ES-7.

12   Testimony of Stephen D. Jenkins, Docket No. 07-0098-EI, In Re: Florida Power & Light Company's Petition to Determine Need for FPL Glades Power Park Units 1 and 2 Electrical Power Plant, January 29, 2007, pp. 8, 14, 26. http://www.psc.state.fl.us/library/filings/07/01362-07/07-0120.ord.doc.

13   For example, the Electric Power Research Institute estimates IGCC capital costs at 20 percent higher than for Super Critical Pulverized Coal. "Super Critical" pulverized coal is a plant that burns hotter than traditional pulverized coal plants, and so emits less pollutants from the stack. See Electric Power Research Institute, *Feasibility Study for an Integrated Gasification Combined Cycle Plant at a Texas Site*, Technical Update, October 2006, p. v.

14   William G. Rosenberg, Dwight C. Alpern, Michael R. Walker, *Deploying IGCC In This Decade with 3Party Covenant Financing*, Vol. I, May 2005 Revision, John F. Kennedy School of Government, p. 2. http://bcsia.ksg.harvard.edu/publication.cfm?program=ENRP&ctype=book&item_id=394.

15   Booz Allen Hamilton, *Coal-Based Integrated Gasification Combined Cycle (IGCC): Market Penetration Recommendations and Strategies*, study for the Department of Energy's National Energy Technology Laboratory, September 2004, p. 52.

16   Electric Power Research Institute, p. 1–7.

17   William G. Rosenberg, Dwight C. Alpern, Michael R. Walker, *Deploying IGCC In This Decade with 3Party Covenant Financing*, Vol. I, May 2005 Revision, John F. Kennedy School of Government, p. 1. http://bcsia.ksg.harvard.edu/publication.cfm?program=ENRP&ctype=book&item_id=394.

18   EPA Final Report, Environmental Footprints and Costs of Coal-Based Integrated Gasification Combined Cycle and Pulverized Coal Technologies, July 2006, EPA-430/R-06/006, p. ES-6. http://www.epa.gov/air/caaac/coaltech/2007_01_epaigcc.pdf.

19   William G. Rosenberg, Dwight C. Alpern, Michael R. Walker, *Deploying IGCC In This Decade with 3Party Covenant Financing*, Vol. I, May 2005 Revision, John F. Kennedy School of Government, p. 2. http://bcsia.ksg.harvard.edu/publication.cfm?program=ENRP&ctype=book&item_id=394.

20   William G. Rosenberg, Dwight C. Alpern, Michael R. Walker, p. 6, footnote 10. http://bcsia.ksg.harvard.edu/publication.cfm?program=ENRP&ctype=book&item_id=394.

The DOE initially estimated the total capital cost for the 600 Megawatt IGCC Mesaba plant in Minnesota at $800 million, but the final cost is currently estimated at $2.155 billion or $3,593 per kW, NOT including carbon capture, transportation or storage.[21] In April 2007, Minnesota's Office of Administrative Hearings rejected the Mesaba plant, finding that:

- neither the project nor the IGCC technology is likely to be a least-cost resource;

- emissions of nitrogen oxides (NOx) and mercury are not reduced significantly, and are not lower than currently available control technology for pulverized coal;

- the technology does not qualify as an "Innovative Energy Project;"

- there's no guarantee of carbon sequestration;

- the plant would cost 9–11¢/kWh, and capturing and transporting the carbon would add at least 5¢/kWh.[22]

In 2006, AEP, the largest electricity generator and coal user in the US, estimated capital costs for a traditional pulverized coal plant at $1,700 per kW; IGCC without carbon capture at almost $2,000 per kW; and IGCC with carbon capture at $2,600 per kW.[23] These costs are far below DOE's estimated capital cost for Mesaba.

GASIFICATION CREATES WATER CONTAMINATION

IGCC more closely resembles a chemical plant than a traditional pulverized coal power plant. Using water to clean the gas creates water contamination problems. Coal gasification wastewater has an average pH of 9.8 (pure water has a pH of 7.0, hand soap has a pH of 9.0–10.0, while household ammonia has a pH of 11.5).[24] The principal contaminant of "process wastewater" is NO3 (nitrate). The Great Plains Coal Gasification plant in Beulah, ND generated 4.83 million metric tons of wastewater in 1988. This plant also produced 245,000 metric tons of gasifier ash, which is removed from the bottom of the gasifier unit. In addition, cooling water is bled from the system to prevent the build-up of minerals that would cause scaling and operational problems. This "bleed" is called "cooling tower blowdown," and the Dakota plant generated 766,000 metric tons in 1988.[25] DOE's IGCC pilot project in Wabash River, Indiana found that elevated levels of selenium, cyanide, and arsenic in the

21   US Department of Energy, *Notice of Financial Assistance for Mesaba*, May 23, 2006. http://www.netl.doe.gov/technologies/coalpower/cctc/ccpi/pubs/2006_program_update.pdf.

22   MPUC Docket No. E-6472/M-05-1993, *In the Matter of the Petition of Excelsior Energy Inc. for Approval of a Power Purchase Agreement Under Minn. Statute 261B.1694, Determination of Least Cost Technology, and Establishment of a Clean Energy Technology Minimum Under Minn. Statute 216B.1693*, dated April 12, 2007. http://www.puc.state.mn.us/docs/calendar/weeklypdf/puc072806.pdf.

23   "The Case for Integrated Gasification Combined Cycle Technology," Presentation to the Michigan Public Service Commission, Lansing, MI, August 22, 2006, by Dale E. Heydlauff, Vice President, New Generation.

24   See http://www.wikipedia.com.

25   EPA, *Report to Congress on Special Wastes from Mineral Processing*, Chapter 5, "Coal Gasification," p. 4–5. http://www.epa.gov/epaoswer/other/mining/minedock/damage/damage.pdf.

wastewater caused a permit violation, and that selenium and cyanide limits were "routinely exceeded."[26] Although IGCC theoretically uses less water than traditional coal plants, the added power demand and reduced output due to carbon capture may not result in overall less water use.

CO₂ CAPTURE

IGCC is being promoted by the coal industry as having the potential to "capture" CO_2, however, studies show that capturing CO_2 reduces plant efficiency and increases water use. According to the Electric Power Research Institute, installation of CO_2 capture equipment has been found to decrease plant output by at least 25 percent,[27] while installation of CO_2 capture equipment increases water consumption by approximately 23 percent.[28]

Additional "capture" costs beyond the plant gate, plus transportation and storage costs, are not factored into the efficiency loss or cost increase.

A July 2006 EPA report estimated CO_2 capture costs at \$24/ton, and says that "widespread introduction" of carbon capture and sequestration technology into the commercial market is "highly uncertain."[29]

CO₂ TRANSPORT

Pipeline costs must be added to total estimated CO_2 capture and storage costs.[30] If stored CO_2 leaks out, the concentrated CO_2 can cause suffocation because it is heavier than air[31]—in 1986, a large release of CO_2 from a volcanic crater, Lake Nyos in West Africa, suffocated and killed 1,700 people. A similar event happened at Lake Monoun in Cameroon. Researchers continue to work on degassing the lakes to prevent another tragedy.[32] Further research is needed on CO_2 migration and seismic shifts from storing large amounts of CO_2 underground.

Pipeline costs for the proposed Mesaba IGCC plant in Minnesota were estimated at between \$25,000 and \$60,000 per inch (diameter of the pipe) per mile plus the cost of repressurization stations to keep the gas flowing.[33] A natural gas pipeline costs

---

26   *Wabash River Coal Gasification Repowering Project Final Technical Report*, DE-FC21-92-MC29310, page 6-14). http://www.osti.gov/bridge/servlets/purl/787567-a64JvB/native/787567.pdf.

27   Electric Power Research Institute, p. v.

28   Id. The EPRI study is the first to evaluate IGCC with CO_2 capture using low rank, high moisture Powder River Basin (PRB) coal.

29   EPA Final Report, p. ES-6, and p. 5.1.

30   Prepared Rebuttal Testimony and Exhibits of Excelsior Energy Inc. and MEP-I LLC, Edward N. Steadman, October 10, 2006, Minnesota Public Utilities Commission Docket No. E-6472-/M-05-1993, p. 42, lines 21–22. http://www.excelsiorenergy.com/pdf/Regulatory_Filings/Docket_E6472_M-05 1993/20061011Rebuttal/Mesaba percent20Docket percent20- percent20EE percent20- percent2025 percent20Rebuttal percent20B.Jones percent202006.10.10.pdf.

31   Prepared Rebuttal Testimony, Edward N. Steadman, October 10, 2006, p. 44, lines 14–15.

32   "Degassing Lakes Nyos and Monous: defusing certain disaster," by Kling, G.W., Evans, W.C., Tanyileke, G., et al., Dept. of Ecology and Evolutionary Biology, U. of Michigan, Ann Arbor, Proceedings of the Nat'l Academy of Sciences of the US, 2005 Oct. 4: 102(40): 14185-90. Epub 2005 Sep 26. http://www.geochemicaltransactions.com/pubmed/16186504.

33   Prepared Rebuttal Testimony, Edward N. Steadman, October 10, 2006, p. 44, lines 25–26.

about \$2–4 million/mile, using a 30 inch pipeline.[34]

CO2 STORAGE AND SEQUESTRATION

CO_2 sequestration differs from "storage" in that it is a more permanent storing of the gas, and must be stored without leaking for thousands of years. We have been unable to safely store solid and liquid radioactive wastes for 50–60 years without leakage. It's unlikely that we'll be able to store a significant part of the world's 28 billion metric tons of CO_2 gas emitted every year without leakage problems. The Minnesota Department of Commerce estimated CO_2 sequestration costs for Mesaba at roughly \$1.107 billion in 2011; and pipeline costs at \$635.4 million.[35]

Carbon sequestration costs are highly uncertain; the National Energy Technology Laboratory states, "the economics of CO_2 recovery are poor in all scenarios...."[36]

A December 2006 DOE Environmental Impact Statement reported that geologic sequestration of CO_2 "is not a reasonable option because [the] technology is not sufficiently mature to be implemented at production scale during the demonstration period for the proposed facility," and it isn't expected to be "technically practicable" for largescale commercial development within the next fifteen years.[37]

A February 2006 presentation on IGCC by Xcel Energy stated that the "wild card" in the IGCC cost equation is CO_2 capture, but no currently operating plants include CO_2 capture.[38] Transport and storage costs must also be included in the total cost of electricity.

An April 2007 MIT study, *The Future of Coal*, states that the US should not increase investment in IGCC or any coal-fired generation that lacks CO_2 capture, and that plants built before CO_2 emissions are capped should not be "grandfathered."[39] The largest CO_2 sequestration project is in Sleipner, Norway, where, since 1996, Statoil has been pumping 1 million tons of CO_2/year into a reservoir beneath the North Sea for enhanced oil recovery, deploying one of the largest offshore platforms in the world. But it would take ten of these projects to store the CO_2 emissions of a single large coal plant.[40]

---

34   See *Oil and Gas Journal*, http://www.ojg.com.

35   Rebuttal Testimony of Dr. Elion Amit, MN Dept. of Commerce, Docket E-6472/M-05-1993. p. 21. http://www.mncoalgasplant.com/.

36   Booz Allen Hamilton, p. 46 (quoting NETL Study by Clayton, Stiegel and Wirner, p. 16).

37   US Department of Energy, Supplement to the Draft Environmental Impact Statement for the Gilberton Coal-To-Clean-Fuels And Power Project, Gilberton, PA, December 2006, p. 3–4; citing CO_2 Capture and Storage Working Group 2002, CO_2 Capture and Storage in Geologic Formations, NCCTI Energy Technologies Group, Office of Fossil Energy, US Department of Energy, January 8, 2002. http://www.netl.doe.gov/publications/carbon_seq/CS-NCCTIwhitepaper.pdf.

38   Xcel Energy PowerPoint presentation, Colorado IGCC Demonstration Project, An Overview of Project Concepts and Objectives, Prepared by Xcel Energy, February 2006, slide 10 of 16.

39   MIT, *The Future of Coal: Options for a Carbon-Constrained World*, 2007, Executive Summary, p. xiv.

40   "The Dirty Rock," by Jeff Goodell, from *The Nation*, May 7, 2007. http://www.thenation.com/doc/20070507/goodell.

EMISSIONS PROFILE NOT GOOD/MORE MERCURY

Mercury emissions per megawatt-hour (MWh) from the proposed Mesaba IGCC plant are 15–27 percent higher than either Supercritical Pulverized Coal (SCPC) or Ultra-Supercritical Pulverized Coal (USCPC) plants. SCPC and USCPC are simply newer types of conventional (pulverized coal) plant technologies that burn hotter and include state of the art pollution control technology.[41]

However, mercury in any amount is one of the most toxic substances known. A 2006 study by the University of Texas Health Science Center reported that for every 1,000 pounds of mercury emitted in Texas counties, there was a 43 percent increase in special education and a 61 percent increase in autism.[42]

Power plants emit more pollutants during start-up than in steady-state operation, and the regulations limiting pollutants generally don't apply during start-up/shut-down. Because gasification plants require about 60 start-up/shut-down events every year (as opposed to 2–3 for pulverized coal), and because it takes a few days for a plant's cold start, pollution emission rates are estimated to increase an average of 38 percent.[43]

CONCLUSION

When total lifecycle costs for coal-fired generation are considered—including coal mining and transportation, power plant construction, CO_2 capture, pipeline construction and transportation, CO_2 storage, coal waste product landfilling, the health effects of air pollution, environmental degradation, and global warming—coal is no bargain. It's just that the coal and utility industry have successfully offloaded these very real costs to citizens, who are "paid" eventually in dirty air, contaminated and acidified water, sick people, and lost lives.

When the currently unaccounted-for, "externalized" costs for coal plants (CO_2 capture, pipeline and transportation costs, storage and sequestration costs, increased risk, liability for explosion or the release of large amounts of CO_2; plus the future cost of global warming, acidified lakes, mercury-poisoned fish, air pollution, asthma, heart attacks, fetal deformities, coal sludge and waste, and the destruction caused by coal mining in our communities), the "higher" costs of renewables aren't so high. We should take NASA scientist James Hansen to heart when he says we should not build one more coal plant, and figure out how to phase out existing ones. Renewables are cheaper.

---

41   Minnesota Pollution Control Agency, Response to Comments on Its Report Entitled Comparison of Nitrogen Oxides, Sulfur Dioxide, Particulate Matter, Mercury and Carbon Dioxide Emissions for IGCC and Other Electricity Generation, p. 3, Docket E-6472/M-05-1993. http://www.puc.state.mn.us/docs/calendar/cal0806.htm.

42   Raymond F. Palmer, Steven Blanchard, Zachary Stein, et al, "Environmental mercury release, special education rates, and autism disorder: an ecological study of Texas, U. of Texas Health Sciences Ctr," Health & Place 12 (2006) 203-209. http://www.seedcoalition.org/downloads/autism_study_UTHSCSA.pdf.

43   Testimony of Stephen D. Jenkins, pp. 17–22. http://www.psc.state.fl.us/library/filings/07/03205-07/03205-07.pdf.

Chapter 28 ▎ Part 7

THE SMELL OF MONEY
Alberta's Tar Sands

Shannon Walsh and Macdonald Stainsby

> *There is no environmental minister on earth who can stop the oil from coming out*
> *of the sand, because the money is too big.*
> —Stéphane Dion, former Canadian Federal Minister of Environment

At Syncrude's Wood Bison Viewpoint, thirty-five km north of Fort McMurray, Alberta, visitors usually first stop to take photos of the carbon-spewing smoke stacks puffing away at the refinery in the near distance before turning their lenses to the grazing bison on "reclaimed" Syncrude land. Syncrude Canada Ltd. is the largest producer of synthetic crude oil in the world, and one of the oldest companies in Alberta's oil patch, producing 111 million barrels of oil in 2007 alone. On a cold afternoon in March, visitors from Ontario, California, Edmonton, Newfoundland, and India pocket their cameras and tread carefully across the deep snow to catch a glimpse of Syncrude's famous imported bison grazing on reclaimed land a stone's throw from the refineries.

The land is not exactly boreal forest, with commercial trees, long grasses, and maintained animals being fed on hay that a local bus driver said he saw being hauled in by truck up Highway 63. The bison, once endemic to the region, have been re-introduced to this patch of reclaimed land with much fanfare. "That's the deal they made with the natives," proclaims an enthusiastic Newfoundlander to his visiting family as they gaze out over the snow at four or five bison casting little black shadows on the white fields, "to put this land back the way it was."

"As long as the buffalo can live here, anything can live here," he explained.[1]

This is ground zero of tar sands development and about as soaked in contradiction as could be expected from what has been coined the largest industrial project in human history—and perhaps the largest environmental catastrophe on the planet right now.[2]

You don't have to look much further than Canada's tar sands to see the petroleum economy spiraling out of control, and with a changing political and economic context for oil production being heralded in, the boom seems to only be changing form.

WHAT ARE THE OIL SANDS?

Alberta sits over one of the largest recoverable oil patches in the world, second

---

1   And the bison may be living better than most. Syncrude provides each bison with its own medical care and veterinarian services, better services than residents than the First Nations community of Fort McKay on the other side of the Syncrude site receive.

2   See research on the "gigaproject" by Oil Sands Truth at http://oilsandstruth.org.

only to Saudi Arabia. Covering 149,000 square kilometers, an area larger than England, the oil patch holds at least 175 billion barrels of recoverable crude bitumen, one of the dirtiest forms of oil extraction in the world.

Unlike conventional ways to recover oil, the tar sands "bitumen" is locked in sand, clay, and silt. The bitumen is a sticky, tar like substance that rests fifty or more meters beneath boreal forest, muskeg,[3] wetlands and river systems. It is an expensive, technologically challenging endeavor to get this oil out of the sand, and it is only in the last few years that it has even become feasible. Industry has invested billions of dollars to develop a massive infrastructure to extract the bitumen out of the sand with methods that continue to be extremely capital-, energy-, and environmentally-intensive. Two extraction processes are the most common: open pit mining, which literally mines the earth for bitumen, and Steam-Assisted Gravity Drainage, known as SAGD, which pumps extremely hot steam deep underground to force the gooey bitumen to the surface.

Both processes use large amounts of fresh water and fossil energy[4] to extract the bitumen, producing more than three times the CO_2 emissions produced by a conventional barrel of oil, and disrupt thousands of square kilometers of boreal forest, fen, and muskeg, creating gigantic toxic dams to contain the post-production wastewater. This equates to more carbon emissions than many countries, with the current tar sands emissions outranking 145 out of 207 nations, sitting between the emissions of New Zealand and Denmark. While mining is an uglier process, leaving a visible scar on the landscape, SAGD is just as environmentally problematic, using twice the energy and almost twice the water than mining processes. Certain SAGD processes now are developing a process of cogeneration—using the coal-like waste remainder from the bitumen for energy, thus increasing emissions even further.

The environmental footprint is huge in relation to water as well: surface-mining operations use between 2 and 4.5 cubic meters of water to produce just one cubic meter of oil. While new processes exist that can substantially reduce water usage, at the moment they are either untenably expensive, producing only small amounts of bitumen, or still in experimental phases. Where gains are made to reduce carbon or water usage in one company or another, the total cumulative impacts continue to rise in all areas as more and more companies and projects come on line.

The cycle is dramatic: on one end an increasingly large amount of water is extracted by an increasingly-large industrial appetite, and on the other end, cumulative carbon emissions quicken global warming and, in turn, water depletion.

The water used by industry ends up filling enormous toxic "tailings ponds": gigantic man-made dams, which store the waste-water collected from the extraction processes. Tailings ponds contain a host of toxic chemicals such as naphthenic acids; Polycyclic Aromatic Hydrocarbons (PAHs); and trace metals such as copper, zinc,

---

3   Muskeg is a type of bogland, the mossy soil in boreal forest. Muskeg often has a water table near its surface and acts as a sponge, holding water back on the land, as well as being a home for many organisms and animals.

4   Primarily natural gas. Left over bitumen they can burn, like coal, and this produces even more carbon.

and iron. The ponds are recycling vats meant to slowly revert water back to a state of non-toxicity. While some of this water is re-used, a large part of it remains standing in the ponds. Current visions imagine that one day the toxins will settle to the bottom of the ponds leaving large artificial lakes speckling the landscape. Visible from space, Syncrude's Mildred Lake Mines tailings pond is currently the largest dam in the world with the exception of the Three Gorges Dam.

Serious environmental worries about the tailings ponds already exist, including the threat of the migration of pollutants into the groundwater and the soil and surface water around the ponds. In 2007, the tailings ponds leaked over 11 million liters per day.[5] In Fort Chipewyan, 250 kilometers downstream from the major oil sands plants, rare cancers, leukemia, lupus, and other auto-immune diseases are on a worrying rise. In a recent study, independent of government and industry, and commissioned by the community, Dr. Kevin Timoney found increased levels of arsenic, PAHs, mercury, and other carcinogenic chemicals associated with tar sands development at dangerously high levels in the soil and water. His report confirms what First Nations elders and community members have long been saying that they have been seeing on the land and in the water, in Forth Chipewyan and Fort McKay, from fish with skin carcinomas and deformities to water levels decreasing and a host of human illnesses.

With advancing technologies and increased expenditure in infrastructure to extract bitumen over the last decade, Canada has supplanted Middle Eastern sources to become the largest foreign supplier of oil to the US, with over a million barrels per day flowing south, 72 percent of which is used for transportation fuels (gas, diesel, and jet fuel). The US has been vocal about seeing Canada as a "friendly" and "safe ally" in keeping North America afloat with the crude oil from Alberta for perhaps another fifty years.

PEAK OIL, CLIMATE CHANGE, AND WATER SCARCITY: AN UNHOLY TRINITY IN THE TAR SANDS

Whether or not we are actually at the summit of Hubbert's Peak—that peak oil moment— whether or not the oil-price bubble finally bursts, what we are probably witnessing is the largest transfer of wealth in modern history.
—Mike Davis

The world consumes 86 million barrels of oil a day—over a billion barrels every 12 days. But very few new oil deposits have been found. For every barrel of oil we now discover, we consume approximately six.[6] The connection between peak oil, climate change, and the oil rush in Alberta is undeniable. The link to capital is both an obvious and complex story to tell.

---

5   Environmental Defence (2008) "11 Million Litres a Day: The Tar Sands' Leaking Legacy" Report. *Environmental Defence*, 09 December.
6   Some believe this number to be closer to nine barrels. See http://www.carbon-cutters.com/Peak_Oil.htm.

While many mainstream environmentalists have welcomed high oil prices in the hopes that it will force market-led solutions to tackle climate change and petrol-economics, it is increasingly clear that rather than the market rising up to develop solutions for climate change, prompted by dwindling oil resources (such as rethinking hyper-consumptive lifestyles), it is advancing in just the opposite direction, attempting to squeeze oil out of the most untenable of regions with gross environmental and human consequences. At the moment we are witnessing what can only be described as the irrational, frantic push of market-forces in their most naked form, precisely at a time where reductions and radical transformation is required.

The tar sands are a case study in the way that the deregulated marketplace so completely spirals out of control. Market-based logics depending solely on self-interest will inevitably come in violent opposition with the very ability of humanity to live. All rational logic has been set aside for the steel arm of the market to generate solutions. While government regulations exist, as the Assistant Deputy Minister of the Oil Sands Division of Alberta Environment Jay Nagendren described, it is the market that directs the Environment Ministry, not the environment. Nagendren explained,

> The premier has said that market forces will dictate the pace of development. So our job is, given that labor forces and finance will decide what kind of conditions need to be set in terms of the cumulative effects, to decide what kind of caps we will have to place on emissions, what kind of restrictions on water use, carbon capture storage, reclamation, tailings ponds, water use, etc.[7]

The role of government to create a resistance to the excesses of capital is clearly not at play in the oil patch. The tar sands presents a gruesome yet succinct reflection of David Harvey's (2006) ideas of uneven geographical development, as it activates the conditionalities around "the material embedding of capital accumulation processes in the web of social-ecological life." What we are witnessing here is a capitalist push towards a total separation between the market's abstract and self-sustaining logic, and the social-ecological realities of our own life-worlds. This disconnect is critically important, we think, at this particular moment in history when the balance between peak oil, climate change, and water shortages hang in a dangerous trinity, effecting the very bare life of most of the planet's population (read here, the expanded impacts on agriculture, food shortages, mass displacement, and migration due to ecological disaster, labor migration to these frontiers of capital, droughts and flooding, effective access to food and safe drinking water, etc.).

This material embedding of capital into our ecological life-worlds is crucially important, especially since many of the environmentalist challenges to climate change use "green capitalist" logics as a frame for post-petrol arguments. When market-utopias take over completely, as we are seeing in the tar sands, its gross excesses become very difficult to curb. The absurdity of reclamation plans in the tar sands, currently approved by government, actually purport to reconstruct entire ecosystems with

---

7   In-person interview, March 2008.

technologies that are still being developed (there is, of course, faith that the market will succeed in developing in some ever-evolving future). They are market-utopias at their most extreme. Boreal forest is "reclaimed" in terms of "equivalent values," which in a recent case has meant that 40 percent of disturbed land must be returned to "commercial forest capabilities," effectively creating a natural environment of harvestable reconstructed commercial forest and artificial lakes. It's an absurdist creation only possible at this point in market-utopian logics.

The truth is that as the world runs out of oil, fresh water is also quickly drying up. Available fresh water represents less than half of 1 percent of the world's total water stock. Many analysts on both sides of the fence, from the World Bank to the Polaris Institute, believe that, by 2025, we will be living in an era of serious water scarcity and water shortages across the globe.[8] The logical incongruity between the pillage of water through the lust for money cannot be more apt. The realities of an impending water crisis impel us to seriously challenge market-led logics within industry and government before it is too late. Green capitalism is most certainly not going to lead us out of what is, ultimately, a market-driven, capital induced crisis.

The tar sands can only be seen as evidence of an untenable state of denial and psychoses around market-based, petrol-energy dependence. Some of the many serious and cumulative human and environmental impacts deserve a brief recounting here:

- Pipeline and refinery projects that cut straight through indigenous land throughout the continent, with serious social, ecological, sovereignty, and health implications for indigenous people, including the construction of the Mackenzie Gas Project, which will bring natural gas from the Arctic straight through unceded Dehcho First Nation territory;

- Health and human impacts of those living in the region of the developments, including the appearance of rare forms of cancers;

- Depleting large amounts of cleaner energy, natural gas, to produce dirty crude, what some call "turning gold into lead";

- Intensive carbon production and adding to climate change;

- Creating new systems of migration of wealth and bodies through trade, resources, and labor agreements, including proposals for thousands of new temporary foreign workers, who will not even be allowed to apply for landed immigrant status;

- Depleting fresh water at a time of increasing fresh water scarcity by drawing off the Athabasca river, whose water system accounts for 25 percent of the fresh water sources in North America;

- Supplying oil for the military industrial project, given that the Pentagon consumes about 85 percent of the oil used by the US government;

---

8   On top of this, as Dr. David Schindler warns, climate change is reducing water levels on the Athabasca River, the equivalent of building a new strip mine along the river every two years.

- · Impacts on fish and wildlife, including the destruction of thousands of hectares of boreal forest and muskeg that acts as an essential "sponge" for water that flows throughout the region.

Perhaps most disconcerting is that most of the tar sands oil ends up as dirty crude, and at the other end of its cycle puffs its way back into the atmosphere out the tailpipes of North American planes, cars, and military vehicles. As Mike Davis (2008) writes, there is a madness to creating a more carbon-intensive process at the very moment when we urgently need to reduce emissions:

> Even while higher energy prices are pushing SUVs towards extinction and attracting more venture capital to renewable energy, they are also opening the Pandora's box of the crudest of crude oil production from Canadian tar sands and Venezuelan heavy oil. As one British scientist has warned, the very last thing we should wish for (under the false slogan of "energy independence") is new frontiers in hydrocarbon production that advance "humankind's ability to accelerate global warming" and slow the urgent transition to "non-carbon or closed-carbon energy cycles."[9]

US president Barack Obama has quickly stepped onto the train of Carbon Capture and Sequestration (CCS), touted by Prime Minister Stephen Harper and Alberta Premier Ed Stelmach as the technological quick fix to the massive carbon emissions issues in the American coal and Alberta oil industries. In fact, Carbon Capture and Sequestration is not a proven technology, is extremely expensive to develop, and has a very limited potential to sequester the carbon emissions that come from tar sands extraction;[10] by some estimates it would only capture 10 percent of tar sands emissions.

"I think to the extent that Canada and the United States can collaborate on ways that we can sequester carbon, capture greenhouse gases before they're emitted into the atmosphere, that's going to be good for everybody," President Obama told CBC chief correspondent Peter Mansbridge. "Because if we don't, then we're going to have a ceiling at some point in terms of our ability to expand our economies and maintain the standard of living that's so important."[11]

But there *is* a ceiling to growth. While this largely unproven technology would bury harmful emissions underground, we would still be left with more exploitation of the tar sands, more depletion of fresh water and natural gas, and more devastation of the boreal forest. As Gerald Butts writes, this technological optimism is like telling our kids "keep smoking—we need the tax revenue. Trust us, we will cure cancer by the time you get it."[12]

It is starkly clear that there is no just and sustainable way to continue living in a petroleum-based economy. The harsh truth remains that the only alternative is a radical rethink of the way that we live, including a serious challenge to capitalism itself.

---

9   Davis, M. (2008) "Living on the Ice Shelf: Humanity's Melt Down," *Tom Dispatch*, 26 June.

10   Butts, G. (2009) "Carbon Capture no silver bullet for tar sands: Only a small portion of greenhouse gases could be sequestered," *The Star*, 27 February.

11   Mansbridge, P. (2009) "Transcript of the CBC News interview with Obama," CBC News, 17 February.

12   Butts, G. (2009) "Carbon Capture no silver bullet for tar sands: Only a small portion of greenhouse gases could be sequestered," *The Star*, 27 February.

But those realities seem to be totally beyond the political will of the Canadian government. Ed Stelmach, Alberta's Premier, has attempted to counter the increasingly negative view of the tar sands, spending $25 million in a "re-branding" campaign. Just as the campaign was being unveiled, hundreds of migrating ducks died after landing on one of the toxic tailings ponds at Syncrude's Aurora North mine site. Usually water canons shoot into the air around the "ponds" to keep birds off, but Syncrude claimed there had been a delay in the installation of the canons after the long winter. Workers in Fort McMurray said ducks dying on the ponds is not a new phenomenon. A former tailings pond worker who wished to remain anonymous, admitted that when she worked on the ponds years ago they were asked to wring the necks of birds who had landed on the ponds and dispose of them in plastic bags.

In February 2009, the government attempted to show it was doing something about the growing environmental concerns in the tar sands, and charged Syncrude for not properly deterring the 500 ducks from landing on its toxic tailings ponds. The charge was filed under the Migratory Birds Convention Act, which has a maximum fine of $300,000, and Alberta's Environmental Protection and Enhancement Act, which has a maximum penalty of $500,000.[13] Pretty small change when you consider that Canadian Oil Sands Trust, the biggest partner in the Syncrude venture, announced $173.6 million in fourth-quarter profits in 2006, up 117 percent from the same quarter the previous year.[14]

CONTINENTAL MARKET-BASED INTEGRATION OF ENERGY: SPP AND NAFTA

While the environmental and human impacts are becoming impossible to ignore, the industry continues to expand the black-gold rush at break neck speed, only slowing to take a breath and absorb the current global recession. Corporate interests aimed at integrating North America's economies and resources have become major proponents in forging this unprecedented push for development. Industry investment into development of the oil sands now totals over $20 billion with $7 billion worth of projects under construction and $30 billion of projects forecast to be completed by 2012. If infrastructure is taken into account, those numbers can increase as high as $200 billion. As Harvey writes, "the circulation of money and of capital have to be construed as an ecological variable every bit as important as the circulation of air and water" (Harvey, 88).

Proposals are afoot to build pipelines that will span the continent—one of Enbridge's pipelines will move 400,000 barrels a day to Illinois by 2011; Kinder Morgan Canada has plans to pipe 300,000 barrels of crude per day from Alberta to Texas, and TransCanada Corp's (TSX:TRP) Keystone pipeline Phase One will move 600,000 barrels to refineries in Illinois and Oklahoma, possibly increasing to as high as 1.1 million barrels.

---

13   Schmidt, L. (2009) "Syncrude charged for duck deaths at tailings pond," *Calgary Herald*, 9 February.

14   CBC News (2006) "Canadian Oil Sands Trust profit jumps 42 percent," *CBC News*, January 2006.

TransCanada Pipeline has also announced its work plan for a crude-oil pipeline through South Dakota.[15] At the same time, the corporate arm is moving further and further north to extract natural gas for these processes. Imperial Oil, Exxon, and TransCanada's gigantic Mackenzie Valley pipeline is still underway, and TransCanada has been awarded the rights to develop an Alaskan gas pipeline. Enbridge plans a Gateway pipeline to the north-central BC coast, which would require supertanker traffic of several hundred tankers a year along the same coastline devastated in March 1989 by the Exxon-Valdez spill. This is only a small sampling of the proposed outlay of pipeline, refinery, and terminal projects across North America. Alongside the pipeline expansions are plans for up to as many as forty new or refurbished refineries to handle tar sands in the continental United States. No refineries have been constructed in the lower forty-eight since the 1970s.

While trade agreements and resource frameworks continue to be a major focus of how this exploitation of natural resources plays out in North America, they also signal a disintegration of the State as such, rapidly creating enclaves and borders around a new kind of capital expansion. Dissolving borders for capital while deepening and entrenching mechanisms of security for bodies and labor that is quickly becoming a hallmark of the tar sands.

Market-driven resource agreements, now being combined with ideas around State energy "security," make national contestations increasingly difficult. The North American Free Trade Agreement's proportionality clause ensures that an average percentage of Canada's energy resources continue to flow south.[16] This guarantees an increasing export of a finite resource. Mexico refused to sign onto this clause, and was exempted, but Canada agreed in order to gain favorable bargaining chips in other areas of trade. Under the clause, Canada must produce the same percentage of export as over the previous three years, worrisome considering that Canada has increased oil exports to the US by 350 percent since 1990. To deepen the irony of a locked-in energy deal with the US, Canada remains one of the only industrialized countries that has not reserved any energy for itself. Gordon Laxer, professor and director of the Parklands Institute at the University of Alberta, argues that Canada lacks a national energy policy that will guarantee energy supplies to some regions of the country in the event of an international energy crisis. Atlantic Canada, Quebec, and some parts of Ontario may have to rely on offshore oil imports from Algeria, Saudi Arabia, and Iraq in the event of shortages. The clause compromises Canada's energy independence while at the same time using a market-based analysis to determine fossil fuel extraction.

In addition, NAFTA's clauses on "national treatment" would confer the same rights over Canada's water resources. The legal, social, and technological precedents being set by the oil sands removal and pipeline expansion beg to be repeated with water.

---

15   Mercer, B. (2009) "TransCanada files plans for its second oil pipeline," *State Capitol Bureau*, 14 March.

16   The proportionality clause originally emerged under the FTA in 1988. For more information about these trade agreements, and specifically the Security and Prosperity Partnership of North America, see http://canadians.org/integratethis/index.html.

A new agreement called the Security and Prosperity Partnership (SPP), further expands NAFTA, ensuring energy security for the United States. Launched in 2005, the SPP extends and expands some the agreements that were troubling in NAFTA in an opaque, undemocratic forum closed to Parliament. Canadian New Democratic Party leader Jack Layton described the process as not simply unconstitutional, but "non-constitutional," held completely outside the usual mechanisms of oversight.[17]

The SPP recommends a "continental energy and natural resources pact" that would create an integrated market place with "streamlined regulatory processes" and "deregulation in each country for cross-border oil pipelines, including a five-fold increase in Canadian tar sands production, and continuing the privatization of energy industries" (North American Energy sector workers meeting, August 2007). As Tony Clarke identifies, Canada is not an energy superpower, but in fact, it has become an energy colony, or energy satellite of the United States.

The North American Energy Sector workers meeting in August of 2007 stated that:

> Through the SPP and the North American Energy Working Group, the governments of Mexico, United States, and Canada have formed an unprecedented collaboration with energy corporations to promote the continental integration of our energy industries and infrastructures....While these working groups bring together government, regulators and corporations at the highest level, they have excluded labour, environmentalists and civil society movements and circumvented the oversight of our elected legislatures.[18]

Rapid, scattered and questionable economic gains, a deepened entrenchment of fortress North America, the dissolution of national borders in order for capital and temporary foreign workers to move across, little to no energy security whatsoever for Canada, and a huge environmental and human catastrophe, leaves the balance between the costs and the gains of this project impossible to reconcile. William Marsden had it right when he titled his book on the tar sands: Stupid to the Last Drop.[19]

While post-petrol energy sources may be inevitable, the "scraping the bottom of the barrel" approach and the almost fundamentalist zeal with which technological and market solutions are being vaunted in the oil patch make it hard to imagine any kind of smooth transition out of the oil crisis. The tar sands represent the crux of where a capitalist madness for oil, driven by a market-economy has led us.

THE ECONOMIC CRISIS & ITS POTENTIALS FOR THE BIG BOYS

On one hand, the economic crisis has slowed down the out-of-control pace of development we were seeing till now,[20] but on another, the crisis is allowing for a

---

17   Falconer, S. (2007) "The Bus to Montebello," *The Hour*, 16 August.

18   North American Energy sector workers meeting (2007) "A Joint Solidarity Statement," *Common Frontiers*, Canada, 18 August.

19   Marsden, W. (2007) *Stupid to the last drop: How Alberta is bringing environmental armageddon to Canada (and doesn't seem to care)*, Toronto, Random House of Canada Limited.

20   In March 2009, Canadian Oil Sands Trust, part of Syncrude, dropped its production estimates from 115 million barrels to 109 million barrels, but they urge investors that they expect an increase in

massive consolidation in power and resources that will make any attempt to slow down tar sands extraction in the future extremely difficult. As Prime Minster Stephen Harper so brashly put it in his first major speech on the recession, the economic crisis is an opportunity.[21]

The apparent slow down can be deceptive if not put in context of the current consolidation of major tar sand operators and the inevitability of peak oil driving oil prices back up. Pairing the economic crisis with market devices set free to self-regulate has already meant the beginning of a massive power grab and consolidation of petroleum companies in the oil patch. As we are seeing in other sectors, the big players are taking advantage of the economic downturn to buy off small and medium players. The massive $19 billion "mega-merger" of Suncor Energy Inc. and Petro-Canada, two of the largest oil companies in Canada, is just one example of how these consolidations are already starting to take shape.[22] The economic "crisis" seems to have given the mega-corporations the ability to rethink their strategy, force out small and mid-sized players, and prepare to gear up once again. The most economically sensible way to get a grip on the oil market at the moment is to buy up smaller producers, rather than looking for new oil reserves. As analyst Ben Dell of Sanford C. Bernstein & Co. explained in a report on potential oil industry consolidation, "Buying an average company, and by extension, its undeveloped reserves, costs $11 (US) a barrel. That is about half of the $21 a barrel, on average, it cost the industry to find and develop new reserves in 2008 through exploration."[23] Exxon Mobil is seen as next in line for a major take over of smaller tar sands operators.

KEEP THE OIL IN THE SOIL: BUDDING RESISTANCES ON THE PATHWAYS OF DESTRUCTION

This is not only about protecting the environment, it is about protecting my people.
—Pat Marcel, elder Athabasca Chipewyan First Nation

There's a sickly smell that hangs around Fort McKay like plumes of yellow smog, a sadness that sticks to your skin, what an Oil Sands Discovery Centre tour guide called "the smell of money." It's easy to wonder how people do it; how they manage to dampen the way they feel when looking out at the ugly visual scar left on the landscape. Talking to people, from riggers to single moms and Tim Horton's employees, it's clear that they just adapt. Like people do everywhere, you become accustomed to a certain level of discomfort, you can close your eyes to terrible things that you know are happening but feel powerless to stop. They are aware of the contradictions in the oil patch, but isn't it impossible, they wonder, to stop this massive machine fueling the planet's oil hunger?

There is complicity to our collective blindness. The consumptive cycle does not function without our active engagement within it. Capitalism is not an abstract

production by the end of 2009.

21   CBC News (2009) "Canada will emerge from slump faster, stronger: PM," CBC News, 10 March.

22   McCarthy, S. (2009) "Suncor's $19-Billion Poison Pill," *The Globe and Mail*, 24 March.

23   Ebner, D. (2009) "With PetroCan gone, who's next?: Cheaper now to buy than dig," *The Globe and Mail*, 24 March.

machine, but it is constructed out of the everyday actions of people everywhere. And their resistances. Simultaneously to the tar sands expansion, resistances are moving, forming, being born, and becoming contagious. While at one end of the spectrum there is a sadness, the bubbling of solidarities and the working out of a strategy is emerging all along the pathways of destruction.

What is most striking are the many average people standing up everyday and joining together through a sense of urgency and injustice in the wake of what once may have been the domain of electoral politics or democratic institutions. Joining together as indigenous and other impacted communities (most often made up of people of color), long-standing activists, Environmental NGOs, disgruntled workers, foreign migrants, and many others, a diverse and eclectic movement is being born throughout North America. Resistance works best when people get involved for their own self-interested reasons—if they look into their own backyards and make the decision that fighting here at home has meaning.

All along the pathway of pipelines and refineries are communities that have already started to mobilize against this massive development. From the Dehcho lands where the Mackenzie pipeline is set to cross through their unceded territory, to the Lubicon Cree who will see the enormous North Central Corridor link up the McKenzie gas pipeline to Fort McMurray, and the multiple communities around the Great Lakes region in Canada and the US who are resisting the expansion and development of tar sands facilities, communities are strategizing and building coalitions for struggles to come. The continent is full of allies.

While the pipelines may facilitate the expansion of tar sands development, they also draw together community-level resistance across the continent. If the struggle continues to emerge from this community level, it can provide more than a simple resistance to the tar sands but a fundamentally revolutionary movement towards self-determination as yet unseen in North America.

At ground zero of this emergent struggle are the residents of Fort Chipewyan, the oldest settlement in Alberta and the home to Dene, Cree, and Métis people. Carbon dating puts indigenous inhabitants here for almost 12,000 years. Almost overnight, the community of Fort Chip has been forced to the forefront of a fight to stop the rapid pace of oil sands development. Over the past year, the community has begun to piece together a government and industry cover-up around the true incidents of toxic contaminants that have been flowing down the river towards them, complicit in the deaths of an increasing number of people in their community. Mobilized across historic divisions, the community has come out fighting at local, national, and international levels. They have no choice. Their lives hang in the balance.

As Athabascan Chipewyan First Nation Chief Allan Adam remarked:
> What they're doing is wrong. Some of our members are thinking that way back home. We are radical. We were radical before I got elected…. We are still radical. Now I have to use it in a different form. Industry and government don't like my approach. But I'm holding them accountable to what is happening to us. The government is going to have to answer our questions.[24]

---

24   In-person interview. August 2007.

Chief Adam is one of the many voices emerging in Fort Chipewyan. He walks slowly back along the pier, clearly grappling with the road that is set ahead of him. "It's been easy for industry to get approvals for new developments from us in the past, but it won't be any more."

As folks in Fort Chip like to say, "the tar sands are downstream from us all." The people of Fort Chip now know what they are up against, but they also know now that they are not alone. They have been the first to step up to the plate. It is now for us all to follow.

FURTHER WEB RESOURCES

- Alberta Energy Utilities Board. Available at http://www.eub.gov.ab.ca
- Athabasca Regional Infrastructure Working group (RIWG). Available at
- http://www.oilsands.cc/
- Blue Planet Project. Available at http://www.blueplanetproject.net/
- Canadian Association of Petroleum Producers. Available at http://www.cpp.ca
- Dehcho First Nations http://www.dehchofirstnations.com/
- Environmental Defense. Canada's Toxic Tar Sands: The Most Destructive Project on
- Earth, Available at http://www.environmentaldefence.ca/reports/tarsands.htm
- Friends of the Lubicon. Available at http://www.lubicon.ca/
- Government of Alberta, Oil Sands, Available at http://www.energy.gov.ab.ca/89.asp
- Greenpeace Edmonton, Stop the tar sands; end our addiction to oil. Available at
- http://www.greenpeace.org/canada/en/recent/tarsandsfaq
- Indigenous Environmental Network (IEN). Available at
- http://www.ienearth.org/energy.html
- Integrate This! Challenging the Security and Prosperity Partnership of North America.
- Available at http://www.canadians.org/integratethis/energy/2007/Dec-13-2.html
- Last Oil Shock. Available at http://www.lastoilshock.com/
- Mike Davis, Welcome to the Next Epoch. Available at
- http://www.tomdispatch.com/post/174949/mike_davis_welcome_to_the_next_epoch
- Mikisew Cree First Nation. Available at http://mikisew.org/
- North American Energy sector workers joint solidarity statement, August 2007.
- Available at http://www.commonfrontiers.ca/Single_Page_Docs/SinglePage_1col_docs/Aug18_07_joint_statement.html
- Oil Crisis. Available at http://www.oilcrisis.com/tarsands/
- Oil Depletion Analysis Centre. Available http://www.odac-info.org/
- Oil Sands Discovery Center. Available at http://www.oilsandsdiscovery.com/
- Oil Sands Truth. Available at http://oilsandstruth.org/
- Pembina Institute, Available at http://www.pembina.org/
- Sierra Club of Canada. Available at http://www.sierraclub.ca/
- Suncor. Available at http://www.suncor.com
- Syncrude Canada. Available at http://www.syncrude.ca
- Tar Sands Watch. Available at http://www.tarsandswatch.org/
- Tar Sands Timeout. Available at http://www.tarsandstimeout.ca/
- To The Tar Sands. Available at http://www.tothetarsands.ca/
- World Water Council. Available at http://www.worldwatercouncil.org/

Chapter 29 | Part 7

NUCLEAR ENERGY
Relapse, Revival, or Renaissance?

Peer de Rijk, on behalf of the World Information Service on Energy

Having been involved for so many years, we are able to recognize and identify trends and developments. And, although the scope of this publication does not make it fit for an analysis of the struggle against nuclear energy and the strategy of the environmental movement, we would like to start by stressing one very important development: even environmental organizations are coming more and more to consider nuclear energy as a possible part of the solution in the fight against climate change. This chapter is not about the more classic, but still very important arguments, against nuclear power (waste, uranium-mining, transports, environmental damage, etc.) but rather shows why nuclear simply cannot be a part of a strategy to combat climate change. However, we nontheless believe the global movements for a more just and sustainable world would be making a tragic mistake if they were to embrace nuclear power just because they are so eager to take action against climate change.

More than fifty years ago, in 1954, the head of the US Atomic Energy Commission stated that nuclear energy would become "too cheap to meter": making electricity with nuclear would soon become the standard and would penetrate into every household and industry. And, at such low costs that no one would even bother to install electricity meters.

Within three months, the first-ever nuclear reactor was grid-connected ... but was in the then-Soviet Union. In June 2004, the international nuclear industry celebrated the anniversary of this first grid connection at the site of the world's first power reactor in Obninsk, Russia, with a conference entitled "50 Years of Nuclear Power—The Next 50 Years."

Real problems for the nuclear industry emerged in the sixties, however, when concerns about safety of nuclear energy first began to be heard in the US. And, while many engineers from abroad were encouraged to work in the US nuclear industry, in order to export the technology and thus increase the market, so too the concerns were exported, traveling very rapidly to Western Europe. In the late 1960s, all European countries had, on the one hand, an active nuclear power program, but on the other hand, the first dissident voices about the dangers and risks were starting to make themselves heard.

By the mid-1970s, those few voices had turned into movements. Public belief in nuclear power was already decreasing, and the accident in Three Mile Island in March 1979,[1] further fed fears and opposition. No matter how much the industry

---

1   The Three Mile Island accident in 1979 was a partial core meltdown in unit 2 (a PWR) of the Three Mile Island Nuclear Generating Station in Pennsylvania. It was the most significant accident in the

tried to make us believe the accident had no health consequences, fewer and fewer people were willing to give them the benefit of the doubt.

By that time, each country with a nuclear program had a massive anti-nuclear power movement. This, in turn, had an enormous influence on the opinions of churches, labor unions, and ultimately political parties, with regard to nuclear energy. In retrospect, one could say that the nuclear industry never recovered from Three Mile Island; more than 110 orders for nuclear reactors were cancelled in the US alone. The last order for a nuclear reactor that was actually built in the US was placed in 1974!

The 1992 World Nuclear Status Report, the first of its kind and published by WISE Paris, Greenpeace International, and the Worldwatch Institute concluded:

> The nuclear power industry is being squeezed out of the global energy marketplace.... Many of the remaining plants under construction are nearing completion so that in the next few years worldwide nuclear expansion will slow to a trickle. It now appears that in the year 2000 the world will have at most 360,000 megawatts of nuclear capacity, only 10 percent above the current figure. This contrasts with the 4,450,000 megawatts forecast for the year 2000 by the International Atomic Energy Agency (IAEA) in 1974.

This turned out to be quite a solid analysis and prediction of developments. At the end of 2008, there were 439 units operating in the world—five units less than at the historical peak in 2002—with a total capacity of 371.7 GW. As there has been very little new building in the past decades, utilities are choosing to increase both capacity and life-time of the existing reactors. This, and the fact that the few new ones were of a much larger capacity than the ones taken off the grid, means that the installed capacity has increased faster than the number of operating reactors.

According to the World Nuclear Association (WNA), in the USA alone 110 increases of capacity have been approved since 1977, some of them up to 20 percent. Not only are upgrades such as this occurring, but also the lifetimes of reactors are being extended. The same trend can be seen in Europe. In December 2008, EDF, the French state-owned utility with fifty-eight nuclear power stations, announced that it wants to extend the lifetime of its reactors from forty to sixty years. This will cost €400 million per reactor, a figure far less than the €4.5–5 billion that a new reactor would cost.

It is very unlikely that the industry will manage to upgrade all the reactors so

---

history of the American commercial nuclear generating industry. It resulted in the release of up to 13 million curies of radioactive noble gases, but less than 20 curies of the particularly hazardous iodine-131. The mechanical failures were compounded by the initial failure of plant operators to recognize the situation as a loss of coolant accident caused by inadequate training and ambiguous control room indicators. In the end, the reactor was brought under control. Full details of the accident were not discovered until much later, following extensive investigations by both a presidential commission and the Nuclear Regulatory Commission. According to the IAEA, the Three Mile Island accident was a significant turning point in the global development of nuclear power. From 1963 to 1979, the number of reactors under construction globally increased every year except 1971 and 1978. However, following the event, the number of reactors under construction declined every year from 1980 to 1998.

that they can run safely for an average of twenty years more than anticipated. Incidents, accidents, and increasing safety requirements have plagued the industry and this situation will not change in the years to come. The current average age globally for nuclear reactors is 23 years, and the average age of the 117 reactors that have been permanently shut down is 22. In the last five years, ten reactors have been shut down—eight in 2006—and only nine have been started up. The capacity has increased in the last years, with an average of 2 GW most recently.

This may sound impressive, however, it is good to put these figures into perspective and compare them with the global net increase in all electricity generating capacity, which stands at 135 GW per year. Now, while it is not in itself good that the production of electricity keeps growing, it is nonetheless encouraging to see where the growth comes from.

World wind energy capacity has been doubling about every three and a half years since 1990. That makes it the fastest growing energy source. By the end of 2006, total world wind capacity was around 72,000 MW, and generation from wind was around 160 TWh. Germany, with over 20,000 MW, has the highest capacity. Denmark, with over 3,000 MW, has the highest level per capita, with wind power accounting for about 20 percent of Danish electricity consumption.

The slightly increased output from nuclear energy will not be sufficient, at least in the short and medium term, to maintain its current 16 percent share of the world's commercial power.

Currently "only" thirty-one countries have nuclear power stations in operation. Out of these, six countries are responsible for almost 75 percent of nuclear energy production. These are the United States, France, Japan, Germany, Russia, and South Korea. Half of these countries are also acknowledged nuclear weapon states.

The decline of the nuclear industry started many years ago. In the 500th issue of the *Nuclear Monitor*, dedicated to victories of the anti-nuclear movement, WISE analyzed the trends and actual situation, declaring that a historic turning point had been reached. "This is the beginning of a trend and it will lead inevitably to the end of nuclear power within a few decades."

And thus, many activists and campaigners from environmental organizations began to believe that the struggle had been won and was over. And, then of course, a new problem emerged: climate change. These two factors led to a huge decline in the number of people willing and able to campaign on nuclear energy issues. "Increasing energy demand, concerns over climate change, and dependence on overseas supplies of fossil fuels are coinciding to make the case for nuclear build stronger. Rising gas prices and greenhouse constraints on coal have combined to put nuclear power back on the agenda for projected new capacity in both Europe and North America," is the oft-repeated mantra of the World Nuclear Association, who continue to forecast a positive future for their industry. Much of this is rhetoric, but it is, nevertheless, effective.

Large international and influential lobby and research bodies keep predicting (or asking for) a large growth in the share of nuclear power. For instance, the OECD: its

"World Energy Outlook" (WEO) 2008[2] asserts the case that only with a tremendous growth in the proportion of nuclear power in the energy mix will it be possible to maintain emissions of greenhouse gases to a level of 450 ppm, a goal that, in any case, the environmental movement is fighting against as it means an acceptance of major climatic changes. Such movements are fighting to have a common goal of keeping emissions under the 350 ppm.

The OECD goal and its opting to solve the problem with more nuclear energy means an enormous predicted and facilitated boost for nuclear power. In order to achieve these goals, the capacity from nuclear power would have to reach 833 GW, with 6,560 TWh of an electricity production in 2030.

Two years before, the WEO rightly observed that "nuclear power will only become more important if the governments of countries where nuclear power is acceptable play a stronger role in facilitating private investment, especially in liberalized markets"—and "if concerns about plant safety, nuclear waste disposal and the risk of proliferation can be solved to the satisfaction of the public."[3]

So, despite massive state support over the course of many decades, and despite spending hundreds of billions of taxpayers' money, nuclear energy has failed to deliver what it once promised. Using the fear of climate change and the legitimate call for action from civil society, activists, NGOs, and the global environmental movement, the nuclear industry is now once again claiming the need for massive state support—notwithstanding all its rhetoric about the free market being the best system to solve the global crises.

They say that, "A global renaissance of commercial nuclear power is unlikely to materialize over the next few decades without substantial support from governments" (9), and that's exactly what they are looking for. With a clever mixture of adding fuel to the fear of climate change, making the case for an inevitable growth of nuclear ("Its happening already, the number of nuclear power stations is growing very rapidly."), and a smart strategy of supporting all international efforts to come to a global treaty on climate change with binding measures, the nuclear industry may well be about to enter an era of massive renewed state support.

In 2007, the secretariat of the United Nations Framework Convention on Climate Change (UNFCCC) published a paper describing possible ways to develop an effective and appropriate international response to climate change. And yes, the reference scenario puts the contribution from nuclear power at 546 GW, while the mitigation scenario states the need for 729 GW from nuclear power plants by 2030.[4]

None of these scenarios will happen without massive incentives. Even the lower figure entails a huge challenge. Building a nuclear power station takes at least ten years and that is only if there is no opposition or other major setbacks on-site. As of

---

2   http://www.iea.org.
3   World Energy Outlook 2006, http://www.iea.org.
4   Source:UNFCCC, "Analysis of existing and planned investment and financial flows relevant to the development of effective and appropriate international response to climate change," 2007 http://unfccc.int/files/cooperation_and_support/financial_mechanism/application/pdf/background_paper.pdf.

the end of 2009 the industry does not have enough educated people, skilled workforce, manufacturing capacity, and available material to achieve these goals.

For the immediate future, the building of new nuclear power plants remains essentially restricted to Asia. Of the thirty-six units listed by the International Atomic Energy Agency (IAEA) as under construction in thirteen countries (as of 12 December 2008), more than thirty are in Asia or the former Soviet Union. A considerable number of these thirty-six officially-listed units have been under construction for more than twenty years—meaning that most of these plants will simply never get grid-connected. The IAEA's PR-machine continues to include the US' Watts Bar-2 reactor (Start date 1972), the Iranian Busher-1 project (start date 1975), and the Russian fast breeder reactor project BN-800 (started in 1985) in their current official statistics for reactors under construction.

IS A RENAISSANCE POSSIBLE?

Imagine that all the problems and public misgivings that plague nuclear power were simply to vanish. And that, from today onwards, there would be real support for nuclear, including sufficiently stable governments the world over that support the nuclear option and would continue to do so in the future. Imagine that possible suitable sites are stepping forward voluntarily, the environmental movements have opted for the largest bailout ever, and—in order to avoid catastrophic climate change—have accepted the building of new nuclear power stations on a massive scale. Would it be possible to even maintain the share of nuclear power that exists in today's energy mix over the coming decades?

For this to happen, it would need eighty new nuclear power stations to be built and brought on-line within the next ten years. That's a new plant every six weeks or so. And, this is assuming that all existing stations would have their lifetimes extended to forty years. Considering that the average age of reactors that have been closed to date is only twenty years, a forty-year-lifetime expectancy might seem optimistic, even though it might seem possible given the progress that has been achieved on the current generation of plants compared to the previous ones.

Just to maintain the current share of nuclear energy, work on building those eighty new plants would have to start tomorrow, or rather, today.

In the slightly longer term, things come into even sharper focus. If we want nuclear power to provide just as much electricity in 20 years time as it does at present, once the first 80 new plants have been built, we would have to build another 200 between 2018 and 2028. This is a completely unrealistic prospect. It is something that was deemed inconceivable even in the glory days of nuclear power, when the industry was flourishing and faith in the atom's blessings verged on hysteria.

Even developments in China or other Asian countries won't fundamentally change the global picture. The economic crisis that hit the world in 2008/9 will certainly have its influence on the once so ambitious plans of China. Officially, the country still wants to increase its nuclear power capacity from about 9,000 MW (9 GW) in 2007 to 40,000 MW (40 GW) by 2020. However, this is now completely out of the question. Only about 10 percent of the additional 31 GW are currently under construction, with five units totaling 3.2 GW having been started in the last three years. Building frequency would have to be more than tripled in order to meet this ambitious goal.

A nuclear-utility-sponsored analysis carried out by the Keystone Centre pointed out that to build 700 GW of nuclear power capacity "would require the industry to return immediately to the most rapid period of growth experienced in the past [1981–90] and sustain this rate of growth for 50 years."[5] The industry organization, WNA, is particularly optimistic, stating: "It is noteworthy that in the 1980s, 218 power reactors started up, an average of 1 every 17 days ... so it is not hard to imagine a similar number being commissioned in a decade after about 2015. But with China and India getting up to speed with nuclear energy and a world energy demand double the 1980 level in 2015, a realistic estimate of what is possible might be the equivalent of 1 1,000 MW unit worldwide every 5 days."[6]

Again, this is not much more than wishful thinking and an attempt to create an atmosphere in which it is impossible to carry on resisting the revival. By releasing highly optimistic figures and scenarios on an almost daily basis, the industry hopes for a situation whereby there is continued concern about climate change, and that no steps are undertaken to solve the problem in a way that does not involve nukes and fossils.

The essential problems of nuclear power remain, and it is very unlikely they will be solved within the coming two or three decades—the period in which we need to realize a swift transition to a truly sustainable energy future.

A report published by Standard & Poor's identifies the barriers:

> The industry's legacy of cost growth, technological problems, cumbersome political and regulatory oversight, and the newer risks brought about by competition and terrorism concerns may keep credit risk too high for even [federal legislation that provides loan guarantees] to overcome.[7]

The few projects currently underway are plagued by cost overruns, technical failures, and unforeseen problems. The European Pressurised Reactor (EPR) is a new reactor design developed by the French company AREVA in co-operation with the German firm Siemens. Serious doubts have been raised concerning the EPR's safety and costs. A study of the EPR's blueprints and experience at the two sites where they are currently under construction (in Finland [Olkiluoto 3] and France [Flamanville 3]) has revealed weaknesses in design, problems during construction phases, and soaring costs. Its backers present it as the only example of an advanced "third generation" reactor; a flagship of the nuclear "renaissance."

The EPR has been promoted as a technology that makes nuclear energy cheaper and more competitive. In 2002, when the decision was made to build an EPR in

---

5   Bradford, et al. "Nuclear Power Joint Fact-Finding," Keystone Center, June 2007.

6   http://www.world-nuclear.org/info/Copy percent20of percent20inf17.html.

7   UtiliPoint International, 21 June 04.

Finland, the government promised that it would cost €2.5 billion and take just four years to build.

The final contract, three years later, put the price at €3.2 billion and construction time was set at four-and-a-half years. Since construction began, less than three years ago, a variety of technical problems have led to a two year delay, extending the construction period to at least six-and-a-half years. The estimated additional cost is €1.5 billion, raising the current price tag to €4.7 billion, almost double the initial estimates. More problems, delays, and cost overruns are likely to occur before the project is completed. The construction contract was signed as a fixed-price, turn-key delivery arrangement from AREVA and Siemens. Extra costs will most likely be borne by the two companies. Nonetheless, AREVA is seeking to claim some of the additional costs from the investor, the Finnish utility TVO. Financing for the Finnish EPR has benefited from state support in the shape of a €570 million loan guarantee provided by the French export agency COFACE. The low interest rates offered by French and German state-controlled banks may be in violation of EU legislation and are the subject of a pending complaint with the European Commission and the European Court.

Referring to the Finnish EPR saga, the International Energy Agency (IEA) warned—already in 2004—against the risk of relying on the new reactor for emission cuts, saying that any delays would inhibit Finland's ability to meet its greenhouse gas reduction targets under the Kyoto Protocol. That risk has become a reality. In August 2007, after twenty-seven months of construction, the project was officially declared to be between twenty-four and thirty months behind schedule and at least €1,500 million over budget. Unlikely to be operational before 2011, Olkiluoto 3 will not be ready in time to contribute to Finland's Kyoto target.

CIVILIAN NUCLEAR PROGRAMS GIVE RISE TO MILITARY ONES

The overall impact of nuclear weapons modernization in existing nuclear weapons states is likely to serve as a substantial encouragement to nuclear proliferation as states such as Iran, with their perception of vulnerability, deem it necessary to develop their own deterrents.

A general consensus exists that we will sooner or later reach peak oil, peak uranium, peak gas, and even peak coal, all leading to intensified competition for supplies. Climate change and the competition over resources are the primary root causes of global and regional insecurity and conflict, including political violence and terrorism. This in itself leads to global militarization. Instability, and thus insecurity, make nuclear weapons more desirable to (democratically-chosen) governments, military elites, and dictatorial regimes because they are considered a relatively cost-effective insurance policy in an uncertain world. More nuclear energy easily leads to more nuclear weapons. Yet, countries supporting nuclear energy as part of the solution to climate change and fossil fuel dependency are at the same time trying to prohibit non-allied countries from manufacturing, or even starting to manufacture, nuclear weapons.

In this sense, Iran's civil nuclear power program is an interesting case. Everything it has done in past years is legal under any international treaty. Yet, the simple fact that it is not considered an ally of the western world and its interests mean that the US and others have been considering a war against Iran. Western countries are now trying to convince the world to accept a regime and treaty that only would allow certain countries to have access to enrichment technology. This is being done through the so-called GNEP proposal announced by Bush in 2006. The list of countries allowed would include only the happy few: China, India, Japan, France, Germany, Netherlands, Russia, the UK, and the US. It would exclude all Islamic countries and all other upcoming economies, who would have to stand in line to buy enriched uranium from those who are privileged enough to have it.[8]

No wonder such suggestions are not acceptable for countries such as Iran. Their line of argument is very simple and effective: Why, if nuclear power is safe and acceptable, can we not have access to the whole nuclear chain and make everything we need ourselves? Why would the technology of enrichment be safe in the hands of some countries and not in our hands? And they are right. There is simply no justification for such a treaty. And, as history teaches, any racist and inherently unjust treaty leads to more insecurity and conflict. Dr. Mohamed El Baradei, Director General of the International Atomic Energy Agency, declared at the 52nd Regular Session of the UN nuclear watchdog's General Conference in Vienna, which took place on October 4, 2008,[9] that every country has the right to introduce nuclear power. Emphasizing that an expansion of nuclear power would create new demand for spent fuel management and waste disposal, he said:

> In the last two years, some fifty member states have expressed interest in considering the possible introduction of nuclear power and asked for Agency support. Twelve countries are actively preparing to introduce nuclear power. Increased demand for assistance has been particularly strong from developing countries, which seek expert and impartial advice in analyzing their options and choosing the best energy mix.

Compared to the 1990s, and especially since "9/11," there is a shift towards a strategic culture that favors the use of military force to secure national interests and protect against vulnerabilities, and an apparent move away from strengthening legal and diplomatic structures for conflict prevention and management. At any rate, this was the situation during the Bush years. Now, it seems that Obama really wants to take another approach. In this context, his call for a nuclear-free world offers some hope, though, of course, it remains to be seen how things will develop. We are again moving towards a world in which the possession (and threat of almost-possession) of nuclear weapons is considered more-or-less normal and accepted as a way to gain political influence.

For example, the UK, France, China, and Russia are all engaging in nuclear weapons modernization programs. Israel maintains a nuclear force, Pakistan and

---

8   *Nuclear Monitor*, issue 653, March 19, 2007.
9   http://www.iaea.org/About/Policy/GC/GC52/index.html#day-5

India are vigorously developing their smaller forces, North Korea probably has a small stock, and the USA is revising its nuclear arsenal for it to be more effective in a so-called "Post-Cold War security environment."

The former US Secretary of State, Henry Kissinger, not known for his pacifism, stated in an article:

> Deterrence continues to be a relevant consideration for many states with regard to threats from other states. But reliance on nuclear weapons for this purpose is becoming increasingly hazardous and decreasingly effective.... Unless urgent new actions are taken, the US soon will be compelled to enter a new nuclear era that will be more precarious, psychologically disorienting, and economically even more costly than was Cold War deterrence.[10]

The question we must ask is whether building new nuclear power plants would make the prospect Kissinger describes more or less likely? The evidence and experience points in one direction: that a new build would push us towards "a new nuclear era."

Much depends upon the health of the Nuclear Non-proliferation Treaty (NPT). At the heart of the NPT is a "Grand Bargain," whereby states that did not possess nuclear weapons as of 1967 agree not to develop them, and states that possess them agree to get rid of them over time. This bargain is breaking down; nuclear weapon states (NWS) have not taken the necessary steps to convince non-nuclear weapons states (NNWS) that they are sincerely committed to nuclear disarmament. Unless NWS take unambiguous steps towards abolishing their nuclear arsenals, more and more states will seek the ultimate deterrent.

This is, in itself, made easier—and is effectively being encouraged—by the choice to build new nuclear power stations. At this moment, it is mainly the usual suspects and big upcoming economies that are making the choice to go nuclear. However, if a stricter post-Kyoto new climate treaty supports nuclear energy as a valid technology ("carbon-friendly technology" will be the marketing strategy and term) we will see a real revival of nukes. And, an even greater revival shall be seen should it be accepted that the big economies (the Annex-1 countries in the Kyoto-treaty, which include all developed countries in the Organisation for Economic Cooperation and Development) are allowed to build nuclear power stations outside of their territories, for instance in the developing world and former Soviet republics, in order to reach their own national targets.

It is a serious mistake to believe that the security consequences of a large expansion of civil nuclear technology and materials can be managed. More nuclear power will lead to more nuclear weapons, more instability, and crises. And it will not help prevent climate change.

Even if the many good arguments do not halt the further expansion of nuclear power, the inherent limitations of resources (money, industrial capacity, building,

---

10   *Wall Street Journal*, January 4, 2007.

staff, and technological development) are quite likely to prevent a real revival. In all probability the industry will die automatically, once faced with its own limitations.

SO, WHY BOTHER?

Because every single new nuclear power station is a disgrace. Not only because it increases dangers and the ever-growing stockpile of plutonium and nuclear waste, but also because it is a major lock-in for the future. It takes away much needed money for renewable energy; it blocks further expansion of decentralized, small-scale solutions; and it postpones the moment when people will have to start making real choices.

As long as we keep building new nuclear power stations, regardless of whether it is just a few or hundreds, we will never be able to choose a real sustainable and socially-just development path. This is why it is so important to keep fighting nukes and coordinate efforts much more between the growing global community of people working in and for the renewable energy sectors, those fighting for climate justice, and the anti-nuclear power movement. Although the nuclear industry may never have the huge expansion it desires, the real limit will only be imposed by the intensity and international nature of struggles against it.

Chapter 30 ▌ Part 7

THE ECOLOGICAL DEBT OF AGROFUELS[1]

Mónica Vargas Collazos

> *Most of us are food producers and are ready, able and willing to feed all the world's peoples.*
> —Declaration of Nyéléni, Forum for Food Sovereignty Mali, Feb 27, 2007

2007 may well pass into the history books as the year in which agrofuels shot to fame. Not only has the media boosted this "alternative" as the way out of the planetary environmental crisis, but it has also received significant incentives from the governments of core countries, which has resulted in an accelerated production of these fuels. A comprehensive and responsible discussion of these issues must take a wide range of considerations into account. We situate our analysis within the paradigm of ecological debt, defined as the debt contracted by the industrialized countries to the rest of the world's countries due to the ongoing, and historically-rooted, plundering of natural resources, as well as the exporting of environmental impacts through the free use of the world's environment. This debt is closely intertwined with the capitalist mode of consumption and production (Ortega, 2007: 20).

A MIRACULOUS SOLUTION

Perhaps one of contemporary globalization's predominant features is that it generates problems that concern humanity in its entirety, and these problems are now starting to gain official recognition. Two global themes were reiterated throughout this year, from G-8 meetings to the World Economic Forum, to United Nations forums: climate change and hunger. In February 2007, after years of intense debate and scorn of even the minimalist goals of Kyoto Protocol, the Fourth Assessment Report of the Intergovernmental Panel on Climate Change (IPCC) finally formally established that human activities are responsible for 90 percent of climate change. Meanwhile, the United Nations Food and Agriculture Organization has stated that more than 850 million people currently suffer from hunger, with a projected 100 million more by 2015. If all the talk from their active proponents is to be taken at face value, it would seem that agrofuels[2] embody the most suitable responses to the twin problems of hunger and climate change. So what does this miraculous solution consist of? The production of biomass-based fuels is currently concentrated in bioethanol and biodiesel. Bioethanol is obtained from products that are rich in sucrose (sugar cane,

---

1   This is a slightly shortened version of an article that was previously published in *Agrocombustibles. Llenando tanques, vaciando territorios*, 2007, Censat Agua-Viva, Bogotá. Permission for translation and reprinting has been obtained from the author, and the article was translated by Kolya Abramsky.

2   We deliberately avoid using the term "biofuels." Instead, we adopt the position taken by the hundreds of peasant organizations that met at the Forum for Food Sovereignty in Nyéléni, which asserts that we are dealing with an industry that constitutes an aggression towards the environment.

molasses from sweet sorghum), from substances that are rich in starch (grains such as maize, wheat, and barley), and also through the hydrolysis of substances that contain cellulose (wood and agricultural wastes).[3] Using modified motors, these fuels can be used to replace gasoline. Biodiesel, on the other hand, is made from vegetable oils (from oil palm, rapeseed, soya, and jatropha) or animal fat. Biodiesel replaces petrol, and can be used either in pure form or as part of a mixture.[4]

Believing that agrofuels do not increase the concentration of CO_2 in the atmosphere—a perception that is currently under fire from many different directions—several countries have passed legislation making their use obligatory for transport. However, the production capacity that this requires is not yet readily available. Preparations are afoot for at least 30 percent of transport fuels in the US to come from agrofuels (especially ethanol) by 2030, which would require an annual production of 227 million liters. The percentage of US maize production devoted to bioethanol increased from 6 to 20 percent between 2000 and 2006, but meeting these fixed targets will require devoting virtually all of its crops to fuel production.

For its part, the European Union has opted in favour of four types of incentives, all of which rely on public resources. These are agricultural subsidies within the framework of the Common Agricultural Policy, tax breaks, requiring transportation fuels to contain at least 5.75 percent biofuels (biodiesel or bioethanol) in their mix by 2010 and double this figure by 2020, and finally the undertaking of pilot projects by public transport companies. On top of all this, the free trade treaty between the European Union and MERCOSUR, which is under negation, is being heralded for the favorable impact it will have on opening up the bioethanol market.[5] In order to meet this future demand, production of the required commodities is taking off in countries that have an abundance of high-quality land, including Brazil, Argentina, Colombia, Malaysia, and Indonesia.

TOWARDS BIO-BUSINESS

All of this opens up some very juicy business possibilities, which explains the fact that large transnational companies are pursuing agrofuels from many different directions (Rulli y Semino, 2007). We are living through a moment of unprecedented convergence between different corporate sectors—petrol, automobile, food, biotechnology, and financial. The very same companies that have made millions in profits through practices that generate climate change,[6] are now set to reap even greater profits through its "mitigation." British Petrol is collaborating with the biotechnology company DuPont to provide the British biobutanol market; ConocoPhillips has contracts with meat producers to produce biodiesel from animal fat or invest in

---

3   Essentially this refers to second generation agrofuels, which are discussed later in this article.

4   For example, diesel qualified with the term B30 indicates that it contains 30 percent biodiesel (GRAIN, 2007).

5   European Strategy on Biofuels, (Brussels, 8.2.2006, COM(2006) 34 final)

6   According to the magazine *Revista Fortune* 2007, the profits of the ten leading transnational companies exceeded €119.691 billion (more than ten times the GDP of the USA). Six of these are petrol companies, three are car companies, and one a leading provider of commodities and foodstuffs.

jatropha crops. Biotechnology companies like Monsanto and Syngenta are intensifying production and research into transgenic seeds, at the same time as Ford, Daimler-Chrysler, and General Motors are preparing to sell over 2 million bioethanol-fueled cars in the coming decade. Wal-Mart plans to sell agrofuels in its 3,800 US shops as part of its standard sales, and companies in the food sector are establishing integrated networks in order to control the entire production chain from seeds all the way to transport.[7]

While it may be crystal clear that agrofuels are a good business, it is far less clear whether or not these energy crops will contribute effectively to the reduction of emissions and to the improvement of living conditions for the most impoverished populations of the planet. In order to answer this question, let us briefly consider some of the consequences of mass production of these fuels.[8]

AGRICULTURE AND CLIMATE CHANGE

Agrofuels are creating a close and peculiar relation between climate change and the worldwide problem of malnutrition at the global level. The large-scale production of these fuels in response to the new demand from core countries is inevitably resulting in a further industrialization of agriculture, and the consequent advance of the deforestation due to soya cultivation in the Amazonia. A report by NASA in 2006 actually established the correlation between the price of soya and the level of destruction of the Amazon rainforest. Similarly, the last twenty years have witnessed Indonesia lose a quarter of its forest cover to palm oil plantations, which have increased in size from 600,000 hectares in 1985 to 6.4 million hectares in 2006.[9]

And so, the idea that boosting agroindustry in order to mitigate the effects of climate change resulting from deforestation is ridiculous. Today's agricultural model is petrol-based, from the production of chemical inputs all the way to the transporting of goods. Furthermore, agriculture and changes in land use (deforestation) count for 14 and 18 percent, respectively, of all greenhouse gas emissions (Stern, 2006). In particular, the conversion of the forests into cultivated lands, the use of nitrate fertilizers, the large scale cultivation of leguminous crops such as soya, and the decomposition of organic wastes all have been identified as responsible for emitting a third greenhouse gas, nitrous oxide. In Brazil alone, 80 percent of emissions come from deforestation caused by the expansion of soya and sugar cane crops. Additionally, it is estimated that the destruction of peat, linked to monocultures, will result in roughly 40 billion tons of carbon being released into the atmosphere (GRAIN, 2007). Finally, according to FAO, rice production is the single human activity that generates

---

7   For an exhaustive examination of the global companies with the largest investments in agrofuels, see: GRAIN, 2007

8   For example, here we do not deal with the close relation between agrofuels and the growth in transgenic crops. Detailed analysis of this question can be found at http://www.etc.group.org, http://www.biodiversidadla.org, and http://www.grr.org.ar.

9   For some, this expansion has meant excellent business. The Malayan business groups Sinar Mas and Raja Garuad are both major players in palm cultivation, biodiesel production, and timber exploitation (Biofuelwatch, Carbon Trade Watch/TNI, Corporate Observatory, 2007).

the largest source of methane—130 million hectares of rice paddies produce between 50 and 100 million tons of methane per year. Thus, we are trapped in a vicious circle, since the FAO has also voiced concern over the negative impacts that climate change has on agriculture and access to food in the poorest countries (FAO, 2007).

RISING GRAIN PRICES AND SPECULATION

According to the Coordination of Agricultural and Animal Husbandry Organizations (Coordinadora de Organizaciones de Agricultores y Ganaderos, COAG), public subsidies for energy crops drive grain producers to devote their land to agroenergy crops rather than animal and human food production. In the core countries, this situation is particularly worrying to the livestock sector. Let us recall that 70 percent of the planet's agricultural lands are devoted directly or indirectly to rearing animals, and the production of animal feed alone requires 33 percent. Cereals represent 55 percent of the production of animal feeds. Thus, taking the Spanish State as an example, of the 30.6 million tons of grains produced and consumed, 23 million are for animal feed (pigs in particular). The other side of the coin is that Spanish production represents just 15 percent of the European total, the European Union being the world's second biggest producer of animal feeds. Cultivable lands are simply not available domestically on a sufficient scale to supply the raw material, and so a large proportion of Europe's grains are being imported from the USA (maize and soya), and Brazil and Argentina (soya) (COAG, 2007).

Recent years witnessed a contraction in grain supplies owing to unstable production, which was tied, in part, to adverse weather conditions. However, demand continues to grow, particularly in the United States, due to increased production of maize-based bioethanol. On the other hand, the continuously rising barrel oil price is having a major impact on the logistical costs related to agricultural production (inputs and transport). In this context, the prices of grains are skyrocketing. This is especially so for maize, which constitutes the grain base in animal feed formulas. At the same time, the production of yellow maize for ethanol use has increased, to the detriment of white maize, which is used for human consumption. This has made the sector an interesting market for speculative capitals. In early 2007, this resulted in the so-called "tortilla crisis."

The United States has embarked on a major program of building bioethanol factories. However, this coincided with a slight reduction of maize production and consequently resulted in a reduction of US stockpiles. These stockpiles represent 40 percent of the world's reserves.[10] This situation allowed the world's most important grain trader, Cargill, to speculate and sell futures in maize to energy companies, alarmingly, resulting in a doubling of the price of maize tortillas in Mexico (Llistar, 2007).[11] As far as the oil-consuming sectors are concerned, an unequal competition

---

10   It is predicted that by 2012 the volume of maize which the US devotes to agrofuels might be double that going to export. This will mean its maize supplies will be reduced and prices will continue to rise (COAG, 2007).

11   Since the signing of the North American Free Trade Agreement (NAFTA), Mexican consumption of this basic good has been chained to US production. Mexico has increased its maize imports from

between cars and human beings is also emerging. Indonesia, which is the world's second largest palm oil producer, is a telling example. Henry Saragih, Secretary General of the Federation of Indonesian Peasant Unions (FSPI), asserts that the rise of agrofuels means that companies such as IndoAgri and London Sumatra now expect to expand their plantations to 250,000 hectares by 2015. Approximately 1.5 million tons of palm are exported to the European Union where they are converted into agrofuels. Meanwhile, people in the producer country, Indonesia, are faced with a shortage of palm oil—a dietary staple (Saragih, 2007).

Faced with this reality, the United Nations Special Rapporteur on the Right to Food has observed that "the production of agrofuels is inadmissable if it brings more hunger and water shortage to developing countries." He went on to recommend a five year moratorium on their production (UN, 2007).

SOCIAL IMPACTS: FROM PLUNDER TO THE DESTRUCTION OF QUALITY OF LIFE

By its very nature, the industrialization of agriculture has proved to be a social failure in several countries. Bolivia, Guatemala, Honduras, and Paraguay present us with a serious paradox: food crops make up a high percentage of the countries' exports, yet malnutrition is taking on a structural character (Gudynas, 2007). Agrofuels have been championed as an alternative source of work that could allow peasants in core and periphery countries alike to increase their earnings and achieve social well being. Yet, in reality, nothing appears further from the truth. On the other hand, the situation in the European Union is still far from clear. Some studies have claimed that 1,000 tons of agrofuels can create between 2 and 8 full time jobs, concentrated especially in refineries and ports (Biofuelwatch, Carbon Trade Watch/TNI, Corporate Observatory, 2007). However, in the periphery countries, which are ultimately set to become the major sellers of raw materials for vehicle fuels, the development of this sector is based on establishing economies of scale and an extremely centralized agro-industrial model where transnational capital and local land-holding elites have increasingly intimate relations with one another (GRAIN, 2007). The inhabitants of the rural communities are becoming ever more expendable and are left with only two options: either to migrate or become agricultural day laborers. Below we will briefly consider a few examples.

The Rural Reflection Group (El Grupo de Reflexión Rural, GRR) emphasizes that the Green Revolution that was implemented in Argentina's countryside, contributed to the population's impoverishment. Thus, in a country that was known as one of the "world's granaries," the National Survey of Nutrition and Health registered in 2006 that 34 percent of children below the age of two suffer from malnutrition and anemia. According to GRR, this phenomenon can partly be explained by the fact that Argentina was converted into a producer of transgenic crops and an exporter of animal fodder, based in large-scale Roundup Ready soya monocultures. In this

---

half a million tons in 1993 to 7.3 million tons (tariff-free) in 2004. The final stages of NAFTA came into effect in 2008, which means that Mexico will become flooded with millions of tons of US maize and beans, raising the possibility of provoking a major social and political crisis.

context, land ownership became concentrated, ruining 400,000 small producers and provoking a rural exodus that swelled the poverty belts in the large cities (Rulli and Semino, 2007). The reality is not very different in Brazil, the world's largest bioethanol producer. The municipality of Ribeirao Preto (São Paulo) is known as the "Brazilian California" due to its technological development in the production of sugar cane. Yet, 30 factories control all the land, 100,000 people (20 percent of the total population) live in *fabelas* (shanty towns), and there are more people in prison (3,813) than there are peasants (2,412) (Vicente, 2007).

During the United Nations Permanent Forum on Indigenous Peoples, which was in session in May 2007, attention was drawn to the fact that indigenous populations are being displaced from their land by the expansion of energy crops. This is contributing to the destruction of their cultures and forcing them to migrate to the cities. In one Indonesian province alone, West Kalimantan, 5 million people have already been forced to leave their ancestral territories (Biofuelwatch, Carbon Trade Watch/TNI, Corporate Observatory, 2007). Thus, the Indonesian peasants stress that the growth of agrofuels threatens to end up eroding their agricultural and food system. Land is concentrated in the hands of a mere handful of large companies, which together own 67 percent of the cultivable land. Palm monocultures have deepened the marginalization of the small producers. In 2006 alone, these plantations provoked 350 land-based conflicts, despite the fact that land reform is enshrined in the Indonesian Constitution and the country's laws. This concentration of land and marginalization of peasants is by no means a new process, as it has been going on since colonial times (Saragih, 2007).

In Paraguay, the advance of transgenic soya and sugar cane monocultures is also giving rise to a frenzied process of investors buying up the best lands. The country devotes 2.4 million hectares to soya production, but is aiming for 4 million in order to fulfill its sale commitments to the European Union. This is a country where 21 percent of the population lives in extreme poverty, 1 percent of the land owners own 55 percent of the land, and 40 percent of the producers cultivate plots that are between 0.5 and 5 hectares. In September 2006, the Supreme Court confirmed that the National Agrarian Reform Institute had illegally sold land to large soya producers. According to the organization Sobrevivencia, approximately 70,000 people abandon the countryside each year after coming under pressure to sell their plots, and, according to various civic organizations, peasant livelihoods and communities are being destroyed in other ways too. This year five people died and seven were injured by the agro-industry's armed guards in the Paraguayan department of San Pedro—one of the zones where the government is promoting ethanol production.[12] In Colombia, Jiguamiandó and Curvaradó, the Afro-descendant communities, experienced an even worse fate: military and paramilitary violence forced them to flee their lands, which were then illegally occupied by the company Urapalma (Redes-AT and GRAIN, 2007b). Risking harsh punishment, some dared to return, only to

---

12   For more information on this see Rulli, 2007 and Biofuelwatch, Carbon Trade Watch/TNI, Corporate Observatory, 2007.

find their homes destroyed, and the previously well-preserved jungle devastated. Oil palm plantations extended as far as the eye could see.

And what became of those who stayed? According to the Brazilian Forum of NGOs and Social Movements for the Environment and Development, the monocultures failed to generate as many jobs as they had promised. In the tropics, 100 hectares of family farming creates 35 jobs, the same area of land devoted to eucalyptus plantations only represents 1 job. In the case of soya it is two, and in sugar cane and palm, ten. In many cases, the cane cutters are only paid if they manage to produce a certain quota, the amount having been predetermined by the company. Needless to say, working conditions are difficult, including the use of agrochemicals without any protective equipment, precarious housing, lack of sanitation services and drinking water, and even child labor.[13]

The populations who live in the vicinity of the palm and soya plantations find their health endangered by the powerful herbicides used. It is estimated that in Malaysia, between 1977 and 1997, an agricultural day laborer died every four days due to poisoning from the herbicide Paraquat. In Argentina, the aerial spraying of herbicides on neighboring soya plantations is causing an alarming number of cancer cases in the Southern province of Santa Fe (Biofuelwatch, Carbon Trade Watch/TNI, Corporate Observatory, 2007).

MEGAPROJECTS AND AGROFUELS

Biodiesel and bioethanol are normally not teletransported from the fields to the petrol tanks, and, in this undeniable fact lies another aspect of the rise of agrofuels that can hardly be described as "bio": the increasing need for integration of infrastructures necessary for their transportation and export. Hence, the need for the, lamentably, resuscitated Plan Puebla Panamá (PPP) and the Initiative for the Integration of South American Infrastructures (IIRSA).[14] These megaprojects consider Latin America's unruly geography to be an obstacle to the extraction of raw materials and the transport of goods. Their mission is to get around it by way of motorway corridors, hydroelectric dams, waterways, electric cables, oil pipelines, etc. And of course, it goes without saying that these projects will bring lucrative profits to companies such as the Spanish Iberdrola and Gamesa (wind park in Mexico), ACS (management of ports and trawlers in Brazil), and even to mostly-unknown consultancy firms such as TYPSA or Norcontrol. And, despite the promises of "local development" (evoking the ideologically bankrupt "trickle down" theory), these megaprojects are in fact harmful, situated as they are on indigenous territories and peasant communities, and traversing zones that are rich in biodiversity.

Although local populations have not been consulted on designing these megaprojects, there has been participation from the Inter-American Development Bank (BID). The BID, which bears considerable responsibility for generating the continent's

---

13   See Biofuelwatch, Carbon Trade Watch/TNI, Corporate Observatory, 2007 and Holt-Giménez, 2007.

14   For more information about the geopolitical dimension of both plans and their social and environmental impacts, see http://www.odg.cat/es/inicio/enprofunditat/plantilla_1.php?identif=582.

debt, is currently promoting agrofuels in several ways. It estimates that Latin America will need fourteen years to convert itself, at the cost of $200 billion, into one of the world's key biodiesel- and bioethanol-producing zones. The BID president himself, Luís Alberto Moreno, codirects a private sector group, the Interamerican Ethanol Comission, together with Jeb Bush (ex-governor of Florida) and Japan's ex-prime minister, Junichiro Kozumi. Thus, the BID supports the expansion of palm plantations in Colombia, and sugar cane and soya in the Brazilian Amazon. In fact, this year the BID's Executive Director approved $120 million for the first stage of a private sector agrofuels project in Brazil, which will go to Usina Moema Açucar and Alcohol Ltda. (São Paulo). The operation forms part of the bank's initiative to develop structures that enable priority debt financing for five bioethanol projects, costing $997 million (IDB, 2007).

On the other hand, it is crucial to ensure that commodities can freely reach the ports, not only those on the Atlantic shores, but also the Pacific, in order to reach Asian markets. Thus, the bank recommends that Brazil spend $1 billion each year on infrastructures over the next fifteen years. It also strives to speed up the IIRSA projects that have been rejected by civil society, such as the Paraguay-Paraná-Plata, the project of improving the navigability of the Río Meta, Ferro Norte (a railway network that would connect the soya states of Paraná, Mato Grosso, Rondonia, and São Paolo), and the Río Madera complex.

The latter, the Río Madera complex, is one of the main projects underway within the IIRSA axis, Perú-Brasil-Bolivia, and is located on the Brazilian-Bolivian border. The project currently consists of constructing two mega-hydroelectric dams in Brazilian territory—in San Antonio and Jirau. These dams are closely linked to the growth of agrofuels, as the hydroelectric power stations will supply the energy to the Brazilian states of Rondonia and Matto Grosso, enabling an expansion of the soya industry. Soya production is particularly important in Matto Grosso, one of the biggest soya producers in the world.[15]

Megaprojects for integrating infrastructures are turning out to be a crucial factor in the transportation of the raw materials for agrofuel production, such as grains. Not only does this entail increasing the external debts of the countries where these plans are being carried out, but it is simultaneously generating a considerable ecological debt, owed by large companies to the local populations. These populations do not participate, nor are they consulted, though they experience major social and environmental impacts from the projects.

SECOND GENERATION FUELS: FROM BAD TO WORSE

Faced with the multiple problems of first generation agrofuels, a new technological response is once again being offered: liquid agrofuels (BtL, Biomass to Liquid), which can be obtained from lignocellulosic biomass such as straw or wood chips. This includes producing bioethanol by fermenting hydrolyzed biomass, as well as agrofuels

---

15   For more information, see: http://www.biceca.org and http://internationalrivers.org/.

obtained by a thermo-chemical process, such as the bio-hydrocarbons obtained by pyrolisis, the forms of gasoline and diesel that are synthetically produced.[16]

The social and environmental impacts generated by the large-scale production of these fuels are, for the time being, relatively similar to those associated with first generation agrofuels. Gathering organic waste from fields requires the use of greater amounts of fertilizers, thus emitting greater quantities of nitrous oxide. Furthermore, the massive harvesting of dead trees will result in loss of biodiversity, given that thousands of species depend precisely on this vegetation waste. This could reduce the forests' capacity to absorb carbon. The other aspect is that the preferred raw material would originate from tree monocultures. The genetics industry is currently research-ing the modification of plants to produce less lignin, in order to facilitate cellulose breakdown and accelerate the plants' growth rhythm. However, releasing transgenic trees into the environment has unknown risks (Biofuelwatch, Carbon Trade Watch/TNI, Corporate Observatory, 2007). Enthusiasts of second generation fuels and tree plantations seem to have forgotten that a forest is not just a collection of trees, but an ecosystem.[17] The World Rainforest Movement reminds us that in Chile tree plan-tations are known as "planted soldiers" (i.e. they are green killers). The plantations occupy massive areas, threatening the inhabitants' traditional sources of subsistence. In Thailand, eucalyptus is referred to as the "selfish tree" because it monopolizes the water necessary for growing rice, the basic peasant subsistence. The monoculture model that has been used by the growing paper industry is being replicated in differ-ent countries, provoking ongoing resistance to its social and environmental impacts.

HUMAN BEINGS, NOT MACHINES

Not only do agrofuels constitute a completely inadequate response to global prob-lems such as global warming and hunger, but the large-scale production of these fuels does not even make any break with fossil fuels, since these are necessary for their production and transportation. Furthermore, agrofuels imply an intensifica-tion of the agro-industrial model, a model that already bears significant responsi-bility for the current environmental crisis and the worsening living conditions of the world's poorest populations. The only beneficiaries from agrofuels are the large business conglomerates, several of which, having participated in the petroleum, au-tomobile, agribusiness, and construction sectors, have already contributed to gen-erating climate change and an unclaimed ecological debt. According to the FAO, the rapid transition towards a greater use of agrofuels could reduce the emissions of greenhouse gases "only if they take into account food security and the environmental consequence" (FAO, 2007). Yet, the FAO proposal places before us an equation that is impossible to resolve, since it is being made within the context of one of the central pillars of capitalist logic, namely the obsession for sustained growth (which itself is not sustainable). Furthermore, its starting point is an over simplistic understanding of the environment and affected populations.

---

16   See Programa del Encuentro Biocarburantes '07 (http://www.iir.es).
17   See the documentary film "Invasión verde," http://www.wrm.org.uy.

This is due to disdain for a key fact: human beings are still not automatons. The planet's millions of impoverished people are not simply machines that require a suitable source of energy. An indigenous leader of the Mixe people (Oaxaca, México) told me that what his community seeks is autonomy. Autonomy is a complex equilibrium that includes concepts such as a community having its own food, hope, decision-making power, thought, language, territory, development path, education, life and death, all of which belong to them. For their part, the Andean communities are fighting for *Suma Qamaña* ("good living") to be introduced into the new Bolivian constitution. Theirs is a territory that, for its inhabitants, is sacred, and where the diversity of nature and its divinities live together with the human species. In Mexico, maize is not simply a basic food staple for the Wixárika, it also has a sacred character. The *milpa*, or cultivated land plot, is like a community where maize, beans, squash, amaranth, and medicinal plants all live together and complement one another (Redes-AT and GRAIN, 2007a). We need to approach dilemmas such as climate change and the contradictions generated by the capitalist system from a position that recognizes humanity's complexity and cultural diversity. In that light, the possible responses are numerous. Indigenous and peasant organizations have given expression to their demands in the all-encompassing and comprehensive concept of food sovereignty. More recently, the concept of energy sovereignty has also been adopted. Popular campaigns around food sovereignty are also beginning to demand a halt to energy crop plantations and a moratorium on EU incentives for agrofuels, and their importation of monoculture-based agrofuels or others that, in other ways, contribute to the ecological debt and threaten food sovereignty.[18]

Let us end this article by underlining a theme that is currently garnering ever greater strength and around which an ever greater variety of ideas for change are forming: degrowth. This is understood as "the need to leave the current economic model behind and break with the logic of continuous growth" (Mosangini, 2007). It is understood as the formulation of an economic, ecological, and socially-sustainable science, which seeks to reground the economy as a subsystem of the biosphere, in respect of its laws and physical limits. An example is the emergence of proposals for production on a local and sustainable scale, organic agriculture, deindustrialization, the end of the current transport model, the end of consumerism and advertising, deurbanization, self-production of goods and services, austerity, and non-market-based exchanges. Such proposals are especially urgent in the core countries. Such initiatives, in a move towards empathy, listening, and collaboration between the different resistances to the capitalist system, will undoubtedly provide a basis from which to responsibly face up to today's global problems in order to recover the possibility of a dignified life for all of us who inhabit the planet.

---

18   See http://www.biofuelwatch.org.uk/ and http://www.noetmengiselmon.org.

BIBLIOGRAPHICAL REFERENCES

Inter-American Development Bank (BID). (2007), "BID aprueba US$120 millones para proyecto de biocombustibles en Brasil," Press Release 25th July.

Barabas, A. (2003), "Introducción: una mirada etnográfica sobre los territorios simbólicos indígenas." In: Barabas, A. (coord.) *Diálogos con el territorio*, Vol. 1, México, INAH.

Binimelis R., Jurado A., Vargas M. (2007), "La trama de los agrocarburantes en el Estado español," *Revista Ecología Política*, n. 34.

Biofuelwatch, Carbon Trade Watch/TNI, Corporate Observatory. (2007), *Agrofuels. Towards a reality check in nine key areas*, TNI-CEO, Amsterdam.

Coordinadora de Organizaciones de Agricultores y Ganaderos (COAG). 2007. *Las verdaderas causas de la subida del precio de los cereales; consecuencias para el sector agrario.* Madrid.

GRAIN. 2007. *Seedling. Agrofuels special issue*, Barcelona.

Gudynas, E. 2007. Bolivia: sueños exportadores y realidades nacionales. *Bolpress*, 2nd June.

Holt-Giménez, E. 2007. "Biocombustibles: mitos de la transición de los agro-combustibles." *Revista electrónica Biodiversidad en América Latina*, 3rd September.

Llistar, D. 2007. "Guerra Norte-Sur: biocombustibles contra alimentos." *Revista electrónica Rebelión*, 19th April.

Mosangini, G. 2007. "Decrecimiento y cooperación internacional." *Revista electrónica Rebelión*, 29th September.

United Nations 2007. *The Right to Food.* Report from the Special Rapporteur on the Right to Food. August. A/62/289.

United Nations Food and Agriculture Organization (FAO). 2007. *Living With Climate Change.* Press Note 10th September.

Ortega, M. (Coord.). 2005. *La deuda ecológica española: impactos ecológicos y sociales de las inversiones españolas en el extranjero.* Sevilla, Spanish State, Muñoz Moya Editores Extremeños.

Redes-AT and GRAIN. 2007a. *Revista Biodiversidad: sustento y culturas*, Montevideo, Uruguay (50/51).

Redes-AT and GRAIN. 2007b. "La fiebre por los biocombustibles y sus impactos negativos." *Revista Biodiversidad: sustento y culturas* (52): 16-20.

Rulli, J. 2007. "Soja en San Pedro-Paraguay. Guardias emboscaron a campesinos por cazar en latifundio." *Revista electrónica de Base investigaciones Sociales*, 29th August

Rulli, J. and Semino, S. 2007. "La génesis de una política agraria. De la OCDE a la producción de biodieseles de soja," article presented in the *Seminario de Expertos sobre biodiversidad y derecho a la alimentación*, Madrid.

Russi, D. 2007. "Biocarburantes: una estrategia poco aconsejable." *Revista Biocarburantes Magazine*, *Estado español*, May.

Saragih, H. 2007. "It's cars versus humans." *The Jakarta Post*, Jakarta, 26 July.

Stern, N. 2006. *Stern Review Report on the Economics of Climate Change*, HM Treasury, London.

Vicente, C. 2007. Interview with Joao Pedro Stedile, leader from MST and Via Campesina Brasil. *Revista electrónica Biodiversidad en América Latina*, 6 June.

CHAPTER 31 ▌ Part 8: Resurrection of the Nuclear
Industry, Its Connection with Global Militarism and
Limited Uranium Supplies

CONFRONTING THE NUCLEAR RESURGENCE
British Government's Maneuvers, EU Policy, and the Nuclear-Fossil Collusion[1]

Sergio Oceransky

NUCLEAR BRITAIN

The UK nuclear industry has been in a terminal crisis for several decades. No nuclear reactors have been planned since Sizewell B entered into operation in 1995, after seventeen years of costly delays caused by grassroots resistance. The last reactor to come into operation before Sizewell B was Torness 2, in 1989, fourteen years after receiving statutory consent from the Secretary of State. Despite Margaret Thatcher's assurance that a new nuclear power station would be built each year under her rule (1979–1990), she and the following Prime Ministers were only able to see through a handful of previously-approved projects, at an agonizingly slow pace and with immense added costs.

The paralysis that followed was caused mainly by a mix of public opposition and lack of economic viability, despite generous governmental support. The Non-Fossil Fuel Obligation (NFFO), created by the Electricity Act 1989, has provided billions of pounds to UK nuclear power generators and forced the electricity Distribution Network Operators in England and Wales to purchase nuclear electricity. According to Pete West of the Severn Wye Energy Agency (SWEA), "from 1990 to 1998, 98 percent of the Non-Fossil Fuel levy was handed over to the nuclear industry. Hansard records from January 1996 indicate Nuclear Electric had received £5.9 billion of public funding from the Non-Fossil Fuel levy during the previous 5 years." The NFFO mandate was later enlarged to include the renewable energy sector in what can only be described as a shameless greenwash operation. Renewable energy producers have received an insignificant share of the funds raised; the NFFO continues

---

1   This is a selection from a previously published piece published as a special issue of the *Nuclear Monitor*, on January 28th, 2008. No. 665. *Nuclear Monitor* is the regular publication of the World Information Service on Energy (WISE) and the Nuclear Information & Resource Service (NIRS). It is reproduced here with permission from both the author and WISE. The article has been divided in two pieces for this book, and another selection is included as Chapter 11, "European Energy Policy on the Brink of Disaster: A Critique of the European Union's New Energy and Climate Package." This original article was written when the new energy and climate policy framework of the European Union was taking shape, as a contribution to the heated debate around it. The debate is over and the EU policy has been passed but the contents of the text are still relevant to discussions on energy and climate policy issues. While there were some important changes in the EU package that was in fact passed, many of the issues discussed here were included in the final package. The complete text of the original article can be downloaded at http://www10.antenna.nl/wise/665/Special/665_Special.pdf.

to be essentially a tool for the channeling of public funds to nuclear power and lately, illegally, to the Treasury, as denounced by the National Audit's Office.

In the last decade, due to the crisis of the nuclear sector and to changes in public opinion, more appropriate tools were introduced, such as the Climate Change Levy for non-domestic energy users (a tax that only renewable energies and cogeneration are exempted from). But the existing orders to subsidize nuclear power under the NFFO Fund, issued in September 1998, will continue in effect until it expires in 2018. In contrast, the Treasury has decided to literally steal the share of the NFFO Fund that should be used for renewable energy. As Oliver Tickell denounced in *The Guardian*, "The Treasury and the Department of Trade and Industry [which controls NFFO funds] justify these payments by claiming the NFFO Fund is 'hereditary revenue of the Crown'—along with income arising from the Crown's traditional rights to treasure trove, swans and sturgeons. Yet despite questioning from MPs, ministers have refused to publish either legal advice or an outline of their legal argument, claiming 'legal professional privilege.'"[2] The NFFO Fund therefore is still, essentially, a funding tool at the service of nuclear reactors.

In spite of such privileges and generous subsidies, the UK nuclear sector came very close to bankruptcy. The UK government created British Energy in order to privatize the eight most modern nuclear power plants in the UK. By the end of the 1990s the company was a stock market favorite, but by 2002 it was in deep financial trouble and approached the government for financial aid. The crisis came as a result of a slump in wholesale energy prices, a failure to obtain tax exemption for nuclear power on the Climate Change Levy, and renegotiations of the nuclear waste processing and power plant decommissioning costs with British Nuclear Fuels plc. The closure of nuclear power plants was avoided when the government made the taxpayers foot the bill (once more) and provided £3.4 billion public money to bail out the company in 2004. At the same time, the Nuclear Liabilities Fund was created by the government to assume the long-term financial liabilities from spent nuclear fuels. The Fund is a mixture of state-funding (estimated between £175 million and £200 million per year) and contributions from British Energy (which is required to provide 65 percent of its profits to the fund), and also acts as a public-funded creditor to British Energy. The Fund therefore established a limit in the private liability for nuclear waste and decommissioning and ensures government-funded profitability to whoever decides to invest in nuclear energy.

In addition to this enormous transfer of public money into the private nuclear sector, British taxpayers are also made to fund the public nuclear sector. As Pete West of the Severn Wye Energy Agency explains, "The £72 billion public liability for clean up of existing nuclear plants refers to the older Magnox reactors that were unmarketable at privatization and therefore still in the public sector. The Nuclear Decommissioning Authority is currently spending £2 billion per year of taxpayers' money on nuclear waste management."

---

2   See http://politics.guardian.co.uk/comment/story/0,1840311,00.html

The combination of all the direct and indirect support measures and subsidies that have been channeled into nuclear power, if applied to the renewable energy sector, would have created the most impressive green-energy-generating capacity in the world, enabling the closure of a substantial amount of fossil fuel and nuclear power plants, and establishing the UK as the global leader in environmental action, sustainable technological innovation, economic and geopolitical independence, and green employment generation. Instead, they went into the nuclear black hole and did not create a single extra kilowatt of installed capacity, a single new job, or a single blip of new knowledge. They just fed the bank accounts of nuclear investors, while the radioactive waste stocks continued growing.

The energy bill announced on the 10th of January 2008 added some more guarantees for the nuclear sector. In a nutshell, the proposed bill assures nuclear investors that the government will pay to resolve crises, will provide even higher indirect subsidies than until now, will cover unexpected costs for handling nuclear waste and reactor decommissioning, and will ensure that new nuclear reactors are built fast, disregarding local opposition and democratic principles where necessary.

The main elements can be summarized as follows:

•   The government promises "greater certainty for investors" through unilateral action to underpin the price of carbon, which becomes the main instrument for indirectly subsidizing nuclear. While the primary tool for this will be the EU Emissions Trading Scheme, the government commits to "keep open the option of further measures to reinforce the operation of the EU ETS in the UK should this be necessary to provide greater certainty for investors."

•   Public money is made available for decommissioning new plants and waste disposal. In theory, operators are responsible for these costs but "if the protections we are putting in place prove insufficient, in extreme circumstances the government may be called upon to meet the costs of ensuring the protection of the public and environment." Section 3.75 of the document indicates that there will be a fixed price for disposing of waste, despite the lack of any plan or strategy in that respect, which means that no one knows what the costs will be. What is clear is that expensive action is required: last year the Royal Society warned that Britain's stocks of plutonium are kept in "unacceptable" conditions and pose a severe safety and security risk. The length of time between starting a new nuclear plant and eventually putting the waste into a geological repository could well be over 150 years. Cost projections in this context are pure speculation, but the public purse will cover all unexpected costs.

•   These provisions limit the long term liabilities of private companies. Private operators do not need to worry about the financial consequences of nuclear pollution, whether caused by waste, accidents, or decommissioning;

the government will meet the costs. Therefore, the proposed bill will facilitate access to loans and capital markets. In addition, the nuclear insurance premium will continue to be grossly undervalued and publicly subsidized (as everywhere else in the world).

•   Other costs associated with the nuclear revival that are likely to be covered by taxpayers include the cost of adapting transmission lines to fit the highly centralized electricity generated by new reactors, security and transport of waste fuel, and protection of nuclear power stations from the effect of tidal surges. A study commissioned by British Energy said that "increases in future surge heights of potentially more than a meter could, when combined with wind speed increases, threaten some sites unless existing defenses are enhanced."

•   The government also announced a planning bill that will make the process of building new reactors quicker and less complex, ensuring that the costly delays in winning planning permission to build Sizewell B will not be repeated.

The reinforcement of an already outrageous and disproportionate level of taxpayers' support to the nuclear sector was the response to the conclusions reached by the 2003 Energy white paper, which concluded that nuclear power's "current economics make it an unattractive option for new, carbon-free generating capacity and there are also important issues of nuclear waste." The Labour Cabinet now claims that there should be no artificial cap (!!) on the amount of nuclear energy generated in the UK. The proposed energy bill will provide all the (apparently non-artificial) subsidies and public guarantees to ensure that the UK remains a nuclear superpower. The Conservatives also assured investors that the political climate will remain supportive of nuclear power in the long term.

The new energy bill presented on the 10th of January is not so new: it has been planned and prepared over the last years. According to Pete West, Tony Blair only agreed to the publication of the 2003 Energy white paper, which specifically ruled out new nuclear power, if there was a review in 2007 including the nuclear option if renewable energies were failing to deliver. Since then, the government applied the worst possible renewable energy regulatory framework and stole funds collected to promote renewable energy, resulting in one of the lowest shares of green power in one of the countries with the largest renewable energy potential in Europe. Last year's consultation into UK Energy Policy, which resulted in the nuclear revival, was so shamelessly manipulated by the Government that Greenpeace successfully challenged it in court: the ruling established that the government had not fairly represented consultees' opinions, which were well argued responses in favor of renewables and energy saving, and against nuclear power. But now, on the basis of policy machinations, resource theft, and illegal and dishonest consultations, the Cabinet misrepresents renewable energy as a marginal component of the energy mix, limited by high costs, public opposition and lack of reliability, and nuclear power as the unavoidable option to secure affordable, safe, and clean energy. This is a showcase example of deliberate and disingenuous hypocrisy.

The uselessness of this nuclear revival is apparent from the Cabinet's own plans, according to which new nuclear reactors cannot begin to be built until 2013 or later, and no new plant would come online until 2018. Past and current experiences, such as the status of the vastly subsidized new reactor under construction in Finland, which is two years late (and £1bn over budget) after just two years' building, indicate that it is unlikely that the first new nuclear plant can open in the UK before 2021. Old reactor closures mean that the share of nuclear power in UK's electricity supply will go down from about 18 percent now to 3 percent by 2020.

Therefore, the new nuclear reactors cannot respond from the energy gap created by the closure of existing ones. They also cannot respond to climate concerns: according to the 2006 report of the Sustainable Development Commission (SDC), reporting directly to the prime minister, replacing all the existing nuclear capacity with new nuclear plants might save 7 million tons of carbon by the late 2020s—equivalent to around 4 percent of total UK emissions. In contrast, as Caroline Lucas (MEP, Green Party) remarks, "the government's own figures show that there is the potential to save more than 30 percent of all energy used in the UK solely through energy-efficiency measures which would also save more money than they cost to implement. Moreover, about two-thirds of the energy used in electricity generation from large, centralized power stations is wasted before it ever reaches our homes, and by itself accounts for a full 20 percent of UK CO_2 emissions."

The nuclear revival is therefore blatantly disconnected to energy or climate considerations. It is purely based on geo-political factors and the quest to maintain a hegemonic position in the world-system, combined with powerful economic interests. One can only hypothesize about the analysis and intentions behind such an irrational policy, since the official reasons will never be publicly disclosed. A later part of this article will do so, focusing especially on the rush for the last remaining (easily usable) uranium stocks, on questions around nuclear proliferation, and on the reasons why powerful economic interests need to delay the (potentially very fast) transition to renewable energies as much as possible.

Of course, the new energy bill was well received by power companies, including French giant EDF (the employer of UK Prime Minister's brother Andrew Brown), German E.on, Centrica (British Gas parent group), and others. EDF was particularly pleased with the commitment to provide a "UK mechanism" for encouraging low-carbon technologies. These corporations have been working very closely with the UK government to shape the EU's energy and climate policy according to the interest of the nuclear and fossil industries.

MAKING SENSE?—THE GEO-STRATEGIC AND MILITARY DIMENSION OF ENERGY POLICY

It is difficult to understand how such an awful energy and climate policy can make sense to policy makers. It is far more expensive, polluting, complex, and risky than a transition to a decentralized 100 percent percent renewable energy system. It will have negative consequences for almost the whole of society, benefiting only a few private interests.

But this is nothing new. The same pattern has repeated itself recurrently in the history of energy policy, since those private interests are perfectly aligned with the geo-strategic and military interests of the state, particularly in hegemonic countries. According to *The Economist* ("Nuclear Power Out of Chernobyl's Shadow," May 6th 2004, print edition), "more than half of the subsidies (in real terms) ever lavished on energy by OECD governments have gone to the nuclear industry."

According to "Federal Energy Subsidies: Not All Technologies Are Created Equal" (by Marshall Goldberg, REPP, July 2000 No. 11), between 1947 and 1961, commercial, fission-related nuclear power development received subsidies worth $15.30 per kWh. This compares with subsidies worth $7.19/kWh for solar and 46¢/kWh for wind between 1975 and 1989. In their first 15 years, nuclear and wind technology produced comparable amount of energy (2.6 billion/Nucl. and 1.9 billion kilowatt-hours/wind), but the subsidy to nuclear outweighed that to wind by a factor of over 40, at $39.4 billion to $900 million.

What we are seeing today is nothing more than the extension of the post-WWII energy policy of "world powers," which served (and still serves) the goal of maintaining military dominance. The relaxation in the nuclear race that followed the end of the cold war is over. The message sent by the differential treatment that the US-UK axis gave to Iraq and North Korea has sent an unequivocal message to the other countries of the world: if you don't have nuclear weapons, you should submit to our domination or be ready for invasion. The neoconservative policies of Bush's and Blair's administrations have left us, amongst other terrible legacies, an irreparable damage to the little credibility that multilateralism still had, the reawakening of nuclear proliferation in peripheral countries, the re-escalation of a military race with Russia, and growing numbers of people ready to die while provoking as much death and destruction as possible in the West. The relation with emerging powers such as China is still quiet, but nobody knows what the future might harbor.

In this context, the only ethical policy choice is working decidedly for complete disarmament, for the complete abolition of all civil and nuclear programs, and for peace, cooperation, and urgent environmental remediation and economic redistribution. Renewable energies play a key role in that process, since they can provide much more energy than we need for affordable prices, and they are present all over the world.

However, the choice of the UK Labour government is clear: it seems to have signed a "state pact" with the Tories in order to marginalize renewable energy and to rebuild a major nuclear capacity. As the *Guardian* reported on 24th of January, "John Hutton, the business secretary, said the UK remained committed to meeting the EU renewable energy target share but insisted that other low-carbon technologies-including nuclear power had to be part of the mix. Battles are likely to develop among the twenty-seven governments over the inclusion of nuclear energy." John

Hutton's report on the EU energy package to the UK Parliament indicates that the UK wants the RE target to include nuclear power: "We set out the framework for our low carbon future in the 2007 Energy white paper. We are already working to implement the domestic measures proposed in that white paper through the Energy and Climate Change Bills currently going through the House. We have also announced decisions on the future of key low carbon technologies such as nuclear power and the development of carbon capture and storage. Having this broad portfolio of low carbon options, alongside renewable energy technologies available to investors, will be essential in moving to a much lower carbon future by 2050."

One of the reasons is the fact that easily usable uranium stocks are getting depleted very rapidly, and the race for nuclear fuel is similar to the race for oil fields. A naïvely optimistic assessment could be that the UK government thinks that by approving many nuclear power plants, uranium will become more expensive and this will discourage other countries from building nuclear power plants. But they know that the price of uranium will not stop other countries from reinforcing their nuclear programs.

We cannot afford to waste time with this sort of policies (and politicians): what we need now, urgently, are intelligent policies based on the common good, rather than on the concentration of power and wealth in the hands of large energy corporations and the state. And if history is of any use, we should be prepared to define and struggle for such policies through grassroots organizing, since the solutions certainly will not come out of the offices of the European Commission, or of any national government.

Chapter 32 ∎ Part 8

JAPAN AS A PLUTONIUM SUPERPOWER[1]

Gavan McCormack

For sixty years, the world has faced no greater threat than nuclear weapons. Japan, as a nuclear victim country, with "three non-nuclear principles" (non-production, non-possession, and non-introduction of nuclear weapons into Japan) and its "Peace Constitution," had unique credentials to play a positive role in helping the world find a solution, yet its record has been consistently pro-nuclear, that is to say, pro-nuclear energy, pro-the nuclear cycle, and, pro-nuclear weapons. This essay elaborates on Japan's aspiration to become a nuclear state, arguing that attention should be paid to Rokkasho, Tsuruga, and Hamaoka, the places at the heart of Japan's present and future nuclear plans, no less than to Hiroshima and Nagasaki, whose names represent the horror of its nuclear past.

The nuclear question in relation to Japan is commonly understood in the narrow sense of whether Japan might one day opt to produce its own nuclear weapons. Prime Minister Kishi, in 1957, is known to have favored nuclear weapons. In 1961, Prime Minister Ikeda told US Secretary of State Dean Rusk that there were proponents of nuclear weapons in his cabinet, and his successor, Sato Eisaku, in December 1964 (two months after the first Chinese nuclear test) told Ambassador Reischauer that "it stands to reason that, if others have nuclear weapons, we should have them too." US anxiety led to the specific agreement the following year on Japan's inclusion within the US "umbrella."[2] Prime Ministers Ohira, in 1979, and Nakasone, in 1984, both subsequently stated that acquiring nuclear weapons would not be prohibited by

---

1   (Editor's note). This article was originally written in 2007, this article is being reproduced here unchanged. Originally delivered as a lecture at Cornell University, 25 October 2007, this paper develops further points made in a chapter of the author's recent book *Client State: Japan in the American Embrace* (London and New York, Verso, 2007). Article posted for *Japan Focus* on December 9, 2007. It is being published with permission from the author. The article is still quite current, and no major development has occurred to change its thrust since writing. Of course the Japanese Government will applaud the Obama initiatives in direction of de-nuclearization, but any actual step to "strip away" the umbrella coverage which the USA offers to Japan would cause an uproar. Indeed, former Finance Minister Nakagawa (close friend and political associate of Prime Minister Aso) has recently repeated his call for a debate on "going nuclear"—which recent expert (Russian) opinion reckons Japan could do in a matter of a single month— so vast are the Japanese plutonium stockpiles and so advanced its rocketry. However, as argued in this article, this is only one dimension of the complex Japanese embrace of the nuclear. The article's argument on the centrality of plutonium to the Japanese state and its future strategies still stands. For more recent articles, the reader is recommended to look at the following solidly detailed reports, both dating from 2009: Emma Chanlett-Avery and Mary Beth Nikitin, "Japan's Nuclear Future: Policy Debate, Prospects, and US Interests," Congressional Research Service, February 19, 2009. http://www.fas.org/sgp/crs/nuke/RL34487.pdf, and Institute for Foreign Policy Analysis, "Realigning Priorities: The US-Japan Alliance and the Future of Extended Deterrence," March 2009. http://www.ifpa.org/projects/japn_ext_deterrence.htm

2   "60 nendai, 2 shusho ga 'kaku busoron' Bei kobunsho de akiraka ni," *Asahi shimbun*, 1 August 2005.

Japan's peace constitution—provided they were used for defense, not offence.[3] In the late 1990s, and with North Korea clearly in mind, the Chief of the Defence Agency, Norota Hosei, announced that in certain circumstances Japan enjoyed the right of "pre-emptive attack."[4] In other words, if the government so chose it could invoke the principle of *self-defense* to launch a pre-emptive attack on North Korean missile- or nuclear- or related facilities.

The former Defence Agency's then parliamentary Vice-Minister, Nishimura Shingo, carried this even further by then putting the case for Japan to arm itself with nuclear weapons.[5] Trial balloons about Japan developing its own nuclear weapons have been floated from time to time. Abe Shinzo, then Deputy Chief Cabinet secretary, remarked in May 2002 that the constitution would not block Japan's development of nuclear weapons provided they were small.[6] North Korea's declaration of itself as a nuclear power in 2005 and its 2006 launch of missiles into the East Sea (Japan Sea) further stirred these calls. Should the North Korean crisis defy diplomatic resolution, and North Korea's position as a nuclear weapon country be confirmed, such pressures would become almost irresistible. Even with that crisis resolved, as now seems increasingly possible, the attraction for Japanese politicians of nuclear weapons as a symbol of great power status has an ominous aspect.

However, I argue that a much broader construction of nuclear threat should be adopted. Japan is simultaneously a unique nuclear-victim country and one of the world's most nuclear-committed—one might almost say nuclear-obsessed—countries. Protected and privileged within the American embrace, it has evolved into a nuclear-cycle country and plutonium super-power. Plutonium is the chosen material on which the future of the Japanese economy is to rest—it is a material that only came to exist because of its destructive potential, so dangerous to humanity that a teaspoon-sized cube of it could kill 10 million people. Today, Japan contemplates, with apparent equanimity, a future in which it accumulates virtual mountains of the stuff.

In general, criticism of Japan tends to concentrate on its past crimes and present cover ups, i.e. on past history. Yet the bureaucratic project to convert Japan into a plutonium-dependent superpower surely concerns the region and the world. And where Japan goes, Asia and the world commonly follow.

WEAPONS

So far as defense policy is concerned, Japan is unequivocal: the core of its defense policy is nuclear weapons. To be sure, the weapons are American rather than Japanese, but their nationality is immaterial to their function: the defense of Japan. The nuclear basis of defense policy has been spelled out in many government statements,

---

3   Andrew Mack, "Japan and the Bomb: a cause for concern?" *Asia-Pacific Magazine*, No. 3 June 1996, pp. 5–9.

4   Statement of 3 March 1999 (quoted in Taoka Shunji, "Shuhen yuji no 'kyoryoku' sukeru," *Asahi shimbun*, 3 March 1999.).

5   "Nishimura quits over nuclear arms remarks," *Daily Yomiuri Online*, 21 October 1999.

6   Yoshida Tsukasa, "'Kishi Nobusuke' o uketsugu 'Abe Shinzo' no ayui chisei," *Gendai*, September 2006, pp. 116–129, at p. 127.

from the National Defense Program Outline (1976) and "Guidelines for US-Japan Defense Cooperation" (1997) to the 2005–06 agreements on the "US-Japan Alliance: Transformation and Realignment for the Future."[7]

So supportive has Japan been of American nuclear militarism that in 1969 it entered secret clauses into its agreement with the United States so that the "principles" could be bypassed and a Japanese "blind eye" turned towards American vessels carrying nuclear weapons docking in or transiting Japan, an arrangement that lasted until 1992.[8] Thereafter, nuclear weapons continued to form the kernel of US security policy, without Japanese demur, but there was no longer any need to stock them in Japan or Korea since they could be launched at any potential target, such as North Korea, from submarines, long-range bombers, or missiles. In 2002, the US articulated the doctrine of preemptive nuclear attack, under Conplan 8022. Conplan 8022-02, completed in 2003, spelled out the specific direction of preemption against Iran and North Korea.[9] By embracing an "alliance" with the US, Japan also embraces nuclear weapons and nuclear preemption.

Japan's position in denouncing North Korea's nuclear program rests on the distinction between its "own," i.e. American nuclear weapons, which are "defensive" and therefore virtuous, and North Korea's, which constitute a "threat" and must be eliminated. Yet logically, if Japan's security—and the security of the nuclear powers themselves—can only be assured by nuclear weapons, the same should apply to North Korea, whose case for needing a deterrent must anyway be stronger than Japan's.

Mohammed ElBaradei, Director-General of the International Atomic Energy Agency (IAEA), criticizes as "unworkable" precisely such an attempt to separate the "morally acceptable" case of reliance on nuclear weapons for security (as in the case of the US and Japan) and the "morally reprehensible" case of other countries seeking to develop such weapons (Iran and North Korea)."[10]

The moral and political coherence of Japan's Cold War nuclear policy depended on the one hand on reliance on the US "Umbrella," and, on the other, on support for non-proliferation and nuclear disarmament under the Non-Proliferation Treaty, but as the US, and indeed other nuclear club powers (Britain, Russia, France, China) made clear their determination to ignore the obligation they entered under Article 6 of the 1970 Non-Proliferation Treaty, and reaffirmed in 2000 as an "unequivocal undertaking," for "the elimination of their nuclear arsenals," the policy was steadily

---

7   To quote only from the October 2005 statement, "US strike capabilities and the nuclear deterrence provided by the US remain an essential component to Japan's defense capabilities ... " Ministry of Foreign Affairs of Japan, Security Consultative Committee Document, "US-Japan Alliance: Transformation and Realignment for the Future," 29 October 2005.

8   Morton Halperin, "The nuclear dimension of the US-Japan alliance," Nautilus Institute, 1999; "Secret files expose Tokyo's double standard on nuclear policy," Asahi Evening News, 25 August 1999.

9   Conplan refers to the global strike plans under which Stratcom (Strategic Command, Omaha) deals with "imminent" threats from countries such as North Korea or Iran by both conventional and nuclear "full-spectrum" options, under President Bush's January 2003 classified directive. (William Arkin, "Not Just A Last Resort? A Global Strike Plan, With a Nuclear Option," Washington Post, Sunday, May 15, 2005.)

10   Mohammed ElBaradei, "Saving ourselves from self-destruction," New York Times, 12 February 2004.

hollowed out. As the dominant Western powers turn a blind eye to the secret accumulation of a huge nuclear arsenal by a favored state (Israel) that refuses to join the
NPT, so they tend to treat Japan too as a special case, extending it nuclear privileges
for reprocessing, partly because of its nuclear victim credentials and partly because
they are well aware that it is Washington's favorite son. Partly, too, perhaps because
of its pacifist constitution.

Over time, like the nuclear powers themselves, once Japan had embraced the
weapons, it paid less and less attention to getting rid of them. Its cooperation with
nuclear intimidation against North Korea contributed to proliferation and brought
closer the time when Japan itself might decide to possess its own weapons. Should it
make that decision, Japan already possesses a prototype intercontinental ballistic missile, in the form of its H2A rocket capable of lifting a 5-ton payload into space, huge
stores of plutonium and high levels of nuclear scientific and technical expertise.[11] No
country could match Japan as a *potential* member of the nuclear weapon club.

Needless to say, countries like Japan that choose to base their national policy on
"shelter" beneath the US umbrella identify themselves with that umbrella's threatening as well as its defensive function. It is a system within which Japan is steadily incorporated, despite the almost total absence of public debate. Japan's leaders appear
to embrace their compliant nuclear status without qualm.

While Japan seems to have no qualms about the nature of the "umbrella" under
which it shelters, the US has spoken plainly about its determination not to rule out
first use of its nuclear force. The Pentagon's "Global Strike Plan," drawn up in response to a January 2003 classified directive from the President, integrated nuclear
weapons with "conventional" war fighting capacity and made clear the reservation of
right of preemption.[12] What that might mean for Korea (and for the region) boggles
the imagination. According to a 2005 study by the South Korean government, the
use of US nuclear weapons in a "surgical" strike on North Korea's nuclear facilities
would, in a worst case scenario, make the whole of Korea uninhabitable for a decade,
and if things worked out somewhat better, in the first two months, it would kill 80
percent of those living within a ten or fifteen kilometer radius, and spread radiation
over an area stretching as far as 1,400 kilometers—including Seoul.[13]

The US—with Japan's support—in March 2003 launched a devastating war on
Iraq, based on a groundless charge that that country was engaged in nuclear weapons
production, had an arsenal of around 7,500 warheads, most of them "strategic" and
more powerful than the ones that destroyed Hiroshima and Nagasaki. It now works
on a replacement schedule to produce 250 new "reliable replacement warheads" per
year, makes great efforts to develop a new generation of "low yield" small nuclear
warheads, known as "Robust Nuclear Earth Penetrators" or "bunker busters" specially tailored to attack Iranian or North Korean underground complexes, deploys
shells tipped with depleted uranium that spread deadly radioactive pollution likely

11   Dan Plesch, "Without the UN safety net, even Japan may go nuclear," *The Guardian*, 28 April 2003.
12   William Arkin, "Not just a last resort: A global plan with a nuclear option," *Washington Post*,
15 May 2005.
13   *Chosun Ilbo*, 6 June 2005.

to persist for centuries, has withdrawn from the Anti-Ballistic Missile Treaty (ABM) and declared its intent not to ratify the Comprehensive Test-Ban Treaty (CTBT), and promises to extend its nuclear hegemony over the earth to space.

Robert McNamara, who used to run the American system, in March 2005 described it as "illegal and immoral."[14] Even though civil nuclear energy cooperation with a non-signatory (especially a nuclear weapons country) contravenes the very essence of the NPT, in 2005 the US also lifted a thirty-year ban on sales of civilian nuclear technology to India, describing it as "a responsible state with advanced nuclear technology." It roundly denounces Iran and North Korea, on the other hand, for their insistence on a right guaranteed for them in Article 4 of the NPT.

Like the US, Japan's non-proliferation policy is contradictory: turning a blind eye to US-favored countries who ignore or break the rules, such as Israel and India, while taking a hard line on countries not favored by the US, such as Iran and North Korea. It is also passive on disarmament, i.e. specifically downplaying the obligations of the US and other superpowers, and because its own defense policy rests on nuclear weapons it is unenthusiastic about the idea of a Northeast Asian Nuclear Weapons Free Zone.[15]

For the past decade the idea of Japan becoming the Great Britain of the Far East has been eagerly promoted on both sides of the Pacific. The nuclear implications of this are rarely addressed, but Britain has long seen nuclear weapons as crucial to its power and prestige. In 2006, the British government declared the intention to renew its Trident fleet, i.e. to rest its defense on nuclear weapons into the foreseeable future. The Japan of Koizumi and Abe sets great store too on the paraphernalia of big power status and it has definitely given consideration to this, as it has to other aspects of the British model.

ENERGY

So much for weapons, but what about energy?

The Japan of "non-nuclear principles" is also in process of becoming a nuclear superpower, the sole "non-nuclear" state that is committed to possessing both enrichment and reprocessing facilities, as well as to developing a fast-breeder reactor.

Japan's Atomic Energy Commission drew up its first plans as early as 1956, and the fuel cycle and fast breeder program were already incorporated in the 1967 Long-Term Nuclear Program. The dream of energy self-sufficiency has fired the imagination of successive governments and generations of national bureaucrats. Trillions of yen have been channeled into nuclear research and development programs. The lion's share of national energy R & D (64 percent) goes, on a regular basis, to the nuclear sector, and

---

14	Robert McNamara, "Apocalypse Soon," *Foreign Policy*, May–June 2005, reproduced in *Japan Focus*, 8 May 2005.

15	For outlines of a "Northeast Asian Nuclear Weapon-Free Zone," see Hiromichi Umebayashi, "A Northeast Asian Nuclear Weapon-Free Zone," Northeast Asia Peace and Security Network, Special Report, 11 August 2005. and Umebayashi Hiromichi, "Nihon dokuji no hokatsuteki kaku gunshuku teian o," *Ronza*, June 2005, pp. 188–193.

additional vast sums, already well in excess of 2 trillion yen, have been appropriated to construct and run major centers such as the Rokkasho nuclear complex.[16]

Nuclear power at present makes a modest and declining contribution to world energy needs—17 percent in 1993, declining to 16 percent by 2003. Just to *maintain existing* nuclear generation capacity globally, it would be necessary to commission about eighty new reactors over the next ten years (one every six weeks) and a further 200 over the decade that followed.[17] To *double* the nuclear contribution to global energy, bringing it to about one-third of the total, a new reactor would have to be built each week from now to 2075.[18] The head of the French government's nuclear energy division, speaking to the April 2006 Congress of the Japan Nuclear Industry Association at Yokohama, estimated that in order to raise global reliance on nuclear power from its present 6 percent to 20 percent by mid-century (i.e., a modest increase), it would be necessary to construct between 1,500 and 2,000 *new* reactors globally.[19] Even such a mammoth undertaking, trebling current nuclear capacity, would still constitute only a modest contribution to solving global energy problems.

NUCLEAR POWER PLANTS IN JAPAN, 2006

At present, there is virtually no sign of that sort of commitment. Of leading nuclear countries, for example, the United Kingdom had more than forty reactors, but closures were set to cut that to a single one by the mid-2020s, and the US, though it had about a hundred reactors, was also expected to decommission many of them during the 2020s.[20] The Bush administration made a determined push to reverse this trend. At present, there are 440 reactors operating worldwide, with 28 more under construction and 30 more promised by 2030 in China.[21] The US has 103, France 59, Japan 55 (29 percent of its power). Despite the near catastrophes at Three Mile Island (1979) and Chernobyl (1986), not to mention Japan's own series of serious incidents, Japan alone has steadily stepped up its nuclear commitment, increasing its number of reactors from thirty in 1987, to fifty-five now, with ten more planned.

Japan, nevertheless, is intent on playing a leading role in pioneering a hitherto unprecedented level of nuclear commitment. Central to the Japanese vision of a nuclear future is the village of Rokkasho in Aomori prefecture. Perhaps more than anywhere, Rokkasho encapsulates Japan's transition, over the past century, from agricultural and fishing tradition—via a traumatic burst of construction and state excesses—to the full embrace of the nuclear state. A remote provincial community— a vast stretch of land over 5,000 hectares and, at that time, still relatively untouched by industrialization—was set aside in 1971 under the *Shinzenso*, or Comprehensive

16   Citizens' Nuclear Information Center (CNIC), "Cost of Nuclear Power in Japan," Tokyo, 2006.
17   "Nuclear power for civilian and military use," *Le Monde Diplomatique*, Planet in Peril, Arendal Norway, UNEP/GRID-Arendal, 2006, p.16.
18   Frank Barnaby and James Kemp, "Too hot to handle: The future of civil nuclear Power," Briefing Paper, Oxford Research Group, July 2007.
19   Quoted in "Genpatsu no seisui wakareme," *Asahi shimbun*, 6 June 2006.
20   Ibid.
21   Michael Meacher, "Limited Reactions," *Guardian Weekly*, 21–27 July 2006, p. 17.

National Development Plan, as one of eleven gigantic development sites, designated to host petrochemical, petroleum refining, electricity generation, and non-ferrous metal smelting on a scale exceeding anything then known in Japan. In due course, the oil shocks and consequent industrial restructuring saw the fading of the dream of an industrial complex. Instead, large-scale oil-storage facilities were set up on part of the site from 1979 on, and the Rokkasho nuclear enrichment, reprocessing, and waste facilities—which took up about one-third of the original site—from 1985. Local government officials had no enthusiasm for the nuclear course, but the deeper they sank into financial dependence the more difficulty they had opposing plans made in Tokyo. An accumulated debt of 240 billion yen was written off with an infusion of taxpayer money in 2000. Until 2005, hopes were high that the International Thermonuclear Experimental Reactor (ITER) might be built there, but that hope collapsed when the project was allocated to France.[22] The prospect in the early twenty-first century was one that nobody in the village dreamed of in 1971—of becoming a center of the global nuclear industry.

Despite the early-twenty-first century Japanese government's mantra of privatization and deregulation, huge sums were poured into nuclear projects that would never have started, much less been sustained, by market forces. While public and political attention focused, in 2005, on the privatization of the Post Office, bureaucrats far removed from public scrutiny, accounting, or debate were taking decisions of enormous import for Japan's future, cosseting the nuclear industry and giving it trillions.

Japan's renewable energy sector (solar, wind, wave, biomass, and geothermal, excluding large-scale hydropower), constitutes a miserable 0.3 percent of its energy generation. There is a planned to rise, over the next ten years, to 1.35 but then a slight *decline* by 2030. By contrast, even China plans to double its natural energy output to 10 percent by 2010, and the EU has a target of 20 percent by 2020.[23] In short, Japan stands out as a country following a course radically at odds with the international community, driven by bureaucratic direction rather than market forces, much less democratic consensus.

THE NUCLEAR STATE—WASTE, FAST BREEDING, AND THE MAGIC CYCLE

The objective set out in the Ministry of Economics, Trade, and Industry (METI)'s 2006 "New National Energy Policy" was to turn Japan into a "nuclear state" (*genshiryoku rikkoku*), with the level of nuclear-generated electricity to be steadily raised, to "between 30 to 40 percent" by 2030 (compared to 80 percent in France, the world's number 1 nuclear country, in 2006).[24] Other reports suggest a goal of 60 percent by 2050.[25] In August 2006, METI's Advisory Committee on Energy Policy pro-

---

22   Tsukasa Kamata, "Huge tract for ITER sits vacant," *Japan Times*, 25 November 2006.

23   Iida Tetsunari, "Shizen enerugii fukyu o," *Asahi shimbun*, 8 June 2004, and "Shizen enerugii nanose," *Asahi shimbun*, 15 April 2007.

24   According to the "New National Energy Strategy" published by the Ministry of Economics, Trade and Industry in May 2006. Keizai Sangyo sho, "Shin Kokka Enerugii Senryaku," May 2006.

25   "Safe storage of nuclear waste," Editorial, *Japan Times*, 25 July 2006.

duced its draft "Report on Nuclear Energy Policy: Nuclear Power Nation Plan."[26] Its "Hiroshima Syndrome" would be put behind it, and inhibitions about safety, radiation, waste disposal, and cost cast to the wind as Japan, the once nuclear victim, sets out to become a nuclear super-state.

Japan's nuclear energy commitment currently does not particularly stand out in terms of its scale, but among non-nuclear weapon states, it alone pursues the full nuclear cycle, in which plutonium would be used as fuel after the reprocessing of spent reactor waste. It is this bid for plutonium super-power status that distinguishes it. Already with stocks of plutonium amounting to more than 45 tons[27]—almost one-fifth of the global stock of civil plutonium of 230 tons[28] and the equivalent of 5,000 Nagasaki-type weapons—it has become "the world's largest holder of weapons-usable plutonium,"[29] and its stockpile grows steadily. Barnaby and Burnie estimated in 2005 that Japan's stockpile on current trends would reach 145 tons by 2020, in excess of the plutonium in the US nuclear arsenal.[30] Japan therefore ignored the February 2005 appeal from the Director-General of the International Atomic Energy Agency (IAEA) for a five-year freeze on all enrichment and reprocessing works, arguing that such a moratorium was applicable only to "new" projects, not ones such as Japan's that had been underway for decades.[31]

Currently (2007), Japan is commencing full commercial reprocessing at Rokkasho. It undertakes with impunity what ElBaradei sees as highly dangerous activity that should be placed under international supervision and strictly limited, doing so in defiance of the international community, but with the blessing of the US. Countries such as Iran and North Korea are told they must absolutely stop doing the same thing (and indeed countries such as South Korea are also blocked from following Japan down the enrichment and recycling path). If Iran and North Korea are a threat to global non-proliferation, then so is Japan. Its forty-five tons of plutonium may be compared with the ten to fifteen kilograms of fissile material that North Korea was accused of illicit diversion in the 1994 crisis (or the maximum of around sixty kilograms it might possess in 2007).[32] The Federation of Electric Power Companies puts the figure of 19 trillion yen on the cost of the Rokkasho facility over the projected

26   Sogo shigen enerugii chosakai, denki jigyo bunkakai, genshiryoku bukai (Subcommittee on Nuclear Energy Policy, Advisory Committee on Energy Policy, Ministry of Economy, Trade and Industry (METI), "Genshiryoku rikkoku keikaku" (Report on Plan to Build a Nuclear Energy Based Nation), draft, 8 August 2006.

27   Frank Barnaby and Shaun Burnie, *Thinking the Unthinkable: Japanese nuclear power and proliferation in East Asia*, Oxford Research Group and Citizens' Nuclear Information Center, Oxford and Tokyo, 2005, p. 17. (Around three-quarters of that is presently being processed in Britain's Sellafield and will be returned to Japan in due course. Eric Johnston, "Nuclear foes want Rokkasho and Monju on UN nonproliferation agenda," *Japan Times*, 2 April 2005.)

28   "Nuclear power for civil and military use," *Le Monde Diplomatique*, cit, p. 17.

29   Barnaby and Burnie, p. 8.

30   Ibid.

31   Mohammed ElBaradei, "Seven steps to raise world security," *The Financial Times*, 2 February 2005.

32   For 2007 estimate, David Albright and Paul Brannan "The North Korean plutonium stock, February 2007," Institute for Science and International Security, 20 February 2007.

forty-year term of its use.[33] That would make it certainly Japan's, if not the world's, most expensive facility in modern history. Experts point out that it would cost very much less to bury the wastes, unprocessed (provided, that is, there is some place to bury them ...), and fear that the actual cost might climb to several times the official estimate.[34] Rokkasho's reprocessing unit is supposedly capable of reprocessing 800 tons of spent fuel per annum, yielding each year about 8 more tons (1,000 warheads-worth) of pure, weapons-usable plutonium.[35] Such a plant—though it would be the only one in Asia—would make little more than a small dent in Japan's accumulated and accumulating wastes, which, in 2006, were estimated at approximately 12,600 tons,[36] not to mention the 40,000 tons of toxic nuclear spent fuel wastes so far accumulated throughout Asia.[37]

As it gets going, Rokkasho is about to release the equivalent of the nuclear wastes of 1,300 power stations.[38] The tritium discharge level will be 7.2 times that of Sellafield in Northern England, recently closed by the British Government. The operation of the Sellafield plant, and the wastes it poured into supposedly deep sea currents for dispersal, led over decades to fish devastation across much of the Irish Sea and leukemia levels in children forty-two times the national average as far away as Carnarvon in Wales.[39] In Rokkasho, the plant operators have secured a permitted level of tritium release of 2,800 times that permitted for conventional reactors, essential to the plant's economic viability, and although it is said to be dispersing its wastes into deep ocean currents, an opposition group scattered postcards into the Rokkasho sea which later turned up right along the Japanese coast, through Iwate, Miyagi, Fukushima, to Ibaraki and Chiba Prefectures.[40]

What, then, will Japan do with its plutonium mountain? To address the general perception that it is the most dangerous substance known to mankind, in the 1990s, it undertook two steps: first, it issued an assurance that it would neither stockpile nor hold more than was necessary for commercial use. From the beginning that pledge

33   Yoshioka Hitoshi, "Genpatsu wa 'kaiko' ni atai suru no ka," *Asahi shimbun*, evening edition, 21 November 2005.

34   Such cost would amount to between one-half and two-thirds of the costs of reprocessing. Yoshioka, cit.

35   Shaun Burnie, "Proliferation Report: sensitive nuclear technology and plutonium technologies in the Republic of Korea and Japan, international collaboration and the need for a comprehensive fissile material treaty," Paper presented to the International Conference on Proliferation Challenges in East Asia, National Assembly, Seoul, 28 April 2005, p. 18.

36   Estimate by Shaun Burnie, Greenpeace International, personal communication, 4 September 2006. For table showing projected spent fuel waste accumulation to 2050, see Tatsujiro Suzuki, "Global Nuclear Future: A Japanese Perspective," September 2006. Nautilus Institute at RMIT University, Melbourne.

37   Michael Casey, "Asia embraces nuclear power," *Seattle Times*, 28 July 2006. US stocks of spent nuclear fuel amounted to 53,000 metric tons as of December 2005, projected to rise by 2010 to between 100,000 and 1,400,000 (*sic*)." (US Department of Energy, May 2006).

38   Kamanaka Hitomi, with Norma Field, Discussion, University of Chicago, 18 April 2007 (Text courtesy Norma Field).

39   Mizoguchi Kenya, "Shuto-ken ni mo yatte kuru—Rokkasho saishori kojo no hoshano osen," *Shukan kinyobi*, 24 August 2007, pp. 14–15.

40   Ibid.

was empty. The stockpile grew steadily because of the many delays to the plans, due largely to the many accidents (including those causing fatalities),[41] cover-ups,[42] and continual budget over-runs that galvanized public opposition to proposed projects.[43] Even if Rokkasho was to function for 40 years, without delays and technical problems, processing without a hitch 800 tons of spent fuel per year, spent fuel volumes will continue to grow. Japan's nuclear reactors are currently discharging each year 900 tons of waste, about 100 more than can be reprocessed by a fully functioning Rokkasho reprocessing plant. This figure is set to reach between 1,200–1,400 tons by 2015 as more reactors are commissioned, which will mean the accumulation of 400–600 tons over and above what can be reprocessed, most of which will remain stored at reactor sites or at proposed regional interim storage sites.[44] That would be added to the current global stockpile of separated plutonium (ca 250 tons)[45] with the gap widening further if, or as, more reactors are built.[46]

Second, the government launched a campaign to persuade the public that there was no need to worry about plutonium. The Japanese Power Reactor and Nuclear Fuel Corporation issued an informational video featuring a character, "Mr. Pluto," who declared that plutonium was safe enough to drink, which he demonstrates, and that there was little risk of it being turned into bombs.[47] When the US Energy Secretary, among others, protested at the video's inaccuracies, it was withdrawn, but the advertising campaign continued.

Until 1995, the plan was to operate fast-breeder reactors, which "breed" (i.e. produce more than they start with) very pure, "super-grade" plutonium. Such programs make little economic sense, since they cost four to five times as much as conventional power plants, and most projects around the world, including the US and UK, have been abandoned on grounds of either safety or cost.[48] The Japanese Citizens' Nuclear Information Center judges that they are "completely incompatible with non-proliferation."[49] Japanese plans were thrown into disarray by the shut-down of the Monju prototype fast-breeder reactor (at Tsuruga, in Fukui Prefecture on the Japan Sea coast) after a sodium leak and fire in December 1995, followed by evidence of

41   Monju experimental fast breeder was shut down from 1995 after leakage of a ton of liquid sodium from the cooling system; two workers were killed, and hundreds exposed to radiation, in a 1999 accident at Tokaimura fuel processing plant when workers carelessly mixed materials in a bucket, causing criticality and near catastrophe; five more were killed when sprayed with super-heated steam from a corroded cooling system pipe in a 2004 accident at Mihama.

42   Plans for large-scale plutonium use in the form of mixed oxide fuel (MOX) collapsed in 1999–2001 when it was revealed by Japanese environmental groups that vital quality control data for fuel delivered to Kansai Electric by British Nuclear Fuels had been deliberately falsified. The effect of this was to galvanize opposition in three Prefectures slated for MOX fuel use—Fukui, Fukushima, and Niigata.

43   Burnie, p. 19.

44   Takubo Masafumi, "Kadai wa New York de wa naku, Nihon ni aru," Sekai, June 2005, 142–51, at p. 151.

45   H.A. Feiveson, Princeton University, Statement at UN meeting, 24 May 2005.

46   Eric Johnston, "Nuclear fuel plant not biz a usual," Japan Times, 10 August 2004.

47   Scientific American (Digital), May 1994.

48   Yoshida Yoshihiko, "NPT o ketsuretsu saseta no wa Beikoku no tandoku kodoshugi," Ronza, August 2005, pp. 154–9.

49   CNIC, "Statement by CNIC and Greenaction about GNEP," 11 July 2006.

negligence and cover-up, and the project was suspended for almost ten years. After years of protest, opponents of the project won a court victory upholding their position that the design of the reactor was flawed. In May 2005, however, the Supreme Court overturned that ruling and upheld the government's decision to proceed. By then, over thirty years, the project had cost already 600 billion yen and had not lit a single lightbulb. Under current government plans, the fast breeder is to be commercialized by 2050—a remarkable seventy years behind its original schedule.[50]

Undaunted, the JAEA has set up, in Tsuruga something called an Aquatom—science museum, theme park, community center—designed to brush off the near disaster and persuade people that this is the future. Display panels explain to visitors that the world has only 40 years of oil left, 65 of natural gas, 155 of coal, and only 85 of uranium for conventional nuclear plants.

> Japan is a poor country in natural resources ... therefore Monju, a plutonium burning reactor, is necessary because plutonium can be used for thousands of years.[51]

Money continues to flow into local Tsuruga projects, including those in welfare and tourism promotion. The spirit of Mr. Pluto is alive and well in Aquatom.

Not only is Monju itself to be resuscitated, but a second reactor, to replace it, is to be built by around 2030, with a cost of "about 1 trillion yen."[52] The bureaucratic dream of energy security for the twenty-first century operates on a higher plane of logic than economics.

Whatever the outcome of the fast-breeder project, the government has also adopted a plan to burn recycled plutonium in conventional light-water reactors, in the form of a plutonium-uranium oxide (MOX) fuel.[53] This process is also several times more expensive than low-enriched uranium fuel and involves much higher risk.

Earlier efforts to start plutonium MOX use in the late 1990s failed. On current plans, Japan's utilities would begin to load plutonium fuel from around 2007–08, but judging by the past, it is likely to take longer, and the gap between the production of plutonium (from both European-based stocks belonging to Japan, and those coming out of Rokkasho) and the ability to load it into reactors will widen further.

The bottom line is that wastes continue to accumulate. Low-level wastes—basically comprised of contaminated clothing, tools, filters, etc.—are held in over 1 million 200-liter drums, at nation-wide reactor sites and at Rokkasho's repository, whose projected eventual capacity is 3 million drums.[54] Forty vast repositories are planned, each 6 meters high, and 24-by-24 meters in length and width, and containing 10,000 drums. Eventually, they will be covered in soil, with something like a mountain built over them, after which they must be closely guarded for at least 300 years, slowly spreading, like giant, poisonous mushrooms or the mausolea of ancient Japanese

---

50   Ibid.
51   Eric Johnston, "Nuclear plants rural Japan's economic fix," *Japan Times*, 4 September 2007.
52   "New fast-breeder reactor to replace prototype Monju," *Asahi shimbun*, 27 December 2005.
53   "Editorial—Pluthermal project," *Asahi shimbun*, 16 February 2006.
54   Hirata Tsuyoshi, "Shinso no kaku haikibutsu," *Shukan kinyobi*, 25 May 2003, pp. 38–41.

aristocrats, across the Rokkasho site. Meanwhile, fluids containing low levels of radiation are being piped several kilometers out into the Pacific Ocean for discharge, the standards for effluent control in place at reactor sites around the country drastically raised (i.e. relaxed) in order to make regular discharges possible.[55]

High level toxic wastes, basically spent fuel, have since 1992 been regularly shipped across vast stretches of ocean to reprocessing plants at Sellafield in the north of England and La Hague, in Normandy, France—each shipment equivalent to about seventeen atomic bombs-worth of plutonium, despite the protests of countries en route and the risks of piracy or hijacking.[56] Once processed, the liquid high level waste is vitrified and put in canisters, each 1.3 by 0.43 meters, which are returned to the Rokkasho site, where they are to be stored initially for 30 to 50 years, while their surface temperature slowly declines from around 500 degrees centigrade to 200 degrees centigrade, at which point it is planned to bury them in 300 meters deep underground caverns where their radiation will further dissipate over millennia. These canisters already more than half-fill their first giant storehouse.

As Japan's reactors reach their "use by" date, they must be decommissioned, dismantled, and the sites cleaned. No one knows exactly what that will cost, but early in 2006, the British authorities calculated £70 billion ($170 billion) for dealing with twenty of their repatriated civil nuclear sites.[57] Whatever the short-term financial inducements on offer from Tokyo, local communities are steadfastly opposed to hosting such facilities, and governors balk at the thought of their prefectures being turned into nuclear dumpsites for literally millennia.

However, the determination of the state and nuclear power industry to press ahead with all possible nuclear developments, and the imperative of doing *something* with the plutonium mountain, constitute powerful, perhaps irresistible forces.

Due to the inadequacy of international nuclear standards, the proliferation hazards associated with reprocessing are greater than most would believe. The best estimates are that a 1 percent loss of fissile materials—or "about a nuclear weapon's worth per month—in such a vast system of uranium and plutonium processing and transport would be impossible to detect.[58] This feeds further uncertainty on the part of Japan's neighbors, especially South Korea and China.

NUCLEAR PARTNERSHIP

In the United Nations, Japan declines to associate itself with the "New Agenda Coalition" (NAC) that came into existence following the nuclear tests by India and Pakistan in 1998, seeking to exert more urgent pressure for disarmament and non-

---

55   Although such discharge only began in March 2006, seawater levels of radioactivity soon rose, sparking protests from the Governor of Iwate Prefecture (into which the currents from Rokkasho flow) and local fishermen. (CNIC, "Active tests at the Rokkasho Reprocessing plant," June 2006; and Koyama Hideyuki, "Sanriku no umi ni hoshano hoshutsu nodo wa genpatsu no 2700 bai," *Shukan kinyobi,* 19 May 2006, p. 5.

56   George Monbiot, "Dirty bombs waiting for a detonator," *The Guardian,* 11 June 2002.

57   Jim Giles, "Nuclear power: Chernobyl and the future: when the price is right," *Nature,* No. 440, 20 April 2006, pp. 984–986.

58   Barnaby and Burnie, p. 9.

proliferation. Japan, however, sees it as too "confrontational"—in other words, too directly challenging the nuclear privilege of the US and the other nuclear privileged powers. For Japan to join NAC, against US wishes, might also have been to weaken the US-provided "umbrella."

While Japan's government and bureaucracy single-mindedly pursues its chosen nuclear superpower path, its embrace of the US tightens, while its distance from Asia widens. In February 2006, Washington included Japan on a short-list of countries for a projected Global Nuclear Energy Partnership (GNEP), a kind of nuclear energy "coalition of the willing" that would include the US, Great Britain, France, Russia, China, and Japan (i.e. the existing nuclear club members, all nuclear weapons states, plus Japan). The world would be divided into "our" states, which can be trusted with weapons (Pakistan, India, Israel) and reprocessing technologies (Japan, and Australia if Prime Minister Howard can have his way) in a system designed to sidestep the existing UN-centered international framework of the 1970 Non-Proliferation Treaty and establish a new cartel to control the production, processing, storage, sale, and subsequent disposal of uranium. Nominally the project is to address global warming and energy needs, but actually it is to address the unsolved problem of nuclear wastes—especially Mr. Pluto—as hundreds of tons of the stuff accumulate worldwide. So, difficult to bury it under Yucca Mountain, why not just use it?

By adopting this project, the US was reversing thirty years of policy banning reprocessing because of the proliferation and cost concerns. It would now sponsor the construction of a new generation of reactors, the reprocessing of spent fuel (something that would become okay when conducted by close allies of the US) and create a boondoggle for companies such as General Electric (and presumably also Japanese companies such as Hitachi) with hundreds of billions of dollars in construction contracts up for grabs. The project would develop a so-called proliferation-resistant recycling and reactor technology, maintain monopoly control over it, and then offer facilities to the rest of the world on a lease basis.[59]

The Japanese government, which has long been negatively disposed towards regional attempts to forge a Northeast Asian Nuclear Free Zone, jumped at this American invitation to join a global nuclear superpower club. Australia too, initially taken unawares by the proposal, soon developed enthusiasm. Prime Minister Howard eagerly sought American advice on his visit to Washington three months later, secured the blessings he sought, and issued a call for a national debate on nuclear energy.[60] Australia could expect to play a key role in such a project, mining, manufacturing, selling, and monitoring it for the duration of its cycle, since it is the "Saudi Arabia" of global uranium (it has the most uranium, with 24 percent of global reserves, although it has thus far chosen to remain a quarry for uranium, not itself processing it).[61] The Prime Minister, along with the Defence, Industry, and Environment ministers, have

---

59   US Department of Energy, "The Global Nuclear Energy Partnership," updated July 2006.
60   Geoff Elliott, "US backs Howard's nuclear vision," *The Australian*, 17 August 2006.
61   Paul Sheehan, "A thirsty world running dry," *Sydney Morning Herald*, 31 July 2006.

all said that Australia should "consider" the option of a nuclear power industry.[62] The global axis of US power evident in its construction of special relationships with the UK, Australia, and Japan would here take on a nuclear dimension.

The major technology it advocates (advanced burner reactor, or ABR) exists only as a theoretical proposition. The principle is the same as the neutron fast breeder reactor (to date a colossal, expensive failure), but without the use of a breeder blanket, which is where the supergrade plutonium is produced. However, the application of a blanket is a simple one compared to the technical challenge of designing a fast reactor to operate reliably. Commercial scale demonstration of the new, American-proposed technology could not be expected for twenty or twenty-five years[63] and the costs are expected to be enormous. The US Energy Secretary indicates that a fund of between $20 and 40 billion will be needed, and implies that a major contribution would be expected from Japan.[64] This requisitioning may, in time, come even to dwarf the levies imposed on Tokyo to fund its Gulf and Iraq wars, sustain the dollar in international financial markets, and feed the missile defense industry. The wastes would still accumulate.

Above all, the partnership is based on positive promotion of nuclear as the core source of future global energy, and it would require public investment of the core countries to flow to the most costly and dangerous option, rather than to true renewables. It goes against the trend of global energy markets. Between 1994 and 2003, global electricity supply increased by 30 percent for wind, 20 percent for solar, 2 percent for gas, 1 percent for coal, and 0.6 percent for nuclear.[65]

There are also serious doubts that the world has enough uranium anyway to follow the nuclear course, even if safety and other issues could be met. John Busby calculates that "primary production would have to be increased 167-fold to match the anticipated global needs exclusively from nuclear power in 2020," and even if nuclear power generation could be doubled—an unlikely proposition—it would be enough to meet only 5 percent of world energy consumption.[66] This uranium shortfall is used by advocates of fast breeder reactors to justify the development of new designs of breeders, despite their failure over the past decades. The agenda of massive expansion, whether of the still-to-be-developed partnership technologies or of the existing light water reactors, is simply fantastic.

Japan, 300 years ago, was a more-or-less sustainable, zero-emissions, and zero-waste society. Under current Japanese government plans, 300 years from now (and indeed for 10,000 years into the future), provided all goes well, the country's northern and eastern regions will be a vast, poisonous complex, over which generation

---

62   Anthony Albanese, "Twenty years on: lest we forget the lessons of Chernobyl," *Sydney Morning Herald*, 26 April 2006.

63   US Department of Energy, p. v.

64   "Kaku gijutsu kaihatsu, Bei 'saidai 4 cho 7000 oku en,' Bei chokan kenkai, Nihon nado no kyoryoku kitai," *Chugoku shimbun*, 17 February 2006.

65   Ian Lowe, "Heeding the warning signs," *The Weekend Australian*, 7–9 September 2007.

66   John Busby, "Why nuclear power is not the answer to global warming," *Power Switch*, 25 May 2005.

after generation, virtually forever, a heavy, militarized guard must be maintained. Whether Rokkasho is to become the representative model of twenty-first-century civilization—and future centuries and millennia—will be determined by the ongoing contest between Japan's nuclear bureaucracy (pursuing the chimera of limitless clean energy, global leadership, a solution to global warming, the maintenance of nuclear weapon defenses (whether American or Japanese)), and the civil society (pursuing its agenda of social, ecological, and economic sustainability, democratic decision making, abolition of nuclear weapons, phasing out of nuclear power projects, and reliance on renewable energy, zero emission, material recycling, non-nuclear technologies). Much depends on the outcome.

In sum, nuclear power is:

• Too slow to constitute a response to the climate change crisis—15–25 years per reactor, and, in the short term, at least it involves actually significantly increasing greenhouse pollution by construction, mining, etc., and is therefore far from being carbon-free;

• Too dangerous and/or too difficult. It rests on some technologies that are unproven, and requires confidence to be sure that highly-poisonous and dangerous materials can be safely managed for millennia, and it is especially incompatible with Japan's earthquake and volcano-prone environment. For example:

(a) Kashiwazaki (Niigata), the world's largest nuclear plant (7 reactors, generating 8,000 MW), was hit by 6.8 magnitude earthquake on 16 July 2007—50 cases of malfunctioning and trouble, including burst pipes, fire, radioactive leaks into the atmosphere and sea; shock more than twice as strong as its design allowed and location was on a fault not hitherto detected. If the country with the world's most advanced scientific and engineering skills could make such disastrous miscalculations, if the nuclear industry could be regularly guilty of malpractices such as data falsification and fabrication, the deliberate duping of safety inspectors, failure to report criticality incidents and emergency shut-downs,[67] could the rest of the world do better?; and

(b) Hamaoka complex in Shizuoka Prefecture (5 reactors, 190 kms SW of Tokyo) sits, like Kashiwazaki, on fault lines, where the Eurasian, Pacific, Philippine, and North American plates grind against each other, in an area where government seismic experts in January predicted that there was an 87 percent chance of a magnitude 8 earthquake within the next thirty years;

• Too irresponsible, bureaucratic, and anti-democratic governments have consistently proven themselves incompetent, have resorted to lying, cover-up, belittling of risk, and imposing their bureaucratic priorities rather than listening to the people (whether in *genpatsu*, bases, or dams); and the

---

67   "Malpractices at Japanese nuclear power plants," Protest Statement by Citizens' Nuclear Information Center, 2 April 2007.

nuclear state can only be bureaucratic, centralized, heavily policed, and non-, if not anti-, democratic;

•   Too expensive. Even the multi-trillions for Rokkasho does not cover all the costs. An equivalent investment in, for example, wind would yield 5 times more jobs and 2.3 times more electricity (almost immediately).[68] And, apart from the costs already mentioned, Kashiwazaki shows that the 6.5 magnitude protection standard for the nation's reactors is inadequate. It is clear that reinforcing to 6.8, or 7.0, will require prodigious outlays also, so far not factored in. On top of this, if the potential costs of a disaster were also factored in, by way of insurance for example, the industry would be un-sustainable. A major quake at Hamaoka would create a disaster potentially dwarfing Chernobyl. 30 million people would have to be evacuated and it might be impossible ever to live in the area thereafter.[69]

The final question is this: Is Japan's drive to become a nuclear super-state compatible with its "Client State" role? The US has always insisted that Japan not be a nuclear weapons state, but, given a forthcoming privileged position within the GNEP, it stands to become a *de facto* nuclear superpower anyway. The Bush administration may be confident that it has locked Japan in to Client State subordination for the foreseeable future, but a considerable potential ambiguity opens up. In the GNEP, more trust is needed, and much depends on continuity of shared identity and role, yet there is, perhaps, diminished certainty about the US ability to ensure that Japan remains forever gripped within the American embrace—dependent. The long-term prospect is for this particular Bush administration policy to diminish the force of its other policies aimed at incorporation and subordination.

---

68   Eric Prideaux, quoting from Greenpeace France's "Wind vs Nuclear 2003," "Atomic power at any cost?" *Japan Times*, 5 September 2007.

69   David McNeill, "Shaken to the core, Japan's nuclear program battered by Niigata quake," *Japan Focus*, 1 August 2007.

Chapter 33 ∎ Part 8

A DIFFERENT PERSPECTIVE ON THE US-INDIA NUCLEAR DEAL[1]

Dr. Peter Custers

The US-India Nuclear Deal was initiated through a framework agreement signed by India's Prime Minister Manmohan Singh and US President Bush in July, 2005. Under this initial statement, India agreed to separate its civilian and military nuclear production facilities, and place all civilian production facilities under the inspection regime of the International Atomic Energy Agency, the IAEA. This nuclear deal, which took three years to complete, officially aims at promoting India's access to uranium and to civilian nuclear technology, through enlarged importation of both. Whereas nuclear energy contributed a reported 2.5 percent of India's energy requirements in 2007, the deal is expected to boost the contribution of the nuclear sector to India's electricity supply—without reducing India's primary dependence on coal. From its very start, the US-India Nuclear Deal has generated huge controversies, both in India and internationally.

This essay discusses the hazardous and wasteful implications of the US-India nuclear deal beyond its implications for the nuclear arms' race in the subcontinent. Most of the key objections against the deal that have been put forward by progressive opponents of the deal in India and internationally, have addressed the fact that it legitimizes India's status as a nuclear weapons' state, and that it will enable India to expand its production of weapons-grade plutonium. Already, India is estimated to possess a sufficient amount of plutonium for the manufacture of at least 100 atomic bombs. Since India reportedly has agreed to place only fourteen out of its twenty-two civilian reactors under the IAEA's inspection regime, it is free to produce, in the remaining eight reactors, another 200 kilograms of weapons-grade plutonium per year.[2] Thus, fears that the controversial deal will enhance the danger of a nuclear conflagration in South Asia appear to be well grounded—even if we leave aside all other interrelated objections that have been raised.

In this essay, the spotlight will not be on India's past and future plans for production of weapons-grade plutonium and nuclear bombs, but on two other major questions. The US-India nuclear deal needs to be fiercely questioned with regard

---

1   This is a revised text of a lecture given at the Jawaharlal Nehru University (JNU), New Delhi, on September 17, 2008. This article has been previously published in *Monthly Review* (US, September, 2009, p.19), *Peace Now* (The Bulletin of the Coalition for Nuclear Disarmament and Peace, New Delhi, March, 2009, p.13), *New Age* (Daily Newspaper, Dhaka, May 24 and 25, 2009), and *The India Economy Review* (May 31st, 2009, p.150). It is being included here with permission from the author.

2   See e.g. Praful Bidwai, "Manmohan's False Nuclear Move," (19 July, 2008, http://www.cndp.org/); also Zia Mian and M.V. Ramana, "Going MAD: Ten Years of the Bomb in South Asia," (29 July, 2008, http://www.cndp.org/).

to its *ostensible* aims, i.e. the vast expansion in the production of nuclear energy. Whereas a more then ten-fold increase in generation of nuclear energy, as foreseen, may help to overcome India's rapidly-growing energy needs—the side-effects in terms of generation of nuclear waste are so ponderous, that from this perspective too, implementation of the deal needs to be preempted. Moreover, as reported briefly in India's national press in September last, when the signing of the deal was being debated—there is a little discussed "*reverse side*" to the nuclear deal, which is the US' additional commercial objectives relating to its arms exports. The US is poised to lobby aggressively, in order to capture a larger share of India's arms imports than it has held up until now.

In order to address these combined issues, this essay utilizes a *holistic view on waste*. In this view, processes of manufacturing that result in military commodities, i.e. in weaponry, basically need to be analyzed as processes that result in waste of economic resources. This, for instance, is the case where economic policy-makers deciding to purchase armament systems do not primarily have in mind security considerations, but macro-economic stimulation of domestic demand for goods. However, this, the production of "*social waste*," generally does not stand alone, but needs to be juxtaposed with the generation of "non-commodity waste" during the same industrial processes. Whereas conventional economics discusses these side-effects of industrial manufacturing under the heading of "externalities"—in this essay the term *non-commodity waste* will be used, whenever reference is made to the ecologically-harmful by-products of industrial manufacturing.[3]

Whereas "*social*" waste and "*non-commodity*" waste are rarely juxtaposed in public debate—the US nuclear deal and its reverse side offer an occasion to do precisely this. As the below cited data on the generation of waste in the nuclear production chain show—the US-India nuclear deal is bound to result in huge quantities of extremely dangerous waste that cannot be sold on the market, but needs to be put aside, at great risks to humans and to our natural environment. Again, the importation of expensive armament systems entails the waste of vast economic resources that could be used towards relieving India's persistent mass poverty, hence should be considered importation of social waste. Moreover, the issues of "social" and "non-commodity" waste can also be posed in relation to the manufacturing of weapons-grade plutonium and atomic weapons, where generation of the two given forms of waste occurs *simultaneously*.[4]

THE NUCLEAR DEAL: IMPORTATION OF NUCLEAR TECHNOLOGY AND IMPORTATION OF US ARMAMENT SYSTEMS

As starting point for my discussion I will take two newspaper articles published in the *Times of India* on September 11, 2007. One of these highlighted the business

---

3   A precursor of the concept of non-commodity waste is the term "discommodities" coined by the Marginalist Jevons, but largely ignored by other economists of his time—and subsequently. See W. Stanley Jevons, *The Theory of Political Economy* (London: Macmillan and Co., 1879), p. 62..

4   For a full discussion, see Peter Custers, *Questioning Globalized Militarism: Nuclear and Military Production and Critical Economic Theory*. New Delhi: Tulika Publishers, 2007.

prospects of the US-India nuclear deal via the sale of nuclear production technology, and via the importation and the construction of nuclear reactors in India. The
second article discussed the aspiration of the US to expand exports of armament
systems to India. To take the article on plans for expansion of nuclear energy production first—it spoke very glowingly about the size of business that will be generated,
mentioning a figure of $40 billion US worth of orders Indian and foreign enterprises
stand to receive, and hailing the deal as a "project" having a financial size of Rupees
2.4 lakh crore.[5] Under the deal, a reported twenty-four light-water reactors will be
imported from abroad and installed along India's coasts(!). India plans to build a
further twelve indigenous nuclear plants, consisting of pressurized heavy-water reactors. At no point in the article are the implications of the nuclear deal, in terms of
generation of additional *nuclear waste*, discussed![6]

In another article published in the *Times of India* on the very same day, the
secondary objectives of the US, which traditionally is not a major seller of military
hardware to India, are described. The article delineates the huge size of India's overall
arms imports. It states that since the Kargil conflict, India has spent a "whopping" $25
billion on imports of weaponry. The country is "poised" to spend another $30 billion
on such purchases over the next five or six years(!). Thus, the US is vying to capture a
whole series of arms orders, which India intends to place on the world-market. India's
import schedule reportedly includes a $170 million plan to buy anti-ship Harpoon
missiles, a Rs 42,000 crore project for the purchase of multi-role combat aircraft, and
197 light utility and observation helicopters worth another Rs 3,000 crore. One deal
that has already been clinched, and has been sent for approval to the US Congress,
is the arms deal—described as India's "biggest ever" with the US—for the purchase
of eight Boeing reconnaissance aircraft, estimated to cost no less than *Rs 8,500 crore*.
At no point in the article is it explained that such lavish spending on arms imports
represents a form of social waste, and that the same financial resources could well be
spent on alleviating the massive poverty that still exists in India.[7]

Officially, of course, the US-India nuclear deal and the listed plans to import
armaments are not interconnected issues. The arms purchases do not directly form
part of the agreement surrounding importation of nuclear technology. And yet it
is probably correct to see the US' hopes to overtake other foreign suppliers of arms
to India as a reverse side of the nuclear deal, as was indeed hinted at in the article
of the *Times of India*. In any case, juxtaposition of the two issues enables us to look
more holistically at the wasteful implications of the Indian government's behavior,
than would a focus on the US-India nuclear deal alone. Hence, below I am going
to address *both* the generation of nuclear waste that will occur in consequence of

---

5   *Lakh* and *crore* are numerical figures commonly used in accounting in South Asia; 1 lakh refers
to 100,000; 1 crore signifies 10 million.

6   Srinivas Laxman, "N-Trade: It's a $40 Billion Opportunity." *Times of India*, New Delhi, Septem-
ber 11, 2008, p. 15; for other estimates regarding the business prospects of the deal, see J.Sri Raman, "How
India's 'Waiver' Has Won," September 9, 2008—http://www.cndp.org/.

7   Rajat Pandit, "In Defence, US Wants to be India's No.1 Partner." *Times of India*, New Delhi,
September 11, 2008, p. 13.

the nuclear deal, and India's arms imports, in order to show the full extent of waste creation that is involved.

THE GENERATION OF HAZARDOUS WASTE IN THE NUCLEAR PRODUCTION CHAIN

Let's take the issue of nuclear-waste-generation first. I do not possess comprehensive data on the nuclear waste that has been generated by nuclear production in India to date, nor am I in a position to give a precise assessment regarding the waste that importation and construction of new reactors will result in. However, the experience of nuclear production worldwide is unequivocal: nuclear waste emerges at each and every link in the nuclear production chain, starting from the very first stage—i.e. the mining and milling of uranium—and up to the stage where nuclear fuel elements are treated in reprocessing facilities. An important source for my own understanding of these issues is the book *Nuclear Wastelands*, written by a group of scientists led by the US-based Indian academician Arjun Makhijani.[8] From this and other sources, I have selected three cases of waste generation, namely: the waste tailings that emerge when uranium is mined and milled; depreciated fuel elements, which themselves are a form of nuclear waste; and the high-level waste that needs to be put aside when former nuclear fuel elements are reprocessed.

Uranium mining is, of course, the very first stage in the whole nuclear production chain. As we know, such mining is also undertaken in India, and would likely be intensified as a result of the US-India nuclear deal. When uranium ore is mined and uranium is prepared and enriched in the process of making nuclear fuel elements, left behind is a truly huge amount of hazardous material, in the form of mill tailings, which contain radioactive substances and are therefore hazardous for humans and nature. Speaking in terms of volume, these tailings reportedly constitute 95 percent of all the nuclear waste that is generated in the nuclear production chain. Among the radioactive substances found in mill tailings are for instance radium-226 and thorium-230, the latter of which has a half-life of 76,000 years, meaning that it will take that many years before half of the radioactivity contained in the thorium will have decayed. In mining uranium and in creating the tailings, capitalist entrepreneurs are not just burdening our children and grandchildren with the consequences of uranium extraction, but entire future generations—for an almost indefinite period of time. The damaging consequences of uranium mining have been well recorded in the US, where nuclear production was started. Here, tailing dams have turned into slurry after downpours of rain. Between 1955 and 1977, a total of fifteen tailing dams broke, and in one such case, the river Rio Puerco was flooded with 94 million gallons of tailing liquids, resulting in contamination of a long stretch of the river.[9]

Another stage in the nuclear production chain known to generate dangerous waste, is when nuclear energy is generated in reactors. The production of nuclear

---

8   Arjun Makhijani, Howard Hu, and Katherine Yih (eds.), *Nuclear Wastelands: A Global Guide to Nuclear Weapons Production and its Health and Environmental Effects*. Cambridge: MIT Press, 1995..

9   Katherine Yih, Albert Donnay, Annalee Yassi, A.James Ruttenber, and Scott Saleska, "Uranium Mining and Milling for Military Purposes," in Arjun Makhijani, Howard Hu, and Katherine Yih (1995), op.cit., p. 121.

energy can be seen as a contribution to human welfare, if purely looked at only from the perspective of energy generation. Yet the hazardous implications of the use of the nuclear fuel rods in the reactors are multifarious. A section of the rods needs to be taken out regularly, as the nuclear fuel elements can be utilized for only three years. Now, in the parlance of economic theory, the fuel elements once taken out are considered "depreciated means of production"; they are presumed to have lost all the value that has been transferred to the new commodity, the nuclear energy. Yet the fuel elements undoubtedly are a form of hazardous waste. Speaking quantitatively, the size of this waste seems small. Yet the radioactivity contained in the spent fuel elements is truly intense, as the radioactive elements present in this nuclear waste include uranium, strontium-90, caesium-137, and plutonium. Of these, plutonium is entirely the outcome of human production, and as such it does not exist in nature. It is known to be the very most toxic substance on earth, its half-life exceedingly long: Plutonium-239 has a half-life of 24,400 years, plutonium-242 as much as 380,000 years. Even *micro-gram* quantities of plutonium, when inhaled by humans, are known to result in *fatal cancers*.[10] Hence, the expansion of construction and utilization of nuclear reactors worldwide is a cause for grave concerns. Each additional nuclear reactor generates spent nuclear fuel rods containing different forms of high-level waste.

The third distinct stage in the chain of nuclear production I wish to refer to, is the stage of *reprocessing*. For decades, policy-makers in the West have tried to make the public believe that they had solved the above-described issue of dangerous waste. They did so by arguing that these fuel rods can well be reprocessed—that they may be treated chemically in reprocessing facilities so as to allow for re-use of the uranium and to pave the way for use of the fresh plutonium for "productive" ends, towards the manufacturing of new fuel elements. Yet it is at the reprocessing stage that problems really pile up. First, it is at this stage that high-level waste comes into being as a distinct category of waste, since the chemical treatment of the fuel rods does not only help to separate out uranium and plutonium, but also results in high-level waste that needs to be put aside. This counts for uranium-236—to be distinguished from uranium-235—incorporated in the fuel elements. Uranium-236, mind you, has a half-life of 24.2 million years. Again, there is the radioactive element Jodium-129 which has a half-life of 15.7 million years. These are time-scales which as humans we are hardly able to imagine, but which make the consequences of nuclear production that much graver. The high-level waste in liquid form put aside after chemical treatment of the fuel rods is commonly stored in tanks.

Now, the risks involved in such storage can be visualized through the accidents that have taken place in nuclear-military production facilities in both the US and the former Soviet Union. It was at the Hanford nuclear complex in the US where the US used to manufacture its military plutonium. There, high-level waste in liquid

---

10   For details on the health and environmental hazards of plutonium production and use, see notably Frank Barnaby, *Nuclear Legacy: Democracy in a Plutonium Economy*. Cornerhouse Briefing Paper No. 2, Sturminster, Newton, UK, November 1997.

form was stored in 117 stainless-steel tanks, each containing half a million gallons of waste. In 1973, a leak was discovered that had caused massive dissipation of radio-activity into Hanford's subsoil.[11] But the most dramatic example of an accident with high level radioactive waste has been reported from the former Soviet Union: In the Cheliabinsk complex, a military-nuclear complex located in the Ural mountains, a tank explosion occurred in 1957. The government of the then USSR suppressed the news of the accident in the name of guarding "state secrets," but Soviet scientists unraveled the accident long before the Gorbachev government instituted an enquiry. Just as in Hanford, the high-level waste from the reprocessing in Cheliabinsk was stored in stainless steel tanks, located in a canyon-shaped area eight meters under the soil's surface. The explosion in Cheliabinsk's tanks resulted in a massive leakage of a reported 22 million *curies* of radio activity—2 million of which were in the form of a plume that reached a height of 1 kilometer above the complex. The explosion and the release of radioactivity destroyed entire eco-systems in the surrounding region. Villages had to be evacuated, rivers and lakes were polluted, and the government had to take draconian measures to contain the danger for the region's ecology.[12]

Above, I have simply summarized data on selected aspects of nuclear waste generation, focusing on waste tailings from uranium mining and milling, on the waste represented by spent nuclear fuel elements, and on the high-level waste that is put aside whenever nuclear fuel rods are reprocessed. Surely, given the risks they represent for humans and for nature surrounding us, there is no way one can belittle the occurrence of multiple waste in the nuclear production chain. Nor can one deny the validity of posing the consequences of the US-India nuclear deal in these terms.

INDIA AS IMPORTER OF WEAPONS SYSTEMS—THE QUESTION OF DISPARATE EXCHANGE

I will now turn to the second form of waste I have identified, namely *social waste*. Here I will focus on the reverse side of the US-India nuclear deal, which is the US' eagerness to expand its arms sales to India. Today, India heads the list of Southern importers of armament systems, replacing, with China, the Middle Eastern oil giant Saudi Arabia. According to a report brought out by the US-based Congressional Research Service (CRS), in 2005 India ranked first among developing nations weapons' purchasers, in terms of the market value of agreements signed to import weaponry. Further, whereas the total value of Southern arms imports in this year was $30 Billion, the value of the agreements concluded by India alone was $5.4 Billion, meaning that India was set to swallow fully one-sixth of the total![13] This data is corroborated by information compiled by the respectable Stockholm-based peace research

---

11   On the leakages of nuclear waste at the Hanford complex, see, for instance, Arjun Makhijani and Scott Saleska, "The Production of Nuclear Weapons and Environmental Hazards," in Arjun Makhijani, Howard Hu, and Katherine Yih (1995), op.cit., p. 44.

12   On the Cheliabinsk catastrophe, see for instance Zhores Medvedev, *Nuclear Disaster in the Urals.* London: Vintage Books, 1980; also see Arjun Makhijani, Howard Hu, and Katherine Yih (1995), op.cit., p. 335.

13   Richard Grimmett, "Conventional Arms Transfers to Developing Nations, 1998–2005." Washington, DC: Congressional Research Service (CRS), The Library of Congress, October 23, 2006).

institute SIPRI. In its 2007 annual report, SIPRI offers comprehensive figures for the value of arms imports by individual Southern states over a period of thirty years. Again, India heads the list of these totals. This of course does not imply that India has been the leading Southern importer in each and every year, but it does signify that the accumulated arms imports of India have been so big over the last decade as to make up for the comparatively "smaller" size of arms imports in earlier decades.[14]

Now, the role that arms transfers between North and South hold in the world-economy can be assessed from either a Southern or a Northern perspective. If looked at from a Southern perspective, one has to reflect on India's arms imports in terms of disparate exchange. The term "disparate exchange" expresses the fact that Southern economies, when importing armament systems from the North, are losers. Whereas they import military commodities that, from a *social* point of view, should be considered *waste*—the Northern states that export the armaments are benefactors, for they directly or indirectly transfer the arms in exchange for raw materials, semi-finished goods and labor-intensive commodities representing wealth. This is indeed a form of international exchange that may be characterized as disparate (as opposed to un-equal) exchange, since there is a qualitative difference between the commodities flowing in parallel between Northern and Southern trade "partners." Although in certain cases the inter-linkages between exported and imported goods are explicit (notably in case of barter agreements where crude oil is exchanged against weaponry)—more generally processes of disparate exchange are less easy to pinpoint, i.e. are *indirectly* interlinked.[15]

To highlight the imperialist nature of this trading mechanism, it needs to be stated that the given trading mechanism was historically instituted by the United States. For when OPEC's oil-exporting countries in the 1970s decided to take their fate in their own hands by insisting on the right to fix the international price of crude oil, the US immediately tried to take advantage of the changing situation. It knew, of course, that increased prices of oil would *inter alia* result in additional dollar incomes for members of OPEC.[16] Hence it feverishly worked to channel such Southern income towards additional Southern imports of weapon systems from the US and other Northern arms exporter—and with success.[17] In the 1970s, leading oil exporters, such as Saudi Arabia and Iran, were easily deluded into buying fighter planes and other expensive weaponry, putting them to the front of the list of Southern

14 For SIPRI's most recent data, see Paul Holtom, Mark Bromley, and Pieter D.Wezeman, "International Arms Transfers." (Chapter 7 of the SIPRI Yearbook 2008: *Armaments, Disarmament and International Security* (Stockholm: 2008), p. 293).

15 An exposition regarding the trading mechanism of *disparate exchange* between North and South is stated in Peter Custers (2007), op.cit., Part 3, Chapter 19: "Unequal Exchange versus Disparate Exchange. A Theoretical Comparison. Succession and Coexistence of Two Imperialist Trading Mechanisms," p. 309.

16 For the views of US State Department officials regarding the implications of the historical price increases decided upon by OPEC in 1973, see Pierre Terzian, *OPEC: The Inside Story* (London: Zed Books, 1985).

17 See e.g. Anthony Sampson, *The Arms Bazaar* (London: Hodder & Stoughton, 1977); and Russell Warren Howe, *Weapons: The Shattering Truth About the International Game of Power, Money and Arms* (London: Abacus, 1980).

importers of weapons systems. Today, as India has emerged as a leading Southern arms importer, the US is eager to expand its arms' sales to India, at the expense of the country's traditional suppliers of arms.[18] And whereas it needs to be assessed whether the exports of social waste from the US will be undertaken at the expense of wealth belonging to India's own population, or rather at the expense of wealth belonging to Indians and other Southern states combined—the arms transfers are bound to represent further cases of *disparate exchange*.

India's massive imports of armament systems can also be analyzed from a Northern perspective. Here we need to highlight the fact that the hegemonic power in the world-system, ever since the days of British imperialism, has used its leverage to export weaponry, as a part of macro-economic policy-making, particularly for the presently-tottering hegemonic power, the US. Ever since the 1960s, the US has used its exports of armament systems as a *replacement* mechanism, as supplement to ensure that American armament corporations are at all times supplied with orders sufficient to protect their production capacity and guarantee accumulation. At the end of the 1980s, for instance, when the US government needed to partly scale down the size of its orders towards monopoly corporations based in the US military sector—it heavily pushed military corporations into expanding their exports. It even employed the second Gulf war, in 1991, towards this end. Moreover, the US Ministry of Defense, the Pentagon, itself embraces the economic logic behind armament exports. This is evident, for instance, from statements contained in its 2006 report to the US Congress, the "Annual Industrial Capability Report" (AICR). As the report states, "Defense exports play an important economic role in strengthening the US defense industrial base"; "about 20 percent (sic) of US weapons systems items are exported ... "; and "sales to foreign customers have frequently been critical to keeping entire production lines open ... "[19] Hence, it is difficult to interpret these sales as necessitated by the US' "security," when the US Pentagon itself admits to its congress that the exports of armament systems represent a leverage for macro-economic policy-making. The combined historical evidence for the past several decades indicates that exports play an active role in solving dilemmas within the US' business cycle, driven as it largely is by military allocations.

JUXTAPOSING SOCIAL WASTE AND NON-COMMODITY WASTE

Let me state my conclusions in brief. As suggested, the US-India nuclear deal should be analyzed in terms of two sorts of wasteful implications. If looked at strictly from a perspective of expanded production of nuclear energy in India—as is the official line of the Indian government—the deal already needs to be severely criticized. In this case, it will undoubtedly result in vastly increased generation of nuclear waste, which, from the standpoint of critical economic theory, is to be considered *non-commodity* waste. Above I have not presented specific data on the waste that India's own

---

18   For India's primary dependence on arms supplies from Russia, see e.g. Paul Holtom, Mark Bromley, and Pieter D.Wezeman (2008), op.cit., p. 300.

19   Office of the Undersecretary of Defense, "Annual Industrial Capability Report" (Washington, DC: AICR—US Pentagon, February, 2006).

production of nuclear energy has generated in the past, but have concentrated on international data regarding the generation of waste at three stages in the nuclear production chain, i.e. the stage of uranium mining and milling, the stage of production in nuclear reactors, and the stage of reprocessing of nuclear fuel elements. These data unequivocally bring out that, in assessing the implications of the US-India nuclear deal, the issue of nuclear waste needs to be taken on board.

Yet if we are to assess the full extent of waste generation implied by the US-India nuclear deal, we also need to reflect on the reverse side of the deal. There needs to be, it seems, greater awareness of the fact that the US does not just intend to use the deal to promote the export of nuclear production technology towards India. The US also is keenly interested in greatly expanding its sales of armaments to India, in view of the fact that India is one of the global South's leading arms importers, along with China. Here again, my data regarding the loss of wealth implied by these deals for India and the South are incomplete. Thus, further research on Indian armament imports should bring out how they express *disparate exchange*. They may lead to loss of wealth for the people of India alone—or ultimately lead to replication of disparate exchange via parallel exports of conventional arms by India to other countries of the global South. In any case, such research would have to focus on the precise way in which foreign currency is generated for payment of these imports, and to make a holistic assessment of the US-India nuclear deal and the mentioned arms deals, we need to *juxtapose* "non-commodity" waste and "social" waste.

Chapter 34 | Part 8

PEAK URANIUM[1]

Energy Watch Group

Any forecast concerning the development of nuclear power over the next twenty-five years has to concentrate on two things: the supply of uranium and the addition of new reactor capacity. Neither nuclear breeding reactors nor thorium reactors will play a significant role in this time because of how long it takes for their development and market penetration.

Taking the uncertainty of the resource data into account, it can be concluded that, sometime between 2015 and 2030, uranium stocks will be exhausted and it will be impossible for production to increase at the rate necessary to meet rising demand, creating a supply gap. Later on, after a few years of adequate supply, production will decline once again, due to shrinking resources. Therefore, it is very unlikely that it will be possible to maintain even present nuclear capacity beyond 2040. If it is impossible to convert all the reasonably assured and inferred resources that exist into volumes of uranium produced, or if stocks turn out to be smaller than the 210 kt (kilotons) that are currently estimated to exist, then this supply gap will occur even earlier.

Uranium resources data indicates that discovered reserves are not sufficient to guarantee uranium supply for more than thirty years. Eleven countries have already exhausted their uranium reserves. In total, about 2.3 Mt of uranium have already been produced, and the remaining uranium reserves are mostly of low quality and concentration, and ever greater amounts of energy are required for their extraction.

At current annual demand, the proved reserves (reasonably assured below 40 $/kg Uranium extraction cost) and stocks will be exhausted within the next thirty years. Likewise, possible resources—which consist of all estimated discovered resources with extraction costs of up to 130 $/kg—will be exhausted within seventy years.

At present, only 42 kt/yr (kiloton per year) of the current uranium demand of 67 kt/yr are supplied by new production. The remaining 25 kt/yr are drawn from stockpiles that were accumulated before 1980. However, these stocks will be exhausted within the next 10 years, making it necessary for uranium production capacity to increase by at least 50 percent in order to match future demand at current capacity.

Important new mining projects (e.g. Cigar Lake in Canada) have recently been beset by problems and delays, casting doubt over whether these extensions will be

---

1   This extract is from the Energy Watch Group report "Uranium Resources and Nuclear Energy" (December 2006), EWG-Series 1/2006. The Authors are Dipl.-Kfm. Jörg Schindler and Dr. Werner Zittel (Ludwig-Bölkow-Systemtechnik GmbH, Ottobrunn/Germany). The complete report is available to download at: http://energywatchgroup.org/fileadmin/global/pdf/EWG_Report_Uranium_3-12-2006ms.pdf

completed in time or even be realized at all.

If only 42 kt/yr of the proved reserves below 40 $/kt can be converted into production volumes, then supply problems are likely to occur even before 2020. If it is possible to convert all estimated known resources up to 130 $/kg U extraction cost into production volumes, a shortage can at best be delayed until about 2050.

This assessment is summarized in figure 1, which summarizes the present supply situation. The production profiles are derived by extrapolating production rates for each country according to its available resources. The large data uncertainty is reflected in the different choices of uranium still available. The dark figure is based on proved reserves (reasonably assured resources below 40 $/kg U extraction cost), the light area above represents the possible production profile if reasonably assured resources up to 130 $/kg U can be extracted. These categories are more or less equivalent to the so called probable reserves.

The uppermost area is in line with resources that include all reasonably-assured and inferred resources. This roughly corresponds to possible reserves. The black line represents the uranium demand from nuclear reactors, which amounted to 67 kt in 2005. The forecast shows the uranium demand until 2030, based on the forecast made by the International Energy Agency in 2006 in its reference case (WEO 2006).

Between 2015 and 2030, a uranium supply gap will arise, when stocks are exhausted and production cannot be increased at the rate necessary to meet rising demand. Only if nuclear breeding reactors operate in large numbers and with adequate breeding rates, can this problem be solved for some decades. But there is no indication that this will happen within the next twenty-five years.

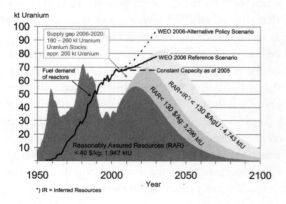

Figure 1: History and forecast of uranium production based on reported resources. The smallest area covers 1,900 kt uranium which has the status of proved reserves while the data uncertainty increases towards the largest area which is based on possible reserves consisting of 4,700 kt uranium.

Possible uranium production profiles, in line with reported reserves and resources, are shown together with the annual fuel demand from reactors. The reserve and resource data are taken from the Red Book of the Nuclear Energy Agency (NEA 2006). The demand forecasts up to 2030 are based on the 2006 scenarios by the International

Energy Agency; a "reference scenario" that represents the most likely development, and an "alternative policy scenario," based on policies aimed at increasing the share of nuclear energy in order to reduce carbon dioxide emissions.

Only if estimates of undiscovered resources from the Nuclear Energy Agency are included, the possible reserves would double or at best quadruple. However, the probability that these figures can actually be turned into producible quantities is smaller than the probability that these quantities will never be produced. Since these resources are too speculative, they are not a basis for serious planning for the next twenty to thirty years.

Nuclear power plants have a long life cycle. Several years of planning are followed by a construction phase of at least five years. Following construction, the reactor can operate for some decades. In line with empirical observations, an average operating time of forty years seems to be a reasonable assumption. About 45 percent of all reactors worldwide are more than twenty-five years old, 90 percent have now been operating for more than fifteen years. When these reactors reach the end of their lifetime, by 2030, they must be replaced by new ones in order for net capacity to be increased.

At present, only 3 or 4 new reactors are completed per year, worldwide. This trend will continue at least until 2011, as no additional reactors are under construction. However, the completion of 15–20 new reactors per year will be required just to maintain present reactor capacity. Today we can forecast with great certainty that total capacity will not increase by 2011 due to the long lead times necessary.

This assessment leads to the conclusion that, in the short term—until about 2015—the long lead times of new reactors and the decommissioning of aging ones, will hinder rapid extension, and that after about 2020 severe uranium supply shortages will become likely. Again, this will limit the expansion of nuclear energy.

As a final remark, it should be noted that, according to the WEO 2006[2] report, nuclear energy is considered to be the least efficient measure in combating greenhouse warming. In the report's "Alternative Policy Scenario" the projected reduction of greenhouse gas emissions by about 6 billion tons of carbon dioxide is primarily due to improved energy efficiency (contributing 65 percent of the reduction), 13 percent due to fuel switching, 12 percent are contributed by enhanced use of renewable energies, and only 10 percent are attributed to an enhanced use of nuclear energy. This is in stark contrast to the massive increase in nuclear capacity stipulated by the IEA and in the policy statements it made when presenting the report.

URANIUM SUPPLY

The definition of uranium resources differs, in several ways, from the reserve classifications used for fossil fuels. The classification into various categories (from discovered Reasonably Assured Resources (RAR) and Inferred Resources (IR) to undiscovered prognosticated and speculative resources) and cost classes (expected extraction

---

2   Quoting the World Energy Outlook of International Energy Agency we refer to the 2006 issue. Data from newer issues of the report does not change the evidence of our report or this article.

cost below 40 $/kg U, below 80 $/kg U, and below 130 $/kg U) gives the impression of a high data quality and reliability. However, the impression is false—at present this quality and reliability does not actually exist. Usually, only "reasonably assured resources" or RAR below 40 $/kg U or below 80 $/kg U extraction cost are comparable with proved reserves regarding crude oil. Other discovered resources (RAR between 80 and 130 $/kg U cost and inferred resources (IR)) have the status of probable and possible resources, while the undiscovered resources are highly speculative, thus making it impossible to use this data in serious projections concerning probable future developments.

At the world level, about 2.3 million tons of uranium have already been produced since 1945. Discovered available RAR are somewhere between 1.9 and 3.3 million tons, depending on the cost class. Estimated additional resources (with lower data quality) are between 0.8 and 1.4 million tons. This is summarized in a table below. The historical assessment shows that discovered resources were marked up in the early years, but after 1980 a substantial marking down occurred (about 30 percent), undermining the credibility of these data. This is discussed later on in the chapter.

The Nuclear Energy Agency also assesses the undiscovered resources within each country and cost class. However, since these are highly speculative (and probably might never be converted into produced quantities), only the aggregated data are summarized in the following table, together with the assessment for discovered resources. It is important to bear in mind that the data quality gets worse as the reader reads from the top of the table to bottom, with the speculative resources having a much larger probability of never being discovered than of being converted into future production volumes.

Resource category		Cost range	Resource [kt]	cumulative	Data reliability
Reasonably Assured Resources (RAR)		< 40 $/kgU	1,947	1,947	high
		40 – 80 $/kgU	696	2,643	
		80 - 130 $/kgU	654	3,297	
Inferred Resources (IR)		< 40 $/kgU	799	4,096	
- former EAR I		40 – 80 $/kgU	362	4,458	
		80 - 130 $/kgU	285	4,743	low
Undiscovered Resources	Prognosticated	< 80 $/kgU	1,700	6,443	
		80 - 130 $/kgU	819	7,262	
	Speculative	< 130 $/kgU	4,557	11,819	
		unassigned	2,979	14,798	

Table 1: *Uranium Resources (Source: NEA 2006)*

The reasonably assured (RAR) and inferred (IR) resources, as well as the uranium already produced, are shown in the following graph. About 2.3 million tons of uranium have already been produced. These amounts are shown as negative values at the left of the bar. RAR below 40 $/kg U are in the range of the uranium already produced. At the present levels of demand for uranium from reactors, about 67 kt/year, these reserves would last for about thirty years. This would increase to fifty years if the classes up to 130 \$/kg U were also included. Inferred resources up to 130 \$/kg would extend the static R/P ratio up to about seventy years (R/P ratio shows the number of years resources last by actual production rate).

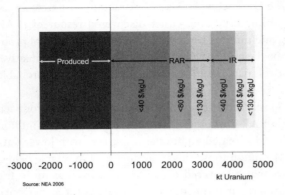

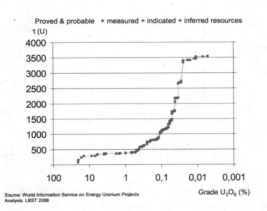

Figure 2: Reasonably assured (RAR), inferred (IR) and resources of uranium already produced

Amongst other criteria, the ore grade plays an important role in determining whether uranium can be mined easily or not. The energy demand for the uranium extraction increases steadily with lower ore concentrations. Below 0.01–0.02 percent ore content the energy requirement for the extraction and processing of the ore is so high that the energy needed for supplying the fuel, operation of the reactor and waste disposal comes close to the energy that can be gained by burning the uranium in the reactor. Therefore, ore grade mining below 0.01 percent ore content makes sense only under special circumstances.

Today only one country, Canada, has reasonable amounts with an ore grade larger than 1 percent. The Canadian reserves amount to about 400 kt of uranium, with highest concentrations of up to 20 percent.

Figure 3: Cumulative world uranium resources (without China, India, and Russia) related to ore grade.

About 90 percent of worldwide resources have ore grades below 1 percent, more than two thirds below 0.1 percent. This is important since the energy requirement for uranium mining is at best indirectly proportional to the ore concentration. For concentrations of below 0.01–0.02 percent, the energy needed for uranium processing—over the whole fuel cycle—increases substantially.

The following figure represents data for about 300 uranium mines listed in the WISE online database (http://www.wise-uranium.org/). It comprises measured, indicated and inferred resources (this is roughly equivalent to RAR + IR data in the previous figure—the difference might be due to some missing data on Russia and China, as well as due to different definitions being used).

The following figure shows the uranium resources and uranium already produced for individual countries. The countries are ranked in the order of volume of uranium already produced. The brown bar on the left shows the uranium already produced, while the different shades of the bar on the right display the different qualities and cost classes of resources. As before, only reasonably assured and inferred resources are included in this figure, since undiscovered resources are deemed to be too speculative to include.

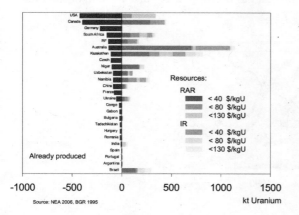

Figure 4: *Cumulative produced uranium and reasonably assured and inferred resources of the most important countries.*

It turns out that twelve countries have already exhausted their uranium resources, having rapidly depleted them over the last decades. These are: Germany, the Czech Republic, France, Congo, Gabon, Bulgaria, Tajikistan, Hungary, Romania, Spain, Portugal and Argentina. It is highly probable that the bulk of the remaining resources are in Australia, Canada and Kazakhstan. Together, these countries contain about 2/3 of these resources below 40 $/kg U extraction cost. However, again, it must be stressed that only Canada contains reasonable amounts of ore with more than 1 percent uranium content. Australia has by far the largest resources, but the ore grade is very low, with 90 percent of its resources containing less than 0.06 percent.

Likewise, in Kazakhstan most of the uranium ore has a concentration of far below 0.1 percent.

The production profiles and reported reserves of individual countries show major downward reserve revisions in the US and France, having passed their maximum production levels. These downward revisions raise some doubts regarding the data quality for reasonably assured resources.

A summary of the history of uranium production in all countries is shown in the following figure. At the bottom are those countries that have already exhausted their uranium reserves. The data are taken from NEA 2006, and for some Eastern European countries and FSU countries, from the German BGR (BGR 1995, with additional data for subsequent years). The figure also includes the uranium demand for nuclear reactors (black line). In the early years, before 1980, uranium production was strongly driven by military uses and also by expected nuclear electricity generation growth rates that eventually did not materialize. Therefore, uranium production by far exceeded the demand of nuclear reactors.

The breakdown of the Soviet Union and the end of the cold war led to the conversion of nuclear material into fuel for civil reactors. This was at least partly responsible for the steep production decline that has occurred ever since the end of the 1980s.

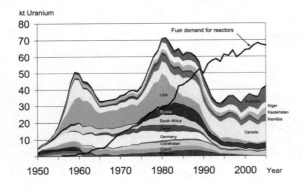

Figure 5: *Uranium production and demand*

At present, production falls short of demand by more than 25 kt/yr. This gap has been closed with uranium drawn from stockpiles. However, consisting partly of stocks at reactor sites, stocks at the mines, and stocks resulting from the conversion of nuclear weapons and the reprocessing of nuclear waste, the total amount of these stocks is very uncertain. In 2002 it was estimated that about 390–450 kt of uranium could come from these sources (BGR 2002). By the end of 2005, these estimations were reduced to about 210 kt of uranium or even less.

In order to ensure the continuous operation of existing power plants, uranium production capacities must be increased considerably over the next few years, well in advance of stocks becoming exhausted. Rising prices and vanishing stocks have led to a new wave of mine developments. Currently various projects are in the planning and construction stage, which, if completed in time, could satisfy the projected demand.

From all the mines that are planned to be in operation for the given time period according to the Nuclear Energy Agency (NEA 2006), about 20 kt/yr of additional production capacity is expected by 2010. This would increase the present capacity from about 50 kt/yr to 70 kt/yr, enough to meet the current demand once the stocks are exhausted.

However, new mining projects very often experience cost overruns and time delays, which will raise doubts as to whether the production capacities can be extended in time. These problems can be observed, e.g. by the development of the Cigar Lake project, which was supposed to produce about 8 kt/yr U_3O_8 (equivalent to 6.8 kt U) starting in 2007. This mine will be the world's second largest high-grade uranium deposit containing about 100 kt proven and probable reserves. Its expected production capacity will increase the present world uranium production by about 17 percent. Therefore its development is a key element in expanding world uranium supply. In October 2008, a severe water inflow occurred, which completely flooded the almost finished mine. At present it is very unclear whether the project can be developed further.

Chapter 35 ▌Part 9: Whither Coal: Expanded Production, Leaving It in the Ground, or Simply Running Out?

BONE AND BLOOD: THE PRICE OF COAL IN CHINA[1]

China Labour Bulletin

In the town of Spring Hill you don't sleep easy
Often the Earth will tremble and groan
When the Earth is restless, miners die
Bone and blood is the price of coal
Bone and blood is the price of coal
—"The Ballad of Spring Hill"[2]

When President Hu Jintaò descended 400 meters into a mine near Datong in the heart of China's coal country on 31 January 2008, he was faced with an acute dilemma. China was in the midst of its worst fuel and transport crisis in five decades—the country desperately needed to bolster coal production in order to generate electricity, to heat frozen homes, and to get the transport system moving again in time for the Spring Festival holiday—but soon after he took office, in 2002, he made a solemn commitment to improve coal mine safety and reduce the number of accidents and deaths in the industry.

President Hu urged the Datong miners to work through the holidays and increase production while making safety their number one priority. He must have known this was wishful thinking. Increased production nearly always carries an increased risk of accidents, particularly when ventilation systems are not upgraded to cope with the attendant gas build-up, as is frequently the case in China. Just three days after

---

1   This chapter is the first section of the March 2008 *China Labour Bulletin Report* No. 6, "Bone and Blood: The Price of Coal in China." It is being reproduced here with permission from the China Labor Bulletin. The full report is available at http://www.clb.org.hk. The second section of the report, which is not included in this excerpt, focuses on the coal mine accident compensation system and the post-accident management and social damage-limitation methods used by local governments. It discusses the new 200,000 yuan compensation benchmark for deaths from coal mine accidents and what this standard has meant for bereaved families in terms of their civil rights. The report uses telephone interviews conducted by CLB's director Han Dongfang with the families of coal mine accident victims and industry insiders to reveal the human face of coal mine tragedies. The interviews illustrate the methods used by government work teams in the wake of accidents, such as controlling the media, pressuring bereaved families into signing compensation agreements, and keeping families apart from each other in order to forestall solidarity actions and deprive them of their negotiation rights. The report further shows how local governments and mine operators make compensation awards contingent on bereaved families signing away their rights to seek further compensation through the court system. The report concludes with policy proposals and recommendations aimed at reducing the number of coal mine accidents, with a focus on the urgent need to give workers a voice and role in the safety monitoring and supervision process.

2   Lyrics from the song composed by Peggy Seeger and Ewan MacColl to commemorate the Canadian Spring Hill mining disasters of 1891, 1956, and 1958.

the president's proclamation, nine miners were killed in a gas explosion at a mine in neighboring Shaanxi province.[3]

The insatiable demand for coal in China has not only led to well-established large-scale mines greatly exceeding safe production levels, it has also encouraged the growth of small-scale illegal mining by unscrupulous business people, eager for profit and unconcerned with the lives of others. Ten days before Hu Jintao's visit to Datong, a group of men attempted to reopen a coal mine near Linfen in Shanxi province closed down three years earlier during the government's safety rectification campaign. Before any coal could be extracted, an explosion ripped through the mine killing twenty people. Li Yizhong, the government official ultimately responsible for coal mine safety, candidly admitted after the event that people had been driven to such reckless behavior by the increased demand for coal resulting from the winter fuel crisis.[4]

This winter crisis has thrown the problems of China's coal industry into sharp relief, but those problems are deep-rooted and systemic and have, according to the government's own statistics, led to the deaths of at least 45,000 miners since the turn of the century. China has relied to a very large extent on domestic coal production to power its sustained and high levels of economic growth over the last decade. Raw coal production shot up by almost 74 percent between 2001 and 2003, to 738 million tons. By 2006, it had increased almost three times to 2.3 billion tons. The Chinese government, which prides itself on "putting people first," is fully aware of the appalling human cost of this rapid growth in coal production and has introduced a vast array of new legislation, regulations, and policy initiatives designed to reduce the number of accidents and deaths in China's coal mines. However, the tangled web of collusion between mine owners and local government officials (upon whom the central government relies to enforce its policies) has prevented Beijing's well-intentioned initiatives from having any significant effect. Accident and death rates have declined from the peak of 2002 when nearly 7,000 miners died, but they remain unacceptably high with 3,786 miners being killed in 2007, according to official figures.

In August 2007, for example, over 12 million cubic meters of water from the flooded Wenhe River poured into the Huayuan coal mine in Shandong, trapping and eventually killing 172 miners. It was the second worst coal mine disaster in the history of the People's Republic of China and came at a time—one year before the opening of the Beijing Olympics—when the Communist Party and government wanted to project a positive image of China to the world. The Huayuan disaster was an unpalatable reminder that, despite the government's efforts over the previous five years to improve mine safety and reduce accidents, China's coal mines remained the world's deadliest.

The *China Labour Bulletin*'s new research report,[5] of which this chapter is an excerpt, identifies the key problems faced by the industry, explores the reasons why

---

3   Reuters. "China mine blast kills nine amid new drive for coal." http://www.reuters.com/article/latestCrisis/idUSPEK132605

4   *USA Today*. "China seeks to improve workplace safety." http://www.usatoday.com/news/world/2008-01-30-chinasafety_N.htm

5   This report is a synthesis of our two recent Chinese language reports on the coal mining industry, *Youxiao de gongren zuzhi: Baozhang kuanggong shengming de bi you zhi lu.* ("Bloody Coal: An

government policy has been so ineffective, and most importantly, reveals the human face of the industry, so often hidden or obscured in official media reports of mining tragedies.

This chapter focuses on the core dilemma faced by the government: increase production or improve safety. It examines the massive safety deficit that exists in the mining industry and examines the government's attempts to narrow that deficit. The chapter points out that the government's mine privatization program, which contracted out thousands of former state-run mines to private operators, has dangerously eroded mine safety, and explains why attempts to close down unsafe mines have failed. It shows how the government's licensing and approval procedures have become an open invitation to corruption, and demonstrates that collusion between mine operators and local government officials is now so widespread and blatant that mine operators openly flout central government directives. Mine owners and local officials conspire to cover up accidents and evade punishment, while the rights and interests of miners are either ignored or blatantly violated. The piece concludes by suggesting that the only effective way to protect the lives and rights of miners is to develop democratically-elected and truly representative workers' organizations that can stand up to the currently overwhelming power of management and safeguard working conditions at the coalface.

COAL MINE SAFETY IN CHINA

According to Chinese government figures, the total number of coal mine accidents increased by over 50 percent, from 2,863 in 2000 to 4,344 in 2002. In response, the leadership in Beijing introduced a series of ostensibly tough measures that it hoped would reduce the number of coal mine accidents: raising the rank of different government departments linked to mine safety; increasing the number of licenses required to operate coal mines; improving the mine inspection and approval system; launching actions against illegal mines; consolidating national coal resources; and closing small mines. But in both 2004 and 2005, the government failed to meet its mine safety targets, as major accidents causing "severe" and "exceptional" loss of life became more frequent and deadly.[6] In 2005, while there was a slight drop in the total number of accidents (3,341 accidents, causing 5,986 deaths), those involving severe and exceptional loss of life reached a peak (58 accidents with severe loss of life, with a total of 1,739 deaths), a 41.46 percent and 77.6 percent increase respectively over 2004. Since the founding of the PRC in 1949, there have been twenty coal mine accidents in which more than a hundred people died; eight of these occurred during the high growth period of 2000 to 2005, and of these eight, six occurred in a thirteen-month period between 2004 and 2005.

Appraisal of China's Coal Mine Safety Management System.") March 2006; and *"Yi ren wei ben"? Meikuang kuangnan yishu tanhua de qishi.* ("Putting People First: A Critique of China's Compensation System for Bereaved Coal miners' Families.") November 2006.

6   China's mine accidents are officially divided into four categories: accidents with exceptional loss of life (thirty or more deaths); accidents with severe loss of life (ten to twenty-nine deaths); accidents with serious loss of life (three to nine deaths); and accidents with loss of life (one or two deaths).

In his 2006 and 2007 New Year's messages, Director of the State Administration of Work Safety (SAWS) Li Yizhong admitted that efforts to prevent accidents causing exceptional loss of life remained ineffective and that serious illegal practices in the coal mining and other key industries persisted.[7] Moreover, there is considerable evidence that local authorities have either concealed accidents or under-reported fatalities to the higher authorities, especially in cases of accidents involving fewer than ten deaths. For these reasons, the real number of mine accidents and casualties in China remains a mystery.

ECONOMIC AND SOCIAL OBSTACLES TO THE IMPLEMENTATION OF COAL MINE SAFETY POLICY

In 2004, raw coal production in China reached 1.95 billion tons, an increase of 228 million tons or 13.2 percent from 2003. In 2005, it reached 2.19 billion tons, an increase of 9.9 percent, and in 2006, it rose by 8.1 percent to 2.3 billion tons. A significant portion of these increases came from mines that greatly exceeded their safe production capacity. Of China's twenty-seven coal producing provinces and regions, twenty exceeded their production targets in 2004, and nineteen provinces did so by more than 10 percent. Three regions—Fujian, Shaanxi, and Beijing—exceeded their production targets by more than 50 percent.[8] Such over-production fatally compromised these mines' ability to ensure safety. According to a survey by the State Administration of Coal Mine Safety (SACMS), China's coal production exceeded 1.7 billion tons in 2003, but that year only 65 percent of that took place in mines (including open-pit mines) that "guaranteed" (*baozhang*) production safety.[9] In 2004, only 1.2 billion tons were produced in mines that met safety standards, with more than 750 million tons produced in mines that failed to do so.[10]

As the following examples illustrate, over-production can be a crucial factor in coal mine accidents, especially when mine ventilation systems are not upgraded to cope with increased production, leading to the build-up of potentially explosive gases. The Daping coal mine in Zhengzhou, Henan province, where 148 people died in a gas explosion on 20 October 2004, had been inspected and approved for an annual production capacity of 900,000 tons. In 2003, the mine produced 1.32 million tons of coal, and from January to September 2004 it had already produced 960,000 tons. Similarly, the Sunjiawan coal mine in Liaoning province, where a gas explosion killed

7 Li Yizhong, "*Yuandan xianci: anquan fazhan, guotai min'an*" ("New Year's address: safe development for national peace and prosperity"), State Administration of Work Safety, 4 January, 2006, http://www.chinasafety.gov.cn/zuixinyaowen/2006-01/04/content_151616.htm; "*2007 nian xinnian heci: yingjie anquan shengchan de 'gongjiannian' he 'luoshinian'*" ("2007 New Year's greetings: making production safety a reality this year"), State Administration of Work Safety, 1 January, 2007, http://www.chinasafety.gov.cn/zuixinyaowen/2007-01/01/content_213605.htm.

8 Lu Baohong, Gao Feng, Liu Jun, "*Woguo kuangnan pinfa chaochan wei huoshou, 1/3 chanliang wu anquan baozhang*" ("Over-production is the chief culprit in most mine accidents in China; 1/3 of the coal output comes from mines that fail to guarantee safety"), *Xinhua Net*, 20 February, 2005, republished in Sina.com, http://news.sina.com.cn/c/2005-02-20/16355150771s.shtml.

9 The official term *baozhang* does not actually mean that mine safety is guaranteed. The death rate in mines that do not exceed safe production capacity also remains disturbingly high.

10 Lu Baohong, Gao Feng, and Liu Jun, op cit.

at least 214 miners on 14 February 2005, had been approved for a production capacity of 900,000 tons, but its actual output in 2004 was 1.48 million tons. The Shenlong coal mine in Fukang county, Xinjiang province, where 83 miners died in a gas explosion on 11 July 2005, had a safe production capacity of only 30,000 tons, but during the first half of 2005 alone it had already produced almost 180,000 tons of coal.[11]

China's coal industry suffers from this fundamental "safety deficit" because investment in mine safety systems and equipment has lagged behind rises in production. After a series of major accidents in state-owned mines during the second half of 2004, SACMS director Zhao Tiechui publicly acknowledged: "State-owned coal mines have run up an extremely alarming safety deficit. We estimate that 51.8 billion yuan will have to be invested in the next three years to clear this deficit."[12] In February 2006, SAWS director Li Yizhong acknowledged that according to the latest surveys and available figures, the "safety deficit" had reached 68.9 billion yuan.[13] This safety deficit is particularly serious in small village-and-township coal mines that invest very little in safety systems and equipment and employ rudimentary mining techniques. According to one estimate, China's small village-and-township coal mines would need to invest at least 8–10 billion yuan to attain even the most basic safety standards.[14] Indeed, most mining accidents occur in small mines that lack basic safety systems and equipment. According to SAWS statistics, at the end of 2004, there were 23,388 small coal pits in China: these accounted for nearly 90 percent of all coal mines and for one-third of total national coal output, but were responsible for more than two-thirds of coal mine deaths.[15]

In recent years, state-owned mines with reasonably effective safety systems and equipment have been unable to satisfy China's rapidly growing demand for coal. There has thus been a rapid proliferation of small coal mines seeking to fill the gap between supply and demand. For example, in 2005, the number of small coal mines in China grew by approximately 38 percent.[16] This rapid growth in the

11   Wang Dalin and Liu Hongpeng, "*Xinjiang Fukang Shenlong Meikuang anquan shengchan cunzai zhongda wenti*" ("The big problem of safe production at the Shenlong mine in Fukang, Xinjiang"), *Xinhua Net*, December 2005, http://news3.xinhuanet.com/newscenter/2005-07/12/content_3211755.htm.

12   "*Shigu pinfa, jingshi zhuanbian zengzhang fangshi shi zhongjie kuangnan genben zhi ju*" ("Upgrading early warning systems is the key to preventing frequent coal mine accidents"), *Banyuetan* (Fortnightly Chats), republished in *Beifang Net*, 1 February, 2005, http://news.enorth.com.cn/system/2005/02/01/000956285.shtml.

13   Li Yizhong, "*Zhongguo you zhongdian meikuang anquan zijin qianzhang 689 yi, ni liangnian buqi*"("China has a serious coal mine safety deficit of 68.9 billion yuan, which will take two years to offset"), *Xinjingbao* (New Beijing Daily), republished on *People.com*, 9 February, 2006, http://politics.people. com.cn/BIG5/1027/4086732.html.

14   "*Woguo difang meikuang anquan shengchan wenti cunzai sanda maodun*" ("Three big contradictions underlying China's problem with local coal mine safety"), *Zhongguo Nengyuan Wang* (China Energy Net), 11 October 2004, http://www.china5e.com/news/meitan/200410/200410110146.html.

15   "*Li Tieying zuo baogao, jiexi meikuang zhong teda shigu pinfa yuanyin*" ("Report by Li Tieying [Vice-Chairman of the 10th NPC Standing Committee]: an analysis of the causes for the frequent occurrence of coal mine accidents with severe loss of life"), *China.com*, 25 August, 2005, republished in *Sina. com*, http://news.sina.com.cn/c/2005-08-25/17057594186.shtml.

16   "*Meitan gongye xiehui diyi fu huizhang Pu Hongjiu da Jingji Ribao jizhe wen*" ("*Economic Daily* interview with Pu Hongjiu, Executive Vice President of China National Coal Association"), *Jingji Ribao*

number and overall production of small mines initially lowered coal prices, but also forced state-owned mines to invest less in safety infrastructure and raise their output beyond designed safe-production capacity. In August 2005, the government began to restructure the mining industry and close down small mines. At the time, the director of the SACMS announced that big state-owned coal mines would rapidly step up production to make up for the loss of output from closed-down mines. The SACMS director believed that, since small mines accounted for only one-third of total coal output, shutting them down would not have too big an impact, as lost production from those small mines could be offset by increasing production in larger coal mines.[17] This put enormous pressure on the state-owned coal mines and has forced them to further exceed their designed production capacity and ignore safety standards, which greatly increases the risk of accidents.

For example, on 27 November 2005, the Dongfeng mine, operated by the Qitaihe branch of the Longmei Mining Group in Heilongjiang province, experienced a coal dust explosion that killed 171 people. The mine's production quota was 480,000 tons, but by October it had already produced 400,000 tons by exceeding its production limit every month that year except in July. On 5 November 2006, a gas explosion at the Jiaojiazhai mine in Shanxi province, which was operated by Xuangang Coal and Power Company, a subsidiary of Datong coal mine group, killed forty-seven miners. By October, this mine had already produced 1.07 million tons of coal, which was not only more than its annual production quota but also the first time in the forty-eight year history of the mine that it had produced more than 1 million tons in a single year.

THE GOVERNMENT'S DILEMMA:
INCREASING PRODUCTION OR REDUCING ACCIDENTS

Beijing's growth-oriented economic polices have created a dilemma: increase production to satisfy the growing economy's insatiable demand for energy or invest in more mine safety systems and equipment to redress the "safety deficit" and reduce accidents.

In poorer regions where coal is the main source of revenue, this is an intractable dilemma for local governments. According to a survey of 100 counties in which coal was a major industry, it accounted for approximately 40 percent of those areas' total industrial output.[18] For example, in Shanxi province, where coal is the most important industry, mining and coking is the main source of revenue for 80 percent of counties. Ten counties around Lüliang in Shanxi have been designated national-level or county-level poverty-stricken counties. In all of them, coal accounts for 70–75

(Economic Daily), 15 December 2005.

17   Liang Dong and Cao Jiyang, "*3 nian guan 1.4 wan ge, shui lai tian xiao meikuang jianchan hou gongxu quekou*" ("If 14,000 small coal mines are closed down over a three year period, who is going to make up for the shortfall in the supply of coal?"), Jingji Cankao Bao (Economic Reference News), 12 December, 2005, republished in *China.net*, http://big5.china.com.cn/chinese/news/1057964.htm.

18   "*Woguo difang meikuang anquan shengchan wenti cunzai sanda maodun*" (Three big contradictions underlying China's problem with local coal mine safety,) Henan Meitan Xinxiwang (Henan Coal Information Network), 14 October, 2004, http://www.hnmt.gov.cn/aqsc/aqsc002/aqsc154.htm.

percent of government revenues.[19] According to SAWS statistics, between 2003 and 2006 there were 15 mine accidents, which killed 155 people, in Lüliang. The Liangjiahe mine in Xixian county, Shaanxi province, where thirty-six miners died in a coal-dust explosion on 30 April 2004, paid the county government more than 1 million yuan in annual taxes, or almost 15 percent of annual revenues. On 7 August 2005, a flood in the Daxing coal mine in a poor mountain area in the town of Wanghuai near Xingning city in northern Guangdong, trapped and killed 123 miners. Until the accident, the mine was the local government's main source of revenue. Zeng Yungao, the owner of the mine, had paid 2.5 million yuan in annual taxes and had given 3 million yuan to local charities and schools. As far as governments in poor regions are concerned, when the central government closes mines for whatever reason, it cuts off the economic lifeline of local communities.

RESTRUCTURING THE COAL MINING INDUSTRY

Prior to 2006, the government applied a "one size fits all" (or "single cut of the knife" in Chinese parlance) approach to coal mine restructuring. Following a major accident, the government ordered all mines in the proximity to stop production and improve safety (*tingchan zhengdun*).[20] The scope of restructuring depended on the gravity of the accident: a smaller accident might only affect a county or region whereas a bigger one could affect an entire province. In August 2005, the government issued two directives to strengthen the restructuring program: "Circular on the Immediate Closure and Restructuring of Coal Mines that Fail to Meet Safety Standards and Operate Illegally,"[21] which stipulated that mines failing to obtain a production safety license were to be closed down, and "Special Regulations by the State Council on the Prevention of Work Safety Accidents in Coal Mines."[22] Faced with growing pressure from Beijing, local governments tried to find ways to follow the directives whilst sustaining their revenues and economic development. In some cases, they simply ignored the directives, while in others they went through the motions of closing down mines. Upon hearing the central government's demands, the vice governor of Fujian province said that shutting down Fujian's small coal mines would cause a collapse of the province's electricity network. He wrote to his subordinates: "The restructuring of local coal mines to improve safety standards ought to be done in accordance with actual local conditions. Whilst mining enterprises need to be urged to take prompt restructuring measures in accordance with national production safety regulations,

---

19   Gao Yu, "*Hai you bi fubai geng weixian de: kan Shanxi meikuang shigu pinfa yuanyin*" ("Even more dangerous than corruption: a look at the causes of the frequent coal mine accidents in Shanxi"), *Jiangnan Shibao* (South China Times), 6 December, 2001, 5th edition.

20   *Tingchang zhengdun* could be translated more literally as "to suspend operations and reorganize" but in the context of the coal mining industry it invariably means to stop production and improve safety measures and equipment.

21   "*Guanyu jianjue zhengdun guanbi bu jubei anquan shengchan tiaojian he feifa meikuang de jinji tongzhi*," issued by the General Office of the State Council, 24 August, 2005.

22   "*Guanyu yufang meikuang shengchan anquan shigu de tebie guiding*," issued by the General Office of the State Council, 31 August, 2005.

the demand for coal also has to be met."[23] As of January 2006, Fujian province had yet to close down any mines or revoke operating permits.[24]

The government's regulations on the suspension of coal mine production were designed to improve safety. They stipulated that after a mine stops production, management must formulate a safety improvement plan, inspect the mine for hidden dangers and conduct safety training for personnel. These procedures would be followed by a local government safety inspection. However in most cases in which production was suspended, these measures were not taken. For example, after a gas explosion at the Yinguangshi mine in the town of Anping near Lianyuan city, Hunan province, that killed sixteen miners on 26 July 2004, the Lianyuan municipal government ordered mines within its jurisdiction to stop production and improve safety. But, according to a journalist who visited the area, the township government and the owner of one small mine had implemented a solution that "satisfied both parties concerned": the township government put the mine's lifting and transportation equipment under lock and key to avoid being held responsible in the event of another accident, while the mine owner sent the miners home to reduce losses while operations were suspended. No review of existing safety standards, formulation of a safety improvement plan, or employee training was carried out.[25]

In some localities government officials turned a blind eye to the continued production of coal during the suspension and safety period. After a gas explosion killed ten miners on 6 January 2004 at the Luobuyuan coal mine in the town of Meitian in Hunan province, the governments of Yizhang county and Chenzhou city declared that they took the accident "very seriously" and instructed the township government to "immediately stop production and improve safety" in local mines. They also organized a team to conduct safety inspections of all coal mines under the city's jurisdiction. However, several mines within Chenzhou municipality, including a number of small collieries in Meitian township, suspended operations during the day while continuing to mine coal at night, which they then sold during the day. In early August, officials from the Hunan Coal Mine Safety Supervision Bureau discovered that in some localities, not only had mines not been shut down, but new ones had been opened.[26]

23   Dong Wei, "*Li Yizhong nuchi xiaomeiyao, anjianjuzhang jiebuliao difang jingji nanti*" (SAWS director Li Yizhong angrily rebukes small collieries but he cannot solve local economic problems), *Zhongguo Qingnian Bao* (China Youth Daily), 7 December 2005.

24   "*Guojia gaiwei tongbao guanbi meikuang qingkuang, yanzhong tuoyan de yao zhuijiu zhuyao fuzeren zeren*" ("Circular by the National Development and Reform Commission on the actual situation surrounding the closure of coal mines and the need to find out who are the main people responsible for the delays"), *Zhongguo Anquan Shengchan Bao* (China Work Safety Herald), republished in SAWS, 20 January 2006, http://www.chinasafety.gov.cn/zuixinyaowen/2006-01/20/content_152949.htm.

25   Huang Maowang and Xiao Jianyong, "*Heise zhi zhong*" ("Coal's burden"), *Hunan Gongrenbao* (Hunan Workers' Daily), republished in Rednet.com, http://people.rednet.com.cn/PeopleShow.asp?Pid=4&id=46172.

26   Li Feixiao and Huang Xiong, "*Xuruo de 'tingchan zhengdun', yanneng shijin kuangnan xuelei? Hunan Lianyuan meikuang diaocha*" ("Sham 'suspensions of operations to improve mining safety.' Who can wipe away the blood and tears caused by mine accidents? An investigation into the Lianyuan mine

RESISTANCE TO THE GOVERNMENT'S COAL MINE CONSOLIDATION AND CLOSURE POLICY

The State Council's "Certain Opinions on Promoting the Healthy Development of the Coal Mining Industry," issued on 7 July 2005, promoted big mining companies and groups, encouraged small mines to restructure and merge into bigger firms, and advocated closing down small mines that were poorly organized, lacked adequate safety, wasted resources, and polluted the environment. On 25 March 2006, the SAWS, the National Development and Reform Commission and nine other government departments jointly issued a guideline entitled "Certain Opinions on Strengthening Coal Mine Work Safety and Standardizing the Integration of Coal Resources," which called for the closure, by the end of 2007, of mines with an annual production capacity of less than 30,000 tons. In June 2006, it was estimated that of the 17,000 small coal mines in China, a third fell into this category.[27] On 28 September, the General Office of the State Council published a guideline issued jointly by the SAWS and 11 other government departments, entitled "Opinion on Improving the Work of Reorganizing and Closing Down Coal Mines" that called for the closure of sixteen types of small mines. SAWS proposed a three-step strategy of reorganizing and closing mines, integrating and upgrading technology and improving mine management, setting a timetable for the closure of 9,887 small mines by mid-2008. In late May of 2007, SAWS published a list of 9,104 coal pits to be closed throughout China, of which 8,884 had already been closed.[28] But the central government's policy of consolidating coal resources and closing down small mines met with considerable resistance from local governments and mine operators. In April 2006, SACMS vice director Wang Shuhe told a reporter that, while the central government had demanded the closure by the end of 2005 of 5,243 small coal mines that failed to meet national standards, most provinces were so slow in carrying out the closures that the deadline had to be extended until March 2006. According to SACMS statistics, by the end of 2006, Shanxi, Sichuan, Heilongjiang, Shandong, and Hunan provinces, in which coal mining is a key industry, had still not completed their designated mine closures. Shanxi and Sichuan provinces only closed down 67 percent and 83 percent, respectively, of the mines slated for closure, and the deadline had to be extended yet again.[29]

To protect their own economic interests, local governments have been liberal in interpreting central government instructions. On 8 November 2006, SAWS issued the "Circular on Coal Mine Accidents with Exceptional and Severe Loss of Life,"[30]which

---

accident"), *Xinhua Wang Hunan Pindao* (Xinhua Net Hunan Channel), http://news.xinhuanet.com/focus/2004-09/06/content_1938527.htm.

27   Chen Zhonghua, Liu Zheng, "*Woguo jiang caiqu 'sanbuzou' jiejue xiao meikuang wenti*" ("China about to adopt a three-step strategy to solve the problem of small collieries"), *Xinhua Net*, 22 June, 2006, http://news.xinhuanet.com/newscenter/2006-06/22/content_4733773.htm.

28   SAWS, http://www.chinasafety.gov.cn/zhuantibaodao/2007mkzdgb.htm.

29   "*Guanbi xiao meikuang zaoyu zuli, jiujing shui zai 'wu gaizi'?*" ("The closure of small mines is obstructed. Who is covering up the truth"), *ENorth.com*, 17 April, 2006, http://news.enorth.com.cn/system/2006/04/17/001283160.shtml.

30   *Guanyu ji qi meikuang teda tebie zhongda da shigu de tongbao* (Circular on Coal Mine Accidents with Exceptional and Severe Loss of Life), SAWS, 28 November, 2006, http://www.chinasafety.gov.cn/2006-11/28/content_206775.htm.

revealed two methods or "technological fixes" (*jishu gaizao*) used by local governments to avoid implementing central government orders. The first method was to substitute mines slated for closure with mines that had long been abandoned or closed. For example, the Changyuan coal mine in Fuyuan county, Yunnan province, in which a gas explosion killed thirty-two miners on 25 November 2006, was placed on the SACMS closure list in early 2006, but instead of closing this mine, the local government closed a mine that had been depleted twenty years earlier. In this way, local governments routinely adjusted the closure list so as to implement as few closures as possible, while ostensibly meeting the target figure.[31]

Another "technological fix" was for local governments, in collusion with mine operators, to go through the motions of inspecting and approving mines for an increase in production capacity so that the mine would no longer be considered a small mine in line for closure. An investigation into a gas explosion that killed 17 miners on 13 March 2006 at the Rongsheng Colliery in Otog Banner, Inner Mongolia revealed that although the mine had a production capacity of 90,000 tons, on 15 January 2005, the Ordos City Coal Bureau approved a "technological fix" to increase its annual production capacity to 150,000 tons. On 10 December of 2005, the coal bureau approved another increase of the mine's designed production capacity to 300,000 tons even though no technological improvements had been made. An investigation by the Inner Mongolia Coal Mine Safety Supervision Bureau revealed that the Rongsheng Colliery had a reserve base of 1 million tons and a service life of less than three years, and therefore under government guidelines should have been closed.[32]

COLLUSION BETWEEN GOVERNMENT OFFICIALS AND MINE OPERATORS

In August 2005, Li Tieying, Vice chairman of the Standing Committee of the National People's Congress (NPC), acknowledged that: "Coal mine accidents that have already been investigated and prosecuted have revealed that corruption was behind almost every accident that caused exceptional loss of life."[33]

Energy shortages and rising coal prices have made coal mining very lucrative. Media reports indicate that, in recent years, net profits varied between 100 and 200 yuan per ton on the Chinese market depending on type and quality. Thus, a small coal mine with an annual production capacity of 30,000 tons can earn net profit of 3-6 million yuan per year, while slightly bigger mines can generate more than 10

---

31   "*Xianchang qinli: '11.25' kuangnan xianchang suojiansuowen*" ("Eyewitness report from the scene of the Changyuan coal mine Accident of 25 November"), SAWS, 28 November 2006, <http://www.chinasafety.gov.cn/2006-11/28/content_206255.htm>.

32   "*Guanbi xiao meikuang zaoyu zuli, jiujing shui zai 'wu gaizi'?*" ("The closure of small mines is obstructed. Who is covering up the truth"), *ENorth.com*, 17 April, 2006, http://news.enorth.com.cn/system/2006/04/17/001283160.shtml

33   "*Quanguo renda changweihui zhifa jianchazu jiu anquan shengchan fa shishi qingkuang zuochu baogao, jianyi xiuding kuangshan anquanfa*" ("Report by the NPC Standing Committee's Law Enforcement Monitoring Group on the actual implementation of the Work Safety Law and proposed revisions to the Coal Mine Safety Law"), *Xinjingbao* (New Beijing Daily), republished on the *Xinhua* website, 26 August, 2005, http://big5.xinhuanet.com/gate/big5/news.xinhuanet.com/newscenter/2005-08/26/content_3405767.htm.

million yuan per year in net profit. According to the official yearbook of Gaoping city in Shanxi province, some 11.5 million tons of raw coal were mined within the city limits in 2003, of which 2.15 million tons were produced by municipal-level mines and more than 8 million by township and village mines. According to a Gaoping city official, 2003 figures show that approximately 7 million tons of raw coal were produced by privately-owned mines that year. Thus, private mine operators made an annual profit of approximately 1.4 billion yaun.[34] The desire for profit makes it difficult for the central government to enforce its policies, thereby enabling small mines to continue operating or even to proliferate.

SAWS director Li Yizhong has identified five types of collusion between corrupt officials and mine operators: Government officials or managers of state-owned enterprises own coal mine shares; officials secretly operate coal mines or protect relatives who operate illegal mines; they flout regulations and abuse their authority to review and approve mines in exchange for bribes from mine operators; they turn a blind eye to or help conceal illegally run mines; and they take part in or tacitly consent to accident cover-ups. The profit motive has driven mine operators and local government officials to collude together, break the law, and obstruct central government directives. Li Yizhong explicitly warned in June 2005 that: "The collusion between government officials and business interests and between officials and mine operators has reached a level that should be taken extremely seriously."[35]

THE CONTRACT SYSTEM

At the end of the last century, in an attempt to turn China's loss-making coal industry into a major source of revenue, the central government turned much of the country's coal resources over to the private sector. The government closed down or contracted out to private entrepreneurs many small and middle-sized state-owned coal mines. The new operators then subcontracted every step of the mining process to labor contractors who hired rural migrants to work in the mines. Some contractors who obtained mining rights immediately subcontracted the mines to other entrepreneurs. Several large-scale state-owned mines have also been contracted out to private operators. For example, sections of the Sunjiawan mine in Liaoning province, where a gas explosion killed 214 miners on 14 February 2005, had been contracted to different mining teams. According to miners who worked in the Sunjiawan mine, the contractors paid the management a fixed sum for a section of the mine and then hired laborers to work that stretch of tunnel.

This complex and intricate system of subcontracting has often hindered accident investigations and made it difficult to find the original contractor legally responsible for the mine. On 3 June 2004, a gas explosion at the Hongda Colliery in Handan

---

34   "*Cong nongming dao qianwan fuweng: toushi Shanxi meikuang laoban baofu shengtai*" ("Rags to riches: a look at the sudden riches of coal mine operators in Shanxi"), *Xinjingbao* (New Beijing Daily), 17 November, 2004, http://finance.sina.com.cn/money/x/20041117/07031159998.shtml.

35   "'*Tai dandawangwei le!*' Li Yizhong nuchi '*guanmei goujie*'" ("'The brazenness!' Li Yizhong inveighs against the collusion between government officials and mine operators"), *People's Net* (hosted by *People's Daily*), 16 June, 2005, http://opinion.people.com.cn/BIG5/35560/3474971.html.

county, Hebei province, killed fourteen miners. According to the investigation team, after the accident the mine managers were unable to say how many miners had been working in the mine because the mining rights had been subcontracted to several operators. The Tianfu Sanhui coal mine in Chongqing, where a coal and gas outburst killed ten miners on 17 October 2003, was a state-owned mine. Three years before the accident, a construction company that was qualified only to dig pits but not to mine coal had been awarded a contract to exploit the mine. At the time of the accident, men who were unqualified to work as miners were working in the pit.

The contracting system has thus facilitated collusion between government officials and mine owners. Once a mining contract is awarded, government officials in charge of issuing licenses, conducting safety inspections and upholding laws invest or become shareholders in the mine either directly in their own names or indirectly through a relative. In exchange for profits, local officials then provide mine operators and contractors with protection from legal scrutiny.

LICENSING AND APPROVAL PROCEDURES

Prospective mine operators currently require six licenses: a mining license, a production license, a business license, a coal mine manager qualification certificate, a coal mine manager safety qualification certificate, and a production safety license. The licensing procedure has made it more difficult for newcomers to get into the coal mining business, but has also created opportunities for rent-seeking government officials. Since July 2005, the government's policy of restructuring and closing down small mines has made it much more difficult to meet the requirements to run a mine, but in practice, this has also served to enlarge the scope for collusion between officials and mine operators. According to media reports, in Shanxi province it is well known that a prospective mine operator will have to pay 5 million yuan to get through the red tape needed to start mining coal. In Taiyuan, capital of Shanxi province, there is a hotel called Coal Tower among whose regular guests is a group of people who specialize in greasing the wheels and taking care of the red tape for people in the coal mining business. A local mine operator told a journalist that on a single occasion he had spent more than 10,000 yuan wining and dining an ordinary government official.[36]

Media reports of coal mine accidents have shown that mines without properly obtained licenses are often still able to remain in business. The investigation of the gas explosion at the Yinguangshi mine in Anping township, Hunan province, which killed 16 miners on 26 July 2004, revealed that the mine did not have all the required licenses, though it was located just 300 meters from the Anping township government office. After a flood at Daxing coal mine near Xingning city in Guangdong province drowned 123 miners, it was revealed that the mine had no mining license or business license and had operated illegally for 6 years. After a gas explosion killed

---

36   Wei Huabing, "*Shanxi kuangzhu zibao heijin beihou de guanshang guanxi*" (Shanxi mine operator reveals the behind-the-scenes connections between officials and businessmen in the oil industry), *Shanghai Dongfang Zaobao* (Shanghai Oriental Morning Post), 1 June 2006, http://news.sina.com.cn/c/2006-06-01/01479082814s.shtml.

twenty-six miners on 28 March 2007, at the Yujialing coal mine in the town of Yip-ingyuan near Linfen in Shanxi, investigators discovered that all six of the licenses the mine needed to operate had expired.

Most accidents, however, occur in mines that have all six required licenses, sug-gesting that safety licensing standards are too low, or that licenses are issued by cor-rupt or negligent officials—or possibly both. Coal mine accident investigators rou-tinely conclude that fully licensed mines either "lacked the conditions to ensure safe production" or "had major hidden safety hazards." For example, the investigation into a gas explosion at the Jinjiangpanhai coal mine in the city of Panzhihua, Sichuan province, that killed twenty-one miners on 12 May 2005, showed the mine had all the required licenses even though it lacked a basic ventilation system. Six months before an underground water leak trapped and drowned sixteen miners on 14 July 2005 at the Fusheng coal mine in the town of Luogang near Xingning in Guangdong, the mine had obtained a production safety license. The investigation showed that the mine had been built next to a limestone cave and there were obvious flaws in its de-sign and choice of location. The Pianpoyuan coal mine in Anshun city, Ziyun county, Guizhou province, where a flood drowned eighteen miners on 15 July 2006, also had all the required licenses. In its ruling on this accident, the Guizhou Provincial Pro-duction Safety Supervision Bureau noted that the Ziyun County Enterprise Bureau had issued a production safety license even though there was a serious safety hazard in the mine. Worse still, the enterprise bureau had failed to close down the mine and impose an administrative penalty even after serious flooding was discovered. In short, one sees a catalogue of incompetence, neglect and corruption in such cases.

MINE OPERATORS OPENLY FLOUT CENTRAL GOVERNMENT DIRECTIVES

The deep-rooted collusion between local officials and mine operators has made many operators increasingly brazen in their flouting of the law. Even in the midst of central government crackdowns, mine operators ignore orders to suspend op-erations and improve safety. Investigations into the gas explosions at the Mengnan-zhuang coal mine in Xiaoyi city, Shanxi province, which killed seventy-two miners on 22 March 2003; at the Baixing mine in Jixi city, Heilongjiang province, which killed thirty-seven miners on 23 February 2004; and at the Xiangyuangou mine in Ji-aocheng county, Shanxi province, which killed twenty-nine miners on 9 March 2005, revealed four common denominators. First, in each instance, prior to the accident the local coal mine safety supervision bureau had found the mine to have a "hidden danger" that could cause an accident. Second, all the accidents occurred after the mines had been ordered to stop production and improve safety. Third, safety super-vision personnel had locked away the mine transportation equipment to make sure the mine would actually stop production. Fourth, the operators had broken the locks to restart mining.

Mine operators and government officials have conspired to create a dense web of collusion, which has made it virtually impossible for the central government to enforce its own laws and regulations. The government has put officials who are guilty of colluding with mine operators in charge of dealing with the mine safety, which is like asking criminals to judge themselves.

COVERING UP ACCIDENTS AND EVADING PUNISHMENT

Mine safety regulations stipulate that if an accident in a small township or village mine run by a private operator causes three or more fatalities, the mine is to stop production and improve safety, and if it fails to do so its licenses are to be revoked. In addition, local government officials may face disciplinary action and lose their jobs. To evade punishment, and in spite of repeated warnings, many mine operators and local officials continue to cover up accidents or lie about the number of victims. SAWS figures, based on surveys and media reports, indicate that in 2006, as many as 89 accidents (including coal mine accidents) involving 204 deaths, were falsely reported or covered up. Of nineteen other accidents with serious or severe loss of life between January and February 2007, six were falsely reported or covered up. In just the five days after 18 March 2007, three major accidents that killed a total of forty-two people were falsely reported or covered up.[37]

When accidents cause only a few deaths, mine operators often try to keep the number of people who know about the incident as small as possible; they transport the corpses to other locations and pay the victims' families "additional" compensation if they promise to keep quiet.[38] But small-mine operators are unable, on their own, to cover up accidents that cause severe loss of life or exceptional loss of life. Such accidents are frequently covered up with the help of officials. The officials are often only too willing to help because they want to protect their careers. For example, after a gas explosion at the Jiajiabao coal mine in Xinzhou city, Ningwu county, Shanxi province, killed thirty-six miners, the mine operators transported the corpses of seventeen of the miners to Inner Mongolia and hid them there. This subterfuge was jointly planned by the director of the Ningwu County Coal Industry Bureau, the mine's chief engineer and the deputy-director of the Xinzhou mine rescue team. The action was approved by both the deputy secretary of the county Party committee and the deputy head of the county.[39] After a flood drowned fifty-six miners at the Xinjing

---

37   Zhao Tiechui (director of the SACMS), *"Zai quanguo anquan shengchan shipin huiyi shang de jianghua"* ("Speech at the National Video Conference on Production Safety"), SAWS, 2 March, 2007, http://www.chinasafety.gov.cn/zhengwugongkai/2007-03/02/content_220849.htm; Jin Guolin, *"kuangnan manbao shijian shangzhou lianfa siqi, guanfang biaoshi jiang jiada chachu"* ("Four coal mine cover-ups last week alone; the government declares that it will step up efforts to investigate and prosecute such incidents"), *Zhongguo Xinwen Wang* (China News Net), republished in Qianlong.com, 26 March, 2007, http://news.qianlong.com/28874/2007/03/26/1160@3744158.htm.

38   Wang Jintao and Li Rentang, *"Meikuang shigu cengchubuqiong, yinman shouduan huayang-fanxin"* ("Constantly changing ways of covering up an endless series of coal mine accidents"), *Xinhua Jiaodiawangtan*, http://www.he.xinhuanet.com/jiaodian; Wang Xiaohong and He Yuanfa, *"Jiekai 10 tiao renming 'sile' heimu—Luoboyuan meikuang shigu diaocha"* ("The inside story of a 'private settlement' over the loss of ten human lives: an investigation into the Luoboyuan coal mine accident"), *Zhongguo Jingji Shibao* (China Economic Times), republished in Sohu Caijing (Sohu Finance), 19 July 2004, http://business.sohu.com/20040719/n221074103.shtml.

39   *People's Daily Online* (English): "Procuratorate blames officials for high rate of coal mine blasts in China." http://english.peopledaily.com.cn/200705/22/enG-20070522_376912.html

coal mine in Zuoyun county, Shanxi, the operator falsely claimed that only five miners had been trapped inside. The township head and Party secretary both knew the
real number of miners trapped underground and aided the mine operator in sending
family members of the trapped miners to neighboring Inner Mongolia in order to
prevent them from speaking to the press.

WHY IS IT SO DIFFICULT TO PREVENT COLLUSION?

The collusion between officials and mine operators has become so entrenched and
widespread that tackling it will require great effort and determination on the part of
the central government. Half-hearted rectification and punitive measures will not be
enough. In areas rich in coal resources, many government officials either manage or
co-manage coal mines. They have formed an alliance with coal mine operators and
established a network in which "one official protects another" and everyone agrees
that "if one is hurt, all are hurt." Yu Yujun, the former governor of Shanxi province
(who was forced to resign in September 2007), said it was not unusual for an illegal
coal mine to have seven or eight "protective umbrellas" (official backers) and that
as there were 3,000–4,000 illegal coal mines in the province, more than 10,000 government officials were involved in illegal mining in Shanxi alone. According to an
industry insider, the governor's figures were low, because they were based on mines
that had been identified by the authorities. In fact, no one has accurate statistics on
the number of illegal mines in Shanxi.[40]

The web of collusion between mine operators and local government officials is a
well-established and widely acknowledged fact. In a speech given at a convention of
leading officials in the coal mining center of Linfen, on 20 December 2007, Shanxi
Provincial Party Committee Secretary Zhang Baoshun explicitly cited "collusion between officials and mine operators and power-for-money deals" as one of the major
reasons for the frequency of mining accidents and for the government's inability to
stop illegal resource extraction over long periods. As one Linfen mine operator told
a journalist from *Liaowang Dongfang Zhoukan* (Outlook Oriental Weekly) the demand for bribes from officials was such a constant that, "I do not fear government officials looking into my affairs. I get worried when they are not looking into my affairs:
because, so long as they are interested in me, I can use my money to square things
with them."[41] Collusion is at the center of the continuing cycle of coal mine accidents
in China. Despite this, central officials continue to place their faith in discredited
and woefully ineffective policies using traditional top-down administrative methods
of issuing orders and entrusting implementation to regional and local governments.
In February 2006, for example, SAWS director Li Yizhong said the effort to reduce
coal mine accidents, "certainly has to rely on local governments and government
departments, public security organs, procuratorial organs and the people's courts,

40   Li Yuxiao, "*Xiaomeiyao liyi geju toushi*" ("A look at the profit structure in small mines"), *Nan-
fang Renwu Zhoukan* (Southern People Weekly), Vol. 26, 2005.
41   Sun Chunlong: "Vice-mayor of Linfen in Shanxi Province humbled, more than 50 coal mine
bosses summoned for disciplinary action" (*Liaowang Dongfang Zhoukan*), on *sohu.com*. http://news.sohu.
com/20080123/n254839797.shtml, January 23, 2008.

as well as disciplinary and supervisory departments, all of which must join forces to enforce the law."[42]

MINERS: THE ONE GROUP IGNORED IN COAL MINE SAFETY POLICY

The vast majority of miners in China today are rural migrants from poor areas and laid-off urban workers. They are paid according to how much coal they produce, have no social security or welfare benefits, and often have to buy their own tools and protective equipment. After a gas explosion killed thirty miners at the Pudeng coal mine in Puxian county, Linfen city, Shanxi province, some of the survivors told journalists that it was only after the coal mine safety supervision department began to investigate the accident that the operator distributed respirators to the miners. They said that if they'd had respirators in the first place there would have been fewer fatalities, because many had died of gas asphyxiation.[43]

Before coal mine accidents occur, gas monitoring devices frequently sound an alarm. Even in small mines without safety monitoring equipment, there are often clear signs down in the pit of an impending accident. But when this happens and miners demand to come up to the surface, operators often refuse to suspend production. This callous attitude has led to many thousands of deaths in China's coal mines. For example, eighty minutes before a gas explosion killed fifty miners on 19 May 2005 at the Nuan'erhe coal mine near the city of Chengde in Hebei province, dangerously high levels of gas were detected, causing the power supply to switch off automatically. The miners had plenty of time to evacuate, but managers refused to suspend production, ordered the power supply turned on, and made the workers continue working underground. In the days leading up to a gas explosion that killed twenty-four miners at the Luweitan coal mine in Linfen, detectors signaled excessively high levels of gas. Safety personnel had urged the management to stop production, and miners had refused to go underground, but the mine boss forced the safety personnel to alter the gas monitor log and threatened the miners with pay cuts or fines if they did not go back down into the pit.

Reckless disregard for miners' lives is also commonplace in state-owned coal mines. For example, the gas explosion that killed 166 miners on 20 November 2004 at the Chenjiashan coal mine, Tongchuan, Shaanxi province, occurred several days after a fire was detected in the pit. Mine operators not only failed to halt production, but they threatened any miner who refused to go underground with a fine or dismissal. The Xuangang Coal and Power Company's Jiaojiazhai mine in Shanxi province, where a gas explosion killed forty-seven miners on 5 November 2006, had a gas monitoring and control system. After the accident, miners reported that the managers had tampered with the gas detectors to raise the level at which the alarm would sound. Prior to the accident, the gas density in the pit had been 4 percent

---

42   See SAWS, http://www.chinasafety.gov.cn/zuixinyaowen/2006-02/28/content_155008.htm.

43   *"Shanxi Linfen Puxian meikuang: anquan guanli xingtong erxi"* ("The mockery of safety management procedures at the Pudeng coal mine in Linfen city, Shanxi province"), CCTV International, 9 May, 2007, http://news.cctv.com/special/C17274/01/20070509/102833.shtml.

above the highest level permissible, but the management made no attempt to deal with the hazard, find out what had caused it, or evacuate miners.

The government's attempts to restructure the industry in order to improve safety standards have largely ignored the voices and interests of those most concerned with the problem, namely the miners themselves. Miners are simply the objects of the government's production safety education, supervision, and reorganization measures. They are denied the opportunity to participate in the process of formulating legislation and policy and are not allowed to play an active role in protecting their own life and safety. Miners' lives are regarded, in effect simply as statistics with which to evaluate the success of government policy; the fewer deaths per ton of coal produced the better. Any attempt by miners to get actively involved in coal mine safety is regarded with suspicion or hostility by mine operators and government officials who tend to see labor organizations as "destabilizing social forces."

Since the beginning of the Industrial Revolution, experience in developed and developing countries has shown, time and again, that workers and their interests must occupy a central position in any occupational safety system. The current Chinese government has proposed an occupational safety system that is supposed to "put people first," but the State Council's 2004 "Decision on Further Strengthening Production Safety" singularly fails to provide for the participation of workers in monitoring safety. This 5,500 character document contains only one sentence that says anything about workers' participation: "Trade unions and Communist Youth League organizations must embrace production safety, give full play to their respective strengths and develop mass activities to promote production safety." Government officials seem more concerned with meeting targets and quotas than the welfare of miners. As long as the number of mining deaths in a certain region does not exceed the quotas set by the government, not only do local officials not have to worry about accidents and deaths on their watch, but can actually cite the number to boast of their achievements.[44]

In is unlikely that the central government will undergo a fundamental shift in thinking in the near future, however CLB has identified a number of practical measures the government could introduce in order to truly "put people first" and increase workers' role in coal mine safety management. Firstly, the government could allow and encourage the establishment of frontline safety monitoring committees in all large and medium-sized mines, with similar teams for small mines, and

---

44   The *State Council's Decision on Further Strengthening Production Safety* (issued on 9 January, 2004) proposed the establishment of a system of production safety targets at the national, provincial, and municipal levels to facilitate the quantitative control and evaluation of production safety conditions. Since 2004, provincial, district, and city governments have issued annual production safety targets and conducted follow-up safety inspections and evaluations. The system comprises seven types of targets: number of accident deaths at the national, provincial, and district levels; number of accident deaths per 100 million yuan of GDP; number of accident deaths per 100,000 persons at the national, provincial, and district levels; number of deaths in industrial and mining enterprises; number of deaths in industrial and mining enterprises per 100,000 persons at the national, provincial, and district levels; number of deaths in coal mining enterprises; and number of deaths in coal mining enterprises per million tonne of coal produced.

federations of safety management committees in areas where many small mines are concentrated. These committees should be empowered to order a work stoppage and mine evacuation when serious hazards are detected. Mine operators who obstruct or refuse to accept such action should be liable to criminal sanctions. This would have an immediate impact on reducing China's coal mine fatality rate.

Secondly, the government should phase out the short-term mine contracting system and replace it with a long- or unlimited-term system that would encourage stable production and a stable workforce. Miners' wages and benefits should also be increased so that they are encouraged to remain in the industry, building up skills and experience that would enable them to better understand and manage safety issues. A higher minimum wage specifically for the mining industry would be an effective first step in raising wages.

Ultimately, the most effective way to guarantee greater worker involvement in safety management would be to allow the establishment of genuinely representative trade unions in the coal mine industry. Currently, the vast majority of unions are effectively controlled by management and do not represent workers' interests, while many smaller mines do not have unions at all. CLB believes that mine unions are essential for proper safety standards to be maintained and respected, and for miners to be treated as human beings rather than as tools for profit. An effective trade union, moreover, could provide a counter-weight to the current monopoly of power held by the mine owners and local officials acting in collusion. As such, it is in the central government's interest to develop grassroots unions and use them as an ally in combating corruption and unsafe work practices in China's mines.

Chapter 36 ▌Part 9

LEAVE IT IN THE GROUND—THE GROWING GLOBAL STRUGGLE AGAINST COAL[1]

by Sophie Cooke

Coal is the dirtiest, most carbon-intensive of fossil fuels, and the biggest historical cause of climate change. If we are to cut our carbon emissions enough to prevent our planet from becoming unlivable for the majority of life on earth, we must stop burning coal. We are at the climate crunch point. It is not just starting to happen; we are now seeing changes from carbon emitted 25–30 years ago. We are on the verge of the "tipping point," the point of no return where global temperatures will trigger other effects that will plunge large amounts of previously locked up carbon into the system and the warming will be unstoppable. We are in the final throes of the last chance to halt catastrophic runaway climate change.

Yet, government and industries fail to take action, as their priorities are to keep industry going and make profit, however short term. Meetings and summits end with promises to make meaningless cuts to which people say "well, it's a start." Investments in "new technologies" mean more industry and money to corporations and little cutting of carbon. Air itself has even been made a commodity to sell on the emerging "carbon market." In the past few years widespread acceptance of the science has meant governments could not keep denying that the climate is changing dangerously and had to be seen to be taking action and providing solutions. From this, a whole swathe of "false solutions" arose, including carbon offsets—a big distraction allowing more emissions, more trade and often opening up more business opportunities to the detriment of local communities and the environment.

Coal is global and so are its effects—from mining to transportation to burning. The first and most affected peoples are indigenous peoples, most often those with the smallest carbon footprints. These are the very people whom the world needs to be learning from, not destroying in order to burn more coal.

All along the coal chain, and all over the planet, people are resisting. With harsher repercussions and different access to media and the Internet, resistance from less privileged nations is reported less and harder to find information about.

Grassroots direct action against coal has been around a long as commercial mining, because, as with all mining, it leaves a swath of destroyed ecosystems and communities. In privileged nations, over the last few years, direct action against the use of coal has grown exponentially. Direct action is about people taking action themselves to stop things they don't want to happen. This is different from lobbying,

---

1   This chapter covers a range of different events and processes in many different countries. Two websites which contain especially useful information on these themes from a range of countries are http://www.risingtidenorthamerica.org and http://www.earthfirst.org.uk.

which uses a range of tactics sometimes similar to those used in direct action, but with the aim to influence policy makers into a "better decision."

Grassroots climate resistance comes in many forms. More recently it has been taken up in such an urgent and widespread way by climate campaigners and eco-activists. Previously, it was often said to be a too big, too difficult, and too un-specific campaign to work on, but now the final urgency of the situation has taken over. People have come together to fight the industries involved in destroying the climate. This focus started with oil, and has since moved on to the dirtiest of the fossil fuels, coal. In the last few years, climate focused anti-coal campaigns have been gaining momentum across the globe.

The movement is expanding, but not in a centrally coordinated way with leaders and mandates that don't fit everyone. Rather it is a movement of solidarity and skill sharing, ideas and support. People are seeing that their struggles and actions are not isolated; they are not alone in fighting a mine or a power station in their community. People are making the connection that the carbon from these places affects us all—all around the globe.

Links are being made more easily in developed nations where activists have greater access to the Internet, share common language, and face much smaller repercussions for taking action. However, in these places work is also being done to see what support can be given to people fighting the coal industry in countries where the consequences of their dissent are much greater.

Groups like the International Rising Tide Network have helped link people working on coal internationally, sharing stories and organizing international days of action on climate change, such as "Fossil Fools Day." Fossil Fools Day '08 and '09 both saw series of international coordinated actions in the UK, USA, Australia, New Zealand, and South Africa, largely focused on coal, but also on oil and gas. Actions ranged from community protests to shutting down work at coal mines and coal plant construction sites.

CLIMATE CAMPS

Combined training, educational, and direct action camps, based around living at an eco-village-style camp, have been springing up around the world. Climate camp, largely born from "Hori-zone," a 5,000-strong direct action base camp and eco-village set up for the G-8 in Scotland in 2005, started up the following year in England at Drax Power Station, the largest point emitter of carbon in the UK. In the following years the "climate camp" idea has spread to sites across the USA, Australia, Germany, Canada, New Zealand, France, Belgium, Holland, Denmark, India, Ireland, and the Ukraine.

These tactics are more useful and face less severe repression in richer nations where individuals have relatively more power and freedom to speak. Many of the camps include invited and financially-supported speakers from indigenous communities affected by coal, and coal activists who face greater state repression in their respective countries. However, even in the countries where camps have been held, state repression is growing. At the UK climate camp at Kingsnorth Power Station in

2008, the area around the camp was shut down and turned into a police state with violence against active peaceful protest, including the battering of teenagers and pepper spraying of a local Member of Parliament. During the violent police repression at the 1st of April 2009 Climate Camp in London on "Fossil & Financial Fools Day," a local newspaper vendor passing by was battered from behind by police inside a cordon and subsequently died.

There is international resistance to all elements of the coal chain, from its mining and transportation to its burning. Here are just some of the coal projects being targeted around the world:

BANGLADESH

Phulbari is the coal capital of Bangladesh and a British company, Global Coal Management, is working to build a huge opencast coal mine there. Local people are opposed to the project, and there has been no genuine public consultation. The proposed mine would divert a river, suck an aquifer dry for thirty years, and evict thousands of people from their homes. The coal would be exported via a railway and port in the Sundarbans, the world's largest mangrove forest. In August 2006, Phulbari witnessed the killing of five people by the paramilitary Bangladesh Rifles during a massive rally against the project. Hundreds more were injured among a crowd of some 50,000 people opposing the mine, which would cover an area containing more than a hundred villages.

VENEZUELA

In Venezuela, the eco-indigenous defense group "Homo et Natura" are accompanying the indigenous and ecologist resistance against the TransGuajira "Poliduct" fossil fuel infrastructure that supports big coal and big oil in the region, the Bolivar or America Harbor, and the rail lines that are part of the coal industry expansion in Zulia.

UNITED STATES

The Appalachian Mountains in the Eastern United States are being literally blown up to mine coal. "Mountain top removal" is a form of strip mining in which the mountains are blasted away, the coal seams plundered, then the spoil dumped into thousands of valleys, leaving unusable toxic land around some of the poorest communities in the US. Local water, homes, schools, and communities are being devastated. Resistance is increasing, and in the last few years Mountain Justice, Earth First!, and Rising Tide have been taking action, targeting mines, power plants, offices, billboards, and the banks that are financing coal. This has included taking over the only bridge into a coal-fired power plant in Virginia, and blockading the construction of new power plants in Florida and North Carolina. The latter resulted in police using tazers and "pain compliance" (when police use torture to get people to submit to them). During direct actions against coal in the US, pepper spray has been used on the faces of people rendered helpless as they've locked their arms onto work machinery. One of the three 2008 North American climate camps was held in the Appalachians to fight the coal industry's stranglehold on the region.

In February 2009, activists from Rising Tide Boston disrupted a lecture at Harvard University being delivered by Arch Coal CEO Steve Leer, who was speaking on the future of "clean coal" technology. The activists enlightened the lecture attendees on the true cost of coal extraction and their involvement in the destructive practice of mountaintop removal. In March 2009, more than 2,500 activists—many willing to risk arrest—successfully blockaded all five entrances to the Capitol Power Plant in Washington, DC.

The summer of 2009 has seen action widespread across Appalachia as part of the continuing campaign to end mountaintop removal. In many places, this involves direct action and civil disobedience leading to mass arrests. In June, activists scaled twenty-story-tall mining machinery to drop a banner that read, "Stop mountaintop removal mining." This was the first time that a dragline had been scaled on a mountaintop removal site. Later the same month, thirty-one people, including NASA climate scientist Dr. James Hanson, were arrested at a Massey Energy facility in West Virginia.

BLACK MESA RESISTANCE

In the Southwestern United States, the indigenous people of the Black Mesa are resisting both coal mining and power plant construction. Black Mesa Indigenous Support says: "In 1974 the US Congress passed a law allegedly to settle a so-called land dispute between the Dineh and their Hopi neighbors. This law required the forced relocation of well over 14,000 Dineh and a hundred-plus Hopi from their ancestral homelands. The "dispute" being settled was, in reality, fabricated by the US government as a way to obtain easier access to strip mine one of the largest coal reserves in North America. The land, known as Black Mesa, is home to thousands of traditional sheep herders, weavers, silversmiths, and farmers. For hundreds of years before Europeans came to the Americas, the Dineh and Hopi existed in balance with each other and with Mother Earth." In thirty years the Black Mesa mine has contributed an estimated 325 million tons of CO_2 to the atmosphere. The expanded mine could potentially contribute an additional 290 million tons of CO_2 to the atmosphere. Lives are being attacked from all sides with the forced relocations; a drop in the water table due to water mining for slurry, which is affecting drinking water springs; and desert plants shrinking by the year from lack of rainfall due to climate change. Over thirty years have passed, and the Elders carry on resisting...

TRANS-PACIFIC SOLIDARITY

The company responsible for much of the destruction in the USA is Peabody Coal, who are now also beginning business in Australia after buying Excel Coal. Local indigenous family members, indigenous members of the United Nations of South Pacific, along with people from two other aboriginal nations and local residents, occupied the offices of Peabody Coal in Newcastle, NSW, in solidarity with the people of Black Mesa in North America. Arthur Ridgeway, who is a traditional owner of the Pambalong area that now makes up Newcastle, said, "Climate change now is a global issue and the global community must now act to protect what is left of our precious environment. These coal companies have no place in our future the way they desecrate the earth and our sacred homelands. Peabody's practices in America have shown that they have no regard for the earth there and they will do the same here. We will not let the desecration of indigenous cultures continue, here in Australia or overseas."

NEW ZEALAND

Over the Tasman Sea, in New Zealand, people have been occupying Happy Valley since January 2006, in an effort to halt development of a proposed opencast coal mine. Solid Energy's mine, a planned ninety-six meter deep pit, would completely obliterate this pristine valley and pollute local rivers with heavy metals and acid mine drainage. The coal produced would create as much carbon dioxide as all of New Zealand's domestic transport. There have been many direct actions carried out, including occupations and banner drops, and the digging up of the front lawn of Solid Energy's HQ. Protestors have twice blockaded railway tracks used to move coal, with two people locked directly onto the rail tracks, and a third hanging 30 meters above the ground from a tree.

AUSTRALIA

In Newcastle, NSW, people are fighting the expansion of the world's largest coal port, which is set to increase its capacity by 60 percent. This resistance has included a string of direct actions including locking onto trains, and obstructing trucks, offices, and construction sites, culminating in the 2008 Australian climate camp. The camp began with an invitation to Aboriginal country by local Pambalong traditional owners. On the lasts day of the camp, over a thousand people marched to the Carrington coal terminal. One climate camper said "The children led the march which meandered along the edge of the coal rail line until we held a five minute silent vigil in front of the massive coal stockpiles. Then, one by one, small groups of people made their way over or under the fence and onto the tracks. By the end of the day, fifty-seven people had been arrested and we successfully halted all coal trains through the Carrington port for the day."

Since then a string of power stations have been targeted with occupations, including in June 2008 when activists locked onto and shut down a conveyor transporting coal into Muja power station in Collie. In March 2009, during "Earth Hour," community climate activists occupied Hazelwood Power Station in Victoria's Latrobe. And, in June 2009, activists locked onto a conveyor belt at Bluewaters coal-fired power station near Collie, Western Australia.

GERMANY

Activists in Germany held their first *"Klima Camp"* in Hamburg, August 2008. The goal was to stop construction of the new HamburgMoorburg coal power plant, which is being built by energy company Vattenfall despite massive local opposition. Resistance against opencast has a long tradition in East Germany, where wanton destruction started in 1923 and carries on to date under the auspices of Vattenfall. Huge areas of the Lausitz (inhabited by the Sorbs, a slavic speaking minority) have been ripped up, many villages destroyed, and ten of thousands of people forcefully resettled. The now partly-destroyed village of Lacoma has been the focus of several longterm occupations, tree protest camps, and sabotage of heavy machinery and pumps over the past decade. Direct action is ongoing. In southwest Germany, local activists have blockaded coal trains for the Mannheim coal power station, as well as access to the power plant itself. In May 2009, several hundred people stormed the plant to protest plans to build an additional plant on the same site.

NETHERLANDS

In the Netherlands, five new coal-fired power plants are in the planning. E.on is making a concerted effort to be the first company to start building next spring. E.on is also planning to build a coal-fired plant in Antwerpen, Belgium. Dutch Earth First! (Groen Front!) have been resisting, with blockades of E.on Headquarters, bike demos, and other protests.

UNITED KINGDOM

Summer 2008 saw the squatting of land near Kingsnorth Power station for the UK climate camp, with days of action, workshops, and skill sharing. Climate campers are vowing to prevent the construction of the first of a string of planned new coal power stations in the UK. An ongoing campaign of direct action civil disobedience to stop construction at the plant itself and target all parts of the chain, including suppliers and contractors, is being planned.

The coal industry continues to extract, transport and burn coal at an increasing rate. There are thirty-three opencast mines in the UK at the moment, with thirty more in planning or development. Using enormous machines, a small number of workers can rip apart the landscape and get to the coal, which means that new mines usually provide very few jobs. Currently 29 percent of the coal used in the UK is mined within its borders—the rest is imported from equally-destructive mines around the world.

FfosyFran opencast coal mine near Merthyr Tydfil, Wales will be the biggest opencast in the UK—and a climate disaster waiting to happen. MillerArgent want to dig up coal which, when burnt, will produce about 30 million tonnes of CO_2 every year. Local people have been opposing the mine for years; more than 10,000 signed a petition opposing the mine, but despite this, the Welsh Assembly, encouraged by Westminster, gave the go-ahead. Since November 2007, campaigners have been helping local residents engaging in civil disobedience on site. The mine has been closed down, more than once, bringing public and media attention to the issue, and supporting the local campaign with rallies and meetings.

On 13 June 2008, twenty-nine climate campaigners stopped a train taking coal to Drax power station in North Yorkshire. Drax is the single biggest source of CO_2 in the UK, and train services delivering coal to the station remained stopped for the day. Up to 100 police officers, some in riot gear, had to be brought in to remove the activists. Protesters used safety signals to stop the train on a bridge overlooking the power station, before they climbed on board and dumped coal onto the tracks. Others used a network of climbing ropes to suspend themselves under the bridge from the train. And in November 2008, an unidentified activist breached the fences of

Kingsnorth power station and shut down a 500 MW turbine, single-handedly cutting UK carbon output by 2 percent.

In July 2008, the anti-opencast squat near Shipley, Derbyshire was evicted. Activists had been occupying the house at Prospect Farm and surrounding trees, and it took about a week after UK won the possession order to get them out of the tunnels. Demolition of the farmhouse was part of UK Coal's plans for a 100 hectare opencast mine removing 1 million tons of coal over 5 years. The plans were opposed by local people and refused by the council, but central government overturned the refusal in spite of planning policy guidelines. When work on the Derby site began with the building of access roads and a plant compound, actions to disrupt work took place, including obstructing a steamroller that was being used to set a tarmac ramp for heavy plant machinery.

In April 2009, 114 people were arrested in a 2 AM police raid on a community center and school in Nottingham. It was "believed that a demonstration was planned at the E.On power station at Ratcliffe-on-Soar." The Ratcliffe-on-Soar coal-fired power-er station is the third largest source of carbon dioxide emissions in the UK and has been previously targeted by activists. Summer 2009 saw actions in Scotland. Scottish Coal were given permission to mine 1.7 million tons of coal from Mainshill Wood in South Lanarkshire, a decision that enraged local residents who have campaigned against this mine for many months; a squatted solidarity camp was set up in the area. Also in July, a group of activists disrupted the operations of Scottish Coal at the Rosewell open-cast coal mine in the Midlothians. Some of the ten activists who stopped work are local residents. They climbed onto digging machinery to prevent work and climbed onto trucks to prevent coal from leaving the Rosewell site. Three UK climate camps were planned for 2009, in England, Scotland, and Wales.

LINKS TO RELEVANT ORGANIZATIONS:

- Rising Tide UK risingtide.org.uk
- Climate Camp UK climatecamp.org.uk
- UK Earth First! earthfirst.org.uk
- Rising Tide North America risingtidenorthamerica.org
- Mountain Justice Summer, North America www.mountainjusticesummer.org
- Rising Tide North America, www.risingtidenorthamerica.org
- Katuah EF! North America www.katuahef.org
- Black Mesa Water Coalition, North America www.blackmesawatercoalition.org
- Black Mesa indigenous support, North America www.blackmesais.org
- Rising Tide Australia http risingtide.org.au
- Friend of the Earth Australia www.foe.org.au
- Save Anvil Hill, Australia www.anvilhill.org.au
- Climate Camp Australia www.climatecamp.org.au
- Homo et Natura www.homoetnatura.org
- Save Happy Valley, New Zealand www.savehappyvalley.org.nz
- Klima camp Germany www.klimacamp08.net
- Phulbari Resistance www.phulbariresistance.blogspot.com
- Groen Front! www.groenfront.nl
- Carbon Trade Watch www.carbontradewatch.org
- Coal mining and local resistance www.minesandcommunities.org
- Indigenous Environmental Network www.ienearth.org

Chapter 37 ∎ Part 9

PEAK COAL

By the Energy Watch Group[1]

When discussing the future availability of fossil energy resources, conventional knowledge has it that an abundance of coal exists globally, thus allowing for coal consumption to increase far into the future. This is either regarded as being a good thing since coal can be a possible substitute for the declining crude oil and natural gas supplies, or it is regarded as a horror scenario leading to catastrophic consequences for the world's climate. However, the discussion rarely focuses on the question: How much coal is really there?

Our report attempts to give a comprehensive view of global coal resources and past and current coal production, based on a critical analysis of available statistics. This analysis is then used to provide an outlook on possible coal production in the coming decades, and concludes that there is probably much less coal left to be burnt than most people think.

POOR QUALITY DATA

The first and foremost conclusion from this investigation is that quality of data concerning coal reserves and resources is poor, both on global and national levels. However, there is no objective way to determine how reliable the available data actually are.

The timeline analyses of data given here suggest that, on a global level, the statistics overestimate the reserves and the resources. In the global sum, both reserves and resources have been downgraded over the past two decades, in some cases drastically.

Even though the quality of data on reserves is poor, an analysis based on these data is nonetheless still deemed meaningful. According to past experience, it is very likely that the available statistics are biased on the high side, and therefore projections based on these data provide an upper boundary of possible future developments.

ONLY RESERVE DATA ARE OF PRACTICAL RELEVANCE, NOT RESOURCE DATA

The logic of distinguishing between *reserves*, which are defined as being proved and recoverable, and *resources*, which include additional discovered and undiscovered inferred/assumed/speculative quantities, is that, over time, production and exploration

---

1   This extract is from the Energy Watch Group report "Coal: Resources and Future Production" (March 2007), EWG-Series 1/2007. Responsible for the report are Dipl.-Kfm. Jörg Schindler and Dr. Werner Zittel (Ludwig-Bölkow-Systemtechnik GmbH, Ottobrunn/Germany). The complete report is available to download at: http://energywatchgroup.org/fileadmin/global/pdf/EWG_Report_Coal_10-07-2007ms. pdf. The extract included here was prepared by Thomas Seltmann at the Energy Watch Group, for the purpose of this book, and they have kindly agreed to include it under the general Creative Commons License. However, the main report is protected by © Energy Watch Group/Ludwig-Boelkow-Foundation.

activities allow for reclassifying some of the resources into reserves. It should be noted that resources are regarded as quantities in situ, of which at most 50 percent can eventually be recovered. In practice, over the past two decades such a reclassification has only occurred in two cases: India and Australia.

In the global sum total, hard coal reserves have been downgraded by 15 percent. Adding all coal qualities, ranging from anthracite (hard coal with high energy content) to lignite (brown coal with lower energy content), reveals the same general picture of global downgradings. The cumulative coal production over this period is small compared to the overall downgrading. Therefore coal production cannot explain the phenomenon.

For global resource assessments, the trend is even more severe: world coal resource assessments have been downgraded continuously from 1980 to 2005, by 50 percent overall.

Thus, in practice, over the past more than two decades, resources have never been reclassified into reserves, despite increasing coal prices.

SIX COUNTRIES DOMINATE COAL GLOBALLY

Eighty-five percent of global coal reserves are concentrated in six countries. These are, in descending order of *reserves*: USA, Russia, India, China, Australia, and South Africa. The USA alone holds 30 percent of all reserves and is the second largest producer. China is by far the largest producer but possesses only half the reserves that the USA possesses. Therefore, the outlook for coal production in these two countries will dominate the future of global coal production (see below).

The largest coal *producers* in descending order are: China, USA (half of Chinese production), Australia (less than half of US production), India, South Africa, and Russia. Between them, these countries account for over 80 percent of global coal production.

Coal consumption mainly takes place in the country of origin; only 15 percent of production is exported, 85 percent of produced coal is consumed domestically.

The largest net coal *exporters*, in descending order are: Australia, Indonesia (40 percent of Australian export), South Africa, Colombia, China, and Russia. These countries account for 85 percent of all exports, with Australia alone providing almost 40 percent of all exports.

The fastest depletion of reserves in the world is taking place in China, where 1.9 percent of the country's reserves are produced annually.

	Largest	2nd largest	3rd largest	4th largest
Reserves 2005	USA 120 Btoe	Russia 69 Btoe	India 61 Btoe	China 59 Btoe
Production 2005	China 1,108 Mtoe/a	USA 576 Mtoe/a	Australia 202 Mtoe/a	India 200 Mtoe/a
Net Export 2005	Australia 150 Mtoe/a	Indonesia 60 Mtoe/a	South Africa 47 Mtoe/a	Colombia 36 Mtoe/a

Table 1: (Btoe: Billion tons of oil equivalent); (Mtoe/a: Million tons of oil equivalent per year)

BEST CASE SCENARIO: GLOBAL COAL PRODUCTION TO PEAK AROUND 2025 AT 30 PERCENT ABOVE PRESENT PRODUCTION LEVELS

Based on the assessment that reserve data may be taken, for all practical purposes, as an upper limit of the quantity of coal that could be produced in the future, some production profiles have been developed.

Our analysis reveals that over the next ten to fifteen years, global coal production may still increase by about 30 percent. This increase will be mainly driven by Australia, China, Former Soviet Union countries (Russia, Ukraine, Kazakhstan), and South Africa. Production will then reach a plateau and will decline thereafter. The possible growth in production until about 2020 outlined in this analysis is in line with the two demand scenarios made by the International Energy Agency (IEA) in its 2006 edition of the *World Energy Outlook*. However, our projection for development beyond 2020 is only compatible with the *IEA alternative policy scenario*, a scenario in which coal production is constrained by climate policy measures, while the *IEA reference scenario* assumes that coal consumption (and production) will continue to increase until at least 2030. According to our analysis, this will not be possible due to limited reserves.

Again, it needs to be emphasized that this projection represents an upper limit of future coal production, according to the authors' best estimate. Climate policy or other restrictions, which may reduce coal consumption (and production), have not been taken into account.

CONCLUSION AND RECOMMENDATION

Global coal reserve data are of poor quality, and appear to be biased towards the high side. The best case scenario offered by our production profile projections suggest that the global peak of coal production will occur around 2025, at 30 percent above current production levels.

A wide discussion on this subject is necessary, leading to better data in order to provide a reliable and transparent basis for long term decisions regarding the future structure of our energy system. The repercussions for the climate models on global warming are also an important issue.

RESERVES AND RESOURCES

A closer look at the historical reserve assessments raises doubts regarding the quality of reserve assessments:

• For instance, China's reported proved reserves have not changed since 1992. For other countries, they have not even changed since 1965.

• In recent years the proved recoverable reserves (as reported by the World Energy Council, London) for other countries—e.g. Botswana, Germany and the UK—have been downgraded by more than 90 percent. Even the reserves of Poland are 50 percent smaller now than they were twenty years ago. This downgrading cannot be explained by volumes of coal produced in this period. The revisions are probably due to better data.

•   Since 1987, India's proved recoverable reserves (as reported by WEC) were continuously revised upward, from about 21 billion tons to more than 90 billion tons in 2002. However, India is the only country with such huge upward revisions.

•   According to the latest assessment by the WEC, total proved recoverable world reserves at the end of 2002 stood at 479 billion tons of bituminous coal and anthracite, 272 billion tons of sub-bituminous coal and 158 billion tons of lignite.

Normally it is argued that reserves are part of the resources. Over time, and with coal prices increasing, more and more resources will be converted into recoverable reserves. This suggests the analogy of an iceberg: of which only the tip is visible, and 90 percent is under water. However, the present and past practice of reporting reserves does not support this view. Many countries have not reassessed their reserves for a long time, and where they have, revisions have been mostly downward rather than upward, contrary to what should be expected.

The estimated resource base should be regarded as a final limit for the amount of coal that ultimately can be recovered. However, in addition to the concerns raised above, the historical assessment of global resources has also revealed substantial downgradings over the last decades. The following figure shows that estimated coal resources have declined from 10 billion tons coal equivalent (~8300 Mtoe) to about 4.5 billion tons coal equivalent (~3750 Mtoe)—a decline of 55 percent within the last twenty-five years. Moreover, this downgrading of estimated coal resources shows a trend supported by each new assessment, so it is possible that resource estimates will be further reduced in future. One can conclude that better understanding and improved information have led to a continuous downgrading. In the following figure, the discrepancy of data for Europe and Asia for 1993 is due to the fact that the former Soviet Union was attributed to Europe in 1993 and to Asia in all the other years.

History of Assessment of world coal resources

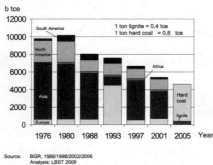

Figure 1: Reported resource assessments by the BGR (German Federal Institute for Geosciences and Natural Resources) since 1976. The physical tons of coal are converted into btce (billion tons of coal equivalent) for reasons of comparison. For comparison, 1 btce = 833 Mtoe.

PRODUCTION

Even though the above discussion is cause for major concern with regard to data quality, for lack of any better data, the most recently reported reserves have been used to assess future coal production. It is very unlikely that recoverable reserves will eventually turn out to be higher than reported. The reasons for this assessment are as follows:

•   As shown above, the resources have been scaled downwards several times since 1980. The most recent reassessment resulted in coal resources that are 55 percent less than they were in 1976.

•   Reserve data have often remained unchanged for many years. In most cases, when the data has been updated it has resulted in revisions downwards rather than upwards.

It is important to bear in mind that if these reserve data turn out to be too optimistic, so too will the derived production profiles. Nonetheless, these projections offer a starting point for further considerations.

The next figure shows coal reserves for the main countries. Reserves of hard coal and lignite are converted into energy units by means of the rough conversion factors as used in BP Statistics: 1 ton of oil equivalent (toe) corresponds to 1.5 tons of hard coal (anthracite and bituminous coal) and to 3 tons of sub-bituminous coal and lignite. Future world production is determined by the production profiles of these countries: USA, Russia, India, China, Australia and South Africa.

The figure also shows coal production in 2005. China is depleting its reserves at an annual rate of almost 2 percent. Therefore, at the present production rate, and if its resources do not turn up as reserves, China's reserves will be depleted in about fifty years. A conversion of resources into reserves has not been observed for almost thirty years. Besides China's special role and production in the "big six" countries, Germany and Indonesia also merit some attention, as they are depleting their reserves at an even faster rate. Germany is the world's largest lignite producer, with a share of about 20 percent of world production.

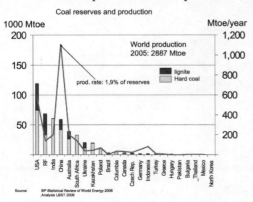

Figure 2:   Distribution of world coal reserves and annual production

Future coal production profiles are estimated by fitting the reported proved reserves to the present and historical production pattern. Provided present trends continue, with China's huge coal depletion rate and its absolute dominance of worldwide production (it is the largest producer by a factor of two), the eventual peak of Chinese coal production will determine the peak of worldwide coal production.

Comparable analyses have been made for each country; a bell-shaped curve is fitted to the historical production data and to the available proved reserves for each. These production profiles do not take into account possible restrictions that may be imposed, such as those on coal quality with respect to pollutants, or policy restrictions due to climate change. These projections represent a future scenario not restricted by political measures. The results are summed up for each region and for each class of coal.

The production data for the different regions are combined in order to give, in the following figures, world production data for bituminous and sub-bituminous coal and separately for lignite. Figure 4 provides a summary for bituminous and sub-bituminous coal. The lower quality sub-bituminous coal is always painted in a darker shade in order to demonstrate the different coal qualities.

The decline rates of future production are reduced by the production of the Former Soviet Union countries in line with their reported sub-bituminous coal reserves. However, it is by no means certain that their reported reserves will ever translate into corresponding production volumes. Some doubts remain regarding the data quality of the coal reserve data for the Former Soviet Union countries, as the last update was carried out in 1998. Therefore, it is probably more realistic to expect the decline following the peak will in fact be steeper than that shown in figure 4.

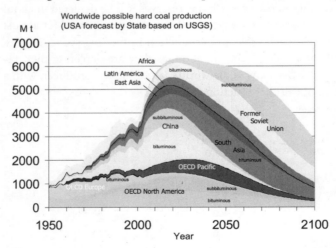

Figure 3: *World production of hard coal (bituminous and sub-bituminous) disaggregated into the ten regions.*

Figure 4 shows the world production of lignite, and to facilitate comparison, the same scale is used as in figure 3. However, the heating value of lignite is much lower than that of bituminous and even lower than sub-bituminous coal. Lignite is predominantly used for domestic heating and to produce power, and is not transported over large distances due to its low energy content.

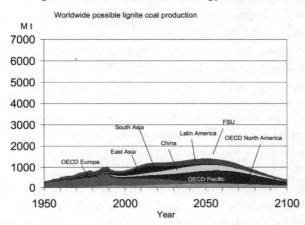

Figure 4:   World production of lignite (bituminous and sub-bituminous) in the ten world regions.

These projected production profiles are based on reported "proved" recoverable reserves (WEC), with the exception of the USA. In the case of the United States, an earlier production forecast made by the USGS (United States Geological Survey) has been used as a guide.

The final figure, figure 5, combines the regional contributions to global hard coal and lignite production and converts them into energy terms. The following factors have been used to carry out the conversion: 1 toe bituminous coal = 1.5 t bituminous coal (For China, South Asia, and Russia the relation "1 toe = 1.6 t" is used); 1 toe sub-bituminous coal = 2 tons sub-bituminous coal; and 1 toe lignite = 3 t lignite.

The figure includes the two scenario calculations from the IEA's *World Energy Outlook 2006*, the "reference scenario" and the "alternative policy scenario."

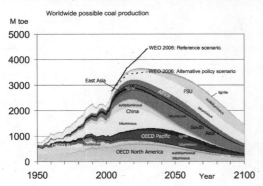

Figure 5:   World coal production in the equivalent of a million tons of oil as calculated in this study based on proved recoverable reserves.

This analysis leads to some important conclusions:

- The production profile of the world's largest producer, China, determines the peak of global coal production.

- The production profiles of China, South Asia, and the former Soviet Union countries are based on resource data that is probably of low quality.

- Apart from the world production profile, regional production profiles are also important. In a world characterized by shrinking supplies of oil (and later gas), coal will attract increased attention. It can be assumed that regional oil and gas supply gaps will initially be closed by using domestic alternatives, probably even by producing fuels from coal. This will have significant consequences for the availability of coal on the world-market (because it will result in reduced amounts being available for export). This is even more so for lignite, since it is not transported over long distances due to its low energy content.

- The IEA's WEO 2006[2] scenarios ("reference scenario" and "alternative policy scenario") are compatible with this supply scenario until about 2020. However, after that, it will only be possible to meet the demand projected in the "alternative policy scenario," as supply will flatten. It will be impossible to meet the demand projected in the "reference scenario," due to supply restrictions.

---

2   Quoting the "World Energy Outlook of International Energy Agency," we refer to the 2006 issue. Data of newer issues does not change the evidence of our reports or this article.

Chapter 38 ▌ Part 10: Agrofuels as the Geopolitical Handmaiden of the Petrol Industry: A Tale of Enclosure, Violence, and Resistance

GLOBAL AGROFUEL CROPS AS DISPOSSESSION

By Les Levidow and Helena Paul

Worldwide, many governments have been promoting biofuel crops as a sustainable way to help avoid climate change and provide energy security—in the name of the common good. Benefits are also claimed for rural development, people's livelihoods, and poverty reduction worldwide. However, biofuel crops have already caused harm in the global South—competition for land use between fuel and food, land seizures, higher food prices, greater agrichemical usage, shifts to monocropping, loss of rural livelihoods, peasants' dispossession from land, deforestation, etc. Beyond the effects of a specific crop, impacts take place at the macro-level as they globally displace production; for example, greater US maize production for biofuels has displaced soya production to Brazil, which, in turn, has increased deforestation by cattle ranching in the Amazon.

Such criticisms were initially raised by NGOs (e.g. Barbara, 2007; Econexus et al., 2007; GRR, 2005; Oxfam, 2007; Semino, 2007), and similar concerns were later taken up by state bodies (e.g. EAC, 2008; EEA, 2008; FAO Media, 2007; RFA, 2008). However, the latter reports have generally described the harm as incidental or contingent—as "negative side-effects." On the contrary, the fundamental problem is agri-industrial systems appropriating resources to produce standard commodities for global markets.

PERSPECTIVES ON MARKET DISPOSSESSION

To elaborate the above argument, this article links two theoretical perspectives: the biofuels market as a global integrated network, and capital accumulation by dispossession. By linking those perspectives, markets can be understood as integrating states in a network of global capital, which reduces their capacity or incentive to protect general livelihoods and environments, especially in the global South. Political and economic elites accommodate the global forces that dispossess communities of resources.

MARKET AS A GLOBAL INTEGRATED NETWORK

Biofuels epitomize a globalization process. Biofuel crops were originally promoted for local or national uses, but greater industrial integration and commodity flows have been recently globalizing biofuels, thus deterritorializing relations between production and consumption. An emerging "global integrated biofuel network" (GIBN) is characterized by greater transboundary flows, weaker influence by states, a homogenization of products and processes, and an integration with analogous networks of fossil fuels, argues Arthur Mol (2007: 302–3). "Overall, there is a tendency towards standardized products that can be detached from the local space of place and be transferred in a globally integrated network" (ibid: 309).

Embryonic small-scale, local biofuel networks are being pressured to integrate into national biofuel regions and then into international commodity flows, as exemplified initially by Brazil.

> Local marginal farmers become increasingly dependent on powerful global players in the GIBN ...
>
> These national biofuel regions result in large-scale monocropping biofuel production and the increasingly centralised, homogenised production and refining of these crops, while local biofuel regions are losing their relevance. Secondly, there is a clear tendency towards the development of a GIBN in which production, trade investment, consumption, control and governance lies beyond the control of nation-states (Mol, 2007: 307; 305–6).

Such systems damage local environmental resources; large-scale, high-input, monocultures degrade soil and water, as well as undermining food availability and affordability for local populations.

Despite those global pressures, many local biofuel regions have significant barriers to agri-industrial cultivation methods. Where land access and cost structures are unfavorable, biofuel crops may be developed as a local energy substitute, especially in peripheral localities that are not well served by conventional fossil-fuel infrastructures (Mol, 2007: 304). Conversely, profitable investment depends upon overcoming those barriers and thus incorporating localities into global value chains.

Under pressure from civil society, government policies may incorporate efforts to address environmental issues, e.g. by monitoring whether biofuel production saves or increases carbon emissions. However, it is much more difficult to mitigate new social vulnerabilities in the global South, given the structural change in power relations between global traders, developing countries, and small-scale farmers, as Mol argues (309–10). Even if governments want to protect local resources and livelihoods from dispossession, they have weaker capacity to exercise effective control.

ACCUMULATION BY DISPOSSESSION

Current dispossession of resources, especially in the global South, has analogies with the "primitive accumulation" that originally turned communal resources into private property. In "the historical process of divorcing the producer from the means of production," entire populations were "forcibly torn from their means of subsistence." In particular, "The expropriation of the agricultural producers, of the peasant, from the soil is the basis of the whole process" (Marx, 1976: 875–76). "The basis" means a prerequisite, even an aim—not simply an ex-post consequence, much less a contingent side-effect.

By analogy, since the 1970s, New Enclosures have attacked a wider range of commons than existed at the beginning of capitalism. This is done by various means, including ending communal control of the means of subsistence, seizing land for debt, substituting migrant labor, turning seeds into private property, etc.

Likewise, extending the original concept, David Harvey (2003) substitutes "accumulation by dispossession" to denote an ongoing process. He draws present-day analogies with early capitalism:

> A closer look at Marx's description of primitive accumulation reveals a wide range of processes. These include the commodification and privatization of land and the forceful expulsion of peasant populations; conversion of various forms of property rights (common, collective, state, etc.) into exclusive private property rights; suppression of rights to the commons; commodification of labor power and the suppression of alternative (indigenous) forms of production and consumption; colonial, neo-colonial and imperial processes of appropriation of assets (including natural resources); monetization of exchange and taxation (particularly of land); slave trade; and usury, the national debt and ultimately the credit system as radical means of primitive accumulation.... All the features which Marx mentions have remained powerfully present within capitalism's historical geography up until now (Harvey, 2003: 145).

Such dispossession remains central to capital accumulation in its recent forms. By analogy to the original enclosures of common land, new strategies seek to enclose broader resources for private use, especially in the face of collective efforts to protect them as common resources. These commons include land, water and knowledge (De Angelis, 2004).

All those dynamics were linked by Karl Polanyi to analyze mass starvation in British-ruled India:

> The catastrophe of the native community is the direct result of the rapid and violent disruption of the basic institutions of the victim ... These institutions are disrupted by the very fact that a market economy is foisted upon an entirely differently organised community; labour and land are made into commodities, which, again, is only a short formula for the liquidation of every and any cultural institution in an organic society (Polanyi, 1944: 159–60).

As another historian has noted, market pressures drove Indian and Chinese peasants into debt as a strategic instrument of dispossession: "Instead of profiting from exchange, they were forced by the market into the progressive deterioration of their conditions of production, i.e. the loss of their property titles" (Medick, 1981: 44).

Through various pressures since then, smallholders have been effectively forced into global commodity markets, thus undermining the earlier basis of food security. As another commons essential for local community needs, forests have been cleared for agri-industrial production. Low productivity is often blamed for food shortages, environmental destruction and deforestation, as if these were essentially technical problems due to extensive methods of land use. Yet the causal relation is often the reverse: intensification through technological development has been decisive for large-scale deforestation (Hecht, 2004: 67; also Angleson and Kaimowitz, 2001).

NATIONAL EXAMPLES

Drawing on above perspectives, we can identify the forces causing agri-environmental sustainability problems and land-use competition from biofuels—namely, agri-industrial monocultures, which opponents call "agrofuels" (Econexus et al. 2007). These systems have several drivers: companies' search for more profitable products; governments' search for export markets, foreign currency and foreign investment, especially as speculation moves from property into agriculture and land; a greater global demand for animal feed and biofuels; and integration of those commodities with other industrial products. Biofuels provide both an incentive and a pretext for grabbing land. Extra incentives come from expectations that agronomic and/or technological changes can increase productivity.

Biofuel production appropriates so-called "marginal" land, yet "indigenous people depend on these now marginal lands for their livelihoods," states an NGO report. With the rise of the global biofuels market, moreover, "The drive to use increasing amounts of marginal land for energy crops will also require more fertilizer use, create more erosion, and further degrade soil fertility, which is essential for food security" (Barbara: 2007: 8, 11). As that quote shows, even some biofuels opponents use the deceptive term "marginal," but for the biofuels market, however, "marginal" means previously unproductive for capital accumulation, thus ignores societal uses. As another NGO report noted,

> [Although] identifying "idle" lands may help bring under-utilised land into production, it may also create risks of dispossession. Where forms of local resource use are perceived as low productivity, land may risk being classified as idle or under-utilised, and therefore available to prospective investors, despite the economic, social or cultural functions it performs for local people (Cotula et al., 2008: 46–47).

Even where smallholders retain access to land, they undergo greater exploitation. They are easily caught in debt traps; often they must borrow funds to buy tools and seeds, as well as basic necessities at a price set by the companies buying the crop. Small-scale producers may become dependent on a large, well-organized company that dominates the local infrastructure.

In those ways, biofuels extend the harm already caused by agri-industrial crop production for animal feed, edible oils, fabrics, etc. This link is illustrated by two main examples below: soya in Argentina, and jatropha in Tanzania.

SOYA MONOCULTURES IN LATIN AMERICA: FROM ANIMAL FEED TO BIOFUELS

A genetically modified (GM) herbicide tolerant crop, patented by Monsanto as Roundup Ready (RR) soya, was designed for applying the broad-spectrum herbicide glyphosate. According to its promoters, the crop allows farmers to spray lower quantities of a benign chemical, thus replacing harmful ones, a claim that has been increasingly disputed. In the USA, for example, herbicide applications have increased on RR soya (Benbrook, 2004). Moreover, the crop-herbicide combination encourages farmers to expand monocultures to more land, especially in the global South.

RR soya has been crucial in expanding soya monocultures in Argentina since 1996—covering more than 15 million hectares just a decade later. Often sprayed from small airplanes or large trucks, herbicide is applied to remove weeds and "volunteer" crops from previous rotations. Large areas are cultivated by direct-drilling machines which apply fertilizer, seed, and pesticide in a single trip. This upscales and simplifies the farming process, often reducing the farmer's need for labor. On their own criteria, these systems have had some success in mass production of a single crop, benefiting some large-scale producers.

Indeed, soya monocultures have caused significant harm to rural communities, local food production, biodiversity, livelihoods, and land access. Farmers are caught between high input costs and low prices for commodities. Land prices and debts have risen. In addition to these difficulties, threats and actual violence have driven people off their land, many fleeing to urban slums. (This section draws on Altieri and Pengue, 2006; Barnett, 2004; Benbrook, 2005; EcoNexus/GRR, 2005; and GRR, 2005.)

General prosperity and nutrition have declined. Mixed farming in Argentina once produced a wide range of staple food products and provided incomes for rural communities. Mechanization and monoculture have greatly reduced the number of jobs. Milk and other foodstuffs now have to be imported into a country that used to produce ten times its own food needs. Hunger and malnutrition have been reported from some regions. Diverse nutritious food production has been marginalized by soya; attempts to replace meat with soya have caused health problems among the urban poor, as Argentine soybeans reportedly contain less protein and amino acids than those from the US, China, and Brazil (Karr-Lilienthal et al., 2004).

Aerial herbicide spraying harms communities. Generations of families, their animals, and crops are made ill—resulting in skin, respiratory, and digestive ailments, and cancers. There is generally no warning and no escape from the spraying. Crops and local biodiversity are lost. Protests are often met with violence.

Forests have been seriously depleted in Argentina; the Chaco Forest previously survived a century of smallholder farming, but large areas had been removed for GM soya by 2004. This removal has led to lower rainfall, more flooding, local climate change and losses of unique biodiversity. Diseases such as leishmaniasis (infecting the skin) have increased in some areas of intense deforestation.

Pest problems have also emerged in the monocultures themselves. The application of huge amounts of a single herbicide induces herbicide tolerance in weeds. By 2002, this had already been recorded in about twelve common weeds in Argentina. As a result of such tolerance, additional herbicides such as atrazine and paraquat are being used to clear the weeds after the harvest.

Likewise, soya monocultures have become vulnerable to disease attacks, such as Asian rust, which has been active in Argentina since 2001. *Fusarium* fungus has also become a threat, requiring farmers to apply fungicides with different equipment and methods. As these problems exemplify, "Excessive reliance on a single agricultural technology, like RR soybeans, sets the stage for pest and environmental problems that can erode system performance and profitability" (Benbrook, 2005).

Soil quality and water resources have also been adversely impacted: After more than a decade of agri-industrial production, often without rotation, soil nutrients need to be replaced and soil structure has been damaged, especially by compaction; glyphosate has adverse impacts on earthworms; yields are not increasing; any further growth in production takes place at the expense of forests, soil quality, and communities that depend on these resources; fertilizer requires energy to produce, some 2 percent globally, and its usage generates N_2O emissions, which may counteract any benefit from biofuel replacing fossil fuels (Crutzen et al., 2007).

In all those ways, RR soya in Argentina has undermined sustainable crop production. Diverse, productive farming systems have been reduced to monocultures, thus adversely affecting biodiversity, crop protection, human health, and rural welfare. Doubts are now being cast on the quality of the crop itself.

Argentina's expansion of RR soya had a political-economic driver, namely that the Menem government undertook a privatization campaign that tripled Argentina's enormous national debt in 1989–99. In parallel, it subsidized investment in facilities for grain transport from agri-industrial areas to ports, as well as for container shipping. Under this government, Monsanto was granted the license to commercialize RR soya. Most soya production there is exported to earn foreign currency in order to service the national debt, especially under political pressure from creditors.

In recent years, biodiesel demand has further driven expansion of agri-industrial soya cultivation in Argentina. It provides a supplementary market for the oil, complementing the animal feed market for the cake:

> Soya biodiesel is not a business to be carried out on a small scale, as cost, running the machinery, the distribution of the forage cake by-product, the size and cost of the overseas freight for exporters, means that the industry can only be taken on by large businesses. As the main producer of soya oil, Argentina is in prime position to satisfy internal and external demand (Semino, 2007: 3).

Soya cultivation has been expanded especially for export, generating conflicts over resources, especially over the commons used by indigenous groups (Valente, 2005).

For a long time in the Southern cone countries, government policies for agricultural mechanization have favored a technology paradigm inappropriate for small-scale family farms. This sector has been thereby marginalized, which has resulted in a rural exodus to urban areas, a trend that is aggravated by biofuel development (Wilkinson, 1997: 40).

JATROPHA IN TANZANIA: CONFLICTING MODELS

As an exception to the focus on food crops, jatropha, traditionally used as hedging to protect fields from livestock, is a poisonous crop being strongly promoted for biofuels. It can grow in marginal areas with little water, and oil from its seeds is used to produce soap and for many other traditional uses, including cooking and lighting. Once the jatropha establish themselves and fertilize the soil, their shade can be used for intercropping vegetables (such as red and green peppers, and tomatoes) for at least the first two years, which can provide additional income for the farmers (Becker and Francis, 2003). Thus livelihoods could improve through land restoration associated with other crops alongside jatropha, according to this scenario.

Proponents of jatropha claim that its use for fuel will not divert resources from food production, yet large-scale jatropha cultivation is already generating conflicts in Africa and Asia. Yields are greatly increased when the jatropha is grown on fertile soil and with more water, so its cultivation may undermine water resources for other uses. Appropriation of fertile land for large-scale jatropha cultivation also displaces food crops, thus intensifying competition for land use among crops and among farmers. As a biofuel source, jatropha poses a stark choice—small-scale production for local needs versus agri-industrial production for global markets.

This choice intersects with wider conflicts over land tenure and local access to commons in traditional forms, which still prevail in many parts of Africa. Financial incentives encourage property claims and new debt burdens, both of which subordinate local production to global markets.

Tanzania faces such a conflict over jatropha. According to an academic study of jatropha prospects there, infrastructural aspects "such as transport, reliable and efficient equipment and its maintenance, and financial support, are seen as important barriers and uncertainties" (van Eijck and Romijn 2008: 322). Conversely, more efficient infrastructure would provide incentives for agri-industrial systems.

> There is indeed a danger that if investment in Jatropha does begin to take off in earnest, the sector could be taken over by big commercial players interested in setting up large plantations. In this scenario, less glamorous but socially useful small-scale projects aimed at energy provision by and for local communities could lose out.

> An influx of large investors could also lead to undesirable competition with food crops. Although Jatropha can grow in hostile conditions, there is increasing evidence that seed yields are sensitive to soil fertility and water availability. Farmers could be induced to become outgrowers for large buyers, converting too much prime crop land to Jatropha cultivation. Poor villagers could also be induced to sell their land to large investors, while it is still unclear whether their short-term gains would constitute adequate compensation for long-term loss of livelihoods and loss of land for food production (Ibid: 324).

In Tanzania's Kisarawe coastal area, for example, biofuels development has been led by UK firm Sun Biofuels PLC, with support from the Tanzanian government (African Press Agency, 2007). Thousands of peasant farmers have been displaced from well-watered, highly-populated land to make way for a jatropha biofuels project, which has appropriated large tracts of fertile land, much previously used or suitable for food production (Edwin, 2007; *The Citizen*, 2008).

Biofuels development threatens other uses of common lands, such as, for example, in a Kisarawe village whose land has been taken over for biofuels:

> Although uncultivated, the land is used by the villagers of Mtamba, principally for charcoal-making, firewood, and collecting fruits, nuts, and herbs … it includes a waterhole which is the only place that they can collect water when it is dry. They also collect clay there to build houses (Bailey, 2008: 22).

Villagers have no formal guarantees for access to the land being appropriated, nor for local employment by biofuels projects. In such ways, prospects for biofuels are causing tensions between requirements of investors and of communities (ibid: 22).

With government consent, foreign companies have been buying up good-quality, well-watered land for jatropha in Tanzania. They have been allowed to develop jatropha according to their priorities, i.e. to produce as much as possible under the best conditions, as soon as possible, for a global market. The drive for higher yield pushes farmers off good-quality land, where they formerly produced food and/or fulfilled other local needs. Such land is being treated as if it were "marginal," as a basis to enclose commons and dispossess communities. Following protest from various critics, the government has recently taken a more cautious approach to changes in land use for biofuels (McGregor, 2008). Tanzania illustrates wider tensions: the search for global markets and foreign investment undermines communities' access to local resources, increases dependence upon unstable global prices, and potentially dispossesses producers.

MORE SUSTAINABLE BIOFUELS?

Amidst controversy over harm in the global South, biofuel proponents emphasize a remedy in technological innovation, especially next-generation crops that would be more efficient and therefore sustainable. An "integrated biorefinery" is being designed for more flexibly processing diverse biomass sources into fuel, feed and/or other industrial products. Crop research seeks genetic changes that can enhance bioenergy extraction—like, for example, integrating energy with feed production, increasing the productive efficiency of crops, broadening their geographical range to "marginal" land, or processing "waste" material. Newly useful resources are called "marginal" or "waste," as if they had no other societal uses; informal uses of resources for local needs are rendered invisible—and readily dispensable.

In such ways, novel biofuels are expected to avoid the sustainability problems of current biofuels. Such ambitious expectations depend on many socio-economic assumptions, including that:

- those problems are caused mainly by inadequate crops, low yield, low productivity, etc.;

- crop production serves finite needs within a given geospatial unit, such that greater yield or production reduces competition among different uses (food, feed, fuel, etc.);

- more efficient resource usage, e.g. by co-producing animal feed with fuel, will enhance sustainable production; and

- biomass "waste" and "marginal land" have no uses other than biofuel production.

Those expectations also depend upon GM crops as a means to address inherent problems of agri-industrial monoculture. Success would depend on further assumptions that:

- genetic modification can increase yield without requiring additional inputs such as agrichemicals; and

- agronomic genes (e.g. herbicide tolerance) can be stacked in ways that reduce losses from weeds or insect pests, while avoiding a further treadmill of pest resistance that would require further genetic and/or chemical solutions.

Such assumptions are contradicted by recent experiences of similar problems. Yield increases have generally depended upon agri-industrial methods and commodity inputs, which eventually generate pest resistance, thus aggravating farmer dependence upon input suppliers (see Table 1).

Most acute in the global South, agri-environmental unsustainability and land-use competition result mainly from industrial monocultural systems—using just a few crop varieties to produce standard commodities for global markets. Development of integrated biorefineries is driven by economic-political forces similar to those that have already caused harm. Likewise, most GM crops are designed for industrial monoculture systems; both are driven mainly by private interests which increasingly converge.

If technically successful, such crops would provide greater financial incentives for shifts towards agri-industrial monocropping, which would aggravate the political-economic pressures that displace food production in favor of animal feed and biofuel. Moreover, GM seeds and agrochemicals take the place of farmers' knowledge and expertise, thus further limiting community use and control of local resources. This effort is part of a long history of multi-national corporations colonizing "a multitude of new spaces that could not previously be colonized, either because the technology or the legal rights were not available" (Paul and Steinbrecher, 2003: 228–29).

Moreover, industrial monocultures threaten actual or potential alternatives that could sustainably fulfill local needs, sometimes including biofuels. Such alternatives provide many kinds of commons: access to land; use of local resources that help to avoid debt traps; forests providing diverse resources such as food, firewood, and grazing areas; the right to save, exchange, breed, and re-use seed; control over what to grow; direct sales to an open local market. Alternatives also provide many benefits, such as:

- Small-scale, locally-focused, diverse agriculture already fulfils nutritional and livelihood needs for hundreds of millions of people in the global South.

- Poly-cultures such as inter-cropping produce a wider range and greater quantity of useful bio-material for local populations than monocultures do.

- Given their biodiversity, such systems may be more resilient to shifts in climate and water resources than large-scale monocultures. For example, push-pull systems of biological control have successfully countered maize pests.

- Farmers develop and exchange their own crop varieties, building in diverse characteristics that can respond to new threats such as climate change, pests or diseases.

- Such systems could produce biofuel in ways that are less socially and environmentally harmful, while still giving priority to local food needs, thus minimizing competition for land use.

	Assumptions	Experience
Land use/ Political economy	Competition for land use results mainly from low yield and so could be avoided by more productive crops, a broader geographical range or more efficient energy-extraction.	Sustainability problems and land-use competition arise from the drive for intensive monoculture feeding global markets. Greater efficiency increases the financial incentives for a shift to agri-industrial monoculture systems.
Markets	Greater production can alleviate competition among diverse uses of biomass, e.g. food versus fuel.	Biofuel crop expansion responds to global markets, whose high prices and greater demands readily consume any extra production or yield, especially given the global linkage between feed and fuel markets.
Diverse uses of biomass	With novel GM crops, plant residues can be used more efficiently by deriving many industrial products from the same biomass, thus minimizing waste and enhancing sustainability.	More efficient use increases economic incentives for monocultural systems to supply biomass for 'biorefineries', feeding global markets for various industrial products, thus displacing local food needs.
'Marginal' land	Crops designed for stress tolerances would have a broader geographical range. Cultivation on marginal land would avoid conflict with food needs.	Agri-industrial production has invaded common land previously used for cultivation or grazing. Such land is called 'marginal', meaning that it had not added value to global markets.
'Waste' biomass	GM techniques can alter plant residues from agricultural fields for more efficient breakdown into biofuel, thus using waste biomass (which otherwise has no use).	So-called 'waste' biomass is essential for soil fertility and nutrients. If removed in large quantities, then this biomass would have to be replaced by chemical fertilizers, whose usage causes direct and indirect harm.

Agronomy	GM herbicide-tolerance/Bt traits help to increase yield by better controlling weeds and pests, thus enhancing agri-environmental sustainability.	Higher yield from current GM crops depends upon intensive inputs – e.g., fertilizers, aerial herbicide sprays – thus polluting soil and water, while generating new pests. Such crops provide a financial incentive for shifting land to agri-industrial systems, not for replacing previous chemical-intensive methods of weed/pest control.
Livelihoods	Through greater efficiency, GM biofuel crops will help to enhance rural livelihoods, especially through greater opportunities for export from the global South.	More efficient/productive crops strengthen the financial incentive to remove common land from other economic uses, while also disciplining, exploiting or even removing labour within new industrial production methods.

Table 1: Novel plants for sustainable biofuel production? Optimistic assumptions versus experience

The above agricultural practices are intensive and efficient in different ways than monocultures that produce a standard commodity. This difference highlights a systemic conflict between local human needs versus "efficiency" for global commodity markets—i.e., between antagonistic accounts of sustainability. Agrofuels create dependence on foreign companies that determine the terms of production and trade.

These issues have become contentious in the EU, in particular, because its ambitious targets for biofuel usage could only be achieved through substantial imports from the global South. As a way to legitimize the targets, the Commission proposed sustainability criteria for any fuels to qualify. Some critics proposed mandatory certification that would include, in its criteria, displacement effects and societal harm, but they were excluded from the 2008 proposal for a Renewables Directive, which narrowly defines the relevant environmental harm and allows economic operators in the fuel chain to arrange their own systems to certify relevant information. Moreover, the Commission shall ensure "that the provision of that information does not represent an excessive administrative burden for operators in general or for smallholder farmers, producer organizations and cooperatives in particular" (CEC, 2008). These terms were accepted by the Council and Parliament. In the name of favoring small-scale producers, the rules favor minimal information and obscure dispossession.

DRIVING DISPOSSESSION, PRE-EMPTING SUSTAINABLE PRODUCTION

Biofuel crops have caused agri-environmental damage, land seizures, higher food prices and competition between food and fuel in the global South. Such harm is widely portrayed as contingent "side-effects," yet they are a result of extending crop monocultures to more land to produce standard commodities for global markets, within an agri-industrial development trajectory.

Even before the rise of a global biofuels market, monocultures were producing crops for animal feed or edible oils. Their production caused systemic harm—e.g. competition for land use, higher land and food prices, labor exploitation, insecure employment, greater agrichemical usage, etc. Various new enclosures deprived rural communities of control over human and natural resources.

These patterns have been extended by agri-industrial production of a few crops—initially for animal feed and edible oils, and, more recently, for biofuels, which opponents call "agrofuels." Through horizontal and vertical integration across the agri-feed fuel chain, industry aims to process the same harvest into multiple industrial commodities as co-products of biofuels. Renewable biological resources are equated with sustainability, understood as an input-output efficiency of resource usage for producing standard commodities.

These agri-industrial systems are driven by the rise of a global integrated bio-fuel network. New markets integrate global capital and states, especially those in the global South that seek foreign investment and export income. As such, these states have a weaker capacity or incentive to protect general livelihoods and environments from agri-industrial development (Mol, 2007).

In the global South, political and economic elites accommodate forces that dis-possess communities of resources, through old and new types of enclosure. Support comes from relatively large landholders, from some smallholders (initially) expecting to gain from contract farming, as well as others seeking access to hitherto common land. Given these incentives and pressures, governments ally with foreign interests against their own people.

Consequently, agrofuels have been thrown onto the fire of pervasive conflicts over land use. Land itself has become a focus for speculative capital, which has been fleeing property and avoiding commodity price falls. Agrofuels exemplify how capital accumulation by dispossession has been extended, within and without the capitalist labor process, since the original "primitive" accumulation of capital.

As an agri-industrial system, agrofuel production links several types of enclo-sures: degraded labor conditions, labor subordination via contract farming, appro-priation of land (often through expulsion), loss of control over production, envi-ronmental degradation through agrichemicals, competition for land use, property rights over seeds, etc. Land is turned into private property, which then operates as capital in global markets. Violence plays several roles in enforcing this change—dispossessing communities from land and labor rights, as well as responding to environmental degradation and resource competition. Moreover, agrofuels expan-sion undermines alternative agricultural systems, the various commons on which they depend, community access to such commons, and community loyalties that sustain them.

In all those ways, the agrofuels project attempts to solve energy problems by intensifying the exploitation of human and natural resources. This means remov-ing resources essential for biodiversity and local needs, thus enclosing commons of many kinds. Agrofuel expansion is driven by attempts to sustain global markets for standard commodities and energy supplies for the global North. Prospects for soci-etal progress lie in collective resistance—which warrants its own study.

REFERENCES

African Press Agency (2007). "Thousands of Tanzanian Peasants to be Displaced for Biofuel Farm," 12 August, http://pacbiofuel.blogspot.com/2007/08/pbn-thousands-of-tanzanian-peasants-to.html, http://tech.groups.yahoo.com/group/biofuelwatch/message/855

Altieri, M. and Pengue, W. (2006). "GM Soybean: Latin America's New Colonizer," *Seedling*, January, http://www.grain.org/seedling/?id=421

Angleson, A. and Kaimowitz, D. (2001). *Agricultural Technologies and Deforestation*. London: CAB.

Bailey, R. (2008). *Another Inconvenient Truth: How Biofuel Policies are Deepening Poverty and Accelerating Climate Change*. London: Oxfam.

Barbara, J. (2007). *The False Promise of Biofuels*, International Forum on Globalization and the Institute for Policy Studies, http://www.ifg.org/pdf/biofuels.pdf

Barnett, A. (2004). "Soya Not Only Destroys Forests and Small Farmers—It Can Also Be Bad For Your Health," Food Magazine, *The Observer* (UK), November: 31–37.

Becker, K. and Francis, G. (2003). "Bio-diesel from Jatropha plantations on degraded land," Department of Aquaculture Systems and Animal Nutrition, University of Hohenheim.

Benbrook, C. (2004). *Impacts of Genetically Engineered Crops on Pesticide Use in the United States: The First Eight Years*, BioTech InfoNet, Technical Paper Number 7, http://www.biotech-info.net/Full_version_first_nine.pdf

Benbrook, C. (2005). Rust, resistance, run down soil, and rising costs: Problems facing soybean producers in Argentina, *AgBioTech InfoNet*, Technical Paper Number 8, http://www.greenpeace.org/raw/content/belgium/nl/press/reports/rust-resistance-run-down-soi.pdf

The Citizen (2008). "Tanzania: Government on Spot over Biofuel Production," 23 July, http://allafrica.com/stories/200807240051.html

CEC (2008). "Proposal for a Directive of the European Parliament and of the Council on the Promotion of the Use of Energy from Renewable Sources." Commission of the European Communities.

Cotula, L., Dyer, N., and Vermeulen, S. (2008). *Fuelling Exclusion? The Biofuels Boom and Poor People's Access to Land*, London: IIED.

De Angelis, M. (2004) "Separating the Doing and the Deed: Capital and the Continuous Character of Enclosures," *Historical Materialism* 12(2): 57–87.

EAC (2008) *Are Biofuels Sustainable?* London: Environmental Audit Committee, House of Commons.

Econexus et al. (2007). *Agrofuels: Towards a Reality Check in Nine Key Areas*, http://www.econexus.info/pdf/Agrofuels.pdf

EcoNexus/GRR (2005). *Argentina: A Case Study on the Impact of Genetically Engineered Soya—How Producing RR Soya is Destroying the Food Security and Sovereignty of Argentina*, EcoNexus (UK) and Grupo de Reflexion Rural (Argentina), http://www.econexus.info/publications, http://www.grr.org.ar

Edwin, W. (2007) "Tanzania: UK firm invests $20m in biofuel farm," *The East African*, 7 August, http://allafrica.com/stories/200708070929.html

EEA (2008). "Suspend 10 percent Biofuels Target, Says EEA's Scientific Advisory Body," 10 April, http://www.eea.europa.eu/highlights/suspend-10-percent-biofuels-target-says-eeas-scientific-advisory-body

FAO Media (2007). "Climate Change Likely to Increase Risk of Hunger," 7 August, http://www.fao.org/newsroom/en/news/2007/1000646/index.html

GRR (2005). *Iguazu CounterConference Against Greenwashing of the Soy Industry*, Grupo de Reflexion Rural, 16–18 March, http://www.econexus.info/publications, http://www.grr.org.ar

Harvey, D. (2003). *The New Imperialism*. Oxford University Press.

Hecht, S. (2004). "Invisible Forests: The Political Ecology of Forest Resurgence in El Salvador," in R.Peet and M.Watts, eds, *Liberation Ecologies*, pp. 64–103. London: Routledge.

McGregor, S. (2008). "Tanzania to Draw Up Interim Guidelines on Biofuel Production," 14 August, Bloomberg L.P.

Marx, K. (1976). *Capital, Volume 1*, Chapter 26)London: Penguin, Pelican Marx Library)

Medick, H. (1981). "The proto-industrial family economy and the structures and functions of population development under the proto-industrial system," in P. Kriedte et al. (eds), *Industrialization before Industrialization*. Cambridge Univ. Press.

Mol, A. (2007). "Boundless biofuels? Between environmental sustainability and vulnerability," *Sociologia Ruralis* 47(4): 297–314.

Oxfam (2007). *Biofuelling Poverty: Why the EU Renewable-Fuel Target May Be Disastrous for Poor People*, http://www.oxfam.org.uk/resources/policy/trade/bn_biofuels.html

Paul, H. and Steinbrecher, R. (2003). *Hungry Corporations: Transnational Biotech Companies Colonise the Food Chain*, London: Zed Books.

Polanyi, K. (1944). *The Great Transformation*. Boston: Beacon Press.

RFA (2008). *The Gallagher Review of the Indirect Effects of Biofuels Production*, London: Renewable Fuels Agency.

Semino, S (2007). "Future Perspectives of the Soya Agribusiness: Biodiesel, the New Market," in J.Rulli, ed., *United Soy Republics: The truth about soy production in South America*. GRR, http://www.lasojamata.org/?q=node/91

Valente, M. (2005). "Argentina: The Environmental Costs of Biofuel," http://ipsnews.net/news.asp?idnews=32959

Van Eijck, J. and Romijn, H. (2008). "Prospects for Jatropha Biofuels in Tanzania: An Analysis with Strategic Niche Management," *Energy Policy* 36: 311–25.

Wilkinson, J. (1997) "Regional Integration and the Family Farm in the Mercosul Countries," in D.Goodman and M.J.Watts, eds, *Globalising Food*, pp. 35–55.

Chapter 39 ∎ Part 10

BRAZIL AS AN EMERGENT POWER GIANT: THE "ETHANOL ALLIANCE"[1]

Camila Moreno

In the new global geopolitics defined by agrofuels and the land grab associated with it, Brazil has played a central role. The Brazilian experiment with biofuels—particularly sugar cane ethanol—is promoted as a global model for sustainable biomass production, but the model is being widely criticized and opposed by the country's social movements and civil society.

Brazil is referred to throughout the world as "the" model of non-food-crop success and competitiveness. Where they stand out is in their more than three decades of domestic use of ethanol fuel, and in the availability of land in what is considered the largest agricultural frontier of tropical arable land already served with infrastructure (financed in great part with generous public money) to serve corporate agribusiness interests. A nationwide effort is being channeled into creating and attracting investments (roads, waterways, dams, irrigation systems, pipelines, refinery plants, tanks, ports, and also credit, research, education, etc.) to guarantee leadership in the new agroenergy era. Petrobras, the Brazilian oil company, plans to lead ethanol export activities in the country, shipping out around 5 billion liters a year by 2012.[2] Bilateral contracts, as with Japan, for example, are promoting the use of ethanol worldwide in a series of industrial and chemical processes.

The leadership of an influential southern country, within the so-called BRIC group (Brazil, Russia, India, and China), emerging as a giant power in the region and becoming a world leading supplier of a new strategic resource can be analyzed on many levels. On the regional perspective, Brazilian expertise and leadership in ethanol has a clear geopolitical counterpart and serious implications for the Latin American region, stemming from what has been dubbed the Brazil-US "ethanol alliance."

The Brazilian government's tireless efforts to establish an international market for ethanol would result in the first internationally-traded agrofuel (currently no official market exists to determine prices and regulations, and global trade of agrofuels has been carried out on the basis of contracts.)[3] The goal of leading this new global market with the first generation of tradable agroenergy commodities, and expanding

---

1   This article is a shorter and modified version of: Moreno, C. & Mittal, A. *Food and Energy Sovereignty Now: Brazilian Grassroots Position on Agroenergy*. March 2008, http://oaklandinstitute.org/pdfs/biofuels_report.pdf

2   http://noticias.terra.com.br/interna/0,OI3364613-EI8177,00.html

3   http://www.ietha.org/ethanol/

its influence and control on the transition to second generation agrofuels (cellulosic ethanol) has been defined as a matter of State. Throughout the world, Brazil is recognized as a country uniquely qualified to lead other developing countries into agrofuels production, exporting knowledge and technology, especially throughout Latin American and Africa. However, we will argue, Brazil is not acting alone in this.

BRAZIL: AN EMERGING POWER GIANT

> Brazil and the United States account for approximately 70 percent of global production of biofuels. Our two countries can and must lead in these areas.[4]

Although Brazil has the largest share of world agrofuels production, with the US, Brazil is the global leader in ethanol exports, and, in 2006, supplied 70 percent of the world's demand. In the 2006/2007 harvest, the production of sugar-cane ethanol was 17.8 billion liters—3.4 billion of which were exported (19 percent of total production) with 56.2 percent being exported to the US, despite the imposed tariff of $0.14 US per liter ($0.54 per gallon). In 2008, Brazil increased production to 26.6 billion liters, an increase of 15.6 percent. This was a historical landmark, in which 5.6 billion liters (21 percent of the production) were exported, 2.8 billion liters of which went to the USA, the largest importer country. However, a considerable part of the Brazilian ethanol exports reaches the United States through the central American corridor, using Caribbean countries as platforms for fiscal reasons, such as Brazilian interests expanding sugar cane production in the region.

In addition, Brazil announced in late 2007, the discovery of a massive offshore oil basin reserve, known as the "pre-salt" (*pré-sal*). This finding puts the country's oil and gas reserves amongst the world's ten largest, turning it into a net oil exporter and the largest proved oil reserve of any non-OPEC country. Discovery of this oil field could boost Brazil's overall reserves and has raised speculation of further discoveries in its largely unexplored offshore oil and gas basins. Initially, Petrobras executives confirmed recoverable reserves of between 5 to 8 billion barrels of oil and gas,[5] but more recent estimates claim around *50 billion barrels* of proved offshore oil and gas reserves.[6] In the context of increasing international demand for oil and gas, led by the growing economies of China and India and accompanied by peaking of production, a decline in output where production has already peaked, and a high price, this latest discovery introduces a new element in the balance of power in any international negotiations that Brazil will engage in from now on, just as it also redefines the terms of previous negotiations.

---

4   Christopher McMullen, Deputy Assistant Secretary for Western Hemisphere Affairs. "US-Brazil Relations: Forging a Strategic Partnership," Remarks to the Brazil-US Business Council, Washington, DC, October 17, 2007. http://www.state.gov/p/wha/rls/rm/07/q4/94355.htm

5   Petrobras executives have said that production can be expected for the period 2012–2013, though many industry analysts say the field's peak production will not occur before 2020. Brazil's big oil find was listed under the "Ten Most Underreported Stories in 2007" by *Times* magazine. December 24, 2007, p. 42.

6   http://www.agenciabrasil.gov.br/noticias/2008/11/07/materia.2008-11-07.4706281756/view

Regionally, this find will bolster the US energy partnership with Brazil, leveraging against the "leftist" governments of other oil rich countries in the continent—Venezuela, Ecuador, and Bolivia—whose political agendas of resource nationalism are not aligned with that of the United States. The "ethanol alliance" suits US interests in isolating Chavéz's influence on the region, strengthening Brazilian leadership in the continent. Meanwhile, President Luiz Inacio Lula da Silva has been emphatic in denying that the offshore oil and gas reserve would alter Brazil's agrofuels policy. He stated that a diversified energy matrix is of utmost importance in the current times and all we could want, as it provides energy security and gives maximum "bargaining power to our country in negotiating its proper position in the new global energy scenario."[7]

BRAZIL AND THE UNITED STATES FORGE AN "ETHANOL ALLIANCE"

Brazilian President Lula's visit to Camp David in March 2007 sealed what has been dubbed the "ethanol alliance" between the United States and Brazil.[8] Aimed at promoting greater cooperation between the two countries on ethanol and biofuels, the agreement promotes a bilateral partnership on research/development; promotes the biofuels industry through feasibility studies and technical assistance; and creates a world commodity market for biofuels through greater compatibility of standards and codes.

At Camp David, both presidents discussed the reduction of agricultural subsidies, which has been the main impediment to the conclusion of the Doha Round of the World Trade Organization (WTO), as well as international standards for foreign trade in ethanol, a technical step to define it as the first commodity in the emerging agroenergy global market.

ETHANOL, ENERGY, AND CLIMATE CHANGE POLITICS

Set in the broader context of a great business opportunity and maintaining US hegemony in the region, the politics of ethanol are creating a solid base for its future by manufacturing favorable public opinion. A starting point for this has been the smoothing over of differences between the Left and the Right—for instance the unusual affinity between Presidents Lula and Bush, despite their ideological differences.

The "ethanol alliance" was able to overcome ideological opposition between the two heads of State—Lula being the furthest left politician ever elected president of Brazil, and Bush, one of the most conservative presidents in recent US history. Their

---

7   Reuters, "Brasil irá investir US $500 milhões em submarino nuclear," October 27, 2007. http://www.alertnet.org/thenews/newsdesk/N03232022.htm

8   "Memorandum of Understanding Between the United States and Brazil to Advance Cooperation on Biofuels," US Department of State, Office of the Spokesman, March 9, 2007. http://www.state.gov/r/pa/prs/ps/2007/mar/81607.htm. Also see "Joint Statement on the Occasion of the Visit by President Luiz Inácio Lula da Silva to Camp David," White House Press Release, March 31, 2007. http://www.state.gov/p/wha/rls/prsrl/07/q1/82519.htm

opposing political views would not get in the way of this new "energy cooperation."[9] This partnership between Brazil and the United States sheds light on how the politics of energy/climate change is defining a new political frontier of our times, and diminishing prior ideological constraints, as if the production of energy had nothing to do with the society that it will be used for.

A clear sign of this affinity, and the emerging tropical leadership, was the announcement, in late June, of Brazil's reengagement in its nuclear energy program. As President Lula framed it: "Brazil can afford the luxury of becoming one of the few countries in the world to master the entire uranium enrichment cycle and, from there, I think we will have far greater esteem as a nation."[10]

Brazil has some of the world's largest uranium reserves, but this announcement did not result in any threat from the US military, and was clearly not a cause for much concern, given that Brazil has historically been a friend and has a record of a good neighbor policy with the United States.

NEW ENERGY DEALS IN A SHIFTING WORLD ORDER

On October 9, 2007, the US Congress unanimously approved bipartisan House Resolution 651 HI, stating that, following the oil shock of the early 1970s, Brazil reduced its energy vulnerability by diversifying its energy sector through sugar-based ethanol. The centerpiece being "cooperation on biofuels," the resolution urges strengthening a strategic partnership between both countries, praising the leadership of Brazil as "decisive," not only as a regional leader, but as a global partner.[11] The Resolution recognizes the strategic relationship between the United States and Brazil, and the wider meaning and importance of the Memorandum of Understanding on biofuels cooperation signed between the two countries in March 2007:

> For years, Brazil has flown *below the radar* in the United States. We never paid much attention to what was happening in the largest country in South America ... we are reaching the end of this period of ignorance and neglect and that we, in America, are finally *waking up* not only to Brazil's importance, but also to how *natural this relationship should be*. Outside of the United States, Brazil is the largest democracy in the hemisphere, Secretary of State Condoleezza Rice has called Brazil "the regional leader and our global partner."[12] (emphasis added)

The use of the military metaphor, "flown below the radar," leaves no doubt that the US conceives of biofuels/ethanol politics as part of its larger Energy Security strategy, aimed at reducing dependence on foreign oil and gas reserves. Even though non-fossil or clean sources of energy are to be progressively introduced (either in a

---

9   Public hearing on United States-Brazil Relations—Rayburn House Office Building, testimony by Paulo Sotero, Director of the Brazil Institute of the Woodrow Wilson International Center for Scholars September 19, 2007.

10   Reuters, "Brazil to Invest $500 Mln. in Nuclear-Powered Sub," July 10, 2007. http://www.alert-net.org/thenews/newsdesk/N03232022.htm

11   House Resolution 651 IH, October 9, 2007. http://www.state.gov/p/wha/rls/rm/07/q4/94355.htm

12   Ibid.

transition forced by oil and gas depletion, escalating costs to pump and transport the remaining reserves, or for the warfare required to further explore fields in certain regions (oil wars)), renewable sources are nonetheless quite far from being an effective substitute for the current dependence on oil, gas, and coal—currently the main energy and raw material matrix of the globalized economy.

A country's agricultural capacity (meaning available arable land and water) to produce biofuels and guarantee a steady supply to international markets, as in the case of Brazil, is becoming an increasingly important factor in negotiating a stronger role in the emerging new world order. Agroenergy fields are already defining a global geopolitical order over southern territories, as part of an "energy security strategy."[13]

BRAZIL, AS A DECISIVE PLAYER

Brazil is paving the way in transforming ethanol into an internationally tradable energy commodity. An improved bilateral relationship is not only necessary and beneficial for Brazilian interests, but US interests as well. The bilateral dialogue is increasingly a two-way street. The United States continues to set the agenda for the international arena; however, Brazil is a decisive player in defining the terms on which that agenda is discussed.[14]

The importance of ethanol as a means to Brazil's rise as a political force in the twenty-first century cannot be understated. Brazil has played a key role in the global promotion of biofuels and the negotiations aimed at developing an international market for ethanol. Even though Brazil's foreign relations policy is based on maneuvering its capacity for biofuel production, for example to gain a permanent seat at the United Nations Security Council, agroenergy is promoted domestically as being "beyond ideology." This has been supported by the oddest political alliances (such as Bush) for the sake of "clean, renewable and thus, peaceful" energy.

The proportion of Brazil's energy matrix met by renewable energy is unmatched. A full 45 percent of the total energy produced and consumed in Brazil comes from non-fossil sources, compared to an average of only 14 percent of renewable sources share in the world energy matrix, and a timid 6 percent average for the OECD countries.[15] Sugar-cane-based fuels (ethanol) and bio-electricity are already second place to oil in the ranking of the largest energy sources in Brazil's energy matrix. Their high moral position on "renewables" comes from the following distribution of energy sources that account for total national supply:

---

13   See "The Geopolitics of Agrofuels," Position paper of the first international meeting of southern organizations to discuss agroenergy and food sovereignty. Quito, Ecuador, June 2007. Available at: http://www.accionecologica.org; in English: http://www.wrm.org.uy/subjects/biofuels/Quito_Manifest.html

14   Brazil's former Ambassador Abdenur, "The Future of US-Brazilian Relations." Meeting held at Woodrow Wilson International Center for Scholars, January 24, 2007. Ambassador Abdenur is a leading member of a generation of diplomats that paved the way towards the opening up of Brazil to the rest of the world during and after the democratic transition of the 1980s … his successful diplomatic career includes serving as Brazilian Ambassador to three of the world's most influential nations—United States, China, and Germany—and as secretary general of Itamaraty, the Brazilian Foreign Ministry.

15   World Resources Institute, 2007.

Oil	36.7 percent
Sugar-Cane (ethanol and co-generation of bio-electricity)	16.0 percent
Wood and other organic feedstocks	12.5 percent
Hydroelectric power plants	14.7 percent
Gas	9.3 percent
Coal	6.2 percent
Other renewables	3.1 percent
Nuclear	1.4 percent

BEN 2008 (National Energy Balance)

What makes Brazil distinct from any other country today is that ethanol/bio-fuels are a whole state project. "Agroenergy" unifies the discourse of several state agencies, from public research to market regulation, under the central coordination of the Chief-of-Staff of the Cabinet, who supervises all Ministries that touch upon the issue. This includes the ministries of agriculture, environment, energy, industry and trade, science and technology. It even includes defense, as *energy* is seen as a matter of *national security*.

To solidify its share in the emerging global clean energy industry, Brazil has adopted quite an aggressive strategy, combining public and private sector interests, part of which includes a strategic regional partnership with the US to its Agroenergy Plan (2006–2011), the most ambitious public policy on agroenergy in the world.[16] The plan was conceived with the goal of consolidating the country's leadership in so-called first generation (biofuels, bio-ethanol, and biodiesel) and to lead the development of second generation cellulose ethanol with important agro-biotech support (seeds and enzymes). Although official figures have been drastically altered since the release of the National Agroenergy Plan (late 2005), from the initial 200 million hectares that were considered "socially acceptable" for the expansion of agroenergy crops (sugar cane, soy, and palm to biodiesel and eucalyptus for energetic forests), down to a more "modest" estimate of 90 million hectares. Out of these, plans are afoot to open 44 million hectares of degraded land just to ensure sugar cane expansion.

ETHANOL AS AN INTEGRATING FORCE IN THE REGION

---

16   The Agroenergy Plan was masterminded by former Minister of Agriculture, Roberto Rodrigues, now with the Interamerican Ethanol Commission. *Plano Nacional de Agroenergia* 2006–2011, p. 51, 2da edição revisada, Ministério da Agricultura, Pecuária e Abastecimento, Secretaria de Produção e Agroenergia. http://www.biodiesel.gov.br/docs/PLANONACIONALDOAGROENERGIA1.pdf.

The Memorandum of Understanding (MOU) signed in March 2007 between the US and Brazil aims to promote greater cooperation on ethanol and biofuels in the Western Hemisphere, including multilateral efforts to advance the development of biofuels in other countries through assistance in building domestic industries. Latin American countries targeted for United States-Brazilian technical assistance and for establishing and/or expanding sugar-cane plantations and mills are the Dominican Republic, El Salvador, Haiti, St. Kitts, and Nevis. According to the official line, the goal is to promote capacity for local production and consumption of biofuels, and to create jobs, reduce dependence on fossil fuels, and spur economic development.[17] However, an examination of the broader forces acting in the region shows that it goes beyond *local* production and consumption.

Brazil's support for ethanol as the driving force of economic and political integration of the Americas is sustained by a determinedly-obstinate President Lula. In addition to the promotion of biofuels, he has also become personally committed to a "visionary" project of unifying the Americas through the sugar cane and biomass industry.

Sharing a similar vision for the region is the non-governmental Interamerican Ethanol Commission.[18] Its members include Jeb Bush (former governor of Florida and brother of former US president George Bush); Roberto Rodrigues, former Brazilian Minister of Agriculture and mentor of the National Agroenergy Plan (prior to joining the government, Rodrigues was the president of the Brazilian Agribusiness Association (ABAG)); and Luis Moreno, president of the Inter-American Development Bank (IDB). The purpose of this Commission is to foster understanding between the public and private sector in order to set specifications and standards, establishing the regulatory framework for the future international ethanol market. The Commission's membership is the true representation of interests behind the ethanol industry and its political leanings.

The Inter-American Development Bank is another actor strongly promoting and financing biofuels production in the region. IDB's April 2007 study, "A Blueprint for Green Energy in the Americas," reports that some Latin American and Caribbean countries have shown "great interest and promise" in the development of biofuels.[19] The IDB study asserts that while the sugar cane harvesting season in Central America is shorter than Brazil's, Costa Rica, El Salvador, and Guatemala have efficient sugar industries and could produce significant sugar-based ethanol. Costa Rica and Guatemala house 44 percent of Central America's ethanol processing factories (2007).

---

17   Christopher McMullen, Deputy Assistant Secretary for Western Hemisphere Affairs, US-Brazil Relations: Forging a Strategic Partnership, Remarks to the Brazil-US Business Council, Washington, DC, October 17, 2007. http://www.state.gov/p/wha/rls/rm/07/q4/94355.htm

18   Other actors who support ethanol/biofuel as a great oportunity for rural development in the region are the regional secretariat of the United Nations Organization for Food and Agriculture (FAO), headed by José Graziano, former minister for the Hunger Zero program; and the Inter-American Institute for Cooperation on Agriculture (IICA), a lead actor in the promotion of the Green Revolution in the region.

19   IDB Report, April 2007.

In the Caribbean, the largest ethanol plants are located in Jamaica and the Dominican Republic. Jamaica has exported the largest amount of ethanol to the United States, most of it reprocessed hydrous ethanol from Brazil.

Benefiting from free trade agreements such as the Caribbean Basin Recovery Act, Caribbean and Central American countries can export ethanol to the US without any tariffs since it does not exceed the agreement's benchmark 7 percent of US domestic production.[20] Under the Dominican Republic-Central America-United States Free Trade Agreement (CAFTA-DR), signatory countries (Costa Rica, the Dominican Republic, Guatemala, Honduras, Nicaragua, and El Salvador) continue to be able to send some share of duty-free exports to the United States under conditions established by the Caribbean Basin Initiative (CBI). Exports from Costa Rica and El Salvador enjoy specific allocations. In the future, these free trade agreements could spur indigenous ethanol production in Central America, which would result in social, economic and environmental problems that are already being experienced elsewhere.

Countries under the Central American Free Trade Agreement, (CAFTA) are the very countries where the IDB is promoting biofuels most strongly: Panama, Honduras, El Salvador, Guatemala, Costa Rica, Dominican Republic, and Nicaragua. Together, these countries account for 700,000 hectares already planted with sugar cane, most of it processed for sugar production. The area under sugar cane cultivation in these countries is expected to jump to 1.05 million hectares, a growth of 50 percent.

On the corporate side, Brazilian agribusiness and industrial conglomerate Dedini has expressed its goal to expand in the Central America and Caribbean region.[21] With cutting-edge expertise in the design and production of industrial infrastructure for the ethanol plants, Dedini is responsible for about 80 percent of national production of ethanol and more than 30 percent of world production, and intends to increase exports, with potential markets in California (via Central America) and the Asian/Pacific region, especially Japan and Korea, and the European Union. Other Brazilian agribusiness groups, supported by foreign investments and regional plans for biofuels, are renting lands, establishing sugar cane fields, and opening new mills in the region. Ethanol, along with heavy infrastructure support (roads, ports, storage tanks, etc.) appears to offer new business opportunities that are being introduced under the rubric of "rural development" programs.[22] Brazil is also helping promote a "biofuel revolution" in Sub-Saharan African countries such as Angola, Mozambique, Burkina Faso, and Congo, among others, by providing technical agricultural assistance. Also, many US organizations and foundations are investing heavily to promote a new Green Revolution for Africa.

Supplying the world-market with renewable energy is becoming the main integrating force in the Americas, quintessentially expressing the terms of the Memorandum of Understanding between the US and Brazil. Catering to the energy security

---

20   Bravo, Elizabeth, "Agrocombustíveis, cultivos energéticos e soberania alimentar na América Latina," *Expressão Popular*, São Paulo, 2007.

21   http://www.dedini.com.br

22   Daño, Elenita, *Unmasking the New Green Revolution in Africa: Motives, Players and Dynamics*, Third World Network and African Center for Biosafety, 2007.

strategy of the United States creates an opportunity for the Brazilian agroindustrial conglomerates to export sugar-cane ethanol and sell technology through this new fuel corridor.

FUELING THE US DEMAND

In January 2007, President Bush announced the US target of reducing petroleum consumption by 20 percent in just ten years, while calling for a seven-fold increase in the current production of over 18 billion liters of ethanol. In the broader context of competitiveness in what is becoming the next global industry—clean power— cellulosic ethanol sets the technological frontier for what is called the "second gen- eration" of biofuels. Commercial availability in the next seven years will rely heavily on the increase of biomass production per hectare, and strong biotech/GMO support for accelerating enzymatic processes, especially fermentation. Cellulosic ethanol re- search, strongly promoted by the new Obama administration, is running to close this gap, specially making use of controversial biotech tools, such as synthetic biology.[23]. The plan for fueling the US energy demand, based around energy security in a near and "green" future relies heavily on biomass dependence for co-generation of fuel and electricity and to boost an entire alcohol-chemical chain in the future (to re- place petrochemical base industries). In other words, just about everything that is produced out of oil and gas will be reproduced from ethylene made out of ethanol. However, this "green" new world depends on securing vast territorial extensions of arable land overseas.

In a transition to a post-oil society, the new agroenergy matrix for the United States' expanded engagement with Brazil is becoming an international issue. Brazil and US confluence on ethanol exemplifies how energy/climate change politics have less to do with environmental concerns than with the emergence of a new power balance and guaranteed energy security. This is likely to have a wider impact in the international arena, as the alliance will come to determine the future world-market for ethanol.

The emergence of an international market of agroenergy commodities, such as ethanol for fuel, and the pressure to ensure supplies, is introducing new corporate actors that are investing heavily in production of agroenergy crops. Since the global adoption of biofuels depends on governmental strategies to ensure "energy security," national biofuel plans and targets are impacting the availability of agricultural land and water worldwide. Despite this impact on global food production and on arable land and water, this agenda is moving ahead without public debate or participation.

The ethanol alliance is an example of how new energy/climate-change politics and cooperation on agrofuels is really about the maintenance of US hegemony and about Brazil's aspirations, supported by its economic and political elites, to become a regional power and a partner of the United States. At the same time, agrofuels are increasing the production of fuel from biomass and consolidating an entirely new

---

23   http://www.etc.group.org/en/issues/synthetic_biology.html

commodity chain of agroenergy products, which is bringing together the strongest corporate sectors: agribusiness and energy.[24]

AGROFUELS AND THE NEW LAND STRUGGLES IN LATIN AMERICA

Within the complex issue of agrofuels and the emerging climate/energy politics the question of land must be considered. The expansion of the agricultural frontier and the change of use of land associated with agrofuel and monoculture production are dynamic and interdependent processes, with wide social and economic impacts, impossible to separate or consider as independent factors.[25]

Brazil's global ethanol leadership and competitiveness requires the burning of sugar cane fields, a grim picture of the plantation of the twenty-first century, where the harshest labor conditions for the migrant labor force coexist with not infrequent occurrences of slave labor. Shocking work conditions—some 500,000 workers who toil from March to November stooped over in the tropical sun harvesting sugar cane to make ethanol—along with emissions from the burning fields are the most widely-publicized effects of the expansion of the plantations.[26]

Expansion of monocultures under the corporate-controlled industrial agricultural system is seen as the main driving force that determines access and control over common natural resources (land, water, forests, biodiversity, oil, gas) and is at the root of nearly all socio-environmental conflicts in Brazil, as well as throughout Latin America. According to the "Resistance to Agribusiness Forum," "The agribusiness model follows the criteria set by the global market, and we are being forced to adopt it as the only means of development and progress for our countries, although it comes with humanitarian and ecological impacts of catastrophic proportions."[27]

Currently, at least 80 percent of Brazilian biodiesel is made out of soy, the increase of which, in Brazil (and neighboring countries of the Mercosur soy complex such as Paraguay, Bolivia and Argentina), is a key culprit of deforestation. The industry is controlled by US corporate giants such as Cargill, Archer Daniels Midland (ADM), and Monsanto. The soy fields are devouring the largest remaining tropical

---

24   According to the United Nations Food and Agriculture Organization (FAO) terminology at its International Bioenergy Platform (IBEP), 2006, p.2: "Bioenergy: energy from biofuels. Biofuel: fuel produced directly or indirectly from biomass such as fuelwood, charcoal, bioethanol, biodiesel, biogas (methane), or biohydrogen. Biomass: material of biological origin excluding material embedded in geological formations and transformed to fossil, such as energy crops, agricultural and forestry wastes and by-products, manure or microbial biomass. Bioenergy includes all wood energy and all agroenergy resources. Wood energy resources are: fuelwood, charcoal, forestry residues, black liquor and any other energy derived from trees. Agroenergy resources are energy crops, i.e. plants purposely grown for energy such as sugar cane, sugar beet, sweet sorghum, maize, palm oil, seed rape and other oilseeds, and various grasses. Other agroenergy resources are agricultural and livestock by-products such as straw, leaves, stalks, husks, shells, manure, droppings and other food and agricultural processing and slaughter by-products." ftp://ftp.fao.org/docrep/fao/009/A0469E/A0469E00.pdf

25   Assis W. & Zucarelli. "De-polluting Doubts: Territorial Impacts of the Expansion of Energy Monocultureshttp," 2007. http://www.natbrasil.org.br/Docs/biocombustiveis/depolluting_doubts.pdf

26   Michael Smith and Carlos Caminada, "Ethanol's Deadly Brew," Bloomberg Markets, November 2007. http://www.bloomberg.com/news/marketsmag/mm_1107_story3.html

27   See complete report on the Resistance to Agribusines Forum, http://www.resistalosagronegocios.info/docs/PoliticalSynthesis-ForumofResistancetoAgribusiness.pdf

forest in the world, and are a root cause of land use change and emissions that worsen climate change. Virgin Amazon forest is cleared to extract and sell valuable tropical wood, followed by burning (and associated damaging emissions) to open new pasture areas for cattle raising—often in former public lands. Brazil's growing role as the world's largest beef producer and exporter is due, in large part, to the illegal and violent dynamics of land acquisition for cattle raising, which are expanding over the Amazon Forest, progressively paved by soy.[28]

Considering the push to expand sugar cane for ethanol and soy for biodiesel, it has been argued that the agrofuels frenzy "is an explosive mixture to industrial monocultures,"[29] characterizing agrofuel production as a major driver of land use change to boost negative environmental, social and economic impacts.

This understanding is especially clear in relation to the southern and tropical countries where the expansion of energetic crops—and the associated land grab—has mainly taken place, usually to supply northern countries' mandatory targets for "renewable" energy. Thus, land acquisition and control emerges as the center of the agroenergy strategy, which implies a new definition of geopolitics that is to secure the international division of labor associated with the production of the new agroenergy commodities, based in the control over arable and mostly tropical lands.

Globally, the land grab associated with the agrofuels boom has risen to something between 15 to 20 million hectares, a figure that only takes into account contracts that were available to the public when portraying the magnitude of the issue was first attempted in 2008.[30] From this perspective, climate and energy politics are inextricably related to a rapid re-configuration of territories, specially affecting the vulnerable populations living on the so called "marginal" lands—that is, land not yet taken by the land market and the industrial agriculture.

Deriving from a family of typical, colonially minded concepts such as *Terra nullius*, or "land belonging to no one," marginal lands in the contemporary imagination play a key role in the discourse of agrofuels promotion, together with its acritical acceptance by public opinion. On the ground, however, land conflicts and territorial disputes related to control of resource-rich territories have recently taken on new dimensions and pose major threats to social movements at the dawn of the "agroenergy era."

The agrofuels "boom" is considered to be a major factor of a "counter land reform" movement taking place, and poses a major obstacle to the still-unfulfilled land distribution and social justice related to a democratic agrarian policy, a condition of any equitable society. On a tide that goes against all the progressive social forces and social movements, agrofuel production and land grabs have been fostering the breakup of rural local economies, displacing peasants and small land owners, and

28   http://issuu.com/greenpeacebrasil/docs/farra

29   Sérgio Schlesinger and Lúcia Ortiz, "Agribusiness and Biofuels: An Explosive Mixture—Impacts of Monoculture Expansion on bioenergy production in Brazil." FBOMS (Brazilian Forum on NGOs and Social Movements for the Environment), 2006. At: http://www.natbrasil.org.br/Docs/biocombustiveis/biocomb_ing.pdf

30   For data and studies on the Land Grab see: http://farmlandgrab.org

concentrating land and political powers associated with territorial control, just as is currently occurring with fossil fuel supplies. In this sense, agrofuel production is clearly defining a new geopolitics, as in the case of the cane ethanol in Latin America, for example.

This framework has guided several analyses from the global south, as expressed and discussed in the "Position Paper of the Global South on Food Sovereignty, Energy Sovereignty and the transition towards a Post-Oil Society." [31] The geopolitical implications of the transition to the agroenergy era—depending structurally on capturing and securing land as the basic condition that makes such a transition possible—cannot be reduced to the pseudo-technical "sustainability criteria," which is the overall frame of the agrofuels debate within, for instance, the European Union so far. It is urgent that public debate around "policy options" addresses the complexity of issues surrounding power relations and structural violence arising from mandatory targets on the increased adoption of agrofuels into domestic markets, considering from the start that they will, by necessity, be produced elsewhere (and on someone else's land).

---

31   "The Geopolitics of Agrofuels." Position Paper of the Global South on Food Sovereignty, Energy Sovereignty and the Transition Towards a Post-Oil Society. Quito, September 2007. http://www.wrm.org.uy/subjects/agrofuels.html

Chapter 40 ▮ Part 10

DYNAMICS OF A SONGFUL RESISTANCE[1]

Tatiana Roa Avendaño and Jessica Toloza

"A single swallow does not necessarily mean that summer is on its way."
—Juan Ventes[2]

Despit the fact that it might appear that our voyage along the length of the South Pacific coast of Colombia came to an end with the final activities in Tumaco, the journey is not over yet. Our "South Pacific Voyage" was a joint initiative of the Process of Black Communities (Proceso de Comunidades Negras, or PCN) and CENSAT Agua Viva, Friends of Earth Colombia (Amigos de la Tierra Colombia), and its goal was to broaden the resistance campaign against agrofuels, called "Filling Tanks, Emptying Territories" (Llenando Tanques, Vaciando Territorios), amongst local communities. Through the debates, discussions, and denunciations arising from the presentations, as well as the warnings about megaprojects that marginalize and bleed the territories, we have been brought face to face with the vestiges of slavery. Such has been the outcome of this campaign for life and freedom in the context of today's marginalization. Like migratory birds, we made our way from port to port, listening to tales of a pained world, aware that the confirmation of the story lay in the lives of the protagonists: peasant men and women. These are the downtrodden victims of injustice, yet they are nonetheless alive with happiness. Together, we built a fraternal fire and shared a small artisanal boat in which we ate together as equals and gently sung ourselves into dissonance. Despite our diverse places of origin (Buenaventura, Bogota, Bahía Málaga, Ladrilleros, Cali, Sala Onda, Guapi, Timbiqui, and Tumaco) and our different professions, we made the journey together in a familial and fraternal spirit. Combining visions and dreams for a single cause, we reclaimed the word, recounting the outrages and injustices of a capitalism whose discriminatory policies and practices are devastating the African population and banishing them from their own territories. Capitalism which, according to Bolívar Echeverría, "implies the alienation of the human subject, and the erosion of its ability to reproduce itself and generate its own ways of being."[3]

This "Pacific" trip through the region began in the Puerto de Buenaventura on 28 September 2007 and ended in Tumaco on the 8th of October. The journey exposed

---

1   This article was published previously in *Agrocombustibles: Llenando Tanques, Vaciando Territorios* (Agrofuels: Filling Tanks, Emptying Territories), published by Censat Agua Viva and the Proceso de Comunidades Negras (Process of Black Communities), Bogotá, Colombia, 2008. It is being reproduced here with permission from Censat and the Author. It was translated from the Spanish by Kolya Abramsky, with assistance from Claudia Roa and Adam Rankin.

2   Member of the South Pacific Voyage, old sailor and peasant from Guapi (Cauca).

3   Bolívar Echeverría. Cultura y barbarie. http://www.bolivare.unam.mx/ensayos/barbarie.html

the permanent state of siege that Afro-descendants face, threatened as they are with the loss of sovereignty, freedom, and their territory by the onslaught of megaprojects. Of crucial importance is agribusiness, especially the oil palm monocultures (originating from Africa) that are being developed in the region.

The multiple grievances and problems witnessed during the trip left us feeling impotent, with a desolate and unpleasant taste in our mouths. Yet, the experience demonstrated the urgent need for activities that improve the communication between these communities and strengthen their abilities to analyze and design local and regional strategies for defending their territory. The campaign seeks to link the entire Afro-descendant population of the South Pacific region with a common understanding and a deepening of the autonomous Plans and Projects for Life, in a way that emphasizes their own capacities to research and acquire knowledge. At the same time, it strives to strengthen their culture and embrace their ancestral wisdom. With this in mind, these communities are concentrating their political efforts on "the ability of humans to make their own decisions about themselves and their ways of living together. This ability is necessarily exercised in a process of acquiring consistency in concrete daily life and in the creation of identities."[4]

Thus, their political perspective serves to reinforce the knowledge of their rights and legal tools; by asserting their ancestrality and culture they are able to cohesively constitute themselves as a threatened people and culture. Alternative proposals are based in appealing to these fundamental aspects. As the popular saying goes, "A single swallow does not necessarily mean that summer is on its way."

The South Pacific is not merely a geographical space—as the inhabitants on the shores of its rivers are fond of saying, it is an entire universe. It is a universe where people still use song to express their feelings and play the marimba to get in touch with their past: "*the devil is … the marimba*" chants the song. And, after feeling and getting to know the South Pacific's coastal and river areas, one might easily imagine that today only one devil exists in the region: megaprojects. The overbearing and indiscriminate presence of these projects is the expression of a development-based logic, characterized by a heavy dose of environmental racism and an indifference to the communities and their cultures. These megaprojects obey a logic that is based in the destruction of natural wealth and the erosion of cultural autonomy. At the same time, communities that are already historically impoverished and degraded in the country's idiosyncratic imagination, are facing displacement.

THE DEVIL MADE HIS APPEARANCE … AND HE WAS ANYTHING BUT A MARIMBA

The word "development" conceals the shadow that megaprojects are casting over the region; the people there refer to it with great caution, just as they might refer to a ghost or an armed man. However, the various organizations and community councils that exist in the region are sounding the alarm.

These development proposals are the products of Colombian governmental initiatives, together with the multi-national financial institutions such as the CAF

---

4   Bolívar Echeverría. *Cultura y barbarie.* http://www.bolivare.unam.mx/ensayos/barbarie.html

(Corporacion Andina de Fomento), the Interamerican Development Bank, and FON-PLATA (Fondo Financiero para el Desarrollo de la Cuenca del Plata). The projects have been drawn-up and implemented without consultation, and do not prioritize the ethno-development projects that the regions' inhabitants have managed to forge around their traditions and visions. Instead, the megaprojects are clearly a strategy aimed at dispossessing and displacing these very same populations. By undermining legislation concerning the *Consulta Previa*[5]—namely Law 70, which was passed in 1993, and Decree 1320, which was issued in 1998—these mega-development projects manage to snatch away Afro-Colombians' right to define their own ways of living that the laws entitle them to.

The Colombian state's interest in territories rich in natural diversity does not come free of charge. Foreign companies and capitals have already mapped out the future of entire communities.

> Rooted in the historical process of capital accumulation, these companies are now developing policies aimed at seizing the peoples' genetic, intellectual and cultural wealth. And, in the name of democracy and civilization, monocultures are being promoted.[6]

These interests do not take the communities into account, quashing and devaluing their beliefs, traditional practices, and labors, and the ground is being laid for a territory void of inhabitants—in other words, no peasants, indigenous people, or Blacks. In the early decades of the twentieth century, legal measures were established to usurp the land from peasants and settlers, and in this case, the Afro-descendant communities specifically. Yet, today, colonial methods still remain intact in their essential features. Now, as in the past, peasants continue to suffer banishment from their land at the hands of large and wealthy landowners—only this time around, these landowners are transnational companies.

That the state instigates eviction and subjugates life to new forms of commodification through the imposed presence of large multi-national companies (the sole beneficiaries of the government's proposals) hinders the existence of viable and peaceful relations between a territory and its inhabitants. This phenomenon of accelerated and unscrupulous extraction of natural wealth, as well as its commodification, is characteristic of the position of southern countries in a globalized market. And, as far as the Black communities of the Colombian South Pacific are concerned, it poses a dramatic and all-encompassing threat to their cultural, biological, and ancestral patrimony.

Throughout the course of the Voyage, the clearest and most evident example of the threat posed by megaprojects encountered was the Deep Water Port in Malaga

---

5   The Consulta Previa is a legal mechanism for consulting the black and Indigenous communities before going ahead with a megaproject.

6   Almendares, Juan. *Reflections on Human Rights, Torture and Cruel, Inhuman and Degrading Treatment and Environmental Justice* [Reflexiones sobre derechos humanos, tortura y tratos crueles inhumanos y degradantes y la justicia ambiental].

Bay in the Valle del Cauca. Not only will this construction impact the local population's right to cultural diversity, territory, and participation, which they are entitled
to under the Consulta Previa, but it will also endanger territories for which collective
titles have already been issued. A group of young environmentalists in the community of Bahía Málaga have initiated an eco-tourism process, which is rooted in a local
community perspective, rather than the typical logic of travel agencies or others who
promote commercial tourism packages which devour landscapes and cultures. To
the contrary, this eco-tourism initiative strives to cherish, reclaim and revindicate
the beauty of the areas' traditions and territory. By doing so, it seeks to raise awareness among visitors to the area so that they will leave with an understanding that
other ways of seeing the world and relating to nature do, in fact, exist. However,
these local ways of life are seriously threatened by the construction of the deep water
marine port, as are their food sovereignty and territorial autonomy, which will end
up being administered by "outsiders."

The region of Gran Patía is also learning about an additional threat: the Waterway
(Acuapista) megaproject, which, together with the Deep Water Port, forms part of the
Archimedes Project. The government's devious approach to implementing this project
has consisted of breaking it down into sub-components and dividing them between
the different municipalities that it will pass through. In this way, the megaproject will
bring together three departments and fourteen municipalities. The Waterway would
traverse the entirety of the region's complex ecosystem of marsh-lands, provoking the
kind of incalculable damage that has already occurred with the Canal Naranjo, which
connects the Patía Viejo river with the Turbia ravine, a tributary of the Sanquianga
river. Built in the 1970s to allow faster transportation of wood extracted from the area,
the canal's construction has accelerated the sedimentation of the Patía river, making
its passage almost impossible. Let us not forget that the rivers are the only means of access and communication for the inhabitants of the Pacific region. Not only would the
loss of the Patía river leave an entire population isolated and marginalized (even more
than they already are), but it would also alter an entire ecosystem and water basin that
has served as sustenance and a cultural reference for decades. The inhabitants still
remember a time when the river was wide and deep. Now all you hear is "*canalete!*,"[7]
the rallying cry for people to set about the task of removing blocks of mud and earth
that are clogging the river. Projects such as the Waterway, and others that form the
Archimedes Project, are being developed within the framework of the Initiative for
the Regional Integration of South America (Iniciativa de Integración Regional para
Sur América, or IIRSA). IIRSA is an attempt to create infrastructure to guarantee the
opening up of new commercial routes, as well as facilitating international trade, a process of pillage brought about by way of Free Trade and Bilateral Investment Treaties.
These infrastructure projects seek to speed up the transport of commodities produced
by large companies and multi-nationals, and result in ever greater degradation and
marginalization of local and regional trade and alliances.

---

7   A *canalate* is a stick that the rafters use when their boats become stuck in the sand due to rising
tides.

Patía comes under the Association of Community Councils of Greater Patía, ACAPA, which was one of the first Associations to receive collective ownership rights to land that was ancestrally occupied by the region's inhabitants. Today 96,000 such titles have been granted, spanning three municipalities: Mosquera, Francisco Pizarro-Sala-Onda, and Tumaco in Nariño. Despite these collective land rights grants, and the black communities' long-standing residence in the region, there have nevertheless been reports of incidents in which land belonging to these collective holdings has been sold to foreigners. This has resulted in the land being exploited through practices that are not traditional to the region, such as extensive livestock grazing.

On the other hand, the municipality of Guapi, situated in the Caucan Pacific region, is being drawn into the dynamic of megaprojects by way of the indiscriminate planting of African Palm in areas that are part of the collectively-held lands. It is not only the old people of Guapi who are worried by these monoculture plantations: the young men and women are also concerned about the threats to their land that are associated with this megaproject that plans to produce African palm for the next sixty years. As much as 15,000 hectares of the Communitarian Council of Lower Guapi's total 23,000 hectares are endangered, and the territority's integrity is in jeopardy due to Salamanca, the palm company that won the concession.

Locals are also concerned about the construction of the small-scale hydroelectric plant at Brazo Seco. They believe that this project will not serve the population's well-being, but rather seeks to guarantee the energy requirements of agribusiness, just as has been the case with other projects in the area. Once again this violates Decree 1320, which was issued in 1998 and concerns the *Consulta Previa*. The Brazo Seco hydroelectric plant also threatens to have a severe ecological impact.

Tumaco is a dramatic case in point: here the Guapireños have had ample opportunity to experience the consequences of producing African palm. Tumaco is the municipality with the largest presence of African palm cultivation in the South Pacific region, and it was here that the sowing began. Today, it is reported that around 40,000 hectares have been planted, compared to only 18,000 in 1998; in less than a decade, the extension of African palm crops in Tumaco has doubled. Meanwhile, the Afrocolombian peasants maintain a traditional culture, based on agriculture that is both varied and sustainable, which has allowed them to turn their land into a microcosmos containing diverse plant and animal varieties. However, according to accounts from people in the area, the oil palm gives rise to nothing but sterility of the soil and a uniform strain of plants that homogenizes the landscape and the territory. Furthermore, it is not even edible! In the words of a woman who attended the meeting of the Communitarian Councils:

> The oil palm is a selfish crop that does not allow for the production of anything else. Those who cultivate it will lose their ability to grow banana, cassava and fruit trees. They won't be able to cultivate anything. Nothing at all. Absolutely nothing. This is why I call the oil palm plantations selfish.

Charo Mina, a leader of PCN, who lives in the United States, and participated in the Voyage, wrote:

> The communities exposed to the cultivation of oil palm in the vicinity of Tumaco have experienced the devastating environmental, social and cultural effects of its presence. Their lands have been expropriated (in many cases violently), their water has been contaminated, and they have lost traditional production practices such as the traditional farming system that is based upon a complex ecosystem combining edible food crops, wood sources and ecological control mechanisms. The monocultures present the Afro-descendant communities with an ethical problem, both in relation to environmental, economic and cultural issues, as well as from a historical perspective. The Colombian government's insistence on imposing monocultures in the collective territories belonging to these communities is an affront to their morality and ethics.

In the mid-1970s palm cultivation was implemented in Tumaco with pressure and coercive and cruel methods. However, since 1999, a new strategy of getting hold of land has been adopted by those promoting palm-oil, a strategy that complements their earlier one. In 1999, Cordeagropaz, the Tumaco Corporation for Agribusiness Development, a public-private entity created to promote so-called "strategic alliances," was created. These alliances have overridden the legal rights of the Boards of the Communitarian Councils by organizing small cultivators of oil palm into business associations that serve to bypass the councils. Cordeagropaz, with assistance from USAID, promotes mediation between the government, banks, and palm companies, and violates the basic rules stipulated in the special ethnicity law. Their alliances seek to intensify the presence of agro-industrial palm plantations in the midst of collectively-held territories, by way of associations that do not have legal decision-making power over the territory. These associations simply express the unequal relations between capital and the local population, where the natives put their lands and their labor at the service of this monoculture, while becoming indebted. Not only are their culture and food sovereignty at risk, but also their actual territory. In order for palm cultivation to be able to expand, the people must vacate their territories.

Thus, it must be understood that the displacement of the black communities due to the government's fervent promotion of megaprojects is an intentional strategy, aimed at weakening the control that these communities have begun to exert since being granted collective land titles and since the establishment of Communitarian Councils. If the Communitarian Councils were to be strengthened and given due recognition as the appropriate governing bodies within these territories, as distinct from merely being grassroots organizations, it would introduce new elements to discussions posed by government policies and the Afrocolombian communities.

AND SO, THE DEVIL ARRIVED WITH HIS DEMONS IN TOW

Numerous policies seeking to integrate the black communities with the rest of the country are based on megaprojects that, in addition to assaulting the ancestral nature of the territories belonging to these communities, also intensify existing conflicts

and threaten the communities. The projects are being generated according to external economic requirements and do not include community consultation though they
involve potential projects on their collective landholdings.

The invasion of illegal crops into various zones of the South Pacific has intensified the armed conflict in these regions. The different sides of the conflict fight
for control over the territory, and the civilian population is caught in the middle.
In the midst of this violence, the government has developed so called "alternative
proposals," which unfortunately have nothing more substantive to offer than further
penetration into territories and displacement of inhabitants. These supposed "alternatives" simply serve to cement hegemonic models that were initially put on the table
by the interests of large-scale capital and multi-national foreign investment, and are
backed up by unjust and unequal trade treaties.

Megaprojects have arisen under the pretext of the Colombian government's
program of eradicating illegal crops. The imposed establishment of oil palm monocultures for the production of edible oils and agro-diesel is turning out to be the
strongest pretext. The communities have suffered the repercussions of the spread of
coca in certain regions of the South Pacific, brought about by outsiders: the indiscriminate glysophate fumigations negatively affects people's health, and harms basic
food-crop production and the territory's biodiversity. Furthermore, the agricultural
products and crops that the government has introduced to replace coca have also
been affected by aerial spraying. One concrete example of this is San José de Tapaje,
a *corregimiento* that forms part of the municipality of Charco.[8]

It is clear that the civilian population is caught in the middle of the armed
conflict, the tranquility of their Pacific homeland suffocated. The communities that
settled on the shores of the Tapaje River have had to sustain the scourge of the armed
groups (both legal and illegal). These armed forces are often stationed very close to
the houses in the community, thus preventing people from exercising their right to
move freely within their own territories and benefit from its natural wealth. This is
in violation of international humanitarian law. After six o'clock in the evening, the
river is a lonesome place, a predatory serpent that inspires terror in all who stumble
upon it.

However, some people are more afraid of being uprooted and future homesickness, than they are of bullets, and so, despite everything, they continue living in
Tapaje. Women, men, old people, and children all continue to bathe in the currents
of the river, continue singing to its waters and have not given up sowing banana,
sugar cane, and hope. Alternative projects manage to survive. One such initiative is
the Association of AfroColombian Women for Peace (Asociación de Mujeres Afro
Colombianas por la Paz, AMAC), a group of women from San José de Tapaje who
has had to resist constant threats against its agricultural and cultural proposals.

The Tapaje River is the epicenter of many problems besides those mentioned
above. When coca and armed groups mix with the civilian population, the result is

---

8   *Corregimiento* is an administrative term for a small populated area that exists within a
municipality.

that the communities and the territories where they live are the most affected. Many families are forcibly displaced.

The displacement suffered by the river communities of Tapaje has changed in important ways recently, and a new category of people has emerged in the process. These are people who refer to themselves as "The Resisters" (*Los resistentes*). In addition to physical dispossession, displacement also has symbolic and psychological aspects. The relations between the inhabitants and their land and its resources undergo profound changes. Of those who stay, children are left with fear in their eyes and women with empty stomachs, but they are not considered displaced peoples, and hence are not prioritized for the government's national assistance program.

"The Resisters" are loathe to abandon their land, referring to it as "their paradise." Aggrieved, they ask themselves why the government fails to offer alternatives to abandoning their homes, and receive only threats and harassment from the different armed groups warning them to "vacate the territory." Bearing the brunt of the violence, they have very few tools at their disposal to continue their resistance. Their main weapons are their culture and the processes of ethno-education that have enabled them to appropriate the territory as their own, by way of love for their traditions and culture. The Resisters have valiantly chosen a life of communion with the land, and these make up the lifeline that they cling to so dearly. The songs, the poetry and the dance are the arms wielded by these men and women who talk to the river and rouse people to clear its channels, giving them the strength to face the bullets that seek to remove them from their homes.

> The displaced and The Resisters alike both have lost their right to freely exercise their culture and social being, owing to their loss of autonomy to freely move, to maintain their traditional crops, to freely exercise their right to organize themselves and to participate politically. They have also lost their right to enjoy themselves and carry out recreational activities. The inhabitants of the Territorio Región of the South Pacific live in a situation of confinement, held hostage in their own territory, kidnapped for what they represent and what they are a part of.[9]

These are the features of the policy of plunder and forced change that is being implemented in the territories that have belonged to the black Colombian communities since ancestral times. There is a sense of being under siege, both from the state and from the transnational companies, whose activities threaten the region's communities and its territories—territories that are recognized as the world's third richest, both in terms of genetic wealth, as well as natural wealth in general.

AND, WITH THE POWER OF TRADITIONAL SORCERY, PEOPLE TAKE ON THE DEVIL

The diversity and culture of an immense lyrical universe is under threat from agrofuel monocultures, as well as the megaprojects that go under the name of "development"

---

9 Comment made by Charo Mina in his report "Colombia's African Diaspora Is the Target of an Extinction Strategy" [La diáspora africana en Colombia está en la mira de una estrategia de extinción].

for the communities. Affecting nature, the geographical landscape, the cultural worlds, the agricultural traditions, and the beauty of a territory that is both friendly and seductive, these initiatives amount to an assault against life.

In many communities, such as Bahía Málaga or San José de Tapaje, people continue struggling for alternatives that will improve the living conditions of men and women alike, and reconcile the communities with their environment and the traditions of their elders.

Finally, the only thing that remains is to recall that upstream we encountered the men and women of a songful resistance. There they were, soaking their clothes and their stomachs in the waters of the river, drinking freshly-made "*biche*" and "*naidy*" juice[10] as they engaged in their daily celebration of life, all the while contemplating the harsh reality of hunger and the indiscriminate spraying of chemicals.

There, the bland color of the skin likens the earth, and it is at this moment when uprootedness and homesickness weigh down on our chests and we feel the burden of those who are unable to roam their territory and freely enjoy their traditions.

Despite the fact that we do not have ancestral and collective lands, that we do not know how to plunge a *canalate* deep into the water, that we do not distinguish between the flavors of *pepa e'pan*[11] and that we do not have a river coursing through our memories, this territory and its people nonetheless opened its heart to us. The women sang us a lullaby and seasoned our palettes with the tasty local herbs, chillangua and chillarán, while the local music reminded us of how arhythmical our feet are under the sound of a marimba. The communities entrusted us to shout to the four winds all the pain and injustice that they are living through in their own lands. And, so, this is how the people in the Pacific live, living as they do in the midst of war, and exorcising bullets and intrusions with prayer and song.

PALM OIL IN COLOMBIA: A TALE OF INTERNATIONAL BACKING, COMMERCIAL NETWORKS AND COMPANIES [12]

The majority of the palm oil produced in Colombia is produced for the national market. In 2005, 85.45% of the oil was sold on the Colombian market as compared with 14.55% in the international market, with 13.229 tons consumed nationally, and 2.253 tons exported.

Unrefined raw materials make up 80% of exported palm products. These are sent to Europe, where they are refined in European plants in order to be re-exported at a later date. Thus, the European market receives the greatest share of exported oil. The main countries recieving exported Colombian palm oil are: Spain, UK, Germany, Holland, and outside of Europe, also Brazil.

---

10   Translator's note: I have been unable to find any English translation for these terms.

11   *Pepa e'pan* is the fruit of the bread tree.

12   This related section was written as part of a report about oil palm in Colombia by Censat Agua Viva, by Irene Vélez Torres, in February 2008. It is a previously unpublished document. This text was translated into English by Kolya Abramsky, with assistance from Claudia Roa and Adam Rankin.

The companies which market palm oil overseas are Colombian national capital and specialize in the palm sector. The two most important exporters are the industrial groups Famar S.A. and Daabon, belonging to the Dávila family. These commercial groupings bring together several international marketers including the international marketing companies Tequendama (owned by the Daabon group) and El Roble (owned by Famar S.A.). Aside from these conglomerates, other companies include Bajirá Industrial, Extraction and Marketing Company [la Extractora y Comercializadora Industrial Bajirá] and the Gradesa International Marketing Company PLC [Comercializadora Internacional Gradesa S.A.].

These marketing companies benefit from favourable credits and taxation arrangements from FINAGRO, the Investment Fund for Peace [Fondo de Inversiones para la Paz] and USAID –The US Agency for International Development.

THE ROLE OF INTERNATIONAL FINANCIAL INSITUTIONS IN PROMOTING AGRO-FUELS

During the period 2006-8, the World Bank increased the funds available for loans in the energy sector by 40%. In a similar vein, the Interamerican Development Bank (IDB) has begun promoting agro-fuels as part of the Initiative for Climate Change and Sustainable Energy which seeks to offer support for clients to diversify their energy matrix. According to the IDB, it will take at least 14 years before Latin America is able to become a large scale producer of agrofuels, for which it will require at least 200 billion dollars. In order to realize this potential, the bank putting its resources into supporting the expansion of African Palm and sugar cane crops.

While the majority of the companies which produce and sell palm oil are national capital, this productive system is nonetheless connected with international capital and its interests. Concretely, it must be stresed that a good part of the loans from which the palmiculturists benefit are loans that the Colombian government has acquired from international financial institutions and are charged to the public treasury.

STRATEGIC ALLIANCES

One of the strategies currently promoted by the Colombian government involves Strategic Alliances. In an official communiqué issued by the Presidency of the Republic on 7th July 2007, it was reported that in the first semester of 2007 18, 500 hectares of palm were sown within the framework of Strategic Alliances. These alliances are led by two key players: the businessman Carlos Roberto Murgas[13] and the

---

13   Roberto Murgas was a functionary of César Gaviria and Andrés Pastrana's governments, and went on to become a key player in Álvaro Uribe's presidential campaign on the Atlantic Coast. Together with César De Hart (president of the Colombian Association of Agricultural Producers [Sociedad de Agricultores de Colombia] and the husband of Martha Pinto de De Hart, the first Minister of Communications in Uribe's government) and Jens Mesa (president of Fedepalma and husband of the current Minister of Communications, María del Rosario Guerra de la Espriella) Murgas formed part of the troika leading the country's agricultural sector. In 1990, Murgas managed the Agrarian Bank [Caja Agraria] for several months during the Gaviria government. He later went on to become president of Fedepalma and the Colombian delegate to the Food and Agriculture Organization (FAO). In 1997 he participated in the presidential campaign of Andrés Pastrana, who subsequently appointed him as his Agriculture minister. The Codazzi refinery, in the department of Cesar, is currently part of his business holdings.

company Indupalma. In 2007 Murgas owned 14,400 hectares, working in a Strategic Alliance with peasants from in regions such as María la Baja, the department Bolívar, North Santander, the region of Catatumbo, the municipality of Tibú and in César. In the period preceding the issuance of the government communiqué, Murgas had received loans for more than 2.25 billion pesos by way of the Rural Capitalization Incentive (RCI). Indupalma, on the other hand, has 4,100 hectares in the Sabana de Torres, Santander. It had recieved handouts of just over 23 billion pesos. These figures showed Murgas to be the biggest player in the Strategic Alliances in 2007.

Murgas is emblematic of the chain of interconnections which exist between public indebtedness, the use of legal instruments such as the RCI to encourage the expansion of these crops, the establishment and imposition of Strategic Alliances which bind the local populations to the palm-based productive system, and the dominance of one single businessman throughout the various phases of production and distribution of palm oil. However, Murgas is by no means the only person within the palm sector's business panorama who exhibits these characteristics. A series of exposés in the country's most representative weekly newspaper provoked a scandal in mid-2007. Incoder, the Colombian Institute for Rural Development, had given out more than 16,330 hectares of uncultivated land in the department of Vichada to 13 close associates of Habib Merheg, a senator from the department of Risaralda. Included amongst the recipients were members of his Legislative Work Unit [Unidad de Trabajo Legislativo], his secretary, lawyer and several directors from the company Cable Unión de Occidente, which Merheg was linked to until 2002. In addition to these lands, the legality of whose transfer is still being disputed, senator Merheg also bought the 2,400 hectare Mirador estate in 2005. The goal of purchasing this land, also in the department of Vichada, was to cultivate palm, a prospect which, in his own words, Merheg found "very emotional".

In general, the type of connections revealed in these specific cases is cause for reflection about the complex web of connections between the companies and promotors of palm in the different stages of production, as well as their relations to the governmental policies which back up the interests of these companies and individuals.

Chapter 41 ▌ Part 10

CALL FOR AN IMMEDIATE MORATORIUM ON EU INCENTIVES FOR AGROFUELS, IMPORTS OF AGROFUELS, AND AGROENERGY MONOCULTURES[1]

Statement issued and signed by diverse organizations in Europe and throughout the world

The undersigned call for an immediate moratorium on EU incentives for agrofuels and agroenergy from large-scale monocultures including tree plantations and a moratorium on EU imports of such agrofuels. This includes the immediate suspension of all targets, incentives such as tax breaks and subsidies that benefit agrofuels from large-scale monocultures, including financing through carbon trading mechanisms, international development aid, or loans from international finance organisations such as the World Bank. This call also responds to the growing number of calls from the global south against agrofuel monocultures,[2] which EU targets are helping to promote.

BACKGROUND

Agrofuels are liquid fuels from biomass, which consists of crops and trees grown specifically for that purpose on a large scale. Agrofuels are currently produced from crops such as maize, oil palm, soya, sugar cane, sugar beet, oilseed rape, canola, jatropha, rice and wheat. Agrofuels are designed to replace petroleum, mainly in road vehicles and trains. Biodiesel and ethanol are the main types of fuel produced. Agrofuels do not include biofuels derived from waste—such as biogas from manure or landfill, or waste vegetable oil—or from algae.

Agrofuels are being promoted by governments and international institutions as a means of reducing greenhouse gas emissions from transport, and improving "energy security," i.e. of helping to ensure regular supplies, stabilise the price of oil and mitigate the impacts of volatile oil prices and possible peak oil. Public support for agrofuels is further justified on the basis of their claimed positive impacts on rural development and jobs in producer countries, promises of "second generation" agrofuels whose production will not compete with the production of food, and assumptions about the availability of large amounts of "degraded" or unused land.

Agrofuels are also being strongly promoted by industry. New corporate partnerships are being formed between agrobusinesses, biotech companies, oil companies

---

1   This statement was originally issued in 2007. As it is a previously published statement, original spelling has been left.

2   For example: "Official Declaration of Chake Ñuhá on the Agrofuels and Environmental Services Traps," Asunción, Paraguay, 24 April 2007; "We Want Food Sovereignty Not Biofuels, signed by Alert Against the Green Desert Network, Latin American Network against Monoculture Tree Plantations, Network for a GM free Latin America, OilWatch South America and World Rainforest Movement," January 2007. http://www.wrm.org.uy/subjects/biofuels/EU_declaration.html

and car manufacturers. Billions of dollars are being invested in the agrofuel sector in a development often likened to a "green goldrush," in which countries are turning land over to agrofuel crops and developing infrastructure for processing and transporting them.

IMPACTS OF AGROFUELS FROM LARGE-SCALE MONOCULTURES

Agrofuels are generally grown as monocultures (including plantations), often covering thousands of hectares. In order to compete in the market, they require government support such as subsidies and tax breaks. Support for agrofuels has to date failed to acknowledge the negative social, environmental and macro-economic impacts associated with this kind of farming.

Forecasts by different UN agencies predict that in the future most agrofuels will be produced in the global South and exported to industrialized countries. Although presented as an opportunity for Southern economies, evidence suggests that monoculture crops for agrofuel such as oil palm, soya, sugar cane and maize lead to further erosion of food sovereignty and food security;[3] threaten local livelihoods,[4] biodiversity,[5] and water supplies;[6] and increase soil erosion and desertification.[7]

Agrofuels are currently being developed within the intensive, mechanised, agro-industrial paradigm, using massive monocultures and inputs of fertiliser and pesticide. There is strong evidence that such agrofuel production will not mitigate climate change but instead may accelerate global warming, as rainforests, peatlands and other ecosystems that are essential carbon stores are being destroyed to make way for plantations. There is also controversy about how much greenhouse gas is generated by the agrofuel production process and whether agrofuels provide any real savings once issues such as fertiliser use (and thus increased nitrous oxide emissions),[8] refining, transport etc., are taken into the equation.

3   C Ford Runge and Benjamin Senauer, "How biofuels could starve the poor," *Foreign Affairs*, May/June 2007, http://www.foreignaffairs.org/20070501faessay86305-p20/c-ford-runge-benjamin-senauer/how-biofuels-could-starve-the-poor.html and Food and Agriculture Organisation, "Food Outlook (Global Market Analysis)" No. 1, June 2007, http://www.fao.org/docrep/010/ah864e/ah864e00.htm.

4   Victoria Tauli-Corpuz and Parshuram Tamang, "Oil Palm and Other Commercial Tree Plantations, Monocropping: Impacts on Indigenous Peoples' Land Tenure and Resource Management Systems and Livelihoods," report to the United Nations Permanent Forum on Indigenous Issues, May 2007, http://www.un.org/esa/socdev/unpfii/documents/6session_crp6.doc and "El fujo del aceite de Palma Colombia-Belgica/Europa acercamiento desde una perspectiva de derechos humanos," HRVE and CBC, November 2006, http://www.hrev.org/hrev/media/archivos/flujoPalma/informe_es.pdf

5   "Agrofuels—Towards a Reality Check in 9 Key Areas," Chapter 4, http://www.biofuelwatch.org.uk/docs/agrofuels_reality_check.pdf

6   "Water for Food, Water for Life: A Comprehensive Assessment of Water Management," International Water Management Institute, 2007, http://www.iwmi.cgiar.org/Press/coverage/pdf/Biofuel percent20crops percent20could percent20drain percent20developing percent20world percent20dry percent20- percent20SciDevNet.pdf

7   Alice Friedman, "Peak Soil: Why Cellulosic Ethanol and Other Biofuels are Not Sustainable and a Threat to America's National. Security," *Energy Pulse*, July 2007, http://www.energypulse.net/centers/topics/article_list_topic.cfm?wt_id=46

8   "Biofuels Threaten to Accelerate Global Warming," Report by Biofuelwatch, April 2007, http://www.biofuelwatch.org.uk/docs/biofuels-accelerate-climate-change.pdf

GM AGROFUELS

Many of the crops currently being used for agrofuels have been genetically engineered (soya, maize, rape). A decade of utilization has revealed that the current range of genetically modified crops have not increased yields or reduced dependence on inputs. However, proponents of genetic engineering in agriculture are already using the threat of climate change to argue for wider use of GM crops and the development of new ones such as GM eucalyptus for agrofuel production. GM crops and trees pose serious risks to biodiversity, ecosystems and the food chain. GM microbes and enzymes being developed as part of cellulosic ethanol research (so-called second generation—see below) could also pose severe risks that have not been researched or even considered by governments.

SECOND GENERATION AGROFUELS

It is being suggested that a "second generation" of agrofuels can be developed that will solve some of the problems posed by current agrofuels, such as competition between food and fuel production. The aim is to find ways (including genetic engineering and synthetic biology) of modifying plants and trees to produce less lignin, engineering the lignin and cellulose so that they break down more easily or in different ways, and engineering microbes and enzymes to break down plant matter. Such high-risk techniques do not challenge the pattern of destructive monocultures designed to feed increasing energy consumption patterns. A moratorium on monoculture agrofuels is needed now, to prevent further damage being done through the over-hasty promotion of agrofuel crops. In the meantime, the promises and potential risks associated with second-generation agrofuels should be fully examined. Whatever the outcome, such fuels will not be available for approximately ten years and decisive action to address climate change is required immediately.

SCOPE OF THE MORATORIUM

The moratorium called for by the signatories will apply only to agrofuels from large-scale monocultures (and GM biofuels) and their trade. It does not include biofuels from waste, such as waste vegetable oil or biogas from manure or sewage, or biomass grown and harvested sustainably by and for the benefit of local communities, rather than on large-scale monocultures. A moratorium on large-scale agrofuels and their trade could favour the development of truly sustainable bioenergy strategies to the benefit of local communities as opposed to the financial benefit of the export-oriented industries.

CERTIFICATION IS NO SOLUTION AT PRESENT

Since public support and targets for agrofuels are being justified for their supposed environmental benefits, a number of different initiatives have been started up to develop "sustainability certification schemes." The undersigned organisations regard certification schemes, whether voluntary or mandatory, to be incapable of effectively addressing serious and potentially irreversible damage from agrofuel production, the main reasons being:

Macro-level impacts such as the displacement/relocation of production to lands outside the scope of the certification schemes cannot be addressed through these schemes.

Likewise, certification cannot deal with other macro-level impacts like the competition with food production, and access to land and other natural resources.

The development of such criteria has to date failed to ensure that communities most directly affected by agrofuel production are included in the discussion and fully consulted from the outset, or to comply with basic procedural requirements ensuring Free Prior and Informed Consent of indigenous peoples whose lands will be affected.

The development of agrofuels is proceeding far more quickly than certification can be implemented.

In many countries, conditions are lacking to ensure the implementation or monitoring of such safeguards, or accountability for those responsible for violating them.

As one certification initiative from the Netherlands, the Cramer Report,[9] says:

> Some of the impacts of biomass production are difficult to assess on the individual company level, and only become apparent on the regional, national and sometimes even on the supranational level. This is true in particular for the impacts caused by indirect changes in land use and is especially important in the themes Greenhouse gas emissions, Biodiversity and Competition between food and other biomass uses. In determining the sustainability of biomass it is crucial to take these macro-impacts into consideration.

At present, there are no concrete proposals for macro-level policy, in addition to certification schemes, that would deal effectively with these macro-impacts.

WHY DOES A MORATORIUM NEED TO BE IMPLEMENTED WITH IMMEDIATE EFFECT?

Despite an increasing number of civil society statements and evidence-based reports expressing concern about the unintended but foreseeable negative impacts of agrofuels and calls to halt their expansion, the agrofuel rush is accelerating. The decision of the high-consumption countries, notably the EU and the US, to introduce significant incentives for agrofuels, such as mandatory targets, publicly funded subsidies and tax breaks, is triggering speculation and investment in plantations and enticing countries in the global South to commit substantial portions of land to agrofuel crop-production.

In the past 18 months, billions of dollars have been invested in agrofuel plantations and refineries and associated infrastructure. In Indonesia, $17.4 billion of investment were pledged in the first quarter of 2007, whilst the government plans

---

9   "Testing Framework for Sustainable Biomass," Final Report from the Project Group "Sustainable Production of Biomass," 2007, http://www.lowcvp.org.uk/assets/reports/070427-Cramer-FinalReport_EN.pdf

to convert some 20 million hectares of land to biofuel plantations. 9–10 million hectares of rainforest are acutely threatened in West Papua alone. In Latin America, the Inter-American Development Bank has announced plans to invest $3 billion in private sector agrofuel projects. Governments in a growing number of countries, including Brazil, Argentina, Paraguay, Ecuador and Colombia, are implementing national strategies to boost agrofuel production that involve financial incentives and investment in and licensing of refineries and infrastructure projects, including new roads, ports and pipelines. Those infrastructure developments will open up old-growth forests and other natural ecosystems to destruction, whilst accelerating the displacement of local communities by expanding plantations. The impacts of this massive, rapidly growing investment in agrofuel expansion will be irreversible and irreparable.

Agrofuels pose a particular threat to tropical forest and wetland ecosystems, as events in Indonesia already indicate. Such forests play a vital role in stabilising climate and creating rainfall. There is evidence that the Amazon rainforest may be approaching a point where deforestation will have reduced the vegetation so much that it can no longer maintain its rainfall cycle, thus threatening much or all of the ecosystem with potentially rapid die-back and desertification.[10] Further destruc-tion of rainforests and peatlands for agrofuels could push the planetary system into accelerated warming, sea level rise and ecological change sooner than fossil fuel emissions alone. If the current rush for agrofuels is allowed to continue while certification and the necessary macro-level policies are developed, the damage such schemes and policies are meant to prevent will already have been done by the time they are in place. The risks of a "wait and see" approach are far too high. The EU should apply the precautionary principle to its approach to biofuels and implement a moratorium.

A moratorium will immediately reduce the demand for crops and trees used as agrofuel feedstocks, thus reversing current increases in commodity prices and putting the brakes on the expansion of monoculture plantations for agrofuels which is threatening ecosystems, food security, communities and the global climate. It will provide time to look at the consequences of large-scale agrofuel production in or-der to make a sound and comprehensive assessment of their socio-economic and environmental implications. This will include assessing the foreseeable impacts of proposed agrofuel targets and ensuring that proposed policies and safeguards are capable of being implemented and preventing the serious negative impacts that are already being experienced. It is essential that civil society, and particularly those most directly affected by the production of agrofuel crops are given a fair chance to assess the impacts of the current promotion of agrofuels. A moratorium on incentives for

---

10   L. R. Hutyra et al, "Climatic Variability and Vegetation Vulnerability in Amazonia," *Geophysical Research Letters*, Vol. 32, L24712, doi:10.1029/2005GL024981, 2005, http://eebweb.arizona.edu/faculty/saleska/docs/Hutyra05_Var.Vuln_GRL.pdf, and also Marcos Daisuke Oyama and Carlos Alfonso Nobre, "A New Climate-vegetation Equilibrium State for Tropical South America," *Geophysical Research Letters*, Vol. 30, No. 23, 2199, doi:10.1029/2003GL018600, 2003, http://www.agu.org/pubs/crossref/2003/2003GL018600.shtml

large-scale agrofuel crop production and a halt to EU agrofuel imports will provide the space required for this discussion.

SIGNATORIES CALL FOR EFFECTIVE MEASURES TO TACKLE CLIMATE CHANGE

Agrofuels have not been shown to mitigate global warming; they actually threaten to accelerate it. The undersigned support urgent cuts in greenhouse gas emissions, based on climate science assessments, which involve a drastic overall reduction in energy use in industrialised countries, strict energy efficiency standards, and support for truly renewable forms of energy, such as sustainable wind and solar energy, as well as the protection of ecosystems and carbon stores.

SOME BRIEF NEWS REPORTS FROM DIRECT ACTION-BASED RESISTANCE AROUND THE WORLD: BRAZIL, UK, GERMANY, AND THE PHILIPPINES[1]

VIA CAMPESINA WOMEN PROTEST AGAINST A CARGILL ETHANOL PLANT IN SÃO PAULO, FRIDAY, MARCH 9 2007

This morning, more than 900 women from Via Campesina occupied the Cevasa sugar mill in the region of Ribeirão Preto, São Paulo state. Cevasa is the largest sugar cane company in Brazil, and was recently sold to Cargill, one of the world's largest agricultural transnational corporations.

The protest is part of a national "week of struggle," under the slogan "Women in defense of food sovereignty." The Ribeirão Preto region concentrates the largest sugar cane industries in the country, which are known for labor violations (including slave labor); since 2004, seventeen rural workers have died in the region due to excessive work. The industry is also responsible for environmental destruction.

The women want to contradict the idea that the production of ethanol can benefit small farmers and protect the environment. They denounce air, soil, and water pollution, and respiratory diseases caused by the sugar cane monoculture. Also, the expansion of this industry creates greater land concentration, and increases poverty and other social problems.

In addition, the protest is against the proposal by the United States government to benefit large ethanol companies in Brazil, which is not in the interest of the majority of the Brazilian population.

Via Campesina women defend another agriculture policy, which gives priority to small farmers, who are responsible for 70 percent of food production in the country. Also, they defend a broad agrarian reform to deal with the serious problem of land concentration.

In order to guarantee food sovereignty, rural workers protest against the visit of US President Bush, and against his proposal to use of the country's resources to deal with the United States' energy problems.

BACKGROUND

In Brazil, beginning in the 1970s, during the so-called world oil "crisis," the sugar cane industry began to produce fuel, which justified its maintenance and expansion. The same was repeated in 2004, with the new Pro-Alcohol program, which principally serves to benefit agribusiness. The Brazilian government began to stimulate the production of biodiesel as well, principally to guarantee the survival and expansion of large extensions of soy monoculture. To legitimate this policy and camouflage its

---

1   Original at http://www.viacampesina.org/main_en/index2.php?option=com_content&do_pdf=1&id=283

destructive effects, the government stimulated the diversified production of biodiesel by small producers, with the objective of creating a "social certificate." The monocultures have expanded into indigenous areas and other territories of native peoples.

In February of 2007, the United States government announced its interest in establishing a partnership with Brazil in the production of biofuels, characterized as the principal "symbolic axis" in the relation between the two countries. This is clearly a phase of a geopolitical strategy of the United States to weaken the influence of countries such as Venezuela and Bolivia in the region. It also justifies the expansion of monocultures of sugar cane, soy, and African palm in all Latin American territories.

CARGILL HQ IN UK BLOCKADED BY CLIMATE CAMPERS, AND ACTION AGAINST AGROFUELS[2]

Since 7:50 this morning, twenty participants in this year's Camp for Climate Action and members of Action Against Agrofuels have been blockading the only access gate to Cargill's European regional head office in Cobham, Surrey. Eight activists have locked onto the gates, closing the site down completely. Agrobusiness giant Cargill is being targeted by the protesters for its role in rainforest destruction and land-grabbing, as well as for profiteering from the food crisis. One member of the group says:

> Cargill is using the boom in agrofuels to expand soya, palm oil, and sugar plantations, displacing communities, food crops, and destroying ecosystems. Destroying rainforests and other biodiverse ecosystems, including healthy soils, is one of the quickest ways of heating the planet. This is why we are blockading the Cargill office two days before the official start of the Climate Camp at Kingsnorth.

According to the UN World Food Programme, 100 million more people are going hungry as food prices have risen by 83 percent in three years. At the same time, Cargill's profits have risen to record levels, going up by 86 percent in just nine months, since the company is profiteering from high food prices.[3] Another group member says:

> For companies like Cargill, agrofuels are an opportunity to make more profits from food, to take over more land from small farmers and communities and to further destroy local food production. This is why we need an immediate halt to government policies such as the mandatory blending of petrol and diesel with biofuels in the UK, as well as an end to "free trade" food and agricultural policies, which solely boost the

---

2   This text is a press release that can be found at http://www.indymedia.org.uk/en/2008/08/405121.html.

3   Cargill owns 25 percent of shares in the UK's biggest biofuel supplier, Greenergy International. They have major investments in US corn ethanol, Brazilian sugar cane ethanol, as well as palm oil and soya, and they are involved in joint ventures to develop GM agrofuels. Cargill is the world's biggest grain trader, the biggest exporter of sugar cane and soybean from Brazil, the biggest soybean crusher in Paraguay and one of the world's five largest palm oil traders. About the harm which Cargill is causing to communities and ecosystems in Brazil, Paraguay and Papua New Guinea, see http://understory.ran.org/wordpress/wp-content/uploads/2007/11/cargill_letter1.pdf.

power and the profits of agribusinesses like Cargill. Instead of agrofuels, we need to have policies that support Food Sovereignty, i.e. which put people's right to food first, and support small-scale, biodiverse, organic farming, instead of energy-intensive agriculture.

HAMBURG, 200 PEOPLE BLOCKADE WORLD'S LARGEST AGRODIESEL REFINERY[4]

As part of the *Klimacamp* (Climate Camp) in Hamburg, over 200 people today took part in a blockade of the world's largest agrodiesel refinery.

The refinery in the Port of Hamburg is operated by agribusiness firm Archer Daniels Midland (ADM). Here, GM soya from South America and palm oil from cleared rainforest in Indonesia is refined for use by German cars. Climate campaigners went to the agrodiesel refinery from two directions and blockaded the two access roads and thus the tankers. One demonstration of cyclists started from the main station in Hamburg and went through the city, past the offices of agribusinesses Bunge and ADM with the subsidiary Töpfer. A second demonstration began in Wilhelmsburg. Rallies were held during stops at a petrol station and outside an animal feed producer.

The climate campaigners protested against the industrial, globalized agricultural model that increasingly displaces production by small farmers. Industrial agriculture is a major cause of global warming, responsible for up to 32 percent, due to massive fertilizer and pesticide use, intensive livestock farming, and rainforest destruction. Industrial agriculture is thus climate killer number one—ahead of the energy sector. On the other hand, climate change also causes enormous problems for small farmers.

Industrial agriculture, with its monocultures, ensures high profits for large landowners. The supremacy of seed and chemical corporations such as BASF, Bayer, Syngenta, Monsanto, Dupont; of agricultural trade companies such as ADM, Cargill, Bunge; of large supermarket chains such as WalMart, Metro, Carrefour; and of food multi-nationals such as Nestlé and Unilever has caused a high degree of dependency of the farming sector on those corporations, and has worsened hunger and malnutrition in the global South.

For a few years, there has been a trend towards agrofuels, and in vast regions, feedstocks for agrofuels are being cultivated. In the US this is mainly maize, in South America sugar cane and GM soya, in Indonesia palm oil, and in Europe oilseed rape. Industrial monocultures hasten the pollution of soil and groundwater. The CO_2 balance of industrial biodiesel production from agrofuels is devastating. During the production of most feedstocks, more CO_2 is produced than during the burning of conventional oil. The trend towards agrofuels has also led to food scarcity and thus bears major responsibility for the enormous global food price rises. ADM is also a major contributor to the rainforest-destroying company Wilmar, which owns over 200,000 hectares in oil palm plantations in Indonesia. In July 2007, environmentalists

---

4   From   http://www.linkezeitung.de/cms/index2.php?option=com_content&do_pdf =1&id=5247

proved that the company was responsible for illegal forest fires in Indonesia. ADM has so far refused to stop rainforest destruction and the eviction of indigenous peoples. Every year, 350,000 tonnes of palm oil are refined in their biodiesel refinery in Hamburg.

MAOIST REBELS ATTACK PHILIPPINE BIOFUEL PLANTATION[5]

Maoist-led guerrillas raided a state-owned plantation used for biofuel production in the central Philippines, the first attack on an alternative energy investment, an army official said on Thursday.

The rebels left leaflets denouncing the operations of a facility producing biofuels from cassava and jatropha, a drought-resistant plant, which competes for crops with food production in the mainly agricultural Southeast Asian nation.

Communist New People's Army (NPA) rebels stormed a jatropha plantation on Negros island on Tuesday, burning equipment and stopping workers from hauling lumber, Colonel Cesar Yano, a brigade commander on Negros, told reporters.

"The workers were not harmed," Yano said.

The rebels oppose the use of food for energy purposes, targeting the 2-billion peso ($42 million) ethanol project because it would plant jatropha trees instead of sugar cane and rice, the traditional staple, Yano said.

Jatropha is considered to be one of the most promising sources of biofuels.

The 10-hectare jatropha plantation in Tamlang valley also sits on what was a rebel stronghold before troops drove the NPA guerrillas deeper into the mountains.

The biofuel plantation is a joint venture between the government and Tamlang Valley Agri Development Corp, a company formed by a local alcohol firm and a political clan related to the finance secretary.

The government has a 35 percent stake in the plantation. There was no immediate reaction from the owners.

The Philippines has been promoting the cultivation of crops suited for biofuels to lessen its dependence on costly imported crude oil. The country imports nearly all of its crude oil needs.

The rebels have stepped up attacks on Negros after an army battalion was removed from the island a month ago and was sent to reinforce troops fighting Muslim rebels on the southern island of Mindanao, officials said.

Manila has been battling Maoist-led guerrillas active mostly in the main island of Luzon and in the central Philippines for nearly 40 years in a conflict that has killed more than 40,000 people and stunted investment in the resource-rich country.

The rebels target mines, plantations, logging and telephone companies to scare foreign investors and raise funds.

---

5   Reuters report issued Thursday 11 September, 2008. Reported by Manny Mogato; edited by Paul Tait.

Section 4:

Possible Futures: The Emerging Struggle for Control of the Globally Expanding Renewable Energy Sector and the Roads Ahead

Energy production and consumption are on the cusp of becoming absolutely central to global political, economic, and financial dynamics. Changes in the energy sector will be intimately intertwined with different possible ways out of the world-economic-financial crisis, and the process is becoming increasingly conflictive. Conflicts exist. They cannot be wished away. Now is not a moment for remaining neutral, but for building a collective realignment of forces. The question as to what kind of long-term broad and powerful coalitions might be built in order to become collectively strong enough to bring about a far reaching and emancipatory transition to a new, and predominantly renewable energy system is of utmost significance.

It is almost certain that the renewable energy sector in general, and wind energy in particular, will experience a massive and rapid global growth in the years ahead. A worldwide struggle over how this growth occurs, and where, and how the fruits of it are distributed is intensifying. Questions of ownership and control are becoming central, as are labor conflicts at the point of production of key infrastructures and raw materials. Territorial conflicts in rural areas rich in renewable energy resources are also key. A political struggle is opening up over who owns knowledge and who receives training in renewable energy, and on whose terms and for what aims. The outcome of these struggles will be the key determinants shaping the type of transition that occurs, and its depth and pace. As "green capitalism" becomes increasingly key to paving the way for a new cycle of global accumulation, "green technologies" are set to become an important site of class struggle, and will be at the heart of attempts to impose a new global deal on workers, both waged and unwaged, across the world.

The institutions of the world-economy are already recognizing this new situation. In addition to the recent Copenhagen debacle, the "timeliness" of these issues can be seen in terms of two other important global institutional developments in the energy sector. In 2008, the International Energy Agency World Energy Outlook anticipated an oil supply crises as soon as 2010 and called for an "Energy Revolution." This date is now already upon us. The International Renewable Energy Agency, IRENA has recently been established and has close to 150 member countries. The Copenhagen spectacle clearly reveals that existing political institutions are completely unwilling to undertake the required changes on the scale and within the time frame necessary to solve the climate-energy crisis. Those few national governments that are in fact willing to push a more emancipatory vision of change are not capable of doing so, while those that are capable are not willing.

As the recent climate conference in Bolivia[1] makes clear, a growing movement exists that believes it will not be possible to solve the twin energy-climate crises within capitalism. The International Energy Agency's call for an Energy Revolution should indeed be taken seriously. However, an anticapitalist one.

The time is ripe for sparking a worldwide energy revolution.

---

1   In April 2010 tens of thousands of citizens concerned with climate change issues met in Cochabamba, Bolivia for the "People's World Conference on Climate Change and Mother Earth Rights."

Chapter 43 | Part 11: Emerging Social Conflicts in the Renewable Energy Sector: The Example of Wind

DENMARK: POLITICALLY-INDUCED PARALYSIS IN WIND POWER'S HOMELAND AND INDUSTRIAL HUB[1]

Preben Maegaard

Contemporary wind technology was born and came of age in Denmark. In the 1980s and 1990s, a dynamic research and production network evolved here, growing at an astonishing pace and resulting in the creation of a new industry. By 2001, there was 2,500 MW of wind power already installed in a country of just over 5 million inhabitants. Wind power on days with strong wind can provide half of Denmark's total electrical need, with normal power quality standards maintained.

However, after the liberal-conservative government took over in 2001, the progressive renewable energy programs, which established Denmark as a world leader in this emerging sector, were abolished. The extraordinary growth in installed capacity came to a standstill. A repowering program, launched by the previous government, resulted in a short term revitalization of the sector in 2002, but did not make a lasting impact.

Since then, Denmark has increased its installed capacity by about zero percent. Only 8 MW was installed in 2006, compared to around 300 MW of annual installation in the 1990s. In 2007, for the first time ever, more installed power was decommissioned than installed: thirty-nine windmills were taken down, while only seven new ones were connected in the whole year. This brought widespread attention to the dramatic situation of the sector.

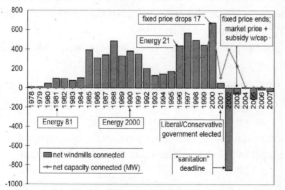

Figure 1: Net windmills and capacity grid-connected by year in denmark. Source: Nordic Folkcenter for Renewable Energy (Data from Danish Energy Authority, Oct. 2007).

---

1   This chapter was previously published in *European Sustainable Energy*, February 2008. It is being reproduced here with permission from the author.

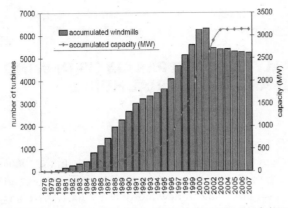

Figure 2: Accumulated windmills and capacity in Denmark (1978–2007): Source Nordic Folkcenter for Renewable Energy (Data from Danish Energy Authority, Oct. 2007).

Even with such poor records in wind power installation since 2001, Denmark is still the leading wind power country on a power percentage and per-capita basis. In 2005, wind power contributed 18.2 percent of Denmark's gross electricity generation, while, in Spain and Germany (the other two wind power giants), it contributed 7.2 percent and 4.4 percent respectively. At the end of 2006, there was 3.136 MW installed capacity in Denmark—with a population of 5.5 million, this is equivalent to 573 Watts per person. In the same year, in Germany, there were 20.622 MW and 82 million people, resulting in 251 Watts per person. On the other end of the scale, in the UK there was 1.960 MW installed, meaning just 32 Watts per person.

THE OWNERSHIP MODEL BEHIND TWO DECADES OF SUCCESS

The most important reason behind such a successful take-off of wind energy in Denmark was the active support and positive involvement of the population. This participatory process started in the early 1980s and culminated in the 1990s, but it collapsed after the liberalization of the energy sector in 1998.

Before liberalization, the strong participation from the population was made possible by a policy based on three principles:

- All farmers had the right to install one turbine on their own land.

- Local residents had the right to become members of wind cooperatives in their municipalities or neighboring municipalities. Exclusive local ownership was the condition to obtain planning permission for cooperative windmills, and there was a limit to the shares that each cooperative member could hold.

- Electric utilities could build large wind farms in agreement with the government.

The absence of financial investors made Denmark's wind sector unique compared to other countries. At the turn of the century, around 150,000 households were co-owners of a local windmill. The ownership model was an integral part of the success of wind energy in Denmark, and the key factor behind the high public acceptance that wind power projects enjoyed during that time. It also enabled a much faster deployment; since large numbers of people were involved in the sector, there was tremendous good will.

In 1992, the role of cooperatives shrank due to a change in planning procedures. In 1998, due to liberalization in the sector, the ownership model changed dramatically; the restrictions on ownership were abolished, and everyone was allowed to own as many windmills as they could get permission for, anywhere in the country. The take-over bids began, resulting in a dramatic decrease in public involvement. At the beginning of th e century, cooperatives were being offered substantial amounts by financial investors to sell off their windmills, sometimes at the price of new ones or more due to the capitalization of re-powering certificates. Investors would then trade those windmills for the right to erect larger ones. As a result, currently about 50,000 households are co-owners of windmills. Consequently, the attitude towards wind power suffered a reversal—now the erection of every single windmill becomes a local problem, and results in bitter conflicts that lead to long delays or cancellations.

CURRENT POLICIES IN DISTRESS

The current wind energy policy, established through a comprehensive political compromise reached by all major parties on March 29 2004, consists of two targets to be reached by 2009. The most important target is installing two offshore wind farms, each 200 MW, which already went through a tendering process. They are supposed to provide power for 350,000–400,000 households. The policy for onshore wind power, disgracefully called "sanitation," foresees the replacement of around 900 medium windmills (of up to 450 kW) by 150–200 new megawatt-class turbines. Around 175 MW of wind power capacity will be decommissioned while 350 MW will be installed, resulting in a net increase of 175 MW.

DONG Energy and E.ON won the tender for the Nysted/Rødsand II offshore wind farm in the Baltic Sea, but the project is shelved and no progress is likely in the foreseeable future. DONG withdrew first, due to lack of economic viability, but also announced its intention to build a 2 x 800 MW conventional coal power station in Lubmin, Germany, at the Baltic Sea. This is one of twenty-five large coal power projects that are planned in Germany, which has not yet implemented a coal-stop of the kind that has been in place in Denmark since 1990. The Danish coal-stop has avoided the emission of millions of tons of CO_2 and (together with the refusal of nuclear power) has enabled the flourishing of wind power and other forms of sustainable energy.

E.ON took over Nysted/Rødsand II and planned to continue without DONG, but, in December 2007, announced that it preferred to allocate its investments to other countries with more profitable tariff schemes than Denmark's. E.ON justified its decision to abandon the project on the grounds of increasing cost of new windmills and other materials. The Danish government is currently investigating the legal aspects of this failure to comply with contractual commitments.

The government had expected Nysted/Rødsand II to become one of the flagships for the 2009 Climate Summit Conference in Copenhagen. Therefore, there is strong political pressure to save the project and see it realized in time for the conference.

The other offshore project is Horns Rev II, which will be located thirteen kilometers north of the existing Horns Rev I. The tender for this project was won by DONG Energy, and will include ninety-five turbines of 2.3 MW supplied by Siemens, plus three experimental turbines of up to 5 MW. The total investment is estimated to be 3.5 billion DKK (470 million EUR), and the annual production is expected to be around 800 GWh. Construction should begin in 2008 and the farm should be commissioned in 2009, in time for the Climate Summit.

The onshore "sanitation" program was justified on the basis of an arguable improvement of the landscape and planning situation. The controversial idea behind this concept is that it is preferable to have a small number of very large windmills than a large number of small and medium-sized ones. However, there is considerable social debate about whether the new megawatt-class windmills, with total heights of up to 150 meters, result in an improvement or worsening of the landscape. The medium windmills are seen by many as an integral part of the Danish landscape, while the large ones lack acceptance and are provoking local resistance.

The "sanitation" program was supposed to be accomplished by 2009, but this timeline is now considered unrealistic. Only a small fraction of the program will be delivered by then, due to constraints and delays in local planning and to growing local protests.

Another step backward in the policy defined in 2004 was the abolition of the purchasing obligation for wind power beyond the minimum amount of full-load hours that receive a guaranteed price. Wind policy in Denmark, the cradle of the successful feed-in system, now establishes that the market should determine everything beyond the minimum full-load hours guaranteed by the law. "Wind power now has to prove that it can compete" the Minister in charge, Bendt Bendtsen, declared in 2002. To blame him might be unfair, as various Danish wind energy experts often publicly claimed that wind power would soon be competitive with conventional fossil energy forms.

AN APPALLING TARIFF STRUCTURE WITH DISASTROUS CONSEQUENCES

The tariff structure in Denmark depends on several variables: on which year the turbine went into operation, how many full-load hours they have already delivered, and whether they are offshore or onshore. The tariff comprises a market power price element and a government subsidy.

Under the present government, a general principle for the tariff scheme has been that the combined tariff for wind energy should not exceed 0.36 DKK (0.048 EUR) per kWh for existing onshore windmills. In the onshore repowering program the price is up to 0.48 DKK (0.065 EUR) per kWh for a maximum of 12,000 full load hours. The tariff conditions for offshore are better: although the price paid—0.518 DKK (0.07 EUR)—is not so high, at least it is guaranteed for 50,000 full load hours.

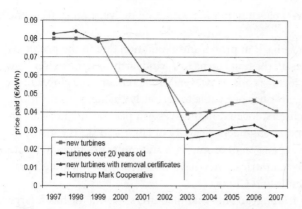

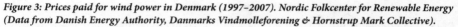

Figure 3: Prices paid for wind power in Denmark (1997–2007). Nordic Folkcenter for Renewable Energy (Data from Danish Energy Authority, Danmarks Vindmolleforening & Hornstrup Mark Collective).

Due to the low prices, some windmill owners have installed brand new mega-watt-class turbines but refuse to put them in operation. They are not interested in connecting them at the existing tariff level, and hope that negotiations about wind power policy currently taking place in the Danish Parliament will result in improved wind power prices. These negotiations have been going on since February 2007 without any agreement. It is impossible to know how long it will still take for politicians to reach an agreement on the future wind energy policy. The Social Democrats have declared that they will not endorse any new policy unless the objectives agreed in 2004 are fulfilled. The government has not, as of February 2008, presented an acceptable solution, so the situation is at standstill. Newly-installed windmills are stalled and their potential contribution to the power supply wasted.

Some wind turbine owners have even declared that they are ready to take down their new turbines and sell them to other countries. They claim that they would obtain a better return doing so than commissioning the windmills, due to the increased prices of new windmills and the low price they would get for electricity. Their arguments have still made no impression on Danish energy politicians.

Offshore projects have been saved from this crisis due to good performance, based on their high capacity of 4,500 full load hours per year. However, the price paid for the electricity is significantly lower than in any other countries.

CONTINUED INDUSTRIAL SUCCESS

Despite all these problems, the Danish wind industry remains in a leading position, even though it must export virtually all its production. Vestas and Siemens (formerly Bonus), in 2006, supplied 27 percent and 7 percent respectively of the global windmill production. More than one-third of the global market is therefore supplied by companies based in Denmark. This is not surprising, since back in the 1990s Denmark controlled 60 percent of the manufacturing. The decline in the market share will certainly continue, as growing numbers of countries develop their own manufacturing capacity in a rapidly-expanding market with long delivery times from the leading manufacturers.

Vestas recently suffered a setback in profits and reputation due to the problems experienced by its V80 turbines, in the first Horns Rev wind farm, and the newer V90 windmills, in British offshore installations. However, Vestas remains the global market leader and so far no competitor has been able to seriously challenge its position.

The two Danish manufacturers, located in the Western part of Jutland, face an acute shortage of skilled labor and engineers. Scarcity of labor and skills constitutes the main constraint for future expansion of production and motivates them to locate additional production in other countries. They have installed fully-integrated production in countries such as India and China.

In addition to these two leading turbine manufacturers, Denmark has a large number of highly-specialized component manufacturers, especially power controls, brake systems, and not least of all blades. The Danish company LM is the leading independent blade supplier, with factories in Denmark and ten other locations in the major wind power countries.

While the two main Danish turbine manufacturers have in-sourced the manufacturing of blades and other components, many newcomers to the wind industry acquire the components from specialized Danish sub-suppliers in order to assure the quality and reputation of their products. In total, the Danish wind energy sector employs 26,000 people. In Germany, which has a population 15 times larger, 60,000 people are employed in the sector.

THE NEED FOR NEW POLICIES AND FRAMEWORK

Windmill manufacturing, one of the fastest international growth sectors, makes a very substantial contribution to the Danish economy. Denmark used to be well-known for its export of bacon, butter, and other agricultural products, but for several years wind equipment has been the main single export item.

This is the primary legacy of the progressive wind energy policies of the 1980s and 1990s. It is high time for the Danish political class to realize their mistake and to put adequate policies back in place. Denmark has hundreds of consumer-owned local energy supply companies for combined heat and power, district heating, and power distribution. New organizational structures and non-profit ownership models to the direct benefit of the involved municipalities may prove to be the most realistic long-term solution for community wind power as well.

Chapter 44 ∎ Part 11

THE SITUATION OF EMPLOYEES IN THE WIND POWER SECTOR IN GERMANY

Martina Winkelmann on behalf of IG-Metall

The wind power industry in Europe is booming. Looked at on an international level, the sector is still in its infancy, but the enormous growth rates and newly developing markets are creating a dynamic industrial sector. Nonetheless, in neither European member states nor other European countries is the full potential being realized. There is still much unused land and many offshore projects that are waiting for the installation of new and improved equipment that will then give the sector further momentum. Both Germany and Europe, as a whole, are leaders in utilizing wind power, but other countries are catching up.

Favorable government conditions have pushed the ascent of the wind power sector which, in turn, has resulted in a rapid increase in employment opportunities. However, to maintain these opportunities in the long term the sector must be cost competitive both nationally and internationally when compared to the more conventional forms of energy production. In order to achieve such an ambitious goal, a good level of union organization and personnel development and training are necessary in the companies.

That presupposes existing co-determination within the factory and comparable and transparent work and remuneration conditions.[1] However, operational practices show a very different picture in the industry: Although there is a lively co-determination culture in many of the suppliers' factories, there are nevertheless many other companies that pay the lowest of wages and exclude co-determination. Here it is abundantly clear that active unionized protection of the interests of the employees is necessary.

The lack of specialists in the field could be a stumbling block in the future and could affect the economic development of the sector.

In the long term, the image of the industry for "Clean Energy" will only be attractive to a potential employee when the working conditions and pay are fair and attractive.

GENERAL POLITICAL CONDITIONS

AIMS FOR THE DEVELOPMENT OF RENEWABLE ENERGIES

In times of rising energy costs and climate change, renewable energies will play

---

1   Co-determination is a practice whereby the employees have a role in management of a company. Co-determination rights are different in different legal environments. In some countries, like the USA, the workers have virtually no role in the management of companies, and in some, like Germany, their role is more important. The first serious co-determination laws began in Germany, where, at first, there was only worker participation in management in the coal and steel industries. But a general law was passed in 1974, mandating that worker representatives hold seats on the boards of all companies employing over 500 people.

an ever-increasing role in the future energy-mix.

There are two aspects that guarantee the future of renewable energies: first, rising oil prices and diminishing fossil fuel resources, and second, the necessity of a vast reduction in greenhouse gas emissions.

To protect the climate, the heads of the European member states and other government leaders agreed on targets to lower greenhouse gas emissions. In Europe, CO_2 emissions are to be reduced by 30 percent by the year 2020. To achieve this Germany would have to reduce the output of carbon dioxide by the year 2020—by 40 percent of the base year 1990. This is possible only with a lasting change of the power supply. In an eight-point plan the percentage of the power supply produced by renewable energies is to be increased to 27 percent (55 million tons), and among other things, measures will be introduced to drastically increase energy efficiency.[2]

The development goals for the renewable energies in Germany and Europe continue to accelerate the growth of the wind power industry.

The goal of the federal government for the development of renewable energy is:

- by 2010, minimum 12.5 percent;
- by 2020, minimum 20.0 percent increase;
- by 2006, already achieved 12.0 percent.

The proportion of primary energy consumption:

- by 2010, minimum 4.2 percent;
- by 2020, minimum 10.0 percent increase;
- by 2006, already achieved 5.8 percent.

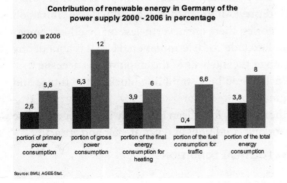

Contribution of renewable energy in Germany of the power supply 2000 - 2006 in percentage

The development of renewable energies in the last few years has meant that goals set for 2010 have already been achieved, and in some cases, surpassed. The federal ministry for the environment pilot study, "Development Strategies for Renewable Energies," states the following is possible:

2   Bundesministerium für Umwelt, Naturschutz und Reaktorsicherheit. BMU Pressedienst Nr. 116/26.04.07—Klimaschutz. Verfügbar unter: http://www.bmu.de/presse/pressemitteilungen_abonnieren/content/39751.php

- 15.5 percent of the power supply by 2010;

- 27.3 percent of the power supply by 2020;

- 8.4 percent of the primary power consumption by 2010;

- 15.7 percent of the primary power consumption by 2020.[3]

In the spring of 2007, European heads of state and governments alike agreed on a formula for climatic and energy policy that included renewable energies.

- The proportion of the primary energy consumption to be increased in all EU member states by an average of 20 percent by 2020.

- The estimated increase in Germany is 16 percent.

- The principal part of the increase has been allotted to wind power energy. However, these quite ambitious goals will only be reached with successful development of offshore wind energy. Already today renewable energies contribute greatly to the reduction of CO_2-emissions (in 2006 a total of around 101.5 million tons). When the wind energy proportion of the power supply is 25 percent, then the CO_2-emissions in Germany would be reduced by approximately 10 percent.[4]

THE GERMAN RENEWABLE ENERGIES LAW

These measures will be supported with the help of a law, the Erneubar-Energien-Gesetz (EEG), which will prioritize the use of renewable energy. With the help of financial aid, the development is to be promoted in power plants fed by renewable sources. The basic idea is that the operators of the supported power plants will be paid a favorable fixed rate for the energy they produce for a certain period of time. This focuses on the production costs of the respective methods of energy production in order to make the plants as economically competitive as possible. The remuneration fixed for the repowered plants decreases annually by a certain percentage, in order to create an incentive for the reduction of costs.[5]

With the development of renewable energies, the dependency on fossil fuel sources such as natural gas, oil, and coal will be greatly reduced, and consequently, the reliance on energy imported from outside Europe.

DEVELOPMENT OF THE WIND POWER INDUSTRY IN GERMANY

ENTERPRISES

Due to various influences, the wind power industry in Europe is currently changing. On the one hand, the wind power plant manufacturers have gone through a consolidation process in the last few years, and on the other hand, new companies have emerged due to acquisitions and partial takeovers.

---

3   Thomas, A. "Windkraft-Industrie 2007, Aktuelle Branchentrends." *IG Metall Vorstand.* 16
4   Ibid, 17.
5   Repowering refers to the replacing of outmoded turbines and other equipment with more up-to-date models resulting in increased efficiency and reduced running costs.

Siemens entered the sector with the acquisition of the Danish company Bonus and the German firm AN Wind. The Indian wind power plant manufacturer Suzlon acquired a majority shareholding in Repower System AG, after a month-long battle for control with AREVA. These are just two examples.[6]

Enercon, who, along with the Danish firm Vestas, are responsible for the production of about 73 percent of wind power plants in Germany, has been the market leader for years. Nordex, Siemens Windpower, and Repower Systems are responsible for a further fifth of the supply.

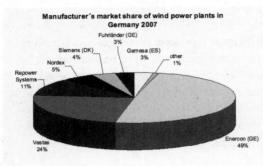

The top ten wind power plant manufacturers are responsible for about 90 percent of sales worldwide. Three of those are German: Enercon, Nordex, and Repower Systems.

While Siemens Windpower produces predominantly in Denmark and the USA, American GE Wind has its own production plant in Germany and has developed a new global wind research center near Munich. In the future, the Spanish Acciona and the Chinese Gold Wind could establish themselves in the top ten.[7]

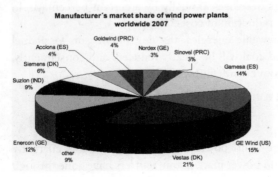

The suppliers contribute significantly to the overall cost simply because the manufacturers then only have to assemble and as a result have less in-house production. Although a number of high-quality components used in the manufacture of plants are produced in other European countries, the vast majority are produced in

6   Thomas, A. "Windkraft-Industrie 2007, Aktuelle Branchentrends." *IG Metall Vorstand*, 9.
7   Ibid, 9–10.

different German locations.

The rapid growth in recent years has resulted in a supply bottleneck, so growth is not governed by demand, but by supply capacity.

OVERVIEW OF TRENDS

Within the last fifteen years, the wind power industry has developed into an important part of the German mechanical engineering sector. The market for wind power plants will also grow substantially in the next years. From this growth, the German wind power industry with its highly technologically-innovative strength can profit further. In addition, the domestic market remains extremely important for the long-term security of Germany as a manufacturing location.

The energy and climate politics continue to propel the worldwide demand for wind power plants. By the end of 2006, more than 74,300 megawatts of wind power had been installed worldwide.[8]

SHORT AND SUCCINCT:

- Accelerated growth of the manufacturers and suppliers of wind power plants.

- A worldwide average of more than 17 percent growth of repowered windpower plants.

- German exports increased from 74 percent in 2006 to over 83 percent in 2007, and have now outperformed the mechanical and plant engineering sectors.

- The German wind power industry accounted for 36 percent of worldwide sales in 2006 (5.63 Billion Euro).

- The number of new installations in the domestic market declined in 2007.

- Predicted increased demand in the demastic market will contribute towards the repowering of windpower plants but at the earliest 2010.

- Germany will install 25.000 MW offshore windpower plants by 2030.

- In Germany employment continues to grow.

- National and international consolidation continues.

- Despite the boom the working conditionen of the employees are less than attractive.

- The climate policy has accelerated the development goal of renewable energies in Germany and in the EU.

- The renewable energies act paper EEG- 2008: will continue to be used as a proven promotional tool.

---

8   Ibid, 5.

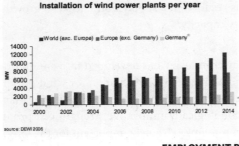

Forecast -
Installation of wind power plants per year

source: DEWI 2006

EMPLOYMENT PROSPECTS

DEVELOPMENT OF EMPLOYMENT

The continuing boom in renewable energies increased the number of jobs in the sector in Germany to 249,000 in 2007[9]—an increase of about 55 percent between 2004 and 2007.[10]

In the renewable energy sector the benefits are seen in both agricultural and industrial production. However, the employees in the wind power and solar energy sectors see the gains only in industrial production.[11]

From the 90,000 workplaces in the wind power sector, about 31,000 are in the direct production of wind power plants and components.[12] The increasing demand for wind power plants in Germany and abroad is responsible for about 10,000 new jobs. The number of workplaces will of course increase if the companies continue to strengthen their position in the export market. For the long term security and development of production, Germany must also develop the domestic market. If they do not, then production could be moved out of Germany, and to the country where the plants will be installed.

If we can further develop offshore wind power energy, then we can secure more jobs in Germany.[13]

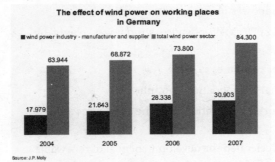

The effect of wind power on working places
in Germany

Source: J.P. Molly

---

9   Schmiade, B. & Becker, F. *Branchenreport Windenergiewirtschaft Europa*, 2008, 28

10   In 2000 there were approximately 60,000 employees; in 2003 approx. 130,000; in 2004 approx. 157,000; in 2006 approx. 214,000; and in 2007 approx. 249,000.

11   Thomas, A. "Windkraft-Industrie 2007, Aktuelle Branchentrends." *IG Metall Vorstand*, 13.

12   Windenergie-Agentur Bremerhaven/ Bremen e.V., Newsletter 08/2008, 9.

13   Thomas, A. "Windkraft-Industrie 2007, Aktuelle Branchentrends." *IG Metall Vorstand*, 13

The increasing lack of specialists in the field is intensifying despite increased training opportunities being offered by the companies.[14]

4.2 STRUCTURE OF EMPLOYMENT

2008	Engineers	Skilled worker	Semi and non-skilled workers	Administration/Management
Wind power plant manufacturer	17.7 percent	52.5 percent	2.2 percent	27.5 percent

Source: Schmiade, B & Becker, F. *Branchenreport Windenergiewirtschaft Europa*, 2008, 39.

STRUCTURE OF EMPLOYMENT IN THE WIND POWER PLANT INDUSTRY IN GERMANY (2007)

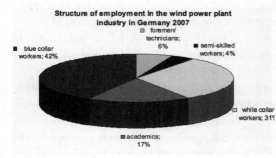

About a third of employees work in the offices, a further half are specialists (skilled/semi-skilled workers, foremen, technicians), and almost 20 percent are engineers. The percentage of skilled and semi-skilled workers is low. Some are not directly employed by the companies but through an employment agency. [15]

The number of job hoppers, who regularly change employer and thus lack any real qualifications in a specific field, has greatly decreased. Most of the new employees are skilled workers in their first job, workers with sector experience, or workers from a similar branch.

As is the case in industry in general, women play only a small role in this branch. In job interviews, the works councils and the employers seem oblivious to the lack of specialists, and of the ways that women could at least fill part of the void. The politicians recognize that women represent a vast reserve of skilled workers, but the employers do little to remedy the problem.[16]

DEVELOPMENT OF WORKING CONDITIONS

Although the working conditions and the codetermination situation in the manufacturing companies are not as attractive as in the suppliers and subcontractors, the conditions are nevertheless better than in the other renewable energy sectors. One of the main reasons why conditions are better in the supply sector is that many of the companies recognize the collective bargaining agreements. Working conditions in the manufacturing sector are extremely varied. In the wind power sector, 42 percent

---

14   Ibid.
15   Ibid., 39–40.
16   Ibid.

of the companies have collective labor agreements, whereas only 15 percent in the solar sector do. In over 53 percent of manufacturers and their suppliers, the workers are represented by a works council. The workers in only 35 percent of the companies in the solar sector have the benefit of a works council.[17]

WORKING HOURS

Market situation, time pressure and shortage of manpower have led to an accumulation of overtime for the employees. The healthy order-books and lack of specialists often results in no alternative for the workers but to work excessive amounts of overtime.

Works agreements on this particular topic are of course in place, but new workplace agreements now realize the urgency of working hours and overtime, in particular. There is not a general picture as far as working hours are concerned. However, one thing they all have in common is that too much overtime is being asked of the workforce.

For this reason the working time account agreements were ever more widely expanded. Meanwhile, hourly accounts with more than 150 additional working hours represent the norm, not the exception. When the workers attempt to balance their account by taking some free time, it is often not possible and the only solution for the employers is to pay the workers for the additional hours. In order for the employers to balance these working time accounts, many agency workers are being hired on short-term contracts.[18]

INCOME

In Germany, most manufacturing companies recognize the collective agreements made in the metal sector, although they often succeed in paying a bit less. When this is the case it leads to conflict between management and workers in the plant. It can be a particular problem when the wage structure is dependant on the location of a plant within an enterprise. Special conflict can develop if the structure differs entirely because of the plant's location—for example, when workers are paid less in a plant because it lies in East Germany. In these situations the goal is not only to increase wages, but first to harmonize remuneration and working conditions. More transparency in wage structures and pay scales could very well lead to more competitive plants and, also, more influential works councils.[19]

WORKS COUNCILS AND CO-DETERMINATION

The data of the study show that there already exists a more stable and structured setup of employees' representatives in the wind power industry than in any other renewable energy sector. Works councils elected by the workers exist in 53 percent of the surveyed enterprises; amongst the suppliers and subcontractors, there is a structure of employee representation that is both long standing and efficient. In comparison, a co-determination policy in the younger manufacturing companies

---

17   Wannöffel, Manfred u.a. "Öko-Branche," *Aufwind,* June 2007, 53.
18   Schmiade, B. & Becker, F. *Branchenreport Windenergiewirtschaft Europa*, 2008, 44.
19   Ibid., 45.

is much less. Here, it is the task of the works councils to manage the effects of operational restructuring and to stabilize their own position in the enterprise. Despite an active co-determination policy in many enterprises we also know of companies that exclude co-determination and the employees' legal right to works councils. The market leader Enercon, in both its national and international locations, plays a rather inglorious role.[20]

As is the case in other industries, the number of employees who are union members is very small. And, on the assembly line, it is less than 30 percent. On the other hand, there are also enterprises in which about 80 percent of the mechanics belong to a trade union. In order to explain these differences, many factors have to be considered such as history and tradition of the enterprise, location, commitment of the union representatives locally, and many others.[21]

APPRAISAL OF THE WORKING CONDITIONS

The excellent growth and development of the industry hasn't translated into attractive nor above-average remuneration and working conditions for the employees. According to the works councils questioned, the conditions are neither attractive nor in fact above average, and in many cases there is much room for improvement. High performance demands, limited income prospects, necessary but often missing education possibilities, and the strong increase of agency workers (over 30 percent) shape everyday life. The necessity for further training is undisputed in the industry, but is limited to mainly the training required to enable the employees to do their daily jobs, and there is often an unwillingness of companies to implement "further training" programs. As far as the highly-qualified workforce is concerned, they are often given much more interesting development possibilities and much more scope and organizational clearance for innovative actions.

In the wind power industry enterprises are often organized into branch associations, however this is more and more rare in associations that recognize collective-bargaining agreements. Blanket or company collective-bargaining agreements covering remuneration conditions exist in 42 percent of the companies surveyed. Collective agreements are valid, particularly in the suppliers of the wind power industry, and other enterprises recognize the general collective bargaining agreements of the metal and electrical industry. Currently, collective bargaining conditions are being strongly contested within the wind power manufacturing industry.

Can organized working and social relations, which provide transparent and accepted work and payment conditions, support the innovative strength of the enterprises and increase the attractiveness of the industry on a long-term basis? The current lack of specialists is still a problem that the industry has to look closely at, and develop solutions for.

The capability of the wind power industry makes sustainable and economical energy a real possibility. Of course sustainability must be considered on a social,

---

20   Thomas, A. "Windkraft-Industrie 2007, Aktuelle Branchentrends." *IG Metall Vorstand*, 15.
21   Schmiade, B. & Becker, F. *Branchenreport Windenergiewirtschaft Europa*, 2008, 47.

economic, and ecological level equally. Therefore it makes sense, also in the wind power industry, to further strengthen the social and works-politics dimensions of sustainability.[22]

SOURCES

Lausitzer Rundschau (2007): Energie sauber—Arbeit unsozial, Lausitzer Rundschau vom 21.11.2007.

Schmiade, B.; Becker, F. (2008): Branchenreport Windenergiewirtschaft Europa. *Arbeitsorientierte Fragestellungen und Handlungsmöglichkeiten.* IG Metall Vorstand Frankfurt am Main, Hans-Böckler-Stiftung, Düsseldorf. 28, 39-40, 44-45, 47

Thomas, A.(2007): Windkraft-Industrie2007, Aktuelle Branchentrends. IG Metall Vorstand, Frankfurt am Main. 5, 9–10,13–17.

Wannöffel, Manfred u. a. (2007): „Öko-Branche" im Aufwind. *Erkenntnisse aus einer explorativen Bestandsaufnahme.* Bochum Juni 2007. 53.

Windenergie-Agentur Bremerhaven/ Bremen e.V. (2008): Windenergie: weltweite Erfolgsgeschichte setzt sich fort. Newsletter 08/2008. 9

ELECTRONIC MEDIA

Bundesministerium für Umwelt, Naturschutz und Reaktorsicherheit. BMU Pressedienst Nr. 116/26.04.07—Klimaschutz. Gabriel: Klimaschutz bedeutet Umbau der Industriegesellschaft 8-Punkte-Plan zur Senkung der Treibhausgas-Emissionen um 40 Prozent bis 2020. Verfügbar unter: http://www.bmu.de/presse/pressemitteilungen_abonnieren/content/39751.php (30.10.2008)

Bundesministerium für Umwelt, Naturschutz und Reaktorsicherheit. Erneuerbare-Energien-Gesetz. Verfügbar unter:
Verfügbar unter: http://www.umweltministerium.de/gesetze/verordnungen/doc/2676.php (30.10.2008)

Bundesministerium für Umwelt, Naturschutz und Reaktorsicherheit. BMU Pressemitteilung Nr. 245/17.09.2007. Erneuerbare Energien geben 235,000 Menschen Arbeit. *Beschäftigungseffekte noch höher als angenommen.* Verfügbar unter: http://www.umweltministerium.de/pressemitteilungen/aktuelle_pressemitteilungen/pm/print/39983.php (30.10.2008)

Wikipedia. Erneuerbare-Energien-Gesetz. Verfügbar unter: http://de.wikipedia.org/wiki/Erneuerbare-Energien-Gesetz (30.10.2008)

---

22   Thomas, A "Windkraft-Industrie 2007, Aktuelle Branchentrends." *IG Metall Vorstand.* 15–16.

Chapter 45 ▌Part 11

FIGHTING THE ENCLOSURE OF WIND:
Indigenous Resistance to the Privatization of the Wind Resource in Southern Mexico

by Sergio Oceransky

Indigenous communities in the Pacific coast of the Isthmus of Tehuantepec (Oaxaca, Mexico) are in the forefront of an emerging global battlefield that will determine, to a large extent, the social and ecological consequences of the transition to clean energy. This bountiful and rebellious region offers the possibility to understand the nature and features of a situation that is likely to become increasingly common in the next decades: the conflict between corporations trying to obtain exclusive access to territories rich in renewable energy sources, and communities asserting control over their territory and defending their resources and livelihoods.

The growing conflict in Tehuantepec is the result of this region being gifted with one of the best wind resources in the world. Thanks to the excellent wind, it is cheaper for large companies to generate wind energy in this region than to buy it from the grid. This has provoked a "wind rush" in which several consortia formed by Mexican and foreign companies are trying to secure exclusive rights on a region endowed with such a great wind.

Although the struggle in Tehuantepec is ground-breaking in many respects, the conflict over the control of RES is not new. The struggle for control over land, water and biomass to generate non-electric forms of energy (food, feed, heat, etc.) is one of the main themes in human history. Regarding the generation of electricity from renewable sources, the use of hydropower has a rich history of conflict. River basins are often under the control of large public or private corporations, which tend to privilege centralized and large-scale dams instead of small-scale decentralized generation, at the cost of (often disadvantaged) local communities. This has resulted in mass displacements and a number of other negative consequences.

However, conflicts around the control of land, water, forests, and other ecosystems are seldom analyzed in connection with the transition to renewable energy. As a consequence, there has not been much discussion about the territorial, economic, and cultural conflicts that are likely to be associated with a shift from fossil fuels to "new" renewable energies (wind, solar, wave, tidal, etc.) as the driving force behind industrial development.

For decades, the analyses and scenarios developed by most renewable energy advocates have assumed that the transition to renewable energy would result in decentralized and community-controlled renewable energy systems. The main reason behind this assumption is that renewable energy sources are decentralized by nature and easy to obtain by anyone with access to technology. In contrast, coal, oil, gas, and uranium are unevenly distributed, the most important deposits are concentrated in relatively few locations and tend to be underground and difficult to extract. Renewable energy technologies have also often been depicted as "technologies for peace," reflecting the assumption that conflicts around the access to energy would be minimal, since all countries have some sources of renewable energy. These overoptimistic assumptions have their roots in the promotion of renewable energy by the anti-nuclear movements of the 1970s, and have been reflected in innumerable texts and depictions of small-scale wind turbines and rooftop-installed solar panels. This has probably contributed to the positive social perception enjoyed by these technologies, but the rapid growth in renewable energy generating capacity that took place in the last decade has proven that the transition to renewable energy is taking a very different shape. It is therefore necessary to develop a more sophisticated framework of analysis regarding this transition, one that transcends the classic green discourse.

We find ourselves at the very beginning of a major change in the energy economy. As Chapter 58, "The Yansa Group: Renewable Energy as a Common Resource," explains, the most significant social, economic, cultural, political, and technological transformations in history were associated with shifts in energy generation: from hunting/gathering to agriculture; from human and animal power for transport and production to the use of wind to cross the oceans; and the use of the steam machine, starting with coal and then adding oil, gas, and nuclear fission, as driver of industry and war. All these transformations have led to increased concentration of power and wealth.

The transition to renewable energy could break this trend. If rural communities have access to the technology, the financing, the training, and the project management skills necessary to undertake their own renewable energy projects, then the energy transition will result in a fairer and more balanced economy. But if transition is undertaken by energy corporations, it will most likely lead to the exclusion of communities from the use of renewable energy sources in their territories, and possibly to their displacement from strategic territories. This will result in tensions and a conflict-ridden, slow and painful change in the energy system, as can already be observed in Mexico.

Whereas the outcome is still uncertain, it is now accepted that transition will happen within a few generations. If the cost of conventional energy sources continues growing and the cost of renewable energy generation continues decreasing, the transition could be undertaken in very few decades. Since we are in such an early phase, it is still possible for communities and social organizations to have a major influence in the future energy economy.

In the near future, communities in many regions all over the world will face a similar situation to the Isthmus of Tehuantepec. A community-based resolution of this conflict is likely to set a powerful precedent, with positive consequences beyond the limits of this beautiful region. It is therefore extremely important to understand and support the struggle of the indigenous peoples of the Isthmus for the collective and democratic control of their renewable energy sources.

This text outlines the emerging struggle in the Isthmus of Tehuantepec, exploring the economic, social, cultural, and political dimensions of the conflict around the use of wind power, a "new" renewable energy.[1] It describes the roles played by the different players involved, paying special attention to the response from the communities affected and the alternatives that some of them are attempting to build.

SOCIO-ECONOMIC AND CULTURAL CONTEXT

The Pacific coast of the Isthmus of Tehuantepec extends from the coast northward approximately 60 km and approximately 60–80 km from east to west. It is inhabited mainly by five different indigenous peoples: Zapotecas, Huaves, Mixes, Chontales, and Zoques, the most numerous of whom are Zapotecas and Huaves. Their territorial rights are recognized, and in almost all cases, collectively organized in so-called *Ejidos* and communities, Mexican legal figures that combine individual land use with collective property. The collective character of some *Ejidos* and communities has been modified through official plans that give more emphasis to private than to common property; however, not all Ejidos and communities have applied these plans.

This is an agricultural region with high-quality land and rich water resources; it is endowed with several important rivers and with the Benito Juárez dam, which provides irrigation to 23,000 hectares. There is also an underground aquifer at a depth of between six and twelve meters, but in some places it emerges at a depth of 1.5 meters. It was a sugar-producing region until the government's sugar policies changed. Today, the main activities are milk production and agriculture. Bettina Cruz Velázquez, specialist in territorial planning and regional development and member of the Assembly in Defense of the Land and Territory of Juchitán, declares that farmers produce three harvests per year in irrigated land and two harvests in non-irrigated land. In the Huave area, a large proportion of the population works also as artisan fishers.

Most of the population lives in poverty, but there is no hunger as a result of food production for self-supply. Lack of access to education is a serious problem. Alejo Girón Carrasco, from the Grupo Solidario in La Venta, remarks that in his community, where the first operative wind farm was built, 76 percent of the population is illiterate. Among those who had the chance to receive formal education, most only completed the third year of primary school. The situation is similar in all the affected communities. As a consequence, *caciquismo* (authoritarian social structures in which the leader commands the community) is still alive: an important part of the population complies with what the local leaders say, especially in communities where political parties have more influence due to the erosion of traditional practices of collective decision-making. In this sense, it seems no coincidence that the first wind farm came into operation in the community of La Venta, which has lost much

---

1   Wind energy is not new; it has been used for millennia to power sea and waterway transportation. However, the generation of electricity from wind only emerged in the late-nineteenth century, and did not reach a significant scale until the 1970s. This article refers only to conflicts around wind resources used to generate power through wind turbines. Fortunately, no conflicts have been registered so far regarding the use of wind for transportation.

of its indigenous inheritance, where the Zapoteco language has been lost, and where political parties have a comparatively larger influence.

STORMY INVESTMENTS

According to an in-depth study of the wind resource of Oaxaca, published in August 2003 by the US National Renewable Energy Laboratory, 33,200 MW of wind generation capacity can be installed in areas with good and excellent wind. If areas with moderate wind resource would also be used, the installed capacity potential is 44,350 MW. These are conservative figures, since they assume an installation density of 5 MW (equivalent to two contemporary wind turbines) per square kilometer, a hub height (i.e. wind turbine height) of 50 meters, and on-shore installations only (i.e. no use of the wind resource in the sea and lagoons). Even though there are small pockets of windy areas in other parts of Oaxaca, almost all of this potential is concentrated in the Isthmus of Tehuantepec.

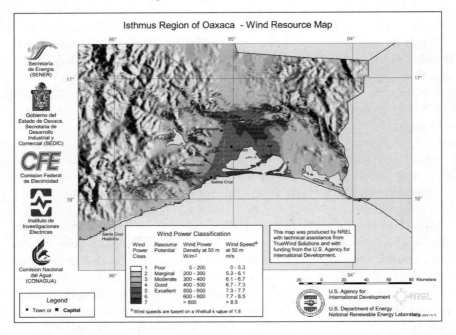

One of the main characteristics of wind in the Tehuantepec region is the high capacity factor (i.e. the amount of hours that wind turbines will be producing power in the year), which is around 50 percent in good and excellent locations. If 33,200 MW were installed in these areas, the total production per year would be around 145,416,000 MW/h. If areas with moderate wind resource are also used, the total production could be up to 194,253,000 MW/h. Increasing the installation density or the hub height would result in a larger production.

The total electricity consumption of Mexico in the year 2008 was 231,400,000

MW/h.[2] Therefore, an intensive use of the wind resource in the Isthmus would produce over half of the electricity consumed in Mexico. This is by far the most significant wind resource of Mexico, and one of the world's richest regions in terms of renewable energy production potential.

Projects to install more than 2,500 MW of wind energy capacity in the region within the next 4 years are already underway or have been approved by the authorities. Out of that total, 1,986 MW will be installed by projects undertaken by private consortia that have applied for, and obtained, power production licenses under the legal formula of production for self-supply. These consortia are formed by foreign (mainly Spanish) companies that build the wind farms and sell wind power, and Mexican and international companies that buy the power. It is a slightly overstretched definition of self-supply.

A growing number of farmers and communities in the region, and local and national environmental and human rights NGOs, oppose these projects. They argue that the wind farm projects were drawn up and are being executed without local consultation or involvement, and that the companies have provided incomplete and/ or incorrect information to land owners in order to obtain abusive land lease contracts. Land owners have presented 180 legal demands in court to nullify the land lease contracts, as well as a case against the former local authority of the *Ejido* where the first operating wind farm was installed. The two private projects that are already underway are facing increasing opposition from the affected communities. Protests in the form of blockades and occupations have been repressed by public forces, including anti-riot police.

LEGAL AND CONTRACTUAL FRAMEWORK

The Law for the Utilization of Renewable Energy and to Finance the Energy Transition (Ley para el Aprovechamiento de Energías Renovables y el Financiamiento de la Transición Energética) was passed in November 2008, as part of a wider energy reform. This reform also included a highly controversial new regime for the use of Mexican oil that was at the center of a mass social and political mobilization. In this context, the renewable energy law received virtually no attention.

The new law does not say much—it only contains some general principles, establishes a fund to finance the transition to renewable energy, and appoints the bodies responsible for developing a renewable energy policy. The Mexican legal framework governing renewable energy is therefore still incomplete. As this text is being written, several legal instruments that complement the law are being prepared, and expected to be passed within the next few weeks or months. They include the Reglament of the Law, a National Strategy for the Utilization of Renewable Energy and to Finance the Energy Transition, a Special Programme for the Utilization of Renewable Energy, and a number of crucially important technical regulations governing the issuance of power production licenses, the rules for the dispatch of energy, the quantification

---

2   Figure published in the website of the Comisión Federal de Electricidad (CFE), see http://www.cfe.gob.mx/es/LaEmpresa/queescfe/Estad%C3%ADsticas/

of externalities associated to different forms of energy generation, and the tariffs to be paid for different forms of energy production. The issues surrounding these legal instruments are explored in further depth at the end of this text.

All the renewable energy projects undertaken and approved so far came into existence in a legal vacuum with regards to renewable energy policy. They were based on the general rules governing electricity production.

In Mexico, the Federal Electricity Commission (CFE) has a monopoly over the transmission grid and over most power generation, due to Article 27 of the Mexican Constitution, which also mandates CFE to generate electricity at the lowest possible cost. However, the Law of Public Service of Electric Energy, approved in 1992, defines five cases in which the private sector is allowed to participate in electricity generation.

According to Dr. Julio Valle Pereña, Director for Promotion of Investments in the Energy Sector at the Mexican Secretary of Energy, private investors have expressed a very keen interest in producing wind energy in Mexico.[3] The public administration has therefore taken measures to resolve the existing obstacles to that investment, creating the conditions in which private projects can take place.

Of the five cases in which the private sector is allowed to generate electricity, the most important one with regards to wind energy is the case of self-supply of electricity, i.e. contracts that allow companies to generate the electricity that they consume. This kind of contract can be obtained by individual companies or by consortia in which some companies sell power and others buy it. The companies can use CFE's grid to carry their electricity, paying a transmission fee. They can use the electricity any time within one year after it was produced. The CFE therefore provides a free energy storage service to private wind energy projects. This has made wind energy projects very attractive to the private sector, since companies pay more for the electricity sold by CFE in peak hours than for the wind energy generated in Tehuantepec.

Under a self-supply contract, if more energy is produced than consumed, the excess production can be sold to CFE, which cannot pay more than the marginal production cost in that node of the network at that moment in time. This option therefore does not make economic sense, given the fact that power can be "stored" in the grid for twelve months. In general, the electricity is consumed during the peak tariff hours (during the day, when the price charged by CFE to power consumers is highest), and the difference between the power energy cost and the peak tariff is the basis for the profitability of these projects.

According to Ramón Carlos Torres Enríquez, Vice-Director for Energy and Environment at the Secretary of Energy, self-supply projects do not decrease the need for other power production infrastructure. If the power providers in the self-supply consortium cannot meet the demand from power consumers, the latter buy energy from CFE, which needs to have an excess in order to cover this potential extra demand.

---

3   Dr. Valle Pereña is now General Director for Research, Technological Development and Environment of the Secretary of Energy. When the interview was conducted, he was Director for Promotion of Investments in the Energy Sector.

The law also allows private investment in power generation for export, without restrictions. About five projects to export to the USA are being studied in Baja California. According to Eduardo Zenteno, president of the Mexican Wind Energy Association, there are projects being studied for an approximate total of 1.500 MW, and some permits have already been released by the Energy Regulatory Commission.

Finally, the law allows private investment in small-scale power generation projects of up to 30 MW. However, the CFE only pays either 90 percent or 85 percent (depending on whether the dispatch has been programed or not) of the marginal cost of power production in that node at that moment in time. This case has been designed for conventional power sources and makes no sense for wind energy, since the payment is insufficient to cover the investment. A small wind farm of the Institute of Electricity Research will enter into operation in the Isthmus under this kind of contract. This project is possible because its main goal is academic, not economic, and receives additional funding from other sources.

In conclusion, the current legal framework for wind energy projects in Mexico is strongly biased towards large players. Due to current laws, the actors in the best position to make use of Mexico's wind resource are corporations that consume large amounts of electricity in peak hours, and power corporations that have sufficient financial resources to cover the high investments required to develop wind farms. The protection of the rights and interests of the communities where the wind resource is located is not regarded at all by the existing legislation.

Private projects receive an important indirect subsidy through the free service of energy "storage" for one year, which allows companies to count all their wind power production as if it had taken place during peak hours, and places on the State the responsibility to install additional capacity to balance the natural fluctuations in wind power generation. Private projects also receive tax incentives; the legislation allows the accelerated depreciation of investments in renewable energy projects, which can be discounted within one year.

INFRASTRUCTURE AND INVESTMENT IN THE ISTHMUS

The main obstacle faced by self-supply projects is the lack of transmission capacity from the Isthmus of Tehuantepec to the center and north of the country, where most of the power will be consumed. For this reason, private wind energy projects were integrated through a process called Open Season (Temporada Abierta), which invited investors to declare the capacity that they wished to install in the Isthmus in order to integrate them into a new 145 km-long transmission line. Open Season started in early 2006 and ended in the middle of 2006. The investors shared the construction costs according to the capacity installed by each one of them. Once their wind farms start operating, they will only pay the variable costs of this new line, but not the fixed costs. In addition, they will pay fixed and variable costs for the existing transmission lines used to transport power to its final destination. The new transmission line is expected to come into operation by the end of 2010.[4]

According to Eduardo Zenteno, president of the Mexican Wind Energy Association, several companies have expressed interest in installing more wind farms in the Isthmus, for a total of 4,000 MW of additional capacity. The viability of these plans will depend on building additional interconnection and transmission capacity.

In addition to self-supply contracts undertaken by private investors, CFE has its own projects in the Isthmus. The first significant wind farm in the region, called La Venta II, was built by a consortium formed by the Spanish companies Iberdrola and Gamesa. The consortium was contracted by CFE to build the farm, which is now owned and operated by CFE. Public investment was taken in order to establish the viability of private projects, and resulted in the construction of a first high-voltage transmission line, which is now also being used by some private projects.

Future public projects will be built and operated by private investors contracted by CFE. The investors will own the infrastructure and deliver energy to CFE in exchange for payment in agreed terms. They will produce energy under the license of Independent Energy Producer, and will be funded by the World Bank's Global Environmental Facility (GEF). A total of 507 MW will be installed under this kind of contract in the Isthmus of Tehuantepec by the year 2012.

Project	Developer	State	License	Capacity [MW]
La Venta	CFE	Oaxaca	Public Service	1.58
La Venta II	CFE	Oaxaca	Public Service	83.30
La Venta III	CFE	Oaxaca	Independent Energy Producer (PIE)	101.40
Oaxaca I	CFE	Oaxaca	PIE	101.40
Oaxaca II-IV	CFE	Oaxaca	PIE	304.20
Subtotal				**591.88**
Eurus	Acciona	Oaxaca	Self-supply	250.00
Parques Ecológicos de México	Iberdrola	Oaxaca	Self-supply	79.90
Fuerza Eólica del Istmo	Fuerza Eólica-Peñoles	Oaxaca	Self-supply	50.00
Eléctrica del Valle de México	EdF Energies Nouvelles-Mitsui	Oaxaca	Self-supply	67.50
Eoliatec del Istmo	Eoliatec	Oaxaca	Self-supply	21.25
Bii Nee Stipa Energía Eólica	CISA-Gamesa	Oaxaca	Self-supply	26.35

---

4   Six private wind farms have been allowed to connect to the transmission line used by CFE for its wind farm in La Venta. In principle, this existing line was going to be used only for CFE projects. The use of this public transmission line has enabled two private projects, developed by two powerful Spanish corporations (giant power corporation Iberdrola and construction and energy corporation Acciona), to start working already. Four more wind farms will be connected to this line in the future. According to Eduardo Zenteno, this arrangement was personally facilitated by the President of the Republic.

Desarrollos Eólicos Mexicanos	Demex	Oaxaca	Self-supply	227.50
Eoliatec del Pacífico	Eoliatec	Oaxaca	Self-supply	160.50
Eoliatec del Istmo (2a fase)	Eoliatec	Oaxaca	Self-supply	142.20
Gamesa Energía	Gamesa	Oaxaca	Self-supply	288.00
Vientos del Istmo	Preneal	Oaxaca	Self-supply	180.00
Energía Alterna Istmeña	Preneal	Oaxaca	Self-supply	215.90
Unión Fenosa Generación México	Unión Fenosa	Oaxaca	Self-supply	227.50
Fuerza Eólica del Istmo (2a fase)	Fuerza Eólica	Oaxaca	Self-supply	50.00
Centro Regional de Tecnología Eólica	Instituto de Investigaciones Eléctricas	Oaxaca	Small-scale production	5.00
Subtotal 2				**1,991.60**
Total				**2,583.48**

PRIVATIZING THE WIND

The access by private investors to the Oaxacan wind resource has taken place in a legal vacuum with regards to territorial and administrative regulations. Given the lack of a proper channel, the initial discussions were undertaken in an informal setting: the colloquia organized by the Fundación para el Desarrollo del Corredor Eólico del Istmo y para las Energías Renovables (Foundation for the Development of the Wind Corridor in the Isthmus and for Renewable Energies).

This foundation played a key role in the process by organizing seven international colloquia on the Isthmus Wind Energy Corridor. These colloquia brought companies and government officials together in order to discuss solutions to the problems faced by the projects and to take further steps. Fernando Mimiaga Morales, secretary of the foundation, explains that the proposal for an Open Season (the process to create a high-voltage transmission line for private investors) came out of the fourth Colloquium (2004) and was taken over by the government.[5] He adds that the colloquia were also useful to make contacts, and for entrepreneurs to visit the area and meet the land owners in the areas that had been allotted to them.

---

5   Other examples are the proposal in the first Colloquium of a Contract of Interconnection for Renewable Energy Sources; in the second Colloquium, of the Action Plan for the Elimination of Barriers to Wind Energy Generation GEF-UNDP-SENER-IIE; in the third Colloquium, the Large Scale Renewable Energy Project, which is the basis for the funding provided by the World Bank; in the fourth Colloquium, the proposal to create a national law for renewable energies; and in the fifth Colloquium, the proposal of Official Mexican Environmental Norm to Regulate the Construction, Operation and Dismantlement of Installations.

Fernando Mimiaga Sosa, Director for Sustainable Energy and Strategic Projects in the Secretary of Economy of the Oaxaca State Government, explains that their business center facilitated the process by which the wind resource of the area was distributed between different companies. The business center proposed an area of operation to each company that showed interest in investing, in order to avoid competition between them.[6] In spite of this, there have been some conflicts, especially with ENDESA, which did not participate in Open Season due to the uncertainties created by the complex takeover of the holding company in Spain. For this reason,

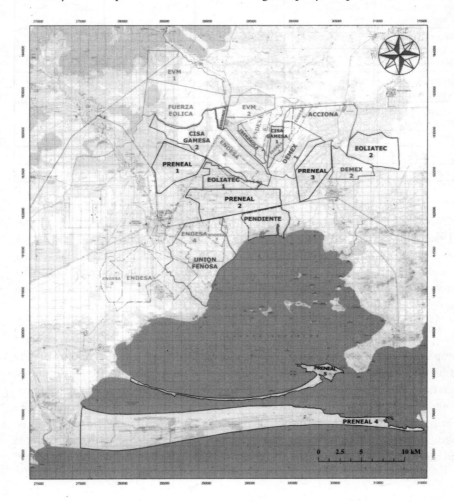

<hr />

6   It should be noticed that Álvaro Velázquez Maldonado, current Director of Operations in Mexico of PRENEAL (one of the Spanish companies that develops projects in the region), worked previously at the business center, where he was responsible for assisting several companies, including PRE-NEAL. It is also remarkable that PRENEAL has an optimal area of operations, having access to more land than any other company, in areas with extraordinary wind resources, including the strips of land that are close to the lagoon.

ENDESA cannot realize projects yet, and some companies are entering some of the eight areas initially allocated to this company. However, this policy has, in general, been successful at avoiding competition between companies.

The resulting division of the territory is reflected in the following map, produced by the Mexican Wind Energy Association.

The territories are marked with the name of the foreign company in charge of energy generation—mainly Spanish companies, including several large utilities. These are the companies that sign contracts to lease the land and are therefore at the center of the conflict with local organizations and communities. The contracts with the national Energy Regulatory Commission are signed by larger consortia of private companies, which include energy consumers as well as producers. The consuming side of the consortia include very large and powerful companies such as Cemex (the largest cement corporation in Latin America) and the Mexican subsidiary of Wal-Mart.

Members of the opposing groups claim that the communities were never informed about the process of territorial division between companies, and were not consulted about it. They add that the colloquia were exclusive meetings of entrepreneurs and government officials where the communities played no role.

Almost all the companies offer the same deal to land owners: 1.5 percent of the gross income that results from energy production, in exchange for the exclusive right over the use of the wind. However, it is not clear from the contracts how the value of this 1.5 percent will be determined. Since the pricing policy within the consortia is internal, the consortia could establish the value on an arbitrary basis. In any case, they will retain 98.5 percent of the gross income resulting from power generation.

The 1.5 percent of gross income offered to local land owners is distributed in different ways, depending on the company, between three groups: one share goes to owners of the land where turbines are erected, another share to owners of land affected by roads or transmission lines, and the rest to land owners whose property is not directly affected by the wind farm, but who have leased their land to the energy companies and foregone the right to use the wind or take any action that could affect the wind (such as planting trees, erecting any kind of construction, etc.). Some companies also offer additional benefits, such as extending the roads that they are going to build for their projects in order to connect geographically isolated communities.

If such a contract would be offered for any other energy source (such as oil or gas), it would be immediately dismissed, particularly in Mexico. However, since the companies involved act as a cartel, and have divided the region in order to avoid mutual competition and be the only bidders, local farmers and communities are not, for the time being, in a position to negotiate a better deal. The companies' attitude, reflected by the president of the Mexican Wind Energy Association, is that they can only accept these conditions or reject the use of their wind. Other alternatives, such as community projects, are, so far, not even considered by any of the players involved.

Federico de la Pisa, from Preneal (one of the Spanish companies planning wind farms in the Isthmus), explains that if a land owner decides not to sign a land lease contract, it has no effect on the developers: the turbines will simply be placed on someone else's land. The only effect will be that those who do not sign will relinquish the income that they could have received, since the wind farm will be built around them, as was done in La Venta II. He adds that the companies prefer that everyone signs the contracts, in order to ensure that no other wind turbines are installed in the area. The companies will pay the same no matter how many people sign the contracts, since it has been decided that all projects will pay 1.5 percent of the production for the land, regardless of the number of contracts signed and the total area covered. If some land owners do not sign, the payment will be distributed among fewer owners, who will therefore receive higher rents. This criterion was decided at the early stages of the wind energy projects, in meetings held between AMDEE (Asociación Mexicana de Energía Eólica, the Mexican Wind Energy Association) and the government. All companies follow it, with slight variations in the distribution of the 1.5 percent amongst the three categories of land owners mentioned above.

Around 1,500 farmers have signed twenty-five to thirty-year land lease contracts that can be automatically extended by the same period, giving companies exclusive access to the wind resource in their land. In exchange, they will receive an extremely low compensation, which bears no relationship whatsoever to the extremely high value of the wind resource in their land. The contracts are strongly biased in favor of the companies; for instance, the company can decide to cancel the contract, but the land owner cannot. The land lease contracts allow the farmers to continue using the land for agricultural purposes, while imposing restrictions regarding construction activities, sale or rent, and often even animal husbandry. Other contracts lease the right to install wind turbines and exclusive access to the wind resource, and place similar constraints on the land owners. Beyond differences in details, they are very unfair contracts, and project developers, opposing groups, and government officials agree that the main reason behind the acceptance of such conditions is the poverty in which local communities live.

The Isthmus is facing an innovative form of privatization of wind, undertaken by way of private contracts between a cartel of powerful and wealthy companies and indigenous peoples suffering economic hardship as a result of centuries of exploitation and neglect. However, some farmers and communities have rejected the deal and are organizing themselves in order to change the terms of the discussion.

GROUNDS FOR OPPOSITION

Members of the communities affected by wind farm projects have generated a still embryonic, but growing, movement to spread more information and foster debate about the projects, promote a collective rejection of the terms offered by companies, and denounce the practices used to obtain contracts. The local groups that compose this movement are not against wind power itself, but against the way that the projects are currently being developed in the Isthmus.

Lack of local and community-based participation is one of the main reasons for the rejection of the current projects. Opposing groups argue that the projects were only discussed between companies and institutions, and that local communities were only seen as providers of land. Since wind is a local resource, and given the great impact that the installation of thousands of megawatt-class turbines will have in their region, they claim that the communities should be the ones deciding how and on what scale this resource should be used, and should participate equally with the other players.

They also argue that the current lack of involvement amounts to a violation of Convention 169 of the International Labour Organisation, signed and ratified by Mexico, which establishes that indigenous peoples have to participate in the formulation, implementation, and evaluation of national and regional development plans that directly affect them. In their view, the only participation that their communities were offered was as contract signatories, under terms decided without them.

Lack of transparency has also generated animosity in the affected communities. Alejo Girón Carrasco, from Grupo Solidario in La Venta, underlines that neither the companies nor the institutions have provided information about the profits expected from the planned investments.

Opposing groups also denounce the use of antidemocratic practices based on *caciquismo* in order to push projects through. Grupo Solidario from La Venta claims that in this community, as in all others, the *caciques* receive a commission for each land lease contract signed by local farmers, resulting in a sudden visible increase in their wealth and possessions.

Members of the opposing groups denounce that the annual rent offered to land owners is an arbitrary amount and is insufficient to compensate for the negative consequences that wind farms have on farmers and communities. According to La Venta's Grupo Solidario, some people signed land lease contracts for La Venta II worth 1,500 Mexican pesos (around $150 US) per hectare, per year,[7] but after pressure from the group, contracts signed later paid 3,000 pesos per hectare per year. A new farm being built in La Venta pays 6,000 pesos per hectare per year, and in other areas with a similar wind resource, apparently 12,000 pesos are being paid per hectare per year—an increase of 800 percent. This contract variation has led many people to conclude that the companies offer as little as possible for the land, and that those amounts have no relation with the value of the wind resource that they receive in exchange.

There is the perception that many projects operate through intermediaries (called "coyotes" by the local population) who keep an important part of the profits. Alejo Girón Carrasco from the Grupo Solidario offers the example of the private project being built in La Venta, which is making payments with hand-written checks signed by a person rather than a company.

Although farming activity can continue once the works have been finalized, it is restricted and sometimes negatively affected in a way that receives no compensation.

---

7   The amounts mentioned refer to the rents that farmers will receive once the projects are in operation. Before the construction work starts, farmers receive a much smaller rent, between 100 and 500 pesos (between $10 and $50) per hectare per year.

For instance, in La Venta II, some roads and the lines of generators were raised, affecting irrigation channels and the natural water flows to discharge rainwater. Although contracts differ from company to company, most restrict the use of the land by the farmers, banning them from planting anything that grows more than two meters, erecting any kind of building, opening wells, etc. Some contracts include a clause that commits farmers "not to install material or animal obstacles" to the wind—a clause that could be used to heavily restrict their activities. In contrast, most contracts grant *usufructo* (unrestricted use rights) of the land to the companies. They often make it impossible for the farmers to rent their land without the company's permission, and give companies a preferential right to buy the land.

Opposing groups also question the legality of contracts signed with individual land owners for land that is communal property, such as in Juchitán, where a very large proportion of the land is communal. Though there is a 1965 Presidential Decree stating that, the legal steps needed to solidify it were not undertaken, and most of the land has been administered as if it was private property (though some areas are still collectively used for grazing and wood collection). The wind energy corporations have signed private contracts with the *caciques* over large areas of common land, treating it as if it was their private property.

Opposing groups claim that, due to the lack of justice towards the communities that own the wind resource, wind farms will contribute to migration to other parts of Mexico and other countries (particularly the USA), and to the influx of external professionals from urban centers. The result will be the disappearance of the existing indigenous cultures, a process that they perceive as territorial displacement by private companies (mostly owned by foreign capital).

According to the Human Rights Centre Tepeyac from Tehuantepec (an organization originally created by the Catholic Church to defend the rights of the local population and, in particular, of indigenous people) and other opposing groups, the companies have distorted the information given to the indigenous population, and in particular, to those who do not speak Spanish. Their contracts were signed with the intervention of interpreters who did not translate the contracts literally, truthfully, or fully. The contracts were not translated into the languages spoken by the communities, even though there is a law of language rights that determines that they must be.

Bettina Cruz Velázquez, member of the Zapoteca community and the Assembly in Defense of the Land and Territory of Juchitán, underlines that the concept of indigenous peoples' development in the region is based on their autonomy and capacity to decide collectively about their future. From her point of view, wind energy projects will erode both aspects, resulting in the loss of indigenous cultural identities that have remained alive in the Isthmus for the last 500 years, despite adversity. She asserts that such a result is probably not incidental but intentional, since the loss of identity is necessary to undertake other kinds of mega-projects in the Isthmus, a region of great geo-strategic interest.

The fact that land lease contracts are valid for twenty-five to thirty years and can be automatically renewed for an equal period is another cause for concern about the future of this territory. The perception of opposing groups is that, after sixty years, there will be no local population left to claim back the land.

All these reasons have led several environmental and human rights NGOs to criticize the wind energy projects in the Isthmus. Mass media published the position of Greenpeace: Cecilia Navarro, communication officer of Greenpeace Mexico, declared, "We do not want corporations to build wind farms that expel communities out of their land. This is not the development that the country needs. We need to develop clean energies together with the communities that own the land, so that they are part of the wind farms and of the decisions."[8]

COMMUNITY ORGANIZATION: LEGAL CHALLENGES AND CULTURAL RESISTANCE

There are groups that reject this kind of wind farm project in almost all affected communities of the Isthmus of Tehuantepec. As a result of their work, around 180 legal demands to nullify the land lease contracts have been presented by land owners against project developers. The juridical argument behind them is that the companies withdrew and manipulated information and acted in a premeditated manner, using the disadvantaged position of farmers in order to obtain larger profits. Many cases are also based on the claim that companies did not provide the contract in indigenous languages, and, in the case of illiterate farmers, did not read the complete contract including the restrictions implied. All the demands were accepted by the court, but not processed. The companies sued were informed (through irregular channels, since the demands were not processed) about the identity of the complainants, and tried to convince the land owners to withdraw the complaint, offering more money—a strategy that was mostly unsuccessful. Then they agreed to nullify the contracts of the land owners who still wanted to get out of their contracts, and to pay all the costs involved (notaries, etc.). Opposition groups believe that companies did this in order to avoid a court case where the legal irregularities involved in getting the contracts signed would be proven.

According to Javier Balderas Castillo, from the Human Rights Centre Tepeyac, the resistance against these corporate projects is still at an early stage and it is not yet a mass movement, since there is not enough information in the communities. His organization has demanded comprehensive information about the wind farm projects since 1995, but never received the information. With the construction and operation of La Venta II, people could see the real impacts, but the movement was not mature enough to confront the situation in an effective manner. It was even more difficult to confront the companies that have been signing land lease contracts for many years. They assess that between 25,000 and 35,000 hectares have already been leased in irregular and unequal conditions.

Bettina Cruz Velázquez explains that the Assembly in Defense of the Land and the Territory of Juchitán was constituted on the basis of rejection of the wind projects, and its members do not accept negotiations with the companies. The Assembly is not against wind power, but against the land grabbing by companies and against

---

8   http://estadis.eluniversal.com.mx/notas/512513.html

the impact that it will have on the life, culture, and territory due to the way in which the projects have been drawn up. They are concerned about how all aspects of social relations will be transformed—for instance the work of women, who play a central role in the Zapoteca culture. These intangible values will be lost due to these projects. They demand complete information, followed by participatory and democratic territorial planning that assures that the impact is minimized and the common benefit as large as possible. The Assembly has few members, about 100 persons who signed contracts in Juchitán, but even though they are a minority, the members are conscious and daring persons. In addition to people who signed contracts, there are people who did not sign, or who decided not to sign due to the work of the Assembly. They have already paralyzed projects in some areas of the Juchitán region, including El Cazadero, where the companies wanted access to 2,000 hectares, but the *Ejido* assembly decided not to approve the project.

There are other examples of the mobilization's impact in communities where no contracts have been signed yet. The *Ejidos* of San Francisco del Mar and San Mateo del Mar, in the Huave region, rejected the wind projects in their respective assemblies. The community of Ixtepec has not signed any contract with private developers, and has decided to undertake a community wind energy project instead (see Chapter 58, "The Yansa Group: Renewable Energy as a Common Resource").

There are two networks that could have a significant impact on the situation in the Isthmus of Tehuantepec. One of them, still taking shape, has the objective of supporting community organizations in the Isthmus in their efforts to get back control of their land. A legal working group was constituted in April 2009, composed by human rights organizations, environmental NGOs, lawyers and academics. This group is working with communities in order to increase awareness about their rights, and to offer legal assistance. The other network is RENACER, the National Community Network for Renewable Energy (Red Nacional Comunitaria por la Energía Renovable), formed in April 2009 by community organizations, human rights organizations, environmental NGOs, etc. This network has three aims: to provide information about the community-based model of renewable energy development that has been successfully implemented in other parts of the world; to help communities to realize renewable energy alternatives, and to work in favor of a legal framework that enables communities to undertake projects on different scales, including large-scale production for the grid.

In August 2009, a forum will take place in Juchitán, organized by the Assembly in Defense of the Land and Territory of Juchitán together with RENACER. The goal of this forum is to give more information to the communities about the situation in the Isthmus and about community-based alternatives. Other objectives are to strengthen cooperation within the region and with organizations from other parts of Mexico, to describe and discuss the new legal and policy framework that is being developed in Mexico in relation to renewable energy generation, and to develop a common strategy and program of work. It is also expected that the forum will give a formal shape to the emerging network to support communities in the Isthmus.

LEGAL AND POLICY DEVELOPMENT

The law passed in 2008 triggered the regulation of renewable energy production in Mexico. This regulation is taking shape now, in the form of several legal and policy instruments.

The law includes two positive provisions: Article 2 asserts that "The utilization of renewable energy sources and the use of clean technologies is of public interest," and Article 21 establishes that affected communities should be consulted on large-scale renewable energy projects and these projects should contribute to their social development. However, these provisions are very vague and offer no guarantee that the communities will have real decision-making power beyond accepting or rejecting ready-made proposals. The real possibilities and role of the different players (state bodies, private companies and local communities) will be set by policy documents and legal instruments that will come out within the next few months. Several of them are already under discussion.

Probably the most important of these policy documents and legal instruments is the Reglament of the Law, which lays out renewable energy policy in more specific terms than the law itself. The current draft of the Reglament is not fair for local communities and does not give them the role that should correspond to them as owners of the land with renewable energy resources. It only considers them as active players in relation to small-scale generation in communities that are not yet connected to the power grid. This is a marginal part of the energy "market" that is of no interest to public bodies or private investors.

In contrast, the draft Reglament of the Law includes a very obscure provision that entitles private projects to get a direct public subsidy (in the form of a payment "for the capacity that they add to the grid") if they are included in the goals set by the Special Programme for the Utilization of Renewable Energy, another legal instrument under discussion. The current draft of the Special Programme for the Utilization of Renewable Energy includes, amongst its goals, the realization of all the large wind farms being built under a self-supply license in the Isthmus of Tehuantepec. Therefore, the passage of the current drafts of the Reglament and the Special Programme would result in these projects, which consist of private generation for private consumption, being entitled to receive a direct public subsidy, in addition to the indirect public subsidies that they already receive in the form of free energy "storage" services and tax benefits.

RENACER has submitted a number of comments to the official body in charge of the legislation, asking to remove this provision and ensure that public subsidies are only available for projects that contribute to the public interest, such as projects undertaken by communities and state authorities. It also demands that communities are considered as legitimate power producers and given the possibility to produce for the national grid. RENACER is not very optimistic about the likelihood that these comments will be taken into account, and will therefore continue advocacy and awareness-raising work on this matter, even after the Reglament and the Special Programme are passed.

COMMUNITY-BASED ALTERNATIVES

One of the topics that will be discussed at the upcoming Forum organized by REN-ACER and the Assembly in Defense of the Land and Territory of Juchitán is the construction of community-based wind farms. The community of Ixtepec has already decided to undertake a community project to sell wind power to the grid, instead of giving up control of its land to private companies. The same idea is being debated in other communities.

The main obstacle faced by this alternative is that most authorities seem to consider communities incapable of undertaking projects beyond a minuscule scale, as reflected by the current draft Reglament of the new renewable energy law. Mexican institutions seem to think that offering private developers everything they ask for is the only way to ensure that renewable energy projects will be realized.

In order to alter this perception and change the course of renewable energy policy, community projects need to be developed and proposed to the authorities and society at large, explaining the range of reasons that make them far more beneficial to society as a whole than private projects. Amongst these reasons is the fact that community projects can stop the process by which the best Mexican renewable energy resources, which will form the backbone of the national energy system, are falling under the long-term control of private (and often foreign) corporations, whose only interest is generating private profit. Another reason is that the contract model proposed community project promoters and results in a long-term reduction in the costs of electricity for all users, and therefore in a large social benefit, in contrast with the privatization of wind resources undertaken by self-supply projects.

RENACER will undertake technical work and awareness raising, and will lobby and advocate to support communities that decide to develop alternative renewable energy projects. It is expected that the joint effort of communities, human rights organizations, environmental NGOs, and other civil society organizations will succeed in providing the conditions under which community projects occur. This would enable the introduction of an alternative to the exploitative corporate projects in the Isthmus of Tehuantepec. Such an alternative would break the deadlock between the two currently existing options (rejection or approval) in the debate about wind energy projects in the region, and foster a conflict-free and community-oriented development of the Mexican wind resource.

Chapter 46 ▌ Part 11

TWO MINI CASE STUDIES:

The End of One Danish Windmill Cooperative, by Jane Kruse[1]

Chinese peasants killed in land conflict over windmills,
an excerpt from a Reuters report[2]

CASE STUDY 1—THE END OF ONE DANISH WINDMILL COOPERATIVE

There is a growing trend for windmills to be owned by individuals, which is a very unfortunate and unfair development.

In January 1988, forty-nine people decided to come together to purchase and install a 200kW windmill in Kallerup in the Thy region of Denmark. The members of this co-op gathered annually in small local restaurants to socialize, receive an annual report for their windmill, and listen to speeches about wind energy and other renewable energy technologies.

But no longer: the windmill has been sold. During its life, it was able to produce enough electricity for 100 families, but the government wants to have even larger windmills, so is giving subsidies to those putting up bigger windmills and decommissioning smaller ones. Because of these subsidies, the cooperative was offered such a large amount of money for their windmill that, in 2005, at the annual meeting they voted to sell it and end the cooperative.

When wind energy was introduced in Denmark it was extremely exciting and popular; it was a frequent topic of discussion between neighbors, colleagues, families, and friends. People felt good about finding ways to tackle environmental issues. Small and medium-sized companies in Jutland jumped quickly at the opportunity to produce windmills and gradually became leading producers in Denmark and internationally.

More than 150,000 families in Denmark invested in windmill cooperatives. The national and local governments did their part to ensure participation by making sure that cooperatively-owned windmills had the right to be connected to the grid, that the utilities would buy their clean energy, and by guaranteeing them a certain price per kWh.

Now it seems the future of windmills is individual owners. Because the ownership is moving away from cooperatives to investors, who are making profits in the millions, more people are starting to protest wind power. What were once beloved windmills are now seen as money-making machines.

---

1   This study was originally published in the Danish newspaper *Politiken* Feb. 9, 2006. It was translated into English by Jane Kruse and Melissa Valgardson, and has been reprinted in this book with permission from the author.

2   This report was posted by Reuters on 24 May 2006.

One of the reasons for this is government wind-energy planning, which dictates the specific coordinates where windmills can be placed. The right for rural farmers and their neighbors to decide where to put up a windmill no longer exists. The owner of the field where these specific coordinates lie is a very lucky person, and often, they are quick to take advantage by putting up large, MW-sized windmills, and sit back to watch the wind blow millions directly into their pockets.

Meanwhile, their neighbor has to see, hear, and even feel the presence of the windmill, living with the change it brings to the landscape, but without the benefits—not paying for electricity—of being an owner. The windmill owner's bank account continues to grow with the 5 million kWh+ of electricity (enough for 1,200 families) being sold each year. With this good fortune, maybe the windmill owner can soon buy the neighbor's house, farm, and who knows what else?

Politicians have to act. Ownership of windmills should only be held by cooperatives or communities, unless, of course, an individual wants a small windmill to cover their own electricity needs. Cooperative and community-owned power ensures that the benefits are shared equally, which is needed for wind energy to regain the support it once had.

AFTERWORD

This great loss of active community involvement was caused by government policy. In the meantime, liberalization and market principles became the new paradigm, with the consequence that, from 2003–2007, installation of new wind power in Denmark was almost 0 MW. It is ridiculous too, that for the right to take down a seventeen-year-old windmill, the cooperative received the same amount of money they paid for it brand new. It was sold October 1, 2005 for 950,000 DKK (130,000 Euro). On top of this, the windmill was perfectly operational and could have gone on producing electricity for many more years.

CASE STUDY 2—WARNINGS, JAILINGS REPORTED IN CHINA PROTEST DEATHS

Chinese city officials have received "serious warnings" and six villagers were jailed after police shot and killed people protesting against the building of a wind farm in southern Guangdong province in December, media reports said on Wednesday.

At least three people died and eight were wounded in Dongzhou village, part of Guangdong's Shanwei city, when police shot villagers protesting against a lack of compensation for land lost to a wind power plant, government officials had said.

Villagers put the number of dead as high as twenty.

"The relevant people who did not do their jobs well and were responsible for the serious incident of lawlessness that happened last December have already been gravely disciplined," China's official Xinhua news agency said on its website (http://www.xinhuanet.com).

"The Disciplinary and Supervision Bodies had made the decision to give the deputy Communist Party chief of the city, Liu Jinsheng, … and the deputy police chief, Wu Sheng, serious warnings," it said.

The Washington-based network Radio Free Asia reported that at least six villagers received jail sentences of between three and seven years in the incident.

"The trial began on May 22, and today, the 24th, all of them received their
sentences," RFA's Mandarin service quoted a relative of jailed villager Lin Hanru as
saying.

Lin and Huang Xijun each got a five year sentence, while Huang Xirang was
jailed for seven years, RFA said in a statement quoting the US-supported agency's
broadcast from Hong Kong.

At least three others got three-year terms, RFA added.

Xinhua said other officials included the deputy mayor, city police chief and head
of the city's construction bureau had received mere warnings.

The Xinhua report stuck by the official line that the unrest was caused by what it
called a small group of lawbreakers misleading the unwitting masses.

Previous reports said the deputy police chief had been sacked and placed under
"criminal detention" for the shooting.

Xinhua gave no further details about the detentions or possible charges.

China is grappling with growing social unrest, fueled by disputes over land
rights, corruption and a growing gap between rich and poor.

Chapter 47 | Part 11

STOP THE PRESSES! VESTAS WORKERS OCCUPY WIND TURBINE BLADES FACTORY AND CALL FOR NATIONALIZATION OF PLANT[1]

WHAT WE'RE FIGHTING FOR

Vestas Blades UK on the Isle of Wight is due to close on 31 July, 2009; 600 jobs will be lost immediately, many more jobs that depend on Vestas will follow. This makes no sense from a green or labor perspective!

The government has just announced a major expansion of renewable energy including wind power. We are calling on Vestas to keep the factories open, saving jobs and offering those who want to leave a better redundancy deal. We are calling on the government to intervene to save jobs at Vestas—through nationalization, if that is what it takes—to show that it is serious about saving the planet.

STATEMENT FROM THE VESTAS WORKERS INSIDE THE FACTORY

As workers at a wind turbine manufacturer, we were confident that as the recession took hold, green or renewable energy would be the area where many jobs could be created—not lost. We were horrified to find out that our jobs were moving abroad and that more than 525 jobs from the Isle of Wight and Southampton were going to be added to the already poor state of island unemployment.

This has sent, and will continue to send, shockwaves of uncertainty through countless families—many of whom are being forced to relocate away from the island.

We find this hard to stomach as the government is getting away with claiming it is investing heavily in these types of industries. Only last week, the government said it would create 400,000 green jobs. How can the process start with 600 of us losing ours?

Now I'm not sure about you, but we think it's about time that, if the government can spend billions bailing out the banks—and even nationalize them—surely they can do the same at Vestas.

---

1   Literally in the last few days before the final manuscript of this book had to be sent to the publishers, workers in the Isle of Wight occupied a Vestas factory, in resistance to the plant's imminent closure, the loss of over 500 jobs, and the planned relocation to the USA. The struggle has received very widespread, and active, solidarity throughout the UK and also in other countries. It is qualitatively a step above the other struggles in the wind industry, both in terms of magnitude and the scope of demands. This is the first worker struggle anywhere in the world that is calling for nationalization of a wind turbine manufacturing plant, or for turning it into a workers' cooperative. As such, the struggle is of international significance. This is an emergency last minute addition to the book, containing news snippets from the Save Vestas campaign website: http://savevestas.wordpress.com.

The people of Vestas matter, and the people of the island matter, but equally importantly, the people of this planet matter. We will not be brushed under the carpet by a government that is claiming to help us.

We have occupied our factory and call on the government to step in and nationalize it. We and many others believe it is essential that we continue to keep our factory open for our families and livelihoods, but also for the future of the planet.

We call on Ed Miliband as the relevant minister to come to the island and tell us to our face why it makes sense for the government to launch a campaign to expand green energy at the same moment that the country's only major wind turbine producer closes.

LATER ON....VESTAS WORKERS BESIEGED BY RIOT POLICE (PRESS RELEASE)

Workers staging a sit-in at the soon-to-close Vestas wind turbine plant on the Isle of Wight are being starved out by police.

The police, many inside the factory and dressed in riot gear, have denied food to the workers who took over the factory offices last night, to protest the closure of their factory. The police, operating with highly questionable legal authority, have surrounded the offices, preventing supporters from joining the sit-in, and preventing food from being brought to the protestors.

Around 20 workers at the Vestas Plant in Newport, on the Isle of Wight, occupied the top floor of offices in their factory to protest against its closure, which will result in over 500 job losses.

Acting without an injunction, on private property, the police have repeatedly tried to break into the office where the protesting workers have barricaded themselves, and have threatened the workers with arrest for aggravated trespass, despite the fact that no damage has been done to the property where the protest is taking place. Police have also forcibly removed people from private property, another action that is of very questionable legality in the absence of a formal injunction.

The officer involved in the latter action was number 3606. The officer who appears to be in charge is 3115.

This heavy-handed response is the latest in a long line of over-reactions to protest by various UK police forces.

PRESS RELEASE OF THE RMT
(NATIONAL UNION OF RAIL, MARITIME, AND TRANSPORT WORKERS)

The occupation of the Vestas wind turbine factory on the Isle of Wight today passed another significant milestone, as the workers held back the scheduled closure date of the facility, and the company wrote to staff this morning confirming that the consultation has been extended indefinitely—a move described by Vestas union RMT as a massive victory.

Vestas had planned to close the factory today, Friday 31 July, but as a result of the occupation, and the global campaign in support of the workforce, they have been pushed back and the extension of the consultation with the workforce means that there is a serious opportunity to draw up a rescue package similar to the one supported by the Scottish Parliament earlier this year, which saved the Vestas factory in Kintyre.

This weekend will see a further show of the strength of the growing support for the Vestas workforce, with crowds from the cancelled Big Green Gathering diverting to the Isle of Wight in what will be another important boost for the Save Vestas campaign.

Tomorrow, Saturday 1 August, there will be a major demonstration in support of the campaign starting at 1pm from St Thomass Square in Newport town center.

RMT has also congratulated Gerry Byrne who took the Vestas protest to the fourth plinth in Trafalgar Square for an hour this morning between 5am and 6am.

Bob Crow, general secretary of Vestas workers union RMT, said:

> The fact that the Vestas campaign has held back the scheduled closure date today is another significant milestone in the fight to save the factory and 625 skilled manufacturing jobs in green energy. The extension of the consultation with the workforce this morning gives us a real chance to work up a rescue plan.

This weekend will see a major demonstration of the growing support for the Vestas campaign, which has fired the imagination of the labor and environmental movements all around the world.

RMT remains deeply concerned as to the well-being of those in occupation, and we will be taking further legal and health advice today. This brave group of workers continues to be denied access to their basic human rights to nutritional food and liquids, and we are making every effort to get supplies through.

CAROLINE LUCAS, MEMBER OF THE EUROPEAN PARLIAMENT, PROPOSES A WORKERS' CO-OP

The Isle of Wight's Green Member of the European Parliament (MEP) is to submit an urgent proposal to the Leader of the Isle of Wight Council for support of a workers' co-op at the Vestas wind turbine manufacturing plant to be established.

In a last ditch attempt to keep the Vestas plant open, Dr. Lucas will call on the government to ensure that:

- The workers of the wind turbine company Vestas are permitted to form a workers' cooperative, and are supported in doing so by the government.

- Financial support (at the very least unemployment benefit) is paid to the workers of Vestas until such time as the proposed workers' cooperative is financially viable.

Dr Lucas commented:

> If the government is serious about tackling climate change, helping to protect the future of UK manufacturing, and safeguarding local jobs, it must act now to keep the Vestas facility open for business.

By submitting a proposal under the Sustainable Communities Act for a workers' co-op, the Council can demand that the government provides the investment and assurances necessary to save this facility—on the basis that it plays a crucial economic and environmental role in the local community.

Failure to keep the Vestas plant open will represent a spectacular failure by the government to match its rhetoric on green jobs with real policy action. It should be seizing the opportunity to create a renewable energy revolution that can see us through a transition towards a more environmentally and economically stable economy. Allowing the Isle of Wight plant to close now would be a massive embarrassment for ministers—and devastating for the Isle of Wight's workers.

Chapter 48 | Part 12: Time to Speed Up! Renewable Energy as a Possible Way Out of the World Economic Crisis?

THE POLITICAL, ECONOMIC, AND ECOLOGICAL REASONS FOR ESTABLISHING THE INTERNATIONAL RENEWABLE ENERGY AGENCY, IRENA
Sharing the Benefits Instead of the Burden[1]

Hermann Scheer, on behalf of the World Council for Renewable Energy

Viewed in a global perspective, it is becoming increasingly clear that the future of energy supply lies with renewable energies. The limits of conventional energy supply—I am talking here about fossil and nuclear energy—are glaringly obvious. Today, world civilization stands at a turning point.

Resources are limited. By now, almost everyone recognizes that oil, gas, coal, and uranium reserves are finite. At the same time, the energy demand of a growing world population is increasing at a faster pace than gains are made in energy efficiency and energy saving. On a global scale, the reserves' curve is declining whereas the demand curve is rising due to the growth of the world's population and the developing countries' thirst for energy. This is resulting in rising energy prices, shortages in national economies, and social problems for ever more countries and their citizens. Access to energy sources has become a global political issue.

The direct costs of conventional energies can only rise, whereas costs for renewable energies can but fall. Renewable energies are infinite and, with the exception of biomass, their primary energy is free. It is the technologies required for the production of energy derived from renewable sources that has to be paid for, not fuels. It is only energy coming from biomass that leads to fuel costs, because it requires work in the agricultural and forestry sector, which must be paid for. The cost of technologies will fall due to economies of scale, as well as the predicted rise in productivity of the deployed technologies, which are still comparatively young. Today's additional costs for renewable energies, if in fact they still exist, are the prerequisite for cost-effective energy in the future, which will be available everywhere on the globe, and for everyone. This promising future is closer than most people who have ignored or underestimated the potential of renewable energies—including some governments, scientists, and those in the conventional energy sector—think.

The second limit of the conventional energy supply system is an ecological one. Even if vast new oil, gas, or coal reserves were to be found somewhere under the Earth's surface, world civilization could not afford to secure and use them. The

1   Based on the keynote speech given by the author at the First Preparatory Conference for the Foundation of IRENA, 10 April 2008, Berlin Germany.

ecosphere's capacity to mitigate damage has already been reached. This means that we have to realize the switch to renewable energies now—even before the known reserves of fossil fuels are depleted. We have a window of perhaps four decades; in other words, we are in a race against time.

Our economy and society are facing the biggest challenge since the beginning of industrialization—and it is not climate change alone. Even if the growing problem of global warming, caused by the past heavy deployment of fossil resources, did not exist, the global energy system would still not stay intact. The problem of the ever-increasing scarcity of energy resources, together with various environmental problems, would remain. From an economic point of view, these are the conventional energy supply system's indirect and external costs, not reflected in energy prices—despite the fact that they will have to be paid nonetheless. Only with renewables will we be able to avoid these costs and free societies from their burden.

For the most part, answering to this challenge is considered an economic burden, a short-sighted assumption that leaves a large imprint on the current energy discussion. The switch to renewable energy leads to several meaningful political, economic, social, and ecologic benefits, which tend to be overlooked if one only considers the microeconomic level, and if only isolated cost comparisons of energy investments serve as the benchmark. On the other hand, using macroeconomic and holistic observations leads to different conclusions.

The macroeconomic benefits are evident, though they cannot, at the same time, be a microeconomic benefit for every participant in the national economy. Smart, informed, and forward-thinking political measures and instruments are mandatory to translate macroeconomic benefits into microeconomic incentives. A very positive example of this approach is the German renewable energy sources law, also known as the Feed-in Tariff Law. There, guaranteed grid-access for electricity produced from renewable sources, a guaranteed feed-in-tariff, and no cap on production all give renewable energy producers high levels of investment security. This law has abolished market barriers so that incentives for investments are stimulated effectively.

It is clear today, but becoming even more clear every day, that renewable energies are the future. However, most countries are not very well prepared for the necessary transition towards them. Just a few years ago, governments throughout the world started to realize that focus must shift to renewable energies and their promotion, and their delayed response is the reason that practical implementation is lagging behind. Many countries already deploy different political and economic approaches to foster the production and use of renewable energy, however, to date, only a few have drafted and implemented substantial and ambitious policies, or have the necessary scientific, technological, and industrial prerequisites at their disposal. Given that renewables have been underestimated for years and have not played an important role in the global energy discussion, this should come as no great surprise.

In the 1950s, the focus lay on nuclear energy. Then, the attitude towards nuclear energy was the opposite of how renewables are regarded today: possibilities were overestimated and the risks were underestimated. Nevertheless, almost all countries oriented their national energy strategies towards nuclear. To support this development, two international institutions were established in 1957: EURATOM in Western Europe and the International Atomic Energy Agency (IAEA), which had a global focus. The establishment of IAEA was welcomed by the UN, but was not founded as a UN organization, and not all UN member states became founding members.

The IAEA's task is more than just preventing the abuse of fissile material; they are also mandated to help governments develop nuclear energy programs, to facilitate technology transfer, and to build human resource capacities. The existence of IAEA, with its staff of roughly 2,000 and annual budget of more than \$250 million, is in itself a valuable motive for setting up IRENA—to establish a balance between nuclear and renewable energies.

IAEA has existed for half a century, and the call to establish an International Renewable Energy Agency was raised twenty-eight years ago—in the framework of the North-South Commission's Report chaired by the former German Chancellor, Willy Brandt. The establishment of such an agency was recommended in the final resolution of the first UN conference on renewable energy in Nairobi in 1981 (Conference on New and Renewable Sources of Energy). Nevertheless, these recommendations were largely unheard. It has frequently been argued that it would be sufficient to mandate existing UN organizations with the promotion of renewable energy.

The focus on promoting renewables internationally has steadily grown in importance. The 1973 oil crisis made it clear for everyone that the oil age would not last forever. To ensure security of supply for fossil resources, the OECD countries established the International Energy Agency (IEA). The IEA is also not a UN Agency—it was called "Club of the Rich." Thus, a third international organization covering energy matters was established—and all three of them have contributed to underestimating renewable energy.

Even though most industrialized nations announced that they would embark on research and development programs for renewable energy after the oil crisis, priority for research and development still lays elsewhere. When oil prices fell back to a lower level in the early 1980s, most countries scaled down their still young renewable energy programs.

On the other hand, the 80s and 90s witnessed a growing unease around nuclear and fossil energies in many societies. Whereas the 80s were characterized—after the catastrophe in Chernobyl—by strong scepticism concerning nuclear energy, the 90s—with the climate reports growing increasingly alarming—took a critical look at fossil energies. However, since many thought that there would not be a realistic alternative to conventional energies, these controversies reached the international energy discussion rather late in the day.

On the other hand, various scientific reports were published that demonstrated the feasibility of using renewables to meet all energy needs: a study by the Union of Concerned Scientists in the US in 1979, a study by the Club de Bellevue, an initiative of scientists from leading French research institutes, or a study focusing on Europe published by the Institute of Applied Systems Analysis in Laxenburg. These examples show that the lack of an international agency for renewable energy helps to explain why these energies have been neglected for so long.

In 1990, the European Association for Renewable Energies, EUROSOLAR—of which I am the president besides my capacity as member of parliament—drafted the first comprehensive memorandum on establishing IRENA and published it widely. At the invitation of Ahmedou Ould-Abdallah, the former energy commissioner of the UN Secretary General, I presented this memorandum at the UN headquarters in New York. Consequently, former UN Secretary General Perez de Cuellar established a task force, the United Nations Solar Energy Group on Environment and Development (UNSEGED). Chaired by Prof. Thomas Johansson, the group concluded that the establishment of the International Renewable Energy Agency was necessary. This proposal was aimed at the Rio-Conference of 1992, where it was expected that the agency would be established. At the invitation of the US senate, the Interparliamentary Conference on the Global Environment, chaired by Al Gore, took place in Washington in 1991. At this conference, I proposed that the conference's resolution should also speak in favor of the establishment of an IRENA, and this proposal was adopted unanimously.

However, opposition, motivated by a range of reasons, meant that not all of these efforts were successful. Existing UN organizations that were partly active in the field of renewables, but with far fewer capacities than the proposed IRENA would have, spoke against the establishment of the agency, as did OPEC states that identified IRENA as a potential rival, those that did not consider renewable energy sources enough to supply the world's needs, and of course, the conventional energy agencies.

However, none of the above critics are able to explain how the global spread of renewables will be supported to the extent that is necessary if not through an agency like IRENA. Therefore, you always have to counter questions like, "Why is IRENA necessary?" or "What is the added-value of IRENA?" with a counter question: "Why should IRENA *not* be necessary if one considers the existence of an International Atomic Energy Agency or an International Energy Agency to be necessary?" Alternatively, "What will we risk if we do not switch to renewables fast enough?"

For many years, in many speeches at international conferences in numerous countries, I have been advocating for the establishment of IRENA. Prerequisite for the founding has always been that one or more governments would take the initiative and build a coalition of like-minded countries that push the establishment of the agency forward. One important milestone on the way towards establishing IRENA was the 2004 International Parliamentary Forum on Renewable Energies, which was hosted by the German Parliament and took place in parallel to the governmental conference "Renewables 2004." Over 300 members of parliament, from 70 countries, took part in the conference, a conference that I was happy to chair. The Final Resolution states: "Promoting renewables requires new institutional measures in the field of international cooperation. To facilitate technology transfer on renewables and energy efficiency and to develop and promote policy strategies, the most important institutional measure is to establish an International Renewable Energy Agency (IRENA), which should be set up as an international intergovernmental organization. Membership would be voluntary, and all governments should have the opportunity to join at any time. The Agency's primary tasks would be to advise governments and international organizations on the development of policy and funding strategies for renewables use, to promote international non-commercial technology transfer, and to provide training and development."

This initiative contributed to the German government adopting IRENA's establishment as one of its policy projects. Today, we are starting with the Preparatory Conference that will lead us to the establishment of IRENA next year.

I am positive that the number of member states will increase swiftly once IRENA is established and has started its activities. IRENA will shorten the way to a global deployment of renewables and will accelerate its pace. We can already be sure today: the Founding Conference of IRENA will be a historic date.

Chapter 49 ∎ Part 12

THE ROLE OF IRENA IN THE CONTEXT OF OTHER INTERNATIONAL ORGANIZATIONS AND INITIATIVES[1]

By the IRENA Secretariat

After intensive preparations involving more than sixty countries, the International Renewable Energy Agency, IRENA, will be founded on 26 January 2009, in Bonn. Mandated by governments worldwide and acting as a global voice for renewable energy, IRENA will act as the main driving force in promoting a rapid transition towards the widespread and sustainable use of renewable energy.

There is a growing consensus that, in order to cope with the challenges of our time, such as increasing global energy demand, global warming, and rising energy prices, the world needs a massive scale-up in renewable energy within a short period of time. This massive increase requires mobilization of human, technological, and financial capacities on a global scale. Thus, IRENA intends to provide global leadership and expertise so as to ensure that renewable energy rapidly reaches its potential.

IRENA will close an institutional gap. It is the first intergovernmental organization to solely concentrate on renewable energy and offer support to industrialized and developing countries alike. IRENA will provide practical advice, and thus support member countries in improving their regulatory frameworks and building capacity. The agency will facilitate access to all relevant information, including reliable data on the potential of renewable energy, best practices, effective financial mechanisms, and state-of-the-art technological expertise.

There are several organizations or initiatives that share IRENA's aim of promoting the use of renewable energy. Amongst these organizations and initiatives are the following: the International Energy Agency (IEA), the United Nations Environment Programme (UNEP), the United Nations Development Programme (UNDP), the United Nations Industrial Development Organisation (UNIDO), the World Bank, the Renewable Energy and Energy Efficiency Partnership (REEEP), and the Renewable Energy Policy Network for the 21st Century (REN21). However, these organizations have different missions, focus, operate with individual mandates, and/or on particular levels (local, regional, global).

In the interest of all countries being members of several organizations, and taking into account the challenge of massive and rapid escalation of renewable energy, unnecessary duplication of work, and thus, wasting of resources should by all means

---

1   This article is an official IRENA document that was written in December 2008 as part of the process of establishing the agency. It is publicly available at http://irena.org/downloads/Role_IRENA_IO.pdf, and is being reprinted here with the permission of Monika Frieling, of the German Environment Ministry (Bundesministerium für Umwelt, Naturschutz und Reaktorsicherheit).

be avoided. There is more than enough work for each organization, requiring optimization of capacities and close cooperation to create synergies.

In order to establish this close cooperation and build the foundation for a trust-based relationship, the initiators of IRENA started, in June 2008, and continued to meet and consult with the other international players. Initial ambiguity gave way to a clear support for the foundation and the welcoming of synergies and cooperation opportunities. All international organizations understand the mission of IRENA and are open to cooperation, and some have explicitly encouraged IRENA to be ambitious in taking a leadership role in the massive deployment of renewable energy. In most cases, opportunities for cooperation have already been identified, and will need further elaboration when IRENA starts operating. The first director-general of the agency will thus be able to expand existing contacts and initiate the next concrete steps for collaboration.

The following paper intends to outline the activities of these international players and the potential options for cooperation with IRENA.

COMPARING THE ACTIVITIES OF IRENA WITH THOSE OF OTHER ORGANIZATIONS

The activities of the relevant organizations and the most interesting fields of cooperation have been identified using the structure of IRENA's initial work program draft of August 2008. This work program is in line with the objectives and activities outlined in the Statute, and was meant to provide a framework for the first phase of IRENA's operation.

As the bold headlines are similar in many organizations, it is important to consider the specific focus, the actual volume, and the approach of the activities, as well as the countries and partners involved.

IRENA

IRENA will act as an international governmental organization, focusing on the promotion of renewable energy and welcoming all UN members to join. Its main tasks are to provide relevant policy advice and assistance to its members upon their request, improve pertinent knowledge and technology transfer, and promote the development of local capacity and competence in member states.

Considering IRENA's tasks, an annual budget of US $25 million and a team of 120 are a realistic approximation of its operating expenses in the first years.

IRENA will be unique in combining:

- a clear mandate from its member governments;

- a worldwide geographical scope, including industrialized and developing countries;

- a complete range of services to support governments in facilitating the use of renewable energy;

- high level expertise in the use of all forms of renewable energy and their integration into energy systems, covering technological, economic, institutional, cultural, social, and environmental aspects;

- an extended network base with research and government institutions worldwide;

- global collection, elaboration, and dissemination of information and knowledge concerning renewable energy; and lastly,

- methods, tools, and networks for promoting experience exchange and accelerating an international learning process.

In addition, IRENA's work will explicitly take into account cultural and social aspects. This is especially important as renewable energies offer opportunities for more decentralized energy supply systems, involving changes in behavior and governance structures. Thus, mutual learning between various economic, environmental, social, cultural, and political conditions is essential for tackling the challenges.

IRENA's initial work program (second draft, August 2008) outlines the future activities of the Agency. The following table summarizes the intended basic tasks and initial activities.

A. Developing a comprehensive knowledge base	• Taking stock of existing knowledge and activities • Developing a reporting system and an extensive database • Developing methodologies for the use of various tools promoting renewable energy
B. Networking	• Cooperating with other organisations, institutions and networks • Consulting with experts from academia and industry
C. Communication	• Establishing an internet-based communication platform • Strengthening international dialogue on renewable energy • Developing a key publication • Building relations with the media
1) ADVISING NATIONAL GOVERNMENTS ON DEVELOPING INTEGRATED APPROACHES TO PROMOTING RENEWABLE ENERGY	• Comprehensive advice on selecting and adapting energy sources, technology and system configurations and organisational and regulatory frameworks • Helping countries to make the best use of available funding • Promoting experience exchange between countries • Assembling a toolbox of successful methodologies and policies
2) INTEGRATING RENEWABLE ENERGY INTO THE URBAN ENVIRONMENT	• Promoting structured experience exchange between countries and cities • Contributing to capacity building • Supporting the development of appropriate local policies
3) STRENGTHENING STRATEGIC COOPERATION IN RURAL AREAS	• Developing integrated approaches for the use of renewable energy in rural areas, while in close cooperation with development organisations, stakeholders and experts • Supporting experience exchange within and between countries and regions • Developing a toolbox of technical, organisational, cultural and political innovations for rural electrification and rural thermal energy supply
4) IDENTIFYING TRAINING NEEDS AND OPPORTUNITIES FOR DEVELOPING RENEWABLE ENERGY	• Facilitating international exchange of experiences and development of coherent methodologies • Developing different approaches for different sources of renewable energy • Contributing to a shared pool of knowledge and methods

5)	INCREASING THE PROPORTION OF RENEWABLE ENERGY USED IN EXISTING ENERGY SYSTEMS	• Identifying and promoting new technical and managerial approaches to the design and management of energy systems and grids appropriate to renewable energies • Producing a series of reference reports describing the current situation, presenting best practice examples and —where useful—proposing standardised approaches and norms.
6)	PROMOTING TECHNOLOGY TRANSFER	• Organising a conference on discussing possible funding mechanisms and strategies for technology transfer • Evaluating specific technology transfer projects in different countries; development of an appropriate assessment methodology
7)	RAISING THE PROFILE OF RENEWABLE ENERGY IN THE ENERGY AND CLIMATE DEBATE	• Developing concepts and policy briefings. • Producing publications, promoting the role of renewable energy in reducing greenhouse gas emissions, combating climate change and guaranteeing the security of energy supply • Developing scenarios for the use of renewable energy • Being present at international climate conferences, observing and informing

IEA—INTERNATIONAL ENERGY AGENCY

The International Energy Agency, IEA, based in Paris, is an autonomous agency linked to the OECD. It was created in 1974 by sixteen OECD countries to ensure energy security after a politically motivated oil shortage and a doubling of the oil price by the OPEC. Today, it acts as an energy-policy advisor to its twenty-eight member countries, all of whom are OECD members, and therefore, no emerging or developing countries are included.

IEA is steered by a governing board. The votes of the member countries are weighted according to their volume of oil consumption. Its objectives are energy security, economic development, and environmental protection. The IEA budget for 2008 is €24.5 million, and it staffs about 190 people.

The IEA has designed arrangements for emergency preparedness, and analyses and monitors developments in the international oil and gas market, undertakes policy analysis and co-operation, collects and processes data ("World Energy Outlook"), fosters energy technology, and, among other things, focuses on energy efficiency and environmental issues. It also produces extensive energy statistics that include some non-IEA countries. The *World Energy Outlook* is the IEA's main publication, quoted as an important reference worldwide. IEA's regular country reports review the respective energy policies.

Since the IEA covers all forms of energy with a traditional focus on conventional energy, only a small—though it is growing—part of IEA activities, is dedicated to renewable energies, strongly supported by targeted voluntary contributions. The International Technology cooperation of the IEA is organized in so-called Implementing Agreements, nine of which concern renewable energy technologies. The agreement on Renewable Energy Technology Deployment (RETD), involving ten countries, is in charge of cross-cutting issues.

IRENA and the IEA fundamentally differ in three regards: IRENA focuses on renewable energy, whereas the IEA covers all energy issues with an emphasis on the conventional energy system, relying on fossil and nuclear sources; IRENA is open to all UN members, whereas the IEA is limited to OECD countries; and IRENA will look beyond the traditional energy supply sector, because renewable energies involve a much larger part of the economy than just traditional fuels (building sector, agriculture, etc.).

Comparing IRENA's work program with IEA activities leads to the following opportunities for cooperation, which have also been discussed at joint meetings:

•   IRENA will dedicate much larger resources to renewable energy than the IEA. Offering in-depth expertise on renewable energy, IRENA can support the IEA in giving renewable energy a stronger emphasis in its cross-cutting activities (e.g. *Energy Technology Perspectives, World Energy Outlook*, statistics etc.);

•   the IEA maintains extended data-reporting mechanisms in order to upkeep its statistics and policy databases. IRENA will have similar needs, but with a different perspective. In order to avoid duplications in member countries and in the organizations it will be important to cooperate on reporting systems;

•   regarding the Implementing Agreements, cooperation should contribute to make best use of the results of the technical and ecological issues and the need for further research.

UNEP—UNITED NATIONS ENVIRONMENTAL PROGRAMME

UNEP is in charge of environmental issues within the UN, and has an extended worldwide network of offices and specialised structures. The UNEP Energy Programme makes up about 3 percent ($12 million) of UNEP's total expenditures.[2] Working with a variety of partners, UNEP helps countries to develop and use tools for analyzing energy policies and programs, climate change mitigation options, energy sector reforms, and the environmental implications of transport choices.

Special attention is given to helping financial institutions improve their understanding of investment opportunities in the renewable energy and energy efficiency sectors. UNEP works with local banks to establish end-user financing mechanisms for renewable energy technologies, and works with the international finance industry (including banks and insurance companies) to lower risks for larger projects and to break down financial barriers. It also provides advice to developing nations' governments on broad policy approaches that will bolster renewable sources of energy. It also supports the creation of an enabling environment for small- and micro-

---

2   http://esa.un.org/un-energy/pdf/un_energy_overview.pdf, p. 30.

businesses in the area of renewable energy. The goal of UNEP's Energy Programme is to bring a longer term, environmental dimension into energy sector decisions.

The UNEP Energy Programme is coordinated by the Energy Branch, located in Paris. The Energy Programme incorporates a wide range of structures that include the URC (UNEP Risø Centre), BASE (Basel Agency for Sustainable Energy), GNESD (Global Network on Energy for Sustainable Development), REED (Rural Energy Enterprise Development), and SEFI (Sustainable Energy Finance Initiative). In addition, regional UNEP offices contribute to the program, mainly focusing on projects and their networks. Work on renewable energies is mostly embedded in more comprehensive approaches. Within UN Energy—a coordination of UN bodies dealing with energy issues—with the Food and Agriculture Organization, UNEP chairs the subcommittee dealing with renewable energies.

Comparing IRENA's work program with UNEP activities leads to the following opportunities for cooperation, as also discussed in a joint November 2008 meeting:

- UNEP's advanced expertise on financing mechanisms by SEFI/BASE can be most valuable in complex policy advice by IRENA, and can eventually help define IRENA's complementary services, meeting the needs of its members in financing renewable energy policies and projects.

- Cooperating with the UNEP energy branch and the multiple projects and networks can contribute considerably to IRENA knowledge base and, in return, can provide updated and well-structured information. Also, cooperation on tools and methods could be most interesting.

UNDP—UNITED NATIONS DEVELOPMENT PROGRAMME

As the UN's global development coordinator, UNDP advocates for change and connects countries to knowledge, experience, and resources to help people build a better life. UNDP supports energy activities that reduce poverty and achieve sustainable development objectives at the local, national, and global level. While annual program expenditures are about US $5 billion, approximately $53 million is devoted to providing access to sustainable energy services.[3] UNDP works mainly at the project level, but is also engaged in strengthening national policy frameworks. UNDP's energy efforts are strongly focused on access to energy and poverty reduction. Renewable energies play an important role, but, together with energy efficiency and conventional energy sources, are only one of several options.

Opportunities for cooperation with IRENA must be further explored while building up IRENA's activities. Discussions can include the following:

- Regarding UNDP's experience in promoting rural energy services and IRENA's in-depth expertise on renewable energy, cooperation in this field may contribute to a stronger role of renewable energy in rural development and poverty reduction.

---

3   UNDP multi-year funding framework report 2007. Program expenditure 2004–2007: 14.3 billion.

- IRENA can offer specialised support and learn from UNDP's wide range of experiences.

UNIDO—UNITED NATIONS INDUSTRIAL DEVELOPMENT ORGANISATION

Established in 1966, UNIDO promotes the creation of wealth and tackles poverty alleviation through manufacturing. "Energy and the environment" (promoting cleaner and efficient use of energy) is one of its three inter-related thematic priorities. UNIDO also facilitates productive activities in rural areas by providing modern and renewable forms of energy, and enhancing the use of renewable energy for industrial applications. The overall budget is US $382 million (2008–2009),[4] only a small fraction of which is devoted to renewable energy issues, most of which is embedded in technical cooperation programs. UNIDO chairs UN-Energy, the coordination of UN bodies dealing with energy issues.

Opportunities for cooperation include the following:

- UNIDO is strongly focused on renewable energies for productive use and industrial application in Africa: cooperation should ensure complementarity;

- coordination and selective cooperation, especially with regard to industrial application of renewables, may contribute to best use of resources in order to promote renewable energy.

WORLD BANK

Among its many tasks, the World Bank Group is one of the largest multi-lateral lenders for energy worldwide (US $4.2 billion). It contains a dense worldwide network of offices with 10,000 staff members. The World Bank's stated objectives in the energy sector are: poverty alleviation, energy security, and climate change mitigation. Renewables play an important role, and the budget is increasing but still accounts for less than 30 percent of the bank's total lending for energy. Within the renewable energy sector, large hydro-power commitments account for more than 60 percent.[5]

The following issues present opportunities for developing cooperation:

- financing mechanisms for renewable energy will be at the center of common interests—meeting the specific renewable energy financing problems and finding ways to convey increasing funds to this sector are issues where cooperation can bring an added value;

- the possibility of cooperating on the World Bank's toolkits for project financing should be discussed; and

- as the World Bank is one of the largest lenders for energy, IRENA can support it to give renewable energy a greater role in the bank's investments in developing countries.

---

4   Annual report 2007.
5   http://siteresources.worldbank.org/INTENERGY/Resources/renewableenergy12407SCREEN.pdf

REN21—RENEWABLE ENERGY POLICY NETWORK FOR THE 21ST CENTURY

REN21 is a global policy network that provides a stakeholder forum for international leadership on renewable energy. Originating at the first International Renewable Energy Conference (Bonn 2004), REN21 connects governments, international institutions, non-governmental organizations, industry associations, and other partnerships and initiatives. With a budget of US $1 million[6] and less than ten employees, REN21 concentrates on policy, advocacy, and exchange; its main product is the annual *Renewable Energy Global Status Report* (GSR). Moreover, since Bonn 2004, it has been involved in the International Renewable Energy Conferences (IRECs) and the monitoring of the corresponding action programs.

REN21 and IRENA share the same vision. Whereas REN21 is a global policy network with broad participation of governments, international organizations and NGO's, but without legal status and limited mandate, IRENA will be a global intergovernmental agency with a clear mandate from its member countries and much larger financial, as well as personnel resources.

Cooperation opportunities, as also discussed in a joint meeting, include the following:

- as a multi-stakeholder network, REN21 will be a valuable partner for IRENA in its aim to involve stakeholders other than member governments and institutions;

- the most important aspect will be to develop close cooperation concerning reporting on renewable energy and the support of international fora for policy makers. Both areas are of vital interest to IRENA and REN21.

REEEP—RENEWABLE ENERGY AND ENERGY EFFICIENCY PARTNERSHIP

REEEP is a Public-Private Partnership launched at the Johannesburg World Summit in 2002. It works with governments, businesses, industry, financiers, and civil society to accelerate the global marketplace for energy efficiency and renewable energy.

The private sector is a key focus for REEEP. Business models, finance facilities, and capacity-building initiatives are developed to enable new enterprises to enter the clean energy sector. All REEEP projects occur in emerging markets and developing countries. It actively facilitates financing for sustainable energy projects and structures policy initiatives for clean energy markets on the ground. REEEP has less than fifteen staff members in eight regional secretariats, and finances its projects through specific fundraising. Together with REN21, it initiated a joint project called REEGLE (Information Gateway for Renewable Energy and Energy Efficiency).

Opportunities for cooperation include:

- A partnership with IRENA for experience exchange;
- cooperation on single projects is conceivable;

---

6   Regular contributions, additional project funding.

- REEEP's worldwide network of partners may be helpful in supporting IRENA's outreach;

- the internet platform REEGLE is an important information source on renewable energy, and thus, cooperation is highly desirable; and

- REEEP has offered to use REEGLE as a common information platform.

SUMMING UP THE COOPERATION OPPORTUNITIES

A detailed comparison of IRENA's structure, tasks, and the draft work program with the ongoing activities of other international organizations, as well as repeated discussions with representatives from most of these organizations led to the following conclusions.

IRENA will play a central role in the context of the international organizations dealing with renewable energy issues by:

- filling a vacant role as a dedicated worldwide inter-governmental organization for renewable energy; providing leadership and expertise for a massive scale-up of renewable energy use;

- expanding and qualifying available resources (financial, human, institutional, expertise, etc.) for promoting renewable energy;

- providing transparency concerning the manifold landscape of ongoing activities and encouraging cooperation;

- offering support to the ongoing activities through in-depth expertise; reliable, comprehensive, and up-to-date information; and experience exchange, tools, and methods;

- strengthening advocacy for renewable energy at all levels; and

- facilitating networking and information exchange.

Good coordination and cooperation offers many opportunities for boosting existing (and introducing new) activities. Strengthening all partners in the joint endeavour to promote renewable energy is the key to IRENA's success.

The following priorities for intensive cooperation have been identified:

- developing and maintaining appropriate reporting systems (with IEA, REN21);

- providing and facilitating fora for policymakers at high levels (with REN21);

- developing and providing easily-used information gateways (with REEEP); and

- improving the access to financing mechanisms (with UNEP).

Overall, IRENA will promote a collective learning process that includes the involvement of all international organizations exchange of experience, the sharing of information on renewable energy activities, and the development of specific formats for IRENA's support of other organizations.

In the context of all international governmental organizations, the role of IRENA will be to lead the global efforts to ensure that there is a sustainable energy supply and that renewable energy contributes to a low-carbon future.

OVERVIEW

IO	Full name	Type	Main renewable energy contribution	Cooperation Priorities
IEA	International Energy Agency	OECD Agency	Renewable energy working party; Policy Database; Reporting System; Analytical work; Reports & workshops	Reporting system methodologies knowledge base coordination on priority issues
UNEP	UN Environmental Programme	UN	Global Network on Energy for Sustainable Development; Single projects; Financing know-how and networks; Analytical work, tools	knowledge base Financing mechanisms experience exchange support for projects
UNDP	UN Development Programme	UN	Energy for Sustainable Development; Rural Energy Programmes ; Huge number of single projects	experience exchange, support for projects
UNIDO	UN Industrial Development Organization	UN	Integrating RE in industrial projects; Single projects	experience exchange, support for projects
WB	World Bank	Bank	Lending, Technical assistance, Policy advice	experience exchange, support for projects; RE Toolkit
REN21	Renewable Energy Policy Network for the 21st century	Network	Renewables Global Status Report; IREC support	Renewables Global Status Report International fora for policy makers outreach to other stakeholders
REEEP	Renewable Energy and Energy Efficiency Partnership	Partnership NGO	Small funded projects worldwide REEGLE database	information platform REEGLE

Structures and renewable energy activities of International Organisations – Cooperation priorities for IRENA

ACCELERATED GLOBAL EXPANSION OF THE RENEWABLE ENERGY SECTOR AS A RESPONSE TO THE WORLD ECONOMIC CRISIS
The Example of Wind

Preben Maegaard

The combination of financial crisis, the world community's need for climate-friendly energy solutions, price stability, and security of energy supply are each sufficient reasons to develop the necessary policies and tools in order to shift towards renewable energy, based on high annual growth rates. Now, with the prospect of an international recession, governments around the world should respond by launching a comprehensive renewal of the energy sector. It seems only natural to transform renewable energy, including wind power, into a new global growth area. National and international initiatives and incentives for using the renewable energy industry are obvious—and urgently needed—as a vehicle for economic growth when a wide range of industrial sectors are in decline, especially the engineering and building industries. Subcontractors in the steel, glass, machinery, ball bearings, and many other component industries will be especially affected by the economic recession, and the consequence of this will be the closure of factories and unemployment.

Accelerated industrial development and proper renewable energy promotion programs can rapidly bring about a shift to an increase in renewable energy to meet the world's need for sustainable power, heat, cooling, and transportation energy solutions. At the same time, it would create thousands of new industries and millions of jobs. The growth of the sector will also create demand for steel, fiberglass, and the many other components of renewable energy manufacturing processes.

The time is right for progressive political initiatives, which in any case are the only medicine capable of combating climate change and safeguarding the welfare of the Earth's inhabitants. This theme will dominate the international debate in the coming decades and will be to the benefit of all countries, not least developing ones. The arguments discussed in this chapter are important political motivations behind an anti-recession policy.

If ambitious programs in renewable energy are launched globally, massive investment in renewable energy products like wind turbines, biogas plants, solar thermal, solar cells, and decentralized cogeneration plants could all become new industrial growth areas in many countries. On the other hand, this option would not be possible in the same way with the conventional energy sector, since mobilizing it to become a motor of growth would give rise to a large number of negative effects, including: increased CO_2 emissions, slow impact (due to the fact that it takes five to ten years to complete centralized power plants), a dependence on imported fuels, and a growing uncertainty about the supply of fossil fuels.

To ensure a rapid transformation, it is necessary to remove the bureaucratic barriers to the extensive use of decentralized energy that exist in some countries. The need for non-traditional forms of technology transfer to countries without their own technological base is urgent, as are comprehensive educational and training programs that apply best practice from the leading renewable countries and institutes. As there is no time to waste, it does not make sense for every country to go through the same failures and experiences that were part of the teething process in the handful of pioneering countries. This would be a complete waste of valuable human and financial resources.

Therefore, appropriate methods and instruments must be made available in the form of non-commercial transfer of technology and training through the international community. Centers and institutes that have, through their pioneering work, paved the way in their home countries, should now commit themselves to the challenge of making their valuable knowledge and experiences available to the 90 percent of the world's countries that have not yet developed renewable energy sectors on any significant scale.

Political, industrial, organizational, and financial transformation from centralized to decentralized energy forms involves significant challenges. The choice of the best solutions requires extensive research, development, testing, and demonstration of new energy solutions, shifting from the use of limited energy resources to the use of renewable energy technology. The sun and wind set no resource constraints for future energy supply. Utilization is the foundation for even more new industries and jobs.

We will see a very high growth, especially in China and India—big countries with significant growth rates and industrialization. Their demands for energy supply for power, cooling, heating, and transport will grow faster than the world average in the coming decade. This development should not be blocked by shortage and high fossil-fuel prices. For them, renewable energy is the most obvious solution. Also, the new-industrialized countries have to realize that they have to respect international climate regulations. China, India, and other Asian countries have often displayed, in contrast to the western countries, an impressive flexibility and readiness to go in different directions. They shall take the opportunity to base their future growth on renewable energy and energy-efficient solutions in all walks of society. Asia has the opportunity to make renewable energy a new growth sector, like the information-technology sector in the 1990s—but the energy sector will become much bigger.

THE WIND INDUSTRY LEADS THE WAY

The wide-scale deployment of renewable energy can be launched quickly. The technologies are already mature, and experience has shown that little time is needed to establish new production facilities. Spain, having introduced renewable energy promotion programs in 1995, represented the world's third largest wind turbine industry in the space of just a few years. Since 2004, China has also become an important wind-turbine producer and may, in a few years, become the world's largest supplier. Many other countries, such as Egypt, Brazil, Turkey, and Pakistan, will manufacture wind turbines within the next few years.

At the beginning of the century, the German wind-turbine industry consumed more steel than its shipyards. After the automotive industry, it was the second largest consumer of steel in the country. With strong growth, as outlined, the wind energy industry is capable of evolving to become the world's largest consumer of steel and many other commodities. This would be a strong stimulus for the world-economy in the anticipated recession.

A global expansion of 300,000 MW in the period 2008–2017 (40 percent growth scenario) would turn wind energy into one of the most important industrial growth areas. This growth can become the cornerstone for the new large industries that manufacture wind turbines, as well as expansion in more traditional sectors.

Wind turbine manufacturers produce turbine components, and design and develop, market and service them. Windmill-specific parts, such as blades, controls, and assemblies, are the wind turbine manufacturers' primary area, however, there are also numerous secondary areas, which are often subcontracted. These include basic materials such as steel, castings, and other metals, fiberglass; epoxy; paint; and raw materials in production. Major subcontracts are components such as gears, yaw systems, ball bearings, generators, electronic components, which mainly come from existing industries. Thus, gear suppliers also sell gears for shipyards, railway equipment, mechanical engineering, etc.

Finally, wind-turbine development creates a demand for transportation equipment, cranes, measuring equipment, security equipment, and many other supplies needed for legal and technical-vocational consultancy, research, certification, and other forms of intellectual services. Investors in wind turbines will be served by a number of specialists with relation to finance, assessment of wind resources, authorization and permits, and their maintenance. Wind energy magazines, trade fairs, exhibitions, conferences, and organizational work are other sectors that generally expand in proportion to the growth of the wind energy industry. Therefore, growth within the renewable energy sector will have a multiplier effect on the economy as a whole.

The cumulative employment effect, with growth levels of 40 percent new wind energy per year, can only be calculated with some uncertainty, however, a Danish example can serve as an indicator of the employment potential globally: The world's largest supplier of wind turbines, Vestas, whose market share in 2007 was 25 percent, delivered about 5,000 MW wind turbines. The company employed 15,000 people, giving a rate of output of 1,000 MW for every 3,000 employees. With global production at 100,000 MW new wind power in 2012, as stipulated above, this would lead to about 300,000 employees in the wind-energy industry worldwide. And, for every person employed in the wind energy industry itself, approximately two are employed in the many secondary industries and service sectors. Inventories and supplies in these areas are uncertain owing to the fact that national statistics do not include employment effects from imported components. Thus, the crane to erect a windmill in the United States can be manufactured in Japan, the steel for the tower may be from China, ball bearings from Sweden, the gearbox from Germany, and the fiberglass and generator can be from Finland, while a Danish wind turbine factory stands as a supplier of the turbine. With an annual production of 100,000 MW from wind energy in 2012, it is estimated that the production of wind turbines will result in the order of 1 million employees. This is about five times more than the number of jobs the industry generated in 2007, and the figure may be even higher because of the uncertainty of the job creation in the many associated industries.

An expansion to 100,000 MW annually requires high levels of conversion in the industry. Perhaps even more challenging, however, is the need for new forms of organization when it comes to ownership and operation of the new decentralized forms of energy. Since it is a regular "new deal," with a special focus on CO_2 reduction from energy production, it can only be done through the interaction between the private and public sectors. Here, the production and installation of wind turbines would be a natural task for the private sector, while the new public infrastructure, which is decentralized in nature, would be a task for existing and new local energy companies. It will lead to the strengthening of the local economy and the local acceptance of wind turbines, which inevitably changes the local visual environment in a very significant way. This is especially true in densely populated countries, where there is opposition to the installation of wind turbines. In unpopulated areas, at sea, etc., alternative forms of ownership may exist and this may be less important if no neighbors are affected.

100,000 MW of new annual wind power capacity will also have a significant impact on the global supply of CO_2-free electricity generation. Lessons from Denmark and Germany, which in 2008 had 20 percent and 8 percent stake in wind power production respectively, show that 1,000 MW of wind power supplies around 2.5 TWh. With an annual generation of 100,000 MW, the new wind generation capacity will contribute with 250 TWh, as compared to the 2006 levels of electricity consumption which was 35 TWh and 600 TWh respectively. The new turbines will not produce less than 250 TWh a year, and actually will produce considerably more because they will be taller than earlier wind turbines. Many will also be installed offshore, where experience has shown that production per MW installed is double that of generation located onshore.

Should these projections seem utopian, it is worth seeing this in relation to the experience of wind turbine development of the world's two largest power consumers, the United States and China, whose electricity consumption levels in 2006 were 2,500 and 1,800 TWh respectively. Nonetheless, when compared to the looming climate crisis and the dwindling fossil fuels, it is necessary to be sober when aiming for the above-mentioned production levels, which, notwithstanding the size of growth in the sector, will still need several decades to replace the current electricity generated from fossil fuels. In addition to meeting existing consumption levels, capacity will have to greatly expand, especially in developing countries that are currently under-served.

In order to achieve much greater use of wind energy, it must be made cheaper. Many countries still reject the widespread use of renewable energy, because the CO_2-neutral forms of energy qualify as a "burden" to the economy. This is seen as important for the businesses and heavy industries that are big consumers of electricity, since even small price differentials affect competition. As a result, many countries are reluctant to introduce wind energy on a large scale. Although work on the commercial use of wind power has been underway since the mid-1980s, wind power is only used on a scale worth mentioning in a small number of countries. As mentioned, in Denmark it provides 20 percent of the country's electricity, and in Germany and Spain the figure is between 6 and 8 percent. In all other countries, wind energy use is negligible in relation to total electricity production. However this is a false argument, since many countries do not consider the externalities of conventional fuels as part of their costs, and, in such cases, the overall economics of renewable energy frequently comes out more favorable than conventional energy solutions.

One obstacle to a higher share in wind power generation globally is, in addition to high electricity prices, the fact that, since 2005, manufacturing capacity within the sector in the well-established manufacturing countries—Germany, Spain, and Denmark—has only just been sufficient to supply the three or four large and stable markets where 85 percent of the global production was installed. This has created a seller's market. However, since there are many more emerging markets with an installed capacity of 1,000 MW or more, which is a kind of take-off point for the development of wind energy on a larger scale, there is a significant need for new manufacturing capacity.

The industry has responded accordingly; the production capacity of wind turbines is greatly increasing, and existing producers are expanding. A remarkable feature of the expansion is that new producer countries, such as the US, China, and India, with very high production potentials, are emerging. In addition, a number of other countries are planning to produce turbines, both for domestic use and export. The MENA (Middle East and North Africa) countries' biggest cable factory, whose headquarters are in Cairo, will turn to wind-turbine production in 2009, and other newcomers are planning production of wind turbines in Brazil, Korea, Iran, and in Eastern Europe.

In particular, China will become important as a producer country. Since 2004, over forty companies have started producing wind turbines of 1 MW or larger. In 2007, two of the top ten wind turbine manufacturers in the world were Chinese, and more are likely to make the list in the coming years. Conversely, European producers can be expected to fall from this prestigious list, despite the fact that they are also expanding. It is important to note that, whereas the European wind turbine manufacturers began as small businesses, which were built up from scratch, the emerging Chinese companies are part of large, well-established industrial groups, located in big cities with plenty of skilled labor and engineers. In Europe, on the other hand, many of the major wind turbine manufacturers are located in sparsely-populated, marginal regions where the shortage of engineers and technical workers is setting a limit to their expansion.

With the growth of existing businesses and the arrival of some fifty new wind turbine manufacturers since 2005, the potential for wind power could lead to high growth rates.

In the following table, it is assumed that growth will not be less than 30 percent per year and is, in fact, expected to be 40 percent. Such high growth levels over a prolonged period are based on a number of assumptions, which are decided politically, financially, and industrially.

Based on the 2007 production figures for wind turbines, this results in the following levels of production:

MW	30 percent annual growth	40 percent annual growth
2007	20,000	20,000
2008	26,000	28,000
2009	33,800	39,200
2010	43,900	54,900
2011	57,100	76,900
2012	74,200	107,700
Total MW	235,000	306.700

By the end of 2007, the grand total of globally-installed capacity was nearly 100,000 MW. If growth rates over the next five years (2008–2012 indexes) are 30 percent, the global installed capacity will rise to around 340,000 MW, or 400,000 megawatts if the growth in each of the years is 40 percent. This corresponds to the growth rates experienced by the photovoltaic industry over the last six years.

Since the financial crisis emerged in the fall of 2008, the share value in the world's largest wind turbine producer, Vestas, has already fallen. In just one month, the shares in the company, which has its own facilities on three continents, fell by 50 percent. Investors feared that already-signed orders for delivery in 2009 and beyond would be cancelled as a result of the uncertainty in obtaining the needed long-term investment for their wind turbines.

Should this situation last for a while, combined with the expansion of new production capacity, it is possible that there will be a very rapid and dramatic reduction in the price of new wind turbines. As a result of demand pressure, from 2005 onwards, the price of wind turbines supplied by the well-known producers has increased by at least 20 percent. With a subdued market and the entry of new production capacity, not least from low-cost countries like India, China, and Egypt, wind turbine prices could drop to well below 2005 levels.

Increased competition and lower prices will make wind power more economically attractive than conventional power-generation technologies. It will help many countries that have either committed to reduce their emissions or that have an increased need for new electricity generating capacity. Consequently, a number of newly-industrialized countries are giving priority to wind power. Not least, coal-fired power production with CO_2 capture and storage (CCS) will face strong competition from the most advanced and economical forms of renewable energy. CCS technology requires huge investment, and results in less efficient coal plants, which creates a comparative advantage for renewable energy.

On the other hand, the perception that investment in renewable energy is risky compared to competing conventional sources, such as coal, oil, and natural gas, is likely to mean that the weakened financial sector will be wary of granting long-term loans. An appropriate response to such a situation would be for international and national agencies and policy makers to introduce new tools for long-term financing of renewable energy in order to assure a constant high growth that can overcome the obstacle of a lack of long-term financing.

At the same time, improved financial instruments must ensure that this move does not simply allow private investors to make use of exorbitantly high pricing in order to lure the most profit-seeking venture capital into an area of technology where products have an operational lifespan of twenty years or more. A too-heavy dependency on offshore wind turbines—where the electricity production cost is over 50 percent higher than onshore wind turbines—will also lead to high prices for renewable energy. This can be seen in the UK, where lack of involvement of local citizen initiatives has blocked the implementation of onshore wind turbines. By 2012, Denmark will also be installing many more wind turbines at sea than on land. This is because the price is guaranteed higher than the corresponding price for electricity produced by onshore turbines.

For non-profit-owned wind turbines, one can use the better price for offshore wind turbines to provide an incentive for municipalities and utilities to establish primary utility and energy storage for electricity and heat with autonomous hybrid plants based on wind, solar, and biomass. Once the turbines have the status of a public utility, many wind sites may be defined by the same principles as high-voltage power lines and other technical facilities that serve the common good. In such cases, the land owner must be given compensation in accordance with normal practice. But this should not be possible for commercial ownership of wind turbines by individuals or companies. In order to obtain local acceptance, especially of the turbines, central utilities also should be prevented from owning local renewable power systems.

With heights up to 150 meters, the turbines are changing the landscape so significantly that this, in itself, will cause protests, which could block the proposed expansion. Therefore, it is important to let the locally-elected politicians decide whether they prefer several smaller turbines instead of fewer very large wind turbines in order to reach a solution that has less visual impact on the local landscape. Such decisions must be taken in the neighborhood where the wind turbines are going to be installed.

Also, it is necessary to debunk the wind turbine organizations' argument that in the future the industry will only be able to deliver very large wind turbines. The international wind turbine statistics show that around 50 percent of all wind turbines set up in 2006 were of 1,500 kilowatts size or smaller. The increased use of wind power must be seen in this context as well. The industry will naturally deliver the products for which there is a market.

The anticipated global wind-turbine production of 100,000 MW added annually, as expected in 2012, will make substantial challenges to the financial sector, regardless of whether funding comes from private investors, local public utilities, or concessionary companies. The cost of 1 MW of new wind power in 2007 was around €1 million, so 100,000 megawatts of new capacity will require the investment of €100 billion in wind energy infrastructure in 2012, with increased amounts in subsequent years.

Is this a prohibitive or even very high amount? One can illustrate the size of investment with a few examples: In 2007, the state budget of Denmark, a country with 5.5 million inhabitants, was around €100 billion. The same year, the oil giants, Exxon Mobil, BP, Shell, etc., had a combined profit—not turn over—of the same magnitude. In comparison with the cost of the war in Iraq, €100 billion is an almost modest amount. Finally, it is trifling compared to the financial packages that the governments of the industrialized countries managed to mobilize in a short time in October 2008 to insure the banks against their impending bankruptcy. Against this backdrop, it is not a question of financial capacity, but rather one of political and organizational commitment. This is what will determine whether or not countries will find serious solutions to the urgent climate and resource problems.

THE NEED FOR COHERENT GLOBAL POLICIES

One of the basic features of global political developments in the coming decades is that they must seek to bring about a shift from fossil fuels to renewable energy. All sectors of society and the economy are penetrated by energy in the various forms of electricity, heat, and transportation. When the renewable energy steps out of its infant shoes in order to become the cornerstone of the future energy structure, it is necessary to implement comprehensive political and organizational decisions.

A systematic transition to renewable energy in primary energy supply will enable the entire spectrum of supply potentials of energy from the sun, wind, and biomass. The cornerstone of this development will be based on creating legislation, planning structures, and financial mechanisms that encouraged decentralized solutions, including, at times, production and consumption on the same site such as for residential use, farms, or individual factories, or even alongside railway tracks to power trains. This is important in order to avoid centralizing the application of technologies that are, by nature, decentralized. However, some forms, such as off-shore turbines, will still be deployed for larger and more centralized use. This is quite different from the model preferred by centralized energy companies which, in order to maintain their market dominance, attempt to turn renewable energies into a predominantly centralized technological form. This would lead to increased costs, through big transmission systems, and these costs can be avoided by keeping the technologies decentralized. Therefore, it would be both politically and organizationally incorrect if we were to favor and lock in, in advance, certain types of ownership of energy technologies.

A transition demands that society, and not least politicians, recognize that, in the future, new organizational and political structures must be established, different from those that have been used since the first energy crises in the 1970s. There are many good results and technologies to build upon, but in future, it should be done in a planned way and with instruments that correspond to the task at hand. Comprehensive reforms should be introduced using the best elements of the policies in the leading renewable energy countries.

Chapter 51 ▌ Part12

ANOTHER CAPITALISM IS POSSIBLE?
From World Economic Crisis to Green Capitalism

Tadzio Mueller and Alexis Passadakis[1]

Tadzio Mueller and Alexis Passadakis[1]

WHAT TO DO IN CASE OF CRISIS?

Things really *have* changed. A little more than a year ago, writing from the left about the necessary and terminal crisis of Neoliberalism was still a very marginal activity, and had that odd feeling of *déjà vu* about it. There is an old joke that says that out of the last three recessions, Marxist economists had correctly predicted fifteen … Okay, maybe it's not the funniest joke, but it is telling: from many a critical perspective, capital(ism) is always in crisis, and all the moves made within "the system" are simply more or less effective attempts to postpone the "final reckoning." Nowadays, though, trying to tell the world that Neoliberalism has entered its final crisis feels rather like carrying coals to Newcastle. The end of a particular era of capitalist accumulation necessarily imposes a particular difficulty on critical analysis: having so often wrongly predicted the downfall of capitalism as the result of a particular crisis, many anticapitalists have developed a tendency to overstate capital's ability to emerge from every crisis stronger, meaner, and more resilient than ever. Obviously, the problem with always saying "capital will win in the end" is the same as with saying "it's going to rain on the washing"—well, yes, if we don't bring the washing in, it probably will. In short, and with the necessary revolutionary pathos: if we don't even believe in the possible (the end of capitalism), the impossible (communism) will surely never come to pass.

At the same time, to not start looking towards the field of force relations that seems to us a very likely outcome of the current crisis, *if and only if* capital and the governments of the world are indeed capable of stabilizing the crisis in such a way that leaves the fundamentals of their power untouched, would be a massive strategic mistake, and would leave us unprepared for what might very well be to come: a brave new world of green capitalism, where what used to be the left wing of global governance (Kyoto, binding environmental regulation, renewable energies, you name it) has moved towards the center. This text, then, is a mix of prediction and anticipation, in order to allow us to act in such a way that might create a *different* future from the one described below. It is about knowing where to strike in the future.

THE TRIPLE CRISIS

THE LEGITIMATION CRISIS

Arguably, the first significant global (as opposed to regional) crisis experienced

1   We are grateful to Kolya Abramsky and many unnamed friends and comrades for comments on earlier drafts of this chapter, as well as for countless discussions that helped us come up with some of the ideas contained here. All remaining errors are, as usual, our own.

by Neoliberalism was a *political legitimation crisis*. Starting in the late 1990s, aided by the spectacular and internationally-visible protests organized by the counter-globalization movement, the institutions of global governance that had been so crucial to the neoliberal project—the World Trade Organization (WTO), the International Monetary Fund (IMF), the World Bank, and others—began to seriously lose public legitimacy. As a result, the WTO negotiations have, since then, been effectively stalled; the IMF was, until the crisis hit, nearly out of business; while the World Bank began to reinvent itself, rather tellingly, as a global "green" bank.[2] All the while the G-8 kept trying to reinvent itself, claiming, at its 2007 summit in Germany, that it was the right institution to solve the climate crisis. At the national regulatory level, too, central institutions, including national governments, were losing legitimacy: in the global North, the disappearance of anything that could be recognized as "Social Democracy" meant that hardly any of the major parties were seen as representing the interests of those disadvantaged by Neoliberalism. In the United States, the author of a recent report on trust in public institutions is quoted in the *Financial Times* as saying that "belief in authority has collapsed," and that "over the last few years the trust between the public and the elites has completely collapsed." As a result, "the public is much less willing to trust corporate leaders' advice on the national economic interest."[3]

To be sure, such a crisis does not necessarily lead to emancipatory political action; it can just as well lead, on the one hand, to apathy and the decomposition of collective political actors, and on the other hand, to an ugly politics of fear and scapegoating. But it also does provide an opening for ideas of social and ecological transformation. The increasingly unequal distribution of incomes and wealth during the neoliberal era, coupled with the non-fulfilment of the free marketeers' central ideological promises (efficient markets, trickle down …) had produced a serious crisis of legitimacy. And *authority* is only stable in the medium term if it is seen as legitimate by its subjects.[4] But the legitimation crisis did not slow down the economics of Neoliberalism much: privatization, commodification, enclosure, they were all proceeding apace. Although here, too, trouble was brewing.

THE ACCUMULATION CRISIS

And so we return to the economic crisis rocking the world-economy, hitting mortgages and banks, food and fuel prices yesterday, international trade and the car sector today, and who knows where and what tomorrow. The *Financial Times*

---

2   For a critical analysis of the World Bank's attempts to "greenwash" itself, cf. Zoe Young, (2002) *A New Green Order? The World Bank and the Politics of the Global Environmental Facility*; on the defeat of the WTO's agenda, see Olivier de Marcellus, (2006) "Biggest victory yet over WTO and "free" trade. Celebrate it!" Available at http://info.interactivist.net/article.pl?sid=06/08/18/0417238&tid. For ongoing critiques of the WTO, the IMF, and the World Bank, as well as materials on the crisis of legitimacy, cf. http://focusweb.org; http://www.ourworldisnotforsale.org; and http://www.brettonwoodsproject.org.

3   Krishna Guha and Edward Luce, (01/10/2008) "failure to lead fuels main street backlash," *Financial Times*. Available at http://www.ft.com/cms/s/0/3c696e88-8f18-11dd-946c-0000779fd18c,dwp_uuid=11f94e6e-7e94-11dd-b1af-000077b07658.html

4   Max Weber, (1964) *Soziologie, Weltgeschichtliche Analysen, Politik* (Stuttgart: Alfred Kroener).

and the OECD (Organisation of Economic Cooperation and Development—the rich countries' club) may have called an end to the crisis, but since none of the structural problems that caused it have been resolved, we probably shouldn't hold our breaths.[5] Among the scramble to find ways out of the crisis, the idea of a new Keynesian "New Deal" has become standard fare in some elite circles. Take Larry Summers, former neoliberal hotshot, now back in power and glory as director of the National Economic Council under Barack "change we can believe in" Obama: "We need to identify those investments that stimulate demand in the short run and have a positive impact on productivity. These include renewable energy technologies and the infrastructure to support them, the broader application of biotechnologies."[6] Just to be clear: a former neoliberal stalwart is proposing Keynesian economic stimulus packages, big government, and all that was evil way back when (he and his friends were in power last time …).

The crisis that is currently rippling through the global economy runs deep. It is a crisis of overaccumulation brought on by the neoliberal attack on global working classes starting in 1970s.[7] The Keynesian deal—high productivity gains in line with high wage deals—was replaced with the, as we know by now, toxic mixture of low wages, easy credit, and lots of cheap goods from Asia/China. In other words: squeeze the working class in the global North (lower wages); globalize the economy and keep wages low in e.g. China (cheap goods); and expand the financial sector to provide cheap credit. Of course, that meant that people got ever deeper into debt, that there was a structural lack of effective demand, that "bubble markets" developed because capital that could not profitably be invested in production went into finance, and so on, and so forth, the litany of causes of the current crisis.

This crisis thus goes right to the heart of capital, and in the words of Lord Stern, the author of a report for the UK government on the possible economic gains to be made from climate change: "We need a good driver of growth to come out of this period, and it is not just a simple matter of pumping up demand."[8] But where to find this new "driver of growth," this magic formula that can kickstart a new round of capitalist accumulation … ?

THE BIOCRISIS

This brings us to yet another crisis, or set of crises, that has not been mentioned yet: the biocrisis. Under this broad heading we summarize those socio-ecological crisis tendencies that arise as a result of the contradiction between the requirements of collective human survival in relatively stable eco-social systems, and the requirements of capital accumulation—or more succinctly put: from the mad idea of going

---

5   Chris Giles et al., (12/05/2009) "Downturn 'bottomed out," *Financial Times*.

6   Larry Summers, (27/10/08), p. 9, "The pendulum swings towards regulation."

7   A "crisis of overaccumulation" occurs when, for any number of reasons (lack of demand for products, market saturation, workers' resistance pushing up production costs so far as to make production unprofitable), too much capital is chasing too few profitable investment opportunities. Capitalists then tend to respond by either trying to open up new markets (e.g. "globalization"), or by bidding up the price of existing assets ("bubbles").

8   Nicholas Stern, (2/12/2008) "Upside of a downturn," http://www.ft.com/climatechangeseries.

ANOTHER CAPITALISM IS POSSIBLE?

for infinite growth on a finite planet. The most prominent of these is no doubt the climate crisis, but further crisis tendencies, all of which stand in a reasonably direct relationship to capitalist production, are (not arranged in any order of importance): loss of biodiversity, lack of access to water, loss of arable land through erosion and desertification, overfishing, destruction of forests, peak oil, and so on.[9]

The concept "biocrisis" thus describes a set of processes that become socially relevant, appear as crises, primarily through the social processes, conflicts, and transformations that are their result—it is about far more than drowning polar bears. For example: desertification in Northern Africa becomes socially relevant insofar as it has concrete effects on human lives, such as displacing people from their habitats and livelihoods, which may, in turn, lead to them attempting to migrate, thus putting pressure on "receiving" countries, which, in turn, creates a very real social conflict as the result of an apparently "ecological" process.

Second, we need to recall that crises are not necessarily bad things from the perspective of capital: crises entail the devaluation of overaccumulated capital and a reduction of overcapacities in industry (i.e. closing down factories), both of which are necessary to restore the profitability of capital assets. Only when capital and industrial capacity is scarce, is it possible to generate profits. And the bigger the bubble—as in the current crisis—the louder and faster it will have to burst; Joseph Schumpeter referred to this as "creative destruction," a kind of radical diet for capitalism.[10]

While serious crises always entail the massive destruction of capital, as well as transformations in the matrix of social power, this destruction of capital is precisely what is necessary for capital(ism) to maintain its innovative, revolutionary power, its famed ability to "constantly revolutionize the means of production," to "melt all that is solid into air," and "profane all that is holy."[11] So, crisis is not necessarily a problem for capital-in-general, and neither is (class) antagonism. The most fascinating example of this was analyzed in the early writings of the Italian *Operaisti*,[12] who argued that the core of Roosevelt's famous New Deal, which contributed significantly to pulling the US economy out of the Great Depression, consisted in *internalizing the cause of one crisis to solve another*. There was, on the one hand, an economic crisis of overaccumulation not unlike the one we are witnessing today—caused by high

9   Millennium Ecosystem Assessment, (2005) *Ecosystems and Human Well-being*, Synthesis, Island Press: Washington, DC.

10   Joseph Schumpeter, (1942) *Capitalism, Socialism and Democracy*, New York: HarperPerennial. Schumpeter was an economist who popularized the term "creative destruction" to describe the regular revolutionizing of economic and regulatory structures and institutions needed to ensure innovation and new "long waves" of economic growth. Crises were seen as a helpful way of sweeping away the old and creating room for the new.

11   Karl Marx and Friedrich Engels, (1978) *The Marx-Engels Reader*—2nd ed., New York and London: Norton, p. 476.

12   *Operaismo* is a marxist tradition of thought as well as a social movement that emerged in the Italian industrial north in the 1960s. In sharp contrast to the Communist Party, which concentrated on gaining state power, they focused on the struggles of workers against their labor conditions and wage labor as such. This struggle is perceived to be the main historical force changing socio-economical relations. Cf. Steve Wright, (2002), *Storming Heaven: Class composition and struggle in Italian Autonomist Marxism*, London: Pluto Press.

productivity gains and high profits coinciding with low wages and a high degree of financialization—and, on the other, a political crisis caused by the sharpening of the class antagonism, powered to some extent by capitalists' fear of the Soviet Union. The trick was to channel this antagonism into struggles for higher wages, which, while not threatening capitalists' control over the production process, generated higher wages thus creating effective demand, or purchasing power, to soak up excess production, *and* forced capital to constantly innovate to maintain productivity growth and therefore high profit rates.[13] It was, in other words, the irreconcilable antagonism between capital and labour that came to drive an entirely new round of capitalist accumulation, a period that would later be seen as the "golden years" of capital.

So here's our argument: while this is by no means a foregone conclusion, the biocrisis is the opportunity that *might* just allow capitals and governments to at least temporarily deal with the legitimation and accumulation crises described above. How? By internalizing the antagonism at the heart of the biocrisis—that between human life and capital—as a driver of a new round of supposedly green accumulation, and a legitimating device for the further extension of governmental authority into the nooks and crannies of everyday life. It is precisely in the political energy surrounding the biocrisis that the potential lies to open up significant new spaces of accumulation through what can be summarized as, from an ecological point of view, the all too slow "ecological modernization" of the economy, as well as structures of governance.

PAPERING OVER THE CRACKS? NATURE AND VARIETIES OF CAPITALIST ECONOMICS

Of course, the question of the relationship between "ecology" and capitalism is not new. In the 1970s, a variety of different conceptions of so-called "green economics" began to appear on the political and intellectual catwalks, usually dressed up to follow the latest intellectual trends. The political transformations of 1989 and beyond put an end that with discussions that aimed for a fundamental conversion of the economy towards a non-market model ("eco-socialism").[14]

The major UN-conferences that took place in the 1990s, especially the Rio conference in 1992, summarized the question of the metabolism between humans and "nature" under the heading of "sustainability." The necessary characteristics of a capitalist economy—profit, private property, growth—thus no longer appeared as structural dangers to the survival of nature and humanity. Rather, the point was to find compromises between the needs of companies and those actors organizing around ecological questions. The desired result was to be a set of standards and guidelines to gently steer the world towards a future that was not going to be much different, just a little better. Remember, the "end of history" had come, but some improvements were still allowed.

---

13   Toni Negri, (1988) *Revolution Retrieved: Selected Writings on Marx, Keynes, Capitalist Crisis and New Social Subjects 1967–83*. London: Red Notes Archive, pp. 9–42.

14   Neil Smith, (1984) *Uneven Development*, Oxford: Basil Blackwell. See also: Thomas Ebermann and Rainer Trampert, (1984) *Die Zukunft der Grünen. Ein realistisches Konzept für eine radikale Partei*, Hamburg: Konkret Literatur Verlag.

However: in the face of the economic, social, and ecological facts created by "globalization" during the 1990s, as well as the ideological dominance of Neoliberalism, "sustainability" increasingly proved to be precisely the hollow concept it had seemed to be from the very beginning, in spite of the hopes it had initially raised in some progressive sectors of the northern middle classes. Green politics turned into "environmental management," into one discrete policy field amongst others, plied by civil society actors and marginal ministries, without ever suggesting the possibility of a fundamentally different society. Accordingly, the current politico-economic rules of the game (liberalization, privatization, market incentives) also structured the field of environmental policy.

With the onset of the second Great Depression in 2007, the old paradigms are crumbling like so many melting icebergs. In light of the increasingly visible struggles around climate change and the defeat of the oil-powered Bush government, a loose transnational coalition of actors pursuing the project of a "green capitalism" is on the move: sectors of liberal parties, international environmental NGOs, green parties, the renewable energy sector, "Silicon Valley," and certain factions of liberal-green financial capital (think insurances).

The concepts developed in the emergence of this project, whether for a (neo-) liberal "green capitalism" or for a somewhat more progressive "Green New Deal" (GND), aim to reconcile capitalist economics with ecology. Where so far the issues meant to make capitalism interesting from the perspective of environmental policy were incentives, "true costs," and flexibility, nowadays the focus is on "growth," which a Keynesian GND is believed to be able to trigger. The idea is that public deficit spending invested in, for example, energy conversion, renewables expansion, research and development, etc., will trigger a massive increase in "green jobs" and, more importantly, a new round of accumulation for the economy as a whole. Thus, it would solve the climate crisis, peak oil, and the world economic crisis all at the same time, killing three birds with one stone. Achim Steiner, Executive Director of the United Nations Environmental Programme (UNEP) summed it up at a press conference where he announced his agency's two-year GND project in October 2008: "The new, green economy would provide a new engine of growth, putting the world on the road to prosperity again."[15] But does it make sense to assume that a project aimed primarily at saving the economy is magically going to save the planet as well?

THE GREEN NEW DEAL: CLOSE, BUT NOT QUITE ...

Alongside attempts like those by the UNEP, the Obama administration, and even the German government to spice up all manner of economic activities by adding a few Keynesian keywords and repackaging them as part of a "green economy," the most significant progressive effort in the field has certainly been made by the Green New Deal Group, a UK-think tank bringing together unorthodox economists, NGO-

---

15   Geoffrey Lean, (12/8/2008) "A 'Green New Deal' can save the world-economy, says UN," *The Independent.* http://www.independent.co.uk/environment/green-living/a-green-new-deal-can-save-the-worlds-economy-says-un-958696.html.

heavyweights, and politicians influenced by the more moderate and reform-oriented side of alterglobalist ideas.[16] It claims to develop a comprehensive answer to the "triple crunch" of financial meltdown, climate change, and peak oil by referring to the lessons of Roosevelt's answer to the Great Depression. Thus the key instruments proposed are financial regulation and deficit spending on renewable energies, leading to the massive creation of green jobs.

So far so good. But the report takes as its point of departure a deeply-dubious analysis: the "triple crunch" is said to be "firmly rooted in the current model of globalization."[17] This is a fundamental misunderstanding, at least if we look at the energy and climate crises: the problems of climate change and peak oil are deeply rooted in what has been called "fossilistic capitalism."[18] The Green New Deal Group's critique stops at the mask of Neoliberalism, thus not seeing the real beast behind it—namely, the madness of trying to, in fact *having to* achieve infinite growth on a finite planet. Which is simply another way of saying "capitalism." Forgetting entirely that the Club of Rome's famous report, *Limits to Growth*,[19] predates anything we might call neoliberal globalization, the report constructs the 1950s and 60s as a "golden age of economic activity,"[20] thus proving that historical memory is indeed very, very short. Forgotten are pesticides, dioxins, road-expansion, and all the other environmentally-destructive infrastructural projects of the post-WWII era. Forgotten is the fact that in order to pacify the class antagonism, Fordism/Keynesianism relied on "externalizing" the costs of this social compromise not only on developing countries and rural areas, but also and especially on "the environment." Forgotten, finally, the fact that this system relied on increased automation to replace and discipline struggling workers, and to reduce the costs of reproducing labor by way of industrialized agriculture—both of these processes required massive amounts of energy, thus greatly increasing total social capital's energy requirements.

Another aspect suspiciously absent in the NEF-program is *labor*. There is a lot of talk about "green jobs" being created. But conditions of labor, wages as a structural factor in the economy, and labor as a social stratum and force are missing, as is the relationship between energy and class struggle in general. It's a New Deal without anyone actually making the deal. A minimum wage and forty hours per week were cornerstones of Roosevelt's program.[21] Trade unions were strong players at the president's table. Without any doubt, this report entails many useful proposals to cut back the power of finance and to fight climate change. However, it evokes an aura of social democratic policy without actually delivering its substance. It uses some catch

---

16   Green New Deal Group, (2008) *A New Green Deal*, London: nef. http://www.neweconomics. org/NEF070625/NEF_Registration070625add.aspx?returnurl=/gen/uploads/2ajogu45c1id4w55tofmpy5 520072008172656.pdf

17   Green New Deal Group, (2008), p. 2.

18   Altvater, Elmar (2005), *Das Ende des Kapitalismus, wie wir ihn kennen. Eine radikale Kapitalismuskritik*, Westfälisches Dampfboot, Münster.

19   Donella Meadows et al., (1972) *Limits to Growth*, New York: Universe Books.

20   Green New Deal Group, (2008), p. 13.

21   Mark Rupert, (1995) *Producing Hegemony: The Politics of Mass Production and American Global Power*, Cambridge: Cambridge University Press.

phrases to pretend it's learned some lessons from history, but floats in a sphere of technocratic nowhere, with policy being made from the top down.

GREEN CAPITALISM: SAVING THE PLANET, OR SAVING THE BOTTOM LINE?

Thus far, we have tried to establish the following: that there is a confluence of crises (we mentioned two, but there are other ones: energy, food, etc.) that is bringing to an end the neoliberal era of capitalism; that there is another crisis, the *biocrisis*, which is as much an opportunity for states and capitals as it is a threat—an opportunity to use the dynamism of these multiple eco-social crises to kickstart a new round of accumulation and legitimate new forms of regulation. We then traced the development of "environmental" discourse from a vaguely critical formation towards a situation where it is at the center of a strategy of capitalist modernization, even if we take as a point of departure one of the more progressive instances of such thinking, namely the proposal for a GND.

But there is still a central question here that remains unanswered: Why oppose this potentially-emerging green capitalism in the first place? The answer returns us to the notion of the biocrisis, to the antagonism between life and capital. This antagonism, we argued, is what might drive the renewed accumulation of green capital, just as the internalized and controlled antagonism between labor and capital drove accumulation in the Fordist era of capitalism. But an antagonism internalized is not an antagonism solved—its role in the scheme relies precisely on its continued existence. Fordism did not solve the class antagonism, it merely internalized it. Neither will green capitalism solve the antagonism of the biocrisis, it will draw energy from it to drive forward that which must always be capital's first and foremost project: the accumulation of *more* capital.

Why is that? Because *money* only becomes *capital* (rather than the stuff we have in our pockets to buy stuff in order to satisfy a concrete want, such as hunger) when it is invested into the production of goods that are then sold in order to achieve a return on the initially invested capital. Or in short: money—production—more money.[22] This process involves a whole range of inputs and requirements, from labor to raw materials, from machines to energy. And historically, although the relative resource intensity of capitalist production might have decreased (i.e., the same product can now be made with less inputs of raw materials), in absolute terms, capitalist production has always been expansive in environmental space, has always required more and more and more inputs—wild-eyed dreams of a capitalist utopia of "immaterial" growth, based on services and the "digital revolution" notwithstanding.[23] Just as the antagonism between labor and capital cannot be solved within a capitalist framework—it is, after all, the very constituent feature of the capital relation—the antagonism between capital and life in relatively stable eco-social systems cannot be

---

22   Karl Marx, (1971) *Das Kapital—Kritik der Politischen Oekonomie. Erster Band*, Berlin: Dietz Verlag.

23   Cf. Bobby Johnson, (4/5/2009), "Web providers must limit internet's carbon footprint, experts say," *The Guardian*, http://www.guardian.co.uk/technology/2009/may/03/internet-carbon-footprint.

solved, because there is a *necessary contradiction* between the *infinite* accumulation of capital and life on a *finite* planet.

But this answer operates on a rather high level of abstraction. Surely, we have to be able to point to a few more concrete aspects of such a green capitalist setup that would make it worth the opposition of radical social movements. In the space that remains, we therefore present a few theses on what the brave new green world envisioned by thinkers from Susan George and Caroline Lucas on the left, to Larry Summers and Ralf Fücks (of the German Heinrich-Böll Foundation) on the right might look like—and why, even on a lower level of abstraction, we remain convinced that there is no way green capitalism can solve the biocrisis. The theses focus, to some extent, on the United Nations Framework Convention on Climate Change (UNFCCC), because we believe that this is going to be a central regulatory instance of green capitalism, comparable to the structural role played by the WTO in neoliberal capitalism, and because we are both active in the emerging global movement for climate justice.

THESES AGAINST GREEN CAPITALISM

• Green capitalism will not challenge the power of those who actually produce most greenhouse gases—the energy companies, airlines and car makers, industrial agriculture—but will simply shower them with more money to help maintain their profit rates by making small ecological changes that will be too little, too late.

• All types of green capitalism fail to acknowledge that the expansive nature of capitalism—its need to grow—will undermine any attempt to reduce its constant imperial demand for more resources. Decoupling growth from energy demand is a myth. Measures for more energy efficiency may lead to "relative decoupling" (less energy per unit produced) but to solve the biocrisis "absolute decoupling" is necessary, which is not possible if the world-economy continues to be bound in an insane logic of growth.[24]

• Because globally, working people, both waged and unwaged, have, over the course of the neoliberal era, lost a significant amount of their power to bargain and demand rights and decent wages,[25] in a green capitalist setup, wages will probably stagnate or even decline, to offset the rising costs of "ecological modernization."

• The "green capitalist state" will be an authoritarian one. Justified by the threat of ecological crisis, it will "manage" the social unrest that will

---

24   In a world of 9 billion people, all aspiring to a level of income commensurate with 2 percent growth on the average EU income today, carbon intensity would have to fall on average by more than 11 percent per year to stabilize the climate, sixteen times faster than it has done since 1990. And by 2050, global carbon intensity would need to be only 6 grams per dollar of output, almost 130 times lower than it is today.

25   It is significant that even in China, held by many on the left to be a place where workers' power was on the rise, wages as a proportion of GDP declined from 53 percent to 41.4 percent. Cf. Jianwu He and Louis Kuijs, (2007) "Rebalancing China's economy—modelling a policy package," *World Bank China Research Paper* No. 7, p. 11. http://www.worldbank.org.cn/english/content/working_paper7.pdf

necessarily grow from the impoverishment that lies in the wake of rising cost of living (food, energy, etc.)[26] and falling wages.

• In green capitalism, the poor will have to be excluded from consumption, pushed to the margins, while the wealthy will get to "offset" their continued environmentally-destructive behavior, shopping and saving the planet at the same time.

• In green capitalism, there is a danger that established, mainstream environmental groups will come to play the role that trade unions played in the Fordist era: acting as safety valves to make sure that demands for social change remain within the boundaries set by the needs of capital and governments, and actually further drive capitalist growth—the more they protest, the more "green technologies" will grow.

• Real solutions to the climate crisis won't be dreamt up by governments or corporations. They can only emerge from below, from globally networked social movements for climate justice, based on the creation of fundamentally different worldwide social relations of production and consumption, and livelihoods.

• As an emerging global climate justice movement, we must fight two enemies: on one hand climate change and the "fossilistic capitalism" that causes it, and on the other, an emergent green capitalism that won't stop it, but will limit *our* ability to do so.

CONCLUSION: OPEN ENDS

This extremely schematic overview of what might be emerging as a new capitalist formation can of course not be the end, but only the beginning of a conversation within emancipatory social movements, as well as the renewable energies sector. Are we going to bet on the green capitalist, market-driven horse? Or are we going for a more fundamental, socio-ecological transformation that can actually deliver both "sustainability" *and* justice? Where exactly the current crisis will lead us, to what political conjunctures, no one can claim to know with any degree of certainty. But for all the misery and horrors the multiple crises we are facing today will bring, they also open up a space for action and struggle—a struggle not just *against* green (or any other) capitalism, but a struggle *for* the constitution of alternatives. It is obvious that we need to enter into a post-petroleum world in the next decades. The way that this transition will happen, and the shape of the world that is to come then, will be determined by struggles that happen now and in the next few years. Things have changed in the last year or so, and will continue to change rapidly. Now is the time to make our moves.

---

26   While these trends are of course contested, we base our assumption about energy prices on the International Energy Agency's 2008 "World Energy Outlook;" and for food on Javier Blas (7/11/2008), "Another food crisis year looms, says FAO," *Financial Times*.

Chapter 52 ▌Part 12

"EVERYTHING MUST CHANGE SO THAT EVERYTHING CAN REMAIN THE SAME"
Reflections on Obama's Energy Plan[1]

George Caffentzis

I am here to speak about the geopolitics of oil, but it has been suggested that I concentrate my presentation on the question: Is President Obama's oil/energy policy going to be different from the Bush Administration's? My immediate answer to this prophetic question will be philosophical: a firm "No" (as is echoed in the title of this talk) and a more hesitant "Yes." The reason for this ambivalence is simple: the failure of the Bush Administration to radically change the oil industry in its neoliberal image has made a transition from an oil-based energy regime inevitable, and the Obama Administration is responding to this inevitability. Consequently, we are in the midst of an epochal shift so that an assessment of the political forces and debates of the past have to be revised and held with some circumspection.

Before I examine both sides of this answer, we should be clear as to the two oil/energy policies being discussed.

The Bush policy paradigm's premise is all too familiar: the "real" energy crisis has nothing to do with the natural limits on energy resources, but is due to the constraints on energy production imposed by government regulation and the OPEC cartel. Once energy production is liberalized and the corrupt, dictatorial, and terrorist-friendly OPEC cartel is dissolved by US-backed coups (Venezuela) and invasions (Iraq and Iran), according to the Bush folk, the free market can finally impose realistic prices on the energy commodities (which ought to be about half of the present ones), and stimulate the production of adequate supplies and a new round of spectacular growth of profits and wages.

Obama's oil/energy policy during the campaign and after his election has the following equally-familiar premise, he presented on Jan. 27, 2009: "I will reverse our dependence on foreign oil while building a new energy economy that will create millions of jobs … America's dependence on oil is one of the most serious threats that our nation has faced. It bankrolls dictators, pays for nuclear proliferation and funds both sides of our struggle against terrorism." In the long-term this policy includes: a "clean tech" Venture Capital Plan; Cap and Trade; Clean Coal Technology development; stricter automobile gas-mileage standards; cautious support for nuclear power electricity generation.

---

1   Originally a chapter on Obama's energy policy was going to be written by Steve Kretzmann, Director of Oil Change International, however, due to his involvement in the Shell trial over Ken Sarowiwa's murder, he was unable to do it. This paper was originally presented at the Geopolitics of Oil Colloquium at Rutgers University, March 4, 2009, and appears in *Turbulence*. It is being reprinted here with permission from both the author and the Turbulence Collective.

The energy policy he outlined in his budget proposal is supportive of a peculiar "national security" autarky (especially when it comes from an almost mythical pro-globalization figure like Obama). Its logic is implicitly something like this: if the US were not so dependent on foreign oil, there would be less need for US troops to be sent to foreign territories to defend the US' access to energy resources. Obama treats oil in a mercantile way, the vital stuff of any contemporary economy (a little like the way gold was conceptualized in the sixteenth and seventeenth centuries), long after mercantilism has been definitely abandoned as a viable political economy. In effect, he is calling for an autarkic import-substitution policy for oil while he is leading the main force for anti-autarkic globalization throughout the planet.

A FIRM "NO"

Obama's paradigm is problematic since it poses the key question of oil policy as a matter of "dependency" and not as the consequence of the present system of commodity production. It does not recognize that oil is a basic commodity; that the oil industry is devoted to making money profits; that the US government is essentially involved in guaranteeing the functioning of the world-market and the profitability of the oil industry (not access to the hydrocarbon stuff itself); and that energy politics involves classes in conflict (and not only competing corporations and conflicting nation states). In brief, it leaves out the central players of contemporary life: workers, their demands and struggles. Somehow, when it comes to writing the history of petroleum, capitalism, working class, and class conflict are frequently forgotten in a way that never happens with oil's earthy hydrocarbon cousin, coal. Once we put profitability and the working class conflict into the oil story, the plausibility of the National Security paradigm lessens, since the US military will be called upon to defend the profitability of international oil companies against the demands of workers around the world, even if the US did not import one drop of oil.

There will be wars fought by US troops aplenty in the years to come, if the US government tries to continue to play for the oil industry in particular and for capitalism in general, the twenty-first century equivalent of the nineteenth century British Empire. For what started out in the nineteenth century as a tragedy, will be repeated in the twenty-first, not as farce, but as catastrophe. At the same time, it is not possible for the US government to "retreat" from its role without jeopardizing the capitalist project itself. Obama and his Administration show no interest in leading an effort to abandon this imperialist, market-policing role as his efforts in Afghanistan, Iraq, and Pakistan, as well as his carte blanche to Israel in its bombing of Gaza, initially indicate.

Thus supporters of the National Security paradigm for oil policy, like Obama, are offering up a questionable connection between energy import-substitution and the path of imperialism. As logicians would say, energy dependence might be a sufficient condition of imperialist oil politics, but it is not a necessary one. This is Obama's dilemma then: he cannot reject the central role of the US in the control of the world-market's basic commodity, at the same time, the inter- and intra-class conflict in the oil-producing countries is making the US' hegemonic role impossible to sustain. Therefore, Obama's oil policy will be quite similar to Bush's.

A HESITANT "YES"

Up until now my argument has been purely negative, i.e., though Obama's oil policy and Bush's are radically different rhetorically, they will have much in common in practice. Obama's goal of "energy independence" will not affect the military interventions generated by the efforts to control oil production and accumulate oil profits throughout the world. These interventions will intensify as the capitalist crisis matures and as the short-term, spot market price fluctuates wildly from the long-term price, and geological, political, and economic factors create an almost apocalyptic social tension.

I do see, however, that there is a major difference between Bush and Obama. The former was a *status quo* petroleum president while the latter is an energy-transition president, i.e., Obama (like Roosevelt in the 1930s and Carter in the 1970s) is in charge of a capitalist energy transition similar to the successful one that substituted oil/natural gas for coal in many places throughout the productive system in the 1930s and 1940s and the unsuccessful one that failed to substitute coal, solar power, and nuclear power for oil/gas in the US of the 1970s. We are in a moment similar to the time when capital began to recognize that coal miners were so well organized that they could threaten the whole machine of accumulation (an experience felt in the British General Strike of 1926 and US coal mining struggle during the 1930s that led to the triumph of the CIO) and had to be put on the defensive by the launching of a new energy foundation to capitalist production, and when Carter despaired of putting the struggle of the oil producing proletariat (especially in Iran) back in the bottle.

In the face of the failure of the Bush Administration's attempt to impose a neoliberal regime on the oil-producing countries, the Obama Administration must now lead a partial exit from the oil industry. It will not be total, of course. After all, the transition from coal to oil was far from total and, if anything, there is now more coal mined than ever before, while the transition from renewable energy (wind, water, forests), in the late-eighteenth century, to coal was also far from total. *Indeed, this is not the first time that capitalist crisis coincides with energy transition, as a glance at the previous transitions in the 1930s and 1970s indicate.*

It will be useful to reflect on these former transitions to assess the differences between Bush's and Obama's oil policies. The different phases of the transition from oil to alternative sources include: (1) repressing the expectations of the oil-producing working class for reparations of a century of expropriation; (2) supporting financially/legally/militarily the alternative energy "winners"; (3) verifying the compatibility of the energy provided with the productive system; (4) blocking any revolutionary, anti-capitalist turn in the transition.

In reflecting on these phases, I note that they offer the kind of challenges that were largely irrelevant to the Bush Administration, since it was resolutely fighting the very premise of a transition: the power of the inter- and intra-class forces that were undermining the neoliberal regime. Consequently, they will provide a rich soil for discussion, debate, and planning in this period. But as the title of my talk was true of the "Firm 'No'" side of my argument in a quite simple sense—the interests of the world-market and the oil/energy companies will be paramount in the deployment of US military power—it also applies to my "Hesitant 'Yes'" side as well, though less directly, since the ultimate purpose of the Obama administration is (*pace* Rush Limbaugh) to preserve the capitalist system in very perilous times. It just so happens, however, that the "everything" that must change is more extensive than had ever been thought before.

The first element in the transition is to recognize that there will be inter-class resistance to the transition from those who stand to lose. Of course, most of the oil capitalists will be able to transfer their capital easily to the new areas of profitability, although they will be concerned about the value of the remaining oil "banked" in the ground. This transition has been theorized, feared, and prepared for by Third World (especially Saudi Arabian) capitalists ever since the first oil crisis of the 1970s. But what is to be done with respect to the oil-producing proletariat? After all, the "down side" of Hubbert's Curve, in a sense, could be seen as a potential payback for a century of exploitation, forced displacements, and enclosures in the oil regions.

The capitalist class as a whole is unwilling to pay reparations to the peoples in the oil-producing areas whose land and lives have been so ill-used. Oil capital's resistance to reparations is suggested by its horror, for example, of paying the Venezuelan state oil taxes and rents that will go into buying back land that had been expropriated from *campesinos* decades ago, and giving it to their *campesino* children or grandchildren. Capital wants to be able to control the vast transfer of surplus value that is being envisioned in these discussions of transition, and without a neoliberal solution it is not clear that it can. Moreover, will the working class be a docile echo to capital's concerns? After all, shouldn't reparations be paid to the people of the Middle East, Indonesia, Mexico, Venezuela, Nigeria, and countless other sites of petroleum extraction-based pollution? Will they simply stand still and watch their only hope for the return of stolen wealth be snuffed out?

We should recognize as far as phase 2 is concerned, that alternative energies have been given an irenic cast by decades of "alternativist" rhetoric contrasting blood-soaked hydrocarbons and apocalypse-threatening nuclear power. But if we remember back to the period when capitalism was operating under a renewable energy regime in the sixteenth through most of the eighteenth century, we should recognize that this was hardly an era of international peace and love. The genocide of the indigenous Americans, the African slave trade, and the enclosures of the European peasantry occurred with the use of alternative renewable energy! The view that a non-hydrocarbon future operated under a capitalist form of production will be dramatically-less polemic is questionable. (We saw an example of this kind of conflict of interest in the protests of Mexican city dwellers over Iowa-grown corn that was being sold for biofuel instead of for "homofuel"!)

As for phase 3, we should remember that an energy source is not equally capable of generating surplus value (the ultimate end of the use of energy in capitalism). Oil is a highly flexible form of fuel that has a wide variety of chemical by-products, and mixes with a certain type of proletariat. Solar, wind, water, and tidal energy will not immediately fit into the present productive apparatus to generate the same level of surplus. The transition will ignite a tremendous struggle in the production and reproduction process, for inevitably workers are going to be expected to "fit into" the productive apparatus whatever it is.

Finally, phase 4 presents the nub of the issue before us: will this transition be organized on a capitalist basis or will the double crisis, opened up on the levels of energy production and general social reproduction, mark the beginning of another mode of production? Obama's energy policy is premised on the first alternative. There are, however, many reasons to call for the negation of this premise that leads to "everything remaining the same." Consequently, we should be investigating with all our energy and ardor the other alternative. Join us.

Chapter 53 ▌ Part 13: Towards a Transition Based on Decentralization, Common Ownership, Dignified Work, and Community Autonomy

SUSTAINABILITY AND JUST TRANSITION IN THE ENERGY INDUSTRIES

Brian Kohler on behalf of the International Federation of Chemical, Energy, Mine and General Workers' Unions (ICEM)

If current patterns of production and consumption must change for environmental reasons, then there will be an impact on employment patterns. Businesses will adapt (with government subsidies), highly-paid executives will gently glide to new positions on golden parachutes, and the environment will presumably improve to the benefit of the general population. Who will pay? Left to the so-called free market, workers in affected industries who lose their jobs will effectively suffer for everyone else's benefit.

The global labor workforce totals over 3 billion workers.[1] An estimated 21 percent of that workforce is engaged in industrial activities that are either directly or indirectly dependent upon energy extraction, production, and consumption.[2] Workers in other sectors, such as agriculture or construction, also rely on energy to fuel production and create jobs.

There is no doubt that, over the next few decades, our current patterns of energy production and consumption will be radically transformed. This is true no matter what energy sources we turn to. It remains true even if technologies like carbon capture and sequestration allow us to continue our reliance on fossil fuels for a time. People working in the energy sector are the most directly affected by the swirling debates around sustainability, and especially climate change. Whether they work in the nuclear, hydroelectric, fossil fuel, wind, solar, or another energy sector, the outcome of that debate will affect their lives and livelihoods, either positively or negatively, but definitely profoundly. Employment in energy-dependent industries and occupations will be affected equally profoundly, if a little less directly.

WHAT IS REALLY MEANT BY SUSTAINABILITY

Sustainable development has been defined as development that meets the needs of the present without compromising the ability of future generations to meet their own needs. It is upon examination of the meaning of the word "needs" that this simple definition becomes more complex.

---

1   *NationMaster*, "World Statistics, Country Comparisons," http://www.nationmaster.com/graph/lab_lab_for-labor-force.
2   *NationMaster*, "World Statistics, Country Comparisons," http://www.nationmaster.com/graph/lab_emp_in_ind_of_tot_emp-labor-employment-industry-total.

Sustainability addresses three broad areas of needs: environmental, social, and economic. To understand how these inter-relate, imagine three puddles of paint on a plate, slightly stirred. The interfaces (social-economic, social-environmental, environmental-economic) are blurred and indistinct, and there is great difficulty in separating one from the other. Within each component exists a myriad of subsidiary interfaces. Sustainability requires an integrative, rather than the traditionally compartmentalized, way of thinking.

If we fail to protect the environment, we will eventually face economic catastrophe and social disintegration. On the other hand, if we consider only narrowly-defined environmental or economic issues in isolation from their social links and impacts, we may destroy cultures, societies, communities, enterprises, and individual working peoples' lives, and have nothing to offer them in return. Balancing and integrating all of these concerns is the essence of sustainability.

Global economic sustainability broadly refers to the smooth functioning of the economic system, including opportunities for growth of the economy in less-developed nations. Employment is our primary means of distributing wealth, therefore a sustainable economy must provide decent work in sufficient quantity to allow people to develop their full human potential. Until the 2008 global economic collapse, business groups had successfully defined their interests as synonymous with economic sustainability. However, recent events have highlighted just how unstable the world's economic system—based on a repeated prescription of deregulation, privatization, and globalization—has become. A sustainable world cannot be built on a casino economy.

An environmentally-sustainable world is ultimately about the sharing of resources and energy. Reliance on non-renewable resources, over-use of renewables, and careless discharges of pollutants have led to a degraded environment. Human beings possess the capacity, which no other species possesses, to render the planet virtually uninhabitable. Without the preservation of the natural environment, neither social nor economic sustainability is possible. Environmental non-governmental organizations (ENGOs) are generally seen as the major advocates for the sustainability of the environment. However, it should be remembered that occupational illnesses and deaths helped identify many of the toxic chemicals ENGOs are now concerned about. Organized labor has long realized that workplace poisons are also environmental poisons, and that ultimately there are no jobs on a poisoned planet.

Social sustainability includes respect for human rights (including labor rights), cultures, and communities. The social dimension values those things that define us as human beings: our creativity, our intelligence (both as individuals and as cultures), our abilities to interact, form families and communities, and care for one another. The labor movement understands that sustainability will never be achieved without addressing the need for fairness, equity, and justice. Business groups have attempted to define the social dimension (and sometimes even the environmental dimension) of sustainable development as a mere subset of the economic dimension.

An article on the *Harvard Business* website clearly expressed this view as follows: "Holding on to an economics-based definition of sustainability helps reconcile broader social interests with the measurement of shareholder value. If we can capture social costs in earnings equations, then we will align social and financial motivations. It would be a loss to let such a useful concept drift into a more emotional definition."[3] We reject this view. Many important indicators of social sustainability can only be qualitative, not quantitative, at best. As Albert Einstein said, "not everything that can be counted counts, and not everything that counts can be counted."

Along with a handful of peace, development, and human rights non-governmental organizations (NGOs) it has been largely the labor movement that has kept the social dimension of sustainability in play.

WORKERS' INTERESTS: CLEAN ENVIRONMENT, GREEN JOBS, JUST TRANSITION

The labor movement will never forget that it speaks for workers. In their interests, unions' positions on global sustainability have historically had three aspects. First, we have tended to take fairly "green" positions on a number of environmental issues, in many cases because they are directly related to the occupational health hazards that our members face. This is especially true in the case of toxic chemicals. Second, we have asked that our governments and our business leaders commit to the creation of decent work, as part of the basic social contract of our society. In the current global circumstance, this provides a strong rationale for our demand that sustainable or "green" jobs be the central part of national and international industrial strategies. Third, we have demanded that workers who are forced out of their jobs for the good of the environment, be compensated. This forms the core of the "Just Transition" concept.

It is worth keeping in mind that labor supports environmental protection and sustainable job creation, even though the main topic of this chapter is Just Transition.

JUST TRANSITION: THE WAY FORWARD

Sustainability is about more than just the environment and the economy. However, neither businesses, environmental groups, nor (sadly) governments have thus far demonstrated much creative thinking on how to manage the social aspects of moving toward a sustainable future.

The only way to manage the social impacts, particularly in regard to employment, of a transition to sustainability without letting the affected workers bear most of the social and economic costs of change is through "Just Transition" programs.

The economic crisis of 2008 has clearly illustrated a failure of the deregulated "free" market. Business as usual has not only created an environmental crisis; it has not even been particularly good at creating jobs in recent years. The areas of greatest economic growth—information technologies, financial services, retail and

---

3   Christopher Meyer, CEO of Monitor Networks, *Harvard Business* blog site "HarvardBusiness. org Voices," http://blogs.harvardbusiness.org/leadinggreen/2008/06/we-need-a-definition-of-sustai.html

food services (for example)—either create relatively few jobs or create relatively low-quality jobs in regards to the wealth they generate. It is not sustainable to try to re-inflate the old economic bubbles. It is time for governments to consider public investments in the public interest, with an industrial strategy aimed at the creation of a sustainable economy.

If the economy up until now has failed to create large numbers of high-quality jobs and led to an ever-increasing disparity between rich and poor, it cannot be considered sustainable. We should consider today's economic crisis an opportunity to re-evaluate the underlying social contracts of society and plan a green industrial strategy that will create large numbers of high-quality unionized jobs. The International Labour Organization (ILO),[4] and the Center for American Progress[5] have both produced interesting analyses that suggest new sustainable, or "green" industries will actually create more jobs per dollar invested than many of today's large industries do.

We need quite radical change in current patterns of production and consumption if we are to avoid sterilizing the planet. The key to getting past the barriers to change is a Just Transition.

Just Transition will never be labor's first choice. Our first choice will always be to determine whether the jobs we have now are, or can be made sustainable. Herein lies a trap. If we are not careful, we can easily become the last defenders of the indefensible, as unions have in the past on issues from clear cutting of old-growth forests, to tetraethyl lead, and as we are in the present on asbestos mining in Canada. Make no mistake, multi-national corporations are eager to have us fight a rearguard action for them while they prepare a soft landing for their managers and stockholders. So, without becoming the last defenders of the indefensible, of course we would prefer to see our present jobs become sustainable. Just Transition is our backup plan.

It would be very much easier to sell sustainability to trade unionists, especially trade unionists in dirty, toxic, or resource-depleting industries if there were excellent examples of Just Transition to point to. There are not. There are some examples that are fairly good, like the programs that were put in place for coal miners and steelworkers in Germany, as these industries contracted there over the last couple of decades. They were impressive in the sense that no unionized worker involuntarily lost her or his job, but nevertheless they failed to create substantial numbers of new jobs in the regions affected. There are no examples of a completely Just Transition, where both workers and communities were fully protected. Until there are, sustainability will be a tough sell to many workers.

The absence of perfect examples does not mean we ought to stop believing in Just Transition, or demanding Just Transition programs from our governments. It

---

4   International Labour Organization, "Green Jobs: Towards Decent Work in a Sustainable, Low-Carbon World," http://www.ilo.org/global/What_we_do/Publications/Newreleases/lang—en/docName—WCMS_098503/index.htm.

5   Center for American Progress, "Green Recovery—A Program to Create Good Jobs and Start Building a Low-Carbon Economy," http://www.americanprogress.org/issues/2008/09/pdf/green_recovery.pdf.

means we have to build those examples and entrench their principles as we progress toward sustainability.

An example of an opportunity for an excellent Just Transition program that we are currently missing is for Canadian asbestos miners. Massive subsidies keep this obviously sunset industry alive. These subsidies could easily be redirected to fund an absolutely first class transition for these workers, their families, and their communities, if we could only find the vision, the consensus, and the will to make it happen. In the energy industries, we know that, in the long run, the world will need to reduce or eliminate its dependence on fossil fuels, even if, in the shorter run of the next several decades, we are hopeful that newer technologies will allow us to reduce the harm of continuing to rely heavily on them. Would it make sense to start planning a Just Transition today?

CLIMATE CHANGE—THE CURRENT TEST

Just Transition applies to much more than the current debate around greenhouse gas emissions, crucial as that debate is. Indeed, forerunners of the Just Transition concept were constructed to deal with threats to parts of the chemical industry, particularly when there seemed to be the prospect of large numbers of toxic chemicals being banned in the 1980s and 1990s.

Presently, the world is (rightly) obsessed with deciding what to do about global warming. The labor movement has had to face up to the realities of climate science. Most unions have done so. The question is not whether to reduce greenhouse gas emissions. The question is when we will do so, and how. These are not two questions, but one. We will take action on greenhouse gases when we believe that we know how to do so.

Resistance to real action is a manifestation of our inability to answer this. Industrialists fear seeing their facilities become obsolete overnight, with the stroke of the regulator's pen. Workers—and the families and communities that depend upon them—fear for their jobs.

In fairness, their fears are not unfounded. Massive change in the way our society operates must take place to preserve the environment and move toward sustainability, even if significant numbers of "green" jobs are created in the process. Like it or not, a transition is coming—and cannot be left to the marketplace. The only way to ensure a Just Transition, is to create structured programs to facilitate it. This means government programs or at least government-guided programs. The so-called free market will not provide a Just Transition. We will not be able to shop our way to sustainability. Surely no one, in the face of our current global economic crisis, can still believe that deregulation, privatization, and contracting out are the answers.

If society wants workers to give up the jobs they have today, those workers will want to know what they will be doing tomorrow—and the answer had better be good. Those who oppose taking action, oppose it because they have no good answer to "the tomorrow question." Additional fear and distrust is being deliberately sown by some industrialists who want workers to fight the battle against Kyoto or its successor for them—not because they think they will win, but to buy time to create their own transition program—a transition program for billionaires and CEOs. The dollar value of each month of delayed action can be calculated, and it is large. When these corporations are ready, labor will find that it has not earned any loyalty from them. Workers will have been busy defending their employers, while their employers will have been busily investing their billions in renewables. When they are ready, these corporations will declare themselves green and leave their workers without jobs, without credibility—the last defenders of the indefensible in the eyes of the public—and without sufficient political power to even negotiate decent severance packages.

However, in the interim, the delaying tactics will have done severe damage—possibly even fatal—to the battle against climate change.

If workers are blackmailed with their jobs, both the environment and workers, who will feel compelled to become the "last defenders of the indefensible," will lose. Therefore workers must not be asked to make this choice.

WHAT WOULD A JUST TRANSITION LOOK LIKE?

Just Transition asks that society consider who benefits from and who pays the cost of implementing measures to protect the environment. To avoid impasse, those costs and benefits must be shared fairly—and not just between countries. Without Just Transition, workers, families, and communities will pay most of the cost of getting to sustainability.

A Just Transition is meant to be an all-encompassing, flexible approach to helping negatively-affected workers. Just Transition is not a suicide pact. It is not merely an enhanced unemployment program. It must keep workers and their unions whole, it must involve workers and their communities in its design, and it must be customized to each situation.

Demands for a Just Transition can be expected particularly to follow efforts by governments to protect the environment. The "visible hand of regulation" will always be experienced by workers in a different way than the "invisible hand of the marketplace." Just Transition programs, therefore, must be an integral part of government policy-making. Sustainability requires that investment in the social infrastructure be recognized as legitimate and necessary for a prosperous future, in the same way that investment in environmental protection or economic development is recognized as legitimate. There is a role for governments beyond legislation and regulation. Governments can provide direct leadership by providing social programs and (at least in some cases) public ownership of selected resources, utilities, and means of production. Surely some resources—such as water, food, and energy—are more than just examples of tradable commodities for financial speculators to gamble on—and access to a fair share of them must be considered a human right.

A Just Transition program is not a traditional labor market adjustment program, which have generally been top-down programs (at least in North America; some European programs have allowed for more worker input) designed to serve the needs and interests of business. A Just Transition program places the needs and interests of workers first.

We propose the establishment of Just Transition funds to provide any worker negatively affected by environmental imperatives with full income, benefits, and educational support until he or she has found comparable work or has made the transition to retirement or self-employment. As examples, a Just Transition program could:

- guarantee a right of first refusal for workers in "brown" jobs being eliminated, to new "green" jobs being created;

- provide income and benefits for each year of service similar to a pension fund;

- provide full income and benefits until normal retirement for older workers;

- provide support for workers wishing to start their own businesses;

- provide educational support, including full tuition and income support;

- provide a wage subsidy to workers who were forced to take a job at lower pay levels so that their total pay would equal that of their old, eliminated job;

- provide redevelopment funds to affected communities;

- provide health care and social services to affected communities, where necessary;

- provide for credible redevelopment and/or transformation ideas for existing industrial sites—even creative solutions that might not otherwise be funded;

- guarantee institutional stability for unions by automatic recognition in new workplaces and jobs created by sustainable or green investment.

Globally, a Just Transition requires that the labor movement pay attention to issues of labor cost parity between the developed and developing worlds, and appropriate legislative and regulatory frameworks (e.g. pollution prevention and forestry practices) that prevent social and environmental (and economic) dumping.

FUNDING A JUST TRANSITION

No doubt many readers would ask how such a comprehensive program could be financed. Even if we ignore the fact that trillions of dollars can apparently be conjured out of nothingness when incompetent or criminal bankers drive their institutions into the ground, there are several options for funding a Just Transition. Certainly a re-direction of presently collected taxes could pay for all of this and more. Alternatively, a "Tobin Tax" (a small tax on international currency speculation) could provide ample funds, while at the same time having a beneficial and stabilizing effect on the global economy.

There is another opportunity for funding adaptation, development, and Just Transition with plenty of money left over. The amount of oil directly consumed by the world's militaries is truly staggering, but if you add in the military contractors, subcontractors, and the so-called "defense industry," it is absolutely obscene. Not only would cutting back on this madness instantly achieve a dramatic and significant reduction in greenhouse gas emissions, but would release truly unimaginable amounts of money for green technologies, development, Just Transition, and all the rest. The labor movement has historically been proud to have links to the peace movements. Perhaps this is the time to renew and strengthen those links. Economic and social development, poverty and inequality eradication, human rights, labor rights, democracy, and environmental protection are all directly linked to demilitarization and peace.

CONCLUSION

Just Transition is an idea. Nowhere in the world has it been fully practiced, although some transitions have been more just than others. As such, there is still room for new ideas to be integrated into the concept. The world spends billions of dollars a year on universities and think-tanks. It would be worthwhile asking some of these to come up with creative ideas on how to structure a Just Transition to a sustainable world.

The ideas contained in this chapter are not solely applicable to the energy industry, but that industry's unique position as both a major source of employment, and a supplier of a basic necessity for development, and life itself, makes it crucial that the right to a Just Transition be recognized for energy workers immediately.

A Just Transition is necessary to defend the social dimension of sustainability. For it is entirely possible to imagine a world in which the economy functions and the environment is preserved, and yet still is profoundly unjust.

Building a sustainable future is a labor issue, and a Just Transition shows us how to get there. There is no future for jobs, unions, or the Earth by pretending that action is unnecessary. Neither is leaving the problem for our children to deal with an option, since the window of opportunity to effectively act may close before they get their chance. Yes, we have a responsibility to worry about jobs and the economy, but there are no jobs on a dead planet.

Chapter 54 ▌ Part 13

KEEPING THE INVESTORS AT BAY
Towards Public Ownership and Popular Acceptance of Renewable Energy for the Common Good

Preben Maegaard

THE NEED FOR ADEQUATE ORGANIZATIONAL MECHANISMS

Today, the necessary technological building blocks for a transition to renewable energy already exist, in the form of decentralized cogeneration plants, wind turbines, large and small biogas plants, solar energy, and various types of biomass for energy purposes. Now, the primary task is to integrate the various forms of renewable energy, sometimes in combination with natural gas, in order to achieve maximum utilization of renewable energy sources and supplies. It is necessary to combine and integrate technologies since no single renewable energy source is sufficient to stand alone.

Until now no country has sought to combine renewable energies into coherent autonomous systems. Instead, they have been attached to the existing fossil-based energy system. One consequence of this is, for example, that wind turbines are periodically shut down when the wind turbines produce too much. There is also an excess capacity of combined heat and power if it coincides with excess. These problems will become increasingly frequent as more wind turbines feed power into the grid, and more CHP systems are utilized. However, a solution can be found by dumping excess wind power into the fuel-efficient heat and power systems, and temporarily shutting down the CHP so that the excess electricity from the wind satisfies the need for heat.

The conflict between renewable energy and conventional power means that there will periodically be a problem of surplus power from the combined supply from wind turbines, solar power, and CHP. The problem need not exist, but is caused by lack of political management and coordination, as well as conflicts of interest. Public ownership can best solve these conflicts associated with intermittent power production. The problem is structural, and requires political solutions with incentives for the wise use of the so-called surplus power, avoiding selling at very low prices to neighboring countries, and establishing major new transmission lines and systems to match the supply peaks, especially when winds are strong.

A comprehensive future conversion to renewable energy requires mobilization of all forms of installations, including both large and small plants. It is not enough to base development on technologies that are currently cheapest, as this could lead to a unilateral deployment of large wind turbines, in particular.

The various renewable forms of energy (solar, wind, biomass) can provide an alternative to fossil fuel when used in combination with one another. None of the renewable energy forms are capable of covering the need for electricity, heat, and transportation if they stand alone. Therefore, there must be a multi-pronged effort involving many kinds of supply systems, energy storage and saving mechanisms, as well as appropriate user-management. A successful conversion involves changing attitudes and habits.

Therefore, we must be careful not to create a legislative framework and conditions that play the different forms of renewable energy against each other. However, this is exactly what happens when the goal of achieving the maximum possible CO_2 reduction from any given investment is set. It is worth bearing in mind positive developments in Germany where, since 1990, laws about differentiated and guaranteed prices have resulted in the development and implementation of a wide spectrum of technologies. By 2007, the choice to embark on a transition to renewables had resulted in 300,000 new jobs and a huge export potential.

The rules for owning and operating renewable energy technologies must therefore be changed. The technologies must be integrated in sustainable, widely-accepted economic contexts as is already normal practice in other public supply services. This requires information and people's extensive involvement, acceptance, and participation. Local communities must be given the right to determine the detailed design and combination of energy from solar, wind, and biomass. The role of the national government is to define and require specific targets for CO_2-neutral energy that must be met by each municipality, goals that cannot be deviated from at the local level.

PUBLIC PROVISION OF RENEWABLE ENERGY FOR THE COMMON GOOD

Europe has a long tradition of public utilities that are responsible for the distribution of water, gas, electricity, and heat. A wide range of public companies exist, some local, some consumer-owned cooperatives, some state corporations, and some limited liability companies. In limited companies (Plc, AG, etc.), the state or municipality may be ordinary shareholders, but sometimes public utilities are run by purely capitalist companies. In such instances, it is normal to have an essential element of governmental regulation concerning control of price and terms of delivery.

The starting premise behind this is that every citizen is a consumer of the utility's products and services. These companies have an effective de facto monopoly status and, although the last years of liberalization were supposed to have created competition in some areas of supply, consumer choice is an illusion. Electricity is electrons. Its supplier (nuclear, coal, the sun) cannot be identified by consumers, and they do not know where the hot water in their heating pipes comes from. There is no difference between natural gas, whether it comes from Siberia, the North Sea, or from Algeria, although there may be several different utilities for the consumer to choose from. Prices are virtually the same and the product is identical.

The supplies of water, electricity, gas, heat, and energy for transportation have in common the fact that they are all daily necessities for domestic consumers, as well as for industry and the public sector and its institutions. Therefore, it is the case in many countries that the same company, often a municipal company, may have supplied all these services for several decades, a system that has generally worked to everyone's satisfaction. People have been able to count on not being deceived or exploited by a monopoly supplier, and local companies were often owned by the people themselves. Electricity prices could vary from municipality to municipality, but only within certain limits. Democratic control over people's representatives, the politicians, has usually guaranteed citizens' interests.

In light of this long tradition in the supply of public goods, it is very relevant to ask why the renewable energy forms, which have become common again in last two decades, have not become a part of the public utilities. In almost all countries, renewable energies are 100 percent capitalist owned and operated. The legal framework that allows for establishing a renewable energy sector even has as its fundamental premise that private companies and individuals who own energy facilities and utilities are legally obliged to:

- let RE-providers be connected to the public electricity supply;
- receive the electricity produced at RE-plants;
- pay a fair and state guaranteed price.

In this way, the roles are distributed and locked up until the political initiative changes the situation. This has been key to the promotion of renewable energy over the last twenty years.

There are two basic models for promoting renewable energy: One is the volume control (quotas and green certificates), and the other price adjustments, mostly known as feed-in tariffs. Of the two models, the feed-in is by far the most successful with 85 percent of all installed wind power based on the feed-in principle. Both models have been introduced to allow non-utilities the opportunity to deliver electricity to supply a system that traditionally has been entirely controlled by companies with monopoly status. However, as will be explained later in this chapter, giving private investors a key role in the development of renewable energy is not an appropriate long-term strategy.

By making municipalities responsible for the shift to decentralized energy supply, based on local wind, solar thermal, solar cells, and biomass resources, municipal and consumer-owned non-commercial companies will have the same need for well-functioning, state-guaranteed tariff systems as do private investors.

It is necessary to state clearly that the entire energy sector has been, and to a large extent still is, based on the underlying view that energy for electricity and heat should come from fossil and nuclear energy, and that using large centralized units is the cheapest way of doing this. Supply companies, whether or not they are publicly owned, have not taken environmental and climatic impact into consideration when choosing their primary energy source. They have also not considered the switch from fossil and nuclear energy to renewable energies a primary concern of public bodies, despite that fact that the switch will result in a better environment and will create new jobs and industries, locally and nationally. Furthermore, such policies could become important as part of an anti-recession strategy in the future.

Neither have public utilities based their choice of primary energy source on the understanding that it is likely that the next few decades will witness a supply crisis for several of the fossil fuels, particularly oil and natural gas, and that the prices of these can be expected to rise sharply. The same can be expected to apply to uranium-based energy. This means that the utilities' long-term investments are not in the citizens' interest, despite the fact that they are often their owners and always their customers.

Because of the supply companies' lack of preparedness for a transition to renewable energy, it is has been left to the private sector—especially private energy developers—to become the new class of proprietors in the renewable energy sector. In order to ensure their investments in wind, photovoltaic, and biogas plants, they have depended on political conditions that have permitted and ensured long-term investment in energy technology and infrastructure. At the same time, in order for an investor to want to engage in capital-intensive energy projects, a revenue incentive must exist. It is not the purpose of this article to focus on the legal framework that best leads to success for renewable energy among private investors, but it is nonetheless important to point out one absolutely clear, empirical fact: guaranteed prices with a long horizon of fifteen or twenty years have an effect that quotas and green certificate models do not have. Three countries in Europe—Spain, Germany, and Denmark—represent 90 percent of all wind energy on the continent, and the sector's expansion in these countries has taken place on the basis of government-guaranteed prices. The result of these policies was that, in 2006, 20 percent of the electricity consumption in Denmark came from wind and 8 percent in Germany, the world's third largest economy. In contrast, countries like the UK and Ireland, with by far the best wind resources in Europe, do not have a wind power development of any real significance.

However, what offered the best political conditions for renewable energy in its pioneer phase is not necessarily the path we must follow in the future if we are to avoid unnecessary conflicts, economic tensions, and inequalities in society. Nor is it the path to follow if we are to create the necessary political and popular base for the continued transition to renewable energies. Just as is the case for other social sectors, ownership will play a crucial role in the renewable energy sector.

There is no doubt that individual private ownership is the right thing when it comes to renewable energy self-sufficiency of each residence or consumer unit. Just as it is neither appropriate nor reasonable to let central transportation companies own and operate people's private transport, it is meaningless to allow the central energy companies to own and operate solar power plants on private people's roofs or the biogas plant belonging to an individual farmer. Wind turbines for personal supply must also be a private matter in the same way as you are responsible for your own house, heaters, bicycles, etc.

However, the situation is completely different when it comes to large wind turbines, solar PV, and other renewable energy facilities, which do not serve individuals but rather supply many households, industries, and institutions. For instance, megawatt-class wind turbines when installed several at a time in wind parks may supply tens of thousands of consumers and require investments of several million euros. Times have changed from the industry's infancy, when wind turbines produced 30–100 kilowatts and mainly supplied individual households and other local stakeholders in the immediate vicinity of the turbine.

The development of renewable energy, based on private investments, cannot be compared to normal competitive commercial enterprise. If prices and other conditions are guaranteed and defined by the state and there is a purchase obligation from the power distributor, the investment risk is very limited. And, if, as is common, a long-term maintenance agreement and production-guarantee agreement with the wind turbine supplier is also signed, the risk is virtually eliminated altogether.

Therefore, it is meaningless to compare this type of wind turbine investment with a normal commercial investment, which almost always involves an active effort and investment risk. Thus, the transfer price is determined by the public, and guaranteed over a longer period of years, leaving a security for the investment to be recovered. This is not common practice in most business activities, where an alternative supplier that is able to offer a similar product will always exist, and so companies must always strive to be competitive. A producer of electricity from wind turbines is not in such a competitive situation; a market for its product is always assured. Rules that are not found in ordinary business activity also apply when it comes to purchasing wind turbine sites. The place for the wind turbine is designated by the government and local authorities, and therefore not the result of traditional commercial efforts.

LESSONS FROM THE DANISH EXAMPLE

Denmark is a pioneer in the use of wind and other renewable energy technologies. By the year 2000, Danish energy politicians had already started applying the brakes on privately-owned wind power. However, they did not distinguish between turbines that were purely objects of financial speculation and those that were owned by private wind cooperatives. This cooperative ownership form, in which the number of shares available to each household was regulated and there was a ceiling on profit levels, was previously widely used in Denmark. With 150,000 households as joint owners of wind cooperatives, it is clear why the expansion of wind power that occurred between 1980 and 2002 (when there was a definitive break in connection with a change of government) had such a broad popular acceptance.

Middelgrunden is frequently mentioned as an example of a successful initiative. This offshore facility lies in the Sound between Copenhagen and Malmö. A Copenhagen electricial company and 7,500 people in the community have established twenty 2 MW wind turbines that provide 5 percent of Copenhagen's electricity consumption. The project was undertaken with wide acceptance, partly because it was a municipal-owned power company and there was broad public participation. Private ownership has not caused problems, as the annual income from being a co-owner of wind turbines is around €1,000, which is a modest contribution to the household economy and similar to its energy costs.

On the other hand, the Danish experience has shown the difficulties in explaining and justifying the right of individuals to benefit from conditions guaranteed by the state, which allow them to reap large fortunes by running utilities that could just as easily be run in the same way as water, gas, electricity, and heating supply have traditionally been run. The problem becomes especially acute when wind turbine suppliers publicly advertise and boast of the privileges of guaranteed prices in the feed-in systems.

Land-use planning, which designates specific sites for wind turbines, accentuates the problem of individual, private ownership. This type of public windmill planning, which has been common practice in Denmark, implies that a land owner (almost always a farmer) has identified a number of locations for wind turbines on the land. This land will therefore gain a very high economic value, regardless of whether the landowners plan to develop and own the turbines themselves or whether they will sell or lease the sites to a wind turbine investor. At times, land may even have a higher price per square meter than land in Manhattan.

However, the farmer or land owner on the other side of the field gap, whose land has not been appointed by the public authority for windmill sites, does not have the same opportunity for capitalization of their land. We cannot respond to requests for labor-free income by awarding sites, set windmills up haphazardly in the country-side, and allow the practice of private windmill ownership to privilege a few owners. That would lead to misdirected and unnecessary speculation of an energy form that is crucial in the transition from fossil and nuclear fuels.

Therefore, a long-term solution must be found to the problem of ownership of large wind turbines and other renewable energy infrastructures.

DEVELOPING NEW FORMS OF COMMON OWNERSHIP TO DEFEND RENEWABLE ENERGIES FROM COMMERCIALIZATION AND SPECULATION

Renewable energy is still young. Technology and tariff aspects have found reasonable solutions, but now we have to develop generally-acceptable organizational structures for decentralized ownership for the common good.

Particularly important here are local cooperatives. These are autonomous associations of persons united voluntarily to meet their common economic, social, and cultural needs and aspirations through jointly-owned and democratically-controlled enterprises.

All over the world, millions of people have chosen the cooperative model of business enterprise to enable them to reach their personal and community development goals. Cooperatives provide 100 million jobs worldwide—20 percent more than multi-national enterprises. The cooperative movement brings together over 800 million people around the world. The United Nations estimated in 1994 that the livelihood of nearly 3 billion people, or half of the world's population, was made secure by cooperative enterprises. These enterprises play significant economic and social roles in their communities. They create and maintain employment, providing income. In addition to being responsible for producing and supplying products and services to their members, they also serve the communities in which they operate.

Cooperatives are autonomous, self-helping organizations controlled by their members. If they enter into agreements with other organizations, including governments, or raise capital from external sources, they do so on their own terms, ensuring democratic control by their members and maintaining their cooperative autonomy.

By putting cooperative principles and ethics in practice, they promote solidarity and tolerance, and as "schools of democracy," they promote the rights of each individual. In many countries, and in a variety of activities, cooperatives are significant social and economic actors in national economies, and contribute to the well-being of entire populations at a national level.

The common-good solution should be given a key role in the decentralized autonomous future-providers of heat and electricity. These will utilize combined solutions of solar, biomass, wind energy, and storage of energy, as no renewable energy solution can stand alone. In Germany, and in Denmark since 1998, wind power is strongly profit-oriented. A new class of green investors is emerging, and they are not residents in the communities where they make their investments. In the early stage of implementation this may be a workable procedure, but for the total transition to a renewable energy society, the local residents must take leadership.

The operation of autonomous renewable energy systems for community supply should ideally be a service that is provided by local cooperatives. Like the supply of water, local district heating, public transport, and other parts of the public infrastructure, cooperative ownership will re-establish the necessary general acceptance of especially big wind turbines at the local level. Like in many other walks of life, decentralized ownership of energy solutions by cooperatives will be seen as generally acceptable the world over.

We can learn from the past 100 years of practice that the state should promote public regulation in favor of local and collective ownership of basic public services. This includes energy. This non-capitalist approach is in line with the promotion of the common good in most democratic societies. Considering the scale and complexity of the transition to a 100 percent renewable energy system, and its urgency, it is only realistic to look for publicly-owned solutions to undertake this task.

For cooperatively-owned wind turbines, public planning with expropriation of the necessary areas for wind turbines will be a normal practice, as is currently the case with power pylons, waterworks, and similar areas of public interest. It is already standard practice to provide monetary compensation when areas are being designated for the common good. This should also be the case in the wind energy sector when they are defined, not as investments for profit, but for the common good. Not-for-profit, traditional cooperatives will fulfill such criteria, and the guild ownership model, as it has been practiced in Denmark for some decades, cannot be considered as cooperatives in this sense. Only *some* of the local residents invest in the guild, such as in the Middelgrunden example described above, and they do so to obtain a private source of income, rather than to supply the community in general with clean, local energy. In contrast, the approximately 200 local, combined heat and power units were established as pure cooperatives. Their board members are elected and do not have individual economic interest in the investment, but work for the benefit of the community that is served by the CHP. Legislation that expropriated areas can only be used for local, public-owned wind turbines must be put in place.

This would be a decisive contribution to attaining local acceptance and make wind energy more competitive.

Legislation must also instruct utilities to buy the electricity from wind turbines at a price determined by the government and guaranteed for a minimum of twenty years. Such laws are used in the most successful wind energy countries. This would be a serious incentive for the local cooperatives to actively be a part of the development of autonomous renewable energy solutions that can offer a complete renewable energy package including wind power.

By making the establishment and operation of large wind turbines the responsibility of local cooperatives or similar forms of public supply, there will be significant savings due to cheaper sites for wind turbines, saved repowering fees, and cheaper long-term financing. This will make wind energy more attractive for the individual community, as well as for the nation. It would also improve supply security, steady energy prices, and secure the fulfilment of international agreements concerning CO_2 reduction.

In Scandinavia, the dominant form of ownership is that which satisfies the common good within the supply of water, district heating, public transport, etc. However, until now, windmill ownership has been private investment, also in the form of cooperatives, guilds, and whatever we call them. People invest in order to make a profit, and therefore the Danish version of windmill cooperatives does not belong to the common-good category.

Public ownership by true local cooperatives will also make wind energy cheaper. Sites for windmills will always be scarce in our part of the world, which results in a capitalized price for land. Together, the repowering certificates and access to a site makes a commercial investment in a new big windmill in Denmark up to 50 percent costlier than if it was installed for the common good of the community by a public company, which happens in many other sectors of society.

It may well prove impossible to implement a renewable-energy community based on the necessary integration of all kinds of renewable energy technologies within a commercialized economy. At the local level, competition between the various suppliers will cause a distorted development. So, in order to avoid capitalization of the feed-in-tariffs and a constant pressure from the investors to get improved tariffs, the best solution is to maintain the feed-in tariffs for common-good investments and offer compensation to the land owners, as has become normal practice for power line compensation.

FINANCIAL SOLUTIONS

Access to know-how is essential. However, as renewable energy is especially capital intensive, it is also necessary for individual countries, especially developing ones, to improve their energy supply without spending too much of their scarce foreign currency on importing energy equipment from the industrialized countries that currently control the technologies, and frequently have prices that make their products prohibitively expensive.

In emerging economies, such as China, Egypt, India, etc., renewable energy supply must be competitive with conventional power plant technology and fossil fuels, for which the countries already have a domestic production. This hurdle can only be overcome if the newly-industrialized countries go about building new RE-industries using their own labor force and wage levels. Otherwise, the expanding sector will simply create more jobs in countries that already have a well-established production, but at costs that are too high for the emerging economies. Such a development would force these countries to fall back on their own already existing high-emission energy forms since this would be the only affordable option.

There are a number of reasons for establishing an appropriate legislative framework to bring about a gradual transition to local public supply of renewable energy and the local acceptance that it requires. Funding schemes and models, similar to those employed by industrialized countries when they established the infrastructure that is a natural prerequisite for modern industrial societies, must be available. Roads, railways, and earlier energy systems were built based on the premise that it was a community task to establish these infrastructures, and to supply water, gas, and electricity to all companies and individuals on terms that were regulated by society. Either companies were publicly owned by the state or municipalities, or they consisted of local consumer companies, which accounted for ownership and operation. Finally, private companies have, at times, taken care of public supply, but, in such cases, concessions have been based on widespread public regulation and legislation to prevent private providers from abusing a monopoly supply. The state provided direct construction grants, guarantees of loans or it was the state companies, which implemented the investment.

Major publicly-owned companies have often arisen as bottom-up initiatives, where many small, local power utilities came together to build large, more efficient power plants and thus the overall transmission network. Such an ownership model was originally found in the German "Stadtwerke" until the political regime took it over in 1935 and put it in the hands of a few large companies. These companies still dominate the German electricity supply today. Denmark is another example: until 2004, the major power plants and the overall transmission networks were owned by local utilities, either municipal or consumer owned. With new legislation in 2004, the major formerly-consumer-owned power plants were sold to commercial companies like Vattenfall and DONG Energy, while the transmission grid, which these local consumer cooperatives also owned, was transferred to a new state company, Energinet.dk, with no financial compensation given. This new ownership pattern of the transmission grid allows for all kinds of producers of electricity (both utilities and non-utilities): large oligopolies, consumer-owned cogeneration plants, and privately-owned wind turbines all have the right to connect and feed into the grid on equal terms. Energinet.dk is responsible for the overall system, ensuring that consumer demand is always met, and they also regulate the level of import and export of electricity. In order to avoid building big and expensive transmission lines in the future, they have an interest in encouraging renewable energy power to be consumed in the vicinity where it is generated.

When it comes to local public utilities, we need to create financing solutions that remove renewable energies, including wind, from private ownership. Past experience shows that there are no problems with financing local public infrastructure and that it can make renewable energy cheaper for society than when private investors are responsible for the development.

CO-OPERATIVES CREATE AND MAINTAIN EMPLOYMENT

Co-operatives provide over 100 million jobs around the world, 20% more than multinational enterprises.

In Canada, co-operatives and credit unions employ over 155,000 people. The Desjardins movement (savings and credit co-operatives) is the largest employer in the province of Québec.

In France, 21,000 co-operatives provide over 4 million jobs. (Source: GNC Newsletter, No 348, June 2007)

In Germany, 8,106 co-operatives provide jobs for 440,000 people.

In Italy, 70,400 co-operative societies employed nearly 1 million people in 2005. (Source: Camere di Commercio d'Italia, "Secondo rapporto sulle imprese cooperative")

In Kenya, 250,000 people are employed by co-operatives.

In Slovakia, the Co-operative Union represents more 700 co-operatives who employ nearly 75,000 individuals.

Co-operatives are significant economic actors in national economies

In Belgium, co-operative pharmacies have a market share of 19.5%

In Cyprus, the co-operative movement held 30% of the market in banking services, and handled 35% of all marketing of agricultural produce.

In Denmark, consumer co-operatives in 2004 held 37% of the market. (Source: Coop Norden AB annual report 2004)

Finnish co-operatives were responsible for 74% of the meat products, 96% of dairy products; 50% of the egg production, 34% of forestry products and handled 34.2% of the total deposits in Finnish banks.

In France, 9 out of 10 farmers are members of agricultural co-operatives; co-operative banks handle 60% of the total deposits and 25% of all retailers in France are co-operatives. (Source: GNC Newsletter, No 348, June 2007)

In Japan, the agricultural co-operatives report outputs of USD 90 billion with 91% of all Japanese farmers in membership.

In Kenya, co-operatives are responsible for 45% of the GDP and 31% of national savings and deposits. They have 70% of the coffee market, 76% dairy, and 95% of cotton.

In Korea, agricultural co-operatives have a membership of over 2 million farmers (90% of all farmers), and an output of USD 11 billion. The Korean fishery co-operatives also report a market share of 71%.

In Kuwait, the Kuwaiti Union of Consumer Co-operative Societies handled 80% of the national retail trade.

In New Zealand, co-operatives are responsible for 95% of the dairy market and 95% of the export dairy market. They hold 70% of the meat market, 50% of the farm supply market, 70% of the fertiliser market, 75% of the wholesale pharmaceuticals, and 62% of the grocery market. (Source: New Zealand Co-operative Association, 2007)

In Norway, dairy co-operatives are responsible for 99% of the milk production; consumer co-operatives held 25% of the market; forestry co-operatives were responsible for 76% of timber and that 1.5 million people of the 4.5 million Norwegians are member of co-operatives.

In Poland, dairy co-operatives are responsible for 75% of dairy production.

In Singapore, consumer co-operatives hold 55% of the market in supermarket purchases and have a turnover of USD 700 million.

In Slovenia, agricultural co-operatives are responsible for 72% of the milk production, 79% of cattle; 45% of wheat and 77% of potato production.

In the UK, the largest independent travel agency is a co-operative.

Chapter 55 | Part 13

TECHNOLOGY FOR AUTONOMY AND SELF RELIANCE
International Technology Transfer for Social Movements

Andrea Micangeli, Irene Costantini, and Simona Fernandez on behalf of the Self-Reliance and Environment Technologies Unit, CIRPS

Technology research for the environment and society is an essential part of research activities applied to international cooperation. Generally speaking, we refer to "technology for self-reliance" for all the processes, structures, and the products aimed at developing a social formation that is rooted in technological principles, instruments and models.

Autonomy and access are two core issues in "technology for self-reliance":

• Access (financial, social, and technical) should be guaranteed to the widest public possible, specifically in the context of disadvantaged situations.

• Self-reliance is the result of the process through which effective capability and social functionality has been built.

People become active participants in their lives without developing a dependence strategy. Self-reliance means building productive social relationships, maintaining relationships that are "non-dependent." Self-reliance processes must guarantee the ecological and social self-reliance of those involved.

Cirps (Self-Reliance and Environment Technologies Unit at Sapienza University of Rome, within CIRPS (Inter-University Research Centre on Sustainable Development) has been working to support these two pillars. In particular, the international cooperation activities and studies run by Cirps focus on a) small social environment, b) low environmental impact energies, c) self production of chlorine, d) disadvantaged work groups, and e) areas that are in permanent crisis or social tension, in both urban and rural contexts.

The motivation for working on this basis lies in a strong belief in the importance of energy and technology in the world. It is easily understood that a big gap exists with regard to technological knowledge. Those in possession of technical knowledge and the means to improve access to such technology have a duty to provide support to those lacking such knowledge and access. At the same time, a focus on self-reliance and autonomy is the only way to avoid establishing a neo-colonial relationship between the actors involved. The issue of "relationship" is extremely important in planning and implementing a project in a crisis-ridden area. The more serious the problem at hand, the wiser the relationships need to be, both locally and internationally, in order to ensure that the beneficiaries successfully achieve autonomy and, consequently, self-sufficiency.

The choice of which project to undertake is an important one, and involves a number of different factors. One of the most important of which is the question of funding. This does not merely concern the issue of the availability of money, but, more importantly, who or what institution is offering it. Cirps considers the ethics of funding one of the cornerstones of its policy; even though Cirps has worked in conflict areas, it has always refused money closely connected to the military, as this without a doubt goes against its principles.

Another crucial aspect concerns the selection of beneficiaries. It is important that the community in which the intervention takes place has already developed its own level of autonomy. This criterion is not defined absolutely and universally, but it is clear that a certain kind of self-organization is necessary in order to establish a good relationship in the territory. Both social movements and organized communities might be considered, but they must accomplish two tasks: they must be representative in the project's area and be willing to implement sustainable changes.

Cirps has been working for many years in the field of self reliance and environment technologies. Over time, it has built up a mature experience in implementing and offering assistance in projects related to water and sanitation and energy sector technologies. In the following pages, some of the principal activities run by Cirps—in the refugee camps, in the field of international cooperation, in the advanced training—are described. The examples include work on renewable energy projects and also those relating to wider technologies, since combining these different technologies is a strategy that is most effective for building self-reliance and autonomy. Of particular importance to Cirps is its interest in promoting activities relating to water in crisis areas, and that decision hasn't been taken randomly. Water is one of the essential resources for human life. Its access must be guaranteed even in places where this is problematic, either for natural reasons (water scarcity) or socially-caused problems (war or isolation). In such situations, technology can be either a tool to guarantee access or to deny it. The projects described below show the kind of intervention that Cirps works on, and the impact it can make on a community. It is hoped that the experience of socially-targeted technology transfer, in its broadest sense, will be of use to those working more specifically in the context of energy in general, and renewable energy in particular, which is the focus of this book.

SELF-RELIANCE TECHNOLOGY FOR THE COMMUNITY: RENEWABLE ENERGY AND WATER PURIFICATION IN ZAPATISTA COMMUNITIES, CHIAPAS, MEXICO

In the field of renewable energy, we have carried out interventions that began with hydroelectric energy production and later continued with the self-production of chlorine in Chiapas, Mexico. This activity, which consists of two projects, is an example of the kind of long-term sustainable development activity that Cirps carries out in conflict areas. The two projects were "Una Turbina per La Realidad" (1997–2001) [A Turbine for La Realidad] and "Cloro Rebelde Zapatista" [Rebellious Zapatistas' Chlorine] (2007–2008).

The area is Chiapas, in the southeast part of Mexico, near Guatemala, and the beneficiaries belong to the La Realidad Community. The projects arose when the Zapatista organization proposed a local study of the community's energy needs and the installation of two plants (microhydro generator, and an on-site electro-chlorination facility).

The project was carried out by CIRPS, an engineering department of La Sapienza University of Rome, LITA (Itinerant Appropriate Technologies Laboratory), and the Italian Zapatista-inspired social movement Ya Basta, and in Mexico it involved a local NGO, Enlace Civil, and above all the community of La Realidad itself. La Realidad has been an active partner in the development, realization, and future management of the system. Public bodies, including Italian mayors and regions, gave financing for the entire project.

La Realidad is a village of about 200 families, with an agricultural subsistence economy based on maize and coffee crops. The use of electricity was quite unknown, with a diesel engine only supplying electricity for the powerful lights of Aguascaliente (where Zapatistas hold meetings with the local community and world civil society) during gatherings, festivals, or other special occasions.

The aim of the microhydro project was to realize a plant for producing energy in a region where nobody in the private sector would carry out such work. It sought to work with and for the community on something that was not extraneous to their skills and culture, giving them the instruments to understand and manage the plant.

In the first phase of the project, men and women were surveyed, in appropriate meetings, in order to find out the energy needs of the community and to propose future applications for electrical power. This identified that the community's main current energy needs were for electric lighting, but it was important to also take into account possible future energy uses, proposed by the community, once a local electric grid had been built. These included a freezer for the small health clinic, machines in the carpentry workshop, and equipment to repair cars and other vehicles.

In 1998, the *encargados*—the people chosen by the community assembly to work on the project—cooperated with Italian volunteers and Mexican engineers to survey and map the land. An area where the small river goes down with some small waterfalls was chosen for the place to intercept the water to the future hydroelectric powerhouse (it involved a drop that measured about eighteen meters).

In the summer of 2002, two courses were held by Italian engineers in the community, one for the community in general, about the opportunities and dangers of electricity, and the other, a more technical one, for the *encargados* so that they would be able to manage the plant in the future. After the courses had taken place, the project could be considered finished. The plant is working well, though not to its full power (30 kW), because the load is not high yet. The community is trained and responsible for managing the plant and solving whatever technical problems may arise in the future. The partners who have worked in the project continue to be in contact with the *encargados* in case of extreme failure. The project's training aspect has been the most important element in terms of making the introduction of the electricity into the village as sustainable as possible. Now that La Realidad has its own power supply, the community itself will decide how to use it.

Following this successful project came Cloro Rebelde Zapatista (Rebellious Zapatista Chlorine). A successful relationship was established with the Governance Council that offered us the necessary human resources both to build the machines on site and to organize the training course. The project successfully improved hygienic and sanitary conditions according to the community's needs, and there was good participation, especially by the women. Last, but not least, it worked within the context of local autonomy and outside of the conventional market, developing a dimension of self-production.

Thanks to OSEC (On Site Electro Chlorination), oxidizing and bactericidal substances do not have to be added to the water in La Realidad. Instead, electrolysis is used to produce the same effect from substances that are naturally occurring in the water. No additional chemicals are required. The process of electrochemical disinfection has several advantages over other processes that are more commonly used to disinfect water. Processes such as chlorination using gaseous chlorine or concentrated hypochlorite solution require additional chemicals. Another advantage of OSEC, which does not require additional use of chemicals, is that the hazards in handling these chemicals are also avoided. Ozonization, and especially ultraviolet irradiation, can be very effective at the point of use, but provide little or no residual disinfection capacity. For an exact adaptation of electrochemical disinfection to the properties of the processed water, it is necessary to know the dependence of the electrolytic active chlorine production rate on the chloride concentration, temperature, current density, and anode material. The essential point in this choice of technology is that no relationships of dependency are built. The self production of chlorine frees the community from the need to request the chemicals that are usually used in water treatment processes. Finally, is the question of the technical impact on the community: technological choice needs to be appropriate for the local situation, in terms of services (maintenance included), local culture, and infrastructure. In this project, these conditions were met by the technology chosen. Machine parts necessary for replacements could be found *in loco*, the fact that the machine was easy to understand allowed a deep penetration of technical knowledge at the local level, and the machine could be installed using local infrastructures.

SOLAR THERMAL AND BOLIVARIAN SELF CONSTRUCTION IN THE VENEZUELAN ANDES

Again relating to solar energy and sustainability, the next project to be considered takes place in Venezuelan Andean communities. Following a path of sustainable development, the Bolivarian government decided to promote and fund a project of the Venezuelan NGO Caribana, which aimed to teach rural Andean communities how to construct solar panels for heating water. This project targets four rural communities in the Venezuelan Andes, in the State of Mérida, the Province Rangel: Gavidia, Mocao, Mixteque, and Mitivivò. These are communities where the problems of poverty and environmental contamination are still present and are getting worse over time. Often local people cannot use hot water because the only water in the area is at freezing temperatures in the water pipes and rivers. The farmers' domestic use of water generates sanitary problems such as a lack of hygiene, osteoarthritis, and the inability to have clean clothes, kitchenware, and living environment.

People use wood or gas to heat water, producing environmental phenomena like deforestation and air pollution. Moreover, deforestation contributes to the erosion of the soil and, consequently, to greater poverty. The Venezuelan government, seeking to resolve this situation in accordance with the new worldwide environmental and social sensitivity, decided to support an innovative project. The purpose was to improve the quality of rural life, using clean energy in a way that would not result in any collateral environmental effects. The local NGO Caribana, active in the field of social and sustainable tourism, tries to promote local culture, social improvement, ecological innovation, and economic development in the Venezuelan Andean zone.

Caribana organized trainings and construction workshops, overseen by Cirps, in order to instruct the farmers on making their own solar panels. To make this project really sustainable, all the necessary materials and components used were those readily available in the surrounding area. After the panels had been built, the NGO would go on to support installation in each community and will continue monitoring to evaluate the impacts on and benefits to the target families. In doing so, the Bolivarian government is trying to develop a new renewable energy culture among its population, enforcing the use of low-impact technologies and using courses and technical training to empower the farmers through skills and education.

SELF-PRODUCED WATER CHLORINATION SYSTEMS, AND AUTONOMY AND RIGHTS FOR DISABLED PEOPLE IN OCCUPIED PALESTINE

Cirps has also run projects in Palestine, relating to the electrochemical production of chlorine in the Gaza Strip, as well as projects around information technologies and disabled people in Jerusalem.

In Gaza, the project's aim was to provide the instruments necessary for guaranteeing water security to the inhabitants of the Bedouin village Um Al Nasser, for irrigation purposes. The intervention was made necessary due to the hard conditions faced by the population (especially the rural one). Once again, we found ourselves in a crisis area, both for natural reasons (shortage of available water) and for artificial reasons (an on-going embargo that continues to isolate the Palestinian people). Working in such a context involved several risks that had to be kept in mind, including a possible escalation of the conflict due to increasing Israeli military attacks and internal clashes, the closing of the Eretz crossing for internationals, delays in carrying out activities, impossibility of accessing areas adjacent to the "Buffer Zone" due to activities from the Israeli army that jeopardized the situation, destruction of work and equipment by the army after the intervention. The closure and isolation of the Gaza Strip makes it necessary to establish a water treatment process that is completely autonomous from external resources. The choice to work in a rural area, and concretely in a Bedouin village, was made in order to cooperate with disadvantaged people. By supporting irrigation, the project was contributing to creating an autonomous agricultural market in the Gaza Strip.

The other project was in Jerusalem. This gives a comprehensive view of two other aspects of technology: information technology and technology for disabled people in crisis areas. As for information technology, it must be related to the interesting scenario of how people react to the difficult question of managing private and public life in the city of Jerusalem. It means an attempt to search for and find social solutions, in a bottom-up dynamic, to a still unsolved issue in the area's conflict, either in accordance with government policies or not. Focusing on citizens is a chance to better understand the real needs of the city and the priorities for coexistence. In particular, the research has a special emphasis on young adults' participation in decision-making processes and in determining their own futures.

As a consequence, the specific aim is to have a comprehensive overview of how communication and information exchange work. Through this study, the existence of spaces for meeting and gathering will be verified, and a dialogue that involves different identities in the same civil society will be had. Focusing the study on young adults' association means researching the new media through which they can establish a dialogue, like the Internet, which must be seen as a new tool for dialogue. Besides real meeting spaces, virtual spaces such as communities, blogs, forums, and discussions should also be taken into account.

Another interesting research project also takes place in Jerusalem. Analysis of the Old City leads to the awareness that lots of barriers exist that divide the city into its quarters—Muslim, Armenian, Christian, and Jewish. The project's aim is to study the situation of disabled people inside each quarter in order to build up a comprehensive vision of how integration is possible. Within this analysis, it will be interesting to see how each community considers disability from a religious point of view and how much it influences the integration of disabled people in their society.

Technology to provide autonomy is the system of processes and products that improve conditions for communities, enabling self-sufficient access to fundamental services. It deals mainly with two issues: community autonomy and individual autonomy, with reference to the bio-psycho social approach as it has been defined by the World Health Organization (2002). The WHO is an important space because, thanks to an accord that was signed in 2001, the medical model (disability as an object to be cured) and the social model (disability as an exclusion tool) are no longer considered two divided issues. This integrated model puts together the personal-medical component and the environmental and social one.

Moving from Jerusalem to the Gaza Strip, the situation gets worst. The struggle of Palestinian people is not only for food, health, and a minimum quality of life, but also to guarantee the same possibilities to the local people. The Israeli border is supposedly open for people with serious illnesses to cross, but how should we define disabled people within this framework? Cirps promoted this project in order to use technologies and education to contribute to the already existing movement for disabled people's rights in the Gaza Strip. Newly-graduated students, disabled and not, will work together with disabled people to improve technology for communication and civil rights.

SUBSISTENCE SOLAR GARDENS, RECYCLING, AND EMPLOYMENT IN THE WESTERN SAHARAN DESERT

Technology plays a major role in the development of the local community. We are working on solar and inclusive gardens to improve Saharawi subsistence agriculture and to promote the use of local energy sources. The project aims to help Sahrawi people develop food self-sufficiency, and is an example of how photovoltaic technology can be used in remote areas: family-run agriculture uses drop-by-drop irrigation and solar energy pumps. As we mentioned before, a project's success is evaluated in terms of sustainability. This one aimed at re-distributing gardens' production to all the families living in the Dakhla camp. More precisely, the central element is the project's economical and technological sustainability. Photovoltaic panels guarantee the availability of water without creating energy costs, which would be high if other methods of electrical generation were used. From a technological point of view, the camp's population has grown accustomed to using them, which has made it easier for the initiative to continue even after the project's conclusion. This is a strong experience and example of self-reliance in the context of a humanitarian situation provoked by occupation and the struggle for independence.

Another project took place in the isolated Dakhla camp, where there is less opportunity for international cooperation. Over the last years, the Sahrawi population involved in the Western Sahara conflict, the Moroccan military occupation, and the mined wall, have requested that Cirps work with them and we agreed.

Cirps proposed income-generating activities for the refugee camps, such as the production of medals made of recycled materials. Although the handiworks are promoted in sports activities and events where prizes are awarded, the main purpose is to build awareness about the Sahrawi people. The Young Sahrawi Medals for Cultural and Sport Events project takes place in "27th February" refugee camp. It deals with projecting and organizing demand for the medals, which come from Italy and other countries. The aim is to promote sports and to offer jobs to young Sahrawi people, with the support of Italian colleagues who advertise the product in Italy and abroad. The intervention strategy is focused on creating new activities linked to a wider consciousness around the Sahrawi question, sport activities, and environmental issues. The workshop produces goods made from recycled materials. Young Sahrawi people reuse the material they find in houses thanks to an awareness raising campaign for correct management of rubbish. The Sahrawi people, as a whole, benefit from the project, due to the wider political visibility they gain in the events where medals are awarded.

CONCLUSION: TECHNOLOGY FOR SOCIAL STRUGGLES

We hope to have given the reader a comprehensive overview of technology applied to cooperation. Presenting specific case studies gave us the opportunity to show how different the application of technology and energy technology can be.

This is especially so, given that our work has focused on the contexts of technology application in crisis areas, which has been a key theme in our choice of which geographical areas to work in. Technology is an adaptable means that can easily be spread all over the world. It relies on knowledge that, since the very beginning, is something to be handed down to the future generations. Furthermore, it is important that it also reaches those living in the current generations, who, for different reasons, have been cut off from it, for the most part not through their own choice. At all times and in all of our projects, our aim is to foster autonomy, and to break the dependent relationship between the so-called "third world" and the industrialized one. Technology must be free to be used, spread, and reproduced, giving a strong impulse to the day-to-day hard lives of people living under oppression.

Chapter 56 ▌ Part 14: Alliances and Conflicts Along the Road to an Anti-Capitalist Energy Revolution

SAVING THE PLANET FROM CAPITALISM
Open Letter on Climate Change in Anticipation of the Poznan Climate Talks, December 2008[1]

Evo Morales Ayma, President of Bolivia

Sisters and brothers:

Today, our Mother Earth is ill. From the beginning of the twenty-first century we have lived the hottest years of the last thousand years. Global warming is giving rise to abrupt changes in the climate: the retreat of glaciers and the decrease of the polar ice caps; rising sea levels and the flooding of coastal areas, where approximately 60 percent of the world's population live; the increase in the processes of desertification and diminishing fresh water sources; a higher frequency in natural disasters suffered by communities across the planet;[2] the extinction of animal and plant species; and the spread of diseases in areas that earlier had been free of them.

One of the most tragic consequences of climate change is that rising sea levels mean that some nations and territories are being condemned to disappear.

The story begins with the industrial revolution in 1750, which gave birth to the capitalist system. In two-and-a-half centuries, the so called "developed" countries have consumed a large part of the fossil fuels that took 5 million centuries to form.

Competition and the capitalist system's thirst for limitless profits are destroying the planet. Under capitalism, we are not human beings but consumers. Under capitalism, in place of Mother Earth, there exist raw materials. Capitalism is the source of the world's asymmetries and imbalances. It generates luxury, ostentation, and waste for a few, while millions throughout the world die of hunger. In the hands of capitalism, everything is turned into a commodity: the water, the soil, the human genome, ancestral cultures, justice, ethics, death … and even life itself. Everything, absolutely everything, can be bought and sold under capitalism. Even "climate change" itself has become a business.

"Climate change" poses a stark choice for humankind: either we continue along the road of capitalism and death, or we embark along the path of harmony with nature and respect for life.

In the 1997 Kyoto Protocol, the developed countries and economies in transition committed to reduce their greenhouse gas emissions by at least 5 percent below

---

1   Translated from the original Spanish by Kolya Abramsky. This version differs slightly from anonymously-translated versions that are circulating on the Internet. Given that it was a freely circulating document, and that it would be very difficult to obtain permission, this chapter is being reprinted without the author's permission.

2   Due to "La Niña" phenomenon, which means that these disasters become more frequent as a result of climate change, Bolivia lost 4 percent of its GDP in 2007.

the 1990 levels. These commitments were based on the implementation of different, predominantly market-based mechanisms.

Rather than being reduced by 2006, emissions of greenhouse gases had increased by 9.1 percent in relation to 1990 levels, thus demonstrating the extent to which the developed countries had failed to honor their commitments.

The market mechanisms applied in the developing countries[3] have not accomplished a significant reduction of greenhouse gas emissions.

Just as the market is incapable of regulating the world's financial and productive system, so too it is incapable of regulating greenhouse gas emissions. On the contrary, the market will only generate big business for financial agents and major corporations.

THE EARTH IS MUCH MORE IMPORTANT THAN WALL STREET AND THE WORLD'S OTHER STOCK EXCHANGES

While the United States and the European Union allocate $4.1 trillion to save the bankers from a financial crisis that they themselves created, programs on climate change receive a mere $13 billion—313 times less than the bankers.

The resources for climate change are poorly distributed; More resources are directed to reduce emissions (mitigation) than are devoted to reducing the effects of climate change that all our countries are suffering from (adaptation).[4] The vast majority of resources flow to those countries that have contaminated the most, and not to the countries that have preserved the environment the most. Around 80 percent of Clean Development Mechanism projects are concentrated in just four emerging countries.

Capitalist logic is nurturing a paradox in which the sectors that have contributed the most to environmental deterioration are the very ones that are benefiting the most from climate change programs.

At the same time, technology transfer and the financing for clean and sustainable development of Southern countries have been just so much hot air, all words and no action.

If we really want to save Mother Earth and humanity, then the upcoming summit on Climate Change in Copenhagen must allow us to make a leap forward. With this in mind, the following are proposals for how the process from Poznan to Copenhagen should look:

ATTACKING THE STRUCTURAL CAUSES OF CLIMATE CHANGE

1) We need to discuss the structural causes of climate change. As long as we do not change the capitalist system for a system based in complementarity, solidarity, and harmony between the people and nature, the measures that we adopt will remain at the level of limited and precarious palliatives. For us it is clear that the model of "living better," of unlimited development, industrialization without frontiers, of

---

3   These are known as the Clean Development Mechanisms.

4   At the present, only one Adaptation Fund exists. It has approximately $500 million for more than 150 developing countries. According to the secretary of the UNFCCC, $171 billion are required for adaptation, and $380 billion for mitigation.

modernity that deprecates history, that is based on the increasing accumulation of goods at the expense of others and nature, is a model that has failed. For that reason, we promote the idea of Living Well, in harmony with other human beings together with our Mother Earth.

2) Developed countries need to reign in their patterns of consumption—of luxury and waste. This is especially so with regard to the excessive consumption of fossil fuels. Subsidies of fossil fuel, amounting to some \$150–250 billion,[5] must be gradually eliminated. It is crucial that alternative energies are developed, such as solar, geothermal, wind, and hydroelectric. Both small and medium-scale application of these technologies is necessary.

3) Agrofuels are not an alternative, because they put the production of foodstuffs for transport before the production of food for human beings. Agrofuels expand the agricultural frontier, destroying forests and biodiversity and generating monocropping. Their production promotes land concentration, deterioration of soils, and the exhaustion of water sources, while at the same time contributing to rising food prices. Furthermore, in many cases they actually consume more energy than they produce.

UNDERTAKING AND ADHERING TO SERIOUS COMMITMENTS FOR REDUCING EMISSIONS

4) There needs to be strict compliance by the developed countries to the commitments that they made to, by 2012, reduce greenhouse gas emissions by at least by 5 percent below 1990 levels.[6] Having failed to adhere to their present commitments, it is unacceptable that the countries that have polluted the planet throughout the course of history are now speaking of larger reductions in the future.

5) New minimum commitments need to be established which hold the developed countries to reducing greenhouse gas emissions by 40 percent by 2020, and 90 percent by 2050, taking 1990 emission levels as the reference point. These minimum commitments must be met internally in developed countries and not through flexible market mechanisms that allow for the purchase of Emissions Reductions Certificates in order to carry on polluting in their own countries. Likewise, monitoring mechanisms must be established for measuring, reporting, and verification. These must be transparent and accessible to the public in order to guarantee compliance.

6) Developing countries, which bear no responsibility for the pollution that has occurred until now, must preserve the necessary space to implement an alternative and sustainable form of development that does not repeat the mistakes of savage industrialization that have brought us to the current situation. A prerequisite to ensure that this process occurs is for developing countries to access finance and technology transfer.

A COMPREHENSIVE FINANCIAL MECHANISM TO ADDRESS ECOLOGICAL DEBT

7) Developed countries must acknowledge the historical ecological debt that they owe to the planet. Accordingly, they must create a Comprehensive Financial Mechanism to support developing countries in implementing their plans and

---

5   Stern, Nicholas, "Stern Review on the Economics of Climate Change," 2006..
6   Kyoto Protocol, Article 3.

programs for adaptation to and mitigation of climate change; innovation, development and technology transfer; the preservation and improvement of their lakes and reservoirs; responding to the serious natural disasters caused by climate change; and the carrying out of sustainable and ecologically-friendly development plans.

8) In order for it to be effective, this Comprehensive Financial Mechanism, must count on a contribution of at least 1 percent of the GDP in developed countries[7] as well as to have at its disposal other contributions from taxes on oil and gas, financial transactions, sea and air transport, and the profits of transnational companies.

9) These financial contributions from developed countries must be in addition to the Official Development Assistance (ODA), bilateral aid, or aid channeled through organisms that do not belong to the United Nations. Any finance that comes from outside of the UNFCCC should not be understood as developed countries fulfilling their commitments under the Convention.

10) Finance must be orientated towards national programs or plans from the different states, rather than towards projects that follow market logic.

11) Financing must not just be concentrated in a few developed countries but must prioritize the countries that have contributed the least to greenhouse gas emissions, those that preserve nature, and/or are suffering the impact of climate change.

12) The Comprehensive Financial Mechanism must be under the aegis of the United Nations, and not under the Global Environment Facility (GEF) and its intermediaries, such as the World Bank and regional development banks. The management of the Mechanism must be collective, transparent, and non-bureaucratic. Its decisions must be made by all member countries, especially developing ones, and not by the donors or bureaucratic administrators.

TECHNOLOGY TRANSFER TO DEVELOPING COUNTRIES

13) Innovation and technology related to climate change must be held in the public domain, not under any monopolistic private patent regime that obstructs technology transfer and makes it more expensive for developing countries.

14) Products necessary for technology innovation and development which are the fruits of public financing must be placed within the public domain and not under a private patent regime[8] in order for them to be freely accessed by developing countries.

15) The system of voluntary and compulsory licenses must be encouraged and improved, so that all countries can access products that have already been patented, quickly and free of cost. Developed countries cannot treat patents and intellectual property rights as something "sacred" that have to be preserved at any cost. The flexible application of the intellectual property rights regime, which is permitted in cases concerning serious public health problems, must be adapted and substantially enlarged in order to heal Mother Earth.

---

7   The figure of 1 percent was proposed by the Stern Review, and amounts to less than $700 billion per year.

8   According to UNCTAD (1998), 40 percent of the resources for innovation and development of technology comes from public financing.

16) It is necessary to recover and promote indigenous people's practices that are in harmony with nature and have proven, over centuries, to be sustainable.

ADAPTATION AND MITIGATION WITH THE PARTICIPATION OF THE PEOPLE AS A WHOLE

17) We must instigate mitigation actions, programs, and plans based on the participation of local communities and indigenous people within a framework of full respect for and implementation of the United Nations Declaration on Rights of Indigenous Peoples. The best way of confronting the challenge of climate change is not through market mechanisms, but rather conscious, motivated, and well-organized human beings who are endowed with an identity of their own.

18) Reducing Emissions from Deforestation and Forest Degradation (REDD) must be based on a direct compensation mechanism, from developed to developing countries. It should be implemented in such a way that it respects sovereignty and ensures broad participation of local communities and indigenous peoples. It requires a mechanism for monitoring, reporting, and verification that is both transparent and public.

A UN FOR THE ENVIRONMENT AND CLIMATE CHANGE

19) We need a world environment and climate change organization to which multilateral trade and financial organizations would be subordinated. Such an organization would promote a different model of development that is simultaneously ecologically friendly and resolves the profound problems of impoverishment. This organization must have effective follow-up, verification, and sanctioning mechanisms to ensure that existing and future agreements are complied with.

20) It is vital to structurally transform the World Trade Organization, the World Bank, the International Monetary Fund, and the international economic system in its entirety, in order to guarantee just and complementary trade, as well as unconditional financing for sustainable development that avoids squandering natural resources and fossil fuels in the process of production, trade, and transportation of goods.

21) In this negotiation process in the run up towards Copenhagen, it is crucial that the participation of our people as active stakeholders at a national, regional, and worldwide level is guaranteed. This is especially important for those sectors most affected, such as indigenous peoples who have always been at the forefront when it comes to defending Mother Earth.

Humanity is capable of saving the planet if we choose to recover the principles of solidarity, complementarity, and harmony with nature, as opposed to the reign of competition, profits, and rampant consumption of natural resources.

Chapter 57 ▌ Part 14

CHARGING RESISTANCE WITH RENEWABLE ENERGY SOURCES
A Solidarity Project with the Zapatista Communities and DIY Wind Generators for Autonomous Spaces

The FARMA Collective

It has been almost 15 years since the Zapatista Army for National Liberation (EZLN) and the indigenous Zapatista communities rose up once again in their 500 year struggle against Western domination and capitalist exploitation. Fighting for a life of dignity and freedom, they have organized their autonomy by satisfying basic needs such as land, health, education, and housing using both "the Fire" and "the Word,"[1] mostly a Word that has echoed in all parts of the world with the sounds of an inspiring rebellion.

Since the initiation of the *Caracoles* (the Conches—the way that the communities are organized in five groups according to their locations) and the *Juntas de Buen Gobierno* (Councils of Good Government—the Zapatista peoples' elected, but instantly recalled representatives) in August 2003, the Zapatista communities have self-organized and managed their own health, education, justice, governance, work cooperatives, and gender equality, always in a manner of governing with obedience to the people and moving forward by asking.

Since 2005, and with the "Sixth Declaration of the Lacandona Jungle,"[2] the Zapatistas have started constructing networks of solidarity against the neoliberal capitalist attack on humanity and nature. Such networks, based on respecting differences and using dialogue, are the "Other Campaign" in Mexico and the "Sezta International" in the world.

Many European collectives have been inspired by the Zapatista struggle and this has led to solidarity projects that helped strengthen Zapatista autonomy or inspired global processes of grassroots networking.

FARMA (Fight for Alternative Renewable Methods and Autonomy) started working as a collective in the autumn of 2006 and set as its primary activities the technical study, fundraising, and finally the construction of a small hydroelectric unit (8 kW capacity) in a seventy-family Zapatista community with a small medical clinic and a school, in Chiapas, Mexico. Apart from that, FARMA organizes workshops for the construction of DIY (Do It Yourself) wind turbines using simple, low-cost materials. These generators are installed in squatted-occupied spaces and social centers in Athens, Greece, where the ideas of equality and self-management are put

---

1   A reference to the book *The Fire and the Word: A History of the Zapatista Movement* by Gloria Munoz Ramirez (San Francisco: City Lights, 2009).
2   http://enlacezapatista.ezln.org.mx/especiales/2

into practice here and now, in order to create the picture of an Another World. In addition, our activities include the organization of various discussions on the Zapatista struggle and renewable energy sources, and we have participated in a series of solidarity actions in Athens. Our collective's main concerns are political solidarity in practice, radical ecology, renewable energy sources, and the right of all people for self-determination, i.e., autonomy.

Electricity can be produced in various ways. In most of them, energy production is regulated solely by financial terms without any care for environmental impacts. The alternatives, which make use of renewable energy sources such as wind, water, and sun, appear to be part of the solution to the problem. However, once these renewable sources are seen as a means of economic benefit and competition, the balance with nature is lost. The result of this attitude is obvious in hydroelectric plants using huge dams, where renewable production is far from friendly to the environment, since it actually destroys local ecosystems. In the context of our activities, it is clear that renewable sources are seen as a means for establishing the autonomy of communities, by empowering them with electricity to be used in schools, clinics, etc., and with minimal ecological disturbance. This way they can be used for decentralized energy production without creating dependencies on centers of power or discriminating between privileged and non-privileged users. This can only be achieved when renewable sources are not controlled by capital as another profitable investment.

Based on this concept, our thoughts traveled over the Atlantic to reach Chiapas, in the mountains of southeastern Mexico, and reached the constantly-evolving Zapatista movement—a movement that is advancing by adapting itself with the needs of the indigenous people, as they are expressed through directly democratic and non-hierarchical processes. Their lives of dignity and self-determination, set the respect of the natural environment as a high priority. On the grounds of supporting the autonomy of these communities in practice and against economic and political power, the idea of constructing a small hydroelectric unit was born and was suggested to the Good Government Councils, who accepted it.

The anticapitalist struggle of the Zapatistas was not restricted to the expropriation of land during the revolt of 1994, but moved forward to the creation of new autonomous structures, that questioned the basis of state control. Now that the people hold the land, they can focus on other aspects of living, such as autonomous schools, clinics, and collective shops.

The Zapatistas, apart from facing constant assaults from paramilitary groups, have to deal with a government policy that tries to break their solidarity and values, trying to buy their dignity through many "aid programs." These include provisions ranging from animals and cement to the supply of electrical energy.

Following the Zapatista point of view: "not to ask from others to do something for us, but to do it ourselves," FARMA began, through Internet research, to discover groups that have built similar small energy projects, gathering knowledge and experience from various parts of the world, such as Thailand, Nicaragua, and Italy. Believing that no matter how specialized and unreachable knowledge may seem, we can obtain it at a level that meets our needs. Against patents, while sharing it and realizing the works ourselves. Without being the "experts," we can access DIY technology on renewable energy sources and spread its application collectively. This DIY concept is the basis for organizing workshops on wind-generator construction by using simple, low cost materials for collectives.

We have learned a lot from our visits to the Zapatista communities. Some things are difficult to describe, but the moment you live it you can feel the change within you. You know you are getting involved with something that is local but simultaneously global, something simple but substantial, so different but familiar, as you learn through solidarity how to resist by being creative with others in equal terms. Feeling solidarity like a bridge that links different pieces of a common struggle, while giving and getting at the same time, sharing experiences and learning from each other. This perspective defined the way we chose to express our solidarity from the beginning to the end of the process.

In this concept, the necessary funding for the construction of the hydroelectric project was raised, without seeking funding from state and capitalist organizations, through self-organized events of solidarity like concerts, parties, and bazaars. The majority of the money was collected through the organization of a concert, where musicians volunteered their efforts and the place was offered free of charge. In the spirit of solidarity and an effort to avoid authoritarian and commercial relationships, we decided not to sell tickets conventionally, but to ask for a voluntary contribution, with a suggested donation of €5. During the event, in readings documentaries, and a discussion, there were references to the repression that the Zapatista movement is suffering on a daily basis. Afterwards, an extensive report on the expenses and the revenue of the concert was publicized, together with a big thank you to everybody that was involved in any way, since the event received great support.

Following the concept of mutual aid and solidarity between communities in struggle, perhaps needless to say, all the time and technical knowledge shared in this project (FARMA participants and other friends and comrades), has been on a voluntary and solidarity basis. No one got paid for their work, and all expenses, such us travel costs, were financed by the people themselves.

AUTONOMOUS AND DECENTRALIZED NETWORKS FOR THE PRODUCTION AND DISTRIBUTION OF ELECTRICAL ENERGY USING RENEWABLE ENERGY SOURCES

Unless we realize that the present economy which is structured upon the ruthless competitive tactic of a dilemma between "expansion or extinction," is a deeply inhumane mechanism, we will falsely tend to put the blame for all environmental problems on technology and overpopulation. We need to see the deeper causes of the problems, namely the globalized market speculations, industrial development and the identification of progress with the interests of corporations.
—M. Bookchin

While trying to describe the histories of our paths, we explored what brought us together on the same struggle—first together in FARMA and then with the Zapatistas.

We all came together from different directions, with different needs, desires, and motivating factors, but we all shared the same questioning of the relationship between humanity and nature. We all brought with us our own small piece of experience, carrying within it the knowledge and inspiration of past movements for social change. Though some of us had engineering backgrounds from university studies, certainly not all did and we nevertheless created on open collective where any person could participate regardless of their level of technical knowledge. All this formed our words and actions.

Informing and taking action against the root causes of climate change were one of our urgent needs, so we got involved with the production, distribution, and consumption of electrical energy. We consider energy, in general, as part of the "commons," part of these things in life that belong to all people, not commodities that can be bought and sold for the production and accumulation of profit. We understand the use of energy as one of the basic human needs and propose that its production and consumption should be carried out in a socially-just manner. At the same time, we understand the authoritarian relationship that we humans impose on nature, which is realized as the senseless and violent exploitation of natural resources and ecosystems. We realize that this relationship stems from the more general idea of exploitation and authority, which is imposed by one person on another. As a result we propose that struggling against hierarchies of all kinds is the only social change that could contribute to the struggle against climate change. This in turn can bring about an idea of social justice that could lead towards a balanced coexistence with the natural world of which we are a part. We also realize that every community, no matter how big or small, has its own way and its own time of doing things. We recognize that the paths are many, with many colors, with many dimensions, and we propose autonomy, synthesis, and respect for the different paths. We realize that socially-just proposals come from the movements, while moving from the bottom and to the left, as the Zapatistas said in the Sixth Declaration. So we are engaged in the struggle for autonomy of all peoples, here and now, through networks of mutual help and solidarity, towards freedom. While trying to stay away from capitalist, state, and authoritarian institutions such as universities and corporations, and the relationships of dependence that they create, we try to appropriate as much technical know-how as we can in order to freely share it and to put it in practice in the everyday lives of people that are struggling for social justice. In this manner we meet with other histories and we move forward together. This is how we came together. This is how we met with the Zapatistas. This is how we met and continue to meet with comrades from near and far. This is how we are moving forward.

Putting all this to practice, we are concentrating on renewable energy sources as one of the most important tools in the construction of autonomy and self-sufficiency. While trying to be independent from the state and multinationals, we have started to satisfy our own needs for energy, based on solidarity and respect for a balanced existence within the ecosystems in which we live in. The idea is to install small-scale renewable energy sources (such as photovoltaic cells, small hydro generators, and small wind turbines) in cooperation with communities struggling for Another World, while trying to share with them the knowledge that we have gathered and contributing to the self-sufficiency of these communities. The installations belong to the communities themselves and are managed by them. The communities themselves specify the production and consumption and issues of quantity and quality according to their needs and with respect to nature. This relationship of coexistence is very strong since the installations are designed according to the natural resources of the region. Our first attempt will be the hydroelectric unit in Chiapas, which scheduled for completion within the next two years. So the goal is to construct autonomous and decentralized solidarity networks for the production and distribution of energy, harmonizing our needs with our natural surroundings. All this goes against the ideas of centralized distribution networks of large-scale energy production units that are managed by economic and political power centers, in order to produce profit.

The technical know-how is shared, with the aim of creating small collectives that have the ability to use the technology according to the needs, desires, and values of their community. This way, decision making is in the hands of the people that are directly affected and not in the hands of some political and economic elite, which would not want to, and could not, know the daily needs of communities all over the planet. Only when technology and science become tools for the self-determination of communities based on solidarity, only then can we consider essential climate action, through local solutions of appropriate technology, in small scale applications. All this we have learned throughout the evolution of our project within the Zapatista communities.

We have seen how the communities decided where an electricity-production project would be more useful to them, and it was in a way that was related to their general principals, not dependant on the "bad government," as they say, and with love towards Mother Earth. We have seen how the community decided, after we informed them about technical issues, which technology would be used. We saw how the community was willing to manage the limited amount of energy that would be produced, collectively and with solidarity. We saw how knowledge can de transferred and become a tool towards autonomy and self-determination, individually and collectively.

A basic advantage of a small hydroelectric installation is its low cost and the simple technical knowledge required to install it: that of a plumber, an electrician, and a builder. Further on, a handmade DIY aproach to the construction of the turbine runner and casing of the 8 kW generator can reduce costs significantly, typically a savings of up to €10,000 compared to products made in the US or Europe. Also, reusing materials such as pipes for small schemes, can reduce costs even further. In this way, when the project is completed and the required know-how has been shared, future installations could be completed and possibly financed by the communities themselves, without a dependence on solidarity collectives, such us ours.

Do-it-yourself approaches to engineering, such as recycling or assembling parts, are not considered in mainstream installations, where appropriate funds are typically found more easily and without asking ethical questions about where they came from. So ingenuity and our collective knowledge becomes our most valuable tool in overcocming economical barriers and patents on knowledge. Of course this has its physical limits and can only work in small-scale applications, unless networking between the small starts to approximate "big."

For this reason, we hope that in the future such projects could be reproduced in nearby communities, as long as they find it useful, by the Zapatistas themselves, and thus giving birth to the first autonomous microgrids. Social change turning small-scale decentralized energy networks into the large-scale energy structures of the future.

DO-IT-YOURSELF WIND GENERATORS, SHARING SKILLS, AND STRENGTHENING AUTONOMY

Having the Zapatista movement as our main source of inspiration made us think that we needed to connect what happens at the other side of the world with the reality in which we live. We have seen that it would not be enough to simply support the struggle against capitalism and the construction of autonomy somewhere else but that it would be necessary to build self-organized procedures where we live, in our everyday life, in a process that aims to strengthen local anti-capitalist social struggles. This is how we got involved with constructing do-it-yourself wind turbines from simple materials. A very plain laboratory was set up in a social center in Athens where many people came and offered their views and hands-on work.

At the beginning it was difficult since we had no manual, not enough knowledge, and no experience of any kind. But we took it step-by-step and tried to share the existing knowledge and create a collective "basket" of knowledge that was growing. Working together and exchanging opinions gave birth to the idea of installing these generators in squats and autonomous spaces in order to achieve autonomy in energy, while minimizing our carbon footprint. By gathering information from the Internet and observing other attempts, we tried to install a DIY turbine at a Prapopoulou Squat, a squatted house in the north of Athens that is used as a social center. As it was one of the first times we actually got to work with tools and materials, it did not have the expected (or desired) results and it ended up with its wings flying all over the place. Nevertheless, it was a liberating experience that showed us which way to go. Our second attempt was a bit more organized, and still included a lot of improvisations, and although it stayed in one piece, it didn't produce much electricity. And then, after one-and-a-half years of having lots of fun as we learned, but without ground-breaking results, we reached that crucial point where things started to come together.

We discovered that the Escanda collective in Spain had a workshop where they were constructing DIY wind generators. There, we built a wind generator that really worked, though only for nine days. Our work there was carried out through a process of free cooperation and equality that made no discrimination of sexes, and accepted no experts that were not willing to share their knowledge in such a way. It was a precious experience that clearly showed that solidarity and skill sharing between collectives from all over the world is crucial, and that the moments where the little steps of every resistance meet can be very inspiring.

During our few years in existence, we never had a very clear view of the path we were on, but we kept moving, sensing new possibilities in the air, and always aiming for radical social change. Like the Zapatistas say: "the path is created while walking." Theory was born from practice and through our needs and desires we expressed the framework of our actions and sensed the new ways that were to be opened.

We have realized that apart from protesting and condemning, it is important to try to be creative and start building today the world we fight to be have tomorrow. Taking knowledge and know-how back from the monopoly of state and capital is an important part of the struggle and helps us believe in our ability to construct little things now, and much bigger later. Creative resistance makes it possible for people to see in action and to live what self-organizing is, to see the benefits and difficulties in practice and not just have theories and abstract ideas about it.

And then an idea that was abstract in our minds for a long time started to take form. It was the idea of getting involved with the way people learn. So, with a group of others, we occupied an abandoned space inside the National Technical University of Athens (NTUA) and created the Freedom Accelerator (επιταχυντής ελευθερονίων/ epitahintis eleftheronion) squat. The basic point of the squat was the creation of an autonomous space of free exploration of knowledge in the heart of the state university, while challenging the university's hierarchical characteristics.

As a result, subjects like libertarian education started to be amongst our interests. Organized workshops for wind generators are starting to take place in a method of equality, trying to abolish power relationships between all the people involved, learning from each other, and seeking to take education in our own hands. The evolution of the squat, that is actually happening as we write, is separate from FARMA and has other ideas as well, like workshops on open-source software, DIY everything, anti-consumerism, land cultivation, etc. Our plan for the future, concerning this project, is to have a workshop of DIY wind generators once or twice a week for students and anybody else who might be interested, as well as to organize some ten-day workshops where people from other places of Greece and other countries can stay, work, and learn how to construct wind turbines in a horizontal process.

To complete the geography and calendar of our very small resistance, we need to travel once again to the other side of the Atlantic. This time to visit self-organized collectives that try to apply appropriate technology while working with indigenous people, communities and squats, always from the left and below, in this so fertile land called Mexico. During our visit to Mexico this summer we made the basis for a good collaboration in future years, as we intend to exchange knowledge and work together on technologies that can support the struggle for autonomy everywhere. So we keep on going ahead step-by-step to the unforeseeable future. There is still a lot of ground to cover with a lot of mistakes to be made, but when longing for freedom, life is never boring or infertile.

NETWORKING MOVEMENTS OF THE GLOBAL NORTH WITH MOVEMENTS OF THE GLOBAL SOUTH

After the Zapatista uprising of 1994 and after the anti-WTO mobilizations in Seattle in 1999, the movements of the global south have started to network with the emerging movements of the global north. Clearly, they fight on the same front against neoliberal capitalist repression in the streets of Genoa, Argentina, North Africa, etc., and it is evident that a new movement that is not interested in taking political power is starting to form from below, creating new structures towards autonomy.

Our paths have crossed with these movements and we have realized that solidarity between all struggles is our weapon, by building local creative resistances that construct our autonomy, along with communication and coordination with these struggles through common networks, as is proposed in the Sezta International that was created in the depths of the Lacandona jungle in Chiapas.

We have seen the movement of the Other Campaign in Mexico, and we have seen the uprisings in Oaxaca and Atenco being violently repressed. We have heard other voices speaking up, in the Indigenous Peoples meeting of the Americas in Vicam in October 2007 and in the Second Meeting of the Zapatista peoples with the people of the world in July 2007. Globally, there is increasing repression of indigenous and rural movements—whether it is the group organizing against the Winter Olympics in Canada; those struggling to keep the oil in the ground in the Niger Delta; the farmers' and indigenous peoples' movements such as the MST (Movimento Sem Terra—Landless Workers Movement) in Brazil, peasants movements in India, and many more as they are expressed through Via Campesina (international peasant movement); the movements in Argentina; and many other known and less-known struggles against the many faces of Neoliberalism and capitalism.

The need to get to know and connect with such struggles is evident as they are part of the global anticapitalist struggle, the fight for our Mother Earth and our right of self-determination as free people. The indigenous struggles of "Tierra y Territorio," for land and territory, for land and freedom, have similar goals to the squatting movements, reclaiming the streets and many more struggles in European history of reclaiming the commons, such as the opposition to the dismantlement of public welfare, and struggles to protect the Earth, such as the antinuclear movement. The meeting of peasants from India with radical youngsters from Europe, for example, who try to find a common ground to plant seeds and share ideas, despite the differences in theoretical analysis, forms of struggle, and culture, has a dynamic worth unleashing.

One of the most important tasks of collectives such as FARMA that travel the world and come face to face with local movements, is to network these local movements with others, while creating a global perspective of a unified anticapitalist struggle. Following the concept "think global, act local" we hope that one day our paths will meet.

FARMA collective, Athens, Greece, November 2008

For more information:
F.A.R.M.A.: www.farmazapatista.blogspot.com
Freedons accelerator squat: www.eleftheronio.org
Prapopoulou squat: www.protovouliaxalandriou.blogspot.com
ESCANDA: www.escanda.org
EZLN: enlacezapatista.ezln.org.mx

Chapter 58 | Part 13

THE YANSA GROUP
Renewable Energy as a Common Resource[1]

Sergio Oceransky, on behalf of Yansa CIC

In the last decade, renewable energy has become increasingly popular. Passionate support for a quick transition to "green energy" is more widespread than ever before, and is gaining ground even in sectors and countries that, even recently, scoffed at the idea that renewable energy sources could contribute in a significant way to advanced energy systems.

Most governments have recently established, or are preparing, specific policies to promote the development of renewable energy. An International Renewable Energy Agency has just been established and will be based in Abu Dhabi, the capital of the United Arab Emirates (of all places!). Large companies such as General Electric and Siemens have bought smaller technological pioneers in order to join the booming branch, and a growing number of banks and investment funds are placing an ever-increasing share of their assets in this sector. The consistent double-digit growth of the last years passed the test of economic recession; as credit dried up for most industries, renewable energy financing slowed down slightly, but continued flowing. All these factors strengthen the increasingly popular idea that the transition to renewable energy constitutes a "silver bullet" that will solve, simultaneously, the environmental, economic, and employment crises. Renewable energy technologies are presented as the gateway to a bright green future.

But something is missing in this rosy scenario: it does not include a serious analysis of the territorial dimension of renewable energy, the most important area of potential conflict in the transition to a sustainable energy economy.

Control over an energy system based completely on renewable energy sources requires, amongst other things, control over very vast territories—a fact that is notoriously absent from most analyses of renewable energy. So are its myriad implications

---

1   This text describes the status and projects of the Yansa Group in late July 2009. The situation is evolving rapidly, as we develop contacts and explore possibilities to improve our plans. The text is a direct result of the work done collectively by the people who have directly or indirectly contributed to the Yansa project. I would like to mention in particular Javier Ruiz, Co-Director of Yansa CIC and one of the cornerstones of the project; T. Díaz, the brain behind Yansa's technological concept; Brooke Lehman, for her contribution towards the establishment of the Yansa Foundation and her wide network of contacts all over the Americas; Olivier de Marcellus, for his wise advice and financial support; Diana Damián, for her participation in the creation of the Yansa Foundation and her amazing network of contacts in Mexico and beyond; Kolya Abramsky for his informed contributions and extensive contacts; Preben Maegaard and Jane Kruse for their permanent inspiration and life-long commitment to the values and ideas on which Yansa as a whole is based; Conrado Moreno for his unconditional support and contribution; and many other amazing persons from diverse countries (Bolivia, Canada, Denmark, Ecuador, Germany, India, Italy, Ireland, Mexico, New Zealand, Spain, Switzerland, Thailand, the UK, the USA, and Venezuela) whose advice, ideas, or volunteer work have contributed to shape this challenging and stimulating project.

for rural communities, for power relations, and for society as a whole.

If the transition to renewable energy is undertaken primarily by rural communities, it may correct the structural imbalance between rural and urban areas brought about by the industrial revolution, which is at the root of most social and environmental challenges faced by humankind. It will be driven (rather than opposed) by communities in areas rich in renewable energy sources, and will, therefore, enable a faster and more effective response to climate change and other environmental problems related to fossil fuels. It will certainly produce a more fair and democratic economy than one where energy oligopolies control immense territories, in addition to controlling energy, a key production factor for all economic activities.

A community-driven transition requires more than good intentions. Rural communities may have access to land rich in renewable energy sources, but they most often lack access to the technology, the financing, the training, and the project management skills necessary to undertake their own renewable energy projects, and therefore cannot make direct use of their renewable energy sources. This puts them at risk of losing control of their land, as is often the case when communities with little power and resources live in areas with strategic importance.

The Yansa Group was established in order to build the financial, technological, and educational resources that will enable rural communities to control and harness the renewable energy sources in their territories, and thereby become a major player in the new energy economy. Yansa's ultimate objective is to contribute to the construction of a commons-based energy system.

THE NEXT GREAT TRANSFORMATION

Energy generation and distribution play a key role in human relations. The most significant social, economic, cultural, political, and technological transformations in history were associated with changes in the way that humanity derives energy from nature. The Neolithic Revolution, which was essentially the change from hunting and gathering to agriculture and animal husbandry as a source of human energy, produced labor and class differentiation, writing, and complex societies and cultures. The use of wind to cross the oceans resulted in colonization and capitalism. Coal and the steam engine brought about the Industrial Revolution. The use of oil changed industrial production, transportation, agriculture, war, and everything else. Nuclear power, though far less widespread than fossil fuels, also had a substantial impact, especially in military and political terms.

We find ourselves at the very beginning of another shift of historic proportions: the return to renewable energy as the primary source of energy for all human activities. The most visible reasons for this change are climate change and fossil fuel depletion, but the main driver will soon be the market. Renewable energy generation becomes cheaper each year, since the technology costs fall and the energy source is free (at least for the time being), while the costs of fossil and nuclear energy tend to rise. Wind energy is already competitive, in market terms, in locations with a good wind resource that are close to appropriate power grids, and it will soon be competitive in many more areas. Solar energy might compete with fossil fuels in only two decades, and perhaps even sooner.

The ownership and power relations that evolve around renewable energy sources will be one of the most important factors giving shape to our future economies, cultures, and societies. Over time, territories rich in renewable energy sources will be of key strategic importance, and bitter conflicts could therefore develop over their control. Fortunately, a positive scenario may be realized if we make use of several factors that can contribute to a community-driven transition to renewable energy.

•   The territorial factor: Most areas rich in renewable energy sources are under the control of rural communities. In many countries, indigenous communities and economically-vulnerable sectors of the population were displaced from the most fertile lands and moved to windy and/or sunny regions that, at that time, were of no interest to more powerful sectors of the population. Many disadvantaged communities, therefore, now have legal control over land rich in renewable energy sources. However, if they lack access to the other factors required for renewable energy production (such as capital, technology, and training), they can only try to negotiate an acceptable deal with companies that have the elements that they miss. But the structural conditions play against them: there are more communities that own land rich in renewable energy sources than companies able to undertake large renewable energy projects, so the latter can often dictate the terms of the relationship. Communities often only have the possibility to accept or reject the deal, or to protest against abusive projects that are undertaken against their will in their territories (see Chapter 45, "Fighting the Enclosure of Wind: Indigenous Resistance to the Privatization of the Wind Resource in Southern Mexico"). Therefore, disadvantaged communities cannot rely on territorial control alone to ensure a fair deal. But territorial control is undoubtedly a very good beginning, and a necessary condition, for a smooth community-driven transition to renewable energy.

•   The technological factor: Most of the technological pioneers in the renewable energy world are moved by social and ecological values. They started working in this sector several decades ago in order to contribute to the construction of a sustainable, decentralized, and fair energy system. Even today, many people who join the sector are moved by the same values. It is therefore possible to mobilize a lot of technological creativity in the process of building the conditions needed for a community-driven transition to renewable energy.

•   The "paradigmatic" factor: There is increasing awareness that the global environmental, economic, and social/distribution crises have common roots in an economic system that is inherently unsustainable, since it requires permanent growth and expansion. Many persons and organizations around the world are looking for innovative solutions that can contribute to create environmentally- and socially-sustainable patterns of production and consumption. This paradigmatic change requires, and should result in, more balanced and democratic power relations. Energy production and consumption is a key part of this equation, on both the environmental and the social side. In addition, everyone agrees on the need for a substantial improvement in the economic conditions and relative power of disadvantaged rural communities. This awareness is also growing amongst persons and organizations that have access to significant amounts of money. As a consequence, the ethical financial sector, where social and environmental impact is just as important as (and sometimes even more important than) financial returns, is experiencing an unprecedented development in size and sophistication. This opens up the possibility to finance the beginning of a community-led transition to renewable energy.

•   The global connections: Over the last couple of decades, the diversity and intensity of communication and information flows have made it possible to connect disadvantaged communities in remote areas, technology experts, experienced trainers, philanthropists, analysts, environmentalists, human rights advocates, legal experts, academics, and many other people who share some basic values and want to work together on a motivating purpose. Complex systems involving myriad collaborative relations are more resilient and creative than monolithic top-down processes. A community-led transition to renewable energy is an obvious candidate for a powerful networking process of this nature.

We stand before a unique window of opportunity to build a fair renewable energy system, and therefore a more equitable and sustainable society. Energy systems do not change very often, and when they do, the period over which the new power relations are defined does not last very long. Once they are established, change becomes much more difficult. Over the next few years, we can collectively give shape to a very positive transformation based on local and global commons. The Yansa Group, which borrows the Brazilian name of the Yoruba goddess of wind, lightning, and passionate change, has been created with the aim of contributing to this process.

LAYING THE FOUNDATIONS: THE GROUP'S STRUCTURE

Making a strong positive impact in the transition to a new energy system is no small undertaking. It requires a number of different areas of activity and expertise, and careful planning of their respective roles. Since Yansa's objective is to contribute to the construction of a commons-based energy system, the financial flows and relationships of control and ownership need to be built with that purpose in mind.

The evolving structure of the Yansa Group will be composed of three different kinds of organizations:

A GLOBAL FOUNDATION

A global foundation, and a number of country-based foundations, that will work with communities who want to make use of their renewable energy sources as a means to build a strong and sustainable economic basis, and who are prepared to help other communities to do the same. The foundation's main area of expertise, and the basis for its work, will be large-scale projects to produce renewable energy for the grid. It will be active on renewable energy for self-supply and on other appropriate technologies for community development, but large-scale production to feed into the grid will be the main priority. This is the core of most energy systems (and will be even more so if liquid fuels are replaced, to a large extent, by electricity as the basis for transport), and for this reason the sector where an alternative community-based model can make the most significant positive impact in society at large.

The foundation will facilitate community training, favoring the direct transfer of knowledge and skills between communities, and complementing it with external experts. This training will encompass different activities aimed at empowering the community as a whole (not only its most trained members) to make collective decisions about the use of its renewable energy sources. It will provide communities with the tools to give shape to their projects in terms of size, siting, compatibility with other uses of the land, etc. Over time, we expect to build up a training network comprised primarily of communities that have already realized projects and are ready to share their knowledge and experience. We expect several of these communities to build dedicated centers for this purpose—or to work in cooperation with existing community-based training facilities, in the framework of long-term cooperation agreements.

If a community decides, on the basis of the training received, to build a large-scale project to sell renewable energy electricity to the grid, the foundation will finance the project and run it in partnership with the community. The profits from the project will be equally divided between the community and the foundation. The part of the profits going to the foundation will be used to finance further community projects elsewhere, while the community's profits will be administered by community-based transparent and democratic structures, following collectively-decided priorities, in order to ensure that it is devoted to strengthening the quality of life, economic opportunities, and environmental sustainability of that community. Renewable energy projects will, therefore, not be an aim unto themselves; they will be part of a broader framework of integral and sustainable community development.

The foundation will also be active in the field of policy advocacy, and on the negotiation of contracts with utilities and public bodies. Together with communities and with coalitions of NGOs, academics, and other allies, it will lobby governments and utilities in favor of policies conductive to community-based renewable energy.

A COMMUNITY INTEREST COMPANY

A Community Interest Company (CIC) that will develop technological expertise, undertake research and development on renewable energy, and manufacture and sell renewable energy equipment. Yansa CIC is developing an innovative wind turbine design, aimed at lowering maintenance costs and fostering the involvement of communities in the operation and maintenance of wind farms. We expect to start manufacturing wind turbines in 2012 or 2013. Once wind turbine production is running smoothly and our technological and productive investments in this field are on firm ground, we will move on to other renewable energy technologies, such as solar and wave energy.

Developing our own technology will make the Yansa Group independent from the relatively few corporations that are at the forefront of R&D and production in renewable energy. It will enable us to adapt the technology to specific wind conditions, and to maximize the degree to which communities get acquainted with the technology, with the objective that they are part of shaping its future evolution.

We expect to make profits by selling turbines in the market, screening our customers to ensure that we are not selling to project developers who displace or abuse local communities. These profits will enable us to supply turbines to community projects in non-commercial terms. Any profits that remain will be donated to the foundation for its work with communities.

Last but not least, the job creation that comes along with manufacturing activities will be helpful when lobbying governments to establish policies conducive to community-based renewable energy. Over time, the CIC will set up manufacturing subsidiaries in different countries, favoring those that offer positive conditions to community wind energy projects.

A LOW-PROFIT LIMITED LIABILITY COMPANY

A Low-Profit Limited Liability Company (L3C) that will serve as a mutual fund to finance the community wind farms supported by the Foundation and the R&D and productive investments made by the CIC.

The capital of the L3C will come from different investors (foundations and other philanthropic organizations, sustainability-oriented investment funds, ethical banks, individual investors) that are interested in making a positive social and environmental impact through their investments, while obtaining a return.

The L3C will provide loans to the Foundation and the CIC, receiving a low but decent interest rate on those loans. It will support the community projects and the technological and productive capacities, without owning them. Over time, we expect the Yansa Foundation and Yansa CIC to build up sufficient assets to be able to have a major role in project financing.

The Yansa Foundation and Yansa CIC will be "owned" by their social objectives. They will not have shareholders, and will not distribute dividends; all their profits and assets will be devoted to fulfilling their aims. Both organizations will have a broad membership, composed of community organizations from all continents, but the members will not receive any dividends or any other form of private profit. Their function will be to ensure that both organizations remain faithful to their objectives. They will meet in general assemblies every few years to appoint the boards of both organizations, and will collectively decide on the strategic line of work for the following years. In turn, the boards will control the work of the executive structures of the foundation and the CIC, and hold them accountable to their mission and mandate.

The L3C will be owned by its investors. It will work as a regular mutual fund, with the important difference that, according to its statutes and to the legal definition of an L3C, its main purpose will be to have a positive social and environmental impact. It will also generate a low return for its investors, with foundations taking most of the risk, and institutional investors obtaining safer terms for their investment. L3C members will retain ownership and management rights in the L3C, but not in the CIC or the foundation.

This financial structure will enable most of the profits to be invested in new community projects, in the further development of technology and productive capacities, and in strengthening the training and sharing capacities between communities. A part of the profits will also be paid to the investors of the L3C, but if the projects are run efficiently it could be only a fraction of the overall returns produced by the group. The rest of this wealth will be turned into common resources. These commons will be built at local level, through the part of the profits from community projects that are invested by the community, and at global level, in two forms: the technological and productive assets of the CIC, and the increased capacity of the foundation to fund further community projects and training facilities/programs.

The relationship between the three parts of the Yansa Group is depicted in the following chart, which also summarizes the group's initial fields of activity. In the future, we expect to work on other technologies in addition to wind power.

Yansa Group Structure

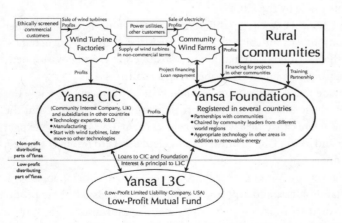

INNOVATION AND PRODUCTION FOR PUBLIC BENEFIT: THE CIC

The Community Interest Company (Yansa CIC) is the only part of the Yansa Group that is already legally established. It was registered on May 1, 2008 in the UK—the only country where this particular legal form exists.

A CIC is a hybrid between a non-profit organization and a company. In most countries, there are strong restrictions on the income-generating activities that not-for-profit organizations are allowed to undertake. These restrictions generally prevent them from generating their own resources by doing business. They therefore depend on external sources of funding (such as grants or donations), or need to have a large enough endowment to enable them to operate on the interest that it generates.

In most countries, organizations that have philanthropic purposes (social, environmental, cultural, etc.) need to register as not-for-profit organizations, since this is the only way to ensure that the organization's assets and activities will be devoted to the purposes with which it was created. If they register a business, their profits and assets are considered private and there is no legal guarantee that they will be devoted to their objectives. But registering as a non-profit makes them dependent on external sources of financing.

A Community Interest Company was created as a legal form in UK legislation in mid-2005, in order to enable organizations with philanthropic purposes to generate their own resources by conducting business. The essential feature of a Community Interest Company is that its activities must be for the benefit of the community. CICs are subjected by law to the so-called Asset Lock, "a general term used to cover all the provisions designed to ensure that the assets of the CIC (including any profits or other surpluses generated by its activities) are subject to meeting its obligations, either permanently retained within the CIC and used for the community purposes for which it was formed, or transferred to another asset locked body, such as, another CIC or charity."[2]

CICs can be companies limited by guarantee or limited by shares. A CIC limited by shares can distribute a limited percentage of its returns amongst its shareholders. A CIC limited by guarantee is a "not for profit" company. Yansa CIC is limited by guarantee: it has no shareholders, distributes no dividends, and is therefore a not-for-profit organization.

The objectives of Yansa CIC are:

1. To promote the development and use of renewable energies and sustainable development generally, alone or in partnership with other organizations;

2. To engage in the manufacture of renewable energy equipment;

3. To work with community-based organizations by:
- providing access to renewable energy technologies and training
- promoting collective and sustainable use of the natural resources for the common good,
- supporting participatory and democratic processes in which communities define their own development models,
- facilitating access to the resources necessary to fulfill these aims;

4. To support the work of non-profit technology centers that produce knowledge in fields of common interest (such as renewable energy,

---

2   Taken from "Community Interest Company Briefing Pack," published by the CIC Regulator (see http://www.cicregulator.gov.uk/CIC percent20guidance/CICBriefingPack2.pdf).

information and communication technology, agriculture, transport, etc.) and thereby promote not-for-profit use of this knowledge, in particular by community-based organizations following existing and new licensing models such as Free and Open Source Software, Open Hardware and Creative Commons; and

5. To raise awareness about the challenges facing our planet and its inhabitants, and about ways to overcome them through positive solutions.

The CIC is currently developing an innovative wind turbine technology platform in Spain, in a region where public authorities offer good incentives. Our design concept is oriented towards high reliability and improved maintenance and repair procedures. Several novel elements should lead to a reduction in the amount of incidents, as well as to a reduction of downtime and repair costs when incidents do occur. Yansa's turbines will not be designed to be the cheapest machines; the aim of the design is to lower the costs per kWh produced over the full lifetime of the turbine.

Another key principle in Yansa's approach to technology development is to design products that are robust and accessible. We will maximize the involvement of communities in Operation and Maintenance and optimize the communication and feedback between the users of the technology and our design teams. For example, some wind turbine manufacturers refuse to share with buyers data produced by the sensors installed in their machines, or refuse to provide sufficient documentation. These and other practices create serious tensions between manufacturers and users, which make improvement and troubleshooting more difficult. We will take a very different approach, based on openness and cooperation with the users and third party service operators throughout the lifecycle of the machines. We therefore expect communities to play a major role in the further development of our technology.

We will also take pro-active steps to avoid early obsolescence of our products. All wind turbines are designed to be operative for at least twenty years, but most manufacturers neglect discontinued lines of products soon after they are out of warranty, making the operation of older turbines very difficult well before the end of their supposed lifetime. We will work with communities to increase the options available as far as spare parts are concerned, in order to minimize the dependency on Yansa in this regard and secure availability throughout the lifetime of the turbines. As one of the members in the CIC team puts it, we aim to design "the AK-47 of wind power."

Once the technology is developed, we will set up manufacturing facilities, possibly in the same region of Spain where we are developing the technology. This region has a long and deeply-rooted history in coal and energy production, but coal mining is being phased out. As a response, the public coal-mining corporation is planning several large-scale wind energy projects. Creating jobs and alternatives in regions where fossil fuels are being phased out is one more dimension of a fair transition process, and the CIC would like to contribute to it.

We are also considering the idea of establishing manufacturing facilities in Mexico, and perhaps manufacturing some components in the US. This will depend on whether these governments establish a policy framework that is conductive for community wind energy projects.

We will sell wind turbines on commercial terms to wind energy developers. We will apply ethical screens in our commercial sales activity, to ensure that we are not providing turbines to projects that displace local communities or violate human rights in any other way. But we will try to reach a large segment of the market that falls within our ethical criteria.

The CIC will supply renewable energy equipment on non-commercial terms to community projects supported by the foundation. Yansa's turbines will be made available to community projects on the most favorable terms possible without neglecting the repayment of debts or the investment in further R&D and production facilities. Any profits that remain from sales in the open market to commercial projects will be donated to the Foundation, after repayment of debts, investment in future activities, and non-commercial supply of equipment.

STRENGTHENING COMMUNITIES AND NURTURING PARTNERSHIPS: THE YANSA FOUNDATION

While the CIC works on technology and manufacturing, and the L3C on financing, the Yansa Foundation is all about communities. Therefore communities and their organizations will play the most active role in the Yansa Foundation.

The Yansa Foundation will be composed of a global body, organically linked with Yansa CIC and a number of national organizations registered in countries where the foundation works. The board of the global foundation will be formed by persons from community organizations from all continents who share the goals and values of the Yansa Group and can mobilize the social energy required for a community-driven transition to renewable energy. The board members of the national foundations will have a similar profile to their respective countries.

In the preparatory meeting of the Foundation, held in September 2008, the Foundation's activities were categorized in four areas of work:

1. Community Projects:
The Foundation will offer support towards the realization of community projects throughout their development:

Outreach: Identifying opportunities for future community projects and outreaching to grassroots community organizations.

Dialogue: Engaging in dialogue around community needs assessments and possibilities, and supporting internal community visioning and assessment processes.

Definition: Supporting communities with defining the specific scope, economic and project plan, technical requirements, and organizational infrastructure.

Realization: Supporting the realization of the community projects through a combination of loans, grants, technical support, education and training.

Long-term Support: Supporting communities in addressing the economic, social, and technological transformation required for long-term project development, and helping to identify future projects.

2. Education and Training:
The Foundation will support participatory and mutual processes of education. These processes will start with communities explaining their political, social and economic realities, as well as their particular expertise, and educational needs and desires. Topics covered may include technology; project management; financial and economic skills; and community participation and organization.

This can be achieved through exchanges, training courses, workshops, fellowships, internships, training for trainers, and regional and global skill-sharing conferences.

The Foundation will place special emphasis on facilitating the education and leadership development of women and other individuals from underrepresented and oppressed groups.

In addition the Foundation will support the development of global, regional, and local learning centers.

3. Technological and Strategic Development:
The Foundation will support the development of technologies that can be freely used in the public domain, which support the principles of the Foundation and the needs identified by communities.

The Foundation will also support networks involved in strategic thinking and technological development in direct support of the Foundation's objectives and educational work.

4. Communication and Information:
The Foundation will support communication and information sharing through conferences, electronic communications, publications and translations, and strategic exchanges.

The Foundation may also support other activities focused on issues related to the Foundation's objectives.

Our practical experience in Mexico, where the foundation is starting its activities, has proven the importance of investing a substantial amount of time and energy in lobby and advocacy work as a first step. Mexico, like many other countries, is now defining its renewable energy policy, and a number of legal instruments that will define the conditions under which renewable energy projects will be undertaken are about to be passed. The Foundation, together with community organizations and NGOs, is promoting a legal and policy framework supportive of community renewable energy projects (see chapter 45, "Fighting the Enclosure of Wind: Indigenous

Resistance to the Privatization of the Wind Resource in Southern Mexico"). We expect that we will need to do similar advocacy work in other countries in the future.

Together with community organizations, human rights and environmental NGOs, academics and other committed individuals, we have formed a network called National Community Network for Renewable Energy (Red Nacional Comunitaria por la Energía Renovable, RENACER). This network comments on the drafts produced by those in charge of producing the legal instruments that will define Mexican renewable energy policy. We are also undertaking activities aimed at informing the communities in areas rich in renewable energy sources about these legal developments, and working towards making their voices and opinions heard by the authorities. RENACER and other local community organizations are organizing a public forum that will soon take place in the Isthmus of Tehuantepec, the region of Mexico with the best wind resources. We are also planning a public information campaign.

In parallel with this advocacy work, the foundation is working in Mexico on outreach and information in Oaxaca and Baja California, and has already established a cooperative relationship with a community that combines all the conditions required for an initial community project: Ixtepec, in the Isthmus of Tehuantepec (Oaxaca). The process in Baja California is in a very promising early phase. Both projects are described below.

INITIAL FOUNDATION PROJECTS IN IXTEPEC AND BAJA CALIFORNIA

Ixtepec is a community in the legal sense. The Mexican Agrarian Law (Ley Agraria) defines the community as the legal figure for rural and indigenous communities that want to keep common ownership and management over their land and resources. The law defines a community as a juridical person and establishes its ownership over the land and a number of protections on that ownership: the right over communal land is "inalienable, cannot expire, and cannot be seized" (inalienable, imprescriptible e inembargable). Ixtepec has decided to remain a legally-defined community, despite the governmental programs offering incentives for the transition to individual property.

The body that administers the commons of a community is the Communal Goods Commissariat (Comisariado de Bienes Comunales). The Commissariat of Ixtepec had already decided that they wanted to undertake a community wind energy project before we had any contact with them, and although they live in an area with a very rich wind resource, they have not signed contracts with any wind energy developer. The Mexican public power utility CFE is building an immense power substation on Ixtepec's communal land that will feed the national electric grid with the energy from all the wind farms that will be built in the Isthmus of Tehuantepec. The Ixtepec Communal Goods Commissariat has engaged in a negotiation process with the CFE, in which they offer to give away their land without payment, in exchange for the CFE supporting them in developing community projects in a number of areas, including wind energy. The CFE has agreed, in principle, and now they are negotiating the details.

The legal protection over communal land means that communities cannot use it as collateral for loans, and as a result it is very difficult for them to get loans of the magnitude needed for a wind farm. Although we are still in the beginning of the process, the Commissariat is, in principle, in great favor of a project in partnership with the Yansa Foundation. In this partnership, the foundation would contribute the wind turbines, and the community the use of the land, and the profits would be divided equally. However, a lot of steps still need to be taken before we can realize this project.

The Yansa Foundation has designed a training program to empower the community of Ixtepec to make a decision about what kind of wind energy project they want to build. The program consists of a series of practical models that will result in different components of the project (the wind map, the environmental impact assessment, the definition of the area where the turbines will be placed, the specific distribution of the turbines, etc.). Community members will be trained through this process on all aspects related to large-scale wind energy production, and on the operation and maintenance of wind farms. The community will be consulted on every major decision, and will have the final word on whether to undertake the project. If they decide to go ahead, we will work to obtain an adequate power purchase contract and all the permits required for energy production. We expect the possibility of the CIC undertaking manufacturing in Mexico will make it easier for Yansa to obtain the contracts and permits required by the public authorities.

The partnership agreement that the foundation will sign with the community of Ixtepec will include terms governing the use of profits generated by the project. The foundation will commit itself to using all funds received from the project exclusively for the advancement of its goals, as mandated by the law, and will invite the community of Ixtepec to be part of the foundation and its work in partnership with other communities. The community of Ixtepec will establish a transparent and democratic governance structure for the management of the funds that it will receive from the project, and draw guidelines and criteria for the distribution of funds in several projects of common interest. This should not be difficult, since the community has functional and democratic governance structures, it has already defined a number of community projects that it would like to undertake in the next few years, and it has very capable members who will certainly be able to produce more common projects for the benefit of the community.

A similar process might take place in Baja California. The conditions there are very different than in the Isthmus of Tehuantepec, so the definition of community will also be different. Ixtepec is a large community with over 30,000 inhabitants and a large number of professionals trained in different areas. The local communities in Baja are much smaller, comprised mainly of fisherfolk, and often do not have legal titles over the land (they have fishing licenses instead). The project that we are discussing in Baja will therefore be regional and will encompass several communities in different areas.

We went to Baja California invited by Comunidad y Biodiversidad A.C. and other organizations from the region that work on community-based marine biodiversity conservation. In the last decades, fish stocks have fallen dramatically due to overfishing, resulting in increased poverty and the loss of biodiversity. As a response, several fisherfolk communities have agreed to the creation of a number of large marine reserves, in order to allow fish stocks to replenish. Once marine biodiversity is restored, it will be possible to resume fishing on a more sustainable basis, in parallel with economic alternatives such as eco-tourism. An immediate and pressing challenge is creating economic alternatives for the fisherfolk communities, whose meager income has fallen along with fish stocks, and who will relinquish a large part of their income due to the creation of the reserves.

In this context, our partners in Baja California see community wind energy projects as a possible way to create alternative livelihoods for fishing communities, allowing for a comprehensive restoration of the previously-magnificent marine ecosystem of this region. Part of the profits generated by the project could be devoted to fund community-based biodiversity projects, in which the fisherfolk (who best know the region and its biodiversity) get paid to look after the reserves, and to the creation of alternative livelihoods based on biodiversity and ecological values. This approach would compensate the environmental and visual impact created by the wind turbines and the associated infrastructure. From an economic point of view it seems feasible, given the fact that the wind resource is extraordinary; there are several locations where there is access to the grid; and in this region, most power plants are fueled by diesel, which means that wind is automatically competitive. The project would contribute to establish a sustainable development path for Baja California, where authorities have identified the ecosystems (particularly marine biodiversity) as one of its main assets.

Soon 501(c)(3) will also be established in the USA, as part of the Yansa Foundation.[3] It will reach out to Native American communities to discuss with them the possibility of working together on community wind energy projects. It will also discuss with community utilities, which still exist in many areas of the US, about the possibility of establishing three-way partnerships in which wind energy is produced in cooperation with Native American communities, with financing from the Yansa Foundation, and fed into community utilities. It will study the possibility of making an impact on the renewable energy plans of the Obama Administration, with a view to shape them in favor of community projects. We will also foster cross-border cooperation between the US and Mexican side of the Yansa Foundation.

We are planning to hold the first proper meeting of the global body of the Foundation in Mexico, in early 2010. At that time, we expect projects to be mature enough to justify the investment involved in bringing leaders of community organizations from places as distant as India, New Zealand, Bolivia, and South Africa. We are looking forward to that moment very eagerly.

This will be the first of what we hope to be a series of inspiring meetings and conferences that will fulfill the communication and information mandate of the foundation. The preparations for a large international event aimed at making a significant global contribution towards grassroots-led efforts, in relation to energy and climate change, are already underway. This event, planned for the second half of 2011, will bring together a diversity of players who often work in isolation on the interconnected fields of energy transition, climate change, rural development, and related issues, such

---

3   A 501(c)(3) is a legal form for a non-profit organization in the United States.

as ownership and control of resources, development and transfer of technology, labor- and gender-related aspects of energy generation, etc. We believe the preparatory process (expected to include several regional meetings) and the event itself will have a deeply-positive impact on the work of all the organizations involved.

FINANCING SOCIAL AND ENVIRONMENTAL IMPACT: THE L3C AND BEYOND

The productive projects of the CIC and the Foundation will require substantial investments. Such large investments are obviously only possible if an attractive return is offered. The Yansa Group will primarily offer a return in the form of social and environmental impact, but it will also offer a financial return.

In order to be able to work with different kinds of investors and attract a sufficiently large pool of capital, we will create a Low-Profit Limited Liability Company (L3C), a new legal form in the USA, which has just been established in a few states (the first one being Vermont, which passed the legislation in April 2008), and is still pending in many others. Though this form is already legal in the whole of the United States, it is only possible to register an L3C in certain areas, but once registered, the company can operate in all fifty states.

An L3C is a hybrid legal structure combining the financial advantages of the Limited Liability Company (LLC) with the social advantages of a non-profit entity. An L3C is basically an LLC that, under its state charter, must "significantly further the accomplishment of one or more charitable or educational purposes," and would not have been formed but for its relationship to the accomplishment of such purpose(s). The charter of all L3C must specify that "No significant purpose of the company is the production of income or the appreciation of property." This does not mean that it is a not-for-profit organization; it is a profit venture, but one where profit making is only a secondary goal.

One of the main advantages of the L3C is that it simplifies the process through which US-based foundations qualify their investments as Program-Related Investments (PRIs). Foundations in the US are legally required to distribute 5 percent of their capital each year for charitable purposes related to their mission and programs. They can do this through grants or through investments that are sanctioned by the Internal Revenue Service (IRS) as PRIs.

Therefore, PRIs offer foundations the possibility to fulfill their legal obligations, support their mission and programs, and generate a modest return in the process. This return increases the amount of foundation capital available for grants or further PRIs, though most foundations do not make use of this possibility because there is a lengthy and expensive process required to qualify.

Because the legal requirements governing L3Cs mirror those for PRIs, which should streamline the process, allowing a foundation's proposed investment to qualify as a PRI. This removes the main barrier that prevents foundations from using this option to fulfill their 5 percent payout requirement.

This innovative legal format appears at a time when most foundations, having suffered severely the consequences of the stock market crisis, are looking for alternative investment opportunities and for asset-building possibilities.

An L3C can structure its capital in different categories, or "tiers," in order to attract investors with different goals. The junior tier is the capital most at risk; investors in this tier have the last claim on the assets of the L3C. In conventional corporations, higher risk is rewarded by higher returns; most usually, shareholders provide junior capital in exchange for co-ownership of the company and the right to receive dividends. In L3Cs, in contrast, foundations provide the junior capital in the form of PRIs. Foundations are legally required to accept a below-market return on the risk that they assume for their investments to qualify as PRIs.

Since PRIs in the junior tier absorb most of the risk, the L3C is able to offer attractive conditions to investors in the other tiers. Senior tier investors will have the first claim on the assets of the L3C. Investing in the senior tier of the L3C is an interesting option for institutional investors (such as green investment funds, ethical banks, and even pension funds) that are looking for safe investment opportunities associated with a positive social and environmental impact. It is also possible to define an intermediate or "mezzanine" tier, designed specifically for conscious investors whose definition of "return on investment" includes social and environmental goals. Mezzanine investors are ready to accept a middle risk position, between the junior and senior trenches, without expecting a higher return in exchange, in order to strengthen the balance sheet of the L3C and help it to attract additional capital in the senior tier.

The risk levels of Yansa projects must be acceptable to the investors in the different tiers of the L3C. The investors retain ownership and management rights in the L3C, and therefore will be part of the decision to provide loans to the projects of Yansa CIC and the Yansa Foundation. These decisions will not be determined by profit considerations, since the profit obtained by L3C members will be limited by its statutes, and will therefore not be larger for projects with, say, a 10 percent return than for projects with a 5 percent return. The L3C should therefore be equally interested in funding all projects that are viable, but will have the possibility to reject projects that are too risky or deemed not to be viable.

The L3C will receive a low interest on the loans it makes to the CIC and the foundation. We expect the projects undertaken by the CIC and the foundation to generate a substantial return beyond the interest that will be paid to the L3C. As a consequence, these two organizations, and the communities that partner with the Foundation, should build substantial assets over time.

The communities will build their common assets in many different forms, according to their own priorities. Options might include cooperative companies and other forms of livelihood and job creation, scholarships and other educational possibilities, improved access to all sorts of services, biodiversity, and environmental restoration, etc.

The CIC will build assets in the form of technology development and production facilities, as well as some reserves to remain operational. Profits generated beyond investment and reserve requirements will be devoted to providing turbines on non-commercial terms, and whatever remains will be donated to the foundation.

Since it will receive 50 percent of the profits generated by the projects undertaken with the communities, over time, the foundation will build a large pool of assets, which will be invested in further community projects. As such, we hope that, the dependency on L3C funds will decrease as time goes on.

The gradual build-up of assets by the Foundation will enable it to have a major impact on the transition to renewable energy. As previous chapters of this book show, during the next two decades, the costs of renewable energy technologies are likely to be decidedly lower than the costs of any other form of energy. As this happens in increasingly large regions (and not only in the ones with the very best renewable energy resources), and for different technologies (not only wind power), we will witness an unprecedented increase in the renewable-energy-generating capacity installed. A large number of communities will be faced with the option of either giving away control over their land and renewable energy resources, or building community projects. A very considerable financing capacity will be required for a community-led transition to renewable energy, which presents a challenge as we cannot count on the relatively-limited number of foundations and ethical investors to provide it. We need to start building up additional capacity now, without depending exclusively on the L3C, or we might find out too late that the window of opportunity for a fair transition in the world energy system is closed.

AN INVITATION TO PARTICIPATE

This process will require a lot of creativity, energy, and money. We hope that enough people and organizations will be inspired by this vision and will contribute to its realization.

On the financial front, we need support in the form of grants and donations to take the projects of the foundation in Mexico (Ixtepec and Baja California) to the stage where they can be funded by loans. They are both very promising, but a lot of discussion, training, and work is still required before we sign the contracts needed to access loans. The Yansa Foundation is looking for grants and donations to finance this preparatory stage, and for the first proper global meeting of the Foundation to be held in Mexico in the first half of 2010.

We are looking for investors for the L3C in order to finance the development of the CIC's wind turbine technology. We need the L3C to raise several million euros within the next months, out of which at least one-third should be junior or mezzanine capital, and the rest senior capital.

We would be very thankful to get probono support in different areas: legal, administrative, production of communication materials (such as videos and publications), etc. We would also welcome help volunteer with translations (especially English to Spanish and vice-versa).

We are always looking for quality advice, and for contacts that can enrich our knowledge and perspectives.

We hope that many people and organizations from all over the world will be inspired by Yansa and become involved in the group.

YANSA IN 2020: A VISION OF CHANGE

Be warned: this part is desire let loose. But it is grounded desire, with a fair chance of becoming reality.

By 2020 we expect the CIC to have a respectable portfolio of wind turbine models, encompassing medium turbines for areas with limited transport infrastructure, a range of large turbines adapted to different wind conditions and grid connection configurations (stand-alone turbines, wind farms with substations, etc.). We may also have developed, or be in the process of developing, a large wind turbine for off-shore wind farms. We expect to have contributed a respectable portfolio of innovations to various aspects of wind energy technology.

We hope to supply a significant share of the wind turbine market, thanks to the reliability of our machines and their competitiveness in terms of cost per kWh produced over their lifetime. This demand will come from commercial customers, as well as from community wind energy projects receiving turbines in advantageous conditions. The CIC will hopefully be amongst the top ten wind turbine manufacturers.

The loans for production facilities in Europe and the Americas might be nearly or fully paid back. We expect the factory in Europe to supply an ever-increasing number of municipal or other publicly-owned projects, which we will target, in particular, through special schemes. We also hope to supply a significant share of the commercial demand related to the EU objectives for renewable energy generation in 2020. We expect the factory (or factories?) in the Americas to supply a large number of projects undertaken by indigenous and rural communities in partnership with the foundation, as well as ethically-screened commercial projects. Hopefully, new markets will be active in other regions with strong rural communities, resulting in production facilities being either already built or under construction in other continents.

We expect that sometime between 2015 and 2020, the CIC will start R&D in other renewable energy technologies, in addition to wind power. The technology choice(s) made, and the approach(es) taken for technology improvement will hopefully reflect a wide consultation and creation process involving a range of end users, engineers, academic institutions, and technology experts, who share the objectives of the Yansa Group and want to take part in this collaborative effort.

We expect the foundation to be active in a large number of countries, and to have established a wide network of partnerships with and between very diverse communities. The first community projects should be running smoothly, selling electricity for six or seven years, covering capital costs (loan repayment rates), and at the same time producing income for community and foundation projects. The first projects should therefore be about halfway through with the repayment of loans, and therefore getting close to the start of the most rewarding period in the lifetime of the projects:

the final six or seven years, when the capital costs have already been covered and the profit margin expands considerably.

On the basis of this income, we expect communities who undertake wind energy projects with the foundation to have acquired substantial collective assets and undertaken projects that result in a noticeable improvement in their quality of life, livelihoods, and sustainability. Their structures and practices will hopefully be strengthened by the influx of common resources to be administered collectively and democratically for the community's public benefit.

In addition to the first community projects in Mexico, by then the foundation will have undertaken a considerable number of training programs with many communities on all continents, and a fair portion of these programs should result in new projects. We expect the first community projects outside of Mexico to take shape before 2015, in several regions around the world. We also expect an increasing share of the educational work to be undertaken, with Foundation funding, by communities that have already developed skills and experience in the wind energy sector, and have developed their own training and skill-sharing facilities and programs.

We also hope that the foundation activities (in combination with the increasing investment capacity of the Yansa Group as a whole) result in legislative improvements offering more conductive conditions for community-based development in several countries.

The foundation board may decide at some point to activate its own technology program. This is only likely to happen after the foundation has built up strong organizational capacities and is generating sufficient income to expand the scope of its work, which in the first years is likely to be focused on training, community project development, and educational/networking/lobbying activities. If it is similar to the ideas that were discussed in the initial foundation meeting in September 2008, the technology program will fund appropriate technology development in a number of fields that are relevant to community life—in particular, with regard to integrated and sustainable natural resource management (agriculture, fishery, water management, forestry, small-scale renewable energy production, soil conservation and restoration, biodiversity, etc.). The Yansa Foundation will support projects in which communities work with experts to identify needs that are not properly covered by available technologies (because the technologies required do not exist, are not efficient, or are unaffordable), and to develop appropriate community-scale solutions. These solutions will be developed in an open-source-based networking process and will be freely available for community-based, not-for-profit applications.

We expect the foundation's networking and educational activities to have contributed to more sophisticated analysis and better collaborative relations between different actors involved in the fields of energy, sustainability, production, technology, human rights, and social relations. Among other things, we expect this to result, at one point or another, in qualitative improvements in Yansa's supply chain. For instance, we hope to be able to increase the amount of recycled metals, and to reduce our demand for the metals provided by mining corporations with awful human rights and environmental records. We also expect, over time, to be able to guarantee good labor conditions, not only for the workers in Yansa's facilities, but also for the workers of Yansa's suppliers.

We expect the foundation to have access to a healthy flow of returns from projects undertaken in partnership with communities, and possibly also to CIC profits. On the basis of these flows, the foundation should have an increasing ability to cover risk for new community projects. Due to the absorption of risk by the foundation, we expect the amount of investors interested in participating in the L3C to grow steadily. This will increase the amount of capital available on low-interest basis for further projects, making it possible for the foundation and the CIC to undertake their projects in very favorable conditions.

By 2020, we expect the membership of the non-profit distributing part of the Yansa Group (the CIC and the foundation) to have grown significantly. We hope that many community organizations from all over the world, representative of different historically disadvantaged sectors (indigenous peoples, peasants/farmers, fisherfolk, women, refugees, etc.) will decide to devote part of their time and attention to follow Yansa's work. Their membership in the CIC and the foundation will commit them to participate in collective processes by which the work will be evaluated, the strategic lines of action will be drawn, and the composition of the boards chosen.

All these community organizations will therefore collectively control (but not own) what we expect to become an extraordinary amount of technological, productive, financial, training, and relational resources. These resources will constitute a global common asset to be used for the fulfillment of the goals of the Yansa Group. We expect this global commons to be the basis for a community-led transition to renewable energies, and therefore to a more fair, equitable, and sustainable economy.

Last but not least, we hope that Yansa will constitute a powerful source of inspiration for similar processes in other fields. The world is likely to benefit from more community-based, commons-building exercises (and commons-based, community-building exercises) aimed at providing constructive approaches to address the global social, environmental, and paradigmatic crisis.

Chapter 59 ▌Part 14

SPARKING AN ENERGY REVOLUTION
Building New Relations of Production, Exchange, and Livelihood

Kolya Abramsky

LINES OF CONFLICT, POTENTIAL COMMONALITIES OF STRUGGLE, AND LONG TERM PERSPECTIVES

This book has sought to show that the process of building a new energy system, based on a large—and possibly even 100 percent—share of renewable energy is not a technical question, but is a profoundly social and political one.

An accelerated transition to a socially- and ecologically-desirable transition will not come about through fate and inevitability, but through deliberate and collective human choices and activities. The many chapters in this book, written by a wide range of players within the global energy sector, have posed the crucial question: Who will actually bring about such a transition and how? Importantly, a transition does not simply entail good ideas winning out over bad ones, but rather that massive numbers of people actively engage in and take control over the production and consumption of energy, in general, and renewable energies, in particular. It will require far larger numbers of people than are currently involved in these issues. This is especially important if the process is to be as far reaching and rapid as possible, and if it is to contribute to a wider emancipatory process rather than giving rise to a chaotic process of social, political, and economic disruption.

People's cooperation is crucial, as the form, pace, and depth of whatever future energy transition occurs will ultimately be determined through their collective activity and labor. However, as this book has sought to show, it will be vital that such cooperation remains a common good in order for it to contribute to self-organized and emancipatory processes that satisfy human needs, rather than being subverted and harnessed for the needs of private profit in the world-market. Another important factor will be our collective ability to find ways out of the economic-financial crisis where we are not pitted against one another, and new hierarchies where the needs of some are satisfied on the backs of others are not formed. For this, it is necessary to correctly identify lines of structural conflict and recognize commonalities of struggle between disconnected and seemingly divergent groups. None of these are small tasks.

For this to happen, and to happen fast, there is an urgent need to build alliances and coalitions between broad social, economic, and political sectors, as well as to mobilize far larger sums of money than are currently available for such processes. Above all, alliances will have to be made with social sectors that are not currently active in bringing about a transition to a new energy system, which will require concrete intentional, strategized, and well-targeted organizational efforts. Understanding and working with very large numbers of people on their own terms, with their priorities and existing structures and movements, will make it easier to develop collectively-defined and implemented short term interventions around clear long-term goals.

There is frequent discussion about leadership in the field of energy in general, and renewables in particular. "Leadership" is a word that is constantly used by many different people and with many different (and sometimes conflicting) understandings. What is this leadership?

The question of leadership is a question about the future; leaders shape the future, opening up certain possibilities, closing others off. Leadership must be understood on two different levels: the first is in terms of the social, industrial, economic, and political sectors (including military when it comes to interstate relations in the existing geopolitical context) involved in financing, building, imposing, and maintaining a collective force that is strong enough to lead. The second refers to individuals within these sectors. It is impossible to tackle the question of individual leadership without first tackling the issue of social forces that may be conducive to leadership.

Crucially, the issue of leadership is not a neutral and natural given, but is the outcome of collective struggle over prolonged periods of time—struggles over who leads, how, and to serve what goals. The different contributors in this book have attempted to shed light on which collective social forces may be conducive to a leadership that will bring about an emancipatory transition to a post-petrol energy system. The pieces have asked which different futures might be possible, and, from there which ones might be more desirable, which ones less.

Not only are established power structures increasingly showing themselves to be ineffective at confronting global problems, they are revealing themselves as a part of those problems. We need to break out of these power relations and create new social relationships on a global scale. In today's crisis-ridden world, which poses the question of multiple possible futures in very stark ways, it is becoming ever-more apparent that, if we are to collectively find a humane and ecologically-sensitive way out, the leadership will be those mass-based and self-organized social powers emerging from below.

RURAL STRUGGLES, ENERGY SECTOR WORKERS AND CITY DWELLERS SPEARHEADING THE TRANSITION

In more concrete terms, there are three broad groups of people who, though they may not realize it, potentially have a lot to gain from technology transfer. Furthermore, they could also become a very significant driving force if provided with access to technical know-how, project management skills, and the right financial and policy instruments. The first group is those living and organizing within rural communities—including paesant, indigenous, and Afro-descendent communities—where the bulk of the world's renewable energy resources are found. These people number, literally, in the billions. The second group is the several million people whose livelihoods depend, either directly or indirectly, on the existing non-renewable energy sector, as well as energy-intensive industries (such as mining, transport, automobiles, industrial agriculture and the export industries, in general). If the transition towards a new and largely renewable energy system is not brought about in an appropriate manner that takes these people's needs into account, their livelihoods may be under serious threat. The third group is urban dwellers, who make up more than half of the world's population. Their struggles for secure and affordable housing, as well as for energy and sanitary services, are becoming increasingly central in the face of the financial crisis, foreclosures, and evictions.

Let us look at the first group, rural communities: Historically, the inhabitants of rural communities have experienced great difficulties in securing autonomous and collective control over their own resources and land (if indeed they do own some land). In many cases, it is rural populations, including indigenous and Afro-descendant communities that bore the brunt of colonialism, slavery, and the highly-unequal social relations that have followed since formal decolonization. These populations often still lack any real decision-making power and ownership over their territories' resources, and face heavy repression when they attempt to gain control of them. In recent years, as agriculture (as well as fishing and forestry) has undergone a process of liberalization throughout the world, through regional and multilateral trade agreements, such as NAFTA, FTAA, WTO, or SADC, it has become increasingly difficult for rural communities to survive, or to do so on their own terms. This is due to the simultaneous pressures of land grabs and concentration, falling agricultural product prices, and the rising costs of agricultural inputs. As several chapters in this book have shown, the struggle for control of the energy sector is making an important contribution in terms of encroaching on possibilities for rural survival.

This is why millions of rural inhabitants, peasants, fisherfolk, indigenous communities, Afro-descendant communities, etc., are organizing in large grassroots organizations from the bottom up in order to bring about fundamental change. Their activity combines addressing the situations of local populations, while denouncing and protesting the global social, political, and economic relations that assure inequality. The Brazilian Landless Labourers Movement (MST), the Karnataka State Peasants Association (KRRS), the Assembly of the Poor (Thailand), and the Zapatistas are just a few of the many important peasant and indigenous organizations that exist around the world. They are all organizing to challenge their subordinate position within the worldwide division of labor and to build autonomous local structures of collective production and decision making, aimed at minimizing their dependence on the world-market and producing for local consumption, based around a concept known as Food Sovereignty.

Such organizations are highly organized within Via Campesina and other global networks. Via Campesina, established in 1993, is an international movement with about 150 member organizations that coordinates peasant organizations of small and medium-sized producers, agricultural workers, rural women, and indigenous communities from Asia, America, Africa, and Europe. Individual member groups of Via Campesina, as well as at times Via Campesina itself, have been key organizations within other important global networks, such as the World Social Forum and Peoples' Global

Action and, most recently, the global resistance process called for by the Zapatistas.

As has been shown in chapters in this book, land is an absolutely central factor here, as one of the main drivers of the transition to a new energy system will be the struggle for control over territories with renewable energy sources. Many rural movements do, in fact, have access to land, so the rural communities who are most well organized are potentially in a very good position to play a key role in this transition process. The importance of the grassroots efforts of the blacksmiths and farmers in Denmark's renewable energy development should not be forgotten.

The second large group of people referred to above are the several million workers (and their dependents and communities) throughout the world who are currently employed, either directly or indirectly, within the fossil and nuclear energy sectors, or other energy-intensive industries. Their structural location means that they will have a key role to play in any shift towards a new energy system, but their livelihoods are also potentially at great risk from such a transformation if they do not have other survival options. Frequently, the concerns of workers in the fossil and nuclear sectors (and to a lesser extent also the energy-intensive industries) and questions of ecological sensitivity and renewable energy have been viewed as mutually opposed and with little room for common ground. Each is seen as having its own valid concern, thus resulting in deadlock.

As has been shown, there is a growing and historically-rooted movement towards what is becoming known as a "Just Transition." Many labor organizations do, in fact, recognize the urgent need to address climate change and to begin a transition to a renewable-energy-based system. The concept of "Just Transition" is based on ensuring that the transition is not carried out at the expense of workers in the existing energy sectors, but rather on their terms and using their skills and knowledge, and that workers are retained where necessary. It also includes the concept of worker-led "clean up" of existing "dirty" sectors, to the extent that such a process is possible.

Labor organizations that currently have a strong policy statement on Just Transition, or similar proposals, include the International Federation of Chemical, Energy, Mine and General Workers' Unions (ICEM); the Canadian Labour Congress (CLC); the Communication, Energy and Paper Workers Union of Canada (CEP); the Just Transition Alliance in the USA; and the Environmental Justice and Climate Change Initiative in the USA; newly-created Confederación Sindical de Trabajadores y Trabajadoras de las Americas (CSA); and the Instituto de Investigación y Estudios Energéticos de América Latina y el Caribe, to name just a few of the more important organizations.

The third group of people mentioned above are urban dwellers and their organizations. Many of the renewable energy technologies can be implemented at the most decentralized level possible, namely the individual residence. This can include, for instance, small-scale technologies, like solar panels for both heating and electricity, wind turbines, biogas digesters, cogeneration systems, energy-efficient building techniques such as passive solar, straw bale, solar skins, and a number of other technologies. If provided with the adequate tools, both technical and financial, these changes are relatively easy to implement by individuals and neighborhoods.

Increasingly, urban dwellers the world over are collectively organizing around the right to affordable and sanitary housing. This is true both in northern countries with relatively stable urban populations but increasingly precarious housing conditions, such as Los Angeles, New York, New Orleans, London, or Berlin, to give a few examples. But it is also true of the rapidly-expanding southern cities, many of which have large populations lacking access to basic energy, water, and sanitary provisions, including Bombay, Rio de Janeiro, and Johannesburg, etc. The economic-financial crisis is likely to increase the scope and importance of urban struggles around housing, and related service provision.

It will be important that these people are involved in the transition, not just because they can benefit from it and are a significant body of people, but because the current model of energy production and consumption creates enormous dependencies for urban populations on rural energy resources, and creates major global inequalities and hierarchies. This situation could potentially become far worse under a renewable-energy-based system, since most energy consumption occurs in cities, but it is rural areas that contain most of the renewable energy resources. It will be impossible to break these dependencies without a very concerted effort and participation from urban energy consumers themselves.

Between them, these three groups of people—rural communities who live in territories rich in renewable energy sources, existing-energy sector workers and those in the energy-intensive sectors, and urban dwellers—comprise the bulk of the world's population and present a considerable social force. This is a truly creative and productive force, and it will be crucial that they are able to tap into and benefit from their own creative power in the coming years. These sectors can surely play a leading role in the transition to a new energy system, including greatly speeding up and deepening the process.

Such movements frequently lack money and other material resources, and this is especially true for rural movements. However, they make up for this with their enormous capacity to educate, mobilize, inspire, and empower people to organize for themselves. Many of them work on the principle of multipliers, which means a small number of trained people may later go on to train a far larger number of people in villages throughout their region, enabling the impact of information, skills, and well-targeted material resources to amplify until it reaches literally millions of people.

The future transition to a new energy system is uncertain. Yet, one thing is certain. Its outcome will largely depend on how it will be brought about, by whom, and on whose terms.

COMPETITION OR SOLIDARITY? TOWARDS AN UPWARD LEVELING BETWEEN WORKERS IN ALL BRANCHES OF THE ENERGY SECTOR

It is becoming increasingly clear that capital is already making every effort to realize a transition to a new energy system on the backs of workers—waged and unwaged—and their communities, both within the energy sector and more generally. As fossil fuel resources become increasingly hard to extract, and as the renewable energy sector expands, these efforts are likely to intensify. This raises the crucial question of how workers in the different branches of the energy sector can relate to each other and avoid being pitted against one another in competition. The relation between workers in the renewable energy sector (the so-called "clean" energies) and those in the fossil and nuclear sectors (the so-called "dirty" energies) is especially important in this regard. In particular, it will be crucial that relationships are built on the basis of solidarity and leveling upwards of standards across the different branches, rather than competition and downward leveling.

The low cost of often highly exploited and repressed labor in the non-renewable energy sector has been an important, though often hidden, subsidy to fossil fuels. This includes Chinese, Indian, and South African coal miners, as well as migrant workers in the Gulf oil states. In Colombia, oil and coal workers live with the constant threat of paramilitary repression, and the country has the highest murder rate of trade unionists in the world. The globalization of China's coal industry, which is accelerating via the World Trade Organization, exacerbates this situation further. As described earlier, several thousand workers die each year in accidents in Chinese coal mines. Effectively, this is a hidden subsidy that renewable energy has to compete with on the world-market. And this has been an important factor (amongst others) in ensuring that fossil fuels remain more competitive in economic terms than renewable energies.

Until recently, the bulk of renewable energy infrastructure production took place in a fairly limited number of high wage countries such as Germany, Denmark, Austria, Japan, the United States, and Spain. The sector has employed comparatively few people, and production has frequently been motivated by strong environmental and ethical or ideological concerns, rather than simply profit. In some countries, such as Denmark and Austria, as well as to a lesser extent in Germany and Spain, cooperatives and other local structures, as well as Independent Power Producers, have owned a significant proportion of infrastructure, frequently with appropriate legal structures to ensure this. These factors have meant that working conditions and wages in the sector have generally been quite good, and there has been a broad convergence of interests between those who own renewable energy companies and the workers within these companies. To date there have been very few cases of industrial unrest within the sector.

As new companies emerge throughout the world, new areas of the world's population are being incorporated into the global division of labor associated with renewable energy, which implies a major restructuring and expansion of the workforce. For the first time, low wage areas of the world-economy are being drawn into the global commodity chains associated with renewable energy in significant ways.

In the space of just a few years, the Indian wind turbine manufacturer Suzlon has become the fifth largest turbine producer in the world, claiming 10.5 percent of global market share in 2007, and China is also rapidly becoming an important wind turbine producer. Since 2004, over forty companies have started producing wind turbines of 1 MW or larger capacity. In 2007, two of the top ten listed wind turbine manufacturers in the world were Chinese companies, and more are likely to make the list in the coming years. Conversely, some European producers can be expected to fall from this prestigious list, despite the fact that major players such as Vestas, Gamesa, Enercon, and Siemens will almost certainly continue to be leading producers. Many other lower wage countries, such as Egypt, Brazil, Turkey, and Pakistan are also set to become wind turbine manufacturers within the next few years. Similar processes are occurring in relation to other renewable energy technologies: China produced 35 percent of the global supply of solar photovoltaic panels in 2007 (up from 20 percent the previous year), most of which were exported to other markets. China already accounts for 70 percent of global production and use of solar water heating systems.

And, although of spurious "renewables" credentials, it is also worth mentioning agrofuels, especially bio-ethanol and bio-diesel, in this context. Brazil, a country with a history of slave-based sugar production and peripheralization in the world-economy, is rapidly becoming one of the key suppliers of sugar, the raw material for ethanol production for the world-market, and the US in particular. Sugar is a low wage/low value raw material sector that is produced for export to high wage consumer countries in the world-economy, where it is then processed into high value fuels. This is the classic division of labor characteristic of core-periphery relations in global commodity chains. A similar story can be told about palm and palm oil, from countries such as Indonesia, Malaysia, and Colombia.

Heralded as the great success of the renewable energy sector, which it undeniably is in some sense, this worldwide expansion to low wage zones of the world-economy also provides an important material basis for the sector to be able to compete much more successfully with the fossil and nuclear sectors. And, just as renewable energy companies are in fierce competition with one another globally, both within this and other branches of the energy sector, so too are their workers (and potential workers) in different parts of the world.

A major concern here is the basic reality that most of the infrastructure and fuel-stocks for renewable energies (such as wind turbines, solar panels, ethanol stocks, etc.) simply do not yet exist on the necessary scale. Given how late a transition to these new energy sources is being left, the shift will have to occur very rapidly as the existing energy regime is quickly becoming increasingly unviable. The unspoken implications of this are that workers in the new energy sectors are going to have to produce the necessary infrastructure very rapidly and under great pressure. There is a strong likelihood that this will necessitate very high levels of worker productivity in order to achieve the desired levels of output in very short time spans. A likely response to supply-bottlenecks will be to build more factories, especially in the current economic climate when there are plenty of unemployed workers who will be willing to work in the expanding sector. Though it would expand the overall output potential for production, more factories do not mean workers will come under less pressure. In fact, it means they will come under even greater pressure, since competition increases between workers and each factory faces greater challenges to survive. Similarly, an industrialization process based on the availability of large pools of unemployed workers throughout the world is a recipe for very bad working conditions and low wages.

However, as the chapters in this book have demonstrated, workers in both high and low wage areas of the world have been quick to resist the roles that they are being assigned in the new division of labor associated with the expanding sector. And, importantly, this is happening both in countries where the sector is well established, such as Germany, and where it is still in the early stages of development, such as the UK. The very real possibility that large-scale and wide-spread worker struggle will develop in significant parts of the expanding sector (as well as the raw materials associated with its global commodity chains, such as sugar or steel) raises important strategic questions, both for workers within the renewable energy sector, and also those in other branches of the energy sector.

As discussed in an earlier chapter of this book, the period preceding the shift away from coal and towards oil was notable for the high level of worker unrest in the coal sector. This created an increasing cost on the sector, and contributed towards making coal increasingly less economically viable. Similarly now, as other chapters in the book have shown, struggles in the oil sector are also making it increasingly less viable. There is an important, though complicated lesson here. Successful struggles by workers (and affected communities) in the oil and coal sector will make these industries more expensive. Paradoxically, the more successful the struggles are, and the more powerful the workers become, the less competitive and viable the sectors become. These pressures make a phase-out of oil increasingly likely, and also contribute towards the acceleration of this process. As such, it is in the interests of not just of the coal and oil workers (and affected communities) that their struggles are successful, but it is also in the interest of those who are actually advocating a phase-out of fossil fuels in the long term. Workers in the oil industry, through successful struggle, may actually undermine their own livelihoods in the long term. This means that it is all the more important to build up renewable energy industries in regions that are negatively affected by phase-out, so that workers and communities there do not lose their livelihoods. In order for this to happen, which is unlikely without targeted and intentional interventions, it is in the interests of both renewable energy advocates and fossil fuel workers to build close and mutually-supportive relationships with one another.

Similarly, the renewable energy sector is relatively new. As such, its workforce is not yet highly organized, either in trade unions or in other forms of organization. However, many of these workers are likely to have previous work experience in the metal and related sectors, so are likely to already have some level of experience with workplace organizing, which can provide a strong basis for organizing in the renewable energy sector. In Germany, the biggest trade union, IG-Metall, is already actively organizing the sector, especially in solar and wind energy. In addition to this, despite the fact that conditions in the various branches are substantially different, it will nonetheless be important that workers in the renewables sector are able to learn and receive assistance from workers and their organizations in the existing energy sectors, which have many decades, or sometimes over a century, of organizing experience.

The above analysis might seem contradictory, and is almost certain to be very difficult to achieve in political and organizational terms, but it may well provide a political basis for upward leveling solidarity between workers in different branches, rather than a devastating downward leveling competition. This will be important to avoid those currently dependent on "dirty" energy and energy-intensive industries being left without livelihoods, and to make sure that the global expansion of the new renewable energy sector is not carried out on the backs of workers. Such cross sector organizing is likely to be especially important in the context of a worldwide economic-financial crisis that implies a generalized assault on workers, and an attempt to divide them across sectors and other hierarchies.

COMMON OWNERSHIP OF ENERGY RESOURCES, TECHNOLOGIES AND INFRASTRUCTURE

As chapters in this book have sought to show, within the energy sector itself, the picture is one of intense struggle. Hardly surprising, given that, in addition to being a highly-profitable commodity, energy is also one of the key means to sustain human life. Struggles over ownership of energy resources, infrastructures, and technologies have been intense in the past, and it is very likely that they will grow more intense in the coming years. Important struggles over the ownership and control of energy production and extraction processes, as well as over access and price, are becoming increasingly central throughout much of the world. This has involved developing a range of different forms of ownership, including community, user, worker, cooperative, municipal, and state, that to differing degrees challenge private ownership and commodification. These struggles have involved broad social sectors: energy users, affected communities, peasants, indigenous peoples, and workers in the energy sectors and more generally. Frequently, those in struggle have faced harsh repression from state and military forces. In many areas, they are literally life and death struggles. Struggles over energy ownership have been at the heart of both foreign military occupations (such as in Iraq), and provide a key material resource basis for wider emancipatory or even revolutionary social processes, such as in Venezuela or Bolivia. These are the struggles that currently define the worldwide energy sector. They are at the heart of the so-called "energy crisis," which is, in no small way, partly a crisis of capitalist control over the sector. These struggles are likely to intensify in the future, and have definitely not already been lost.

This is true for fossil fuel reserves, such as oil (in Nigeria, Iraq, Ecuador, Venezuela, and Colombia) and gas (in Bolivia). In Colombia there are early rumblings of an imminent struggle for nationalization of the country's coal resources. It is also true in relation to electricity generation and distribution infrastructure and pricing, such as in South Africa, France, Germany, the Dominican Republic, India, South Korea, and Thailand (again, to name just a few examples). Similarly, there is a worldwide process of resistance to the privatization of forests, one of the main sources of non-commercial biomass fuels, which meet the domestic energy sources for approximately 2 billion people worldwide. Women, the people who mainly collect and process these fuels, are at the heart of such resistance, especially in Africa, Asia, and Latin America.

Importantly, such struggles are also intensifying in relation to the globally expanding "new renewable energy" sector. Since the 1970s, many pioneering initiatives in renewable energy had a strong emphasis on cooperative and local control. This has included farmers' wind energy cooperatives in Denmark; citizen energy projects in Germany (including cooperatives, buying local grids, and all-women initiatives); a worker-owned cooperative in Spain that was successful in becoming one of the important producers of wind turbines for the world-market, and was a member of the Mondragón industrial cooperative group. These local and democratic ownership structures mainly emerged in northern countries, the major pioneers of the new renewable energy technologies in this period. However, there have also been some interesting examples in southern countries, such as in Nepal with micro-hydro, Argentina and wind, and India in relation to household and village level biogas digesters. Collective and locally-controlled renewable energy infrastructure played a significant role in China's rural energy development during the early years of the Chinese revolution—a very different story, which there is not time to go into here.

Such processes, which emphasize a democratic and participatory community-controlled development of renewable energies, and contribute to the ability of the inhabitants of the territories rich in these energy resources to build a somewhat autonomous and empowering development path, are now frequently undermined. This is occurring because of threats posed by private investors, companies, and free trade agreements, all with the full support of national (and international) policies aimed at undermining previous forms of democratic and participatory control.

As chapters in this book have shown, the question of ownership and control over the territories rich in renewable energy resources is also becoming key. In Mexico, indigenous communities are being deceived so that the country's wind resources (amongst the best in the world) can supply electricity to major multi-national companies, such as Walmart. In China, peasants have been killed by police as they protested inadequate compensation for wind turbines installed on their land. In Denmark, rural wind energy cooperatives are finding it increasingly hard to compete with private investors and are being taken over.

Another important question relates to the ownership of knowledge and renewable energy technologies. Despite some very initial murmurings of an open source technology and non-commercial technology transfer movement arising in the renewable energy sector, inspired by the open source computer software movement, such a process is still virtually non-existent, and most research and technology transfer operates within the context of national and international patent regimes.

Significantly, important labor struggles are also emerging in the renewable energy sector, especially in relation to the production of the raw materials for agrofuels. This includes sugar in Brazil or Colombia; palm in Colombia, Indonesia, and Malaysia; and soya in Argentina and Paraguay; amongst others. In Germany, a leading country in the production of wind and solar energy infrastructure, the major trade union, IG-Metall, is organizing workers against poor working conditions where the infrastructure is produced. So far, these struggles are more centered on working conditions, rather than ownership, but there are a few exceptions to this. In Indonesia, workers in the palm plantations have also taken steps to take over the mills. And, in the very days when this manuscript is being completed, what is likely to prove to be a historic turning point in the wind industry, is unfolding in the UK. The country's only wind turbine component manufacturing plant (owned by Vestas, the world's largest producer of wind turbines) currently faces closure, and the lay off of 600 workers. The workers have occupied the plant, and are demanding its nationalization. They have been met with both widespread social support and riot police. The issue remains unresolved.

A discussion of ownership is relevant to three key questions: How, and at what pace will the existing reserves of fossil fuels be used (or not used)? How will the remaining fossil fuels be used (or not used) in a coordinated, planned, and minimalist manner in order to build renewable energy infrastructure as quickly as possible, so as to produce the necessary reductions in carbon emissions to prevent irreversible climate change? And, last but by no means least, according to what priorities and by whom will these decisions be made?

Of crucial importance is the question of who owns the fossil fuel reserves (and associated technologies and infrastructures), as this determines who makes decisions concerning their use (or their non-use) and the priorities on which these decisions are based. In many countries where a significant shift towards renewable energies has already occurred, the process has been mainly based on public policies. The market is now becoming an increasingly important driver in a small number of countries, and will certainly become more so in the future, as inevitably the cost curves of renewable energies and fossil fuels converge, and a world-market in the sector is built up. However, it is becoming increasingly clear that "national" economies do not really exist—in the renewables sector or any other. They are always mere subsections of the world-economy as a systemic whole, and it is this worldwide division of labor that needs be analyzed. Furthermore, while certain countries have made significant shifts towards renewable energy, a wider process of transition, involving a significant phase-out of fossil fuels on a global-scale has not yet occurred.

It is far from clear that a rapid and smooth global transition towards renewable energy and away from fossil fuels and nuclear energy will even be possible if this process is based on the idea of building a world-market, which might be an unreachable illusion that will provoke immense human suffering. Also, it is important to bear in mind that, while some cost reduction in the renewable energy sector will be driven by technological developments, others will be caused by reducing labor costs. If a competitive downward leveling of workers across different branches of the energy sector is allowed to become a driving factor in the reduction of costs, it will almost certainly have a very destructive impact.

There are three major reasons to believe that some kind of common, worker, community, cooperative, public, or state ownership of fossil fuels (as well as their associated infrastructures and technologies) might play a crucial role in providing a political and economic basis for shifting away from these fuels, collectively and rapidly. These forms of control and ownership, which, despite having important differences between them (especially in the degree of democratic participation that they are based on) nonetheless share certain important considerations and aspirations. They would make it possible:

- to use the world's remaining fossil fuel resources in a rational, co-ordinated and collectively planned way, rather than in the wasteful way in which the competitive market logic allocates resources. This is vital if the transition away from oil and coal and towards renewable energy is to be rapid and orderly, rather than prolonged and chaotic;

- to speed up the transition process by asserting collective control and establishing a political decision-making process regarding whether these resources are used or not used, in order to collectively plan an intentional and comprehensive phase-out, in accordance with collectively-agreed priorities and pace;

- to put the economic revenues from the rent of these resources under common control for common benefit during the period when they are still in use. This will allow these revenues to be used either for broadly-defined collective social needs, or, more specifically, to finance a rapid transition towards renewable energy. It can also provide a cushion against some of the more disruptive aspects of the process—a process that requires large sums of money.

As the chapters in this book have shown, struggles are already occurring throughout the world against the privatization of energy resources and technologies, especially in the oil, gas, and electrical sectors, and in favor of some form of common, collective, or public ownership. Frequently this has involved harsh state repression and also foreign military occupation. These struggles are far from trivial concerns, and many people have already lost their lives. However, it is likely that these struggles will take on an even greater urgency, intensity, and centrality in the coming years. Given that coal is becoming increasingly important once again, it is also likely that ownership struggles over coal, which are not currently a major issue, are likely to heat up. In Colombia, early discussions of such a strategy are already afoot within the main coal miners' union. And, while some may view this as a paradoxical perspective, given that these fuels are undeniably carbon emitting, it is nonetheless almost certain that such struggles can make a vital contribution to building a global collectivity that is strong enough to bring about a rapid and lasting transition towards renewable energy, that is as least-socially-destructive as possible.

Of course, common or public ownership will almost certainly *not* guarantee any of these outcomes. It is no panacea. For instance, in Venezuela, the main driver towards the expansion of renewable energy is actually the oil industry, and it seems that the country will certainly not stop extracting oil in the near future, but rather will use wind energy to do so. Common or public ownership of energy resources (fossil or renewable) and their associated infrastructures and technologies cannot be understood as blueprints to be implemented from above by policy makers. They are not theoretical models or predictions. If we are ever to see such ownership structures become the dominant form of ownership, they will be the outcome of lengthy and complex struggles, led by grassroots social movements against capitalist relations within the energy sector (and more generally), with both users and workers in the sector playing a key role in these struggles. It will be important to create political spaces that are broad enough to include these struggles.

On the one hand, it might not appear very realistic to expect the fossil fuel economy to be collectivized within any useful timeframe, except in countries where there is already a strong grassroots social and political mobilization process, as is the case in Latin America. On the other hand, however, it is also worth bearing in mind that the struggles over energy, which are strong in this region, have not simply come about because of the existing high levels of political mobilization, but have themselves been a key contributing factor to those high levels of political mobilization. It is quite likely that struggles over control and ownership of energy can play an important role in deepening a process of social struggle more generally, as success in these struggles would provide an important material basis from which to support wider goals of social transformation.

THE GLOBALLY-EXPANDING RENEWABLE ENERGY SECTOR: EMERGING CONFLICTS AND THE NEED FOR NEW ALLIANCES

A rapid global expansion of the renewable energy sector is already underway and is likely to continue for many years to come. According to the REN21 2007 Global Status Report, annual global growth has been between 20 and 60 percent depending on the technology concerned. At least until early last year, before the economic-financial crisis kicked in, demand for renewable energy infrastructure was far outstripping supply.

However, the sector's expansion is taking a form that was not widely predicted by many in the field. The peaceful image of inherently decentralized renewable energy technologies is giving way to a reality of intensifying social and economic conflict. The dominant strands within the renewable energy sector have been very slow to acknowledge, let alone come to terms with, the immense conflicts that are emerging around the control over land, water, forests, and other ecosystems with an abundance of renewable energy resources, and the related labor conflicts that are emerging. Those who are seriously working to promote an accelerated transition to a new global energy regime based on democratic, participatory, and decentralized access to renewable energies remain a small minority of highly-committed individuals and small organizations that are swimming upstream against the wider renewable energy sector.

The broad alliance that has characterized the sector's evolution to date is coming under great strain and is starting to crack. Tensions are rapidly emerging between ecologically-motivated concerns and the profit motive, between small producers and large producers, between small and large energy consumers, between companies and workers, between producers and consumers, and (perhaps most importantly) between commercial and non-commercial energy use (and technology transfer). The question of ownership and control is also becoming central.

Renewable energy, as with other energies, is not an idea but a material reality, existing in complex and continually-evolving global commodity chains. These chains exist within, are shaped by and, in turn shape the capitalist world-economy. Crucially, as the chapters of this book have sought to show, the fact that renewable energy is a commodity means that it is also a site of struggle. There is a struggle over how, where, and by whom surplus is *produced*, and a struggle over how, where, and to whom it is *distributed*. And, last but not least, is the struggle over why it is produced in the first place, a struggle that is intensifying as the sector expands.

A dominant approach to international renewable energy technology transfer, as exemplified by the newly-created International Renewable Energy Agency (IRENA), is to identify "best practice" mechanisms and then to look for appropriate political and institutional ways that these practices can be replicated and transferred around the world. Some of these "best practice" approaches have indeed been very good in terms of their ecological and social desirability. As shown in different chapters, certain experiences of renewable energy have simultaneously resulted in a high level of renewable energy capacity and use, and also shown a path of community empowerment, autonomy, and energy sovereignty—at least at the local level. Until now, however, the problem has been that these "successes" have only occurred in a tiny handful of countries, despite the fact that they are certainly worthy of replicating around the world. The hope is to find a process to facilitate conditions for a far-reaching and, above all, rapid "global take-off" of the sector. As such, despite its limitations, IRENA is undoubtedly the most progressive international body devoted to renewable energies.

This "take-off" approach, however, is eerily reminiscent of earlier debates surrounding "industrialization take-off" based on "modernization theory" and the whole host of "development" strategies and policies that followed. This approach suggested that, with a heavy dose of patience and through implementing the appropriate policy measures, all countries of the world could, at some point, industrialize and "catch up" with the "most advanced" ones. Such a perspective is, of course heavily flawed and has been completely discredited through the actual course of history. Angola never did "catch up" with the USA, and the USA would have to decline beyond the realms of our wildest imaginations for such a thing to ever occur within the context of capitalist relations. This is not to say that *some* countries will not catch up or at least substantially close the gap. This may well happen, especially with the restructuring of the world-economy. However, what will definitely not happen is that *all* countries will catch up. The "level playing field" of development is, in fact, profoundly uneven—it has never been level, and it never will be. Furthermore, some countries and regions of the world are "underdeveloped" precisely because others are "developed." The underdeveloped world and developed world are not independent of one another, but hierarchically related, mutually shaping each other.

Yet, it seems that some of the most progressive, internationalist, non-commercial, and forward-thinking wings of the world's renewable energy sector, many of whom have been instrumental in establishing IRENA, are in the process of forgetting the very important lessons of this experience. As with modernization theory, the "best practice" strategy for expanding the renewable energy sector is rooted in a two-fold understanding, both of which are false. On the one hand, it assumes that nation states are autonomous units. On the other, it assumes that currently existing inequalities (in this case, in the global energy system and related technologies) can actually be solved through simply expanding the current system so that the number of "renewable energy losers" is reduced, and the number of "renewable energy winners" increased. Implicit here is the view that it is only a matter of time and careful application of the right procedures (this time, in the realm of renewable energy) before the losers are able to catch up with the winners, and equality (or at least relative equality) can prevail.

At a general level, inequalities in global technology transfer are linked to structural features of the world-economy and its flows of labor, capital, raw materials, and knowledge. Technology transfer does not predominantly happen through a process of global agreement to disseminate "best practices," but through industrial competition and restructuring, which includes class struggle in the worldwide division of labor (which implicates workers in some countries in the exploitation of workers in others). It is dependent on wage differentials between different places. And, just as "underdeveloped" and "developed" zones of the world do not exist independently of one another, but are connected through a hierarchical relationship, so too are "hi-tech" and "low-tech" ones. The world-economy needs "low-tech" zones as the pillar on which "hi-tech" ones can actually exist. Within the context of actually-existing social relations, the model that expands technology until it is universally distributed is simply not achievable. This does not necessarily imply that it is impossible for certain technologies (in this case, renewable energy technologies) to be distributed on a much more even basis throughout the world, but simply to say that such an effort would involve an uphill struggle against wider systemic dynamics, and would require a conscious effort to do so and to obtain the necessary means for allowing it to happen.

A crucial issue here is the manufacture of the means of production. In the case of renewable energies, this means wind turbines, solar panels, storage systems, wave generators, refineries and fuel-stocks, and many other types of equipment and their component parts. An important question will be how the division of labor associated with the production of these means of production will develop in the coming years. This will be one of the key factors in determining whether the sector is able to really spread worldwide, or whether it will remain located in just a small number of centers of production.

If, as seems to be rapidly becoming the case, manufacture of these means of production remains under monopoly (or oligopoly) control, the rest of the world will have no other option but to import from these countries (and companies) at high cost, or to pay expensive licensing fees to work their way around patent mechanisms. The other side of this equation is likely to be that at least some of the countries that do not produce the means of production needed by the sector will be assigned a different role in the division of labor. Certain countries are already rapidly being assigned the role of producers of raw materials for export onto the world-market at low prices. This includes steel, sugar, palm, vanadium, silicon, lithium, and many other materials necessary for manufacturing renewable energy infrastructure and storage mechanisms. A good portion of these are associated with extractive industries—a sector that frequently involves poor labor conditions, ecological degradation, and displaced populations. Furthermore, the fact that many of these commodities are produced in low-wage zones of the world-economy and then imported to high wage ones means that they are traded on the world-market on terms that benefit the importing countries to the detriment of those exporting. This process is known as unequal exchange. Already a small number of countries—including Brazil, Argentina, Tanzania, Indonesia, Malaysia, and Colombia—are becoming key raw material providers in the global commodity chains related to sugar, palm, soya, and jatropha.

The global flows of knowledge, raw materials, money, and labor that shape the sector are undergoing a far-reaching and highly-uneven restructuring. The division of labor, workforce, and market associated with the renewable energies sector globally is still relatively small and young compared to most other global industries. The long term evolution of the global workforce, market, and ownership structures within the industry is still a very open question; it could develop in many different directions. However, this outcome will not be determined either by chance or by good intentions, but rather by the outcome of struggles for control of the sector. It will be a struggle that places states and companies in competition with one another. It will also be a competition between workers (both waged and unwaged) and their communities. Much depends on the degree to which technology transfer is a commercial process or a non-commercial one, which itself will only be determined by the outcome of struggle. Five key factors that decide the outcome of the transition will be: the struggle for territorial control over areas rich in renewable energy resources; the ability to create a skilled worked force; the struggle to control workers in the sector, and their struggles against being controlled; control over the knowledge and technology; and access to the necessary capital.

Since we are in such an early phase, it is still possible for communities, social and workers' organizations to have a major influence in shaping the future renewable energy economy. There is a great need to understand and support the emerging movements working for the collective, autonomous, and decentralized control of the expanding sector. It is still very small relative to other energy sectors, and the bulk of the renewable energy infrastructure remains to be built. As such, the next years offer a window of opportunity to ensure that a significant share of the sector can, in fact, come under common ownership and benefit emancipatory social processes. However, time is short, and unless appropriate, globally-reaching interventions are made very soon the window will likely be quickly closed.

As the book has sought to make clear, a transition to renewable energies might well be carried out on the backs of communities who live in territories that are rich in renewable energy sources, and workers who produce the necessary infrastructure. This is already leading to new forms of exclusion, dispossession, violence, and exploitation, or at best the draining of these resources for use elsewhere. The current expansion of the world-market is an attack on rural communities throughout the world. Whereas fossil fuels and nuclear energy resources are found in a small number of locations, renewable energy resources are broadly spread throughout much of the planet, giving increased strategic importance to large parts of the rural world. This means that the quest for renewable energy could result in a new and perhaps unprecedented landgrab by companies and investors, which would create the potential for even more extreme patterns of displacement and appropriation of land than other forms of energy have done.

This is already occurring with alarming rapidity and brutality due to the rapid global expansion of agrofuels produced for trade in the world-market (rather than for local community-controlled consumption). To a lesser extent, it is also occurring in relation to wind. In particular, the dependency of urban areas (where large quantities of energy are consumed) on rural ones (who produce it) is becoming an increasing point of conflict. Therefore renewable energies, in addition to offering emancipating possibilities for constructing autonomous and decentralized energy systems, also represent a new threat for rural communities (especially indigenous and Afro-descendent), making them increasingly vulnerable to loss of control of their territories and even displacement.

As described in these pages, struggles over territory, labor, and ownership, are all becoming central in shaping the global expansion of the renewable energy sector. A transition, predominantly based on the collective and democratic harnessing of renewable energies, has the potential to result in a significant decentralization of energy production and equalization of access. Communities and individuals could assume greater control over their territories, resources, and lives enabling an emancipatory social change that is based on the construction of autonomous relations of production, exchange, and livelihood. This is especially so for rural communities, which, in theory at least, are ideally located to benefit from renewable energies and to lead the way, since they are richest in natural resources such as wind, sun, biomass, rivers, seas, animal wastes, etc. And this can happen astonishingly fast if communities are given the appropriate tools.

For this reason it is very important that rural communities with rich renewable energy resources have access to the necessary tools so that they can collectively decide on the use of their resources. It is also crucial that urban and rural communities are able to collectively develop solutions that satisfy people's basic needs on the basis of collaboration and cooperation, rather than through a conflictive process that pits rural and urban inhabitants against one another. Similarly, workers in the energy and energy-intensive sectors and urban dwellers also need to be given the appropriate tools. But none of this is likely to happen "spontaneously"—it will require well-organized and well-reasoned interventions.

Given the inadequacy and slow pace of the transition to date and the new conflicts that are emerging, it is reasonable to ask whether the alliances that have been useful in building up the renewable energy sector in the past are likely to continue to be the most effective or if new ones are needed. It is always tempting to follow the path that is easiest in the short term, which frequently involves maintaining the status quo and not alienating current allies, but, in the long run, it makes more sense to clarify where structural conflicts of interest or possibilities for convergences of interests actually lie. Only on this basis can coherent long term strategies—and hence short term interventions—be developed.

One of the important terms that renewable energy advocates frequently use to describe themselves is "independent." Independence allows for an uncompromising and focused effort to achieve the goal at hand, subordinating all other concerns to it. Independence means being free from ties to convention for convention's sake. There must be flexibility to try new approaches; if necessary, to break from existing approaches if they are no longer appropriate and helpful. And, above all, independence means the ability to define the goals of one's own work and to decide without external interference who to work and not work with. This clearly includes deliberately taking sides in political conflicts, if warranted.

The transition to a new energy system, whether it is renewable based or otherwise, is an inherently political process. It involves real material conflicts of interest and struggle, not just differences of opinion that can be settled around boardroom tables. As these struggles become increasingly visible and gain importance, it is becoming more urgent that those in the renewable energy sector actively and overtly take sides in certain conflicts. The claim to neutrality, which large parts of the renewable energy sector espouse, is a false option. Ultimately, neutrality involves shying away from the responsibility of developing a political analysis as the basis for future action, which may involve taking specific sides in conflicts.

It will be important to assist movements getting access to skills, money, and infrastructure through a process of non-commercial technology transfer, since without this, efforts will go nowhere. These efforts are already underway, but they are still very much the exception. Non-commercial technology transfer is especially important to the rural energy worker, energy-intensive sector workers, and urban dwellers' movements. Above all, there is a need to confront and break with the patent regimes that, with backing from international institutions such as the World Trade Organization, ensure monopoly—or at best oligopoly—control of knowledge that urgently needs to belong in the common domain. People in the renewable energy sector need to defend common ownership of this knowledge and devote their efforts to developing open source research and technology development for common use.

Renewable energy events—conferences, seminars, and training programs—are frequently (though not always) very expensive and completely beyond the means of most grassroots movements, and they have to be made accessible. The sector's dominant strategy of having high-prestige high-budget events in five-star hotels needs a serious rethink. It is both elitist and wasteful. In the long term, the crucial point is not so much getting grassroots movements to the kind of elitist corporate events that are currently happening (which are quite often irrelevant to the needs of movements anyway), but rather for movements to be able to have their own autonomously-organized events, on their own terms and to satisfy their own needs. However, in order to be able to do this, they also need to be able to participate in existing events, at least in the short run, so that they can obtain the necessary contacts and know-how.

As this book has sought to show, leadership in an emancipatory transition process is unlikely to come predominantly from structures from above, like governments, multilateral institutions and agreements, or corporations. It is more likely that autonomous movements, self-organizing from below in order to gain greater control and autonomy over their own lives, will lead the way. This is not to say that state regulation is unimportant; it is completely essential in order to secure a legal and institutional framework (as well as financial support) conducive to a grassroots process led from below. However, the regulatory process is very unlikely to be the driving force of the changes, but rather a necessary process that enables wider changes. Furthermore, it is highly unlikely that emancipatory regulation that is strong enough to be effective could even come about without major pressure from below.

And, finally, it will be crucial to get a serious discussion about capitalism going within the more progressive elements of the renewable energy sector. Despite the fact that the renewable energy sector is expanding within capitalism, there is a strong reluctance to talk about this, or the possibility of building alternative social relations that go beyond capital. In addition to talking about technical processes related to energy production and consumption, there is a need for a wider debate about how, and for what purposes, wealth is produced and distributed in society, and how people's subsistence needs are met.

Should some people in the renewable energy sector ask such questions, they would inevitably be met with hostility and organized resistance from certain others, especially those who seek to privatize and monopolize the fruits of renewable energy for their own benefit. Resistance is likely to come at the political, economic, financial, and industrial levels. This is only to be expected. And, to put it crudely, now that the sector is able to stand on its feet, the dominant commercial interests in the sector do not need those who are pushing for a less commercialized and more democratic use of the resources and technologies. Rather than shying away from conflict and espousing a false unity that we are "all working on renewable energy together," now is a time for identifying potential allies as part of a wider political struggle. In fact, it is time to go one step further, and to actively encourage a split within the complacent and politically-timid renewable energy sector. The renewable energy world is facing its moment of truth, and there is no running away from political responsibility and struggle. And while there is a danger that some avenues will almost certainly be shut off, should such a split occur, the possibility of other ones opening up is immense. That will be important in building a new reality that is likely to mainly involve people who are currently not involved in the renewable energy sector, with some (perhaps only a minority) of the people who are already are.

SPARKING A WORLDWIDE ENERGY REVOLUTION

Building a new energy system based around a greatly-expanded use of renewable energies could make an important contribution to the construction of new relations of production, exchange, and livelihood that are based on solidarity, diversity, and autonomy, and are substantially more democratic and egalitarian than those that currently exist.

However, substantial changes in the system of energy production and consumption will require substantial changes in production and consumption relations at a more general level. Faced with the fact that forced and chaotic degrowth have seemingly had a far greater effect, in terms of emissions reductions, than years of regulation, it seems that an urgent question facing emancipatory social and ecological struggles is how to collectively and democratically construct a process of planned rapid and broad degrowth, based on collective political control and democratic and participatory decision making over production, consumption, and exchange at the widest level possible.

Of particular importance to building a new energy system are the major energy-intensive industries, such as transport; steel; automobiles; petrochemicals; mining; construction; the export sector, in general; and industrialized agriculture. These are also some of society's key means of generating wealth and subsistence.

However, it is very difficult to imagine that it will be possible to bring about a rapid and far-reaching change at the pace and scale that is necessary (both for social and ecological reasons) unless these key means of generating and distributing wealth and subsistence are under some form of common, collective, participatory and democratic control, decision making, and ownership. Furthermore, it will become increasingly important to find ways of ensuring that they are used to meet the basic needs of all the world's population, rather than the profit needs of the (currently existing) world-market and the select few workers and communities who are able to reap these benefits. Leaving the process to the logic of accumulating profit in the world-market is likely to be both far too slow and also immensely socially disruptive and brutal. In other words, as well as bringing these sources of wealth under some form of collective control, there is also a need to decommodify them as much and as quickly as possible.

However, following years of market-led reforms and unprecedented concentrations of wealth and power, we are still very far from this reality. This is true both in concrete terms and also in terms of our collective aspirations and strategic approaches. Dominant political strategies for achieving change are firmly rooted in discussions of how to achieve minor regulatory reforms (at best including state ownership), rather than a more fundamental shift in power relations pertaining to structures of production, ownership, and control. This is currently true even in many progressive and radical circles.

Consequently, an urgent task is to discuss what kind of short term interventions might help make such a political agenda more realistic to achieve in the near and medium-term future. It is not a new discussion. In the past, collective ownership, management and control of key means of production (either worker, community, cooperative, or state) have been at the heart of radical proposals for social struggle. Furthermore, radical critiques of existing state communism, socialism, social democracy, and their respective bureaucracies did not lie in a rejection of collective ownership of key means of production as such. Instead, their critique was based on a strong critique of the fundamentally-limited nature of state ownership as a model for democratic, participatory, and self-organized social change from below, based on an understanding that state control is simply a modified form of private ownership and capitalist class relations.

The current situation concerning struggles over ownership and control of the energy sector has been the subject of much of this book. Let us now briefly review ownership struggles in areas of production (and reproduction) that are important in relation to a transition to a new energy system.

Land is one of the most basic elements of subsistence for humans throughout the world, and is also essential for capital accumulation generally, and in relation to the energy sector specifically. It is a key means of production and also reproduction of human life. Collective ownership and decommodification of land is still at the heart of many, if not most, rural and indigenous struggles throughout the world today. It is in these struggles that there is perhaps the clearest political discourse. Throughout the world, large numbers of rural communities (including indigenous peoples and Afro-descendant communities) are struggling against the negative social and environmental consequences of energy extraction, infrastructure, and transportation, regardless of the energy source in question. These are communities who have been struggling for many years against the impact of fossil fuels, nuclear energy, and large-scale hydro, and communities who are, in some cases, resisting the negative impacts of the recent arrival of new renewable energy technologies. Particularly important in this regard is the emerging worldwide resistance to agrofuel production, an energy source that scarcely merits the name "renewable." Another important area of emerging struggle is in relation to territorial conflicts related to energy infrastructure projects occurring within the framework of Kyoto's Clean Development Mechanisms (CDMs). Territorial conflicts are also becoming increasingly important to other renewable energy technologies, especially wind. Territorial conflicts relating to mining and the processing of minerals and other raw materials necessary for building renewable energy infrastructure and storage mechanisms are also likely to become of central importance.

The outlook for ownership and decommodification struggles in the energy intensive industries, such as cars, aviation, transport, or tourism, is much more pessimistic. Importantly, the dominant strategic discourse from major organizations in these sectors is equally pessimistic. Ownership struggles have, by and large, already been lost to the extent that they are more-or-less nonexistent. For the last many years, most struggles in these sectors have revolved around demanding certain reforms in the production and work process, as well as improved user access. However, little space remains open for serious struggle (or even discussion) for major changes to patterns of ownership and control.

At the more radical end of ecological critique, there is frequently discussion about the need for a profound change in production relations. However, while such ecological discourses are often very strong in rural struggles, the organizations and collectives with such perspectives frequently lack the social base necessary for bringing about a wider change. In particular, they have little capacity to contribute to serious debate within trade union and other worker organizations within these sectors (and sometimes lack even the will, as many trade unions, especially in northern countries, have been integrated into the structures of capitalism). On the other hand, the dominant "green" discourse, though often well connected to trade union organizations working on "sustainability" from a worker perspective, hardly talks about ownership of the key means of production. Most campaigns from this broad group of organizations are pushing for change within the existing framework of social relations, and do not have a systemic critique of the existing model of economic development. Finally, though it is important to point out that significant differences exist between trade unions in northern and southern countries, the dominant trade union discourse of workers in these sectors favors tripartite bargaining, "decent work," and social peace, based around regulating production for private profit in an expanding world-market.

However, the economic-financial crisis offers an opportunity to reopen this old discussion, since the old model of Keynesian class compromise and stabilization of struggles aimed at changing ownership patterns of the key means of production is dead, and in all probability is unlikely to be resurrected. Starting with the economic and financial collapse of Argentina in 2001, factory occupations and self-managed industrial production and exchange returned to the radical landscape in an important way. In the wake of the current worldwide financial and economic crisis, a wave of factory struggles including worker occupations has spread around the world, including in the US, the UK (including, as mentioned above, in a wind turbine component plant), South Korea, and numerous countries in Eastern Europe. Such struggles are largely defensive, related to redundancy conditions, rather than proposing a new model of ownership, production, and control, and are still on a very small scale. Notably, the Detroit car factories have virtually been left to go under, rather than being taken over by workers and communities and converted into renewable energy production plants. Yet, even the head of the United Autoworkers Union made a fleeting and cautious reference to worker occupation of the plants, albeit way too little, way too late. Yet, this is a rhetoric that has not been used in such places for many decades. These are small processes, but nonetheless of great importance. The industries in crisis are some of the key energy-intensive industries, such as cars and steel, which are especially relevant to the issue of energy transition and a worker-community-led conversion processes.

The stark reality is that we are very far from bringing about the kind of change in production and consumption relations that is needed to solve the climate/energy crisis. We may never be in a position to do so. However, if we are even to imagine avoiding a socially- and ecologically-disastrous process of climate change and enforced change, it will be important to at least pose the question of how this might come about. Until we face up to this, efforts at a far reaching change in the world's energy system (and, consequently, on climate change) are unlikely to go very far. The task of collectively taking over the key means of production and decommodifying the major productive processes are immense. We are certainly not ready for it now.

What is both possible and long overdue, however, is taking some initial steps to deepen a long term strategic debate about how, and for what purposes, wealth is produced and distributed in society, and how people's subsistence needs are met, as part of a shift to a new energy system. Through a process of debate, we will hopefully be able to slowly develop collective interventions that contribute to these goals, so that in the medium term, as the economic-financial and ecological crises deepen, we might be able to do what is not possible now, and collectively plan production and consumption, based on a class struggle that brings together workers (both waged and unwaged), communities, users of energy, and the energy intensive sectors across the hierarchically divided division of labor. This will be an important step towards bringing about a profound democratization of how wealth is produced and distributed throughout society. Furthermore, *unless* the discussion on production is reopened, it is very likely that the "solutions" found to the economic-financial crisis will be extremely authoritarian.

While common or public ownership of either fossil or renewable energy sources, or energy-intensive industries will almost certainly not guarantee a larger process of emancipation, it nonetheless could offer an important material basis from which to pursue such wider changes. And, in this regard, much work remains to be done in order to collectively appropriate the skills, money, and infrastructure necessary to ensure that movements are able to bring large sectors of the energy sector, both renewable and otherwise, under collective autonomous control.

The International Energy Agency has called for an "Energy Revolution." Well, let them have one. But let's make it an anti-capitalist one.

In less than fifteen years, a number of highly active, imaginative, visible, and above all effective global anti-neoliberal and anticapitalist networking processes have come into existence. They are based on a coming together of people who are engaged in wider social, political, economic, and cultural struggles aimed at collective emancipation from oppressive and discriminatory relationships that are characteristic of today's world-economy. Great efforts have been made to create common global spaces for simultaneously denouncing inequalities and coercive centralized power structures, while seeking to construct alternative social relationships based around solidarity, diversity, and autonomy. In particular, the following organizational processes stand out has having played an important role: the World Social Forum, Peoples' Global Action, Via Campesina, Indymedia, and the different global initiatives of the Zapatistas. Although these processes, each with their own different internal dynamics and slightly differing political perspectives, have all been very significant, they are merely the tip of the organizational iceberg.

In the little over fifteen years of their existence, these initiatives have had very rapid and far-reaching successes. They played an enormous role in strengthening communication and opening up an ongoing conversation that seeks to build common, yet diverse, political perspectives between large numbers of different and fragmented social struggles in many different countries. Importantly, great attention is paid to principles of self-organization, autonomy, diversity, and non-hierarchical organizing, which creates a very fertile context for many new organizations, networks, and collectives to spring up. Another of their successes is that in just a few short years, they achieved the seemingly impossible; in the midst of a triumphalist, post-Cold War, capitalist rhetoric, they dared to denounce capitalism. Furthermore, their efforts were so successful that they rapidly plunged the system and its major global institutions into a crisis of legitimacy. Institutions such as the World Bank, International Monetary Fund, the World Trade Organization, World Economic Forum, and G-8 are increasingly unable to hold their summits without major protests, immense security costs, and harsh media criticism. The same is true for summits relating to multilateral and bilateral free-trade agreements. These institutions are not just facing a crisis of legitimacy, but also deep existential crises.

And now, with the onset of the economic-financial crisis, which these globally networked struggles predicted and anticipated, there is a far wider questioning of both the inevitability and desirability of capitalism's continued existence. Increasingly, the crisis is developing a political dimension as legitimation structures are coming under greater strain. This raises the urgent necessity for these global networking processes to enter into their next phase, a phase that is proving to be extremely complex and difficult. Movements are faced with the task of moving beyond the exchange of information and protest coordination toward building long term autonomous and decentralized collective relations of production, exchange, and consumption. In particular, the financial-economic crisis reveals the urgently necessary, but enormously difficult, task of massively reducing our dependence on financial institutions. The financial institutions, which in any case only offered security to small numbers of people throughout the world, now offer it to even fewer. And it is becoming increasingly hard to survive in the world of waged work. However, the process of collectively disengaging from these structures and creating alternatives is a huge task, especially in the core capitalist countries where people's daily lives are so intertwined with this world.

It will only be possible to break our dependence on money if we are able to build major capacity in the non-commercial and mutual support-based satisfaction of our basic needs. Especially important in this regard are food, energy, water, shelter, health, education, social security, and pensions. This will be necessary in order to reduce our dependence on waged labor. For this to happen, it will need to reach a far greater collective capacity than currently exists. And, paradoxically, this means that movements will have to be able to access large sums of money, infrastructure, skills and knowledge, as well as many other sources of wealth—again, on a far greater scale than movements are currently able to muster. In a nutshell, it will require a concerted worldwide effort to acquire key means of generating wealth and sustaining life.

In order to build collective self-reliance in a way that does not create long term dependencies, a twin-pronged approach necessary. On the one hand, there is the need to struggle to greatly expand the provision of collective social wealth by public institutions, namely local governments, states, and regional or multilateral bodies and agencies. This can take the form of demanding public funds and an increasing share of public wealth, as well as access to interest-free and unconditional loans that could enable movements to buy and run collectively-controlled and non-commercial sources of wealth generation in the areas described above. It will be necessary to create strong enough mobilizations and enough pressure on national governments and international institutions to force these concessions. However, rather than being understood as final goals, such strategies are merely stepping stones towards building autonomous capacity, and it will be necessary to fight for such resources in ways that maintain autonomy and avoid cooptation.

And, on the other hand, it will be necessary to use these resources to contribute to a broader and more long term process of social reconstruction and transformation based on a fundamental shift of power relations from the grassroots upwards. In particular, as has already been discussed in relation to energy specifically, it is becoming increasingly important to once again place the seizure of the key means of production (and reproduction) at the heart of revolutionary strategies, both with and without compensation.

Again, this is a monumental task, one that will not occur without strong social mobilization and struggle. However, it is a process made much more possible and realistic by the massive bankruptcies and devaluation of capital that the crisis entails, as it has left a trail of abandoned buildings, companies, and other pools of social wealth that are deemed "non-competitive" and hence useless. And, crucially, if these are not taken over, collectivized, and moved outside of the commercial sphere, they will be bought up on the cheap and will fuel a new round of socially- and ecologically-disastrous capital accumulation. Furthermore, the construction of new social relations along the above lines is also likely to be crucial in order to avoid disastrous authoritarian "solutions" to the financial-economic and political crises.

Given the centrality of energy, as both a key means of production and key to human subsistence, all of the above suggest that a major priority should be to advance a deep and long-term convergence of two global processes that have, until now, for the most part been developing in relative separation from one another. These are the global anti-capitalist networks described above, and those pushing for the construction of a rapid and smooth shift towards a new, and global, predominantly-renewable energy system. Such a convergence is likely to offer important opportunities for grassroots organizations to strengthen their abilities to collectively contribute to shaping and leading the changes in the global energy system that are currently underway, and that will, without a doubt, accelerate and intensify in the coming years. It would also make an important contribution towards finding an emancipatory way out of the ever-worsening economic-financial crisis.

There is only a small gap between the slogan of the global anticapitalist networks, "Another World is Possible/Other Worlds Are Possible," and the slogan "Another Energy is Possible/Other Energies are Possible." The clear associations made by the global anti-war movement between war and oil interests in the Middle East and further afield has had an important impact. On the one hand, it has greatly delegitimized multi-national oil companies throughout the world and reaffirmed the urgent need to move away from this source of energy. And, on the other, oil workers in Iraq have powerfully asserted their ability to shape the course of world-impacting events and have been at the heart of international anti-war networking among movements. There is clearly only a very fine line between the mass global anti-war movements that emerged around the war against Afghanistan and Iraq (and most recently Lebanon and Gaza), and a mass global movement for a new global energy system.

Until now, such a convergence has been quite slow, scarcely given any formal recognition within either process, and largely spontaneous and uncoordinated. Arguably, the last few years have seen the convergence slowly get underway, although it is still based on a very small number of as-yet not very influential people and organizations who are active in both areas. For the most part, the two sets of organizations and their projects still have very little in common.

The world-wide mobilizations around the Copenhagen COP 15 summit offered the opportunity for an important acceleration and deepening of this convergence process, and throughout the mobilization process movements were able to collectively develop coherent common perspectives leaving a variety of Climate Justice networks which continue to function after the summit itself. However, the mobilization process faced significant difficulties, and the points of disagreements and divergences were large, despite broad consensus on many key issues. Despite the difficulties, these world-wide mobilizations provide much scope for optimism, and should be seen as the beginning of a next phase, in which the convergence process is made much more far reaching, accelerated and explicitly acknowledged, as an intended goal of mobilizations. Until Copenhagen, the dominant approaches on climate change had vested most of their energy on promoting regulatory reforms, rather than on more fundamental changes in the social relations on which constitute the world-wide division of labour, the capitalist world-economy. This is true for the majority of governments, multilateral institutions and also large sectors of so-called "civil society". Importantly, also included here are the major national and international trade unions and their federations, as well as a whole range of social and environmental Non- Governmental Organizations (NGOs). However, the problem at hand is one of production, and the reproduction of lives and social relations. It is not simply a problem of regulation.

The failure of Copenhagen showed the failure of the regulatory approach, and this is an important change to register. Despite the patent inadequacy of pushing for a regulatory approach, efforts in this sphere will almost certainly continue to be pursued in the coming years, as governments from major powers and international institutions attempt to rebuild faith in the regulatory approach. As the legitimacy of this approach lies in tatters, increasingly not just in relation to climate change, but also in relation to the "solutions" offered to the world economic-financial crisis, efforts to rebuild the COP process on climate change are likely to seek to contribute to shoring up legitimacy. This may well still be possible, at least in the short term, and in certain, predominantly northern, countries where the effects of climate changes are less immediately visible and directly impacting on people's lives than they are in southern countries.

Consequently, movements need to be very wary of being pushed back onto the terrain of regulation, as this approach is likely to result in a disempowering demobilization process in which the main message is to trust political and economic leaders, rather than to self-organize for a long term process of struggle. However, movements throughout the world are currently still extremely ill-prepared for the conflicts involved in the transition to a new energy system. Without an urgent change of course, we will rapidly move towards a capitalist driven transition to a new energy system, with all the new forms of enclosure, violence and exploitation that such a transition is certain to entail.

Such a global convergence is unlikely to be easy, and a number of potential obstacles exist. In particular, there is an urgent need for these different groups to break down their unfamiliarity with each other, as well as perhaps a certain degree of suspicion and distrust of each other's struggles and choice of tactics and long term strategies, especially with regard to choices between confrontation or lobbying existing power structures. This will only happen through becoming informed about each other's work, building respect based on diversity, and learning to support one another wherever possible.

The seeds are clearly there, should we choose to plant them. The potential for convergence could almost certainly create a formidable mass social force for positive and rapid change. Should such a convergence take off in the coming years, this would make it possible to develop common actions and proposals on a far bigger scale than has been possible until now.

A common political basis for such a convergence will have to emerge through a lengthy collective discussion, based on the different input of the many different struggles and organizational initiatives already underway in the field of energy. This will be a very challenging process, as it will be important to create a framework that is simultaneously broad enough for many different struggles to feel comfortable participating in, but narrow enough to clearly define a certain political orientation and long-term course of struggle.

The following points are intended to contribute to this process.

1. *The need for rapid and extensive reductions in CO_2 emissions is non-negotiable, and affected communities and workers must lead the discussion of how to bring about this change.* A crucial question concerns the meaning of "clean energy," and the extent to which it is possible to "clean up" existing "dirty" energy and energy-intensive industries. To the extent that is possible, it will be important that it is brought about in such a way that is empowering for affected workers and communities (who, after all, are the ones who know the industries better than anyone else), rather than at their expense. And, to the extent that "clean up" is not possible, dislocated workers and communities will need to be protected and provided with opportunities to create alternative livelihoods. Similarly, international compensatory mechanisms will have to be developed to avoid unfair penalization of particular countries whose main source of national revenue may be the revenue that comes from selling "dirty energy." In particular, the question of ecological debt and reparations is crucial, since people in different regions do not share equal responsibility for climate change.

2. *Managing resource scarcity collectively and fairly.* The question of peak oil starkly exposes the need to develop ways of collectively managing scarcity in a fair manner that avoids very destructive power struggles and exacerbating already-existing growing inequalities (especially in relation to class, race, gender, and age) and a forced imposition of austerity measures on people. Solutions must actively strive to avoid pitting different communities and workers, both waged and unwaged, in different regions of the world against one another, so as to ensure that capital pays the costs, not labor. Failure to do so is almost certain to result in the transition being carried out on the backs of these workers and their communities. If emancipatory movements are unable to force capital to shoulder the burden, it is likely to prove immensely divisive and destructive.

3. *Collective efforts must be taken to ensure that the globally-expanding renewable energy sector contributes to a positive shift in power relations, and does not provide a new basis for exploitative ones.* Renewable energy has enormous potential to allow communities increased control of and benefit from the natural resources that exist in their territories. Conversely, there is also the danger that new structures of inequality, domination, hierarchy, and marginalization may arise. Such problems have been characteristic of the fossil and nuclear energy system, and there is a danger that a new energy system could reproduce and further exacerbate these problems. It will require coordinated and intentional action to avoid these scenarios.

4. *Energy sovereignty and autonomy as a basis for reducing energy and fuel dependency, and energy-related inequalities in the world-market.* There is an urgent need to simultaneously take steps towards equalizing access to energy, and to also reduce the structural dependency that high-energy-consuming

regions have on regions that are net exporters of energy. This is important in order to move towards overcoming the unequal and coercive global power relations on which this situation is based and, in turn, reinforces. These problems will only be resolved through communities being able to exert greater collective control over the energy resources that are both produced and consumed in their regions.

5. *Finding energy and climate solutions that contribute to, and speed up, a wider process of long term emancipatory social change in the face of the current world financial-economic and political crisis.* Substantial changes in production and consumption of energy will require substantial changes in production and consumption relations at a more general level. The process of building a new energy system, based around a greatly expanded use of renewable energies, has the potential to make an important contribution to the construction of new relations of production, exchange, and livelihood that are based on solidarity, diversity, and autonomy, and are substantially more democratic and egalitarian than those that currently exist. Furthermore, the construction of new social relations along the above lines will also likely be crucial in order to avoid disastrous "solutions" to the financial-economic and political crises.

It will be important to greatly strengthen our collective capacity for exchange and mutual support of different struggles in defense of livelihoods, rights, and territories related to the global energy sector. Similarly, it will be necessary to strengthen our capacity for exchange and support of struggles in defense of common/collective/cooperative or public ownership and control of energy resources, infrastructures, and technologies.

Parallel to this, there is the need to build solidary, upward-leveling relationships between workers in different branches of the energy sector and avoiding downward-leveling competition between them. And, linked to this, is the challenge of developing long-term collaboration and cooperation initiatives in non-commercial renewable energy technology transfer, open source technology research, education, training, and grassroots exchanges. For this to happen, two factors are key: building up human resources and also strengthening and increasing collective capacity to raise funds.

Finally, as part of a shift to a new energy system, there is the very urgent task of deepening a long-term strategic debate about how, and for what purposes, wealth is produced and distributed in society, and how people's subsistence needs are met.

Like so many worthwhile tasks in today's crisis-ridden world, none of this will be easy. Yet, the stakes are high. Failure to construct alternatives rooted in new relations of production, exchange, consumption, and livelihoods is likely to have disastrous effects. But there has perhaps never been a better chance to do so, and building a new, renewable energy-based energy system has the potential to make an important contribution to this process.

Then, and only then, it might just be possible to begin seriously discussing an energy revolution, and a whole lot more.

LIST OF CONTRIBUTING ORGANIZATIONS
AND INDIVIDUALS

ORGANIZATIONS

China Labour Bulletin (Hong Kong/China) (CLB) is a non-governmental organization that promotes and defends workers' rights in the People's Republic of China through the development of democratic trade unions in China, the enforcement of domestic labour laws, and the full participation of workers in the creation of civil society. In addition, CLB seeks the official recognition, in China, of international standards and conventions providing for workers' freedom of association and the right to free collective bargaining.

Corporate Watch (UK) is part of the growing anti-corporate movement springing up around the world. It is a research group supporting the campaigns that are increasingly successfully forcing corporations to back down from environmentally-destructive or socially-divisive projects, and dragging the corrupt links between business and power, economics and politics into the spotlight, against the resistance of the complacent, corporate-led mainstream media.

Energy Watch Group (Germany) Energy policy needs objective information: The Energy Watch Group is an international network of scientists and parliamentarians. The supporting organization is the Ludwig Bölkow Foundation. In this project, scientists are working on studies, independent of government and company interests, concerning: a) the shortage of fossil and nuclear energy resources, b) development scenarios for regenerative energy sources, as well as c) strategic deriving from these for a long-term secure energy supply at affordable prices. The scientists collect and analyse not only ecological, but above all economic and technological connections. The results of these studies are to be presented not only to experts but also to the politically-interested public. Objective information needs independent financing; a bigger part of the work in the network is done unsalaried. Furthermore, the Energy Watch Group is financed by donations that go to the Ludwig Boelkow Foundation for this purpose. More details can be found on our website: http://www.energy-watchgroup.org

Environmental Rights Action/ Friends of the Earth Nigeria (ERA/FoEN) is a non-governmental advocacy organization founded on January 11, 1993 to deal with environmental human rights issues in Nigeria. ERA is the Nigerian chapter of Friends of the Earth International (FoE), the world environmental justice federation of seventy-seven autonomous members from almost as many countries, campaigning to protect the environment and to create sustainable societies. ERA hosts the secretariat of Oilwatch International, the global south network of groups and communities resisting destructive crude oil and gas activities.

Fight for Alternative Renewable Methods and Autonomy (FARMA) (Greece) A collective of political solidarity with the Zapatistas, which is mainly involved with radical ecology and renewable sources of energy. The main goal of the collective is the installation of a small hydroelectric unit in a Zapatista community that will satisfy the needs of the local people, and also sharing the acquired technical knowledge of the project with them, as part of the process of constructing their autonomy. The money required for this project has been raised by solidarity events within the Greek anticapitalist movement, rather than from state and capitalist organizations. At the same time, in the mainframe of constructing autonomy here and now, FARMA organizes workshops on D.I.Y (do it yourself) wind generators in self-organized spaces, and has already installed two of them in squats in Athens, Greece and now wants to share the acquired knowledge with other autonomous spaces of the world. http://www.farmazapatista.blogspot.com, farma@riseup.net.

Focus on the Global South is a non-profit policy analysis, research and campaigning organization, working in national, regional and international coalitions and campaigns, with social movements and grassroots organizations, on key issues confronting the global south. Focus was founded in 1995 and is attached to the Chulalongkorn University Social Research Institute (CUSRI) in Bangkok, Thailand. It has country programs in the Philippines and India.

IG Metall (Germany) is the biggest trade union in the world. As of November 2008, it had 2.3 million members. Its headquarters are in Frankfurt. It has seven regional offices and about 170 local offices.

Integrated Sustainable Energy and Ecological Development (INSEDA) (INDIA) INSEDA is the national Indian organization of grassroots NGOs involved in the promotion of renewable energy programs. Its special focus is the implementation of biogas development in rural areas, as well as promotion and transformation of selected Indian villages into model-cum-demonstration eco-villages, jointly with its member NGOs, as part of a broader effort to build process-oriented, people-centered, sustainable human development. Its work focuses on the poor, weaker, marginalized, and other vulnerable, and deprived sections, including women in rural communities. INSEDA is a membership organization, at present having about fifty Indian NGOs as its members. All the members are committed to the promotion of low cost affordable renewal energy technologies and have fairly well developed infrastructure at the grassroots level to implement developmental projects/programs.

The International Federation of Chemical, Energy, Mine and General Workers' Unions (ICEM) is a global trade union federation representing more than 20 million workers in the energy, mining, chemicals, rubber, pulp and paper, glass, ceramics, cement, environmental services, and other industries. Its headquarters are in Geneva, Switzerland. (www.icem.org)

Midnight Notes and Friends is a collective with thirty years of reflecting and writing on capitalism, crisis, and class struggle. Their writings can be found at http://www.midnightnotes.org

Observatorio de Multinacionales en América Latina (OMAL) (Spanish State) is a project of the Association Peace With Dignity. Its purpose is to denounce the social, environmental and cultural impacts of the presence of Spanish transnational companies in Latin America. OMAL carries out on the ground research, writes reports, and has a public access database of articles and news items on its website (www.omal.info). It also participates in meetings, public talks, and related campaigns in order to contribute critical analysis of multi-national companies.

Observatorio de la Deuda en la Globalización (Spanish State, Catalunya) is a co-ordinating network of individuals and organizations dedicated to research-activism. The research is centred around the problem of North-South relations and on the creation of debts between communities due to contemporary globalization processes. It studies the mechanisms associated with the financial debt (external debt) of the poorest periphery countries with the countries of the Core (the Spanish State in particular) in order to be able to make consistent, precise, and well-documented denunciations whenever it is necessary. At the same time, it analyzes a number of other "external debts" contracted by core countries toward countries of the periphery. This includes ecological debt, historical debt, and social debt. The network is coordinated by, and has its core team at, the Cátedra UNESCO de Sostenibilidad at the Polytechnic University of Catalunya. (**www.odg.cat**)

Oilwatch: The Oilwatch network was born out of the need for communities affected by the oil industry to develop global strategies and to build support for their processes of resistance in the struggle against these operations. The activities of Oilwatch include the exchange of information about the operations of oil companies in each affected country, their operational practices, and also the exchange of information about different resistance movements and international campaigns against specific companies. Oilwatch strives to contribute to a heightened environmental consciousness, at the global level, by exposing the impacts of the oil industry on tropical forests and their local populations, while at the same time establishing the relations between the sector's activities and the destruction of biodiversity, climate change, and unpunished human rights violations. The international secretariat of Oilwatch is located in Nigeria.

Public Services International Research Unit (PSIRU) researches the privatization and restructuring of public services around the world, with special focus on water, energy, waste management, and healthcare. It produces reports and maintains an extensive database on the multi-national companies involved. This core database is financed by Public Services International (PSI), the global confederation of public service trade unions.

Saving Iceland is a campaign against heavy industry and large energy projects in Iceland from a deep ecological and anarchist perspective. It was founded in 2004 from a call for international support for direct action opposing construction of the Kárahnjúkar dam in the Icelandic highland wilderness. This dam is the largest hydroproject in European history and has the sole aim of providing energy to the aluminium industry.

Self-Reliance and Environment Technologies Unit at "Sapienza" University of Rome (Italy). This unit, operating within Research Centre on Sustainable Development (CIRPS), works on two basic pillars: A) Access (financial, social, and technical) should be guaranteed to the widest public possible, specifically in the context of disadvantage situations; B) Self-reliance is the result of the process through which effective capability and social functionality is built. In particular, the international cooperation activities and studies run by CIRPS focus on small social environments, low environmental impact energies, self production of chlorine, disadvantaged work groups, and areas that are in permanent crisis or social tension, in both urban and rural contexts. CIRPS has been working for many years in the field of self reliance and environment technologies. Over this time, it has built up a mature experience in implementing and offering assistance in projects related to water & sanitation and energy sector technologies. http://www.cirps.it

World Information Service on Energy (WISE) (coordination office in Netherlands) is a global network of grassroots initiatives and action groups against nuclear energy. It was founded in 1978 and since then has acted as an information switchboard. Its main aim is to support and empower grassroots initiatives all over the globe and to help them effectively fight nuclear power. This is mainly done through gathering, analyzing, and distributing useful information. WISE has published the *Nuclear Monitor* twenty times each year since 1978.

Yansa CIC is a non-profit-distributing Community Interest Company created in 2008 in order to contribute to a community-led transition to renewable energy. Its field of activity is the development and manufacture of renewable energy equipment, starting with wind turbine generators. It is part of the Yansa Group, which includes other organizations focused on project development with communities and ethical financing.

INDIVIDUALS

Nnimmo Bassey (Nigeria) is a human/environmental rights activist. He is the executive director of the Environmental Rights Action (ERA)—Nigeria's foremost environmental rights advocacy group, and chair, Friends of the Earth International—the world's largest federation of grassroots organizations fighting for environmental and social justice. Bassey is a member of the international steering committee of

Oilwatch International. He is also a practicing architect in Nigeria, as well as a published writer and poet.

Patrick Bond (South Africa), a political economist, is senior professor at the University of KwaZulu-Natal School of Development Studies, where he directs the Centre for Civil Society (http://www.ukzn.ac.za/ccs). Patrick's recent authored and edited books include *Climate Change, Carbon Trading and Civil Society* (UKZN Press and Rozenberg Publishers, 2008); *The Accumulation of Capital in Southern Africa* (Rosa Luxemburg Foundation, 2007); *Looting Africa: The Economics of Explotiation* (Zed Books and UKZN Press, 2006); *Talk Left, Walk Right: South Africa's Frustrated Global Reforms* (UKZN Press, 2006); and *Elite Transition: From Apartheid to Neoliberalism in South Africa* (UKZN Press, 2005).

George Caffentzis (USA) is a member of Midnight Notes Collective; co-author of *Midnight Oil: Work, Energy, War, 1973–1992*; and author of *No Blood for Oil—Energy, Class Struggle and War 1998–2004*; and Professor of Philosophy at the University of Southern Maine. He can be reached at caffentz@usm.maine.edu.

Sophie Cooke is a climate and coal campaigner, who, for the last several years, has been working with groups and direct action campaigns all over the world, including Rising Tide national and local anti-coal campaigns in the UK, North America, Australia, and New Zealand. She has also spent many years on the road running workshops, trainings, and events in hundreds of locations globally.

Dr. Irene Costantini holds a Linguistic and Cultural Mediation Degree, and is currently pursuing a masters degree in Oriental Studies. Her research area is the Arabic world, culture and development. She took part in the cooperation project "Chlorine Self-production plant for effluent water to irrigate in Gaza Strips." Currently she is carrying out research on "Information Technology to promote a dialogue within Jerusalem," and works on technology from a social and cultural point of view as an applied tool to develop social awareness.

Peter Custers (Netherlands) is a theoretician on arms production and an international campaigner. For many years he has been working to support class struggles waged by landless peasants, garment workers and other sections of the oppressed in Bangladesh and South Asia. He is a member of International Development Economics Associates (IDEAs) and of the Euro-Memorandum Group, and is author of *Capital Accumulation and Women's Labor in Asian Economies* (Sage, New Delhi, India/Zed Books, London, UK, 1997) and *Questioning Globalized Militarism: Nuclear and Military Production and Critical Economic Theory* (Tulika Publishers, New Delhi, India/Merlin Press, London, UK, 2007). He can be reached at antimil@hotmail.com or at http://www.petercusters.nl.

Claire Fauset (UK) is a researcher for Corporate Watch and a climate change activist. Claire is author of "What's Wrong With Corporate Social Responsibility?" and "Technofixes," a critical guide to large scale technological solutions to climate change, including biofuels, hydrogen, nuclear power, and carbon capture and storage. The report highlights the way in which a "techno-fixated" approach undermines efforts towards real solutions.

Dr. Simona Fernandez holds a political science degree, and is currently pursuing a masters degree in the same field. Her work deals with International Cooperation in the Arab World. She took part in the cooperative project, "Chlorine Self-production plant for effluent water to irrigate in Gaza Strips," is currently carrying out research titled, "Social Research on Disabled people's autonomy in Jerusalem." Her work on technology is from a psycho-social point of view with a particular focus on issues relating to people with disabilities.

Marc Gavaldà (Spanish State, Catalunya) holds a bachelor's degree in Environmental Sciences. He is a researcher and makes documentary films about petrol-related conflicts in Latin America. Author of *La Recolonización* (The Recolonization) (2003), *Viaje a Repsolandia* (Travel to Repsolandia) (2005), *Repsol YPF un discurso socialmente irresponsable* (Repsol YPF a Socially Irresponsible Discourse) (2007). Producer of the documentary *Patagonia Petrolera* (2008).

Erika González (Spain) has a degree in Biology. She is currently a researcher at Observatorio de Multinacionales en América Latina (OMAL) – Peace With Dignity Her research pertains to the impacts resulting from the activities of Spanish multinationals in Latin America, with a particular area of focus on energy. She is coauthor of the book *La energía que apaga Colombia. Los impactos de las inversiones de Repsol y Unión Fenosa (The Energy Which Is Switching Off Colombia: The Impacts of Repsol and Union Fenosa's Investments)* (Icaria, 2007) and the *Atlas de la energía en América Latina y Caribe* (Energy Atlas for Latin America and the Caribbean) (Paz con Dignidad, 2008).

David Hall (UK) is director of Public Services International Research Unit (PSIRU), and is responsible for its work. He specializes in water, energy and healthcare, and the design and maintenance of the PSIRU database and website. Before joining PSIRU he worked at the Public Services Privatisation Research Unit, which developed a database on privatization for the UK trade unions. He had previously worked for trade union research units, and as a lecturer in higher education. He has written books on public expenditure and labor law. Contact: d.j.hall@gre.ac.uk

Ewa Jasiewicz (UK) is a freelance journalist and solidarity activist based in London. She spent 9 months living in occupied Iraq working with Iraqi unions including oil workers, as well as time in Palestine supporting communities of resistance. She is involved with the international Hands Off Iraqi Oil campaign and "Naftana"—the UK support group for the Iraqi Federation of Oil Unions.

Tom Keefer (Canada) is PhD candidate in Political Science at York University where he is researching the political economy of oil and energy. He is an editor of the anti-capitalist journal *Upping the Anti* <www.uppingtheanti.org> and is also active in prison solidarity organizing and support for indigenous struggles in Southern Ontario. He can be reached at tkeefer@yorku.ca.

Brian Kohler, BSc, cCT, MCIC has over thirty years of experience in occupational health and safety, environmental, and sustainability issues. He is currently responsible for developing policies and services in these areas for the International Federation of Chemical, Energy, Mine and General Workers' Unions (ICEM), a global federation of labor unions.

Jaap Krater (Netherlands) has published numerous articles on heavy industry, hydro and geothermal energy in Icelandic, Dutch, and UK media. He recently organized the conference "Global Consequences of Heavy Industry and Large Dams" in Reykjavik. Jaap has a history of involvement with ecological direct action and is a former spokesperson of Saving Iceland. He is particularly interested in how technology influences our perception of the natural world and will now be researching values of biodiversity and social-ecological resilience in the South Pacific. jaap.krater@groenfront.nl

Jane Kruse (Denmark) is Director of Education and Training Programmes at the Nordic Folkecenter for Renewable Energy. She has been active for many years in the Danish wind turbine cooperatives movement and was the former Chairman of the Hornstrup Mark Windmill Cooperative. She is also former Chairperson of the Danish Renewable Energy Association, and a member of the Sydthy Municipality Board.

Nancy LaPlaca (USA), J.D., is part of a national movement working to de-bunk "clean" coal and carbon sequestration, which provides energy consulting to political candidates and intervenes at the Colorado Public Utilities Commission on these issues: (1) clean energy over fossil fuels, (2) more realistic natural gas fuel cost increases (3) future availability and cost of natural gas, (4) discount rates, and (5) including externalities, such as water use, health effects from fossil fuels and global warming damage.

Les Levidow (UK) is a Senior Research Fellow at the Open University, where he has been studying agri-environmental issues. A long-running case study has been the agbiotech controversy, details of which can be found at the Biotechnology Policy Group webpages at http://technology.open.ac.uk/cts/bpg.htm He can be contacted at L.Levidow@open.ac.uk.

Preben Maegaard (Denmark) is the director of the Nordic Folkecenter for Renewable Energy, Senior Vice President EUROSOLAR, The European Association for Renewable Energy, Founder and President-Emeritus of World Wind Energy Association, and chairperson World Council for Renewable Energy. For many years he has been active in renewable energy, both in Denmark and internationally. He has organized and participated in numerous national and international seminars, workshops, and conferences in the field, and has authored or co-authored numerous reports, books and articles.

Esperanza Martínez (Ecuador) is coordinator of Oilwatch's campaign to keep oil underground. Founded in 1996, Oilwatch is a network of resistance to the petroleum sector's operations in tropical countries, with members in Africa, Asia, and Latin America. The Ecuadorean organization of Oilwatch is called *Acción Ecológica*, and has sustained anti-petrol struggles for many years.

Gavan McCormack (Australia) is emeritus professor at Australian National University, coordinator of Japan Focus (http://japanfocus.org), and author, most recently, of *Client State: Japan in the American Embrace* (New York, 2007) with Japanese, Korean, and Chinese editions in 2008.

Andrea Micangeli, Ph.D (Italy) is the coordinator of the Self-Reliance and Environment Technologies Unit at "Sapienza" University of Rome, within CIRPS (Research Centre on Sustainable Development). Since 1994, he has been working to promote sustainable development both in Italy and in socially unstable contexts. At the moment, he is running and coordinating projects in different countries, including:

> Sahrawi Refugee Camps, Algeria, sustainable agriculture with drop to drop irrigation system, recycling dump material, and promoting sports, thanks to the production of medals sold at sports events.

> Gaza Strip, technical assistance and training course for the installation of (self-producing chlorine plant) to implement the use of effluent water for cultivation.

> Mucuchies, Venezuela—promoting a solar panel system for sustainable development.

> Chad, reforestation project and data collecting for Gis.

> Italy—a project aimed at providing solar energy through solar panels to fifteen prisons.

His experience also includes providing drinking water in Basra, Iraq; mycrohydroeletric production in Chiapas; implementing autonomy for disabled people in Afghanistan and East Europe. He teaches in the Faculty of Engineering and Psychology, where he sits on the professorial board of six master courses in renewable energetic technologies and in project management for cooperation and international crisis.

Alejandro Montesinos Larrosa (Cuba) is a writer, publisher, and journalist, who is a mechanical engineer and holds a masters of science (1987), as well as a bachelors and masters degree in journalism (1994). He is director of the publishing house CUBASOLAR and the magazines *Energía y tú* (Energy and You) and *Eco Solar*, and author of *Matrimonio solar* (Solar Marriage) and *Hacia la cultura solar* (Towards a Solar Culture).

Evo Morales (Bolivia) is the president of Bolivia [Note: this contribution is an open letter that Morales wrote for the Poznan climate change talks at the end of 2008, which circulated widely via email and the internet. Unlike the rest of the chapters, it has been reprinted without the author's knowledge.]

Camila Moreno (Brazil) works on social and environmental impacts of biotechnology and agribusiness expansion in Brazil and Latin America. She is a PhD candidate at UFRRJ/CPDA and is a member of the Political Ecology working group of the Latin America Council of Social Sciences, CLACSO. She can be reached at cc_moreno@yahoo.com

Conrado Moreno Figueredo (Cuba) is Professor at the Centre for the Study of Renewable Energy Technologies (CETER), at the Technical University of Havana, member of the Cuban Academy of Sciences, doctor in Technical Sciences, and the Cuban coordinator of the World Wind Energy Institute. He is the author of *Diez preguntas y respuestas sobre energía eólica* (Ten Questions and Answers about Wind Energy).

Tadzio Müller (Germany) lives in Berlin, and after many years of being a counter-globalist summit-groupie, is active in the emerging climate-action movement. Having escaped the clutches of (academic) wage labor, he is currently writing a report about "green capitalism" for the Rosa Luxemburg Foundation, and otherwise doing odd translation jobs. He is also an editor of *Turbulence: Ideas for Movement* (www.turbulence.org.uk).

Raymond Myles (India) is a renewable energy and socio-technical specialist with about thirty years of experience in the field. He has been involved in the systematic promotion of biogas technologies, with focus including its low-cost household use. He has authored and co-authored three manuals on three different low cost Indian household biogas plans. His work has involved planning, research, design, training, as well as hands-on implementation and coordination of rural-oriented field technologies. In 1995, he founded INSEDA, Integrated Sustainable Energy and Ecological Development Association. For the last seven years he has been actively promoting implementation of sustainable, energy-based, community-oriented, eco-village development program, in partnership with INSEDA's grassroots member NGOs within the country. He is also active in different international processes around non-commercial technology transfer, including INFORSE (the International Network for Sustainable Energy), which he co-founded, and now acts as South Asian regional coordinator for the group.

Trevor Ngwane (South Africa), a community activist, is secretary of the Soweto Electricity Crisis Comittee, and previously worked for the Anti-Privatisation Forum, Alternative Information and Development Centre, and several trade unions. He was the subject of a film, *Two Trevors go to Washington*, and a *New Left Review* interview by Tariq Ali, following his expulsion as leader of the African National Congress regional branch in Soweto, for opposing Johannesburg water privatisation in 1999. He formerly tutored sociology at the University of the Witwatwatersrand, and has begun a masters thesis on strategy/tactics and the conflict between socialist and autonomist politics in Soweto.

Sergio Oceransky is one of the founders of the Yansa Group, and Co-Director of Yansa CIC. His diverse professional experience ranges from work with social and environmental organizations to management positions in corporations. He has lived and worked in several European, Asian, and Latin American countries, undertaking collaborative projects. Previous to his involvement in the Yansa Group, he was Coordinator of the World Wind Energy Institute.

Alexis Passadakis (Germany) studied political science and global political economy in Berlin and Brighton. Alongside the biocrisis, his research areas are world trade and privatisation of public services and goods. In the last few years, he has worked with different NGOs and is currently member of Attac Germany's coordinating council. He is active in the emerging climate justice movement and co-organized the first German climate action camp in Hamburg in 2008. Alexis lives in Berlin.

Helena Paul, co-director of EcoNexus, has worked on the protection of indigenous peoples' rights and tropical forests, agricultural biodiversity, oil exploitation in the tropics; patents on life and genetic engineering since 1988. She is co-author of *Hungry Corporations: Transnational Biotech Companies Colonise the Food Chain* (Zed Books, 2003). She can be contacted at h.paul@gn.apc.org.

Peter Polder (Netherlands) is thirty-three, and is a longtime activist within the Dutch branch of the EarthFirst! network. He currently works as an energy analyst. He has also extensively researched the interaction between the climate- and peak oil crisis.

Bruce Podobnik (USA) is Associate Professor of Sociology at Lewis & Clark College (Portland, Oregon). He is the author of *Global Energy Shifts* (Temple University Press), and co-editor of *Transforming Globalization* (Brill Academic Press).

Pedro Ramiro (Spain) holds a doctorate in Chemical Sciences. Since 2005, he has been a researcher at the Observatorio de Multinacionales en América Latina (OMAL)—Peace With Dignity. His research pertains to the impacts resulting from the activities of multi-national companies from Spain and other countries, and his particular area of focus is on energy companies and Corporate Social Responsibility policies. He is coauthor of the book *La energía que apaga Colombia. Los impactos de las inversiones de Repsol y Unión Fenosa* (The Energy That is Switching Off Colombia: The Impacts of Repsol and Union Fenosa's Investments) (Icaria, 2007).

Klaus Rave (Germany): Since 1995, he has been a member of the Executive Board of Investitions bank Schleswig-Holstein, Germany, and, from 1997, the vice president of the European Wind Energy Association (EWEA)—the world largest renewable energy association." Should read "Since 1995, he has been a member of the Executive Board of Investment bank Schleswig-Holstein, Germany, and, from 1997, the vice president of the European Wind Energy Association (EWEA)—the world's largest renewable energy association.

Peer de Rijk (Netherlands) started working on nuclear energy issues in 1980 as a volunteer-activist in several small grassroot groups. Between 1990 and 1996, he was an energy campaigner at Milieudefensie (Friends of the Earth Netherlands), and from 1997 to 2000, he was a nuclear energy campaigner for WISE. Since 2000 he has been executive director of WISE.

Tatiana Roa Avendaño (Colombia) is an environmental activist, member of Censat Agua Viva, and Friend of the Earth Colombia, and member of the International Commitee of the Oilwatch network. She has supported a range of resistance processes such as the struggle of the U'wa people, the Process of Black Communities of Colombia, and writes regularly for the magazine *Ecología Política*, published in Barcelona, Spain.

Miriam Rose (Iceland) has been closely involved with Saving Iceland for three years, after completing a BSc in Environmental Science at Sussex University. She has recently left Iceland and is now studying at the Centre for Human Ecology in Scotland, where she is particularly interested in community resilience and empowerment, with specific reference to the situation in Iceland.

Kristina Sáez (Spain) has a degree in Biology. She works as a researcher at the Observatorio de Multinacionales en América Latina (OMAL)—Peace With Dignity, where she researches Spanish transnational companies in Latin America, with a particular focus on the energy sector. She is the coauthor of *Atlas de la energía en América Latina y Caribe* (Energy Atlas for Latin America and the Caribbean) (Paz con Dignidad, 2008).

Hermann Scheer (Germany) is president of EUROSOLAR, the European Association for Renewable Energy, General Chairman of the World Council for Renewable Energy (WCRE), and a member of the German parliament since 1980. For years, he has been an author, policy innovator, and global leader in the field of renewable energy.

Macdonald Stainsby (Canada) is a writer, hitch-hiker, and social-justice activist, based in Edmonton. He can be reached at mstainsby@resist.ca.

Jessica Toloza (Colombia) is an anthropology student at the National University of Colombia. She is a video maker and participated in the production of a range of videos in Colombia. Currently she is specializing in visual anthropology.

Monica Vargas (Bolivia/Spanish State) is a sociologist and social anthropologist, currently doing research at the Observatori del Deute en la Globalització (Catedra UNESCO de Sostenibilidad, Polytechnic University of Cataluña, http://www.odg.cat), where she is responsible for the research on ecological debt, and is researching for her doctorate. She is the author of several articles on ecological debt, agrofuels, food crisis, and food sovereignty, and is technical advisor for the Campaign "No te Comas el Mundo" (Don't Eat the World!) (www.noetmengiselmon.org). Coordinator of a monograph *Agrocombustibles: ¿Otro negocio es posible?* [Agrofuels: Is Another Business Possible?] (2008, Barcelona, Editorial Icaria, forthcoming), and author of *Nunca Más un México sin Nosotros* (Never again a Mexico Without Us) (2001, México, INAH-Conaculta). She can be reached at monica.vargas@odg.cat.

Shannon Walsh (Canada) is a filmmaker and writer currently finishing her first feature documentary, *H2Oil*, about the human and social costs of oil sands development.

She can be reached at shannon@loadedpictures.ca. The website of "H2Oil" is http://www.h2oidoc.com.

Dale Jiajun Wen is an activist scholar, originally from China. She is a senior scientist at Action 2030 Institute (www.action2030.org), a NGO think-tank focusing on long term policy strategies for sustainable development. Dale can be contacted at dale.wen@gmail.com.

Martina Winkelmann (Germany) was born in Dortmund, in Nordrhein-Westfalen. She is currently living in Wiesbaden, Germany, and has been employed at IG Metall since 1991. Since 2002, she has been based at their main headquarters in Frankfurt, and since 2005, she has worked in the Crafts Plant Policy SME/ Mechanical Engineering Sector, where she is responsible for four branches: the wind and solar energy sectors, the automation & robotics sector, and the heating industries, for the whole of Germany. She is also responsible for co-determination in SME, for new concepts to increase works councils in SMEs.

ABOUT THE EDITOR

Kolya Abramsky has worked for over a decade with a range of grassroots social and environmental organizations from around the world. His work has included education, international mobilizations, publications, and translations. Since 2003 he has focused on energy. He is currently coordinating Towards a Worldwide Energy Revolution. This global process aims to bring people from a wide range of organizations and struggles together to build long term alliances in order to prepare for an anticapitalist transition process to a new energy system. The coalition of organizations involved in convening this event come from Colombia, Ecuador, India, Netherlands, Nigeria, South Africa, and the UK. In 2008–2009, Kolya was a visiting fellow at the Institute of Advanced Studies in Science, Technology and Society, in Graz, Austria, where he received the Manfred-Heindler Award for Energy and Climate Change Research. In 2008, he received a Masters degree in sociology from the State University of New York, Binghamton. In 2006 he was coordinator of the Danish-based World Wind Energy Institute, an international effort in non-commercial renewable energy education, involving different renewable energy centers from around the world. Originally from the UK, Kolya has lived in several European countries, as well as in the US. He is the editor of the book *Restructuring and Resistance: Diverse Voices of Struggle in Western Europe*.

Support AK Press!

AK Press is one of the world's largest and most productive anarchist publishing houses. We're entirely worker-run and democratically managed. We operate without a corporate structure—no boss, no managers, no bullshit. We publish close to twenty books every year, and distribute thousands of other titles published by other like-minded independent presses from around the globe.

The Friends of AK program is a way that you can directly contribute to the continued existence of AK Press, and ensure that we're able to keep publishing great books just like this one! Friends pay a minimum of $25 per month, for a minimum three month period, into our publishing account. In return, Friends automatically receive (for the duration of their membership), as they appear, one free copy of every new AK Press title. They're also entitled to a 20% discount on everything featured in the AK Press Distribution catalog and on the website, on any and every order. You or your organization can even sponsor an entire book if you should so choose!

There's great stuff in the works—so sign up now to become a Friend of AK Press, and let the presses roll!

Won't you be our friend? Email friendsofak@akpress.org for more info, or visit the Friends of AK Press website: http://www.akpress.org/programs/friendsofak